EcoProduction

Environmental Issues in Logistics and Manufacturing

Series editor

Paulina Golinska, Poznan, Poland

About the Series

The EcoProduction Series is a forum for presenting emerging environmental issues in Logistics and Manufacturing. Its main objective is a multidisciplinary approach to link the scientific activities in various manufacturing and logistics fields with the sustainability research. It encompasses topical monographs and selected conference proceedings, authored or edited by leading experts as well as by promising young scientists. The Series aims to provide the impulse for new ideas by reporting on the state-of-the-art and motivating for the future development of sustainable manufacturing systems, environmentally conscious operations management and reverse or closed loop logistics.
It aims to bring together academic, industry and government personnel from various countries to present and discuss the challenges for implementation of sustainable policy in the field of production and logistics.

More information about this series at http://www.springer.com/series/10152

Mitsutaka Matsumoto • Keijiro Masui •
Shinichi Fukushige • Shinsuke Kondoh
Editors

Sustainability Through Innovation in Product Life Cycle Design

Springer

Editors
Mitsutaka Matsumoto
National Institute of Advanced Industrial Science and Technology (AIST)
Tsukuba
Japan

Keijiro Masui
National Institute of Advanced Industrial Science and Technology (AIST)
Tsukuba
Japan

Shinichi Fukushige
Osaka University
Suita
Japan

Shinsuke Kondoh
National Institute of Advanced Industrial Science and Technology (AIST)
Tsukuba
Japan

ISSN 2193-4614 ISSN 2193-4622 (electronic)
EcoProduction
ISBN 978-981-10-0469-8 ISBN 978-981-10-0471-1 (eBook)
DOI 10.1007/978-981-10-0471-1

Library of Congress Control Number: 2016950433

Printed on acid-free paper

This Springer imprint is published by Springer Nature
The registered company is Springer Science+Business Media Singapore Pte Ltd.

Preface

Rooted in mass production, mass consumption, and mass disposal, our societies worldwide are facing constraints on the availability of material resources and fossil fuel resources. Innovation in product life cycle design is one of the critical elements for realizing a circular economy and sustainable society. Product life design is critical for enabling reduction of material input, overall energy input, and product system waste and emissions of industrial systems in our societies.

This book consists of the selected papers presented at the EcoDesign2015 international symposium – the 9th International Symposium on Environmentally Conscious Design and Inverse Manufacturing – which was held in Tokyo December 2–4, 2015. The EcoDesign symposium, since the first symposium in 1999, has been leading research and practices of product life cycle design, and eco-design of products, services, production systems, supply chains, consumption, economies, and society. The symposium has been providing a distinguished platform for professionals from academia, industry, and policymaking to share knowledge and discuss the contemporary topics in the area with people from various backgrounds and regions in the world.

In EcoDesign2015, over 200 studies were presented. The topics for the symposium came from the state of the art of the studies on eco-design and included case studies from various industries and regions. The presentations were delivered by worldwide researchers, especially those from Asian and European countries. Increasing numbers of researchers from emerging economies in Asia also participated in the symposium in 2015. These indicate that eco-design is becoming a global issue in our society, and the importance of this symposium has been increasing even more.

Of the 200 presentations in EcoDesign2015, the papers for 70 presentations are contained in this book. For the papers for which the authors preferred having a peer-review, full-paper peer-reviews were conducted. The 70 chapters in the book are the papers that were accepted as the results of the peer-reviews. The topics of the chapters reflect the topics presented and discussed in EcoDesign2015. They consist of product development for sustainability, design for sustainability in emerging economies, business design for sustainability, sustainable production and material recovery, strategy for a sustainable society, eco-innovation strategy, eco-design of social infrastructure, and sustainability assessment and indicators. All these topics are critical for enabling innovation in product life cycle design.

We would like to express our appreciation to all the contributors, supporters, and participants of EcoDesign2015. For the publication of this book, we especially appreciate the executive committee members who cooperated in the peer review of the papers. As many as 59 members cooperated in that review.

Tsukuba, Japan — Mitsutaka Matsumoto
Tsukuba, Japan — Keijiro Masui
Suita, Japan — Shinichi Fukushige
Tsukuba, Japan — Shinsuke Kondoh

Cooperating Reviewers

- Kazuhiro Aoyama (The University of Tokyo, Japan)
- Casper Boks (Norwegian University of Science and Technology, Norway)
- Hsiang-Tang Chang (National Kaohsiung First University of Science and Technology, Taiwan)
- Jahau L. Chen (National Cheng Kung University, Taiwan)
- Otmar Deubzer (Fraunhofer IZM, Germany)
- Keishiro Hara (Osaka University, Japan)
- Hidetaka Hayashi (EcoDeNet, Japan)
- Hiroyuki Hiraoka (Chuo University, Japan)
- Tomonori Honda (AIST, Japan)
- Allen H. Hu (National Taipei University of Technology, Taiwan)
- Winifred Ijomah (University of Strathclyde, UK)
- Masahito Inoue (Meiji University, Japan)
- Keiichi Ishihara (Kyoto University, Japan)
- Yasushi Itoh (Fujitsu FIP Corporation, Japan)
- Takashi Iwamoto (Keio University, Japan)
- Toshihiko Iwamoto (Tokyo University of Information Science, Japan)
- Hyung-chul Kim (Ford Motors, USA)
- Jong-min Kim (Korea National Cleaner Production Center, Korea)
- Yusuke Kishita (AIST, Japan)
- Paivi Kivikyto-Reponen (VTT, Finland)
- Hideki Kobayashi (Osaka University, Japan)
- Masakazu Kobayashi (Toyota Technological Institute, Japan)
- Michikazu Kojima (Institute of Developing Economies, Japan External Trade Organization, Japan)
- Hitoshi Komoto (AIST, Japan)
- Tsai-Chi Kuo (Chung Yuan Christian University, Taiwan)
- Ki-ho Lee (Korea National Cleaner Production Center, Korea)
- Mattias Lindahl (Linkoping University, Sweden)
- Dan Lockton (Helen Hamlyn Centre for Design, UK)

- Nozomu Mishima (Akita University, Japan)
- Masahiro Miyo (Fujitsu Corporation, Japan)
- Yuji Mizuno (The Institute of Applied Energy, Japan)
- Jiro Nakamura (NTT Advanced Technology Corporation, Japan)
- Osamu Namikawa (Hitachi, Ltd., Japan)
- Nariaki Nishino (The University of Tokyo, Japan)
- Tomomi Nonaka (Aoyama Gakuin University, Japan)
- Shinichi Okamoto (Tokyo University of Information Science, Japan)
- Shuji Oowada (Waseda University, Japan)
- Harald E. Otto (Universita Politecnica delle Marche, Italy)
- Sergio A. Palleroni (Portland State University, USA)
- Tomohiko Sakao (Linkoping University, Sweden)
- Julie M. Schoenung (University of California Davis, USA)
- Li-hsing Shih (National Cheng Kung University, Taiwan)
- Shana S. Smith (National Taiwan University, Taiwan)
- Bin Song (Singapore Institute of Manufacturing Technology, Singapore)
- Sittha Sukkasi (MTEC, Thailand)
- Erik Sundin (Linkoping University, Sweden)
- Shozo Takata (Waseda University, Japan)
- Kenji Tanaka (The University of Tokyo, Japan)
- Tomohiro Tasaki (NIES, Japan)
- Koji Tokimatsu (Tokyo Institute of Technology, Japan)
- Kazutoshi Tsuda (Osaka University, Japan)
- Jui-Che Tu (National Yunlin University of Science and Technology, Taiwan)
- Haruhisa Uchida (Tokai University, Japan)
- Werasak Udomkichdecha (MTEC, Thailand)
- Sharif Ullah (Kitami Institute of Technology, Japan)
- Yasushi Umeda (The University of Tokyo, Japan)
- Michinori Uwasu (Osaka University, Japan)
- Renee Wever (Delft University of Technology, Netherlands)
- Winco Yung (The Hong Kong Polytechnic University, Hong Kong)

Contents

Part VI Eco-innovation Strategy

Part VII Eco-design of Social Infrastructure

Part I
Product Development for Sustainability

Investigating Types of Information from WEEE Take-Back Systems in Order to Promote Design for Recovery

Louise Lindkvist, Natalia Alonso Movilla, Erik Sundin, and Peggy Zwolinski

Abstract Waste electrical and electronic equipment (WEEE) recovery facilities have been set up for the last decade to promote a circular economy. Their activities focus on the reuse, remanufacturing and/or recycling of products. Currently, little information reaches designers regarding the requirements that these facilities have on product design. Therefore, most products are not designed to be properly recovered. The aim of this paper is to explore the nature of product life-cycle information from recovery organisations that could be shared in order to improve resource efficiency. The focus is on how information exchange can benefit the end-of-life phase of forthcoming designed products. Two levels of information have been identified, macroscopic and microscopic. Our study is illustrated with a detailed analysis of the French WEEE compliance scheme and an in-depth analysis of an IT remanufacturing facility in Sweden. Based on the case studies, we have identified current and potential information flows between different stakeholders that could benefit design for recovery.

Keywords Product recovery • WEEE • Product life-cycle information • Information exchange

1 Introduction

A circular economy promotes sustainable development by stimulating reduction of waste and enhancement of resource efficiency [1]. The quantity of waste electrical and electronic equipment (WEEE) produced worldwide in 2014 is estimated at 41.8 million metric tons, and it is predicted to continue growing at a rate of 4–5 % per year [2]. In order to address this, different take-back systems have been established

L. Lindkvist (✉) • E. Sundin
Division of Manufacturing Engineering, Department of Management and Engineering, Linköping University, Linköping, Sweden
e-mail: louise.lindkvist@liu.se

N.A. Movilla • P. Zwolinski
G-SCOP Laboratory, University Grenoble Alpes, Grenoble, France

M. Matsumoto et al. (eds.), *Sustainability Through Innovation in Product Life Cycle Design*, EcoProduction, DOI 10.1007/978-981-10-0471-1_1

throughout Europe to ensure proper management of WEEE [3, 4]. This management is assumed by recovery facilities, formally known as treatment operators, that perform different operations such as collection, handling, shipping, sorting, storage, transport, trading, treatment and preparing for reuse of WEEE [5]. According to the WEEE directive, the minimum quota of recycling is 4 kg per capita and year [6]. In countries such as Sweden and Holland, these rates are about 16 kg per capita and year; this means that there is a lack of strong driving forces to collect and recycle more WEEE, at least from a legislative perspective, which in turn means that more economic drivers are needed to increase the WEEE recycling in these countries [7].

One factor that is influencing the recoverability of a product and its components is the design [8]. Thus, in order to increase recoverability and resource efficiency, it is important that original equipment manufacturers (OEMs) take into consideration the recovery phase at the design stage [9]. However, OEMs are not always involved in treatment operations [10], resulting in a lack of knowledge on the recovery strategies and processes that their products follow. Therefore, OEMs cannot adopt the potential requirements that treatment operators have on product design. Design and recoverability aspects are decoupled in many cases, and the treatment centres' operations are not often considered in the design stage [11].

A close cooperation between OEMs and the different stakeholders from the take-back system would offer designers a better understanding of the expertise of treatment operators. Proper information exchange becomes indispensable for the success of such cooperation [12]. That way, expertise in the treatment operators can be utilised to enhance the recoverability of used products.

The aim of this article is to explore the nature of product life-cycle information from recovery operators that could be shared in order to improve resource efficiency.

2 Theoretical Background

2.1 Recovery Scenarios

Take-back systems have been established throughout Europe to implement the extended producer responsibility (EPR) principle [3]. Known under the name of EPR schemes or compliance schemes [3], they are in charge of the management of different waste streams, i.e. temperature exchange equipment (TEE), large equipment (LE), small equipment (SE), display appliances (DA), lamps [13] and, since 2015 in some countries, photovoltaic panels (PV).

Recovery can be defined as the process of obtaining valuable products, components or materials from waste [14]. The main recovery strategies are reuse, remanufacturing and material recycling [15]. If these strategies cannot be realised, products are incinerated or disposed of. Since these strategies do not allow recovery

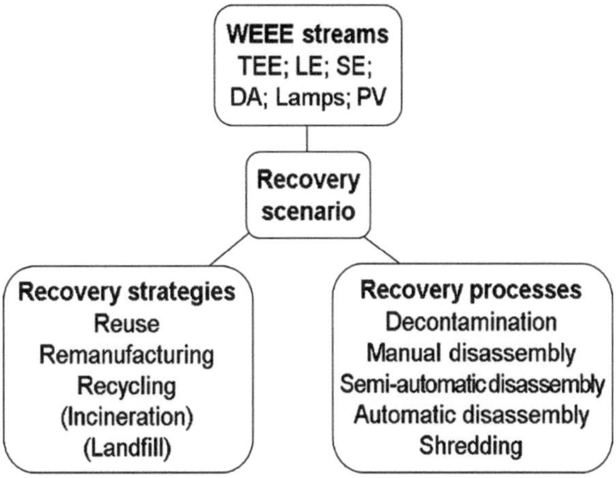

Fig. 1 Variables influencing recovery scenarios for the preprocessing stage

of the materials, we will not consider them as recovery strategies. The first phase of the recovery process generally consists of the separation of components and materials (preprocessing), followed by a second phase in which the different products, components and materials are recovered (post- or end-processing) [12]. In this article, we will focus on the information exchange of the preprocessing phase. The operations that WEEE undergo during the preprocessing phase primarily focus on the following recovery processes: decontamination, disassembly (manual, semiautomated or automated) [15] and shredding [16]. To our understanding, the combination of WEEE streams, recovery strategies and recovery processes results in different recovery scenarios that each require different design decisions (see Fig. 1).

Not all the combinations of WEEE streams, recovery strategies and processes are feasible. For instance, it is not possible to reuse a product through shredding or to recycle DA through shredding (due to their content of hazardous substances). There are some strategies that determine the type of process; i.e. reuse of products is nowadays mostly carried out through manual processes.

2.2 *Product Life-Cycle Information*

A sustainable material life cycle requires information exchange between the actors in the product life cycle [12]. It is important that the mind-set when designing products changes, from being focused on manufacturing and use to also including activities later in the product life cycle, such as service and remanufacturing. In fact, product designers should determine the end-of-life strategy in the design phase [17].

In Fig. 2, a simplified illustration of the product life cycle and the potential information flows from the end-of-life (EoL) phase are marked with red-dotted

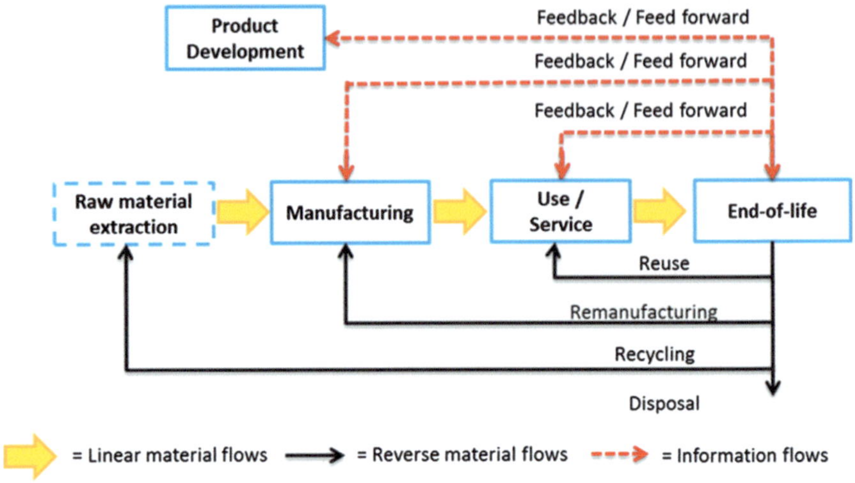

Fig. 2 A simplified illustration of the product life cycle with its material and information flows

arrows. In order for designers to know that their design lived up to expectations, information feedback is commonly used, foremost from customers/users. However, different stakeholders have their specific functions and processes and thus requirements on the product design. With that perspective in mind, regarding information feedback from the entire product life cycle in the design phase seems to be the natural next step in order to improve the development of sustainable products and services.

Design strategies that address increased recoverability of products include, e.g. Design for Disassembly (DfD), Design for Recycling (DfR) and Design for Remanufacturing (DfRem).

Kroll et al. point out that in addition to good design methods like DfR, feedback to the designer is also needed. However, DfRem requires nondestructive disassembly and reassembly of the product back to its original condition. In the recycling scenario, the product will not be returned to its original condition. On the other hand, nondestructive disassembly of the product could increase the amount of material separated and recovered and therefore lead to a more sustainable recycling process. Further, time for disassembling components could be saved if the design allowed for easy disassembly [18], as some products are manually disassembled at some recovery operators.

Ideally, the product and its components should be designed for the intended recovery scenario.

3 Types of Information from the WEEE Take-Back Systems

In order to know what type of information is needed, we have asked ourselves the following questions: What information do we need in order to identify the different recovery scenarios? Within one recovery scenario, what information will be useful for designers in order to improve resource efficiency? Subsequently, we have distinguished two main types of information that we will refer to as macroscopic and microscopic related. Macroscopic information provides a holistic view of the EPR scheme, while microscopic information provides detailed information about the activities that occur in the different recovery scenarios that should be fed back to design and other stakeholders of the product's life cycle.

In order to find the right answers, it is important to ask the right research questions and to state the right research hypothesis. Stating the right question will allow researchers to identify what methods are necessary to retrieve the information and to interpret them. The following subsections explain what questions to ask in order to know what information is included in the macroscopic and microscopic levels.

3.1 Macroscopic Information

The current recycling performance can be improved through a holistic vision on material flows, business models and design [12]. Macroscopic-related information will allow designers to gain some insight on the EPR scheme in a comprehensive way. On one hand, it is important to identify the recovery scenarios that are being carried out at a given time (past, present or future) and at a given place (country, continent or world). On the other hand, characterising the treatment operators allows understanding of why they carry out the different recovery scenarios.

From a design perspective, different operators and scenarios will have different design requirements. Generic design guidelines could be valid for some types of treatment operators, but not for all of them. OEMs should implement design requirements specific to the operators that will recover their products in order to improve resource efficiency. The main questions that need to be answered to obtain product life-cycle information at a macroscopic level are shown below.

Regarding the identification of recovery scenarios:

- What are the waste streams, recovery strategies and recovery processes that different operators carry out? Or, stated more simply, what are the different recovery scenarios that treatment operators realise?
- How many recovery operators carry out one or more recovery scenarios?
- If they carry out more than one recovery scenario, which are they? Is there a link between the different recovery scenarios? For instance, do recovery facilities

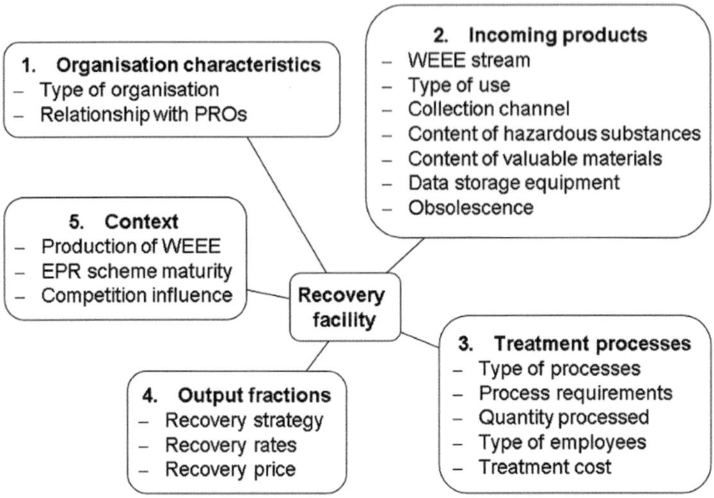

Fig. 3 Variables influencing treatment operators at a macroscopic level

realise several recovery scenarios under the same waste stream, recovery strategy or recovery process?

Regarding the characterisation of treatment operators:

- Why do treatment operators realise the different recovery scenarios?
- What are the variables influencing the different recovery scenarios that they carry out?

A qualitative study of the French WEEE compliance scheme has allowed us to develop a model containing the variables that induce products to be treated by one treatment operator or another. We have classified them into five categories (see Fig. 3). A more detailed description of the different variables is providied by Alonso Movilla and Zwolinski [14].

After identifying the variables that influence the different recovery scenarios, it is possible to formulate more specific questions, like:

- What is the influence of the quantity collected on the recovery processes? What is the influence of the collection channel on the recovery process?
- What is the influence of type of actor on the recovery strategies? What is the influence of PROs on the recovery strategies?

Designers are able to increase their level of understanding of the recovery phase whenever they know what recovery scenarios the different treatment operators realise and why they realise them, meaning what are the variables that influence them. A case study on the identification of the recovery scenarios of the French WEEE take-back scheme is shown in Sect. 4.1.

3.2 Microscopic Information

Microscopic information is the information that focuses on the treatment itself, e.g. what operations are realised, what components are removed and for what purposes (material recycling, component reuse, remanufacturing, etc.) and what tools are used. There is a need to have access to information about the particular recovery scenarios that treatment operators carry out in order to adapt the product design accordingly. This information can be divided into three main categories: detailed knowledge about the recovery process, detailed knowledge about the requirements of the original product and knowledge about the requirements of the recovered product.

Different treatment scenarios have different process steps. The generic ones are collection, preprocessing (that might include de-pollution, manual disassembly and shredding) and end-processing [19]. Remanufacturing includes more steps and operations than classical reuse or recycling processes. For that reason, we will focus our analysis using the remanufacturing recovery strategy as our focus.

Remanufacturing includes inspection, cleaning, disassembly, reprocessing, storage reassembly and testing [20]. These process steps are often not automated. Due to the high variability of the state of the incoming cores, skilled workers are needed in the remanufacturing process. Efficient remanufacturing requires access to the drawings and specifications of the original product. If such valuable information is missing, re-engineering strategies need to be applied [21]. The exact characteristics of the individual remanufacturing process are determined by the features of the product. For instance, some parts with profiles exposed to a lot of grease can be challenging to clean. Other components may be more difficult to separate and may require the development of specific separation technologies.

Products that are recycled go through a number of process steps before they end up as individual fractions of materials (Fig. 4). The WEEELABEX standard [5] asks operators to provide information about their processes and fractions.

Further, knowledge about the requirements of the product design for the intended recovery scenario is essential in order to achieve a more efficient recovery process. The RemPro-matrix (Table 1) highlights design features that have an impact on how efficient the remanufacturing process steps will be [20].

Not many products are designed for remanufacturing (e.g. [22]). However, since many studies show that remanufacturing is indeed good for the environment [23], increasing the amount of products that are remanufactured is one relevant approach to tackle the environmental issues and reduce the environmental impact.

Efficient recycling also relies on ease of separation of components and materials. In the disassembly process, information about the separation processes at the recovery centres is often not known, and optimal separation technologies would require information exchange between the OEM and the recovery centres [12].

A remanufactured product should ideally go through multiple life cycles and thus be used again. According to Lindkvist and Sundin [11], there are many requirements on products intended for remanufacturing, including:

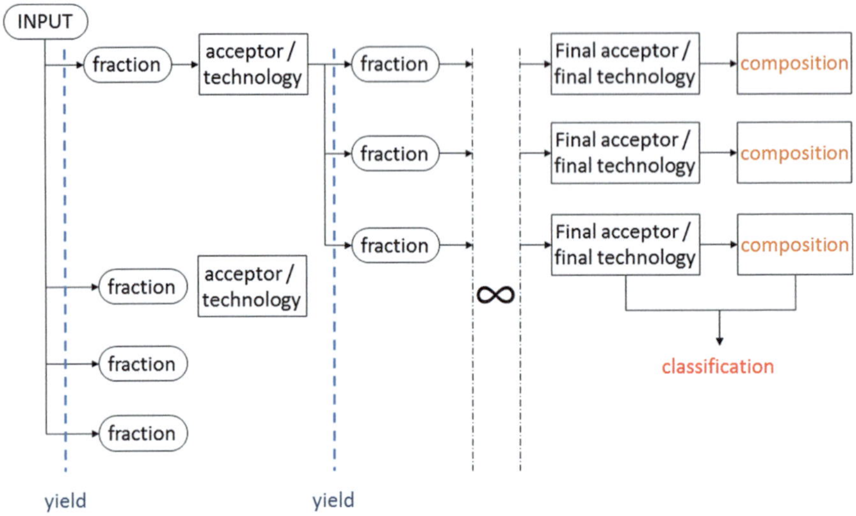

Fig. 4 Input, fractions and technologies from the WEEELABEX standard [5]

- Cleaning aspects
- Component quality
- Component quality of purchased parts
- Disassembly qualities
- Finish/surface issues
- Material selection
- Packing aspects
- Standardisations sought for
- Verification aspects
- Weak component analysis

These qualities relate back to the design of the products, and designers are able to adapt the product design accordingly if they know the characteristics of the specific recovery process steps and the requirements that those process steps induce on the product design. In order to explore this, some questions can be asked:

- What does the recovery process look like?
- What properties of the product determine the success of the recovery process?
- What product features are beneficial and non-beneficial for the recovery option?

Answering these questions should guide what information to feed back to product design. A case study of an IT remanufacturer in Sweden where these questions are explored is presented in Sect. 4.2.

Table 1 Generic steps of a remanufacturing process and desired product properties as described by Sundin [20]

Product property	Remanufacturing step						
	Inspection	Cleaning	Disassembly	Storage	Reprocess	Reassembly	Testing
Ease of identification	X		X	X			X
Ease of verification	X						
Ease of access	X	X	X		X		X
Ease of handling			X	X	X	X	
Ease of separation			X		X		
Ease of securing						X	
Ease of alignment						X	
Ease of stacking				X			
Wear resistance		X	X		X	X	

4 Case Studies

4.1 French WEEE Compliance Scheme

In this section we identify the recovery scenarios performed by the treatment operators of the French WEEE compliance scheme in 2012. The data was collected by the French Environmental Protection Agency (ADEME) and, although confidential, was shared with us. From the identified 196 treatment plants, 169 answered the survey. Of these, 151 correctly reported the questions used in our study, resulting in a response rate of 77 %. Treatment operators provided information about the recovery processes and strategies they carried out on the different waste streams. The different options given by the survey are as follows:

- Waste streams: TEE, LE, DA, SE or lamps
- Recovery processes and strategies: product reuse, component reuse, manual disassembly, mechanical treatment on WEEE and mechanical treatment on decontaminated WEEE

Since the strategies and processes are merged, we have made various assumptions: product reuse and component reuse are carried out through manual operations; manual disassembly, mechanical treatment on WEEE and mechanical treatment on decontaminated WEEE are done for recycling purposes. In order to simplify the analysis, we have grouped product and component reuse under the strategy "reuse" and mechanical treatment on WEEE and mechanical treatment on decontaminated WEEE under "mechanical" recovery processes.

This analysis focuses on determining what recovery scenarios are carried out by treatment operators. The results are shown in Table 2.

Table 2 Recovery scenarios performed in France in 2012

ID	Waste stream	Recovery strategy	Recovery process	No.
1	TEE	Reuse	Manual	28
2	TEE	Recycling	Manual	8
3	TEE	Recycling	Mechanical	12
4	LE	Reuse	Manual	28
5	LE	Recycling	Manual	27
6	LE	Recycling	Mechanical	28
7	DA	Reuse	Manual	19
8	DA	Recycling	Manual	63
9	SE	Reuse	Manual	23
10	SE	Recycling	Manual	55
11	SE	Recycling	Mechanical	28
12	Lamps	Recycling	Mechanical	4

Another part of our analysis focuses on knowing how many different possible combinations of recovery scenarios treatment operators incorporate. Our analysis included the following three variables:

- Waste streams: Up to 17 different combinations of the five waste streams were found. This means that some operators recover just one type of waste stream, while others recover two, three, four or even five streams.
- Recovery strategies: Some operators reuse the products, others recycle them, and others perform both.
- Recovery processes: The processes the operators use can be manual, mechanical or both.

There are 48 possible combinations of these 3 variables at the treatment operators. Table 3 shows the number of treatment operators that carry out the different combinations of recovery scenarios, waste streams, recovery strategies and recovery processes.

The results are represented in the bubble charts of Figs. 5 and 6. Figure 5 shows the number of treatment operators (bubble size) that carry out the different combinations of waste streams (X-axis) and recovery strategies (Y-axis). For instance, there are 18 operators whose activities focus on the reuse of TEE and LE. Figure 6 shows the number of operators (bubble size) that carry out the different combinations of waste stream (X-axis) and recovery processes (Y-axis). In Fig. 5 we observe that 22 operators carry out manual processes on DA and SE. It would have been desirable to show the results in a 3D bubble chart. However, in order to facilitate visualisation, the results are shown in the two mentioned figures. Thanks to Table 3, Figs. 5 and 6, we can draw the following conclusions:

- Of the 12 identified recovery scenarios, recovery facilities performed 5 at the most, with 1 or 2 being the most performed. Most operators treated 1 or 2 different waste streams at a time.
- Recycling was the most performed recovery strategy, followed by reuse.
- Manual treatment was the most performed recovery scenario, followed by mechanical treatment and mixed processes.

Waste streams were quite mixed within the operators, although the most performed together were TEE with LE (white goods) and DA with SE (probably brown goods that needed manual disassembly for separating hazardous substances and valuable materials).

4.2 Swedish IT Remanufacturing Organisation

The remanufacturing case company is a small company that launched its business about 10 years ago. It started as a result of frustration from the two founders over how computers were unnecessarily scrapped after use. The idea was to establish a treatment process so that laptops and other IT products could be reused at least once; to solve this, they set up a remanufacturing process. As an independent actor, the company buys used equipment and sells remanufactured equipment. Since the

Table 3 Number of treatment operators performing the different combinations of recovery scenarios, waste streams, recovery strategies and recovery processes

No. of recovery scenarios	No. of treatment operators	No. of waste streams	No. of treatment operators
1	50	1	52
2	54	2	66
3	25	3	22
4	16	4	10
5	6	5	1
Type of recovery strategies	No. of treatment operators	Type of recovery processes	No. of treatment operators
Reuse	30	Manual	97
Recycling	103	Mechanical	37
Both	18	Both	17

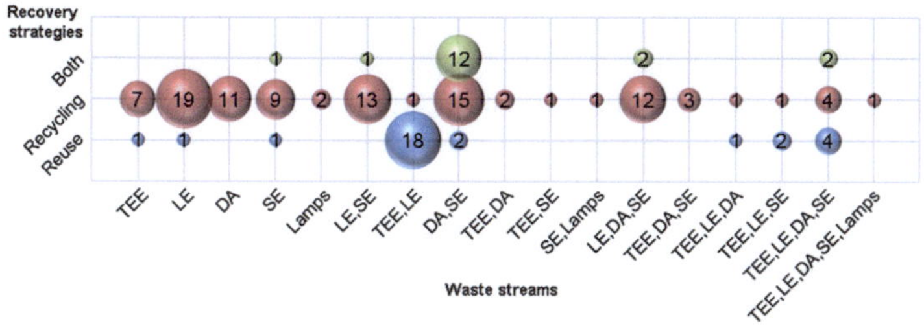

Fig. 5 Number of recovery facilities per waste stream and recovery strategies in France in 2012

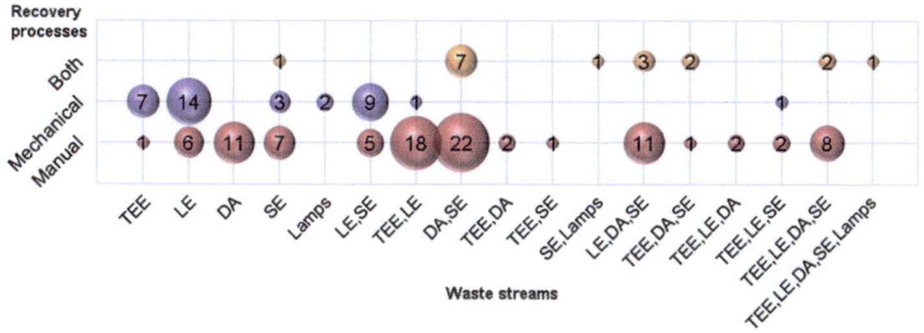

Fig. 6 Number of recovery facilities per waste stream and recovery process in France in 2012

goods are to be returned to useful life once remanufactured, the information stored in the hardware needs to be deleted in a secure manner. Some applications require completely erasing all traces of information through a high-security process.

The remanufacturing process today is not as comprehensive as desired by the remanufacturer. Initially, there is a manual inspection and a classification of the core quality. Then the core goes through some steps where original identifications are erased, the software is updated and the information on the hard drive is erased. The battery's condition is verified, and if it is not up to standard, it will be replaced by a cannibalised battery. The remanufacturer has limited opportunities to remanufacture the exterior of the laptops and other electronic equipment. Design trends are sometimes in conflict with long use or with use that does not damage the delicate surfaces that are attractive to the consumer at the moment of purchase but impractical in the long run. The company has long lists of design improvements, for laptops primarily, that would greatly facilitate remanufacturing and also make more products remanufacturable. At present, many IT products are unnecessarily scrapped by the company because the exterior does not live up to the customers' high standards. The customers require the product to perform and look like new and a lower price is a must, while environmental aspects are not that prioritised in the Nordic market. Exterior requirements are not that strict in the Eastern European market, which allows more products to be recovered.

However, being independent, the company has no influence over the design of the product. This means that some models are more easily remanufactured than others, and the products have some features that benefit remanufacturing and others that hinder it. For instance, some laptop covers have a "self-healing" coating. This means that if there are minor scratches on the surface, they are easily erased once the surface is polished. On the other hand, some covers are easily damaged. For instance, the trend with white screens proved to be very impractical as the white turned yellow after exposure to the sun.

There are quite a few design alterations that would facilitate the company's remanufacturing process. Some examples of microscopic information that could be useful in the design process are "standardise AC adaptors on all laptops", "minimise number of screws used to assemble the product", and "allow easy access to memory card". The company sees potential in at least doubling the times that the product can be remanufactured if its design was better adapted accordingly. It has plenty of feedback to product design, but being independent, no channel to provide it through. In other value chains, there are also many barriers for such feedback to reach the designers [11].

5 Discussion

Regarding the case study on macroscopic information (Sect. 4.1), we observed that not all the waste streams, recovery strategies and processes identified in Sect. 2.1 have been documented in the 2012 French survey on WEEE [24]. Photovoltaic panels, remanufacturing, automated disassembly and semiautomated disassembly should be included in future surveys.

We have based our macroscopic model and case study on qualitative and quantitative analyses of the French compliance scheme. It would be desirable to carry out some in-depth analysis of other take-back systems and, if possible, have information on the European WEEE take-back systems as a whole, to gain a more complete picture of the recovery situation of WEEE. For example, in the Swedish take-back systems, there is strong civic support in environmental protection leading to high collection and recycling rates [7]. However, there is not a strong drive to collaborate with stakeholders within the Swedish take-back systems [7], and little information is transferred back to OEMs from the WEEE collectors, recyclers and remanufacturers since the information is normally not collected and demanded.

The retrieval of recovery-related information can be challenging nowadays. It requires organisation, time and the involvement of different stakeholders. Figure 7 shows the existing and desirable information flows that we have identified. There are three organisations gathering information from recovery facilities: PROs, national agencies and the WEEELABEX organisation. However, no feedback related to the recovery practices is fed back to them. For the French national agency, ADEME, the information is confidential, but the type of information gathered is public (see [24]). However, for PROs and the WEEELABEX organisations, both the information and the type of information gathered remain confidential. In order to make information publicly available, transparency could be included as a minimum legal requirement in the Waste Framework Directive [3].

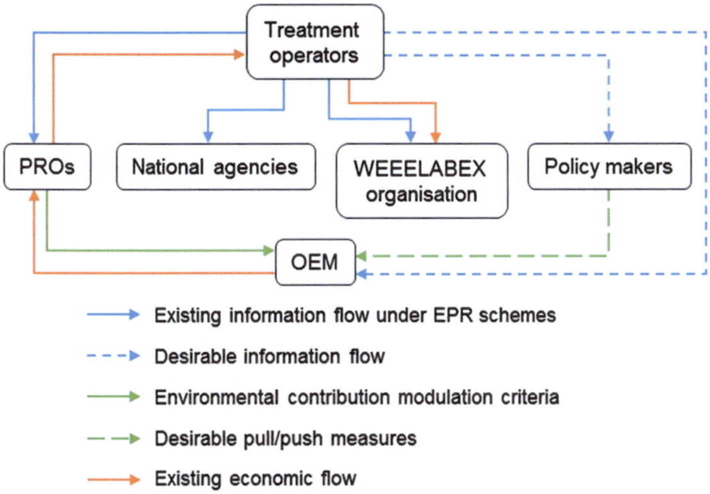

Fig. 7 Existing and desirable information flows

Policy measures could be seen as good incentives to improve the design of products [25]; hence, policymakers should be able to access the information. In an ideal situation, OEMs will work together with recovery facilities to promote the scenarios that allow improvement of the most resource-efficient solutions.

During the design phase, different design strategies to enhance the environmental performance from a life-cycle perspective, specifically ecodesign strategies, should be applied. If the OEM does not consider the recovery phase during the design phase, by its own initiative, certain regulations or laws could help stimulate sustainable development. Design for recovery requirements could, for instance, be integrated into policy initiatives – push measures (e.g. the Ecodesign Directive) and/or pull measures (e.g. the EU Ecolabel).

To share and search for information is essential. In the case of remanufacturing, the problem might be that no one asks for the information feedback from remanufacturers. If the OEM is not remanufacturing, an independent third party like a national or international agency could be the one demanding access to information that could help improve the design of the products, similar to the WEELABEX system for recycling centres.

Information needs to be interpreted and documented to eventually be turned into design knowledge. The knowledge about the recovery scenarios should eventually lead to integration in ecodesign methodologies. Thus, it is important to know not only the information flows but also the format of the feedback. The information could be transferred in many formats such as via documents, databases, digital pictures, recordings, animations, etc. The digital information age facilitates the transfer of data. However, only digital information may not be sufficient. Ideally, dialogue between the parties will help increase the mutual understanding of their respective processes and particular needs. However, sharing information can be a highly sensitive subject, in conflict with intellectual property rights and so on. If handled by a third party, such aspects can be regulated and the information used to benefit the circular economy, where all parties are able to thrive while co-creating resource-efficient product life cycles.

6 Conclusion

In this study we wanted to highlight the importance of information sharing between treatment operators and product design in order to improve design for recovery.

The case studies show examples of what information can be obtained from recovery organisations, both from a macroscopic and a microscopic perspective. Currently, the OEMs have limited access to information from the recovery organisations, and thus there is potential for increasing the information sharing between the actors. If the design was adapted for the intended recovery scenario, that should lead to more resource-efficient products.

Future studies will focus on how to integrate information from the recovery organisations in the design phase.

Acknowledgements The authors would like to thank the French Environmental Protection Agency (ADEME) for sharing its database on WEEE treatment operators. The authors further acknowledge VINNOVA, the Swedish innovation agency, for financing the case study presented in this paper. Finally, we would like to extend our gratitude to the respondents for their contributions.

References

1. Ellen McArthur Foundation (2013) Towards the circular economy, economic and business rationale for an accelerated transition
2. Baldé CP, Wang F, Huisman J (2015) The global e-waste monitor – 2014. United Nations University, IAS – SCYCLE, Bonn
3. European Commission – DG Environment (2014) Development of guidance on Extended Producer Responsibility (EPR)
4. European Parliament and Council (2012) Directive 2012/19/EU of the European Parliament and of the Council of 4 July 2012 on waste electrical and electronic equipment (WEEE), Official Journal of the European Union, pp 38–71
5. WEEE Forum (2013) WEEELABEX standard on treatment V10.0
6. European Parliament and of the Council (2003) Directive 2002/96/EC of the European Parliament and of the Council of 27 January 2003 on waste electrical and electronic equipment (WEEE), Official Journal of the European Union L 037, 13/02/2003, pp 24–39
7. Lee HM, Sundin E (2012) The Swedish WEEE system – challenges and recommendations. In: 2012 I.E. International Symposium on Sustainable Systems and Technology (ISSST), pp 1–6
8. Mathieux F, Froelich D, Moszkowicz P (2001) Development of recovery indicators to be used during product design process: method, potentialities and limits. In: Proceedings EcoDesign

2001: second international symposium on environmentally conscious design and inverse manufacturing. pp 281–286
9. Mizuno T, Kunii E, Fukushige S, Umeda Y (2012) Recyclability evaluation of LCD TVs based on end-of-life scenarios. In: Dornfeld DA, Linke BS (eds) Leveraging technology for a sustainable world. Springer, Berlin, pp 179–183
10. Ramani K, Ramanujan D, Bernstein WZ, Zhao F, Sutherland J, Handwerker C, Choi JK, Kim H, Thurston D (2010) Integrated sustainable life cycle design: a review. J Mech Des 132 (9):091004–091004
11. Lindkvist L, Sundin E (2015) Assessing barriers for available life-cycle information feedback transfer to product design. Proceedings of ICoR 2015, Amsterdam, to be published
12. Vanegas P, Peeters JR, Cattrysse D, Duflou JR (2011) Information exchange requirements for cradle to cradle implementation in an industrial setting: EoL treatment of flat screen televisions, presented at the Life Cycle Management, Berlin, 2011
13. Froelich D, Sulpice D (2013) The "Eco-design" guide of French WEEE compliance schemes
14. Alonso Movilla N, Zwolinski P (2015) Design for recovery: applying multivariate statistics to define groupings of French WEEE Pre-treatment operators. Proceedings of ICED 2015, Milano. To be published
15. Duflou JR, Seliger G, Kara S, Umeda Y, Ometto A, Willems B (2008) Efficiency and feasibility of product disassembly: a case-based study. CIRP Ann Manuf Technol 57 (2):583–600
16. Williams JAS (2006) A review of electronics demanufacturing processes. Resour Conserv Recycl 47(3):195–208
17. Fukushige S, Kunii E, Yamamoto K, Umeda Y (2011) A design support system for scenario-based lifecycle design. pp 919–927
18. Charter M, Gray C (2008) Remanufacturing and product design. Int J Prod Dev 6 (3–4):375–392
19. Sundin E, Bras B (2005) Making functional sales environmentally and economically beneficial through product remanufacturing. J Clean Prod 13(9):913–925
20. Sundin E (2004) Product and process design for successful remanufacturing, Linköping Studies in Science and Technology, Dissertation No 906 Linköping University, Linköping
21. Kurilova-Palisaitiene J, Lindkvist L, Sundin E (2015) Towards facilitating circular product life-cycle information flow via remanufacturing. Procedia CIRP 29:780–785
22. Hatcher GD, Ijomah WL, Windmill JFC (2011) Design for remanufacture: a literature review and future research needs. J Clean Prod 19(17–18):2004–2014
23. Sundin E, Lee HM (2012) In what way is remanufacturing good for the environment? In: Design for innovative value towards a sustainable society. Springer, Dordrecht, pp 552-557
24. ADEME (2014) Inventaire 2012 des centres de traitement de déchets des équipements électriques et électriques (Inventory 2012 of WEEE treatment operators)
25. Ardente F, Mathieux F, Recchioni M (2014) Recycling of electronic displays: analysis of pre-processing and potential ecodesign improvements. Resour Conserv Recycl 92:158–171

A Framework for Sustainable Product Development

Daniel Kammerl, Damian Schockenhoff, Christoph Hollauer, Dominik Weidmann, and Udo Lindemann

Abstract Although consumers are increasingly critical to include environmental and social aspects in their purchase decision, environmental friendly product design is not widespread, and for many products there are no sustainable alternatives. Additionally, sustainability is a very complex issue which overtaxes many companies forcing them to leaving it behind. The concept for a framework presented in this paper is a first step towards supporting sustainable development by making this complex task tangible (problem layer), projectable (goal layer) and accessible (action layer).

Keywords Product development • Sustainability • Framework

1 Introduction

The consumption of products is a measurement for wealth and prosperity of a society. Today, most consumer products are designed to last from cradle to grave: Limited raw materials are sourced and transformed into a product which is used and eventually landfilled or incinerated. This wasteful and unsustainable way of consumption is responsible for the pollution of the ecosystem and the depletion of natural resources. Thus, sustainable growth can only be built on eco-friendly, sustainable products that use fewer resources, do not pollute the environment and avoid the generation of waste. Although consumers are increasingly critical to include environmental and social aspects in their purchase decision, environmental friendly product design is not widespread, and for many products there are no sustainable alternatives. Methods and tools for sustainable development are often considered a burden by companies. This is caused by the fact that additional capital and time are needed for applying those approaches, and the motivation for developing sustainable products is often based on governmental incentives, legislative conditions or marketing benefits. It is aggravated by the fact that most of the applied methods, e.g. life-cycle assessment, come into operation not until the late phases of

---

D. Kammerl (✉) • D. Schockenhoff • C. Hollauer • D. Weidmann • U. Lindemann
Institute for Product Development, Technische Universität München, Munich, Germany
e-mail: kammerl@pe.mw.tum.de

M. Matsumoto et al. (eds.), *Sustainability Through Innovation in Product Life Cycle Design*, EcoProduction, DOI 10.1007/978-981-10-0471-1_2

the development process because they require a large information base, but changes in this phase yet are very expensive. Additionally, sustainability is a very complex issue which overtaxes many companies forcing them to leaving it behind [1].

1.1 Motivation

The sooner sustainability efforts take place in the development process, the less they cost [2]. In order to develop a sustainable product with a minimum of financial expenses, sustainability has to be systematically planned and implemented straight from early phases of the development process.

However, in practice sustainability efforts concentrate on the late phases of product development and marketing [3]. Although companies consider sustainability efforts a burden in early phases, they willingly spend a lot of money in late phases, e.g. on marketing campaigns. But those efforts do not make the product itself more sustainable.

The integration of the attribute sustainability in early phases of the product development process is a possibility to create more sustainable products that do not harm the ecological nor the social environment, without imposing additional efforts on the companies. This challenge is a motivation for this paper.

1.2 Goals

The main barriers for sustainable product development are additional financial efforts and intricacy. The goal of this paper is to present an approach which reduces both of these barriers. Therefore, existing concepts are analysed and discussed. On this base a new concept for sustainable product development is suggested. The final approach should empower companies to integrate sustainability in early phases of the development process and to systematically plan and implement the product attribute sustainability.

1.3 Research Methodology

In the beginning, an initial descriptive study was conducted. For this purpose an extensive literature study in the field of sustainable product development was applied exploring the field of methods and approaches in this area. In the following prescriptive study, a conceptual framework for designing sustainable products has been developed. The proposed solution concept is an extension, adaptation and integration of existing concepts, similar to a design-for-X guideline. For evaluation of the solution concept, a second descriptive study was conducted. In this context, the framework has been evaluated by using it for the development of a mechatronic product and rating it by criteria defined by experts in the field.

2 State of the Art

In this section existing frameworks, methods and approaches for developing sustainable products are presented. They are the result of the literature study.

2.1 Frameworks

In order to improve the ecological properties of products and services, there are several general frameworks for their development presented in literature. For example, Beucker et al. [4] describe their concept of “Integrated Product Policy”. Within this framework, all life-cycle phases of a product or service are considered at the same time, and thus one may take suitable measures for the improvement of sustainability aspects at exactly those moments of the life cycle, where it is most effective.

Within this context, a possible mean for assessing the environmental impact of a product may be the so-called “integrative environmental controlling” approach [5]. Thereby, one may use an enterprise resource planning system, i.e. a software system, for the efficient planning of a company’s resources such as financial and working capital as well as human resources.

Another prominent framework is the Cradle2Cradle (C2C) concept by McDonough and Braungart [6]. The basic idea of this concept may be easily tied up: a world without waste. The authors propose that most environmental problems, such as global warming, deforestation, air pollution and the generation of huge amounts of waste, may be traced back to the behaviour of the recent Western civilization. To avoid this, the C2C principle supposes to keep existing material in closed circles. Therefore, one may distinguish between two principle circles or kinds of material: while natural materials are biologically decomposable and may be led back to the biological circle, synthetically produced resources have to be kept within closed technical loops.

Finally, the Framework for Strategic Sustainable Development is to be mentioned. This framework was developed by the Swedish non-profit organisation “The Natural Step” and offers a method for establishing a sustainable society [7]. The basic idea of the framework is the so-called backcasting: first, one creates an imaginary ideal of a future situation and secondly, the required means for reaching this ideal state are derived.

2.2 Methods and Approaches

In addition to the previously described general frameworks, there also exist more specific methods for developing sustainable products and services. Their general purpose is to support social development towards more sustainability awareness [8]. One of those frameworks is the Guideline for Improvement of Product Development by McAloone and Bey [9] which focuses on guiding the user through the developing process step by step and thereby inspiring him for intense reflection of sustainable solution alternatives. Similarly, Panarotto and Törlind [10] propose sustainability innovations workshops for the same purpose. Those workshops enable developers to create ideas playfully and thereby take the customer into their focus within a multidisciplinary team.

Furthermore, there are various methods that combine existing development tools with the idea of sustainability. For example, the whole-system and life-cycle method (WSLC method) by Faludi [1] enables the combination of classical system thinking with a life-cycle analysis and brainstorming techniques. Besides, there exist several QFD-modifications [11–13] as well as the so-called BioTRIZ method by Vincent et al. [14], which are all variations of their well-known and established "paragons".

Lastly, in addition to those exclusively sustainability-oriented approaches, the IPP Bayern method [15] combines environmental thoughts with the economic interest of a company. Its goal is to improve the interactions between a product and its surrounding along the entire product life cycle. Within IPP Bayern, suitable methods for sustainable product development shall be transferred especially to middle scale as well as small companies in order to support them in developing ecologically friendly and sustainable products while keeping in mind their economic ambitions.

2.3 Conclusion

As the previous section shows, various frameworks and approaches do exist that shall help development teams in bringing up sustainable product and service solutions. Nevertheless, the existing approaches are quite vague and do not offer the possibility of supporting the actual designer in his task of integrating the aspect of sustainability into the product.

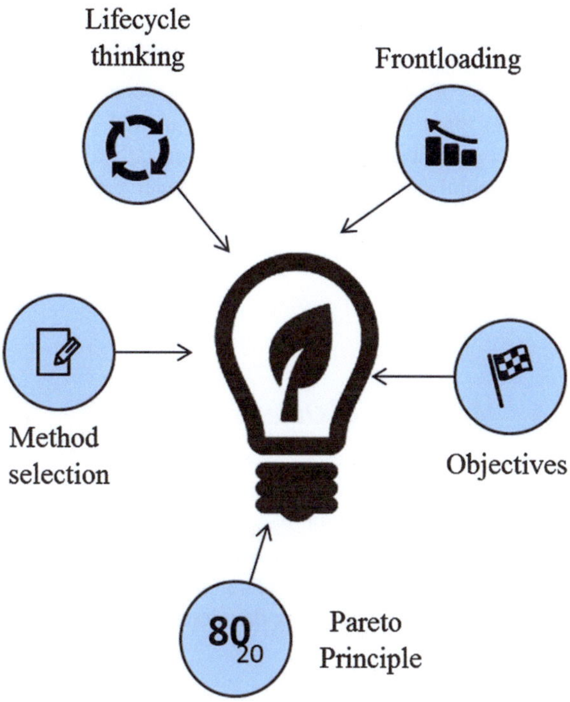

Fig. 1 Challenges for a method supporting the development of sustainable products

3 Challenges

Based on the literature study, challenges for a method supporting the development of sustainable products were deduced. They are illustrated in Fig. 1 and described in the following.

3.1 Life-Cycle Thinking

A sustainable product does not show any negative impact on ecological and social systems. This does not only include the utilisation phase but also the impact from mining raw materials, the chain of production, the supply chain and the use phase and finally the end of life. Developing sustainable products means to consider the entire life cycle of a product.

3.2 Front-Loading

Engineers often neglect sustainable aspects during the concept phase and only consider them when the product is launched. The rule of ten says that the costs of product modifications increase exponentially along the life cycle. At the same time, fundamental decisions about the product properties are being made in the early development phase [2]. Front-loading of the sustainability activities seizes this issue. It means to prepone sustainability effort to earlier stages of the development process.

3.3 Modelling of Objectives

Target planning is the source for successful work in product development [16]. A guided search for solution needs clear, verbalised and accepted goals by all parties [17]. Goals often arise the problem or the situation. However, they are not always obvious. This is why methodical support is essential for verbalising and finding goals, especially in complex development projects [17].

3.4 Selection of Methods

Methods need to be applied according to the situation and in a flexible, productive way. Therefore, methods for certain objectives of sustainability have to be singled out. The operator needs to be assisted with choosing and adapting the right method and guiding the designer in using the method. Often these methods are quite superficial, and the operating instructions are insufficient.

3.5 Pareto Principle

Compared to the challenges during product development, objectives of sustainability seem to be less urgent and important. Although the environmental consciousness of the customers increases, it is still the price playing the major role in the purchase decision [18]. Measures aiming for sustainability need to be cost-efficient and effective in order to lead to a positive cost-benefit ratio. The Pareto principle states that often only 20 % of the causes are needed to reach 80 % of the effects. According to this principle, small adaptions of a product may already reduce the negative impact on ecological and social systems considerably.

4 Framework

Taking into account the challenges described in the former chapter, we propose a framework for supporting the development of sustainable products.

Just as within the Three-Layer-Model of Giapoulis [19], the planning of the development starts with the consideration of the entire system and the definition of strategic problem fields (problem layer) on the abstract problem layer. On the basis of this problem definition, specific sustainability goals are formulated within the goal layer (goal layer). Therefore, the desired sustainability characteristics are chosen from a list of different criteria, and their desired characteristics are specified. As every sustainability property may be achieved by certain methods and design principles, the goal layer also specifies the operative activities. The respective methods are applied on the action layer, and thus, in this layer the results are generated, e.g. in the form of product ideas, product functions, sketches or instructions. Figure 2 shows where and how the five control levers touch within the Three-Layer-Model.

4.1 Problem Layer

The problem layer describes the holistic and strategic view onto the topics sustainability and product development within a company. Here, the goal is to improve the understanding of the problem and the system in order to enable successful development. A lack of understanding or a wrong understanding of the greatest sources of unsustainability may lead to wrong sustainability goals and thus endanger the original goal, which is the development of a sustainable product. To avoid that, there are several guiding questions with hints for their answering offered on the problem layer. Discussing these guiding questions is the basis for a successful development of a sustainable product. The problem layer provides the description of a sustainable and an unsustainable product life cycle and supports the designer in finding possible weak points by comparing the current state to a negative or the optimal state.

4.2 Goal Layer

On the goal layer, the sustainability goals of the regarded product are defined. Modelling the goals enhances the developer's understanding of the problem, gives a certain direction to the search for solutions and allows to rate the solutions as well as the project success. However, the product property "sustainability" is quite vague and subjective. The sustainability of a product is often rated by means of the respective CO_2 emissions, its energy efficiency or its content of contaminants.

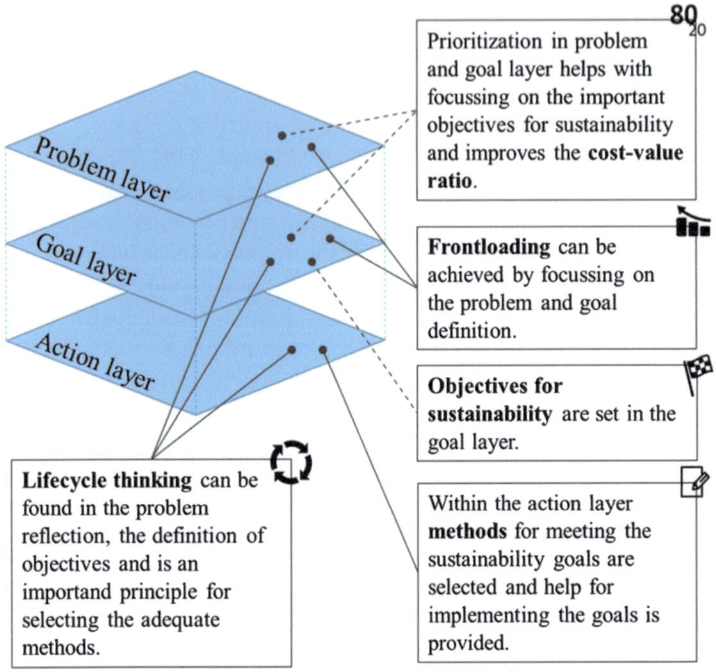

Fig. 2 Challenges addressed in the three layers

However, a product that has a poor energy efficiency may nevertheless be sustainable, if the respective energy exclusively originates from renewable energy sources. In the same manner, a very energy-efficient product may be quite unsustainable, if there is the need for a great amount of energy from fossil resources for the provision of the required raw material. In order to enable a structured and comparable goal planning, the product property "sustainability" is thus described with the help of measurable criteria. The goal layer helps in refining the problems and turning them into goals. For this purpose possible sustainability goals taken from literature are linked to the problems and are combined with measurable criteria. The designer is supported in selecting the goals. So, the goal layer is the link between problem layer and action layer.

4.3 Action Layer

On the path from the idea to the final product, the action layer contains all activities that are required to fulfil the previously defined goals. Development takes place on the action layer after having identified all negative influences of a product and having defined the respective sustainability goals on the goal layer. A possibility for reaching goals systematically is the use of methods. The action layer's purpose is to offer a selection kit, which suggests adequate methods and design principles to the developer in dependency of the respective sustainability goals. This kit includes a set of operating instructions telling the designer how system from business model up to the product structure should be designed. So, finally the methodical supports the designer in eliminating the sustainability problems and achieving the desired goals.

5 Use Case

A Frappuccino machine was developed within a student research project at our institute. The machine is modular and was derived by the analysis of customer demands. The customer is able to configure the functional range of his individual machine by selecting the desired units. The units needed for a Frappuccino are the coffee unit, the ice crusher and the syrup unit. Further units are only optional. Thus, the project focused on the ice crusher and the syrup unit. To augment the sustainability of the first prototype depicted in Fig. 3, the framework was applied.

As a first step, the impacts on ecological and social systems by the machine were identified at the problem level. This was carried out by collecting the information in the model of an unsustainable product life cycle by means of presentation cards. The most important deficiencies derived from this method were the high energy consumption of freezing and boiling water, the material intensity and the poor recyclability.

In the second step, the so-called objective level, the criteria material, energy and recycling were selected based on the identified deficiencies. The goals were to reduce energy and material consumption as well as the amelioration of recyclability.

In the third step, the solution space is generated. The Frappuccino machine delivers the ingredients espresso, ice, milk, and syrup in one unit and mixes them in a separate unit. Any additional element provokes higher energy and material consumption and thus higher costs. Furthermore, the recyclability is decreased. The functionalities "mixing ingredients" and "crushing ice" can be carried out by the same unit. Instead of mixing espresso, crushed ice and syrup in a separate unit, one could also deliver uncrushed ice, espresso and syrup to a milling unit. This unit crushes the ice and mixes the ingredients at the same time. As a consequence, the mixing unit could be omitted, reducing energy and material consumption as well as cost.

An idea to reduce energy consumption could also be deduced: The functionality "freezing water" uses energy and causes disadvantageous heat. By contrast, the functionality "boiling water" requires heat. If the machine generated heat in one unit and needed heat in another unit, one could save energy by smart conducting the heat.

On the action layer, a DfX guideline for integral construction was suggested, which helps combining the two modules and implementing the first idea. The

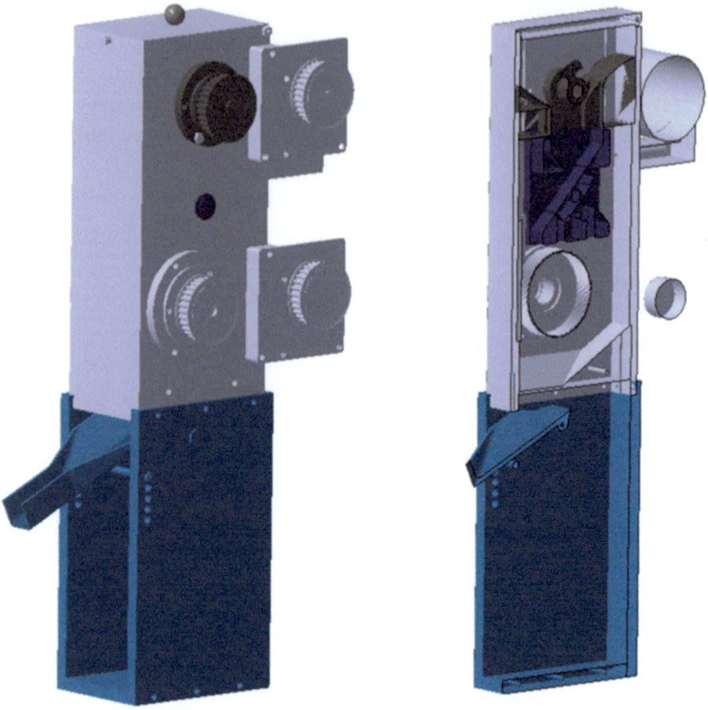

Fig. 3 CAD drawing of ice crusher module

second idea of redirecting the energy flows can be realised by systematically changing the course of the pipes.

6 Conclusion

Within product development, most sustainability properties are defined, and out of it arises an essential responsibility for the redesign of conventional products into sustainable ones. But methods for sustainable product development are often considered as a burden by companies because of additional costs and development time. Additionally, the methodical support often is superficial and lacks of specific operating instructions. Sustainability efforts mainly concentrate on the late development phases, such as product marketing, or the rating of products via life-cycle assessments. However, the efficiency of these subsequent actions is limited according to the rule of ten. The aim of this article is to support the development of sustainable products within companies by setting the focus on the early development phases.

At the beginning, a first descriptive study was carried out. On the basis of an extensive literature study, methods and frameworks regarding sustainable product

Table 1 Strengths and weaknesses of the suggested framework

Strengths	Weaknesses
Understanding and goal modelling of the property "sustainability"	Use in industry is not yet proven
	Insufficient consideration of the supply chain
Support for the creative search for sustainable solutions	Lack of aid for selecting of goals
Incorporation of established methods	Single layers not yet linked sufficiently with each other

development were selected and analysed. Upon the identified strengths and weaknesses of these concepts, five challenges for a method supporting the development of sustainable products were derived. Within the following prescriptive study, the concept for the Three-Layer-Model for developing sustainable products was proposed based on these challenges. On the problem layer, an example of a model of a sustainable product life cycle shows how a company can handle sustainability issues systematically. On the goal layer, the product property "sustainability" is divided into sub-goals. This allows for an early planning and a subsequent rating of the respective sustainable product properties. Furthermore, every sustainability goal can be expressed by means of measurable indicators. The action layer suggests methods and design principles in order to reach the sustainability goals that are previously defined on the goal layer.

The results were discussed in consultation with experts within industrial and academic environments and thereby acknowledged positively. The results are summed up in the following (Table 1):

7 Outlook

A first step towards a support for designing sustainable products was made. In further research the three single layers have to be combined so they interlock better. Besides, a better support in selecting the goals has to be provided. Also the method tool kit has to be amplified and filled with further methods for ecodesign. A very important point that has to be dealt with in the future is the deduction of action instructions based on ecodesign methods for product designers for incorporating sustainability into the product. Finally the approach has to be applied in industry.

References

1. Faludi J (2015) A sustainable design method acting as an innovation tool. In: ICoRD'15–research into design across boundaries, vol 2. Springer, New Delhi, pp 201–212
2. Ehrlenspiel K, Kiewert A, Lindemann U, Mörtl M (2014) Kostengünstig entwickeln und konstruieren. Kostenmanagement bei der integrierten Produktentwicklungen, 7th edn. Springer, Berlin. ISBN 978-3-642-41958-4
3. Pure Strategies (2014) The path to product sustainability. A Pure Strategies Report. http://info.purestrategies.com/path-to-product-sustainability. 26 Feb 2014
4. Beucker S, Bierter W, Fichter K, Noack T, Springer S (2006) Nachhaltigkeitskonzepte für Innovations-prozesse. Fraunhofer-IRB-Verlag, Stuttgart. ISBN 3816770401
5. EU-Kommission (2014) Integrierte Produktpolitik. http://ec.europa.eu/environment/ipp/home.htm. 6 Feb 2014
6. McDonough W, Braungart M (2002) Cradle to cradle – remaking the way we make things. North Point Press, New York. ISBN 9780865475878
7. The Natural Step (2015) The Natural Step's website. http://www.naturalstep.org/. 10 Mar 2015
8. Byggeth S, Broman G, Robèrt KH (2007) A method for sustainable product development based on a modular system of guiding questions. J Clean Prod 15:1–11
9. McAloone T, Bey N (2009) Environmental improvement through product development. A guide. Danish Environmental Protection Agency, Kopenhagen. ISBN 978-87-7052-950-1
10. Panarotto M (2011) Creative methods for sustainability driven innovation. Master's Thesis
11. Masui K, Sakao T, Kobayashi M, Inaba A (2003) Applying quality function deployment to environmentally conscious design. Int J Qual Reliab Manag 20:90–106
12. Rahimi M, Weidner M (2002) Integrating Design for Environment (DfE) Impact Matrix into Quality Function Deployment (QFD) process. J Sustain Prod Des 2:29–41
13. Dong C, Zhang C, Wang B (2003) Integration of green quality function deployment and fuzzy multi-attribute utility theory-based cost estimation for environmentally conscious product development. Int J Environ Conscious Des Manuf 11:12–28
14. Vincent JF, Bogatyreva OA, Bogatyrev NR, Bowyer A, Pahl AK (2007) Biomimetics: its practice and theory. J R Soc Interface 3:471–482
15. Lindemann U (2003) IPP Integrierte Produktpolitik. Druck+Satz Grafex, Feucht
16. Lindemann U (2009) Methodische Entwicklung technischer Produkte – Methoden flexibel und situationsgerecht anwenden. 3. korrigierte Aufl. Springer, Berlin. ISBN: 978-3-642-01423-9
17. Haberfellner R (2012) Systems Engineering. Grundlagen und Anwendung, 12th edn. Orell Füssli, Zürich. ISBN 978-3-280-04068-3
18. Gassmann M (2014). Den Deutschen fehlt das grüne Gewissen, http://www.welt.de/wirtschaft/article131589087/Den-Deutschen-fehlt-das-gruene-Gewissen.html. 29 Mar 2015
19. Giapoulis A (1998) Modelle für effektive Konstruktionsprozesse. Shaker, Aachen. ISBN 978-3826535543

Reducing Conflicts of Interest in Eco-design: The Relation of Innovation Management and Eco-design in the Automotive Sector

Therese Elisabeth Schwarz, Kerstin Schopf, and Astrid Arnberger

Abstract Product development lately has to deal with economic, technical, consumer-related and environmental issues. Environmental concerns owed to changing climate, local natural catastrophes and lack of resources are pushing into the foreground. Eighty percent of a product's environmental footprint is determined in the design phase. Eco-design, integrating environmental matters while improving the environmental performance, is familiar to industries but rarely applied, hemmed by various barriers and obstacles. Conflicts of interest emerging in the R&D phase impair considering and enforcing a switch to eco-design. Conflicts are produced by different legal and normative frameworks, aims of corporate departments or product requirements. A closer look at two life-cycle phases (product development and end of life) helps understand eco-design in automotive battery development.

Keywords Eco-design • Future waste • Innovation • Management • Batteries

1 Introduction

Keeping an eye on life cycles turns the R&D phase of products and associated processes into a challenge to companies and their suppliers. Characteristics of emerging product lines also cover environmentally compatible design, besides economic efficiency and functions. Requirements of target groups and the framework of standards force companies to improve their competitiveness on the market. To handle this, innovation management is integrated in the company's strategic framework. But eco-design is not. Why not?

T.E. Schwarz (✉) • K. Schopf • A. Arnberger
Chair of Waste Processing Technology and Waste Management, Montanuniversitaet Leoben, Leoben, Austria
e-mail: therese.schwarz@unileoben.ac.at

M. Matsumoto et al. (eds.), *Sustainability Through Innovation in Product Life Cycle Design*, EcoProduction, DOI 10.1007/978-981-10-0471-1_3

2 Eco-design in Strategic Management

Johannson [1] describes eco-design as reducing the environmental impact of a product's life cycle while preserving or even improving other attributes like price and function. So the aim of eco-design is to streamline a product in various directions.

The purpose of this paper is to discuss the extensive introduction of the principles of eco-design in strategic and innovation management. Basic methods applied in both management areas are reviewed and transferred to the automotive battery production sector, used as a case study.

Strategic management, being a part of business economics, is closely tied to product policies and the main goals of a company. It aims at an encompassing view of a company, while eco-design strives for an encompassing view of a product. Various methods used in the field will be considered and discussed in the following, focussing on eco-design.

An important tool of strategic management is SWOT analysis. Four main targets are addressed: strengths, weaknesses, opportunities and threats [2].

A company's strength, as it emerges from supporting eco-design, might be identified as a public image of taking environmental care and setting goals for sustainable product design, ranging from the senior management down to individual projects. It aims at a cooperative relationship with stakeholders and a detailed internal and external information exchange concerning environmental topics [1]. Some internal or external drives for eco-design may be found in almost every sector: reducing costs, better quality, the company's image, committed staff, environmental benefits, corporate governance, requirements of the market and social responsibility [3].

This multi-stakeholder and multidimensional process implies various conflicts of interest between departments, product characteristics and methods. The 'eco-design' topic may be interpreted by departments and their teams in different ways (weakness). Conflicts emerging from product development are the greatest challenge to designers. Eco-design recommendations have to be analysed for safety, quality and proper functioning of different product life stages. Any conflicts should be discussed by experts or intended tools be applied to support decision-making and prioritisation. As a consequence, familiarity with eco-design features, rules and goals is an essential starting point for promoting these innovative approaches.

Opportunities for effective product development are the individual legal and normative provisions that establish a framework for supporting activities in eco-design and promoting companies that are already practising them. Legal regulations are applied to ever more product groups, making the producer accountable not only for matters related to end-of-life topics but also for the extraction, treatment and usage of raw materials. The Directive 2005/32/EC emphasises various energy-relevant matters like conserving energy by stand-by processes, lighting, external power supply, etc. This includes the organisation, such as taking responsibility for environmental data and safe handling of products by importers, benchmarking or CE labelling [5]. Other product group guidelines for eco-design have been derived from this directive. The European Directive 2006/66/EC covers the responsibility of the 'batteries and accumulators' product group [4]. These and other frameworks have inspired manufacturers to take recycling and end-of-life costs into account which they have to pay in advance. The European Union intends to reduce the overall consumption by 365 TWh – a full 12 % of the EU's total power consumption in 2009 [6].

Eco-design is often considered a 'welcome secondary benefit'; therefore, buyers do not want to pay for its more expensive development. Developing 'green' products has remained a market niche of the automotive industry for the last few decades. Concerning references on management, 'Porter's five forces' [7] emphasise the threats companies are exposed to, affecting both the market share and the corporate environment. These four threats create opportunities for flexible and innovative companies: (1) Threat of new market entry, (2) substitution, (3) buyer and supplier power and (4) competition. Battery-powered mobility and especially battery development are very competitive markets although only a few battery-powered vehicles have been sold so far. This competitive environment is stimulated by battery manufacturers entering the automotive market (threat of new market entry).

Another phenomenon is substitution. Products that do not excel against competing products tend to drop out of the market. With regard to batteries, cell chemistry is their essential characteristic. Lithium-ion batteries are favoured by R&D against other chemistries because of their higher intensity and current. Modern compounds like lithium-cobalt-oxide (LiCoO 43 %), lithium-nickel-manganese-cobalt-oxide (NMC 32 %) and lithium-manganese-oxide (LMO 11 %) are widespread [8]. Metal-air chemistry and different kinds of alternative fuels (fuel cell, natural gas, etc.) will be closer considered in the future [9]. As a result, focusing on the development of only one power-generating system limits the access to future automotive markets.

The balanced scorecard is a tool for measuring the performance that has been developed by Kaplan and Norton [15]. Its purpose is to assess four different performance categories and to combine them in a possible future analysis. The financial view involves different business indicators relevant for the shareholders. Other parameters of the tool cover customer and corporate views. Corporate processes and customer satisfaction are analysed. The 'organisational learning' category gives an outlook on the future and ties to social innovation. The non-market view and the sustainability assessment that have been added to the original parameters at a later stage include further characteristics to obtain an encompassing insight into the company's performance. Figure 1 shows how the environmental parameter is integrated in the present categories and allows analysing economic and other quantitative factors [16]. Concerning eco-design, the goals of product development should be entered in the individual categories. A non-market-related goal might be to reduce greenhouse gas emissions of the next generation of batteries by 10 %, or a financial goal might be to decrease cost by saving on material and energy during the production. This is the foundation on

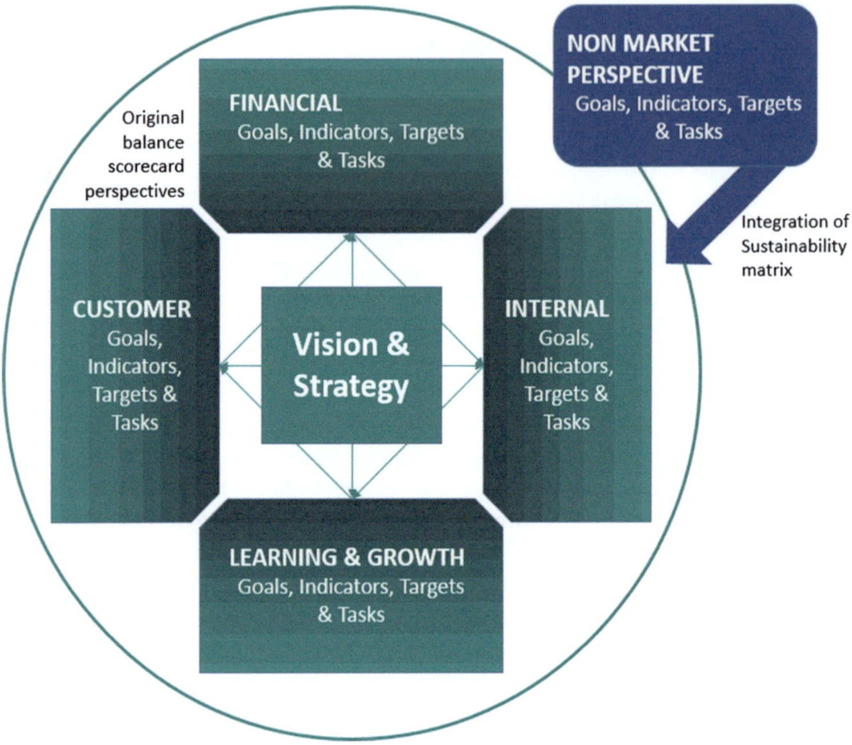

Fig. 1 Balanced scorecard with integrated sustainability (According to Ref. 16)

which concise goals, indicators like energy-saving targets and tasks for implementation, are defined. The balanced scorecard with added sustainability shows options for how to introduce environmental goals in management, much like other business targets.

Senior management and PICs should accordingly define environmental goals for the company and its projects. The environmental matters should be integrated in the corporate strategy and be treated and rewarded like any other business subjects. On the project level, environmental milestones support commitment and priorisation. A workshop should be run by an external moderator and involve an eco-design expert lobbying environmental tasks. Another important matter is the multidimensional and interdisciplinary thinking of all project members and their commitment to innovation and change [1].

References reveal that environmental issues and the supplier chain management are changing courses as new performance indicators, control systems and data requests are implemented [12].

Buyers and final-product manufacturers significantly affect the development of products. The original equipment manufacturers (OEM) in the automotive industry ask suppliers (of all levels) to report the compositions down to their elements. Parts data, construction plans and other relevant information are uploaded to the

International Material Data System (IMDS) [10] that is synchronised with the Global Automotive Declarable Substance List (GADSL) [11]. Recovery concepts, life-cycle data and measures to reduce energy (fuel) consumption are required. Such control mechanisms force companies to establish data collection systems and to consider all life-cycle phases (which is another opportunity for eco-design).

Collaborating with suppliers and experts on environmental concerns can help achieve an interactive development of products beyond a mere specification sheet. The main task is to inform suppliers about criteria and activities needed for eco-design and to simplify and customise the tools for everyday use [1]. Designing an environmentally compatible product implies collaborations with stakeholders and the supplier management as an essential part of components design. Skills and experience of corporate staff should reduce any direct environmental impact that may occur during the life cycle and resulting problems with designing. Evidently, the starting point of the design process in the early development phase should be the smallest unit. Deciding on materials, shape, processes and functions is basic to the very beginning of designing [1].

A couple of interviews and workshops have helped identify the teams that are relevant for decision-making and the design process of a battery pack. It is essential to understand the design process applied in the automotive industry or battery development which is divided into four phases.

1. A-phase: Ideas for the product are collected and initial workshops on the basic customer demands and functions held. A prototype is developed and constructed but often out of parts that have not been obtained from the final supplier, and the ultimate material, shape, etc. may not yet have been defined. The purpose of this prototype is to plan the functions.
2. B-phase: This prototype helps shape ideas in detail. In this phase, meetings with suppliers and decisions about forms and materials are obligatory. A second prototype that includes most of the final properties is produced.
3. C-phase: The process is determined and will be streamlined. Different parts of the battery system are produced, applying mass production procedures, and initial practical use tests are performed.
4. D-phase: The complete battery is made in batch production with all product characteristics. The characteristics are streamlined and additional concepts developed, like after sales or recycling concepts.

Especially in the A-phase, the knowledge and expertise of project engineers, simulation directors, constructers and electric engineers contribute significantly to improve the product. The technical draughtsmen and industrial engineers provide gateways from planning to implementation and should be trained in eco-design methods. Two phases emerge from eco-design decisions: concept and layout design [20]. Innovation management takes a closer look on the corporate procedures of product design.

3 Eco-design in Innovation Management and Product Development

Handling innovation is a basic parameter of developing sustainable products, besides strategic decision-making or market and product policies. From the innovative point of view, factors like technology push or market pull are crucial for making decisions. Every company unit has different methods of how to set, reach and assess its goals. The ability to interact and collaborate is required in any corporate structure. According to Cai and Zhou [13], internal and external drives are creating integrative powers, encouraging eco-innovation. Eco-innovation is defined by the OECD as the result of two characteristics: an emphasis on the environmental impact and the development of products and processes with their life cycle standing in the foreground and another on social innovation and new institutional structures [14].

Innovation management distinguishes incremental and radical innovation. Incremental innovation means inventions related to an existing product that provide little innovative capacity. Some parameters are modified but the functions and basic characteristics remain the same. The second type of innovation is more fundamental: any product characteristics, functions or even target groups may be altered. Usually, these radical innovations are not market pushed: the customers cannot specify their needs and expectations with regard to properties of the new product [17]. Robert Cooper [18] describes the Stage-Gate model that introduces innovation management in common product development. Six different stages have been defined that are separated by gates, each of which is charged with reviewing the previous development process and providing exit options. Repeated reviews of the innovation process could make eco-design a factor in the gate's decision. Including eco-design criteria in the concept will focus on transparent decision-making and the active participation of various interdisciplinary employees. On the other hand, the Stage-Gate process supports the submission of eco-design criteria and goals to other PICs and also forces project members to introduce eco-design into early stages. Eco-design should in fact be integrated in the primordial designing process. Decisions on environmentally compatible products are made in meetings, workshops and the proper designing process (drawing, materials and components). Discussing and informing about life cycles on the approach might contribute here, even though eco-design goals should be isolated and taken very serious by the designers.

Various design areas and product characteristics, like those listed in Table 1, can be integrated in the Stage-Gate process to introduce eco-design criteria into the development phase; see Fig. 2.

Identifying eco-design issues may look easy but is hard to achieve. Topics that cover the main issues of eco-design are suggested by Luttropp and Lagerstedt [19] in an extended checklist; see Table 1. Many other checklists and qualitative tools can help identify critical points at early development stages but often seem to be very unspecific.

Table 1 Criteria for eco-design actions and associated design phase [19]

Eco-design areas in product design	Automotive design phase
Reducing material (weight), energy and resources, material substitution (quality and purity) or surface treatment	A, B, C
Choosing non-toxic and nonhazardous substances	B
Increasing reuse/remanufacturing, recovery, any kind of waste treatment, design for long life and disassembly	B, C
Enforcing the use of renewable materials and the reduction of fuel consumption/transport	B, C
Reducing components and joining elements or excipients	C, D
Promoting structures for easy housekeeping, upgrading and repairing (modular)	A, B, C
Involving life-cycle stages, impact and costs, circular economy (closed loop)	A, B, C, D
Clarifying and identifying materials and streamlining processes and transport	C, D
Adding functions to a product/selling the service	A, B
Enabling the consumer to correctly use and dispose of the product	C, D

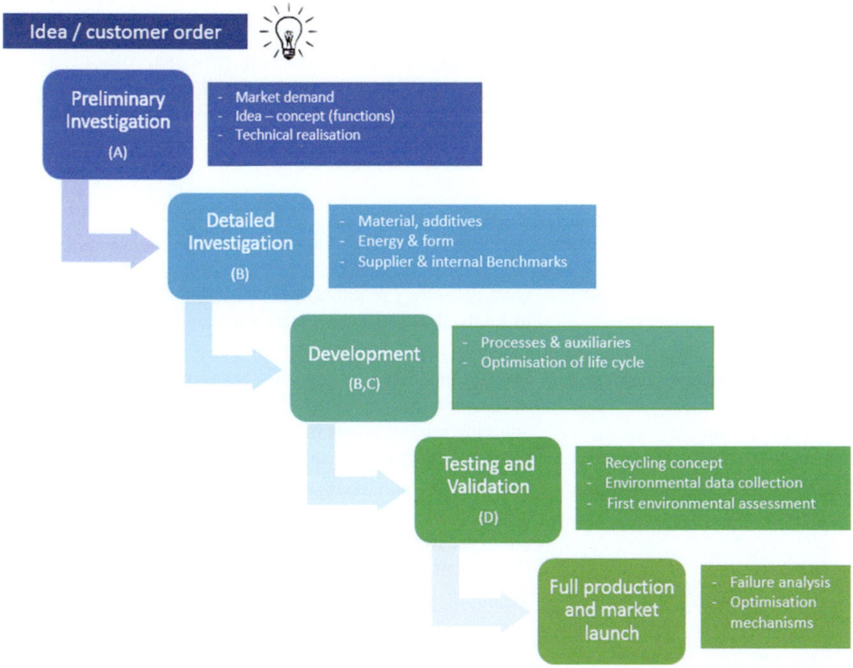

Fig. 2 Stage-Gate process and eco-design (According to Ref. 18)

Hence, methods like life-cycle assessment (LCA) or failure mode and effects analysis (FMEA) are effective tools for detecting environmental hazards on low levels. Even complex product parts can be examined and an encompassing view of the life cycle and any failures, hazards or obstacles be identified before a specific concept is accepted.

Checklists and internet tools, however, can be easily adopted by designers. Obviously, industry tools should be intuitive and flexible yet more precise and customisable than many of the present software tools are. Eco-design tools and checklists need to be customised to match the company's activities and to cover relevant topics and examples.

Even though the eco-design tool is intuitively structured, it ought to be supported by company members. Training employees in eco-design does not prevent conflicts from arising in the development process, and these have to be identified. Potential areas of conflict can be safety/health, quality, mass production, product functions, design issues, legal/normative framework, consumer requirements, financial criteria and marketing or other eco-design criteria. There is often a trade-off, but sometimes, especially when normative and legal requirements interfere, eco-design recommendations have to be withdrawn.

Concerning the design of vehicle batteries, these conflicts of interest may arise from different causes. Looking at the first development phases (A, B), conflicts of interest emerge either from technical solutions, from organisational requirements or from quality parameters. At later stages of the design process, other factors increase: financial aspects, shaping, presentation design, customer requirements, etc. Table 2 shows selected examples of conflicts.

Table 2 Selection of eco-design recommendations and conflict areas

Eco-design rules/criteria	Conflict area examples
Reducing structures – similar joints	Quality – safety requirements warrant, due to different kinds of joints, that production staff will assemble the product correctly and not suffer any injury
Purity of material – facilitating recovery	Product functions – composites are commonly used because of their low weight, stability and other characteristics, but applying several layers means that recovery is difficult and the material should be separated from other components
Streamlined structures – reducing functions	Decisions by purchasing/marketing – purchasing may want to reduce overall costs and decline a supplier with a perfect but expensive component or marketing may request to add a component that improves the look of the product
Using renewable materials	Technology and functions – if renewable materials do not allow meeting the technical requirements
Using non-toxic substances	Mass production and functions – many process steps require toxic substances as adhesives
Design for reusing, remanufacturing and recovering	Safety and functions – components and product parts have to be solidly fixed

4 Eco-design in the End-of-Life Phase

Products that include hazardous and/or a quantity of different substances and have a long life cycle (causing more future amounts of waste) are defined hereafter as 'future waste'. These product groups may have complex structures that prevent their submission to material recovery treatment. A few examples are vehicle batteries, solar panels, wind turbines, etc. [21].

The European Union is setting higher quotas for the recovery of 'future waste'. The end-of-life vehicle Directive 2013/28/EU by the European Parliament and Council [22] stipulates reuse and recovery targets. Since January 2015, at least 95 % of the average mass of all end-of-life vehicles per year must be recovered or reused. According to the current legislation, a removed battery is classified as recovered. Incineration does not constitute material recovery here but energy recovery. The stipulated recovery quotas for the whole car define percentages of material and energy recovery.

These legal constraints place a stronger focus on battery parts. The recycling quotas for different battery types, covering material recycling without incineration, are listed [23]:

- Sixty-five percent for lead-acid batteries
- Seventy-five percent for nickel-cadmium batteries
- Fifty percent for other battery types

The European Regulation 493/2012 defines the recovery of batteries. A clear definition of the battery parts included in the calculation should follow, however [24].

Latest high-tech innovations are more complex than before, due to an increasing number of included functions, substances and design requirements. Battery packs for cars, motorbikes or bikes consist of different modules of power cells, electrical components, cables, housing, cooling and many other parts.

The main parts of a battery are the cells – according to the size and task of the battery, there may be less than ten cells or more than 8000. Different compositions and shapes (prismatic, cylindrical or pouch shaped) demand different requirements from eco-design. This paper is not meant to discuss the design and properties of cells; however, this part contributes a lot to environmental loads, especially in the production and recycling phases.

When designing a product that ought to be recoverable, the proper materials have to be decided about. Metals have a high economic value and contribute much to a battery pack (housing, cells). The main problem with recovering metals is how to purify the input fraction or at least to obtain a composition that the industry accepts. Like, a merger of aluminium and steel implies a loss of value and difficult separation. Another important group of components are electronic and electric parts: besides metals, polymers like polyamide contained in electronic and electric equipment are substantial secondary raw materials that should be recovered to reach the recycling quota. Furthermore, different output fractions (electrolyte, critical raw materials,

Table 3 Material compositions and component groups of the battery systems of different electric vehicle types (Data: battery producers)

Material	PHEV	Motorbike	Bike
Aluminium	16.1 %	26.2 %	10.7 %
Steel/iron	13.4 %	1.3 %	2.1 %
Copper	2.7 %	1.6 %	3.2 %
Cells	58.9 %	60.7 %	71.0 %
Electronic parts	Included in other categories	1.9 %	7.2 %
Plastics	2.3 %	7.6 %	5.7 %
Others	6.6 %	0.7 %	0.0 %

etc.) cannot be treated to obtain high-quality fractions for battery-to-battery recycling.

Table 3 shows different materials and components and their composition. The following data have been collected from various research projects performed by the Montanuniversitaet Leoben. The size of the battery and its number of cells determine the vehicle type; however, decisions about the materials are distinct from each other.

Integrating pouch cells (covered with a plastic envelope) reduces the share of metals. On the other hand, manufacturing processes may also predetermine the use of materials. Welding, adhesive bonding or screwing demand distinct excipients. Altogether, the materials of a battery produce direct and indirect emissions.

The chemistry of cells is not the subject of this paper. Labelling its chemistry, plastics and metals would increase the recyclability and help prevent quality and safety problems emerging from recovery.

The varying constructions, shapes, numbers and compositions of batteries may mean trouble for a common eco-design or recovery concept. Table 4 compares different kinds of batteries and their characteristics.

Various battery constructions impair the recovery of secondary raw materials and energy. Considering how the product may be disassembled for recovery can help understand the problems of repairing, reusing and remanufacturing and even provide important suggestions for future product generations.

The International Disassembly Information System (IDIS) [25] shows disassembly methods for different cars. The main purpose is to remove plastics, batteries and other parts that may interfere with subsequent treatments like shredding. How to disassemble the batteries after separation is not discussed in any detail, though. To cover this subject, various research projects managed by the Montanuniversitaet Leoben have aimed at developing an improved disassembling concept for selected batteries. It had been essential to apply a bill of materials (BOM) and different tools for separation, testing and recording (hammer, screwdrivers, voltmeter, cameras, etc.). First, every step has been recorded by two cameras taking video and image frames, including background data like time required, tools, the sequence of disassembly and the safety equipment. The disassembly experiments have been assessed in (1) a presentation with description, photos and videos of each step; (2) a

Table 4 Battery characteristics (Data: battery producers)

Comparison of different battery types							
	PHEV 1	PHEV 2	EV 1	EV 2	E-motorbike	E-bike 1	E-bike 2
Battery technology (chemical)	LFP	LFP	NMC	LMO	NMC	NK	NK
Weight [kg][a]	100	120	250	290	25	4	4
Modules	4	3	5	48	3	1	2
Cell type	Metal can	Metal can	Pouch	Pouch	Cylindrical	Cylindrical	Pouch
Number of cells	88	120	180	192	360	70	20
Battery layout	Under the boot	Under the boot	Under seat and floor	Under seat and floor	Under seat	Under seat	Under seat

NK not known, *PHEV* Plug-In Electrical Vehicle, *EV* Electrical Vehicle, *E-* Electrical
[a]Rounded values

spreadsheet containing information on disassembly levels, separation procedures, joints, masses, etc; and (3) a disassembly tree showing the sequence of disassembly. Labelled disassembly points in the battery pack may warrant safe and quick disassembly.

Another important eco-design factor is the procedures involved. Not only production but also recovery procedures have to be considered. Note that manufacturing steps like coating facilitate the production but impair the recovery. A full separation of the materials may not be economically feasible, while a blend of materials cannot be sold on the raw material market. This evidence suggests that eco-design will be successful only if the best compromise is identified and accepted.

5 Discussion and Outlook

Eco-design accesses a life-cycle view of products and includes non-technical and noneconomic issues. Eco-design goals and rules need to be integrated in strategic and innovation management. Combining management methods with eco-design highlights difficulties and obstacles while showing benefits of the idea. This study emphasises a combination of methods and shows its effects on automotive battery production. The purpose has been to define obstacles that hamper the introduction of eco-design and to develop concepts for how to avoid these by using original management tools.

Eco-design in R&D processes reveals conflicts of interest deriving from product characteristics or company departments. These conflicts should be taken into account when eco-design is introduced. Using the adapted checklist to reduce conflicts of interest helps identify the relevant teams and emphasises topics that should be discussed in workshops. Conflicts of interest in different eco-design products will have to be discussed in case studies to derive a complete checklist and a tool to overcome these obstacles. In the case of battery development, the complex product characteristics will often adversely affect eco-design processes.

The Montanuniversitaet Leoben has tried in various projects to make the industry familiar with these efforts. Disassembly experiments and the life-cycle assessment of batteries have shown how product characteristics, structures and processes can be improved. Eco-design may be introduced in companies once available tools are flexible, customised and intuitive and once sufficient know-how and commitment are available (especially on the senior management level).

References

1. Johannson G (2002) Success factors for integration of eco-design in product development. Environ Manag Health 13(1):98–107

2. Harvard Business Review (2007) SWOT analysis. https://hbr.org/product/swot-analysis-i-looking-outside-for-threats-and-opportunities/an/5528BC-PDF-ENG. Accessed 19 May 2015
3. ADAM –European Union projects: innovation and EcoDesign in the ceramic industry. In: ED-IC Ecodesign manual – investigation of the motivating factors for eco-design. URL: https://www.adam-europe.eu/prj/5887/prd/1/4/Tool%201_Motivating%20factors.pdf. Accessed 18 Mar 2015
4. European Union (2006) Directive 2006/66/EC of the European parliament and of the council on batteries and accumulators and waste batteries and accumulators. In: Off J Eur Union nr. L266/1 from 26 Sept 2006
5. European Council (2005) Directive 2005/32/EC on establishing a framework for the setting of ecodesign requirements for energy-using products. In: Off J Eur Union nr. L191/29 from 22 July 2005
6. European Union: adopted eco-design implementing measures to reach energy savings by 2020. URL: http://ec.europa.eu/enterprise/policies/sustainable-business/ecodesign/product-groups/index_en.htm#h2-2. Accessed 17 May 2015
7. Porter M-E (1979) How competitive forces shape strategy. Harv Bus Rev
8. Avicenne (2012) The worldwide battery market. URL: http://www.kigeit.org.pl/FTP/PRCIP/Literatura/063_The_worldwide_battery_market_2011_2025.pdf. Accessed 12 June 2015
9. Armand M, Tarascon J-M (2008) Building better batteries. Nature 451:652–657
10. International material data system (IMDS). URL: http://www.mdsystem.com/imdsnt/startpage/index.jsp. Accessed 24 May 2015
11. Global automotive declarable substance list (GADSL). URL: http://www.gadsl.org/. Accessed 24 May 2015
12. Srivastava (2007) Green supply-chain management: a state-of-the-art literature review. Int J Manag Rev 9(1):53–80
13. Cai W, Zhou X (2014) in the drivers of eco-innovation: empirical evidence from China. J Clean Prod 79:239–248
14. OECD (2009) Sustainable manufacturing and eco-innovation: framework, practices and measurement. In: Synthesis report. OECD, Paris
15. Kaplan RS, Norton DP (1992) The balance scorecard – measures that drive performance. Harv Bus Rev
16. Möller A, Schaltegger S (2005) The sustainability balanced scorecard as a framework for eco-efficiency analysis. J Ind Ecol 9(4):73–83
17. Strebel H. (2007) Innovations- und Technologiemanagement, 2nd edn. Facultas wuv, UTB
18. Cooper RG (2008) Perspective: the stage-gate idea-to-launch process – update, what's new, and NexGen systems. J Prod Innov Manag 25:213–232
19. Luttropp C, Lagerstedt J (2006) EcoDesign and the ten golden rules: generic advice for merging environmental aspects into product development. J Clean Prod 14(2006):1396–1408
20. Eco-Design Workshop with industry partners, 2015-06-22 in Graz, Austria
21. Pomberger R, Ragossnig A (2014) Future waste – waste future. Waste Manag Res 32(2):89–90
22. European Union (2013) Directive 2013/28/EU of the European Parliament and of the council on end-of-life vehicles. Off J Eur Union Nr. L 135/14 as of 17 May 2013
23. European Parliament and the Council (2006) Directive 2006/66/EC on (waste) batteries and accumulators. URL: http://eur-lex.europa.eu/legal-content/DE/TXT/PDF/?uri=CELEX:32006L0066&from=de. Accessed 19 June 2015
24. European Commission (2012) Regulation on Directive 2006/66/EC on detailed rules regarding the calculation of recycling efficiencies of the recycling process of (waste) batteries and accumulators. URL: http://eur-lex.europa.eu/LexUriServ/LexUriServ.do?uri=OJ:L:2012:151:0009:0021:en:PDF. Accessed 19 June 2015
25. International Dismantling Information System (IDIS). URL: http://www.idis2.com/. Accessed 24 June 2015

Computer-Aided Design for Semi-destructive Disassembly

Shinichi Fukushige, Yumi Shiraishi, and Yasushi Umeda

Abstract Disassembly is a fundamental process for component reuse, remanufacturing, and material recycling of all assembled products. In the last decades, design methodologies for disassembly that simplify products' structure, fastener, and disassembling process have been widely explored. However, the shape of components to be recycled or discarded need not be maintained in the end-of-life process. This means that we can separate valuable and/or hazardous components more rapidly by deforming or breaking surrounding components designed for the deformation or destruction process. This article proposes a computer-aided design method for such semi-destructive disassembly. In the detailed design phase of the method, linear shape features that enable to break a product into desired shape of chunks are added to the product geometric model. The semi-destructive process aims at extracting reusable, recyclable, or hazardous components more efficiently than manual disassembly with higher quality than shredding. The system supports a designer in determining the location of the linear features called split lines.

Keywords Design for disassembly • Semi-destructive disassembly • CAD

1 Introduction

In Japan, manual disassembly and machine shredding are cooperatively used in recovery processes of end-of-life (EoL) products [1, 2]. The manual disassembly is a process suitable for extracting valuable or unshreddable/hazardous components before shredding, though it is often a labor-intensive and costly process [3]. In particular, manual operations are inevitable in secure recovery of functional components for reuse and remanufacturing.

S. Fukushige (✉) • Y. Shiraishi
Department of Mechanical Engineering, Osaka University, Osaka, Japan
e-mail: fukushige@mech.eng.osaka-u.ac.jp

Y. Umeda
Department of Precision Engineering, The University of Tokyo, Tokyo, Japan

M. Matsumoto et al. (eds.), *Sustainability Through Innovation in Product Life Cycle Design*, EcoProduction, DOI 10.1007/978-981-10-0471-1_4

However, prior studies have shown that the cost of EoL treatment heavily depends on the time required to disassemble a product [4]. The economic feasibility of disassembly can only be guaranteed if such manual operations are drastically simplified in modern EoL process [5, 6]. Therefore, numerous efforts have been made on the research on design for disassembly (DfD) [4, 7–9] to improve the ability of products to be disassembled at the design stage. These studies mainly focus on the simplification of product structure and disassembly process, assuming that all fasteners of disassembled components are removed.

In this direction, disassembly embedded design (DED) [10] and design for product-embedded disassembly (PED) [11] are promising approaches. DED is an idea of embedding separation features inside a product during manufacturing phase and activating them at disassembly phase. The features remove multiple components with one disassembly action. In PED, locator features of components, instead of conventional fasteners (e.g., screws), are designed to constraint relative motion among the components in a product. The combination of the locators enables "self-disassembly" triggered by removing small number of fasteners. However, the design of the separation features and locators that allows simultaneous disassembling by a trigger has many difficulties in determining the shape and layout of the components [4, 12]. Furthermore, such additional feature design may increase the manufacturing time and cost.

Generally speaking, a product can be disassembled in the reverse order of assembly sequence [13]. However, we do not need to maintain the shape of components to be recycled or discarded in the EoL processes [14]. This implies that we can separate valuable and/or hazardous components (we call such components "targets") more rapidly by deforming or breaking components to be recycled or discarded. This leads to the idea of "semi-destructive" disassembly, in which target components are disassembled without any injury while other components can be broken in order to increase efficiency of the disassembly process.

In this direction, active disassembly (AD) [4, 6, 15, 16] is an emerging technology in which specific external triggers deform fasteners simultaneously and disassemble a product efficiently by using the characteristic of smart materials (e.g., shape memory alloys). At present, this technology is not widely used due to the cost and durability of the fasteners over the time [12].

Cutting and sorting by robots are other possible ways of effective dismantling [17], although it is currently limited because of the wide variety of returned products, which requires a great capacity of intelligence and flexibility in operations [13].

Therefore, a design approach for supporting destructive disassembly by manual operation is addressed in this paper. In other words, semi-destructive disassembly is positioned in the middle of disassembly and full destruction (including shredding and smashing) and combines strong points of both of them.

This paper proposes a computer-aided design system for designing "split lines" that enable to break a product into desired shape in the EoL process, like labels and caps of PET bottles as shown in Fig. 1. The split lines are shape features added to the surrounding components of the targets in the detailed design phase. The

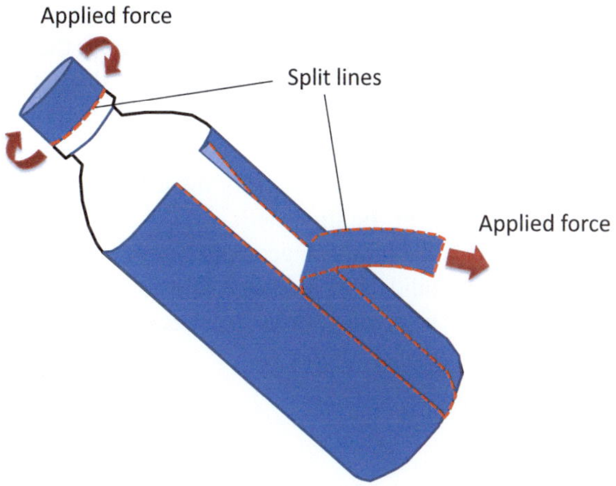

Fig. 1 Example of split lines (PET bottle)

destruction process by split lines releases the targets from geometric constraints in the product. The result of a case study on an air conditioner shows that the redesign of the product shape for semi-destructive disassembly reduces the number of steps needed for extracting the heat exchanger for the aluminum recycling.

2 Computer-Aided Design for Semi-destructive Disassembly

2.1 Outline of the CAD System

The proposed system assumes that a solid model and an EoL scenario [3] of a product are provided. The scenario describes EoL treatment for each component, including manual disassembly, machine shredding, sorting, remanufacturing, final treatment, and so on.

At the outset of the method, the designer identifies target components to be preferentially extracted from the product in the EoL process to recover the components' value, by reusing, remanufacturing, material recycling, or removing hazardous materials, before shredding or smashing the product.

In many cases, such target components are covered by other components and geometrically constrained. We view such structure as an analogy to "kernel and shells" in which the surrounding components (shells) obstruct the extraction of the target (kernel) from the product. Without disassembling the shell components, the targets can be retrieved from gaps or holes of the shells (we call such a space a "window"), which can be made by breaking the shells in line with the split lines. These linear shape features enable to break each shell by applying force to designated points of the shell surface. Each split line divides a shell into two fragments and the peeled one makes a window.

In a detailed design phase, the designer adds a split line to each shell, and the system supports the designer in determining the location of the lines on the geometric model. The rest of this section describes the details of the system: Sect. 2.2 outlines a shell classification algorithm. We assume that the target components are surrounded by multiple-layered shells. In the first step, the system classifies all components except for the targets into layers of the shells. And we call the outermost layer the first shell.

Section 2.3 describes a geometric processing algorithm in the next step for generating a candidate of split line (we call it "baseline"). To assist the designer's decision in this step, the system visualizes a feasible domain to be split and suggests the candidate on each shell.

2.2 Shell Classification

Formally, components included in the first shell S_1 of a product P are defined as follows (see Fig. 2): Given a ray r emanating from any point s in a target component T and intersecting with a component C, if a point p in $r \cap C$ exists such that any point q ($\neq$p) in $r \cap P$ is closer to s than p, then C is a member of S_1. Where $r \cap C$ and $r \cap P$ denote intersection segments between the ray and C and P, respectively. This definition means that when a target moves to infinite distance in the 3D model, the last components intersecting with the target are classified into the 1st shell.

By applying the same classification algorithm to other components, the components located in the outermost layer except for the first shell are classified into the second shell. In the same way, all components are classified into shells from the first to the last shell. Figure 3 depicts a two-dimensional example of the shell classification. Note that shells do not always enclose a target entirely in this method (see the second shell in this example). We look this deficiency space as a window of the shell.

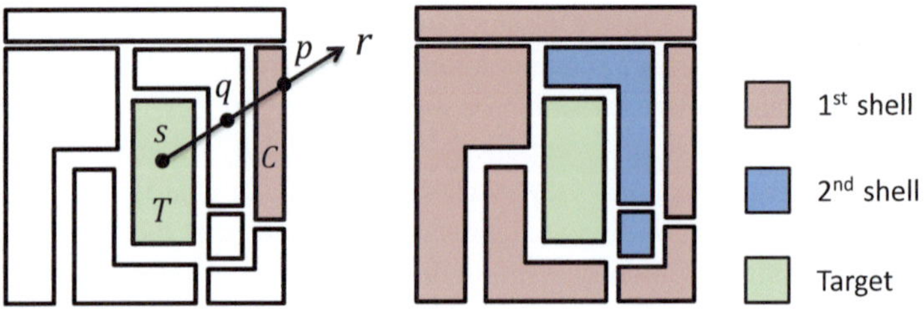

Fig. 2 Shell classification

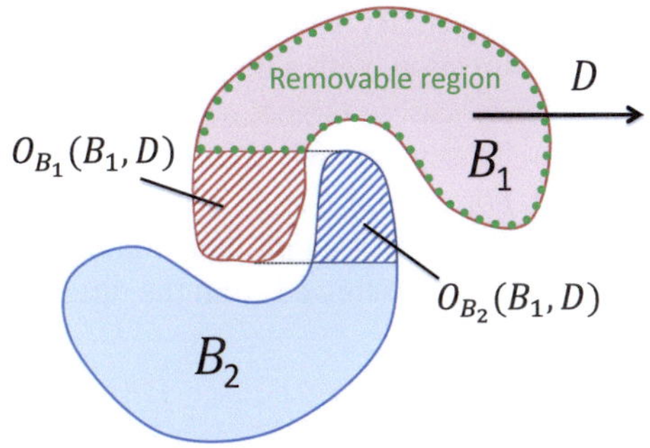

Fig. 3 Removability between two rigid bodies

2.3 *Generation of Split Line Candidates*

2.3.1 Definition of "Removability" Between Rigid Bodies

First, we define removability of rigid bodies. For the sake of simplicity, this article considers parallel translation of the bodies without rotation. Let B_1 and B_2 denote two rigid bodies in a three-dimensional space (see Fig. 3). Let D be the direction of relative movement of B_1 from B_2.

When B_1 moves in the direction of D, the cumulative set of intersection between B_1 and B_2 is defined as an obstacle region $O_{B_1B_2}(B_1, D)$. If this region is an empty set, B_1 is defined as "removable" from B_2 with respect to D.

$O_{B_1B_2}(B_1, D)$ is composed of two subregions $O_{B_2}(B_1, D)$ and $O_{B_1}(B_1, D)$, a part of B_2 and B_1, respectively. $O_{B_2}(B_1, D)$ is the shadow of B_1 projected on B_2 in the direction of D. In order to remove B_1 from B_2 toward the direction of D without breaking B_1, $O_{B_2}(B_1, D)$ must be cut apart and removed from B_2.

On the other hand, $O_{B_1}(B_1, D)$ is the shadow of B_2 projected on B_1 in the backward direction of D. To remove B_1 from B_2 without breaking B_2, $O_{B_1}(B_1, D)$ must be cut apart from B_1. Therefore, $\overline{O_{B_1}(B_1, D)} \cap B_1$ is defined as the "removable region" of B_1 in which, by splitting this region from $O_{B_1}(B_1, D)$, the remaining part can be removed toward the direction of D.

2.3.2 Setting of Extraction Directions to Targets

After the shell classification, the designer sets an extraction direction D_T to each target component T. With this direction, the obstacle region $O_P(T, D_T)$ of the target is calculated by using an intersection between the solid models of the target T and the product P.

The process of the semi-destructive disassembly can be formalized as the sequence of removing each obstacle region by breaking and peeling the shells. When the designer changes the direction, the obstacle region moves within the product model. By using this feature, the designer interactively searches for a feasible direction, with a help of the system's visualization of the obstacle regions.

In many cases, targets are fixed with other components by fasteners (e.g., screws). To separate the targets securely, accessibility to the fasteners should be ensured. For example, when a printed circuit board (a target component) is screwed to a chassis, the extraction direction of the target should be determined so that the screws can be unscrewed from the direction.

2.3.3 Calculation of Removable Region

Let us here focus on the relationship between a shell S_i and its inner layer shells (the outer layer shells are ignored here). The window of each shell should be large enough to remove the obstacle region $O_{S_i}(T, D_T)$ of the target (see Fig. 4). Likewise, the fragments of the inner shells, which will be divided and peeled, are removed through the same window of S_i. The split line should be added to the shell so as to encircle the obstacle regions $\cup_{j>i} O_{S_i}(F_j, D_j)$, where F_j is the peeled fragment of the inner shell $S_j (j > i)$ and D_j is the direction of removing F_j as shown in Figs. 5 and 7.

Furthermore, since shells are, in most cases, geometrically engaged by each other, shell fragments cannot always be peeled off due to the interdependent constraints between shells. To avoid such inter-shell engagement, the system identifies a removable region on each shell. As defined in Sect. 2.3.1, $\overline{O_{S_i}(S_i, D_i)} \cap S_i$ is the removable region of S_i toward an extraction direction D_i (see Fig. 6). When a split line is added to the inside of this region, it is guaranteed that the fragment divided by the line can be removed to the direction as shown in Fig. 7.

The system visualizes the obstacle regions of the target and the inner shell fragments. To remove them, the removable region of S_i should include the obstacle regions entirely. The designer adjusts the removable region by steering the direction D_i on the geometric model of S_i.

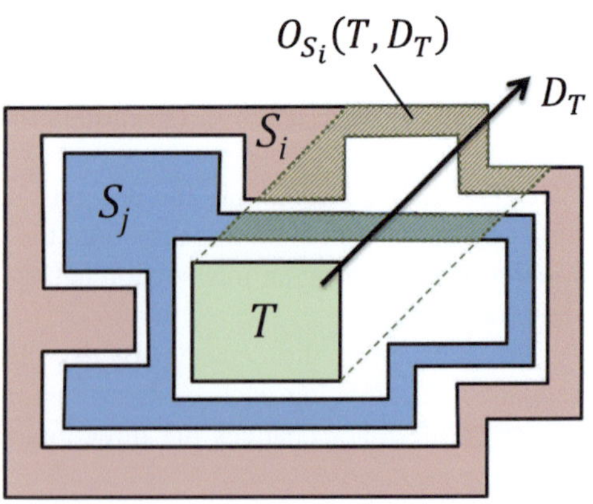

Fig. 4 Obstacle region of the target T

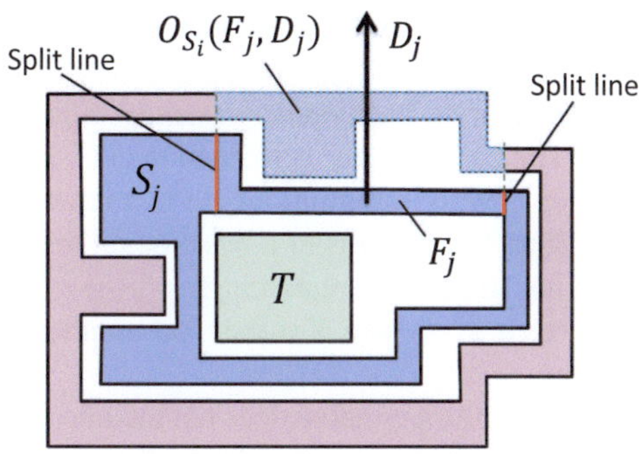

Fig. 5 Obstacle region of the inner shell fragment F_j

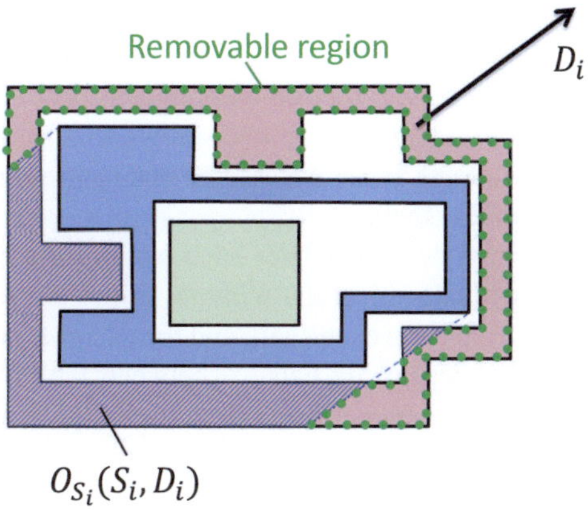

Fig. 6 Removable region of the shell S_i

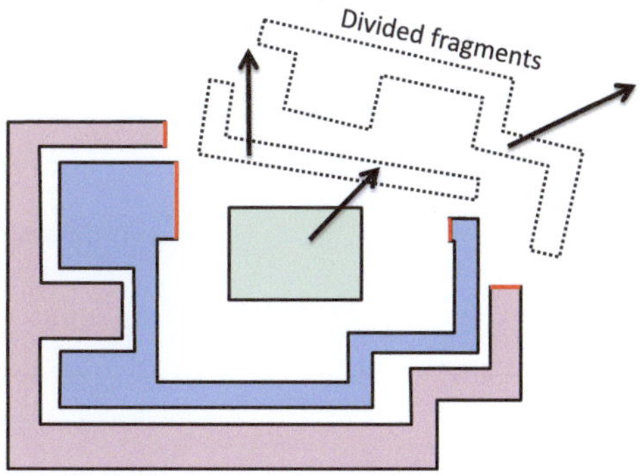

Fig. 7 Opening the windows by splitting

This calculation is applied to each shell from the most inner one to the first in a stepwise manner. The extraction direction D_T of the target is unchanged during this calculation for all shells, while the direction D_i for removing each shell is arranged individually.

2.3.4 Generation of Baselines as Candidates of a Split Line

Next, the system generates baselines on each shell as hopeful candidates of a split line. The baselines are generated to encircle obstacle regions and each combination of screws with minimum length in a removable region. The broken lines shown in Fig. 8 are the patterns of a baseline on the shell when the window enclosed by the baseline includes screws ranging from zero to four. The number of the baseline patterns is 2^N, where N is the total number of screws located within the removable region.

Preferably, a window does not include any screw. In such a case, the window can open without unscrewing after splitting the shell. However, no unscrewing is not always possible due to mechanical or aesthetic reasons. Here, "unscrewing" is a representative example of a laborious disassembly operation. The method deals with other fasteners such as snap-fit in the same manner.

In the figure, the red-colored area denotes a pattern of a window including the obstacle region and the two screws with minimum length of baseline (boundary of the window). Here, we assume that shorter lines are preferable owing to the cutting cost of the shell.

The system visualizes each pattern of a baseline in the order of the number of screws included in the window from zero (equal to the boundary of the obstacle region) to N on each shell, and the designer selects a baseline and modifies its shape by considering the balance between the efficiency of breaking and other design aspects such as performance, reliability, functionality, aesthetics, and manufacturability of the product. The main focus of this study is not on such detailed mechanical analysis and aesthetic design but on support for designers in deriving feasible candidates of split lines based on geometric information. For supporting the modification of baselines, the system indicates the removable regions, obstacle regions, and various patterns of baselines.

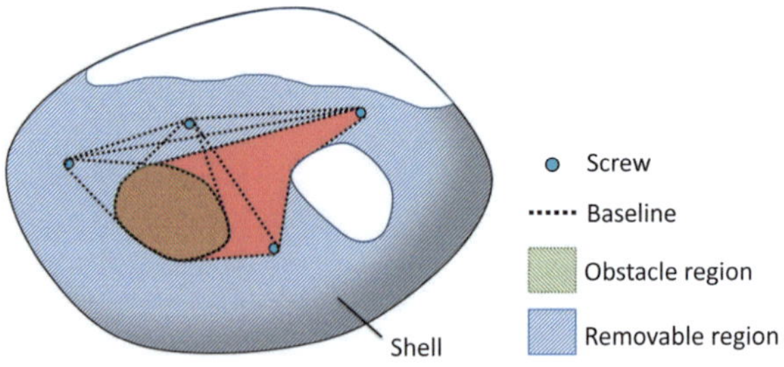

Fig. 8 Pattern of a baseline

3 Case Study

As a case study, we applied the proposed method to a 42-in. LCD TV (with LED back light). By disassembling the product and surveying the components and fasteners, we made a geometric model as a base model. Based on the current EoL treatment of flat panel TVs in Japan, the electrical circuit boards are set as the targets to be extracted before shredding, aiming to recycle precious materials, such as gold, silver, and platinum.

The prototype system based on the proposed method imported the geometric model and classified all components other than the targets into shells. The back cabinet, the front panel, and the panel frame were assigned to the first shell S_1, and the target was located just under the third shell composed of the board cover and the main chassis (see Fig. 9).

Next, we examined six candidates of extraction direction on the 3D model, which are perpendicular to the front, back, right, left, upper, and under faces of S_1. Among them, we selected the back face direction D_T that minimizes the obstacle region of the targets. Figure 9 shows the segmentation result of S_1 with respect to the selected direction, where the dark gray area on the back cabinet is the obstacle regions of the targets and the peeled fragments of the second shell, and the deep blue area is $O_{S_1}(S_1, D_1)$, namely, the other area (light gray area) in the first shell is the removable region.

The red area highlights a pattern of the baseline generated on the 1st shell, which encloses the obstacle region and four screws with minimum length. We selected and modified this baseline to simplify the line shape for easy breaking.

Figure 10 shows the result of the semi-destructive disassembly of the TV with the split lines, where the divided fragments and the targets are extracted in the same direction after splitting the back cabinet and removing the board cover as a peeled

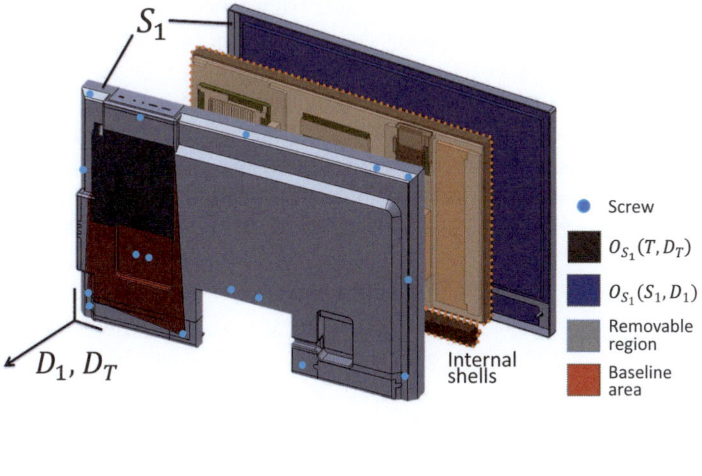

Fig. 9 Segmentation of the first shell

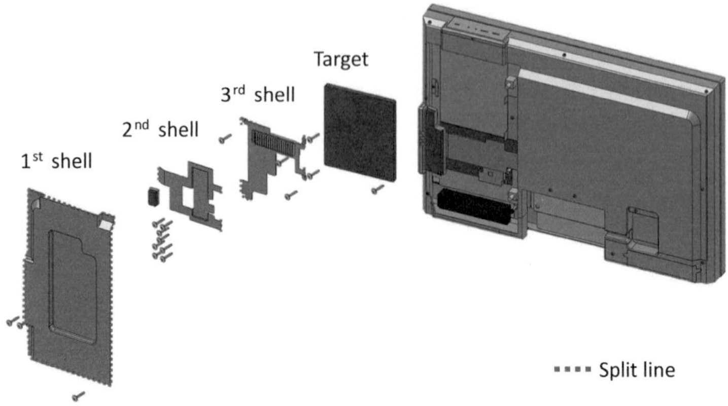

Fig. 10 Result of dismantling with split lines

Table 1 Comparison of steps needed for the dismantling between the current design and the new design with split lines

Operation type	Original design	New design
Unscrewing	33	17
Splitting shell	0	4
Removing fragment	5	5
Total	38	26

fragment of the second shell. This solution presumed that the shells are broken by applying concentrated force by using, for example, partial destructive tools [18, 19]. In this case, the shock of the destruction is insignificant for the printed circuit board recycling.

Finally, we determined a disassembly sequence by evaluating the efficiency of dismantling. In this case study, we evaluated the sequence with a simple method in which the number of elemental operations (steps) needed to extract the circuit boards was counted, such as breaking shells, removing fragments, and unfastening screws.

Table 1 compares the evaluation results between the disassembly of the original design and the semi-destructive disassembly of the new design with the split lines. The number of steps was reduced by 32 % from the original, which requires the disassembly of almost all components to reach the targets. The new design reduces the number of unscrewing operations from 33 to 17 by destructing the shells along the split lines.

4 Discussion

The case study illustrated the support process of the proposed system in adding split lines to the product. The system generated removable and obstacle regions on each shell by using geometric processing algorithms and derived the baselines as feasible candidates of split lines on the 3D model of the product.

In modifying the detailed shape of the baselines, mechanical analysis including mechanics of materials might be needed for ensuring the functional integrity of the product. Generally speaking, split lines weaken the stiffness and rigidity of the product. Shells should be divided only when designated amount of forces are applied to designated points. The designer should find out a well-balanced solution that keeps the functionality of the product in the use phase and makes split lines broken easily in the dismantling phase.

Besides the split lines determined in the design phase and the shells split in the EoL phase in the case study, various approaches can be applied for splitting the shells. For example, determining feasible lines to be cut in the EoL phase and cutting them by a circular saw are plausible applications of the method.

While the proposed method did not assume the changes of product structure and fastener location, the cooperation of design changes in the structure, fastener, and shape of components with split lines is a promising approach to increase the efficiency drastically. In the case study, for example, if the boundary between the cover and the chassis was changed to follow the split line, the efficiency of cutting the shell increased. Incorporating such design changes into the proposed method is one of our future issues.

The shells were divided and peeled individually in the case study. In some cases, multiple shells can be divided and removed at a time. Such a combined division is critical for reducing dismantling time. The extension of the method for combining split lines of multiple shells is another future issue.

For evaluating the semi-destructive disassembly sequence, we used a simple measurement of counting the steps needed for the sequence. For more accurate estimation of operation time and costs at design phase, employing the concept of the Work Factor (WF) method [20] is a plausible approach. In the case of the semi-destructive disassembly proposed here, the length of split lines, materials, and shapes of divided components, as well as tools used for splitting, should be included in the time estimation formula of WF.

5 Conclusion

This paper presented a design support system for semi-destructive disassembly. The system aids a designer to determine the location of split lines by calculating the feasible regions of a product to be split and by generating hopeful candidates of the split lines. The case study demonstrated that redesign of an LCD TV for semi-destructive disassembly reduces the number of operations needed to separate the electrical circuit boards. The evaluation of the mechanical performance and functional integrity of the redesigned TV is an important future work. Other future works include the extension of the proposed system, such as the division of multiple shells at a time; the cooperation of design changes in the structure, joining, materials, and shape of components with split lines; and the optimization of the semi-destructive disassembly sequences.

Acknowledgments This work was partially supported by the Asahi Glass Foundation.

References

1. Masui K, Hata T, Umeda Y (2007) Current status and challenges on environmentally conscious design in Japanese manufacturers. In: Proceedings of the 5th international symposium on environmentally conscious design and inverse manufacturing (EcoDesign 2007), Tokyo, No. B1-3-4F (CD-ROM)
2. Peeters JR, Vanegas P, Duflow JR, Mizuno T, Fukushige S, Umeda Y (2013) Effects of boundary conditions on the end-of-life treatment of LCD TVs. Ann CIRP 62(1):35–38
3. Mizuno T, Kunii E, Fukushige S, Umeda Y (2012) Recyclability evaluation of LCD TVs based on end-of-life scenarios, In: Proceedings of the 19th CIRP conference on life cycle engineering. 179–183
4. Duflou JR, Seliger G, Kara S, Umeda Y, Ometto A, Willems B (2008) Efficiency and feasibility of product disassembly: a case-based study. Ann CIRP 57(2):583–600
5. Mathieux F, Recchioni M, Ardente F (2014) Measuring the time for extracting components in end-of-life products: needs for a standardized method and aspects to be considered. Procedia CIRP 15:245–250
6. Willems B, Dewulf W, Duflou JR (2005) Design for active disassembly (DfAD): an outline for future research. In: Proceedings of IEEE 2005 international symposium on electronics and the environment. pp 129–134
7. Boothroyd G, Alting L (1992) Design for assembly and disassembly. Ann CIRP 41(2):1–12
8. Kopacek B, Kopacek P (1999) Intelligent disassembly of electronic equipment. Ann Rev Control 23:165–170
9. Westkamper E, Feldmann K, Reinhart G, Seliger G (1999) Integrated development of assembly and disassembly. Ann CIRP 48(2):557–565
10. Masui K, Mizuhara K, Ishii K, Rose C (1999) Development of products embedded disassembly process based on end-of-life strategies. In: Proceedings of the 1st international symposium on environmentally consious design and inverse manufacturing, Tokyo, pp 570–575
11. Takeuchi S, Saitou K (2008) Design for product embedded disassembly. Stud Comput Intell 88:9–39
12. Carrell J, Zhang HC, Tate D, Li H (2009) Review and future of active disassembly. Int J Sustain Eng 2(4):252–264
13. Langerak E (1997) To shred or to disassemble? Recycling of plastics in mass consumer goods. In: Proceedings of the 1997 I.E. international symposium on electronics and the environment. pp 63–68.
14. Wakamatsu H, Matsuishi M, Morinaga E, Arai E (2010) Disassembly support system for used products considering destruction of their parts. Service robotics and mechatronics, Selected Papers of the International Conference on Machine Automation ICMA2008 (2010), pp 37–42
15. Choido J, Jones N (2012) Smart materials use in active disassembly. Assem Autom 32(1):8–24

16. Hussein H (2003) Active disassembly – the end of electronic life as we know it. Mater World 11(11):20–22
17. Vongbunyong S, Kara S, Pagnucco M (2013) Application of cognitive robotics in disassembly of products. Ann CIRP 62(1):31–34
18. Seliger G, Basdere B, Keil T, Rebafka U (2002) Innovative processes and tools for disassembly. Ann CIRP 57(1):37–40
19. Mouton Christof (2014) Design and optimization of a flat panel dismantling line. Master Thesis, Katholieke Universiteit Leuven
20. Karger DW, Bayha FH (1987) Engineering work measurement. Industrial Press, New York

Potential of Common Methods to Integrate Sustainability Requirements in the Product Development Process: A Case Study

Maike Kosiol, Dominik Weidmann, Daniel Kammerl, and Udo Lindemann

Abstract Sustainability is a process of change, which ensures present generations to meet their needs without preventing future generations from meeting their necessities. Products have a huge impact on sustainable development since product properties determine the product's environmental impact. This contribution analyzes the potential of commonly known and established methods to support developers in designing sustainable products. Possibilities to integrate sustainability requirements, as well as thereby arising difficulties, are considered. The development of an ice crusher is the use case for the analysis.

Keywords Sustainability • Development process • Case study • Methods • Sustainability requirements

1 Introduction

The core ideas of sustainability are saving the climate, saving our standards of living, and creating a livelihood for future generations [1]. These topics are influenced by the lifestyle of the human population. The population expands rapidly and interacts with its environment by consuming products and services. This is inseparably linked to an increased energy demand, resource exploitation, and environmental pollution including exhaust gases [2]. In the long run, these impacts affect societies by diseases or climate change [3]. The development and mass production of technical everyday products once enabled the so-called consumer society. Decisive product properties affecting all phases of a product's life are already defined in the product development process [4]. The product life cycle covers all stages from the first idea over the development, production, distribution, usage, and finally to the disposal or even beyond [5]. To meet the demand of designing sustainable products along the entire life cycle, the focus is to be set on the product development phase.

M. Kosiol • D. Weidmann (✉) • D. Kammerl • U. Lindemann
Institute of Product Development, Technische Universität München, München, Germany
e-mail: weidmann@pe.mw.tum.de

M. Matsumoto et al. (eds.), *Sustainability Through Innovation in Product Life Cycle Design*, EcoProduction, DOI 10.1007/978-981-10-0471-1_5

In literature numberless methods are available, which support developers to achieve specific objectives in context of sustainability. A number of authors analyzed the applicability of such methodologies. They figured out that many tools are not used in industrial development processes. Tools which might be predestined to integrate sustainability aspects are not immediately applicable, due to their complexity, the required time to apply them, budget, personnel, or the lack of knowledge about how to use them [6].

This paper focuses on methods which are already commonly known and established in product development. The goal is to analyze how these methods can, in addition to their primary application background, support developers in designing products that meet the requirements of sustainability at the best possible rate.

This paper has the following content:

The second chapter is about sustainability in literature, especially in the content of the product development process. The third chapter describes the research methodology. The fourth chapter analyzes the potential to integrate sustainability aspects to methods, which were applied within the scope of a use case. Chapter 5 draws a conclusion and gives a brief outlook.

2 Sustainability in Product Development

For our current understanding, 1987 is said to be the birth year of sustainability. In this year, the Brundtland Commission took place, which published a definition of sustainability that is still valid today: "Humanity has the ability to take development sustainable to ensure that it meets the needs of the present without compromising the ability of future generations to meet their own needs" [7]. The commission describes sustainability as a continuing process of change which reaches for harmony and enhancement in the interaction of resource utilization, technological development, institutions, and investments [7]. The Brundtland report describes sustainability as a broad field, and consequently, many discussions about sustainability focus on particular aspects.

Sustainability is commonly subdivided into three dimensions: economy, ecology, and society. They represent the multitude of sustainability requirements [8]. A product is torn between these three dimensions, which have partly conflicting claims. Figure 1 shows the interdependencies: economy, ecology, and society demand requirements for the product. Additionally, interactions between the dimensions exist; e.g., economy provides workplaces for society; the environment is the basis of life for the society and provides raw materials which are used in the economy.

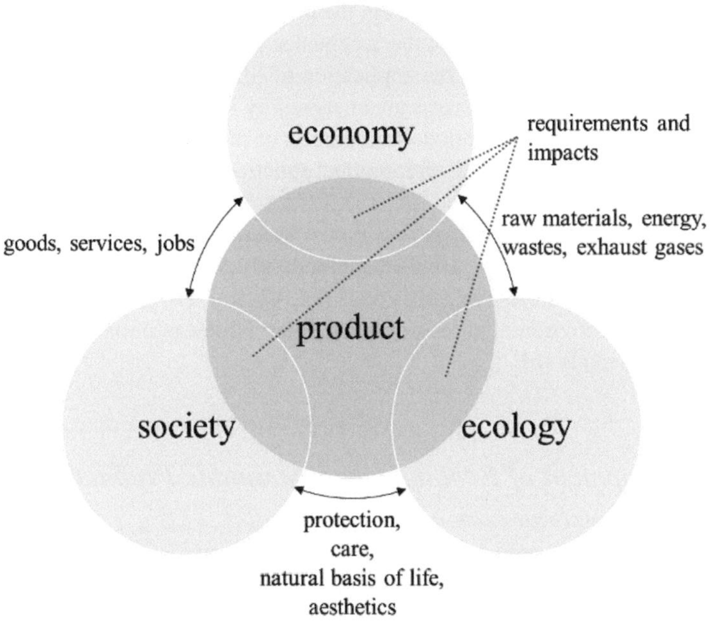

Fig. 1 A product in the area of conflict of sustainability dimensions (Modeled after Ref. [8])

2.1 Development of Economically Sustainable Products

Literature with a focus on the economical sustainability of products isn't widely spread, because this is a self-motivated goal for companies with a long-range plan [8].

2.1.1 Objectives

The development of economically sustainable products means to create successful products which are in line with the market and the customer. Further goals are reducing the amount of goods, such as costs and time, and increasing quality and efficiency.

2.1.2 Methodological Tools

Using methods can lead to better quality, efficiency, and saving of time. Directives and standards serve to design products that are in line with the market. Checklists support the identification of sustainability requirements [9]. Another useful tool is based on value analysis (VA) with the goal to design or redesign a product at low cost. Therefore it is analyzed which functions the consumer is willing to pay for.

Only these functions are implemented in the product, and in doing so, environmental benefits are obtained. Based on the acceptance of the product by consumers, this method was enhanced through the application of life cycle costing (LCC). From a life cycle perspective, product costs are analyzed by integrating environmental cost with the internal cost. The continuous evolution of this method toward sustainability led to the Eco-VA (assesses each product function from the customer, environmental, and cost perspective) and to the LCECA (mathematical model that enables an isolated calculation of environmental cost independent of the product component). Life cycle planning is a recent method, which combines quality function deployment (QFD), life cycle assessment (LCA), and TRIZ. It aims to integrate quality, cost, and environmental aspects in a systematic way into the early stages of a product's life cycle [6].

2.2 *Development of Ecologically Sustainable Products*

Currently products are produced and disposed at an alarming rate. As much as 75 % of material resources used in products and their manufacture are disposed back to the environment as waste within a year [10]. The statistic speaks about so-called throwaway products, which cause challenges toward resource extraction, waste collection, growing landfills, and the arising environmental influences.

2.2.1 Objectives

Ecologically sustainable product development focuses on products that cause less influence on the environment throughout their entire product life cycle compared to previous or competing products [11]. Approaches to achieve this objective are, for example, the usage of regenerative energies [12] and assuring the emission-free usage [13]. Another approach is the consideration of the products' end of life already in the early design process by taking decomposition, material separation, and reuse into account [9]. However, it is important to maintain, or better, to increase the product incentive for the customer. Therefore an additional ecological value is a core issue [14].

2.2.2 Methodological Tools

Most approaches in connection with sustainable development are searching possible solutions concerning the environment. Numerous tools and methods exist to support ecological goals. A multitude of laws and quasi-statutory texts and directives as the VDI 2243 [4], the 2005/32/EC [15], or the 2002/96/EG [16] give practical advice and contain goals and limiting values. Integrated product policy (IPP) aims for constant improvement of products with a focus on their impact on mankind and environment over the entire life cycle [17, 18]. Methods like environmental accounting or life cycle assessment (LCA) deal with the evaluation of the product's ecological performance [19]. A wide range of tools allows quality function deployment (QFD) to be applied in order to consider environmental requirements during the early stages of product development [20]. The purpose is to check if the product corresponds to the customer's requirements, including environmental requirements. Further methods often combine the House of Quality (HoQ) of the QFD with LCA-based techniques: e.g., House of Ecology (HoE) deals with environmental requirements instead of quality requirements, E-QFD combines LCA and HoQ in order to evaluate alternative design proposals, and Readiness Assessments for Implementing Design for the Environment Strategies (RAILS) focuses on supplying tools for selecting environmental improvement options that could be carried out [6].

2.3 Development of Socially Sustainable Products

The product development's role in designing socially sustainable products was already formulated in the 1970s. Products which were aligned to the actual human needs should be developed. They should generate positive social aspects instead of contributing to the erroneous trend of society, which means every deviation from the environmentally optimal behavior with minimal impacts on the environment [14]. In literature judgmental terms are often used in the context of products with a social focus. Expressions like "socially acceptable" assume inherently a damage [8].

2.3.1 Objectives

In the design process, social aspects can especially be included in the requirement analysis. It is within the responsibility of the product development to gather information about the society's needs or about personal preferences. Further areas in connection with the social dimension are work and product safety [8].

2.3.2 Methodological Tools

For the social dimension, far less comprehensive methodical support is available in contrast to both previous dimensions. The few existing approaches often focus on one specific issue, like "design for the third world"; "design for teaching and training devices for the retarded, the handicapped, and the disabled"; or "design for elderly people" [8]. Evaluating approaches like the social life cycle assessment often use indicators defined by the United Nations Environment Program [21]. Indicators can be obtained by a listing of the Committee on Sustainable Development

[22], the standard SA 8000 of the Social Accountability International [23], the norm DIN ISO 26000 [24, 25], or [26].

Failure mode and effect analysis (FMEA) is primarily used to identify, assess, and prevent failures related with product safety. The adapted environmental impact and factor analysis (EIFA) considers environmental issues related to reducing, reusing, and recycling [6]. E-FMEA identifies and assesses potential environmental impacts rather than potential failure and its influence on inventory data life cycle and human-related causes [6, 20, 27].

2.4 *Sustainability Relevant Design-for-X Approaches*

The three previously considered sections can be completed by Design-for-X strategies. These strategies are listed separately in this section because of their huge number and importance. Many Design-for-X strategies cannot be assigned to one specific sustainability dimension although every approach addresses a particular issue that is caused by properties of a product or affects others. Each approach gives practical advice in guidelines and has a particular objective, which is represented by X [28, 29]. Figure 2 shows some particularly crucial approaches in connection with sustainability. They are structured according to the dimension they can mostly affect.

Design for Sustainability serves as a melting pot for more detailed approaches of the three dimensions of sustainability. Especially Design for Upgradability and

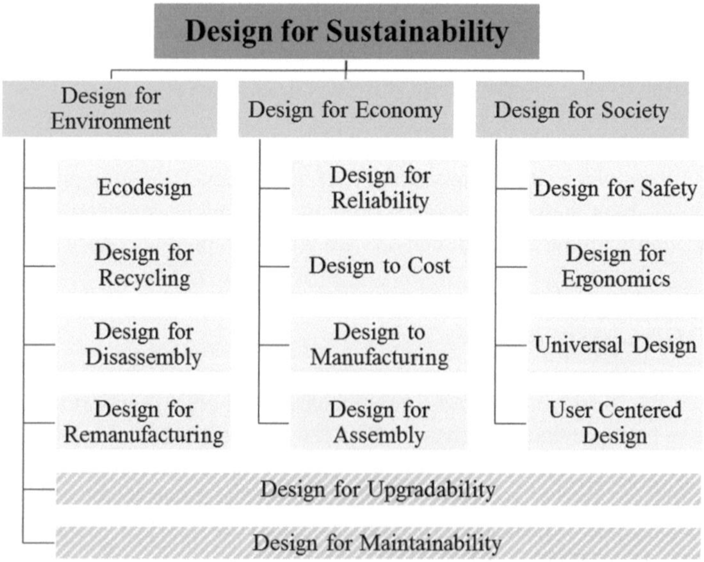

Fig. 2 Overview on central design-for-X approaches with focus on sustainability

Design for Maintainability target at life extension of technical products which affects all three dimensions:

- Design for Upgradability: foreseeing of further use phases by enabling the addition of new functions [3]
- Design for Maintainability: taking measures to preserve and recover a product's initial state [30]

Design for Environment is about minimizing the impacts on the environment during the product life cycle as much as possible, e.g., by considering a realistic mix of recovery and disposal [31]. For example:

- Ecodesign: includes the integration of ecological and economical aspects into product development by considering the entire life cycle [11]
- Design for Recycling: economically restoring material and energy to production processes that accrue by producing and consuming products [19]
- Design for Disassembly: simplifying the process of dismounting [30]
- Design for Remanufacturing: functional or withdrawn products can be renewed through a series of industrial operations [32]

Design for Economy is used as an umbrella term for strategies that try to optimize products with a focus on the economic dimension. For example:

- Design for Reliability: reduction of unplanned interferences and component failures and thus increasing of reliability [8]
- Design to Cost: contains approaches to reduce costs [33]
- Design to Manufacturing: designing suitable for production [33]
- Design for Assembly: suitability for mounting a product [3]

Design for Society develops products that support the improvement of social difficulties [8]. For example:

- Design for Safety: minimization of risk potential by the work with products and machines [8]
- Design for Ergonomics: adaption of technical systems to humans in order to avoid health impairment by using the product as planned [30]
- Universal Design: designing products that can be used by as many users as possible, including people with disabilities [8]
- User-Centered Design: designing interactive products fit for purpose [8]

3 Research Methodology

The product development process contains all steps from the product idea to the generation of manufacturing documents [34]. Procedure models help to deal with the complexity of design processes. Such models in combination with methods can support sustainability studies and hand out advice for decision-makers [35]. The

Association of German Engineers (VDI) provides an established procedure model to develop and design technical systems and products by following seven steps in the directive VDI 2221 [36]. We chose this model because it is commonly known and refers to specific resulting documents like requirement specification. This paper analyzes one method representative for each of the first four process steps to determine the potential of the methods to integrate sustainability requirements in the early design phase.

The first step clarifies the problem definition. Identified requirements are documented in the requirement specification. In the second step, functions and their structure are determined and illustrated in the function modeling. In the third step, solution principles for the functions are identified. Reverse engineering is a method which provides possible effective structures that need to be analyzed in this step. The fourth step structures the solution principles based on various criteria. The weighted points rating evaluates solution possibilities in a quantitative and qualitative manner. More methods have been analyzed within the framework of this project, but would go beyond the scope of this paper.

The presented methods were integrated in a real development project, whose focus was not especially on sustainability aspects, but on cost, time, and effort. Consequently, the selection of the methods was based on close to reality conditions and to cover an as large as possible part of the development process.

Analysis and evaluation of the methods are carried out in the framework of a student project in cooperation with a manufacturer. This development project followed the steps of the briefly introduced procedure model and especially analyzed the creation of conclusion documents. The focus of this paper is not on the detailed description of the methods themselves but on their possibilities to integrate sustainability requirements. The theoretical findings have been evaluated within an interdisciplinary team of five students by practical application of the methods in a use case: the goal was to design an ice crusher add-on for a fully automated home coffee machine with a special focus on sustainability (Fig. 3). The ice crusher has to be integrated in an existing coffee machine in order to produce a frappé with just a single push of a button.

4 Case Study: Ice Crusher

4.1 Analysis and Discussion of Applied Methods

4.1.1 Requirement Specification

The requirement list basically structures requirements systematically. It helps to comprehend responsibilities and changes within the entire development process.

The method serves as a structured documentation of all kinds of requirements that have to be considered during the development process. Each line contains one requirement which is specified by the following columns entering information as an

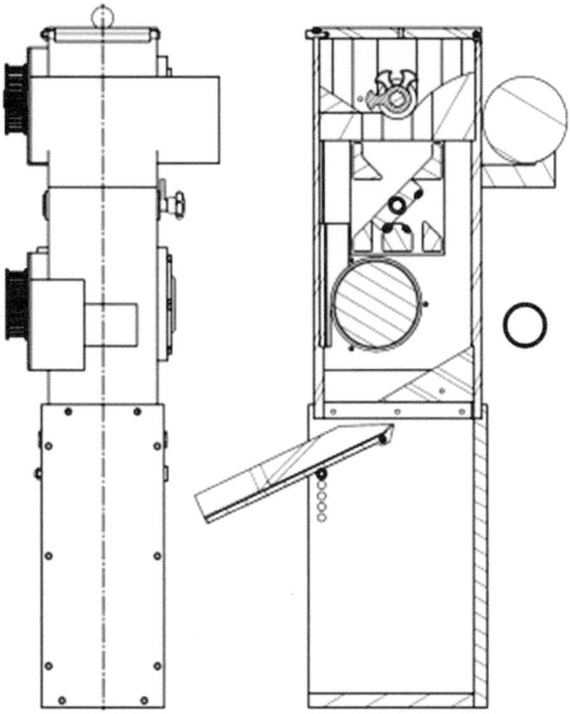

Fig. 3 Use case: the development of an ice crusher add-on for a fully automated home coffee machine

identification number, characteristic, characteristic value, origin, explanatory note, responsible person, and information concerning the revision tracking in the list (Fig. 4).

The requirement list can integrate relevant sustainability issues. It is possible to divide the list into different phases of the product life cycle. In the use case, the requirement list was structured into the phases: product development process, manufacturing, use phase, application, and after use phase. Developers' thoughts are guided to often forgotten phases like disposal. Within these main phases, we created further subgroups, e.g., in the phase manufacturing: material choice, production, and quality.

As shown, sustainability criteria can be easily integrated into a requirement list. However, it is challenging that requirements are not yet defined or known in the early stages of the development process. At least limiting values for sustainability requirements might not be available. Other requirements are not yet evident, which leads to another challenge: the requirement specification does not principally prevent forgetting key aspects of sustainability. Finally, a high number of laws, standards, guidelines, and directives exist, which provide requirements with relevance for sustainability, but this flood of information could overcharge developers. Because of the scale and the complexity of this topic, it is not possible to take every thinkable sustainability requirement into account. Iteration loops with a specific

			Characteristic Values							Revision Tracking		
No.	**Description/ Name of Requirement**	**Short Form**	**min.**	**exakt**	**max.**	**Unit (phys.)**	**Origin**	**Date**	**Responsible Person**	**What? Why?**	**Who?**	**Date**
...	...	...	...	...	...	...	...	...	...	...	...	...
							End-of-life Stage					
15	**Disassembly**											
15.1	guarantee accessability for removal tools	-					VDI 2243	28.10.14	MK			
15.2	standardize connection elements	-					VDI Report No. 1570	29.10.14	MK			
...	...	...	...	...	...	...	...	...	...	...	...	...
16	**Recycling**											
16.1	recycling rate for small household appliances	R		80		weight-%	Directive002/96/EG of the European Parliament & Council	13.02.03	JL			
16.2	place reusable modules easily separable	-					Product Design Suitable for Recycling of Kahmeyer & Rupprecht	18.06.05	MV			
...	...	...	...	...	...	...	...	...	...	...	...	...

Fig. 4 Requirement specification

focus, respectively, could support developers. However, developers have to focus on some core sustainability requirements.

4.1.2 Relation-Oriented Function Modeling

Functional models reduce the complexity of a system and thus increase the understanding of developers.

At the beginning of the modeling, it is important to define a clear objective and a suitable level of abstraction. A system boundary has to be set to separate the considered system from its environment. Technical functions are described by the combination of a substantive with a verb. The relation-oriented modeling differs between harmful and beneficial functions. The causalities between those functions are relations. Based on the main function of the system, further beneficial and harmful functions are identified, which are required to realize the main function.

Focusing on sustainability, this method offers the possibility to especially analyze harmful functions. Solution-neutral functions can be added in order to prevent harmful functions or at least to weaken them, and specific problem formulations can be derived. Harmful functions are unwanted functions, which influence the system negatively. In the context of functions, the focus can be set on the numerous aspects of sustainability, e.g., created emissions, corrosion of materials, used energy, and further more. The harmful functions can address various life cycle phases. It is also possible to derive problem formulations in the background of socially sustainability by considering noise pollution, the handling of hazardous substances, or of heavy objects. By marking harmful functions, the developer can

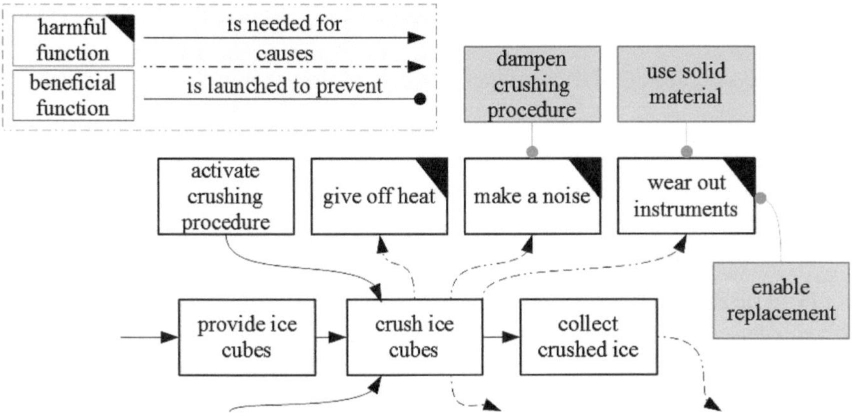

Fig. 5 Extract from the relation-oriented function modeling

concentrate on them. Their setting in the model enables analyses to avoid them, as they can be considered immediately in the appropriate context of their interactions. Possible consequences of modifying harmful functions can directly be noted.

Figure 5 shows an extract of the model which was developed in the framework of the ice crusher project. In this context all functions, which are functionally unwanted or against the principles of sustainability, were named harmful. We explicitly integrated functions to avoid harmful functions, e.g.,"enable replacement" in favor of "wear out instruments."

In our opinion this method offers a great potential to assimilate sustainability criteria into the early stages of the product development process. However, the results of this method still have to be questioned critically. Models always display a reflection of the reality. Several assumptions support the abstraction of the reality. Models and their outcome are always limited in their significance, and they should not be the sole basis for decisions. Another handicap of the relation-oriented function modeling is that it usually considers only few phases of the product life cycle. Primarily it considers functions of the use phase of the product. However, there are other types of function models that look at material flows and flow of energy that occur during all phases of the product's life. Consequently, they are predestined to be used in the context of ecological aspects of sustainability.

4.1.3 Reverse Engineering

Reverse engineering generates ideas for the current design process by analyzing existing products. This method allows developers to gain an extensive insight into the state of the art. They get new impulses for requirements [37] and data about the product that can be used to quantify the characteristic values.

Typically, products with an outstanding position in a certain area are chosen. The purchased product is analyzed in detail, so the developer could create a copy of the product by himself. Therefore the product is fully disassembled – intellectually or physically.

The disassembling of the product is a central aspect in the context of sustainability because it is the basis for recycling, remanufacturing, maintenance, etc. Thereby the following questions can be considered:

- Which and how many tools are required?
- Is it easy to find connection points?
- Is it possible to separate different materials?

In the framework of the development project, two different ice crushers were purchased, which cover the price segments cheap household appliance and professional gastronomic business. The ice crushers were analyzed with the focus on sustainable product design. The disassembling without detailed knowledge about the appliance made us aware of essential sustainability aspects. This is exactly the starting situation in recycling processes. It was hard to detect connection elements that protracted the disassembly process. Another identified deficit was the usage of different connection elements, so different tools and frequent tool changes were necessary. We noted positively that almost every used material could be separated without any damage.

Reverse engineering is a cost-intensive method, because of the purchase of the product, and the analysis can take a lot of time and expert knowledge. A critical aspect of reverse engineering is the developer's fixation on the seen solution. This might prevent innovative and creative solutions.

4.1.4 Weighted Points Rating

If different solution concepts are generated, they have to be evaluated. The weighted points rating can be the basis for the decision-making. This method can consider a huge number of aspects, while the comparability between the different solution alternatives is guaranteed. Developers deal with the specific characteristics of each possible solution and finally present a ranking.

But first of all, evaluation criteria have to be defined. This should be a well-balanced system of criteria in order to not overestimate specific aspects. Sustainability requirements can be easily integrated in this method. The list of requirements can be an orientation for choosing criteria. Fiksel et al. [38] provide, e.g., criteria that aim at sustainability. They provide indicators to evaluate a product, structured into the three dimensions of sustainability (e.g., recycling revenue (economic), life cycle energy (environmental), or illness avoided (societal)). In the development project, we considered aspects of all three dimensions, e.g., costs (economic), noise emission (societal), or resource requirements (environmental). The evaluation of different concepts is based on the degree of fulfillment of the evaluation criteria. A value function or the relation to the remaining alternatives helps to transfer the degree of fulfillment of each criterion into point values. Besides each evaluation criterion can be weighted (e.g., 1, lower importance; 3, average

Evaluation Criteria:	Cost		Safety		Noise Level		Energy Con-sumption		...	Sum		Ranking	
Weighting	-	*3*	-	*9*	-	*3*	-	*9*	...	unweighted	weighted	unweighted	weighted
Solution 1	2	6	1	9	2	3	1	9	...	6	27	2	3
Solution 2	2	6	2	18	1	3	1	9	...	6	36	2	2
Solution 3	3	9	3	27	3	9	3	27	...	12	72	**1**	**1**
...	...	...	...	...	...	...	...	...	...	...	...	...	...

Fig. 6 Weighted points rating

importance; 9, high importance compared with the other criteria). At this point the focus can be set on sustainability criteria. Both the choice of criteria and the weighting should be done in an interdisciplinary team to minimize subjective influence on the evaluation. The point values are multiplied with the weighting and finally added line-by-line (Fig. 6, every line contains one solution alternative and its evaluation of the several criteria). The ranking is the result of the added point values, whereas the solution with the highest point values is the best evaluated solution regarding the present task.

Various sustainability requirements can be easily integrated into the weighted points rating. Due to intercultural differences, the specification of sustainability aspects might be challenging, and interdependencies between the evaluating criteria can cause overestimations of specific criteria.

4.2 *Final Discussion*

The analyzed methods offer potential to integrate aspects of sustainability. However, none of the presented methods consider every aspect of sustainability. Consequently, a set of methods, which integrate sustainability aspects into the product development process, is necessary, and they are already available.

However, resources such as budget, time, available employees, and manufacturing possibilities limit the application of sustainability into product design. Furthermore, those available resources are often fully used up to realize the basic functions of the product which has priority in terms of products that are in line with the market.

The use case showed the complexity of the thematic, but also pointed out that sustainability criteria can be easily integrated in common methods. If background knowledge can be provided and if one just spares a thought to the thematic, the more sustainability requirements are considered, the more complex and time lasting is the process. Developers deal with interdependencies, which can't be met all concurrently. In the context of the use case, we need several iterations to get over the narrowness of thinking and working capacity, which restrict the simultaneous integration of all interacting functions.

5 Conclusion and Outlook

In the context of product design, many methods exist to integrate sustainability requirements. Those tools vary widely in their complexity, quality, and the time which is required to apply them [6]. Methods that necessitate high-application effort are often not used in companies. Therefore it makes sense to integrate sustainability criteria in such methods that are already widely applied in the product development process.

Case studies integrating sustainability aspects are mostly theoretical examples, without the backing of a product design company [6]. This use case was executed in cooperation with a world-leading company specialized in domestic appliances. However, as it was an academic-based research, the exchange was not as intensive as it could have been within a company. We recommend further use cases in the framework of a company.

This contribution analyzes four established product development methods concerning their potential to integrate sustainability aspects. All four, requirement specification, relation-oriented function modeling, reverse engineering, and weighted points rating, are highly interesting in context of sustainability integration.

Many other methods exist, which are predestined to integrate sustainability aspects to the product development process, such as scenarios [39, 40] and portfolios [35, 41, 42]. As the thematic is so extensive, a product in the context of sustainability will always remain a compromise capable of improvement [29].

As a further step toward sustainable products, we recommend to measure their sustainability objectively. It is not sufficient to integrate as many sustainability requirements as possible without knowledge about their effects.

In conclusion, the spreading and application of sustainability requirements in the product development process will only take place in the long term if it promises personal, financial, or scientifically success [8].

References

1. Peters S (2011) Materialrevolution. Nachhaltige und multifunktionale Materialien für Design und Architektur. Birkhauser, Basel, p 6
2. Himmel W (2004) Vom Auspuff zum Lenkrad. In: Schwarz EJ (ed) Nachhaltiges Innovationsmanagement. Springer, Wiesbaden, pp 69–82
3. Ponn J, Lindemann U (2011) Konzeptentwicklung und Gestaltung technischer Produkte. Systematisch von Anforderungen zu Konzepten und Gestaltlösungen, 2nd edn. Springer, Heidelberg, pp 275–278
4. VDI 2243 (2002) Recycling-oriented product development. VDI-Verlag, Düsseldorf
5. Feldhusen J, Grote KH, Nagarajah A, Pahl G, Beitz W, Wartzack S (2013) Vorgehen bei einzelnen Schritten des Produktentstehungsprozesses. In: Feldhusen J, Grote KH (eds) Pahl/Beitz Konstruktionslehre, 8th edn. Springer, Berlin, pp 227–438

6. Bovea MD, Pérez-Belis V (2012) A taxonomy of ecodesign tools for integrating environmental requirements into the product design process. J Clean Prod 20:61–71
7. United Nations (ed) (1987) Development and international economic co-operation: environment. Report of the World Commission on Environment and Development, Geneva 4 August 1987, Oxford University Press, Oxford, pp 15–38
8. Hanusch D (2011) Die Entwicklung nachhaltiger Produkte – Theoretischer Anspruch und Möglichkeiten zur praktischen Umsetzung in der sozialen Dimension. In: Fortschritt-Berichte VDI Reihe 1 Nr. 413, VDI Verlag, Düsseldorf, pp 25–146
9. Behrendt S (1997) Life cycle design. A manual for small and medium sized enterprises. Springer, Berlin, pp 31–60
10. Jaafar IH, Venkatachalam A, Joshi K, Ungureanu AC, De Silva N, Rouch KE, Dillon OW Jr, Jawahir IS (2007) Product design for sustainability: a new assessment methodology and case studies. In: Kutz M (ed) Environmentally conscious mechanical design. Wiley, Hobolen, pp 25–66
11. Hermenau U (2008) Einleitung. In: Abele E, Anderl R, Birkhöfer H, Rüttinger B (eds) EcoDesign. Von der Theorie in die Praxis. Springer, Berlin, pp 1–12
12. Rosemann B, Meerkamm H, Trautner S, Feldmann K (2000) Methodische und instrumentelle Unterstützung zur Optimierung der Kreislaufführung technischer Produkte. In: Verein Deutscher Ingenieure (ed) Recyclingorientierte Entwicklung technischer Produkte 2000. Management komplexer Zielkonflikte, Fellbach. VDI Verlag, Düsseldorf, pp 27–42
13. Franze HA (2000) Integration von Recyclingaspekten in den Produktentwicklungsprozess. In: Verein Deutscher Ingenieure (ed) Recyclingorientierte Entwicklung technischer Produkte 2000. Management komplexer Zielkonflikte, Fellbach. VDI Verlag, Düsseldorf, pp 17–26
14. Abele E, Anderl R, Birkhofer H (2005) Environmentally-friendly product development. Methods and tools. Springer, London, pp 95–180
15. European Parliament, European Council (ed) (2005) Directive 2005/32/EC establishing a framework for the setting of eco-design requirements for energy-using products and amending Council Directive 92/42/EEC and Directives 96/57/EC and 2000/55/EC of the European Parliament and of the Council, Brussels 6 July 2005
16. European Parliament, European Council (ed) (2003) Directive 2002/96/EG on Waste Electrical and Electronic Equipment (WEEE), Brussels 27 January 2003
17. Frerichs G, Venturini, P (2001) Commission of the European Communities (ed) Green paper on integrated product policy, Brussels
18. Bayrische Staatskanzlei (ed) (2001) Umweltpakt Bayern 2011. Nachhaltiges Wirtschaften im 21. Jahrhundert. Bayrische Staatskanzlei, Munich
19. Giudice F, La Rosa G, Risitano A (2006) Product design for the environment. A life cycle approach. Taylor & Francis, Boca Raton, pp 83–104
20. Großmann J, Hansen S, Melzer L, Günther J (2008) EuP-Implementierung. In: Abele E, Anderl R, Birkhöfer H, Rüttinger B (eds) EcoDesign. Von der Theorie in die Praxis. Springer, Berlin, pp 13–46
21. Benoît C (ed) (2009) United Nations environment programme guidelines for social life cycle assessment of products. Belgium
22. United Nations (ed) (2007) Indicators of sustainable development: guidelines and methodologies, 3rd edn. United Nations, New York
23. Social Accountability International (2014) Social accountability 8000 – international standard, 4th edn. New York
24. Lawn P (ed) (2006) Sustainable development indicators in ecological economics. Edward Elgar, Cheltenham
25. DIN (ed) (2011) DIN ISO 26000 Guidance on social responsibility. Beuth, Berlin
26. Bossel H (1999) Indicators for sustainable development: theory, method, applications. International Institute for Sustainable Development, Winnipeg

27. Fritz C, Müller K (2008) Umweltgerechte Energieversorgung für mobile Kommunikationsgeräte. In: Abele E, Anderl R, Birkhöfer H, Rüttinger B (eds) EcoDesign. Von der Theorie in die Praxis. Springer, Berlin, pp 83–113
28. Feldhusen J, Grote KH (2013) Die Hauptarbeitsschritte des Gestaltungsprozesses. In: Dies (ed) Pahl/Beitz Konstruktionslehre, 8th edn. Springer, Berlin, pp 465–478
29. Ehrlenspiel K, Meerkamm H (2013) Integrierte Produktentwicklung. Denkabläufe, Methodeneinsatz, Zusammenarbeit, 5th edn. Hanser, Munich, p 354
30. Pahl G, Beitz W, Feldhusen J, Grote KH, Heusel J, Bronnhuber T, Hufenbach W, Helms O, Schlick C, Klocke F, Dilger K, Müller R (2013) Gestaltungsrichtlinien. In: Feldhusen J, Grote KH (eds) Pahl/Beitz Konstruktionslehre, 8th edn. Springer, Berlin, pp 583–752
31. Wolf B, Weißmantel H (2000) Recyclingorientierte Entwicklung – Ein Schritt auf dem Weg zum umweltgerechten Produkt. In: Verein Deutscher Ingenieure (ed) Recyclingorientierte Entwicklung technischer Produkte 2000. Management komplexer Zielkonflikte, Fellbach. VDI Verlag, Düsseldorf, pp 191–200
32. Bras B (2007) Design for remanufacturing processes. In: Kutz M (ed) Environmentally conscious mechanical design. Wiley, Hoboken, pp 283–318
33. Ehrlenspiel K, Kiewert A, Lindemann U, Mörtl M (2014) Kostengünstig Entwickeln und Konstruieren. Kostenmanagement bei der integrierten Produktentwicklung, 7th edn. Springer, Berlin, p 119
34. Feldhusen J, Grote KH (2013) Der Produktentstehungsorozess (PEP). In: Dies (ed) Pahl/Beitz Konstruktionslehre, 8th edn. Springer, Berlin, pp 11–24
35. Grunwald A, Kopfmüller J (2012) Nachhaltigkeit. Eine Einführung, 2nd edn. Campus, Frankfurt, p 104
36. VDI 2221 (1993) Systematic approach to the design of technical systems and products. VDI-Verlag, Düsseldorf
37. Lindemann U (2009) Methodische Entwicklung technischer Produkte. Methoden flexibel und situationsgerecht anwenden, 3rd edn. Springer, Heidelberg
38. Fiksel J, McDaniel J, Spitzley D (1998) Measuring product sustainability. In: J Sust Prod Des, 6 July 1998, pp 7–18
39. Niemann J, Tichkiewitch S, Westkämper E (eds) (2009) Design of sustainable product life cycles. Springer, Berlin, pp 152–158
40. Peleš S, Warnecke G (2000) Transformation von Recyclinganforderungen in der Produktentwicklung – Methoden und Praxisbeispiele. In: Verein Deutscher Ingenieure (ed) Recyclingorientierte Entwicklung technischer Produkte 2000. Management komplexer Zielkonflikte, Fellbach. VDI Verlag, Düsseldorf, pp 67–82
41. Zhao S (2013) Integriertes Produktmanagement von ökonomischen und ökologischen Produkteigenschaften – Eine EcoDesign-Methodik für kleine und mittlere Unternehmen. In: Fortschritt-Berichte VDI Reihe 1 Nr. 419, VDI Verlag, Düsseldorf, p 26
42. Weaver P, Jansen L, van Grootveld G, van Spiegel E, Vergragt P (2000) Sustainable technology development. Greenleaf, Sheffield, pp 184–203

Part II
Design for Sustainability in Emerging Economies

Perspectives on Sustainable Product Design Methodology Focused on Local Communities

Hideki Kobayashi

Abstract The next generation of eco-products, known as sustainable products, must address social concerns as well as satisfy economic and environmental criteria. One of the social aspects of a product is its contribution to the local community. The main driving force of global economic growth currently comes from emerging countries, such as China and India, and it will move to developing countries in the future. To support the development of sustainable products in these countries, a research agenda is presented for establishing a locally oriented sustainable product design methodology. Also, a concept of an extended function-structure analysis method based on a combination of reverse engineering and field observation is proposed. In this method, the scope of analysis is extended in terms of space and time to understand comprehensively the explicit and implicit site-specific characteristics of a product and its life cycle.

Keywords Sustainable product • Locally oriented design • Reverse engineering • Appropriate technology

1 Introduction

Eco-design aims to enhance usage, emotional, and economic value and reduce the environmental burden throughout the entire product life cycle and is becoming increasingly important in addressing global environmental problems. Accordingly, many eco-design methodologies based on life cycle thinking have been developed [1–3]. Some methodologies have already contributed to producing eco-products. Recently, the comprehensive concept of sustainability has become more focused. In academia, sustainability science has been introduced, which examines the dynamic and complex relationship between nature and society [4, 5].

In product development, eco-design focuses on technical innovations and is assessed based on functional, economic, and ecological criteria. Moreover, the concept of design for sustainability (DFS) has been proposed which is more

---

H. Kobayashi (✉)
Department of Mechanical Engineering, Osaka University, Suita, Japan
e-mail: kobayashi@mech.eng.osaka-u.ac.jp

M. Matsumoto et al. (eds.), *Sustainability Through Innovation in Product Life Cycle Design*, EcoProduction, DOI 10.1007/978-981-10-0471-1_6

Fig. 1 Scope of eco-design

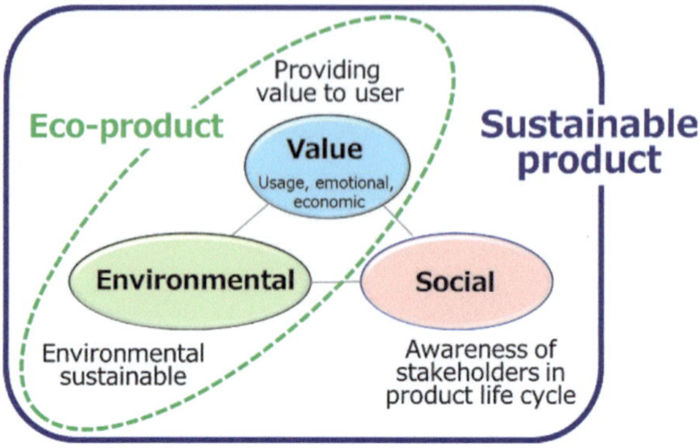

comprehensive than eco-design [6, 7] and is useful to develop the next generation of eco-products, called sustainable products. In this paper, sustainable products are defined as being socially responsible as well as eco-efficient (Fig. 1). Eco-products and sustainable products share the life cycle thinking strategy, in which all stages of the product life cycle are considered. In DFS, engagement of stakeholders is emphasized, which is particularly important for community goods [7]. The local community is also regarded as a stakeholder to be considered in the social life cycle assessment (S-LCA) [8].

The local characteristics of a product and its life cycle have been overlooked in the mass production and mass consumption paradigm, which is to sell products according to global standards. It is more cost efficient to produce a large number of homogenous products. Currently, the driving force of global economic growth is shifting from developed countries to emerging countries such as China and India. It is expected that some of this driving force will shift to developing countries. Thus, product development that meets specific local needs is important for global manufacturing companies and local communities.

In developed countries, eco-design guidelines and product environmental evaluation methods based on general principles or theories are used to design eco-products. Reducing the number of functions in a product or lowering its price is an easy way to develop products for emerging or developing countries. However, these approaches often fail because of the differences in climate, historical background, language, education, social infrastructure, and national character between developing and developed countries. Products must be affordable in developing countries, although this is just one of the necessary conditions. Although currently there is little systematic eco-design for local community, it is becoming increasingly important.

In this work, a research agenda is presented for establishing a locally oriented methodology of sustainable product design. In addition, an extended function-structure analysis method is proposed based on the research agenda. Related work and outstanding issues are summarized in Sect. 2. In Sect. 3, a research agenda is proposed for a locally oriented sustainable product design methodology.

A new design method is described in Sect. 4, and concluding remarks are provided in Sect. 5.

2 Related Work

2.1 Globalization and Glocalization

The global industrial product market depends on mega-trends in technological development. Technology-driven product development is called "product-out." A product-out approach for local communities is globalization based on a competitive product with a common specification worldwide, such as Shimano cycle gears. Globalization of a competitive product creates a telecoupling or teleconnection between local consumption and global land use. For instance, 92 % of the final domestic consumption in Japan is teleconnected to land use overseas [9]. Because the aim of product design for globalization is different from locally oriented design, it is beyond the scope of this study.

Another product-out approach, known as "glocalization," is to localize an original product for a specific region, which efficiently enhances product variety [10]. Glocalization is a portmanteau of globalization and localization. Glocalization is achieved by minor changes or customization of an existing product. Figure 2 shows a television for ASEAN countries [11]. The sales point of the product contains an integrated battery and a radio-frequency (RF) booster because power supply is unstable and reception is weak in those countries. This combination of a localized product with advanced technology is a typical example of glocalization.

Fig. 2 Toshiba's power TV with an integrated battery and a new RF booster [11]

Some design methods can be applied to glocalization. Design for mass customization (DFMC) is a product development methodology for meeting diverse user preferences in a cost-efficient way [12]. Similarly, design for variety (DFV) [13] and product family design methodologies [14] are also suitable. However, they may not be useful for supporting product development in heterogeneous markets, such as developing countries, considering that few successful cases have been reported.

2.2 Field Observation and Local Development Team

Many personal activities with a product depend on the local community and are a result of geographical, climatological, geopolitical, historical, educational, and cultural factors and the community's infrastructure. A popular "market-in" approach is field observation where a skilled local observer or analyst identifies a gap in the market in terms of price, infrastructure, required performance, or user education level. Because field observation is independent from other techniques, it can be easily applied to product development for an emerging or a developing country, although it is relatively expensive.

Field observation can be also used in combination with other approaches, such as glocalization (Fig. 3), and it is a basic element of locally oriented design. However, the success of field observation depends on the person who analyzes the market. Thus, a more systematic, reproducible method is required.

GE Healthcare launched the MAC400 ultra-portable electrocardiograph in India at one-third of the price of similar imported systems [15]. The MAC400 was developed by a local cross-functional team (CFT). As a result, the MAC400 has been accepted in a broad range of markets in 194 countries, and it has sold particularly well in EU countries. This is a well-known case of reverse innovation [10]. Although product development by a local CFT is effective, it is not possible for every manufacturer because of the constraints of organizational management costs.

Fig. 3 A combination of glocalization and field observation

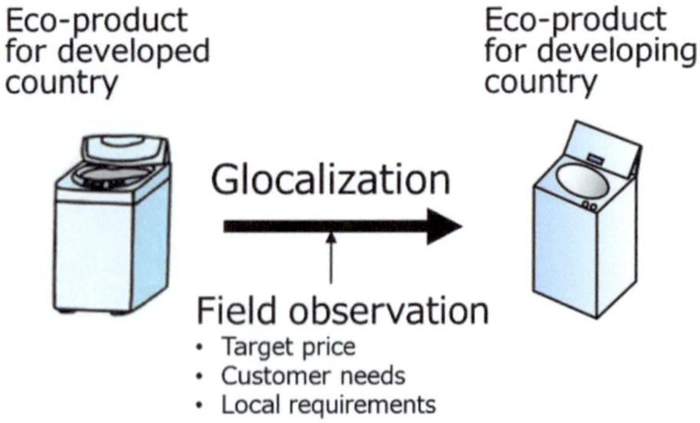

2.3 Appropriate Technology and Co-design

Appropriate technology (AT), which provides sufficient performance at an affordable price, is a new method, first detailed in the book *Small is Beautiful* [16]. In emerging and developing countries, power supply infrastructure is unreliable, making low-power appliances more important than in developed countries. A food storage device, called the Mitticool, that does not use electricity was developed for the Indian market (Fig. 4) and is an example of AT [17]. The Mitticool is inexpensive and it can keep vegetables fresh for about 5 days. The cabinet is made of local clay with no hazardous materials and can be processed safely at the end of life (EOL). It can be produced by local manufacturers, benefitting the local community by increasing local revenue and decreasing the risk of infection and disease for families at the bottom of the pyramid (BOP).

An AT design methodology that is a co-design process between engineers and members of the local community was proposed [18]. Figure 5 shows the concept for

Fig. 4 Photograph of the Mitticool food storage device, which does not require electricity [17]

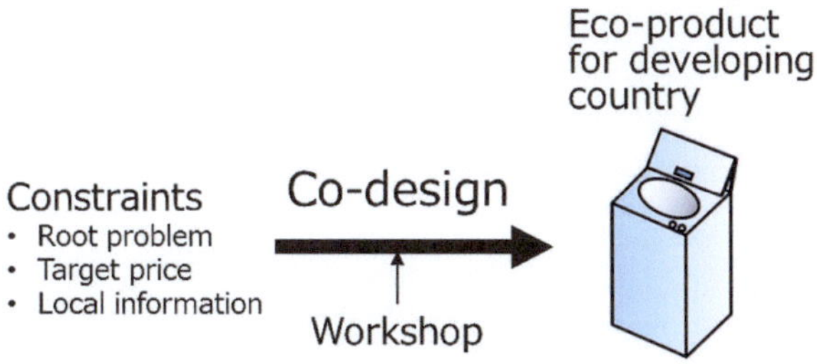

Fig. 5 Co-design approach

a co-design approach, in which a workshop is held at project milestones. A gatekeeper, who connects local people with engineers, plays an important role. The design quality depends on the gatekeeper.

Current co-design approaches are empirical rather than systematic. Therefore, a more systematic, reproducible methodology is desirable. Another critical problem encountered in co-design is the difficulty of sharing information among project members, which can result in gaps in understanding. For example, this was observed among Japanese, Cambodian, and Vietnamese people when they were using English as a communication language. Even if all project members understand English well, their styles of communication will be slightly different because of differences in their cultural backgrounds.

2.4 Reverse Engineering and Redesign

Reverse engineering is a systematic product-out approach for product development [19]. The approach consists of steps including investigation and function analysis through product teardowns and comparing with customer requirements. Based on a reference product, a designer modifies the product function and structure and adds a new technology to meet customer needs (Fig. 6). For example, Samsung uses a combination of field observation by local specialists and reverse engineering of advanced products [20]. Reverse engineering an excellent local product as a reference eco-product could help identify important local requirements. However, this assumption has not been verified.

A key of reverse engineering is function-structure analysis (FSA). FSA is a fundamental method for engineering design. Various FSA methods have been proposed such as [21], which means it is important to consider and select an adequate FSA method for the design objective.

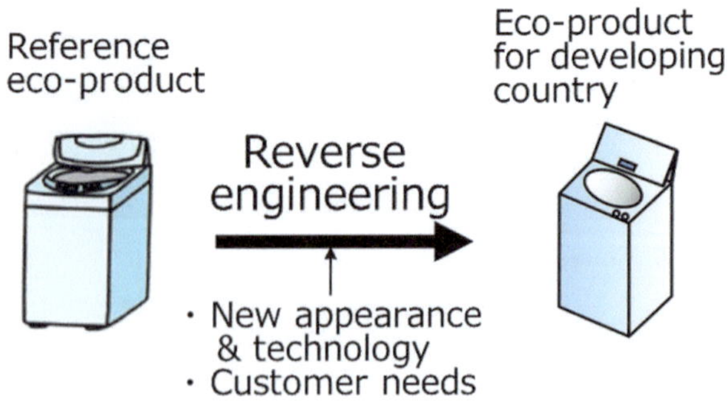

Fig. 6 Reverse engineering approach

2.5 Environmental Problems with Products in Developing Countries

Environmental awareness is essential for developing sustainable products. There are specific environmental problems in product life cycles in developing countries. The first is the lack of databases of environmental burden related to product life cycles. Life cycle assessment (LCA) is a key evaluation method for eco-design. The International Organization for Standardization LCA consists of the following four steps: goal and scope, life cycle inventory (LCI), life cycle impact assessment (LCIA), and interpretation [22]. In the LCI analysis, the total input and output of energy and materials throughout a product life cycle are estimated, and they are converted to environmental burden, such as carbon dioxide emissions, using the LCI database. However, official comprehensive LCI databases have not been developed for emerging or developing countries. To address this problem, a native LCA for selecting an available LCA dataset for a national context has been proposed [23]. However, there is a limit to how precise the estimation of life cycle environmental performance can be.

The second problem is pollution control by waste treatment of harmful materials such as heavy metals. The EOL treatment of products is often inadequate in developing countries. For instance, in China e-waste is recycled via parallel formal and informal processes [24]. In these countries, the risks caused by an informal life cycle process cannot be eliminated. Restricted harmful materials might be used in a product for developing countries, which could be harmful to workers and the local environment at its EOL stage. Therefore, informal life cycle processes must be considered in locally oriented design.

The third problem is that environmental priorities may be different in developing countries compared with developed countries. As discussed in Sect. 2.3, low-energy consumption and renewable energy technology are more important in developing countries than in developed countries. However, reducing carbon dioxide emissions is a low priority. Therefore, attractive eco-technology for product users is different.

3 Research Agenda

Based on the research survey in Sect. 2, the existing approaches are shown in Fig. 7. The locally oriented design methodology should be developed so that it is positioned in the systematic "high reproducibility" and "market-in" part of the figure. The following research agenda toward this goal is proposed.

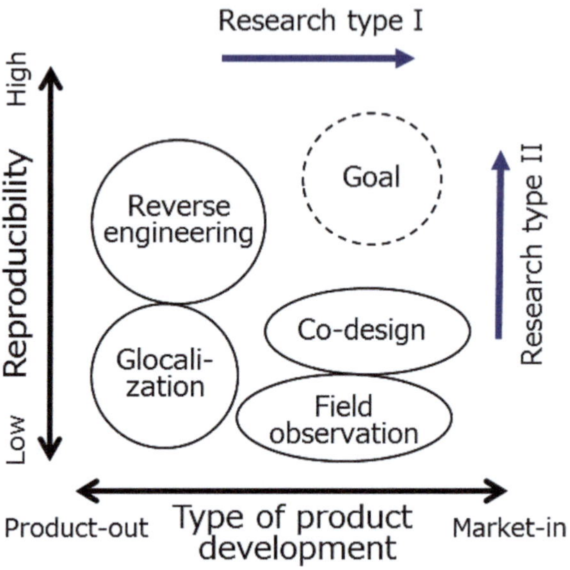

Fig. 7 Locally oriented product design approaches

3.1 Glocalization Approaches

Glocalization will remain as the major approach in global-wide manufacturing companies because the method is easy. However, this approach is the weakest in Fig. 7. The following steps can be taken to improve some of the current weaknesses:

- Use of existing design methods, including DFMC and DFV
- Development of an original product design method for glocalization

3.2 Field Observation Approaches

Field observation is an analysis method and is independent of the design approaches. The following improvements are suggested:

- A procedure to reflect the field observations into the design process systematically
- An engineering communication support tool to connect observers and engineers

3.3 Co-design Approaches

Although the co-design approach is a market-in method, it is not cost efficient or highly reproducible. The following improvements should be considered:

- Development of support for design collaboration to minimize the communication gap caused by cultural differences, even if all members are from developed countries or developing countries
- Selection and education of gatekeepers to lead the co-design process between local people and engineers
- Development of a management method for design projects involving diverse team members

3.4 Reverse Engineering-Based Approaches

As mentioned in Sect. 2, reverse engineering is a current systematic approach. The following research is necessary to improve this approach to address more market-in considerations:

- Improvement of function analysis to handle a broader range of information, including explicit and implicit site-specific requirements
- Integration of a reverse engineering methodology for considering product life cycle such as that in Ref. [3] and "market-in" methods

3.5 Common Technologies

Two common technologies can be developed to enhance research types I and II (Fig. 7). In type I research, local fabrication with additive manufacturing (AM) is a possible route. For manufacturing systems, the term "local" has a specific meaning. Traditionally, a large number of industrial products are produced under precise quality control from a large, efficient factory. AM technology, which is often referred to as "3D printing," is widely used from prototyping to generating complex components for the healthcare, aeronautical, and automotive industries.

Local fabrication using AM is also expected to be used as a tool for sustainable design [25] and could allow recycling and manufacturing to be done in the local community [26]. AM requires a local supply chain rather than a global supply chain, avoiding teleconnections between production and consumption [9]. For local AM fabrication systems to create sustainable products focused on local communities, the following research should be conducted [27, 28]:

- Evaluation of durability, recyclability, disassemblability, and toxicity of products made by AM
- Case studies of LCA and S-LCA of AM-based manufacturing systems

In research type II (Fig. 7), knowledge management (KM) systems are useful. Currently, there are few studies of locally oriented design or AT. Aoki et al. proposed the universal, personal, and customizable (UPC) platform for

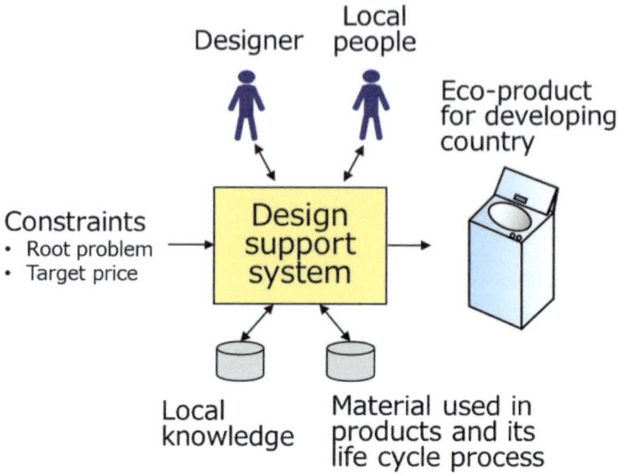

Fig. 8 Design support system for KM

designing and fabricating AT [29]. A system based on the UPC platform supports whole design and fabrication process, including design, prototyping, field testing, and feedback from the field. They highlighted the importance of nebulous knowledge in AT development. Tamura et al. proposed a design methodology for locally oriented manufacturing as an approach for accomplishing social sustainability [30]. It supports locally oriented design using a product model, its value chain, and a checklist.

To reduce environmental burden, it is not necessarily essential to develop an LCI database for developing countries at this time. Rather, a database of materials used in a local product and its life cycle are more important for reducing environmental pollution throughout the product life cycle. The following areas should be addressed (Fig. 8):

- Collection of product life cycle data for local products, particularly at the *EOL* stage
- Development of a database of the materials used in local products
- Development of local knowledge, including alternative fabrication methods and uses
- Establishment of a design support system using feedback from local people

4 Extended Function-Structure Analysis

For locally oriented sustainable product design, an extended function-structure analysis (EFSA) is proposed which is based on a combination of reverse engineering and field observation (Fig. 9).

The EFSA focuses on some expansions of the traditional FSA. The first is spatial expansion to represent site-specific constraint at the product usage stage. Here, it is assumed that products used in daily life are related to other products and services,

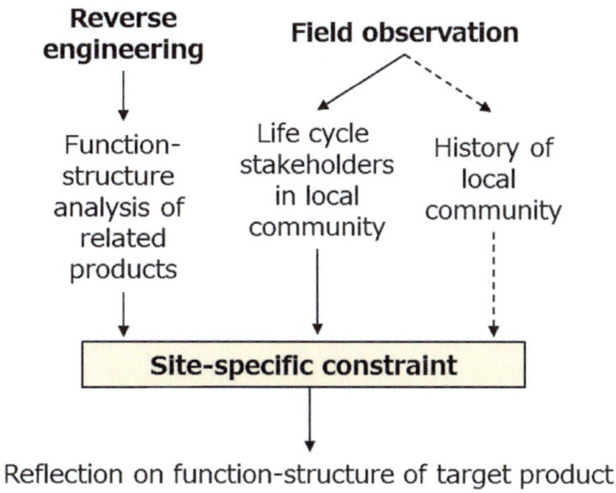

Fig. 9 Schematic of an extended function-structure analysis

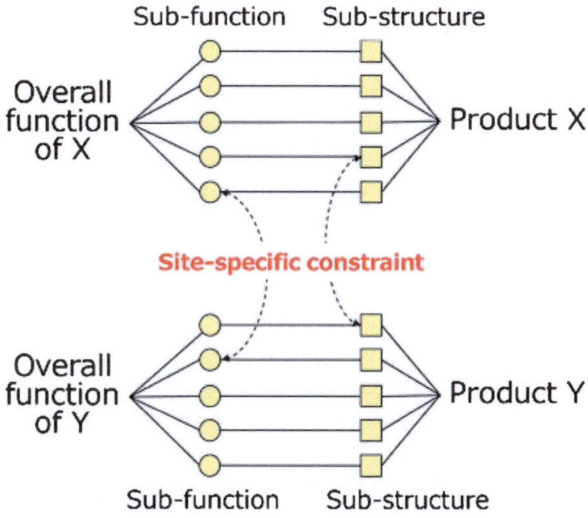

Fig. 10 Spatially extended function-structure

such as housing and electricity infrastructure. Reverse engineering approach using FSA is applied for this purpose. A subfunction or a substructure of product X might be linked to that of product Y via common site-specific constraint (Fig. 10).

The second is life cycle time expansion to represent the requirements of stakeholders of a product life cycle process in local community, particularly manufacturing and EOL. For example, if local treatment for an EOL product is required explicitly, it affects selecting materials for products.

The third is retrospective time expansion to represent a historical reason for the current site-specific constraints. A sort of historical analysis may be useful to understand why different product architectures exist in different locations. Field observation supports the second and third expansions.

Then acquired site-specific constraints reflect on FSA of a target locally oriented product.

5 Concluding Remarks

In the EFSA, the existence of common site-specific constraint is assumed. We can feel it when we look at whole products of an average family [31]. Therefore products used in daily life at the same site seem to share common implicit and explicit constraints. For example, weak electric infrastructure encourages nonelectric or ultralow-power appliances, and weak water infrastructure fosters utilization of rainwater and purification of water. In Ref. [30], a checklist plays an important role for designing locally oriented product, which includes local information of a product. However, the effort for making checklist is large and its quality is fluctuated by an analyst. The EFSA method is useful to collect checklist information comprehensively.

In Sects. 2 and 3, hybrid approaches combining field observation and glocalization or reverse engineering were discussed. It is easy to combine these approaches because field observation is independent from glocalization or reverse engineering. The EFSA is one of those approaches. Other hybrid approaches are also possible, such as combining reverse engineering and co-design. Up to now, the two approaches have not overlapped. To integrate the approaches, integration of the design processes is necessary and this is left as a topic for future work.

In general, locally oriented sustainable product design is important for developed countries as well as developing countries. Developed countries and their local communities do not necessarily have similar cultures. For example, European countries, the USA, and Japan have different cultures. Locally oriented sustainable product design can be used for customization among them, too.

In this paper, a research agenda was proposed for establishing a locally oriented sustainable product design methodology. It provides comprehensive perspectives on this research field and suggests new research ideas. Based on the research agenda, a concept was proposed for the EFSA method to represent site-specific information more explicitly, by expanding the FSA of a single product on the space and time axes. Future work will include studies based on the research agenda and verification of the proposed EFSA method with case studies.

Acknowledgments The author thanks Ass. Prof. Shinichi Fukushige of Osaka University for valuable discussion.

References

1. Ramani K, Ramanujan D, Bernstein W, Zhao F, Sutherland J, Handwerker C, Choi J, Kim H, Thurston D (2010) Integrated sustainable life cycle design: a review. ASME J Mech Des 132:091004
2. Umeda Y, Takata S, Kimura F, Tomiyaama T, Sutherland J, Kara S, Herrmann C, Duflou J (2012) Toward integrated product and process life cycle planning – an environmental perspective. CIRP Ann Manf Tech 61(2):681–702
3. Kobayashi H (2005) Strategic evolution of eco-products: a life cycle planning methodology. Res Eng Des 16(1–2):1–16
4. Kates R et al (2001) Sustainability science. Science 292:641–642
5. Komiyama H, Takeuchi K, Shioyama H, Mino T (eds) (2011) Sustainability science: a multidisciplinary approach. United Nations University Press, Tokyo
6. Arnette A, Brewer B, Choal T (2014) Design for sustainability (DFS): the intersection of supply chain and environment. J Clean Prod 83:374–390
7. Spangenberg J, Fuad-Luke A, Blincoe K (2010) Design for Sustainability (DfS): the interface of sustainable production and consumption. J Clean Prod 18:1483–1491
8. Dreyer L, Hauschild M, Schierbeck J (2006) A framework for social life cycle assessment. Int J LCA 11(2):88–97
9. Yu Y, Feng K, Hubacek K (2013) Tele-connecting local consumption to global land use. Glob Environ Chang 23:1178–1186
10. Govindarajan V, Trimble C (2012) Reverse innovation. Harvard Business Review Press, Boston, USA.
11. Toshiba (2010) Power TV. https://indahelektronik.wordpress.com/tag/toshiba/. Accessed 6 Nov 2015
12. Tseng M, Jiao J (1998) Design for mass customization by developing product family architecture. Proceedings of DTEC DETC98/DFM-5717. ASME
13. Martin M, Ishii K (2002) Design for variety: developing standardized and modularized product platform architectures. Res Eng Des 13:213–235
14. Jiao J, Simpson T, Siddique Z (2007) Product family design and platform-based product development: a state-of-the-art review. J Intell Manuf 18:5–29
15. University of pennsylvania (2010) 'Reverse innovation': GE makes India a lab for global markets. http://knowledge.wharton.upenn.edu/article/reverse-innovation-ge-makes-india-a-lab-for-global-markets/. Accessed 6 Nov 2015
16. Schumacher E (1973) Small is beautiful: economics as if people mattered. Blond & Briggs, London, UK.
17. The Hindu (2012) http://www.thehindu.com/features/metroplus/society/refrigerator-sans-electricity/article3896858.ece. Accessed 6 Nov 2015
18. Sianipar C, Yudoko G, Dowaki K, Adhiutama A (2013) Design methodology for appropriate technology: engineering as if people mattered. Sustainability 5:3382–3425
19. Otto K, Wood K (1998) Product evolution: a reverse engineering and redesign methodology. Res Eng Des 10:226–243
20. Itohisa M, Igari E, Yoshikawa R (2007) Reverse engineering development process in Samsung electronics. University of Tokyo MMRC discussion paper 165 (in Japanese)
21. Pahl G, Beitz W (1988) Engineering design – a systematic approach – the design council
22. ISO14040 (2006) Environmental management – life cycle assessment – principles and framework. ISO
23. Silvestre J, Lasvaux S, Hodkova J, Brito J, Pinheiro M (2015) Native LCA – a systematic approach for the selection of environmental datasets as generic data: application to construction products in a national context. Int J LCA 20:731–750
24. Chi X, Streicher-Porte M, Wang M, Reuter M (2011) Informal electronic waste recycling: a sector review with special focus on China. Waste Manag 31:731–742

25. Diegel O, Singamneni S, Reay S, Withell A (2010) Tools for sustainable design: additive manufacturing. J Sus Dev 3(3):68–75
26. Heyer S, Seliger G (2011) Open manufacturing for value creation cycles. Proc EcoDesign 2011:109–114
27. Kohtala C, Hyysalo S (2015) Anticipated environmental sustainability of personal fabrication. J Clean Prod 99:333–344
28. Huang S, Liu P, Mokasdar A, Hou L (2013) Additive manufacturing and its societal impact: a literature review. Int J Adv Manuf Tech 67:1191–1203
29. Aoki S, Akaishi M, Hori K (2012) UPC Platform for design and fabrication of appropriate technology. Proceedings of the 26th Anal Conf JSAI 1/1-R-4-3 (in Japanese)
30. Tamura T, Kobayashi H, Umeda Y (2015) Proposal of design methodology for local oriented manufacturing in developing countries. Procedia CIRP, to appear.
31. Menzel P, Mann C, Kennedy P (1995) Material world: a global family portrait. Counterpoint

Proposal of a Design Method for Local-Oriented Manufacturing in Developing Countries First Report: Problem Description and Knowledge Representation

Tomoyuki Tamura, Hideki Kobayashi, and Yasushi Umeda

Abstract One of the largest bottlenecks of social sustainability is disparity between developed and developing countries. For decreasing the disparity by raising the quality of life in developing countries, various products and services are expected to be diffused. Conventional products designed for developed countries are not fit to developing countries because of the differences between "locality" including culture, economy, and other situations. We should focus on locality of developing countries in the product design. In this study, we propose "Local-Oriented Manufacturing (LOMan)." LOMan is a concept to encourage designers to focus on locality at manufacturing and use stages. The objective of this research is to propose a design methodology for LOMan. Especially, this paper clarifies the problem of LOMan by describing the results of case studies and a field survey. Based on these results, this paper then proposes "Local-Oriented Manufacturing map (LOMmap)" as a method of knowledge representation for the LOMan design. As a result, we found two kinds of information are needed for supporting LOMan design. One is the information on the local circumstances. The other is the information on influences on a product life cycle caused by the local circumstances. LOMmap supports a designer in determining product specification for LOMan design by representing the two kinds of information.

Keywords Design • Sustainable manufacturing • Social sustainability • Appropriate technology • Local-oriented manufacturing

---

T. Tamura (✉) • Y. Umeda
Department of Precision Engineering, The University of Tokyo, Tokyo, Japan
e-mail: tamura@susdesgin.t.u-tokyo.ac.jp

H. Kobayashi
Department of Mechanical Engineering, Osaka University, Osaka, Japan

M. Matsumoto et al. (eds.), *Sustainability Through Innovation in Product Life Cycle Design*, EcoProduction, DOI 10.1007/978-981-10-0471-1_7

1 Introduction

One of the largest bottlenecks of social sustainability is the disparity between developed and developing countries. The main disparity is economic disparity, which causes other disparities. A measure to decrease the disparity is to raise the quality of life (QOL) in developing countries by, e.g., diffusing various products and services. However, economy, law, culture, and other situations in developing countries disturb diffusing the products made for developed countries. The situations have different characteristics area by area. "Locality" means such characteristics inherent to the areas where the product is manufactured and used. The current product design for BOP business focuses only on decreasing price of product [1]. Conventional products designed for developed countries cannot be diffused in developing countries because of the differences between the locality in developed countries and that in developing countries. Therefore, we should focus on the locality of developing countries in designing and manufacturing products.

For applying technology to developing countries, the concept of "appropriate technology" was proposed [2, 3]. Advanced technologies in developed countries do not always bring the best results in developing countries because there are some problems caused from the constraints of the economic situations, culture, environment, and other characteristics of developing countries [4]. Therefore, technology should fit to the characteristics of developing countries.

For developing products fit to the developing countries, we need design methodologies in addition to the appropriate technology [5]. Sianipar et al. proposed a design methodology for AT that supports describing and evaluating design results. However, the methodology does not clarify connection between locality and product. Clarifying the connection will be a support to develop the products suitable to locality.

Such design methodologies do not exist.

In this study, we propose "Local-Oriented Manufacturing (LOMan)" as an approach for achieving social sustainability. We focus on the design stage. The design stage is important for LOMan because design decides all characteristics of a product. The LOMan design denotes the design of product that fits to the locality. The objective of this research is to propose the methodology for the LOMan design. In this paper, we propose a method of knowledge representation to determine the specification of the product based on LOMan. The companies may have already collected data and executed market research for local situations. But they did not succeed in LOMan at least in Japanese electric manufacturers. We observed that they failed to structuralize the collected data and to transfer the extracted knowledge to the designer correctly. We, therefore, propose the method of knowledge representation to represent them to the designer in this paper. The rest of this paper is organized as follows: Sect. 2 proposes the idea of LOMan. Then, we clarify the problem of LOMan by describing the results of case studies and a field survey in Sect. 3 and Sect. 4, respectively. Based on these results, Sects. 5, 6, and 7 propose a method of knowledge representation for supporting the LOMan design. Then,

Sect. 8 describes an example of the use of the design support tool. After describing the results in Sect. 9, Sect. 10 concludes this paper.

2 Local-Oriented Manufacturing

Currently, each process of product life cycle is executed in different areas, for example, the area where the product is designed and manufactured is different from the area where the product is used, whereas each area has its own locality in, for example, lifestyle, culture, economy, institution, and legislation. Such locality influences on the product design and each process of the product life cycle.

Figure 1 shows an image of LOMan. Locality influences on the product life cycle including manufacturing, remanufacturing, and use stage. At the same time, the products influence on the locality. In LOMan, the product should be fit to the influences. Generally speaking, the product designed and manufactured for one area is not suitable for other areas because of the difference in the locality. Methods for supporting the LOMan design and manufacturing are needed.

LOMan is the concept for designing and manufacturing a product so as to be suitable for locality. LOMan focuses on the locality of the areas for manufacturing, product use, and other life-cycle processes at design and manufacturing stages. For designing a product and manufacturing it based on LOMan, some methodologies are needed, for example, design for LOMan, supply chain management for LOMan, and manufacturing planning for LOMan. We focus on the design stage in this paper; we aim at proposing a design methodology based on LOMan to support a designer in designing a product to be suitable for the locality.

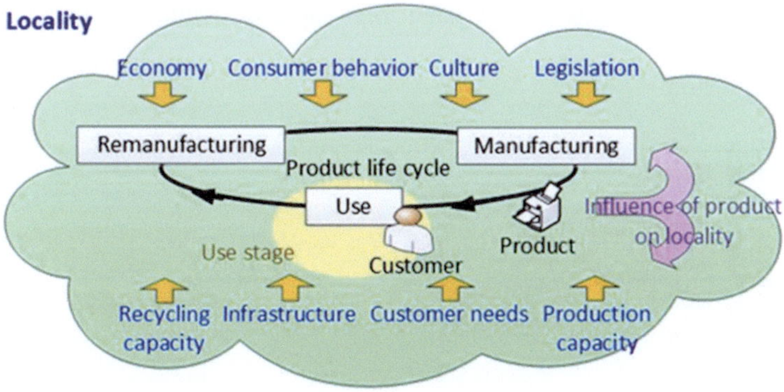

Fig. 1 Image of LOMan

3 Case Studies

By comparing the products for developing countries with those for developed countries, we investigated the influences of the locality where the product is used.

We chose refrigerators and vacuum cleaners for the comparison because they are typical consumer electrical appliances, and therefore, they may reflect the area where the product is used.

3.1 *Influences of the Economic Conditions on Refrigerators*

We investigated the influence of economic conditions on characteristics of the product. We used GDP per capita as an economic condition (x-axis of Fig. 2) and the average total capacity of the refrigerators as a characteristic (y-axis of Fig. 2). Figure 2 shows the correspondence between the GDP per capita and the average total capacity of refrigerators. Since the correlation coefficient between GDP per capita and the average total capacity of refrigerators is 0.707, there is a correlation between them, whereas the average total capacity of refrigerators in Malaysia and these facts imply that some hidden factors such as culture and lifestyle besides economic conditions influence on the average total capacity of the refrigerators. For the LOMan design, the product should be adopted to these factors.

3.2 *Refrigerator*

We compared a refrigerator for Indonesia with that for Japan. We chose a refrigerator for Japan that has almost same total capacity as that for Indonesia. The

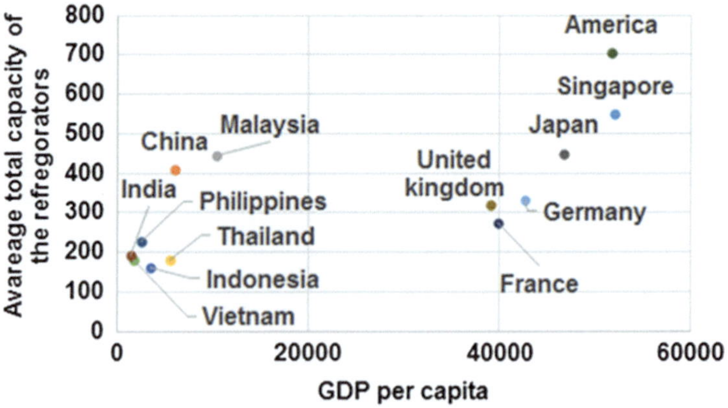

Fig. 2 Diagram of GDP per capita (US$) [6] and the average total capacity of refrigerators (L) [7]

Table 1 Specifications of refrigerators for Indonesia and for Japan

	For Indonesia	For Japan
Price ($)	142	380
Type	GR-Y188	NR-B175W
Manufacturer	Toshiba	Panasonic
Number of doors	1	2
Total capacity (L)	160	168
Color	Purple	Brown
W*D*H (mm)	607*526*1100	480*586*1293
Energy consumption (W)	75 (Compressor)	57 (Compressor)
		120 (Defroster)
Refrigeration room capacity (L)	133	124
Freezer capacity (L)	27	44
Refrigerant	R134a	Isobutane

refrigerator for Indonesia is for a family and the refrigerator for Japan is for a single. Table 1 indicates their specifications. Figure 2 indicates that the total capacity of this refrigerator for Japan is smaller than the average total capacity of the refrigerators in Japan, whereas the total capacity of the refrigerator for Indonesia in this case study has almost same total capacity as the average total capacity of the refrigerators in Indonesia.

First, we compared their specifications (see Table 1). There are three main differences in Table 1. The first difference is in their prices. Price of the refrigerators for Japan is higher than that for Indonesia. It is assumed that this difference mainly comes from the difference of the door numbers. The second difference is in refrigerants. Isobutane has lower global warming potential than R134a, but it is flammable. It is assumed that this difference came from the difference of environmental legislation between these countries. The third difference is in energy consumption. When the compressors are only used, the energy consumption of the refrigerator for Japan is smaller than that for Indonesia. If the defroster is used in the refrigerator for Japan, its energy consumption is higher than that for Indonesia.

Then, we compared their characteristics not expressed in the specifications. The comparison revealed two typical differences. The first difference is in material indication. The plastic cabinets used for the refrigerator for Japan indicate their materials, for example, PS, whereas the refrigerator for Indonesia does not indicate materials. The difference in the environmental legislation in Japan and Indonesia might cause the difference of the material indication. The second difference is their top boards. The top board of the refrigerator for Japan is heat resistant, but the top board for Indonesia is not. In Japan, a microwave oven is often put on a small refrigerator. The difference in usage environment may cause this difference.

From this case study, we confirmed the influence of the locality on the refrigerator. Legislation, usage environment, and economy are example of the elements of the locality. Information on them is needed for the LOMan design.

3.3 Vacuum Cleaner

We compared vacuum cleaners for Malaysia with one for Japan. Table 2 shows specifications of the vacuum cleaners. The comparison of the vacuum cleaners shows, for example, the difference in the power controller on the horse (see Fig. 3). The vacuum cleaner for Japan is controlled by switches. Those for Malaysia are controlled by the hole and the slider. The air goes in through the hole. The amount of the airflow is controlled by changing the size of the hole by moving the slider. The switches of the vacuum cleaner for Japan have four functions, namely, switching the brush on/off in the head, saving energy, changing the power, and switching on/off. The power control by the switch needs smaller area than the power control with the slider, and the switches can have some functions in addition to the power control, whereas having many functions may increase the cost and some of them may not be used by the user. User can adjust the power with the size of the hole.

For the LOMan design, information on the preference of the consumer in the area is needed.

Table 2 Specifications of vacuum cleaners for Malaysia and for Japan

	For Japan	For Malaysia	For Malaysia
Price ($)	334	107	130
Type	EC-PX600	CVC-PH2000CH	VCDC20AV
Manufacturer	SHARP	Cornell	Samsung
Power	850 W (max)	1800 W	2000 W
Weight (kg)	4.3	7.26	5.6

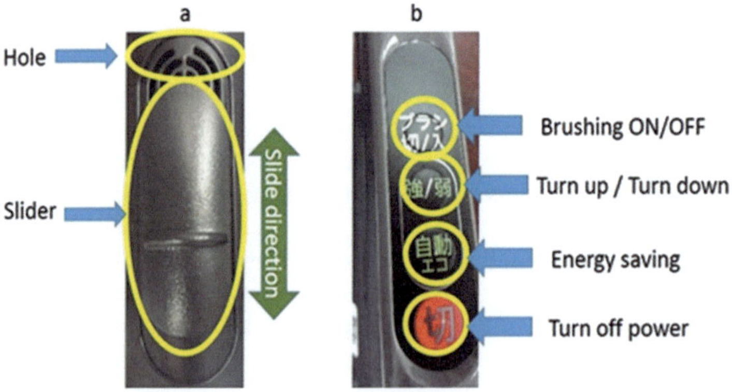

Fig. 3 (**a**) The controller of the vacuum cleaner for Malaysia; (**b**) the controller of the vacuum cleaner for Japan

4 Field Survey

We visited electric shops in Malaysia and looked for the locality of Malaysia in vacuum cleaners and refrigerators.

In Malaysia, two doors, 300–400 L refrigerators with 84–112 US $ are the most selling. In addition, freezers with 200 l capacity for the family are also selling well. The food culture may influence on the choice of the refrigerators, for example, they stored large amount of foods in a freezer.

Samsung advertised their inverters in their refrigerators as "Digital Inverter" in Malaysia. Japanese refrigerator manufacturers did not advertise in such a way. By such advertisement, people who do not know the inverter in refrigerator well may get an impression that inverters used in Samsung refrigerators are special. The impression raises the brand image of Samsung in Malaysia.

5 Framework of the LOMan Design

From the case studies and the field survey, we found the relationship between the locality and a product and its life cycle (see Fig. 4). The locality influences on the product at each process of product life cycle. On one hand, if the influences are good for the user, we should enhance the influences; on the other hand, if the influences are bad, we should decrease the influences. In the LOMan design, a designer should find out the influences of the locality on the product and, if needed, add countermeasures for the influences at design stage. Such countermeasures should be embedded into product design and product life cycle.

For supporting the LOMan design, the following three kinds of information are needed:

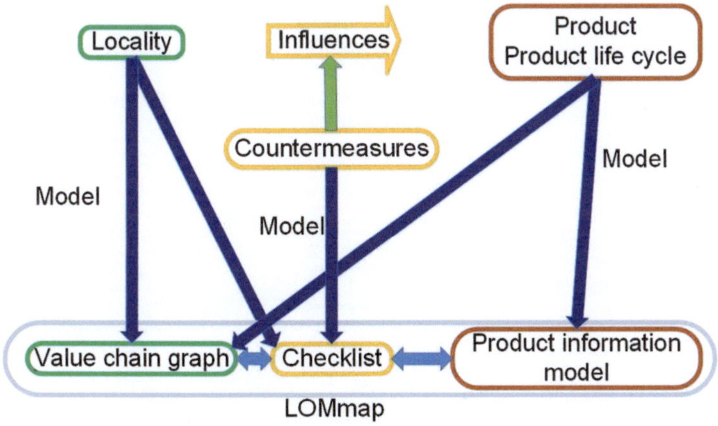

Fig. 4 Requirements for the LOMan design

1. Information on the product life cycle and the areas where the processes of the life cycle are executed, for example, products are sold in Malaysia.
2. Information on the potential influences of the locality on a product life cycle and potential candidates of countermeasures for them; the influences of the floor on the head mechanism of the vacuum cleaner material are an example. To use hard rubber and collect dust is an example of countermeasure.
3. Information on the product, for example, the structure of the vacuum cleaner.

6 Proposal of Local-Oriented Manufacturing Map

6.1 Structure of LOMmap

We propose "Local-Oriented Manufacturing map" (LOMmap) as a knowledge representation scheme for the LOMan design. LOMmap supports a designer in determining the specification of a product for the LOMan design by representing the three kinds of information described in Sect. 5. Figure 5 represents the construction of LOMmap, which consists of a value chain graph, checklists, and a product information model and support tools.

The value chain graph represents the product life cycle and the area where each process of the product life cycle is executed. The checklist represents the potential

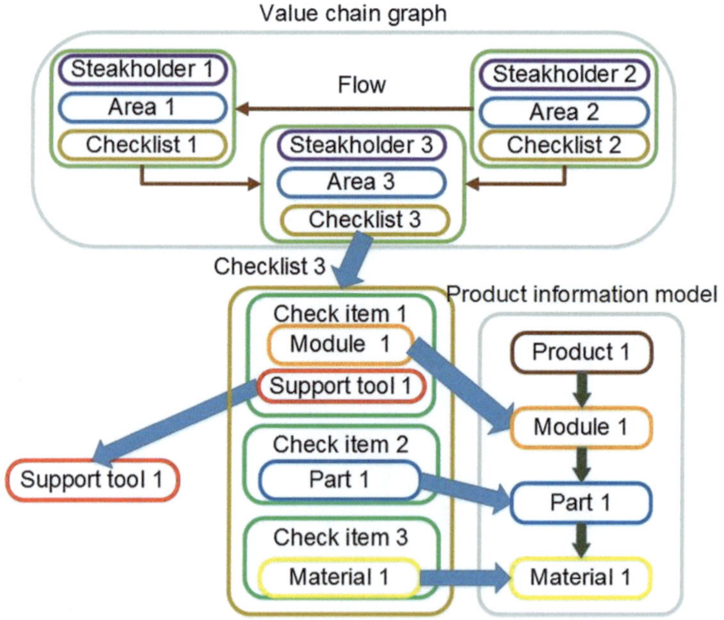

Fig. 5 Construction of LOMmap

influences of the locality on a product and potential candidates of countermeasures to them. The product information model represents the information on the product.

LOMmap is used at the design stage. LOMmap represents one type of product for an area where the product is used. If the supposed area is changed, another LOMmap should be constructed.

6.2 Value Chain Graph

The value chain graph represents stakeholders and the areas where the processes of the product life cycle are executed in the form of a graph (see Fig. 6). Each node represents a stakeholder of the product such as a customer and the area where the stakeholder concerns the products, parts, and materials used in the products, and other resources such as money and electricity. Each arc represents the flows of them. Each node and arc in the value chain graph have a checklist.

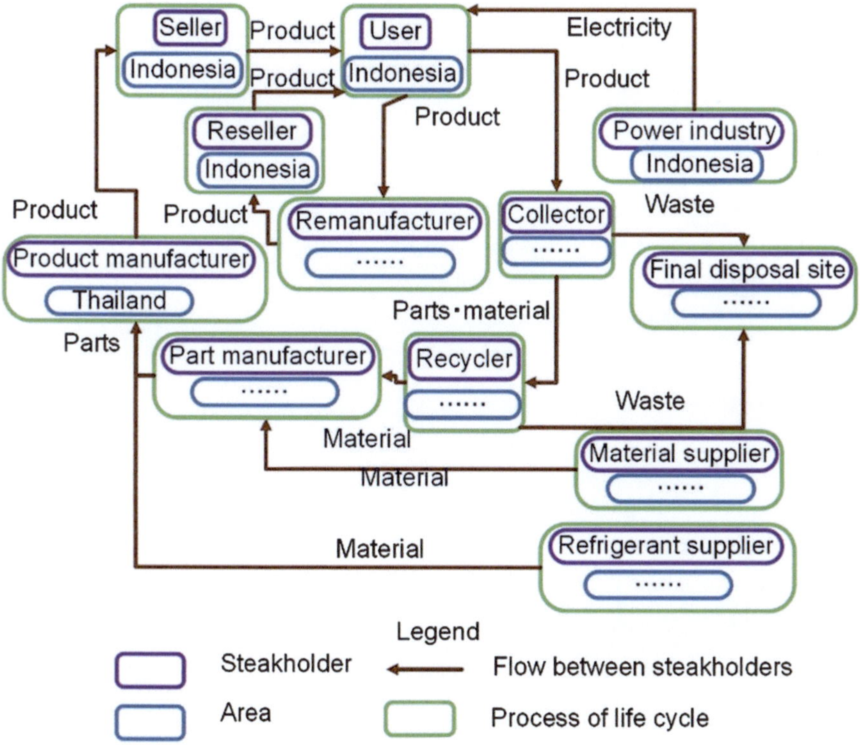

Fig. 6 Value chain graph of the refrigerator for Indonesia (hypothetical)

6.3 Product Information Model

The product information model represents the structure of the product (see Fig. 7). The model consists of a product, modules, parts, and materials of the product. Each element of the model is connected with check items that refer to the component. Each node has the information on attributes of the element.

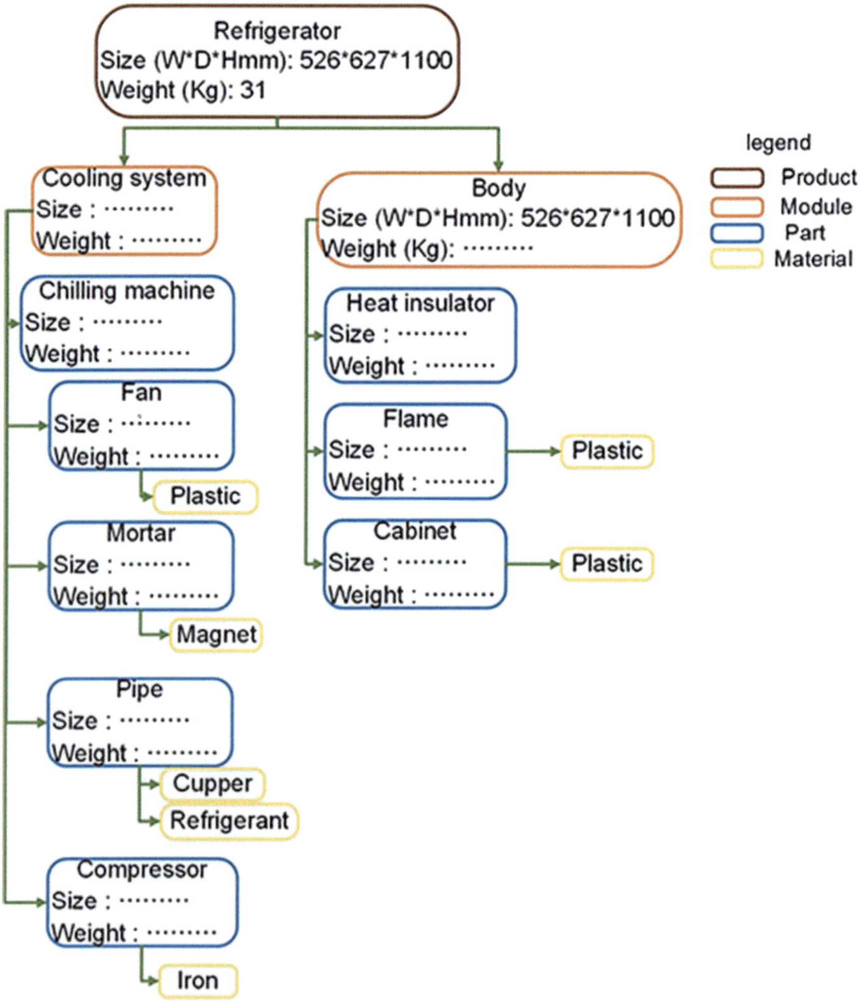

Fig. 7 Product information model of the refrigerator for Indonesia (hypothetical)

6.4 Checklist

A checklist consists of check items. Each check item represents the potential influence and the potential countermeasures and has information described below (see Tables 3 and 4):

- Requirement: This item represents the requirement for the LOMan design.
- Aspect: This item represents categories of the check item, such as law, culture, and economy.
- Targeting area: This item represents the area where the process is executed.
- Element of the value chain graph: This item represents the element of the value chain that has the check item.
- Influence: This item represents the influence of the locality on the product.
- Notes for influence: This item explains the detailed information on the influence.
- Local information: This item represents reference information on the targeting area.
- Element of the product information model: This item represents the elements of the product information model influenced by the locality.
- Support tool: If the designer needs more additional information, a support tool is made and connected with the check item, for example, the tool that shows the prices of the vacuum cleaners in the world.
- Relevance to the product: This item represents the degree of importance of the check item to the product design. This item is represented by a point from 1 to 5. The high number shows the high relevance.
- Idea: This item represents ideas of countermeasures.
- Action: This item represents some detailed solutions for realizing each idea.
- Point of action: This item represents how appropriate the action of the check item is by a point 1–5. The higher point shows higher appropriateness.
- Priority of action: This item represents the priority of the action. This item is calculated by multiplying the relevance to the product and the point of the action.

7 Use of LOMmap

This section describes the process for determining specifications of a product by using LOMmap.

First, the designer determines the product to be designed and the targeting area. Then, experts prepare LOMmap before starting the design process. In this study, we suppose the designer who does not know well the locality of the areas.

Table 3 Example of the check item [1]

<table>
<tr><td colspan="2">Item</td><td colspan="3">Details</td></tr>
<tr><td colspan="2">Requirement</td><td colspan="3">Does the refrigerant obey the environmental regesration in the area ?</td></tr>
<tr><td colspan="2">Aspect</td><td colspan="3">Law</td></tr>
<tr><td colspan="2">Targeting area</td><td colspan="3">Indonesia</td></tr>
<tr><td colspan="2">Element of the value chain graph</td><td colspan="3">Product manufacturer</td></tr>
<tr><td colspan="2">Influence</td><td colspan="3">The environmental registlation regulates the refrigerant available for the refrigerator.</td></tr>
<tr><td colspan="2">Notes for Influence</td><td colspan="3">Regulation on the alternative chlorofluorocarbon for the refrigerator is different in each area. Each alternative chlorofluorocarbon has different global warming potential. There are some flammable alternatives.</td></tr>
<tr><td colspan="2">Local information</td><td colspan="3">Using R134a for the refrigerator in not prohibited in Indonesia</td></tr>
<tr><td colspan="2">Element of the product information model</td><td colspan="3">Refrigerant</td></tr>
<tr><td colspan="2">Support tool</td><td colspan="3">N/A</td></tr>
<tr><td colspan="2">Relevance to the product</td><td colspan="3">4</td></tr>
<tr><td></td><td>Idea</td><td>Action</td><td>Point of action</td><td>Priority of action</td></tr>
<tr><td>1</td><td>To use the refrigerant satisfying the minimum requirements for the regulation</td><td>To use R134a</td><td>3</td><td>12</td></tr>
<tr><td>2</td><td>To use the more environmentally conscious refrigerant</td><td>To use Isobutane</td><td>4</td><td>16</td></tr>
</table>

Table 4 Example of the check item [2]

<table>
<tr><td colspan="2">Item</td><td colspan="3">Details</td></tr>
<tr><td colspan="2">Requirement</td><td colspan="3">Does the refrigerant obey the environmental regesration in the area ?</td></tr>
<tr><td colspan="2">Aspect</td><td colspan="3">Law</td></tr>
<tr><td colspan="2">Targeting area</td><td colspan="3">Indonesia</td></tr>
<tr><td colspan="2">Element of the value chain graph</td><td colspan="3">Product manufacturer</td></tr>
<tr><td colspan="2">Influence</td><td colspan="3">The environmental registlation regulates the refrigerant available for the refrigerator.</td></tr>
<tr><td colspan="2">Notes for Influence</td><td colspan="3">Regulation on the alternative chlorofluorocarbon for the refrigerator is different in each area. Each alternative chlorofluorocarbon has different global warming potential. There are some flammable alternatives.</td></tr>
<tr><td colspan="2">Local information</td><td colspan="3">Using R134a for the refrigerator in not prohibited in Indonesia</td></tr>
<tr><td colspan="2">Element of the product information model</td><td colspan="3">Refrigerant</td></tr>
<tr><td colspan="2">Support tool</td><td colspan="3">N/A</td></tr>
<tr><td colspan="2">Relevance to the product</td><td colspan="3">4</td></tr>
<tr><td></td><td>Idea</td><td>Action</td><td>Point of action</td><td>Priority of action</td></tr>
<tr><td>1</td><td>To use the refrigerant satisfying the minimum requirements for the law</td><td>To use R134a</td><td>3</td><td>12</td></tr>
<tr><td>2</td><td>To use the more environmentally conscious refrigerant</td><td>To use Isobutane</td><td>4</td><td>16</td></tr>
</table>

7.1 Preparation of LOMmap

LOMmap is made for one type of product for a targeting area. For constructing the LOMmap, we assume the value chain graph is prepared by an expert who knows well about the product life cycle and the areas where each process of the product life cycle is executed. The expert might be a different person from the designer. We assume that the company collects the check items including local information, ideas, and actions for various targeting areas and constructs knowledge base for them. We assume that the designer prepares the product information model by describing the structure of the reference product, which might be a previous generation of the same type of the target product or a competing product model by a competitor.

An expert constructs the LOMmap. The expert connects each check item to an element of the value chain graph and an element of the product information model. The expert is a different person from the designer and the expert of the value chain graph. The expert should know well the product life cycle and the product which is the same type of the target product.

7.2 LOMan Design by Using of LOMmap

We model that the designer uses the LOMmap for designing a new product as follows:

1. With the prepared LOMmap, the designer understands the current product life cycle from the value chain graph.
2. The designer chooses an element of the value chain graph and checks its checklist.
3. In the checklist, the designer checks each check item as follows:
   - 3.1 The designer understands the requirement, the aspect, the local information, the influence on the product, the notes for the influence, and of the element of the product information model by reading through the check item. If the support tool is connected with the check item, the designer uses the support tool and perceives the additional information. In other words, the designer checks the green items in Table 3.
   - 3.2 The designer sets the relevance of the check item to the product (i.e., the blue item in Table 3).
   - 3.3 The designer checks for the ideas and the actions (the green items in Table 4) as the candidates of the solution for the check item. The designer may add new ideas and actions to the check item.
   - 3.4 The designer determines the point of action for each action. And the LOMmap calculates the priority of actions (the blue items in Table 4).

4. The designer repeats the stages 3.1–3.4 for all elements of the value chain graph.
5. The designer determines the specifications of the product by examining the actions in the order of descending priorities of the actions. According to the specifications, the designer redesigns the target product and modifies the product information model and value chain graph of the redesigned target product.
6. The expert modifies checklists for changes in LOMmap.
7. The designer repeats stages 1–6.

8 Example

We constructed a tiny LOMmap for a refrigerator made for Indonesia. As an example of a check item, we chose the influence of environmental legislation on the choice of the refrigerant described in Sect. 3. We constructed the value chain graph shown in Fig. 6, the product information model shown in Fig. 7, and the check item described in Tables 3 and 4, respectively. In Fig. 7, we described size and weight as attributes. In Tables 3 and 4, the items in green color represent the information for the designer, and those in blue color represent the items that the designer determines.

The designer uses the LOMmap as described in Sect. 7. The designer looks for the information from the green items in Table 3 and sets the relevance to the product. Then, the designer sets the point of actions. As a result, the action "Using Isobutane" has the highest priority of action. The designer determines "to use isobutane" to the specification of the refrigerator.

9 Discussions

By using LOMmap, we succeeded in representing the relationship between locality and its influences on the product life cycle. The value chain graph represents the stakeholders and areas of the product life cycle. The checklists represent influences of the locality on the product life cycle. The product information model represents the structure of the product. LOMmap is for the designer who does not know well the locality of the targeting area where the product is used. By using the LOMmap, the designer understands influences of the locality on the product life cycle and decides the actions for adapting the product to the locality. By executing the action, the specification of the product is expected to become suitable for the locality.

On the other hand, we can point out several issues:

- Some methods for collecting information on the locality influence the product life cycle, and the ideas and actions from the previous products for the checklists are needed.

- Solution candidates designed by the designer may be contradictory with each other. For example, Sect. 8 chooses the isobutane as the refrigerant for the refrigerator. However, isobutane is flammable and this causes the fire risk. The method to assess and solve the contradictory is needed.

10 Conclusion

This paper proposed "Local-Oriented Manufacturing (LOMan)" for achieving social sustainability. LOMan is the concept to focus on the locality at manufacturing and use stages. From case studies to a field survey, we clarified the issues for the LOMan design.

We proposed LOMmap as a method of knowledge representation for supporting a designer in determining specification of a product in the LOMan design and a designer who does not know well the locality of the targeting area.

LOMmap consists of the value chain graph, the product information model, and the checklists. LOMmap represents requirements for the LOMan design in the targeting area by the relationship among these three items. This structure is effective to represent the locality and its influences on the product life cycle.

Future works include:

- Generally speaking, solution candidates designed by the designer may be contradictory with each other. Some support tools for solving these contradictions are needed.
- We are planning to execute a practical case study of the LOMan design and verify advantages and drawbacks of LOMmap.

Acknowledgments This research is partially supported by public interest incorporated foundation "JKA" through its promotion funds from Keirin race.

References

1. Yoshitaka (2013) Institutional changes and BOP business: sustainability and dispersed knowledge. Japan Academy of International Business Studies, Tokyo
2. Shumacher EF (1973) Small is beautiful: a study of economics as if people mattered. Blond & Briggs, London
3. Tanaka N (2012) Appropriate technology and alternative society. Iwanami syoten, Tokyo (In Japanese)
4. TEITEL (1978) On the concept of appropriate technology for less industrialized countries. Technological Forecasting and Social Change, Amsterdam
5. Sianipar, Yudoko, Dowaki, Adhiutama (2013) Design methodology for appropriate technology: engineering as if people mattered. Basel, sustainability

6. The world bank, World Development Indicators GDP (current US$), URL: http://data.worldbank.org/indicator/NY.GDP.MKTP.CD, 1818 H Street, NW Washington, DC 20433 USA (202) 473-1000, 2010
7. JETRO (2010) A survey of well-selling products and price of services in Asia countries and areas. URL: http://www.jetro.go.jp/jfile/report/07000299/asia_kakaku.pdf. JETRO, Tokyo (In Japanese)

Environment-Community-Human-Oriented (ECHO) Design: A Context-Appropriate Design-Thinking Process for the Well-Being of Individuals, Communities, and the Local Environment

Sittha Sukkasi

Abstract This work builds upon the user-centric "design-thinking" methodology to form environment-community-human-oriented (ECHO) design, a process that strives to create solutions that not only meet the needs of the potential users but also create positive experiences and meaningfully influence their communities and the environment. As important as the users, the environment and communities are also key design considerations and target beneficiaries of the design outcomes. ECHO design was applied to solve the lack-of-safe-drinking-water problem in under-resourced communities. The resulting solution was an integration of products and services, consisting of an inexpensive, easy-to-use-and-maintain, aesthetically pleasing, and environmentally friendly water-disinfecting device; a model to fit the use of the device into the local daily routines, skills, resources, communities' cultures, social conducts, spending habits, health understanding, and environmental settings; and a business model aiming to sustain the use of the product, health-oriented mind-set, and positive long-term impacts on the individuals, communities, and the environment.

Keywords Design thinking • Human-centered design • Environmentally conscious design • Solar water disinfection

1 Introduction

Design has been practiced in many disciplines, subjected to various interpretations, and given different definitions. It can be described as "a plan for arranging elements to accomplish a particular purpose" [1]. Designers seek to "match human needs with available technical resources within the practical constraints of business" [2]. Design can also be considered a problem-solving process for finding solutions

---

S. Sukkasi (✉)
MTEC, National Science and Technology Development Agency, Pathumthani, Thailand
e-mail: sitthas@mtec.or.th; sittha@alum.mit.edu

M. Matsumoto et al. (eds.), *Sustainability Through Innovation in Product Life Cycle Design*, EcoProduction, DOI 10.1007/978-981-10-0471-1_8

Fig. 1 Remote controls [3]

Fig. 2 Ticketing service of a mass transport system

that meet human needs. The resulting solutions are typically in forms of products and/or services. Some designs, however, might result in solutions that can deliver the required functionalities to satisfy the users' needs, but fail to further create positive experiences for or meaningful impacts on the users. For example, the remote controls in Fig. 1 might satisfy the users' basic needs: to be able to operate devices, e.g., televisions or cable boxes, from a distance away from the devices. However, many of the existing remote controls, like those in Fig. 1, are unintuitive, difficult to master, and sometimes even frustrating to use.

Another example is the ticketing services of some urban mass transport systems that primarily have ticket machines that only take coins. Consequentially, many passengers who do not have the right amount of coins have to line up at the customer service windows to buy tickets or, in some systems, to exchange for coins first so that they could then go use the ticket machines, making the experience significantly and unnecessarily stressful during rush hours (Fig. 2).

Such products or services fail to create positive user experiences often because the designers (1) did not comprehensively understand the root problems and/or (2) did not exhaustively explore potential solutions before selecting and implementing one. Consequently, the delivered solutions might solve only the symptoms of the real problems, leaving the users not completely satisfied or with a negative experience.

This paper presents a design process that strives to create solutions that meet the needs of and create positive experiences for the potential users, as well as meaningfully influence the users' communities and the natural environment. A case study that used this design process is also discussed.

2 Design Thinking

Design thinking is a methodology that is intended to avoid the shortfalls such as those in the abovementioned examples. Instead of trying to solve the given problems right away, design thinking involves first determining the right problems to solve and then exploring a wide range of potential solutions to deliver the right one [4]. Design thinking is user centric, aiming to not only meet the needs of the users but also improve their experiences and lives [2]. It is about bringing key design constraints—technical feasibility, business viability, and desirability to users—into a harmonious balance. Design thinking is not a rigid design process, and there are several variations of design thinking. While they may differ in detailed descriptions, they all involve similar elements: divergent-convergent thinking, iterative and nonlinear processes, exploration, experimentation, and, most importantly, user-centric mind-set.

One variation of design thinking is the British Design Council's Double Diamond model of design (DDMD). Based on 25 design methods, DDMD consists of four stages: *discover* and *define* phases for determining the right problem and *develop* and *deliver* phases for providing the right solution [5]. Figure 3 illustrates the phases of DDMD. The *discover* phase is meant for the designers to keep their perspectives wide, notice new things, and gather insights. In the *define* phase, the designers narrow down the actable possibilities and identify the fundamental problem. Then, the designers create, prototype, test, and iterate through various possible solutions in the *develop* phase. Finally, the refined solution is finalized and produced in the *deliver* phase. The *discover* and *develop* phases are divergent in nature, as a number of possible ideas are created, while the *define* and *deliver* phases are convergent, as they involve refining and narrowing down to the best idea.

Another design-thinking process is human-centered design (HCD). It involves iteration of four successive activities—observation, ideation, prototyping, and testing—until a satisfactory solution is obtained (Fig. 4) [4]. The process ensures that people's needs are met, that the resulting product is understandable and usable, that it accomplishes the desired tasks, and that the experience of use is positive and enjoyable. HCD can take place within the diverge-converge process of DDMD.

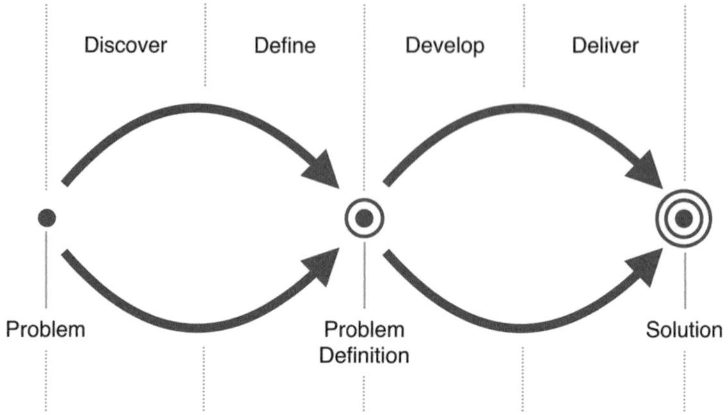

Fig. 3 Double Diamond model of design (DDMD) (Adapted from Ref. [5])

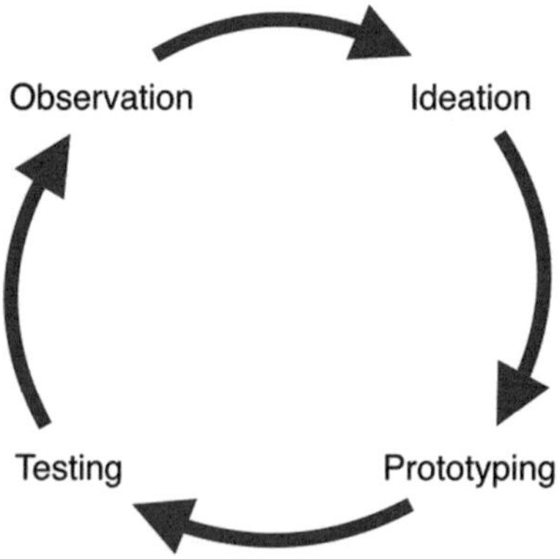

Fig. 4 Iterative cycle of human-centered design (HCD) (Adapted from Ref. [4])

3 Environment-Community-Human-Oriented (ECHO) Design

In many scenarios, focusing on the lives and experiences of the potential users alone is inadequate, and the well-being of the environment and local communities needs to be considered as well, in order to achieve improvement that is effective and sustainable in the long term. In many settings, solutions that are developed solely for the benefits of human, without proper regard to the environment and local communities, could even do more harm than good. Effective solutions in such cases need to be not only user centric but also context appropriate.

This work builds upon design thinking to form environment-community-human-oriented (ECHO) design, a process that strives to create solutions to meet the needs of and create positive experiences for the potential users, as well as meaningfully influence the users' communities and the natural environment. In ECHO design, the

Fig. 5 ECHO core mind-set

environment and communities are key design considerations, but not merely for the benefits of the users' experiences; the environment and communities themselves are also the target beneficiaries of the design outcomes. The interrelations between the users, their communities, and the environment form the core mind-set around which the design goals and activities are oriented (Fig. 5). The subtle difference between the user-centric and environment-community-human-oriented mind-sets may seem trivial but can actually create profound impacts on the design activities and resulting solutions.

Built upon design thinking, ECHO design is, thus, not a rigid process. Indeed, the ECHO mind-set can be flexibly adapted to any variations of design thinking, including DDMD and HCD. The ECHO mind-set can also be used in conjunction with the tools and activities that are typically employed in product design, development, and engineering processes, such as design for assembly, technical specifications, mock-up modeling, etc. Figure 6 illustrates one possible form of ECHO design, in which the ECHO core is integrated into a design process that employs tools from both design thinking and product engineering.

All the tools and activities in the process in Fig. 6 connect to the ECHO core. For example, the *observation* activity is conducted on not only the potential users, their activities, and their capabilities in the actual domains where the solutions will be used; the designers have to also observe the communities of the potential users and their social, cultural, and economic settings, as well as the natural environment and its conditions. The *insights* are gained not only for the sake of the users, e.g., to understand their needs better, or to deliver a product that satisfy the users more effectively. Rather, insights that would lead to solutions that ultimately benefit the communities and environment are needed as well. During *creative thinking* and *brainstorming*, in general design processes, creating numerous ideas without regard for constraints is advised. ECHO design further encourages generating ideas that are in the social or environmental aspects as well, even if they may not be strongly linked to the users. Researching social and environmental issues in the local context can provide new perspectives on the problems. Similarly, creating new ideas from the social and environmental points of view could lead to novel solutions that would have not been realized had the focus been on the potential users only. In *prototyping* and various types of *model making*, mock-ups are created to learn, test, and further refine ideas, which could be of potential problems or solutions. Prototypes, models, and the ideas they embody are tested in ECHO design with respect to not only how

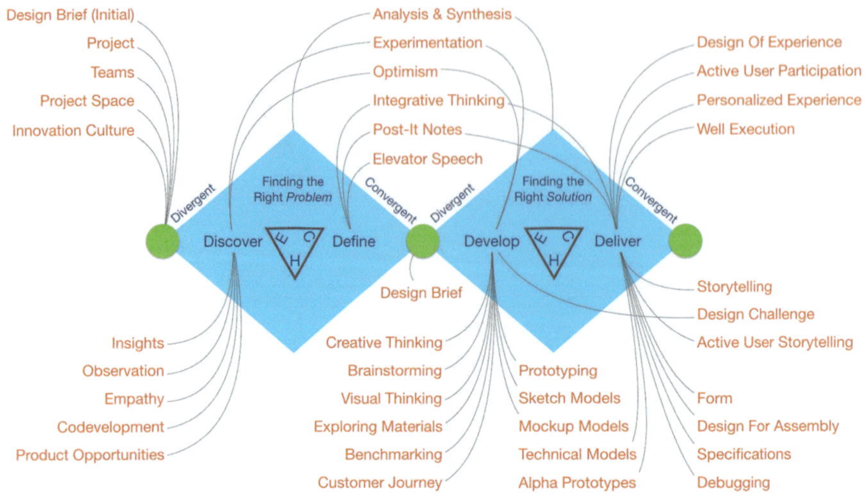

Fig. 6 ECHO design with tools employed from both design thinking and product engineering

well they satisfy the users' needs and experience but also how positively they impact the environment and communities.

ECHO design, like other variations of design thinking, is iterative and nonlinear. Through iteration, the problems can be well defined, and ultimately, well-refined solutions can be developed and delivered to achieve the well-being of individuals, their communities, and the environment.

4 Applications of ECHO Design

ECHO design has been applied by the UpWater project [6] to tackle the problem of lack of access to safe drinking water in under-resourced communities near the Thailand-Myanmar border.

4.1 General Problem Description

Globally, more than 700 million people still lack access to water that is safe enough to drink [7]. Many of them have to rely on surface water and other water sources that are usually contaminated with pathogens that cause diarrheal diseases, which kill as many as 1.5 million people per year, the majority of which are young children in developing countries [8]. The problem affects people in resource-poor areas the most. Over 80 % of the people who lack access to clean drinking water live in rural areas. Moreover, more than one third of people who live in poor countries do not have access to clean drinking water, while almost everyone in rich countries does [9]. These inequities make the problem particularly challenging, as they necessitate an effective, context-appropriate solution that is designed for under-resourced settings of rural and poor areas.

4.2 Discovering Insights into the Users and Problems

First, still during the *discover* phase of ECHO design, in order to gain *insights* into the problems, the UpWater project team set out to identify communities that comprised representative users who could benefit from potential solutions that the project could deliver, i.e., communities whose members were facing the problem of lack of access to safe drinking water. Several communities near the Thailand-Myanmar border in Sangkhlaburi district in Kanchanaburi Province of Thailand were identified as an appropriate starting point. The team then tried to gain as many insights as possible into the potential users themselves, their communities, and the environment (Fig. 7). After several iterations of interviews, *observation*, and *empathizing*, many valuable insights were gained.

The potential users in the communities were mostly Burmese migrants, who had immigrated to Thailand and settled their families in the area since many years or, for some people, decades ago. They did not live in refugee camps, with access to international aids. Instead, they lived in houses that they had built on land that they did not own. Typically, the land belonged to a local temple or local Thai people who allowed the migrants to live there with the conditions that they provided labor for the landowners' nearby rubber plantations or rice farms. They are legally recognized as migrants in Thailand but did not have rights and privileges of permanent residents or citizens. Their incomes were typically low. Most earned their living by working as day laborers and selling non-timber forest products. Those who were more entrepreneurial owned small vegetable or animal farms or convenient stores. Most families had many children, who received basic education at local schools. It could be empathized that the transient nature of the migrants' livelihood (i.e., their limited legal rights and their lack of secure jobs and permanent residences) would have significant impacts on some of the people. Such impacts manifested in many observable forms, such as the temporary nature of their houses.

The sources of drinking water for most of the potential users were wells, small streams, and springs. A small group of people bought drinking water in 20-l containers from local stores. Water samples from those sources were collected and tested, and the results showed that they all contained *Coliform* bacteria that could cause diarrhea. Further discussion with the potential users revealed that they had limited understanding of water safety and diarrheal diseases. Most families drank water without treating it first. Most did not realize that even though the water from the local natural sources may be clear and seem free of contaminants, it still contained pathogens that might not do much harm to the adults but could cause fetal diarrhea to young children.

Fig. 7 Potential users, their communities, and the environment in UpWater project

In addition to gaining insights into the potential users and their problems, the team also observed and tried to understand the communities of the users as well. Four different languages (Thai, Mon, Karen, and Burmese) were spoken in the communities, but most people were fluent in only one or two of them. In each community, the households were typically close, in terms of both space and personal relationship. The members of many households were parts of extended families. Most households in the communities did not have electricity. Some had bought and installed solar panels for lighting and television. Neighbors often gathered at the households with televisions. The community members usually helped one another with construction, farming, and other labors. Many people were members of their community's saving group, which had been helped organize by a local development agency called Pattanarak Foundation. The saving group of each community helped stabilize the members' financial future through saving, provide loans, and make some products that were otherwise expensive or difficult to purchase, such as rice, fertilizers, and animal feeds, available locally at fair prices.

Furthermore, the team studied the natural environment around the communities and found that it predominantly comprised forests, rice farms, and rubber plantations. The forests were abundant with food and non-timber products. During the winter/dry season (around November–February), the temperature could be cold enough to necessitate multiple layers of clothes, and the areas around the household often became dry and dusty. It rained heavily and regularly during the rainy season. There was usually plenty of water in the natural sources except sometimes in the dry season. Sunlight was abundant, sometimes even in the rainy season. The overall natural environment was rich but fragile and at risk of being overexploited.

4.3 Defining a Design Brief

Based on the gathered insights, a clear problem definition that represented the needs of the potential users, their communities, and the environment could be formulated. *Integrative thinking* was employed to incorporate the multiple needs and problems that the team had discovered into a coherent *design brief*. Clearly, there was a need for a solution that could remove pathogens from the drinking water and lower the diarrheal risks for the potential users. However, that alone would not be enough. Simply providing a water treatment to the users would solve one basic need, but it might not create positive experience or meaningful impacts on the users, the environment, and communities in the long term. The solution would need to be inexpensive to obtain, implement, and maintain. It would also need to be easy to use and not require any special knowledge from the users or put extra burden on the limited resources of the environment. The solution also needed to be fit for the way of life of the users and communities. In addition to providing clean drinking water, the solution would also need to impart clear understanding on water safety, hygiene, and related diseases to the all stakeholders in the communities. Moreover, continuity of such understanding and commitment to maintain drinking water safety in the communities needed to be established and fostered. Furthermore, the local environment and natural resources needed to be preserved and, if possible, enriched.

4.4 Exploring and Developing Potential Solutions

Several potential solutions were explored, designed, and developed for different parts of the problem. With respect to water treatment, many potential treatments were explored. Since it was discovered earlier that the water from the local sources contained pathogens but was free of other contaminants, the options could be narrowed down to those that were primarily meant for water disinfection. The aforementioned needs, such as affordability, accessibility, ease of use, ease of maintenance, etc., eliminated many of the remaining options, such as chlorination, leaving two practical options: boiling and solar water disinfection (SODIS). Further exploration discovered that globally only about 20 % of the populations in low-income countries used point-of-use water treatment methods [10]. By far the most popular method was boiling, which in many places could exacerbate existing problems of overuse of locally available fuels, smoke, and poor indoor air quality, which was a major cause of respiratory infections and other health risks such as burning.

Solar water disinfection (SODIS) stood out as an attractive potential solution, as it was simple, affordable, and yet effective. To disinfect water with the SODIS protocol, the user fills the water into a clear container and exposes it to sunlight for 6 h (or approximately one day) on a sunny day or consecutive 2 days if the sky is cloudy [11]. SODIS had been scientifically proven [12, 13], internationally recognized [14–16], and used by many people worldwide [17].

Recently, a new SODIS container design had been developed (Fig. 8). It could disinfect water much more effectively than the original SODIS protocol, yet still maintain its simplicity, affordability, and portability [18]. The method would be able to capitalize on the abundance of sunlight in the environment of the target users, while helping reduce the burden of firewood collection from the nearby forests. SODIS would be proposed as the main water treatment method, while boiling would be promoted alongside it for use during the rainy season.

In addition to water treatment methods, options for education campaigns and training materials were also explored. A number of training materials for water safety, hygiene, and related diseases already existed. Relevant topics were selected and further refined. Additionally, new graphical illustrations were made for all the relevant topics. The illustrations were made relatable to the audience by using drawing of local clothes, objects, and places. Written words were intentionally left out from the graphical materials so that they could be used with all four of the locally spoken languages. To ensure that the education campaign can effectively reach the potential users, local representatives were selected and trained to be field agents. The education activities were designed and planned to be carried out in the evening at the households of the target audience. This would allow the people to be in their comfortable environments and more receptive to the lessons, and more people would be available to attend after returning from their work during the day.

Another part of the solution that was explored, designed, and developed was a plan to involve pilot users from the communities. The water treatment devices

Fig. 8 Improved SODIS [18]

would be provided and integrated into their daily routines. The pilot users would help provide feedback on the usability of the water treatment and hopefully become the agents of change for their communities. This part would leverage the tight-knit nature of the communities. It was assumed that some of the neighbors would be intrigued and follow the lead of the pilot users to adopt the water treatment and safe practice as well, ultimately leading to the overall health betterment of the communities.

The last piece of the solution was designed and developed to help ensure that the solutions would have a sustainable impact. Different distribution models were explored. It was concluded that providing the water treatments to the potential users like a charity would unlikely survive in the long term. Instead, a sustainable distribution and business model was designed. The local saving groups could become the promoters and distributors of products or components related to the water treatments. The new water treatment device was already designed such that no special tools, skills, or materials would be required to produce it. The products could be sold to the members of the communities at prices calculated to be fair based on the costs of materials and labor, as well as typical incomes and expenditures of the people in the communities. Using local people as vendors would generate economic benefits on top of health benefits for the communities and would ensure that products are sold through peers, a model that had often been more effective than NGOs or businesses. Ultimately, the local safe drinking water campaigns should be able to sustain themselves without relying on outside funding.

In summary, the solution that was designed and developed was an integration of several components, consisting of a water treatment method that was inexpensive, compact, easy to use and maintain, easy to understand, and environmentally friendly; a campaign to promote better understanding of drinking water safety and improve the overall health of the communities; and a distribution and plan to ensure long-term positive impacts on the users, their communities, and the environment.

4.5 Implementing the Solutions

The *delivery* of the solution to the potential users was part of the design. Since the ultimate goal was the well-being of the users, their communities, and the environment, the focus of the delivery was not on the water treatment device. Rather, the team spent a lot of time on the education part of the solution. After the local promoters were confident that the potential users had better understanding and awareness on the issues, the water treatment devices were offered to those who were interested. By then many potential users understood the value of the device and viewed it as a tool that could help them achieve the health betterment that they wished for rather than as another charitable donation.

The lead users spurred interest of their neighbors, many of whom followed suit and adopted safe drinking water practices. Many users appreciated the simplicity of SODIS and found it less time-consuming than boiling. Several people also realized that they preferred the taste of water that was solar disinfected to that of boiled water, which smelled of smokes from the firewood. Some people appreciated that they could save their firewood for cooking instead. Consequently, a significant amount of firewood collection from the forest was avoided, reducing burden on the environment.

The local saving groups were presented with the business plans to have them continue the distribution and promotion of the water treatment. Some of the groups already expressed interest.

4.6 Iterations and Improvement

Several insights were gathered as the team proceeded through the ECHO design process. These insights led to the following adjustments to the design and, ultimately, improvement on the solution.

Through trial usage in real use conditions over an extended period, several issues with the design of the SODIS in Fig. 8 could be observed. For instance, if the bags were not hung to dry after each day of use, they would look old and dirty quickly. Also, if the bags were still wet by the next morning, the outer and inner layers would stick together, reducing the effectiveness of the air insulation of the device. To overcome those issues, several new design alternatives were explored, prototyped, and tested. Eventually, the final design, named SODIS X, was developed (Fig. 9). It was much simpler, easier to use, to clean, and to dry, and looked much more refined, which the potential users greatly appreciated.

The education part was also adapted to be a module in training courses that Pattanarak Foundation regularly gave to local people and visitors from other areas. The impact of the campaign could be expanded to a greater audience. A special version of the education campaign was also created to tailor to schoolchildren. New materials, such as coloring sheets, were developed. The lessons were integrated into

Fig. 9 SODIS X

several subjects in school, including science, art, social study, language, and nature conservation.

With regard to sustainable betterment, another distribution and promotion channel was developed. In addition to local business by the community saving groups, other future development training campaigns, such as sustainable food production, would also incorporate the safe drinking water promotion as well.

5 Summary

Built upon design thinking, ECHO design is a design process that strives to create solutions that meet the needs of and create positive experiences for the potential users, as well as meaningfully influence the users' communities and the natural environment. As important as the target users, the environment and communities are also key design considerations and target beneficiaries of the design outcomes. The interrelations between the users, their communities, and the environment form the core mind-set around which the design goals and activities of ECHO design are oriented.

ECHO design was applied to solve the problem of lack of access to clean drinking water in poor communities. The resulting solution was an integration of products and services, consisting of an inexpensive, easy-to-use-and-maintain, compact, easy-to-understand, aesthetically pleasing, and environmentally friendly water-disinfecting device; a model to fit the use of the device into the local daily routines, skills, natural resources, household practices, communities' cultures, social conducts, spending habits, basic health knowledge, and environmental settings; and a business model aiming to sustain the use of the product, health-oriented mind-set, and positive health, economic, and environmental impacts in the long term.

Acknowledgments This work is part of UpWater project ("Breakthrough Point-of-use Water Treatment and Sustainable Health Betterment for Under-resourced Communities"), which is supported by the *Stars in Global Health* program of *Grand Challenges Canada* (grant number S6 0562-01-10).

References

1. Eames Office LLC (2015) Design Q & A text. Retrieved 2015, from http://www.eamesoffice.com/the-work/design-q-a-text/
2. Brown T, Katz B (2009) Change by design: how design thinking transforms organizations and inspires innovation, vol viii. Harper Business, New York, 264 pp
3. Jared, Corin (2006) Remote controls. Retrieved 2015, from https://flic.kr/p/d39sd
4. Norman DA (2013) The design of everyday things, vol xviii. Basic Books, New York, 347 pp
5. The Design Council (2015) The design process: What is the double diamond? Retrieved 2015, from http://www.designcouncil.org.uk/news-opinion/design-process-what-double-diamond
6. Sukkasi S (2014) UpWater. Retrieved 2015, from http://www.upwater.org
7. WHO, UNICEF (2014) Progress on sanitation and drinking-water – joint monitoring programme update 2014. WHO Press, Geneva
8. Prüss-Üstün A, Bos R, Gore F, Bartram J (2008) Safer water, better health: costs, benefits and sustainability of interventions to protect and promote health. WHO Press, Geneva
9. Breu F, Guggenbichler S, Wollmann J (2013) World health statistics 2013. WHO Press, Italy
10. WHO, UNICEF (2011) Drinking water equity safety and sustainability: thematic report on drinking water 2011. WHO Press, Geneva
11. Swiss Federal Institute of Aquatic Science and Technology (2002) Solar water disinfection – a guide for the application of SODIS. Swiss Centre for Development Cooperation in Technology and Management, St. Gallen
12. Acra A, Karahagopian Y, Raffoul Z, Dajani R (1980) Disinfection of oral rehydration solutions by sunlight. Lancet 2:1257–1258
13. Swiss Federal Institute of Aquatic Science and Technology (2013) SODIS: scientific publications. Retrieved 2015, from http://www.sodis.ch/methode/forschung/publikationen/index_EN
14. Swiss Federal Institute of Aquatic Science and Technology (2006) Eawag: Red Cross prize for Sodis. Retrieved 2015, from http://www.novaquatis.eawag.ch/media/20060624/index_EN
15. UNICEF (2009) Soaps, toilets, and taps – a foundation for healthy children. United Nations Children's Fund, New York

16. WHO (2007) Combating waterborne disease at the household level. World Health Organization, Geneva
17. McGuigan KG, Conroy RM, Mosler HJ, du Preez M, Ubomba-Jaswa E, Fernandez-Ibañez P (2012) Solar water disinfection (SODIS): a review from bench-top to roof-top. J Hazard Mater 235/236:29–46
18. Sukkasi S, Terdthaichairat W (2015) Improving the efficacy of solar water disinfection by incremental design innovation. Clean Technol Environ 17(7):1–15

Persuasive Design Aid for Products Leading to LOHAS Considering User Type

Li-Hsing Shih

Abstract This study proposes an aid for generating new design that persuades users to more bodily movement for both health and energy harvesting, leading to LOHAS (lifestyle of health and sustainability) by using a behavior model that suggests users' motives and ability are keys for successful persuasion. The work consists mainly of three parts. Firstly, a questionnaire survey is conducted to gather users' basic information and users' motives and ability in conducting exercise and using products that require bodily movement. Regression models are built to identify potential user's motives and lack of ability based on their information. Secondly, a domain model for case-based reasoning is proposed that explains how design and technology can help persuade users to perform target behavior including moving upper limbs, lower limbs, and whole body. A case library contains more than 90 cases with information of five groups of attributes of the domain model. In the third part, a six-step procedure is proposed based on the concept of case-based reasoning to find useful suggestions on product design from retrieved cases by specifying target users and target behaviors. An illustrative example is presented at the end to demonstrate the proposed approach.

Keywords Case-based reasoning • Design for sustainable behavior • LOHAS • Persuasive design

1 Introduction

Designing products that influence more sustainable behavior is a current focus of many studies, instead of traditional focus on reducing environmental impact and energy use in material extraction and manufacturing stages. Example literatures on design for sustainable behavior (DfSB) include Lilley [1], Wever et al. [2], Lockton and Harrison [3, 4], and Pettersen and Boks [5]. The essence of the studies is establishing relationship between users' behavior change and design principle so

L.-H. Shih (✉)
Department of Resources Engineering, National Cheng Kung University, Taiwan, Republic of China
e-mail: lhshih@mail.ncku.edu.tw

M. Matsumoto et al. (eds.), *Sustainability Through Innovation in Product Life Cycle Design*, EcoProduction, DOI 10.1007/978-981-10-0471-1_9

that useful suggestion on design could be made for achieving target sustainable behaviors. But explicitly describing the cause-effect relationship is not easy since behavior change involves various interacting psychological and social factors. Shih and Hsin [6] tried to overcome the difficulty of describing relationship between design and behavior change based on two ideas: (1) the relationship may be easier to describe if we narrow down the scope and focus on certain type of product and (2) instead of establishing explicit rules to describe the relationship, case-based reasoning may work by properly retrieving past successful cases and getting useful suggestions from past experience. Shih and Hsin [6] and Shih and He [7] intentionally narrowed down the scope and focused on a type of products requiring more human bodily movement. The reasons of selecting this type of product design are twofold: (1) bodily movement could serve as a source of energy harvesting (or called human power) that helps reduce conventional energy consumption and (2) bodily movement often brings benefits in health as well as rehabilitation. Persuading users to more bodily movement could generate clean energy and get healthier body and could also be recognized as a mean leading to a lifestyle of health and sustainability (LOHAS), which was first introduced by Ray and Anderson [8] and often used to define a particular market segment focusing on sustainable living and ecological initiatives.

A case-based reasoning (CBR) approach that extracts knowledge from past cases to provide useful suggestions for new product design is proposed to facilitate DfSB for the underlined product. Concepts of case-based reasoning including setting domain knowledge model, retrieving, and reusing cases are utilized to extract knowledge and experience from existing cases (Aamodt and Plaza [9], Aha [10]). In this study, relationship between behavior change and useful design principles is highlighted via a case library that includes product cases leading users to more bodily movement. Several research such as persuasive technology proposed by Fogg [11, 12] and Fogg and Hreha [13]; persuasive system design (PSD) by Oinas-Kukkonen and Harjumaa [14]; and design with intent (DwI) by Lockton et al. [4, 15] were integrated and extended in building the domain knowledge model which eventually defines the information recorded in case profiling. Shih and Hsin [6] had shown some results of the above ideas. They proposed a domain knowledge model that includes a novel category of target behaviors, design principles, design patterns, and applicable technologies that work with design principles leading to behavior change. A case library was constructed accordingly where each case was expressed as a set of attributes of the domain knowledge model. Designers could therefore get useful suggestions on applicable technologies, design principles, and patterns from retrieved cases as target bodily movement are given as input of conditional search.

In this study, two additional key factors of a behavior model proposed by Fogg [12] including users' motives and ability are deliberately included in the domain knowledge model. Inclusion of users' motives and lack of ability expands and strengthens the power of the domain knowledge model in describing the relationship of behavior change and product design, especially when target users are known. The reason of including these two factors is twofold: (1) by knowing who target users are and their motive and lack of ability could be identified, and (2) by knowing target behavior and target users with identified motive and ability, more suitable cases could be retrieved and more useful suggestions of design principles and applicable technologies could be found. Naturally, the inclusion increases the workload in case profiling, while why and how target users may use the case product requiring prior analysis and is then stored as case knowledge. The advantage of doing so is that more proper cases could be retrieved and more useful design suggestions could be found since motives and lack of ability of target users are identified as additional information in searching for "similar" cases.

The research work includes three parts. Firstly, a questionnaire survey is conducted to gather local users' basic information (about who the users are) and users' motives and ability in conducting exercise and using products that require bodily movement. From the responses of the survey, regression models were built to identify user's motives and lack of ability by knowing user's basic information like age, gender, habit of exercise, and so on. Secondly, a new domain model for case-based reasoning is proposed that explains how design and technology can help persuade users to perform target behavior including moving upper limbs, lower limbs, and whole body. There are five groups of attributes in the domain model including target behavior, design principle, design pattern, applicable technology, and newly added user's motive and ability. A case library contains more than 90 cases with information of the five groups of attributes. Finally, a searching engine is written based on the concept of case-based reasoning to find useful suggestions on product design by specifying target users and target behaviors, while motives and lack of ability of the target users could be identified based on the results of regression analysis. Following sections will explain detail of the work and present an illustrative example.

2 Statistical Models for Identifying Users' Motives and Lack of Ability

Since products that require bodily movement are of interest, potential users' attitude, behavior, and habit regarding exercise and bodily movement are investigated as well as their motives and lack of ability in using products requiring bodily movement. A questionnaire containing questions about people's age, gender, education, vocation, time, and frequency of doing exercise, motive, and lack of ability were presented to local people to gather information about who the users are and their motives and lack of ability when they use products that require bodily movement. Statistical model based on regression analysis is built to describe the relationship between user information (independent variables) and their motives and lack of ability (dependent variables).

There were more than 45 questions containing four groups of questions:

1. Respondents' basic (demographical variables) information such as age, gender, education, and vocation.
2. Exercise habit including health condition, frequency, and average time for weekly exercise.
3. Motives for using products requiring bodily movement. A typical question is like: do you agree providing audio device (like playing music) is important in using product that requires bodily movement? Respondents can rate level of agreement on the statement with Likert 5-point scale.
4. Lack of ability. A typical question is like: if the product have some device to help you reduce complicated operation, would you be more willing to use it?

Questions in (1) and (2) were used to identify who the users are, while questions in (3) and (4) include various measures of users' motive and lack of ability for using product that require bodily movement. The measures were summarized into ten types of motives and five types of lack of ability as shown in Tables 2 and 3. The detail questionnaire design could be found in Sun [16].

The questionnaire survey was conducted in January of 2014 in Taiwan. More than 290 responses were collected, while 250 responses were identified as effective responses. Some statistics of the respondents are gender (male, 51 %; female, 49 %), age (under 25, 40 %; 25–40, 21 %; 40–60, 35 %; above 60, 4 %), self-assessed health condition (bad, 3 %; alright, 38 %; good, 44 %; very good, 15 %), frequency of exercise (none, 14 %; one to two times per week, 48 %; three to five times, 30 %; more than six times, 8 %), and average time per exercise (less than 30 mins, 47 %; 30–60 mins, 37 %; more than 60 mins, 16 %).

Regression models were built with 7 independent variables measured in question groups (1) and (2) and 15 dependent variables measured in question groups (3) and (4). For each motive or ability, a regression model was built and tested for significance. For example, M1 stands for the first motive: pursuing visual pleasure (ranging from 1 to 5 where 5 means the strongest) and seven independent variables (x1–x7) including age, gender, education, vocation, health condition, frequency, and time of weekly exercise. The regression models can be used to express relations between levels of the motives or lack of ability and types of target users. In other words, by having target users' information such as age, gender, frequency, and time of weekly exercise, their level of motives and lack of ability can be estimated via the regression models.

After significant testing for the 15 regression models, regression models for M2, M3, M4, M6, and M7 passed the significant tests, while regression models for five abilities (A1–A5) passed significant tests. To use these regression models, one can substitute independent variable values describing target users to obtain the level of motive and lack of ability. For example, one of the regression models is like

$$\mathrm{M2} = 3.317 + 0.580^{*}\mathrm{x1} - 0.183^{*}\mathrm{x2} \quad (1)$$

where M2 denotes the motive of audio pleasure, x1 denotes gender (x1 = 0 denotes male; x1 = 1 denotes female), and x2 denotes age. Other independent variables did not pass significant t-test in this regression model. Once target user is identified with gender and age, the regression model can help estimate the level of motive M2. The regression coefficients of model (1) indicate that female and young people significantly prefer product providing more audio pleasure like music playing. Please ensure that all figure and tables are cited in the body text consecutively. Figure and tables are placed near the cited point.

3 Knowledge Extraction from Cases with Domain Knowledge Model

In case-based reasoning, a domain knowledge model is needed to describe the relationship between target behavior and applicable design principles and extract past experiences from cases. The model contains knowledge to answer questions like what kinds of design would influence user behavior leading to bodily movement and what types of design or technology used would be more effective in persuading users for using the new product.

To establish the domain model, this research begins with reviewing relevant literatures and defining the scope. The second step is to propose a domain model with case attributes later used in building a case library. The third step is to cross-check the proposed domain model with collected information of existing cases. After the three steps, a domain model with five sets of attributes that describe the cases is defined, including:

1. Target behavior involving bodily movement
2. Motives and lack of ability of users
3. Design principles modified and extended from the list in PSD
4. Design patterns modified and extended from the list in DwI method
5. Technology used together with design for effective persuasion

The following are detail discussion of the five sets of attributes contained in the proposed domain model:

1. Target behaviors. The idea of Behavior Wizard by Fogg and Hreha [13] was referred herein to present target behavior while Shih and Hsin [6] suggested it composed by two axes: three types of behavior changes and three moving parts of user's limbs that were specifically proposed for the underlined product design. Three types of behavior changes adopted from Behavior Wizard were chosen including (a) users try out new behavior, (b) users re-perform beneficial actions, and (c) users increase behavior intensity. Since the essence of this study is to persuade people into more bodily movement, it is crucial to identify which parts of the body exercise may be classified into upper, lower limb, or whole-body movement. This classification provides a simple and more objective way on viewing the different bodily movement and assists in retrieving cases in further applications. This target behavior is depicted with a 3×3 matrix in Table 1.

Table 1 Target behavior matrix

	Do new behavior	Do familiar behavior	Increase behavior intensity
Upper limb			
Lower limb			
Whole body			

Table 2 Motives of users for having more bodily movement

M1. Visual pleasure	M6. Accomplishment
M2. Audio pleasure	M7. Fashion
M3. Physical excitement	M8. Cooperation or friendship
M4. Learning	M9. Competition
M5. Health	M10. Sustainability (energy generation)

Table 3 Ability needed for more bodily movement

A1. Saving time
A2. Less physical effort
A3. Less brain cycle
A4. Less social deviance
A5. Turning behavior to routine

2. Motives and lack of ability of users. Fogg [12] stated that motive, ability, and trigger are three factors for behavior change. Since more understanding of users' motives and lack of ability could help choose effective design principles and patterns in DfSB, these two factors were included in the domain model as well as case profiling. For each case, why (motives) and how (ability) target users use the case product are described in the case profile. In retrieving cases, if potential users' motives and lack of ability are known via prior investigation (as shown in Sect. 2), more appropriate cases and useful suggestions can be found. For the underlined-type product, ten types of motives and five types of lack of ability were proposed and shown in Tables 2 and 3.
3. Design principles: The 28 design principles presented in persuasive system design (Oinas-Kukkonen and Harjumaa [14]) provided a checklist in case knowledge extraction. Since this study only focuses on the specified type of product design, not all design principles would be useful. After checking with collected cases, 17 out of 28 design principles were found appearing in the 98 cases.
4. Design patterns. The 101 design patterns from DwI method (Lockton et al. [4, 15]) were examined and inspected with case products related to this domain. Forty-three out of the 101 design patterns were used in the 98 cases, while additional six design patterns were added by this study. The six additional design patterns include full recording, virtual reality, energy feedback, virtual rewards, system praise, and public exhibition, which were useful in persuasive design for the underlined product. For example, "energy feedback" means there

is energy generated by human bodily motion and usually the energy can be used by users. "Public exhibition" means that product could be placed in public places like airport or rail station on purpose since some users are more willing to try it because other people can see them using the product.

5. Technology is another major ingredient in persuasive technology. Fogg [12] pointed out that technologies can play roles in three ways, as tools, media, and social actors, thereby presenting various functions and influencing behaviors differently. Applicable technology was included in the domain model because it often works together with design means in persuading people to behavior change. Therefore it is important to sort out the technologies that are commonly used in underlined products to understand their effect. Eighteen modern technologies, mainly information and communication technologies, used in the collected cases were summarized including video display, recording, audio player, health condition detecting, vibration, motion detector, internet connection, energy harvesting, etc.

4 A Case Library

A case library is built up by case products that are carefully gathered to present actual examples of the designated type of products; therefore these cases must be chosen and conform the following criteria:

(a) Case product goal must relate to health or sustainability (human power energy harvesting) domains; characteristics in these cases somewhat fit the idea of lifestyle of health and sustainability (LOHAS), examples include healthcare systems, exercise and fitness, and energy harvesting.
(b) The behaviors performed in using these case products must include human bodily movement, which is defined by the World Health Organization as any body movements produced by skeletal muscles that require energy expenditure.
(c) Each case has applied modern information and communication technologies.

A total of 98 cases were collected for the case library, and the major sources of these cases are websites like Yanko designs, Gizmag, and Inhabitant that introduce new green-designed product, fitness products, and other ingenious designs of modern designers. Other sources of cases are reviewing literature of different research that matches the scope of this study. These 98 cases varies differently from health to sustainability perspective like fitness or "exergaming" products, physical therapy, healthcare systems for the elderly, and energy harvesting from human motion.

Besides basic information like name, picture, and source, each case profile must include information of the five sets of attributes mentioned in Sect. 3. The followings are some notes in establishing case profile.

1. Which part of bodily movement is involved in the case? Upper body, lower body, or whole body? And what kind of behavior change?

2. What users' motives are expected by the designer of the case for using the case product? What kind of ability for behavior change is the case product enhancing? To answer these questions, one needs to know who the target users of the case product are. If there are multiple groups of target users, all possible motives and abilities should be included in the case profiling.
3. Check which design principles are used in the case product. Detail descriptions of 28 candidate design principles can be found in Oinas-Kukkonen and Harjumaa [14]. For the 98 cases, 17 design principles were found useful.
4. Check which design patterns are used in the case products, while detail descriptions of 101 candidate design patterns can be found in Lockton et al. [15]. After observing the 98 design cases, 43 out of 101 design patterns were found useful, while additional six design patterns were proposed by this study.
5. Check what kinds of information and communication technologies are used in the case products.

To make an appropriate case profile, authors carefully collected product description and sometimes tried out the product before profiling and extracting information bearing in mind the domain model. After case profiling, indexing was conducted to transform the descriptions into indices denoting different attribute levels. Indices break down the descriptions of attributes into codes and help sort out the differences between cases, which is very important for case search and retrieval.

Another worth noting issue is that these cases were collected from websites, market commercials, and product brochures. It is difficult to verify how successful the case products actually persuade users, let along knowing effects to various types of users. This may be seen as a drawback of the proposed method applying the case library since the effectiveness of the retrieved knowledge applied in new design cannot be verified.

5 An Aid for Retrieving Useful Cases and Providing Design Suggestions

In this section, an approach for generating design concept is proposed based on concepts of case-based reasoning, while an information system based on data management software ACCESS was written to aid the process, making case retrieval and proposing suggestions easier and faster.

Product designers may begin by identifying their design requirement through specifying target behavior with the 3×3 behavior matrix (Table 1), depending on which parts of the body are required to be active and what kinds of behavior flavors are expected. If target users are specified, prior analysis could be conducted to find corresponding motives and lack of ability to use the new product so that retrieval of useful cases could be more effective. With specified target behavior and target users, suitable cases could be retrieved for providing past experiences in design principles and design patterns. When multiple cases are retrieved, frequently used

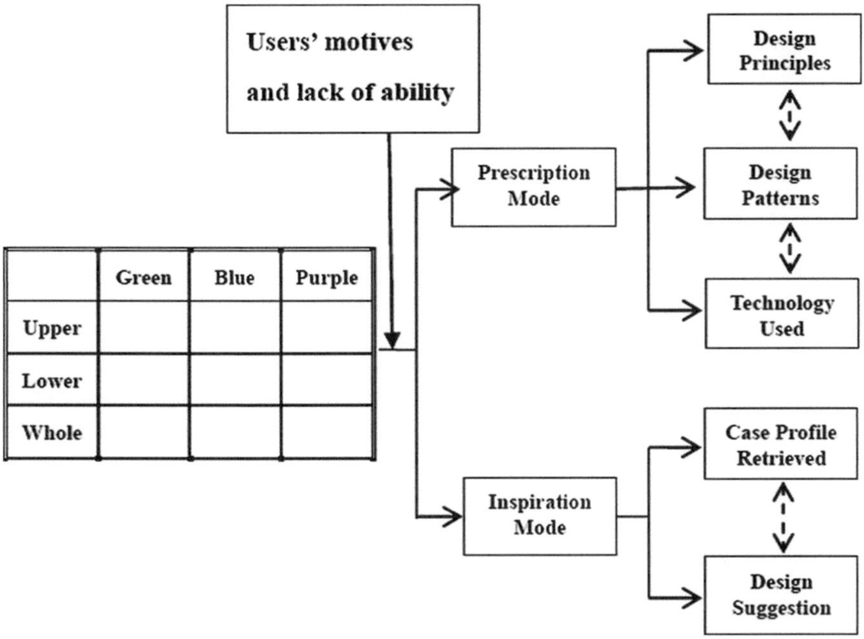

Fig. 1 Application flowchart of the proposed method

design principles and patterns for particular target behaviors are identified and presented. Technologies that often cowork with design principles and patterns are also identified. Designers can refer directly to retrieved case profiles or get inspired by the design suggestions including applicable technology, design principles, and design patterns. Please refer to Fig. 1 illustrating the idea of applying the proposed approach. In inspiration mode, retrieved case profiles could provide inspiration, while in prescription mode, suggestions on design principles and patterns and applicable technologies appeared in retrieved cases are summarized and deliberately presented for stimulating new design ideas.

To summarize, a six-step procedure is proposed to use the case library and to find useful suggestions on design principles, design patterns, and applicable technologies:

1. Specify goal of the project either an existing product that needs new idea to improve or a completely new product starting from conceptual design.
2. Specify target behaviors including expected part of the body moving and behavior flavors. In other words, see which cell in Table 1 (a 3×3 matrix) applies to describe the target behavior.
3. Conduct prior analysis on target users to find their potential motives and lack of ability in performing the target behavior. This is not easy and may require a lot of effort and detail consideration in a case-by-case basis. Section 2 tried to reduce the working load of prior analysis to some extent and established in advance

several regression models to estimate levels of motive and lack of ability of target users described with seven independent variables (x1–x7) including age, gender, education, vocation, health condition, frequency, and time of weekly exercise.

4. Use results of (2) and (3) as input conditions to search for similar cases and retrieve case profiles and indices that include design principles, patterns, and technologies. Relax minor condition if no case could be retrieved and conduct the search again. The information of design and technology contained in the retrieved cases may provide inspiration in getting new design concepts. Other information of retrieved cases like product description and picture may be helpful in inspiring ideas.
5. When multiple cases are retrieved in the search, summarize the frequency of appearance of each design principles, design patterns, and technologies used. More frequently used design principles, patterns, and technologies should be recommended with priority. This is a simple and logic way to highlight useful suggestions out of multiple retrieved cases.
6. Find ideas of product design using combinations of recommended design principles, design patterns, and applicable technologies. An information aid based on data management software ACCESS was written to help conduct steps (4) and (5), making case retrieval and summarizing useful suggestions easier and faster.

As mentioned at the end of Sect. 4, detail study of the effectiveness of using the proposed approach is not included in this study. The proposed procedure cannot guarantee the generated ideas are better. This is also a drawback of using the concepts of case-based reasoning since old experience (retrieved knowledge) may be useful but may not perfectly suit for new problem. Designer needs to make judgment after new ideas are generated.

6 An Illustrative Example

To illustrate the approach, an illustrative example is presented herein. The following are the results of conducting the proposed six-step procedure with the case library:

1. For illustration, suppose one wants to develop a new product to help elder women in Taiwan keep lower body moving and healthy.
2. Target behavior is taking lower body movement assuming that users want to increase intensity of familiar exercise rather than try new exercise. A good example of familiar exercise is riding a bicycle. So, the problem may be specified as to find innovative ideas to persuade elder women to more indoor cycling.
3. Assume target users are elder women who seldom exercise. Assuming users' motive of using the new product is "physical excitement" and also assuming they lack all five abilities in Table 3. In other words, they want to make it easy,

simple, less physical effort, easy to get used to, and so on. Regression models developed in Sect. 2 could be used to find significant motives and lack of ability when more information (seven independent variables) of the target users are provided.

4. With given information from the above three steps, one can conduct conditional search for cases meeting the attributes above. Seven cases out of 98 cases were retrieved. They are Wii Fit Rhythmic Game, Wii Fit Jogging, Wii Fit Snow Skiing, Piezoelectric Harvester shoe, POWER Leap Power Floor, Piezoelectric Floor in Tokyo Station, and FitWet. One can look over the detail case profiles of the seven to get inspiration or go to the next step.
5. Highlight design principles, design patterns, and ICT technologies in the seven retrieved cases and conduct statistical analysis. Find the frequency of appearance of each of them in the seven cases, highlight the most frequently appeared ones. For example, Table 4 shows the top three recommended design principles, patterns, and technologies from the seven retrieved cases where the number in parenthesis denotes the number of times they appeared in retrieved cases.
6. With the suggestions of design principles, design patterns, and applicable technologies, one may get inspiration from each individual suggestion or combinations of them to obtain new design concepts. For example, inspired by the suggestions in Table 4, new product concepts may emerge and contain the following ideas:

   (a) Let the pedal vibrate as a reminder whenever users stop cycling (use design pattern “real-time feedback” and technology “vibration”).
   (b) Let pedal generate heat in winter to keep users’ feet warm (use design principle “rewards”). This is important since local senior users believe keeping feet warm is essential for a healthy body. Designer may add a pair of boots attached with pedals that can warm users’ feet.
   (c) Turn the cycling into energy generation source that provides electricity for music playing, but music may stop once cycling stops (use technology “energy harvesting” and design principle “tunneling”).
   (d) Let users know her health conditions such as blood pressure and calories consumption by setting sensors on the pedals (use design pattern “summary feedback” and technology “health condition monitor”).

Table 4 Frequently used design principles, patterns, and technologies in retrieved cases

Design principle	Design pattern	ICT technology
Tunneling (7)	Real-time feedback (4)	Energy harvesting (4)
Reduction (6)	Summary feedback (4)	Visual display (4)
Rewards (6)	Energy feedback (3), Matched affordance (3)	Health condition monitor (3), Vibration (3)

7 Conclusion

A case-based approach is proposed to aid persuasive design for a specified type of products that persuades users to conduct more bodily movement. A domain model describing cause-effect relationship between design and behavior change is established and helps extract knowledge from past cases, thereby facilitating finding useful design suggestions. Two key factors including users' motives and ability are deliberately included in the domain model which expand and strengthen the power of the domain model in describing the relationship of behavior change and product design, especially when target users are specified. A case library that includes more than 98 cases containing experience of persuasive design is constructed so that cases meeting searching conditions can be retrieved for design suggestions. For each case, information specified in the domain model such as target behaviors, users' motives, users' lack of ability, design principles, and applied technologies are extracted and stored in the case library.

A six-step procedure is recommended for step-by-step execution with the established case library. Conditional search is conducted where target behavior, user's motive, and ability are input conditions, and cases meeting these conditions are retrieved. Target users' motives and lack of ability can be obtained in advance by using regression models that include seven independent variables describing target users. When there are multiple retrieved cases, frequently used design principles, patterns, and technologies are recommended with priority. One can either get inspiration directly from case profiles of retrieved cases or devise new design concept based on the recommended design principles and patterns and technology. A computer program based on ACCESS is written to promptly conduct the conditional search, retrieve useful cases, and identify design principles and applicable technologies for facilitating product conceptual design. An illustrative example is presented at the end demonstrating application potential of the proposed approach.

Acknowledgment Author would like to thank Ministry of Science and Technology of Taiwan for the financial support. (MOST 103-2621-M-006-002)

References

1. Lilley D (2009) Design for sustainable behaviour: strategies and perceptions. Des Stud 30 (6):704–720
2. Wever R, Kuijk J, Boks C (2008) User-centered design for sustainable behaviour. Int J Sustain Eng 1(1):9–20
3. Lockton D, Harrison D (2008) Design with intent: persuasive technology in a wider context. Persuasive technology: third international conference, Oulu, Finland, proceedings. Springer, Berlin, pp 274–278
4. Lockton D, Harrison D (2010) The design with intent method: a design tool for influencing user behaviour. Appl Ergon 41(3):382–392

5. Pettersen IN, Boks C (2008) The ethics in balancing control and freedom when engineering solutions for sustainable behaviour. Int J Sustain Eng 1(4):287–297
6. Shih L-H, Hsin H E (2012) Persuasive design for LOHAS products with physical activity. Electronics Goes Green International Conference, Berlin
7. Shih L-H, He W-S (2013) Persuasive design aid for products leading to happy, healthy and sustainable lifestyle. International conference. on innovation, communication and engineering, China
8. Ray P, Anderson SR (2000) The cultural creatives: how 50 million people are changing the world. Three Rivers Press, New York
9. Aamodt A, Plaza E (1994) Case-based reasoning: foundational issues, methodological variations, and system approaches. AI Commun IOS Press 7:39–59
10. Aha DW (1998) The omnipresence of case-based reasoning in science and application. Knowl-Based Syst 11:261–273
11. Fogg BJ (1998) Persuasive computers: perspectives and research directions. Proceedings of the SIGCHI conference on human factors in computing systems. ACM Press/Addison-Wesley Publishing Co, Los Angeles, pp 225–232
12. Fogg BJ (2009) A behavior model for persuasive design. Proceedings of the 4th international conference on persuasive technology. ACM, Claremont, pp 1–7
13. Fogg BJ, Hreha J (2010) Behavior wizard: a method for matching target behaviors with solutions. LNCS 6137:117–131
14. Oinas-Kukkonen H, Harjumaa M (2009) Persuasive systems design: key issues, process model, and system features. Comm Assoc Inform Syst 24:485–500
15. Lockton D, Harrison D, Stanton N A (2010) Design with intent: 101 patterns for influencing behavior through design. UK: http://danlockton.com/dwi/Main_Page
16. Sun J J (2014) Persuasive design system development for LOHAS products considering users types. MS thesis, Dep. of Resources Eng., National Cheng Kung Univ., Taiwan

Preliminary Research on the Perception and Implementation of Sustainable Supply Chain in Indonesian Companies

Jessica Hanafi, Helena Juliana Kristina, and Ogi Y. Poernama

Abstract The implementation of free trade agreements around the globe has forced companies to differentiate their business to stay competitive globally. Companies are expected to be sustainable, fulfilling all aspects of the triple bottom line. Therefore, a sustainable supply chain management is a challenge to be implemented. This study serves as a preliminary research to discover the perception of sustainable supply chain management among Indonesian companies and its current implementation. The research was conducted through questionnaires and in-depth interviews to representatives of the companies. The industrial respondents are varied to different types, from multinational companies, state-owned companies to local companies. Results suggested that local companies have yet gain understanding of the benefit of being sustainable. Only state-owned and multinational companies have good perception on the issue and just starting to implement the tools of sustainable supply chain management to reduce costs.

Keywords Sustainable supply chain • Sustainable perception

1 Introduction

Free trade agreements, globalisation and online markets have made competition among industries unlimited to their regions. People from different countries can directly compare products from other manufacturers from other countries. This exposed market results in increased competition among companies and throughout the supply chain. To stay in competition, a sustainable supply chain is warranted. Nowadays, a company and its supply chain is expected to be sustainable, fulfilling all aspects of the triple bottom line. The pressure and demands come from all the stakeholders, internal and external. Customers expected safe and environmentally friendly products; the government and NGOs require companies to be concerned about the society and the environment. The management and shareholders insisted on increased profit. These demands embodied into a sustainable supply chain

J. Hanafi (✉) • H.J. Kristina • O.Y. Poernama
Department of Industrial Engineering, Universitas Pelita Harapan, Tangerang, Indonesia
e-mail: jessica.hanafi@uph.edu

M. Matsumoto et al. (eds.), *Sustainability Through Innovation in Product Life Cycle Design*, EcoProduction, DOI 10.1007/978-981-10-0471-1_10

management (SSCM). Carter and Rogers [1] defined SSCM as a strategic achievement and integration of social, environmental and economic objective in an organisation through systemic coordination of key business processes between organisations to improve long-term economic performance and value chain of individual organisation. In a sustainable supply chain, the long-term risks of diminishing resources, energy cost fluctuation, product obligation, pollution and waste management can potentially be reduced [2].

In 2010, the Indonesian government launched the Master Plan for Acceleration and Expansion of Indonesia's Economic Development. It aims to catapult Indonesia into a high-income nation with a GDP of USD 14,250–$15,500 in 2025. This is expected to be achieved by developing nationwide six economic corridors with the main goals to increase value adding and expanding the value chain for industrial production processes and increase the efficiency of the distribution network; encourage efficiency in production and improve marketing efforts to increase national competitiveness and strengthen the national economy; and strengthen sustainable global competitiveness towards an innovation-driven economy [3]. These goals are aligned with the objectives of sustainable supply chain. Therefore, to achieve the master plan objective, a sustainable supply chain for Indonesian industries is essential.

The objective of this paper is to acquire information on the perception and implementation of SSCM-related activities in Indonesian companies. This study serves as a preliminary research of SSCM on selected companies and the current implementation of SSCM in these companies.

This paper is outline as follows. First, a literature review on the aspects of SSCM is explained. Then, the method applied in conducting this study is described. Indicators on measuring the sustainability of the supply chain are explained in this section. The perception and implementation of SSCM are evaluated and analysed in the subsequent section. Finally, the last section will conclude the findings of the study and provide a future outlook of this topic.

2 Sustainable Supply Chain

The three basic constituents of supply chain management are managing material flow, relationship management, and managing information flow [4]. Similarly these constituents accord with sustainable supply chain management. Supply chain considers the product from the initial processing of raw material to delivery to the customer. With the sustainable supply chain, these activities also include product design, manufacturing of by-products, by-products produced during product use, product life extension, product end of life and recovery processes at end of life [5, 6]. Therefore, a sustainable supply chain can be considered to consist of sustainable procurement, sustainable manufacturing, sustainable distribution and sustainable marketing. By capturing these activities, a sustainable supply chain can be achieved.

Sustainable procurement captures three aspects of the triple bottom line. The social aspect deals with assuring ethically sound supplier-buyer relationships, establishing codes of conduct, preventing child labour and considering minority-owned suppliers. Environmentally, procurement activities should facilitate recycling, reuse and resource reduction and connect environmental sourcing to strategic topics to minimise risks [7]. After the selection of materials in sustainable procurement, these materials are then processed sustainably through sustainable manufacturing.

Global Environmental Management Initiative (GEMI) described four levels of classification in strategic sustainable procurement, starting from compliance to implementation of sustainable procurement to company strategic function (total quality approach) [8]:

1. First level: Compliance. The buyer evaluates supplier's performance based on its compliance to environmental, health and safety regulation. Preferences will be given to suppliers with matching environmental policies and standards to the company.
2. Level 2: Implementation and system development. A system exists to evaluate potential suppliers' environmental policies. Suppliers who do not qualify the environmental requirements of the buyer will be removed from the list.
3. Level 3: Integration into general business functions. Supplier selection models are integrated with environmental priorities evaluation system. All business units are involved and coordinate the evaluation of suppliers.
4. Level 4: Total Quality Approach. Corporation gives preference to suppliers who accept and implement International Chamber of Commerce principles on sustainable development. Company collaborates with suppliers to identify and implement appropriate improvements in the corporation and supplier's environmental management system.

Sustainable manufacturing consists of ten principles, namely, safe product and packaging, satisfactory service, waste minimisation, removal of hazardous substances, conservation of energy and material, safe workplace, efficient working condition, continuous improvement of employees, beneficial to surrounding society and sustainable economy for the company [9]. O'Brien [10] also adds that organisation must be lean and clean where re-engineering, Kaizen and company's metrics must address sustainable issues. A product does not end with the customer but a concept of product life cycle must be applied to the whole manufacturing system.

Sustainably conducted procurement and manufacturing leads to sustainable distribution of products. Sustainable distribution considers both facility for distribution supply chain and also the transportation. Since distribution in the sustainable supply chain also includes end of life, reverse logistics becomes an important part of a sustainable distribution. Related to distribution supply chain, the environmental aspect of a sustainable distribution considers the energy consumption, resource consumption and green house gas emission from the logistic activities. Human toxicity potentially becomes the social metrics for this set of activities [11].

The usage stage of a product is affected by sustainable marketing practices. Sustainable marketing is concerned with sustainable development and offers marketers a holistic approach to make the product [12]. Sustainable marketing is expected to promote sustainable consumer behaviour and offer suitable products, with the aim of economic and environmental sustainability, where consumer behaviour and consumption patterns need to be considered.

3 Method

The research was conducted through questionnaires and in-depth interviews to representatives of the companies. Six companies were selected by non-probability sampling to represent multinational companies, state-owned companies, licensee of multinational brand, supplier to multinational companies and local companies. Two types of local companies are represented, which are medium- and large-sized companies.

3.1 The Questionnaire

The questionnaire consisted of two parts. The first part aims to understand the perception of the company on the importance of SSCM activities to the company. The questionnaire used five-point Likert scale measurement. The second part of the questionnaire aims to understand the perception of the company on the implementation of SSCM activities in the company. This questionnaire is measured by using nominal scale. The questionnaires are based on Pagell and Wu [13] which generally discuss on company policy regarding sustainability, using global list of companies' best practice on sustainability as a benchmark.

There are five key topics in the questionnaire, which are orientation, awareness and same vision on sustainability, ensuring supplier continuity, reconceptualisation of the supply chain, supply chain management practices and performance measurement [13]. Details of the key topics are described as follows.

3.1.1 Orientation, Awareness and Vision on Sustainability

This consists of general attributes to support SSCM, managerial orientation and the role of social and environmental responsibility in decision-making process. These attributes include how sustainability fits the business model, proactive stance/ organisational commitment, internal supply chain integration, touchstone value/

guardrail usage in decision making and daily conversation about sustainability and integrate environmental efforts into the entire organisation.

3.1.2 Ensuring Supplier Continuity

The key topic in this attribute is on how far a company will do to develop a long-term relationship with their suppliers. Ensuring suppliers' continuity is an important aspect of a sustainable supply chain [14]. A long-term relationship with the supplier will benefit both parties. The buyer will have raw materials with guaranteed quality, less risk of delayed shipment and price reduction to decrease production cost. As for the supplier, long-term relationship with the buyer will reduce the risk of losing customers. Attributes in this key topic are decommoditised inputs, supplier development, reducing supplier risk, supply development to improve other chains, continuity of suppliers and transparency.

3.1.3 Reconceptualisation of the Supply Chain

The reconceptualisation of the supply chain discusses about changing what the supply chain do, transforming the supply chain into a closed-loop system and reconceptualising who is involved in the supply chain [15].

3.1.4 Supply Chain Management Practices

This key topic discussed the implementation of supply chain management best practice in the world. Among those practices are conducting supplier selection based on the goals of sustainability; collaboration with supplier to achieve sustainable goals; supplier certification based on social and environmental activities; involving in developing supplier activity related to improving supplier performance in social and environmental aspect, which is important to the supply chain; and improving traceability. Included in this key topic also are reduction in fossil fuel and carbon emission, reverse logistics and demand for information on the product produced by the supplier.

3.1.5 Performance Measurement

It is difficult to measure the sustainability performance of an organisation. Therefore, this attribute focused on awarding employees who are able to guide and motivate other employees to achieve the goal of sustainability [16].

3.2 In-Depth Interview

A semi-structured in-depth interview was then conducted to some representatives from each company to understand the actual condition in that company. The questionnaire and interview results were then analysed and compared. A comparison between theory and reality and the actual practices of SSCM between these companies were developed. The interview was conducted to production, warehousing, distribution and marketing divisions. The interview aims to know the understanding of the company and the respondents on the topic and how sustainable activities might have been conducted in the company. Some examples of the interview questions include company policy on sustainability, understanding of product life cycle concept and its implementation, strategies on carbon emission reduction and energy savings and company policy on distribution system and reverse logistics.

4 Results and Discussion

4.1 Company Perception

The questionnaires aimed to capture the perception of company on the importance of SSCM activities throughout the company and their perception on the implementation of the activities (Fig. 2.1.1). Some companies may not have the right understanding of SSCM activities; therefore in the second part of the questionnaire, the implementation index is based on their perception.

The Likert scale from the first part of the questionnaire is used to calculate the perception index of importance of SSCM activities. The value of each key topic are then totalled and divided by the maximum value of the topic. Therefore, the organisation, which rate the entire key attributes as very important (Likert scale = 5), will achieve a maximum index of 1.

The second part of the questionnaire, as mentioned previously, uses nominal scale. The perception of company implementation on SSCM activities is converted into weighted score. The value is 1 for questions answered "Yes", 0.5 for "Limited application", and 0 for "No". The average score of all the questions is called perception index of SSCM implementation. Table 1 shows the perception index on SSCM activities in the selected companies.

The result shows that MNC companies achieve the highest index on the perception of importance of SSCM activities in their company and also the highest perception of implementation. The total perception index is calculated by multiplying the importance rating and the implementation index. The importance rating serves as a weighted score towards the implementation. Therefore, the companies who perceive SSCM activities as important but there is no implementation in the

Table 1 Perception index of SSCM activities in sample companies

Type of company	Index of importance	Index of implementation	Total perception index
MNC	0.9	0.85	0.77
State owned	0.85	0.82	0.70
Vendor to MNC	0.69	0.62	0.43
MNC brand licensee	0.54	0.54	0.29
Local (medium size)	0.78	0.25	0.20
Local (large size)	0.51	0.12	0.06

company will have low score, such as in the case of medium-sized local company presented in this study.

The total perception index shows that multinational company (MNC) has the highest index, followed by state-owned company, vendor to MNC, MNC brand licensee, medium-sized local company and large-sized local company. It is obvious that the MNC will be directed by their overseas parent company and state-owned company is directed by regulation from government. Similarly, vendor to MNC would have strict requirements from their buyer, as classified in the GEMI four-level supplier strategic performance mentioned above. The next section will compare these perception to the actual implementation based on the in-depth interview findings.

4.2 Actual Implementation

An in-depth interview was conducted to each of the selected companies to find out the actual implementation of SSCM activities in the company. Four divisions, namely, manufacturing, warehouse, distribution and marketing, were interviewed. The perception of sustainability varies according to the type of companies. Local companies have limited or no knowledge on sustainability, while some companies perceived sustainability as green products. The interview revealed that the Indonesian market still only considers price and appearance of the product and sustainable issue has not been voiced as requirement from customers. Although some companies may have already known green materials for their products, these companies may opt not to use these materials due to economic reasons. For companies with exposures to multinational companies, industrial policies are mostly related to improvement in cost and production efficiency and waste minimisation. However, for local companies, even manufacturing concepts, such as lean manufacturing or JIT, have not been implemented due to economic reasons. In some applicable industry where recycling is possible, recycling process is already implemented.

In distribution aspect, sustainability has not yet been addressed. However, most companies implemented regular maintenance for vehicles, planned routing for deliveries and emission control. The industrial respondents added that government regulation and policy has to be developed into sustainable development by

Fig. 1 Actual implementation of SSCM activities

conducting campaigns and raising awareness in society. Incentives such as lower importation tax for green materials are one example.

An analysis based on the understanding of the concept and actual implementation of the activities is performed. Figure 1 shows a radar chart representing the activities. Out of the four divisions evaluated, warehouse division has the most sustainable activities compared to other divisions. Four out of six types of companies studied have shown sustainable warehousing activities with implementation index above 0.80. These companies are MNC, state owned, MNC brand licensee and vendor to MNC. Three types of companies show sustainable activities in manufacturing and marketing divisions with implementation index above 0.8. Distribution division has the lower implementation with only one type of company with implementation index of 0.8. The rest of the companies studied have implementation index lower than 0.60.

Most of the companies have used technology to manage the warehouse and to improve the productivity and effectiveness of the warehouse. Efficient material handling is applied and carbon emission in the warehouse is minimised. Meanwhile in the distribution division, although most companies have used vehicles which have passed the emission test, the company do not have policy regarding distribution using outsourced companies.

The average index for actual implementation is calculated by dividing the total index by the number of attributes related to SSCM activities. The types of company with the highest index are MNC, state owned, vendor to MNC, MNC brand licensee, medium-sized local company and large-sized local company, respectively. The MNC shows implementation index higher than 0.9, while state owned and vendor to MNC have index higher than 0.7. The local companies show index lower than 0.5, which means that they have not conducted activities supporting SSCM. The result of the actual implementation is aligned with the perception of the company.

5 Summary and Future Work

Sustainable Supply Chain has not been perceived as important among companies in Indonesia. Out of six types of companies selected to participate in this study, only multinational company and state-owned company have already perceived SSCM as important. Both of these companies also have a good perception on the implementation of SSCM activities in their company. However, when interviewed regarding actual implementation, it was found that the vendor to MNC and MNC brand licensee actually already have some activities supporting the sustainable supply chain, related to their compliance to the multinational company and cost reduction strategy. This preliminary research shows that there is a gap of understanding on sustainable supply chain management and its benefit to local companies.

With the emergence of local companies in Indonesia, it is important to raise awareness of sustainable supply chain management to stay competitive globally. Future work will focus on developing strategies and framework to develop sustainable supply chain management for Indonesian companies, especially for small and medium enterprises.

Acknowledgements The authors would like to thank the Ministry of Research and Technology Competitive Grant No. 108/K3/KM/2015 and the Japan World Exposition 1970 Commemorative Fund for providing financial support for young researchers from Asian countries to present at the EcoDesign 2015 symposium.

References

1. Carter CR, Rogers DS (2008) A framework of sustainable supply chain management: moving toward new theory. Int J Phys Distrib Logist Manag 38(5):360–387
2. Srivastava SK (2007) Green supply-chain management: a state-of-the-art literature review. Int J Manag Rev 9(1):53–80
3. (2014) MP3EI : masterplan percepatan dan perluasan pembangunan ekonomi Indonesia, 2011–2025. In: Luky Eko Wuryanto, Imron Bulkin (eds). Jakarta: Kementerian Koordinator Bidang Perekonomian : Kementerian Perencanaan Pembangunan Nasional/Badan Perencanaan Pembangunan Nasional
4. Preuss L (2009) Addressing sustainable development through public procurement: the case of local government. Suppl Chain Manag: Int J 14(3):213–223

5. Linton JD, Klassen R, Jayaraman V (2007) Sustainable supply chains: an introduction. J Oper Manag 25(6):1075–1082. doi:10.1016/j.jom.2007.01.012
6. Bevilacqua M, Ciarapica FE, Giacchetta G (2012) Design for environment as a tool for the development of a sustainable supply chain. Springer, London/New York. Retrieved from https://books.google.com/books?hl=en&lr=&id=nAM5J10-nbMC&oi=fnd&pg=PR5&dq=design+a+sustainable+supply+chain+bevilacqua&ots=kCkpkZ6svs&sig=F1PjyMq2RFGdi5sFvmNwFKbep18
7. Schneider L, Wallenburg CM (2012) Implementing sustainable sourcing—does purchasing need to change? J Purch Supply Manag 18(4):243–257. doi:10.1016/j.pursup.2012.03.002
8. Hamner B (2006) Effects of green purchasing strategies on supplier behaviour. In: Greening the supply chain. Springer, London, pp 25–37. Retrieved from http://link.springer.com/chapter/10.1007/1-84628-299-3_2
9. Lowell Center for Sustainable Production (nd). Retrieved 28 Jun 2015, from http://www.sustainableproduction.org/
10. O'Brien C (1999) Sustainable production–a new paradigm for a new millennium. Int J Prod Econ 60:1–7
11. Bouchery Y, Ghaffari A, Jemai Z (2010) Key performance indicators for sustainable distribution supply chains: set building methodology and application. Cahiers de recherche, 8. Retrieved from http://www.researchgate.net/profile/Asma_Ghaffari/publication/229014792_Key_Performance_Indicators_for_Sustainable_Distribution_Supply_Chains_Set_Building_Methodology_and_Application/links/0c9605281e9345973f000000.pdf
12. Reutlinger J, others (2012) Sustainable marketing: the importance of being a sustainable business. Retrieved from http://www.theseus.fi/handle/10024/50565
13. Pagell M, Wu Z (2009) Building a more complete theory of sustainable supply chain management using case studies of 10 exemplars. J Supply Chain Manag 45(2):37–56
14. Liker JK, Choi TY (2004) Building deep supplier relationships. Harv Bus Rev 82(12):104–113
15. Gladwin TN, Kennelly J, Krause T-S (1995) Shifting paradigms for sustainable development. Acad Manag Rev 20(4):874–907
16. Sroufe R (2003) Effects of environmental management systems on environmental management practices and operations. Prod Oper Manag 12(3):416–431

Analysis of User Needs for Solar Cooker Acceptance

Robert Wimmer, Myung Joo Kang, Chaipipat Pokpong, and Adeshir Mahdavi

Abstract Many researches were conducted and proved that a solar-based cooker can provide high temperature sufficient for cooking foods. However, solar-based cookers have not been widely used in households yet. This paper aims to investigate the reasons for the low acceptance of solar cookers on the part of the households, based on an analysis of the user needs related to cooking behavior, in order to suggest design criteria of a solar cooker which is more likely to be accepted by households. Questionnaires, observation, and interviews were used to clarify user needs in Thailand and Austria. A prototype was used as a tool to bridge user needs and stove capability. The result of this study focuses on (1) user expectation and (2) compatible design in practical use and building infrastructure.

Keywords User needs • Solar cooker • User acceptance • Kano's model

1 Introduction

This paper presents the research and development work done for a solar cooking stove. It is crucial to gain market acceptance for the solar cooker because it is powered by a renewable energy source without conversion into electrical energy or liquefied petroleum gas (LPG) consumption. It is perceived to be a very new and unknown household device by the majority of consumers which results in low acceptance.

Many research and development projects have proved that several solar cooker designs can provide sufficiently high temperature for general cooking twice a day [1]. Nevertheless, actual and practical use of this kind of device in households has been very rare. The main question which initiated this research was why a solar

R. Wimmer (✉) • M.J. Kang • C. Pokpong
Center for Appropriate Technology, Vienna University of Technology, Vienna, Austria
e-mail: rw@grat.at

A. Mahdavi
Department of Building Physics and Building Ecology, Vienna University of Technology, Vienna, Austria

M. Matsumoto et al. (eds.), *Sustainability Through Innovation in Product Life Cycle Design*, EcoProduction, DOI 10.1007/978-981-10-0471-1_11

cooker is not widely used in households, although it meets the technical requirement of certain temperature for general cooking.

A firewood cooker is a low-investment device to install in a kitchen. There are 2.5 billion people who use biomass for cooking worldwide, but 1.5 million people were killed by the smoke from open fire and the use of traditional cookers [2].

The primary user need related to traditional open firewood cookers is cooking foods. Yet, since the new types of solar-based cookers have already shown that they can cook foods, there must be other reasons why these solar-based cookers have not immediately replaced the firewood cookers. Thus, user needs related to a solar cooker itself, as well as users' pre-use, during-use, and post-use experience and expectations, need to be investigated. This study discovers actual user needs and explores how to implement them in a solar-based cooker.

2 Methodology

This study explores the user needs by using basic methods such as literature review, questionnaires, interviews, and user behavior observation to gain insights. The study focuses on two different climate conditions: Austria represents the temperate zone and Thailand represents the tropical zone. It is very interesting to know the user insight in both different climate conditions. The data were analyzed by using Kano's model to clarify user needs. Then, a prototype of a solar cooker was developed considering actual user needs so as to evaluate user acceptance. The prototype was tested in Austria to explore the maximum heating performance in an area with less sunlight availability.

2.1 Literature Review

The literature review was used to get a better understanding of cooker principles as well as general user needs such as ISO standard [3], patent [4], and academic articles.

2.2 Questionnaires

Two hundred sets of questionnaires were distributed to cooking stove users in Austria and Thailand. The questionnaire comprised two parts. The first was to find out basic information such as personal data of the users, their type of cooker and preferred design attributes. The top ranking of the users required a design attribute in the first questionnaire, and it was investigated in more detail in the second questionnaire. The second part was to explore the users' opinion about preferred design attributes on their existing cooker by rating to gather qualitative data. For each question, the participant selected one of multiple choices as follows: like, expect, neutral, tolerant, and dislike, describing their satisfaction.

2.3 Observation

The purpose of the observations was to gain a better understanding of how users interact with their cooker.

Six users were selected; three females and three males were asked permission to visit and observe cooking behavior in their kitchen. Those kitchens should have sufficient space for 2–3 people to do the observation. The participants were asked to cook a simple omelet for two persons. The actions of the participants were monitored in seven stages: (1) goal establishing, (2) planning, (3) action sequence, (4) cooking performance, (5) perception, (6) interpretation, and (7) comparison [5]. User activities were recorded by photographs and video as well as note taking. The camera was set on a tripod at 1.5 m above the ground. It was usually located about 2 m away from the user. The observations took about 5–20 min or until the users completed their task. The standing point of the observer should not be close to the cooking zone, but it should clearly see all user activities as much as possible. The observer should consider kitchen layout, action space, and lighting direction. The observer should not interfere with the user by asking, talking, moving, or making the participant feel uncomfortable during the activities. If the observer needs to move the position to see the participant's activity from a different point of view, he or she should move slowly and quietly. The observer can talk to the participants before an activity starts to make the participants feel comfortable and relaxed. Figure 1 shows the observation procedure during the data collection.

2.4 In-Depth Interviews

Interviewing is one of the most efficient methods to gain user data [6]. This method strongly relates to the observation stage. The interviewees were the same participants who were observed. They were interviewed one by one immediately after they had used their cooker. Prepared questions might not cover all actions, so the interviewer might have some additional questions that relate to unexpected behavior during the observation. The interviewer asks short, direct questions concerning the user's action to get more information; however, the interviewer should provide for a relaxed interview. The answers were collected by note taking, voice recording, and video recording. The questions aim to get deeper insights and clarify some interesting keywords from the users.

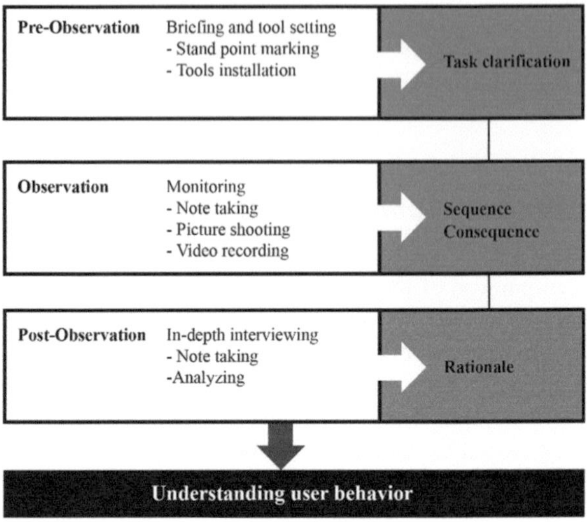

Fig. 1 Observation procedure chart

2.5 *In-Depth Interviews*

The user needs for cooking stove usability were analyzed through the Kano model to clarify the relationship between user requirements and user satisfaction. The Kano model is a user need classification tool which corresponds to user satisfaction. If the user demands are met, the user will be satisfied; if not, they will be dissatisfied. The user will be delighted when a cooking stove has attractive design attributes. However, when an attractive design is expected as basic requirement, user satisfaction will decrease. The initial survey results showed 11 design attributes that users need for comfortably using a cooking stove. The needs were classified into five groups in the Kano model to define need levels for using a cooking stove: (1) must-be quality (M), (2) one-dimension quality (O), (3) attractive quality (A), (4) indifferent quality (I), and (5) reverse quality (R) (see Table 1).

An example function (positive) question: If your cooker can be used anytime, how do you feel? The answer could be "Expect." The result in the second row can then vary depending on the answer from the dysfunctional (negative) question. Suppose the user answers "Dislike" to the following question: if your cooker cannot be used at nighttime, how do you feel? Then the result of this feature will be "M." It means a cooker must be able to work both daytime and nighttime. However, an extremely contradictory answer from the user needs to be questioned. For instance, a user answers "Like" to the positive question "if the cooker is of big size, how do you feel?" but then also answers "Like" to the negative question with the same attribution that asks "if the cooker is not of big size, how do you feel?" The result will be "Q" (questionable). It means this answer is not reliable.

Table 1 Five-level classification of Kano's model

Customer requirements		Dysfunctional (negative) questions				
		1 Like	2 Expect	3 Neutral	4 Tolerate	5 Dislike
Function (positive) questions	1 Like	Q	A	A	A	O
	2 Expect	R	I	I	I	M
	3 Neutral	R	I	I	I	M
	4 Tolerate	R	I	I	I	M
	5 Dislike	R	R	R	R	Q

Q questionable

2.5.1 Must-Be Quality (M)

The first group refers to basic attributes that, when missing in a design, users will absolutely be dissatisfied with. The temperature performance is of highest importance for using a cooker. The users immediately refuse to use insufficiently high temperatures for their cooking. However, when temperature can reach a sufficient level for cooking, it does not increase user satisfaction. The user considers it as a basic need regarding a cooker.

2.5.2 One-Dimension Quality (O)

The one-dimension quality refers to a design attribute of cooking stoves that makes users satisfied or dissatisfied when it is not fulfilled. User satisfaction increases proportionally with a better performance. As an example, a cooking stove that can quickly reach the expected cooking temperature will satisfy a user more than a slower cooker design.

2.5.3 Attractive Quality (A)

The third group consists of quality attributes that influence user acceptance of a new design. An attractive quality attribute can add user satisfaction to a cooking stove, but it does not cause dissatisfaction when the cooking stove does not have this attractive quality. For example, aesthetic appearance can add more value to a cooking design, but it does not impact user satisfaction as long as the cooker is still properly working.

2.5.4 Indifferent Quality (I)

An indifferent quality attribute refers to a quality attribute that can be either positive or negative for user satisfaction. For example, a cooker on-off switch is a critical issue in identifying a design direction. Even though a switch that's easy to turn on and off can help a user control the cooker, it can be harmful for children by unintentionally switching on the cooking stove.

2.5.5 Reverse Quality (R)

The reverse quality attribute group is similar to the indifferent quality group as it might satisfy one user group while it also dissatisfies another user group. For example, a high-technology cooking stove provides a precise temperature for cooking, but users need to spend a lot of time for setting it right. Another group prefers a simple cooker design.

3 Finding and Discussion

3.1 Type and Performance of Solar Cookers

3.1.1 Panel Solar Cooker

A panel solar cooker (Kyoto box) is the most economical and easy-to-make solar cooker. It is made from a foldable polypropylene reflector and a clear acrylic cover. This cooker looks like a box with a reflector inside the panel wall and a black-painted cooking vessel. The foldable design provides a portable feature and is easy for installation. A black-painted vessel can absorb the heat from sun radiation much quicker than a light color-painted vessel. "Maria Telkes created a wooden box oven in year 1959 which can reach temperature up to 225 °C" [7]. Figure 2 illustrates a panel solar cooker (Kyoto box) with four reflectors.

3.1.2 Parabolic Solar Cooker

There are various designs of parabolic shape to concentrate the sun radiation to heat up a cooking vessel that is placed in the middle of a focus spot. This solar cooker type is more difficult to make and has higher costs than a panel solar cooker. "A parabolic solar cooker project in Somalia found that it is only kind of solar cookers acceptable to locals as their cooking speed was comparable with that of charcoal" [8]. A sun tracking system can help the device to maximize the temperature for cooking. This type of solar cooker is usually used for outdoor cooking and can harm

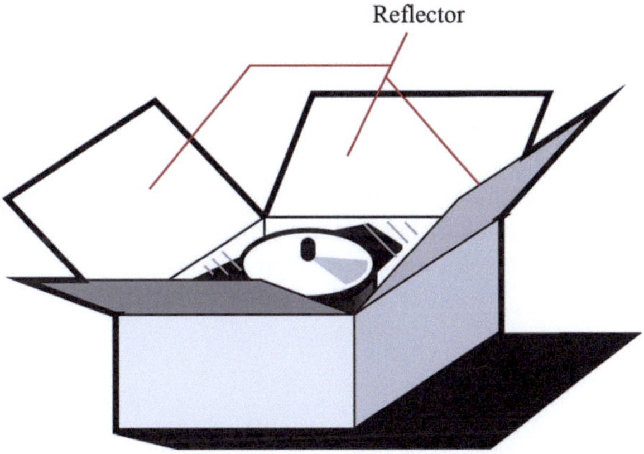

Fig. 2 Panel solar cooker

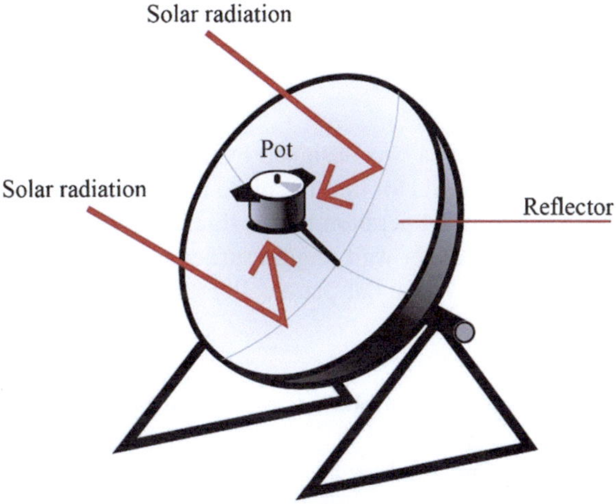

Fig. 3 Parabolic (dish) solar cooker

the user's eyes during cooking. Figure 3 illustrates the characteristic of parabolic dish solar cooker.

3.1.3 Indirect Solar Cooker Station

Flat Plate Solar Collector

This type of solar collector is economical in both design and production. It uses the thermosiphon principle for hot oil circulation in the system. It does not require any mechanical part or electric pump. A solar collector box prototype was built from wood, black-painted copper tubes, and clear glass. This solar collector box was topped with double glass with an air gap in between to protect the skin from burning by accidentally touching the glass surface. The copper tube was painted in black color to maximize heat absorption from the sun radiation. The solar collector was set at an angle of 30°–40° to the ground. There are four reflector panels to reflect more sunlight into the solar collector box. Reflector films were applied on foldable polystyrene foam panels which were placed on the edges of the solar collector box. The reflector can be folded and adjusted to get maximum sun radiation or to close the solar collector box. Those reflector panels can be used as a protection for the glass on the solar collector when it is not in operation.

Thermal Storage

The thermal storage tank can keep the hot temperature of the thermal oil in the system when sun radiation is not available, such as on a cloudy, rainy day.

Cooking Area

The cooking area was designed by using a double-wall cooking vessel with a gap between the walls for thermal oil to run through. The thin wall is more sensitive to fluctuation in temperature than a thicker wall. However, a thick wall takes longer time than a thinner wall to gain sufficient heat for cooking. Figure 4 illustrates the section views of the flat-plate collector with indirect cooker station.

3.1.4 Prototype of Solar Cooker and Solar Fridge

The prototype combined a modified LPG refrigerator and a solar cooker. These devices share the heat sources from the solar collector and heat storage tank. A thermal storage tank was made from steel oil barrels with 50 l capacity. The storage tank was filled with round stones to reduce the oil volume in the heat transport system. The tank was covered completely with insulation material to keep the heat inside the storage part.

The pipeline was designed by keeping it straight and minimizing joints as much as possible to improve the hot oil flow rate since there is not much pressure from the thermosiphon. A curve in vertical direction might cause a problem due to air bubbles blocking the hot oil circulation flow. The pipeline also has an overflow container to prevent oil from spilling when it is getting hot and increasing in volume. There are two possible ways to connect the pipeline to the heat source: (1) the solar refrigerator or cooking stove can use direct heat from the solar collector when sun radiation is available, and (2) those appliances can use the heat from the storage tank when sun radiation is not available.

Figure 5 shows the exterior appearance of flat-plate collector with indirect solar cooker prototype.

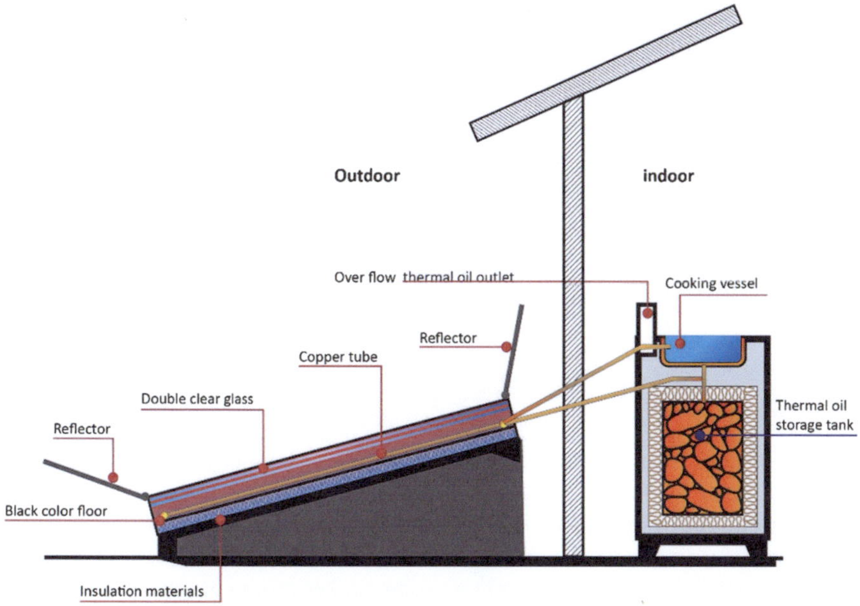

Fig. 4 Technical concept for a flat-plate collector with indirect cooker station (Adapted from Dr. Schwarzer's patent)

Fig. 5 Flat-plate collector with indirect solar cooker prototype

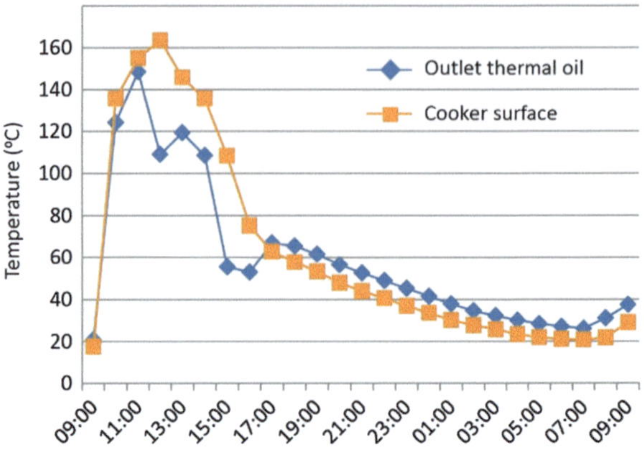

Fig. 6 Testing result of adapted indirect solar cooker

The advantage of this solar cooker design is that users can cook inside the building.

The prototype can produce heat on a cooking vessel surface up to 165 °C which is sufficient for simple food cooking, warming, and pasteurizing. Figure 6 presents the performance temperature of indirect solar cooker prototype.

3.2 *User Needs and Acceptance Criteria for Using a Solar Cooker*

The results of this survey and literature review revealed 11 attributes that a user needs for comfortably using a cooker.

3.2.1 Cooking Temperature Requirements

Temperature performance should be between 75 and 232 °C for safe cooking. Cooking time depends on the temperature level and food characteristics. Cooking time can be shortened by adding higher temperature. Those different temperatures can kill germs for safe foods consumption. Table 2 shows general cooking requirements between 75 and 190 °C only to make cooked foods. The higher range 177–233 °C is used for baking.

3.2.2 Fast Reaching of High Temperature

Users prefer to shorten their cooking time. In the last 10 years, the average use of a cooker was about 45–60 min per cooking time. The survey showed that 76 % of the participants spent only 20–30 min for using a cooker. They need a

Table 2 Minimum temperature for food safety requirements

Category	Foods	Cooking methods	Temperature
Meat	Beef, pork, lamp, turkey, chicken, duck, seafood	Varying	75[a]
		Deep frying, steaming	175–190[b]
			100<
Soup	Water, stock	Boiling	100<
Bakery	Bread	Baking	218–232[c]
	Cake		177–190[c]
	Cookies		177–205[c]

[a] [9]
[b] [11]
[c] [10]

high-performance cooker to achieve this goal. The level of user satisfaction from using a solar cooker directly correlates to fast reaching of high temperature for their cooking. A small-scale solar cooker takes longer time to produce sufficient temperature for medium range cooking. The results of the prototype testing show that the prototype of solar cooker takes about an hour to reach 75 °C which is the minimum temperature for food safety.

3.2.3 Precise Control

A user needs to know the current temperature for cooking. An interaction between displays and control deign is very important for user perception. The survey reveals that 67 % of participants do not know the cooking temperature during their cooking. There are three methods to perceive the current cooking temperature: (1) users monitor a flame characteristic of gas cooker, (2) users look at the lighting signal from an electric cooker to predict cooking temperature from their experience, and (3) users use a knob position and graphics to indicate a cooking temperature level. A solar cooker needs to show the current temperature level or preferred temperature to help control the cooking temperature. The results of the surveys also reveal that it is sufficient to know the temperature range for cooking. Three temperature ranges were suggested by the participants that are sufficient for cooking: high, medium, and low heat. The temperature from the solar cooker is controlled by the flow rate of thermal oil in the system. In this case, the user needs efficient control and display design for better perception.

3.2.4 Easy Temperature Controlling

The on-and-off switch shows conflicting requirements from two user groups. Some users want to have a switch that is easy to turn on and off, but many users don't want to have an easy-to-turn switch because of children and for security reasons. Therefore, a good ergonomic design can help users to control the switch and thus temperature during their cooking. A good relation between displays and control can help users to easily control the cooking temperature. A good grip on a control knob should consider: shape, size, movement direction, position, color, and nonslip materials.

3.2.5 Prompt Use

The results of the survey show that the participants use their cooker at least twice a day. They need 20–30 min per cooking time. A cooking stove should have a sufficient energy input and backup system for that basic need. The literature study found that a solar cooker can provide sufficient heat for cooking at least 6–7 h [12].

The heat storage tank of the solar cooker can keep the hot temperature oil for cooking when sun radiation is not available.

3.2.6 Safety

Safety is a basic standard requirement for any household appliance. The solar cooker does not produce smoke or flames which could cause health problems and fire accidents. Users can be harmed by using a cooking stove because of heat and physical sharp edges. Product designers should avoid using ambiguous cooking zoning that causes skin burning from touching. Warning graphics or interface designs are needed to clearly indicate a hot zone on a cooking stove surface. A cooker design should avoid small corners and gaps which are difficult to clean. A sharp edge can also harm users when they clean a cooker surface. Users need a warning system, with both sound and visual signals, for security reasons.

3.2.7 Number of Burners

The survey results show that all participants would like to have more than one burner. Eighty-five percent of those have four burners on their cooker. It clearly indicates that users need a multi-burner with different sizes for their cooking. Users can cook with two burners at the same time to save cooking time. However, users have negative feedback on a cooking stove that has more than four burners.

3.2.8 Easy to Clean

Forty-one percent of the participants clean their cooking stove every day after cooking. One fourth of the participants are not satisfied with their cooker because it is very difficult to clean. The interview indicated that users need a dirt-free surface during their cooking. The users are satisfied with a flat and smooth surface on their cooker because it is comfortable for removing dirt stains.

3.2.9 Energy Saving and Alternative Energy Used

The feedback from the survey shows that the participants would like to use alternative energy input for their cooker to reduce electricity load and to increase self-sufficient living by decreasing fossil energy consumption.

3.2.10 Aesthetics and Appearance Design

There is a broad range of definitions among participants for what makes an attractive physical appearance of a cooker. For the most part, it can be assumed that the design should be compatible with furniture and the surrounding environment. However, there are some users that prefer a dominant design to make the cooker different from a conventional design.

3.2.11 Durability and Maintenance

The users are satisfied with their cooker if it has a long working life with regular maintenance. The participants expect that their cooker should work at least 5 years. Increase in satisfaction is directly linked to working life of the cooker.

Table 3 presents the percentage of user acceptance keywords in different preferred design attribute of the solar cooker.

4 Summary

Based on the comparative analysis of the research conducted with conventional cooking stoves, the results of this study showed that solar cooking stoves have not been widely used in the residential sector because of three main reasons: the solar cooking stove can only partly respond to basic attributes that users expect to have in their cooking stove, particularly timing, such as prompt use and fast temperature responding.

Most of the prototypes in research do not have a user-friendly appearance. A solar cooking stove still needs further ergonomic design and aesthetics to increase user satisfaction.

A solar cooking stove requires an extra infrastructure in a conventional building. A solar thermal energy system installation is not associated with a number of products that use medium temperature. However, sharing the infrastructure with other household appliances might increase investment value.

Table 3 Needs classification of cooking stoves according to Kano model

Needs	Frequency 1	Frequency 2	Frequency 3
1. Temperature performance	M (43 %)	O (41 %)	A (8 %)
2. Fast reaching of high temperature	O (38 %)	A (35 %)	M (15 %)
3. Precise control	M (47 %)	O (37 %)	A (12 %)
4. Easy to control (switch on/off)	I (45 %)	O (22 %)	R (19 %)
5. Continuous use	M (46 %)	I (17 %)	O (15 %)
6. Safety – non toxic	M (60 %)	O (25 %)	A (12 %)
7. Number of burners	I (46 %)	O (41 %)	A (10 %)
8. Easy to clean	A (57 %)	O (34 %)	M (8 %)
9. Energy saving	O (49 %)	M (32 %)	A (21 %)
10. Aesthetic appearance	I (28 %)	A (23 %)	O (21 %)
11. Durability	O (61 %)	A (23 %)	M (11 %)

M must be, *O* one dimension, *A* attractive, *I* indifferent, *R* reverse, *Q* questionable

An accepted solar cooking stove design must consist of four attributes as follows: temperature performance, precise control, prompt use, and safety. There are four attributes that can increase user acceptance of a solar cooker: fast reaching of high temperature, energy-saving ability, and ease of use. In addition, an energy infrastructure system in a building can be adapted for supply and can be shared with other household appliances to reduce building integration costs. The new design and development of a flat-plate collector with indirect solar cooker shows potential to solve some problems by adding heat storage tank into the system so that users can use it during nighttime as well as share the energy resource with other household devices. There is a potential to develop and improve user acceptance of solar cookers by a hybrid design with conventional energy resources.

References

1. Mirdha US, Dhariwal SR (2008) Design optimization of solar cooker. Renew Energy 33:530–544
2. Bruce N, Perez-Padilla R, Albalak R (2002) The health effects of indoor air pollution exposure in developing countries. World Health Organization (WHO), Geneva
3. ISO standard, EN 60332-2-9/A12, household and similar electrical appliances-safety-, November 2007
4. Schwarzer, Klemens. Germany patent DE 4142119 A1. 24 June 1993
5. Donald N (2013) The design of everyday things. Basics book, New York. ISBN 978-0-465-05065-9
6. Monthalee S (2003) Product design for innovation and reverse engineering. Technology Promotion Association (Thailand-Japan), Bangkok
7. Thirugnanasambandam M, Iniyan S, Goic R (2010) A review of solar thermal technologies. Renew Sust Energ Rev 14(1):312–322
8. Mac Clancy J (2014) Solar cooking: why is it not yet global? Food Cult Soc 17(2):301–318
9. Foodsafety (2014) Foodsafety, from keep food safe: http://www.foodsafety.gov/keep/charts/mintemp.html. Accessed 24 Jan 2014

10. Degrave (2010) http://www.degraeve.com/reference/cake-baking-temperatures-times.php. Accessed 15 Sept 2012
11. Canolainfo (2007) http://www.canolainfo.org/recipes/index.php?page=21. Accessed 24 Apr 2012
12. Wilson DG (2012) Development and optimization of a thermal-storage solar cooker. http://img3.wikia.nocookie.net/__cb20141110184045/solarcooking/images/0/01/Thermal_Storage_-_David_Gordon_Wilson_2013.pdf. Accessed Sept 2015

Sustainable Renewable Energy Financing: Case Study of Kenya

Tabitha A. Olang and Miguel Esteban

Abstract The present work examines the views of renewable energy stakeholders with regard to what they deem as sustainable methods of renewable energy financing in the Kenyan market. It also touches on the preferred source of renewable energy and the factors that need to be considered for their successful implementation. The study made use of online surveys, key informant interviews and site visits to collect data. The findings are intended to be used to inform decisions regarding sustainable financing solutions that are suitable for the Kenyan market. The key finding is that although the Kenyan government has taken a rather hands-off approach to the solar energy market, this market has been able to thrive in the private sector in the recent years.

Keywords Feed-in tariffs • Political risk • Power purchase agreement

1 Introduction

Sustainable development is defined as development that satisfies the needs of the present without compromising the ability of future generations to satisfy theirs [1]. Electricity generation is clearly crucial to the sustainable development of emerging economies as it plays a major role in the alleviation of poverty and constitutes the engine of socio-economic development. Moreover, energy is one of the most important aspects to be considered when establishing the interaction between the technological, economical and political landscape of a given country or region.

Excluding South Africa, sub-Saharan Africa has 37 gigawatts (GW) total installed electricity capacity, which is less than half of that in the United Kingdom, as of 2012 [2]. Moreover, the lack of plant maintenance and skilled personnel and weaknesses in transmission and distribution systems are some of the reasons why most of the installed capacity is not even available for generation. Hence, the region has the world's lowest electricity penetration rate (32 % overall), with rural areas

T.A. Olang (✉) • M. Esteban
GPSS-GLI, The University of Tokyo, Kashiwanoha, Japan
e-mail: tapaulie@gmail.com; esteban.fagan@gmail.com

M. Matsumoto et al. (eds.), *Sustainability Through Innovation in Product Life Cycle Design*, EcoProduction, DOI 10.1007/978-981-10-0471-1_12

having an energy access rate of only 16 % [3]. The region therefore needs to install approximately 7 GW of new-generation capacity annually in order to meet the increasing energy demand and support rapid economic growth [4]. In particular this need to increase installed capacity is particularly important to countries with rapid growing economies, such as Kenya, which has recorded a steady economic growth of 5.7 % in 2013 and a projected 6.5 % in 2015 [5]. Hence, this country will be used as a case study of the problems facing the sustainable financing of renewable energy in emerging economies.

Kenya is rich in potential sources of renewable energy, namely, wind, biomass, hydro, geothermal, biogas and solar resources [6]. However, according to The World Bank, only 23 % of the Kenyan population have access to electricity as of 2010 [7]. In 2008 Kenya identified energy as one of the enablers to its long-term development programme, known as Kenya Vision 2030 [8]. Kenya ratified a new constitution in August 2010, which started a process of regional devolution where there are now two levels of government, namely, those at the national and county levels. The Ministry of Energy and Petroleum (MOEP) anticipated a sharp rise in electricity demand as the new county governments operationalised, implying that various economic activities would spring up at the county levels. Specifically, energy-intensive industries, such as mining, production of iron and steel, irrigation schemes, agro-based industry, operation of petroleum pipelines and electrification of designated railway lines, were some of the activities that will likely result in a sharp increase in energy demand.

The Kenyan Government through the MOEP [9] therefore proposed a roadmap known as the 5000+ MW programme in 2013 to increase the country's generation capacity from 1664.1 to 6700 MW by 2016. This aggressive 5000+ MW programme aims to mainly achieve its target by a sharp increase in four types of electricity production, namely, geothermal (1646 MW), natural gas (1050 MW), wind 630 MW and coal 1920 MW. New capacity will be developed by government power utilities (KENGEN) and independent power producers (IPPs) under a private-public partnership framework. The necessary transmission will be developed by the Kenyan government (KETRACO). The project is mainly financed by the World Bank, which has invested over $650 million in the Kenyan energy sector in recent years [10]. The programme's focus is on reducing the cost of electric power by over 40 % for all end users [9]. As of November 2014, Kenya's installed capacity was 2294.82 MW, as shown in Fig. 1 [11].

The energy sector institutional structure previously had all its functions concentrated between the MOEP and Kenya Power and Lightening Company (KPLC). However, reforms in the sector led to the unbundling of functions, specifically generation (KENGEN), transmission (KETRACO), distribution (KPLC) and oversight and policy functions (ERC). Other key stakeholders in the Kenyan renewable energy sector include Kenya Renewable Energy Association (KEREA) and one emergency power producer known as Aggreko (which currently have an installed capacity of 30 MW). KEREA is made up of 29 private sector members who are dedicated to facilitating the growth and development of renewable energy business in Kenya [12]. Currently, six IPPs are operating in Kenya, accounting for a total

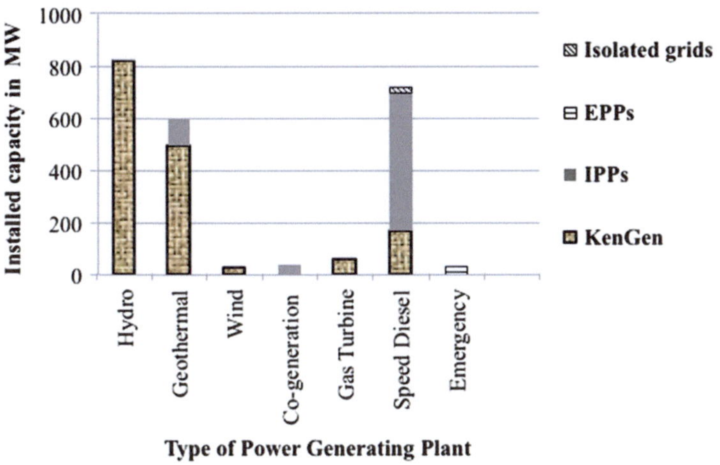

Fig. 1 Kenya's installed capacity as of November 2014

installed capacity of 725.32 MW, which is 31.63 % of the country's installed capacity (see Fig. 1).

The National Energy Policy (2004) was the first Kenyan policy to mandate the promotion of private sector investments in renewable energy. The Energy Act (2006) established the Energy Regulatory Commission (ERC), which promotes development and use of renewable energy technologies. One way of promoting the development of renewable energy technologies was by encouraging potential IPPs to carry out feasibility studies on renewable energy sources. In a bid to safeguard the time and resource investments made by private investors while undertaking feasibility studies and to boost renewable energy development in general, the MOEP proposed to set feed-in tariffs (FITs) for electricity generated from renewable energy sources. This led to the enactment of the feed-in-tariffs Policy (2008). This policy has undergone two revisions and currently the feed-in-tariffs policy (2012) is in operation. As a result of it, large geothermal power plants have been built, and it is significant to note that Kenya ranks eighth with regard to countries with the highest geothermal energy-installed capacity [13]. For example, the Olkaria 1 Power Plant is situated in the Rift Valley in Kenya and has an installed capacity of 185 MW.

However, despite the government enthusiasm for geothermal energy, this form of energy is one of the many available to the country. Despite this, to the authors' knowledge, no research has been carried out on the actual perception of private sector practitioners with regard to renewable energy financing in Kenya. Hence, the present work attempts to investigate how appropriate Kenyan government policy is from the point of view of the private sector in Kenya. The reason for doing so is that it is clear that a better understanding of various stakeholder perceptions with regard to the future of renewable energy in Kenya would help to ensure that equity in the consultation processes can be achieved. When mismatches are identified, consultation can help to build consensus, increasing the chances of successful deployment by making sure that effective policies are put in place.

2 Methodology

2.1 Data Collection

The study used three methods of data collection, namely, structured online surveys, key informant interviews and site visits. Structured online surveys were distributed via email to 43 energy experts and practitioners in Kenya in the month of February 2015. The initial contact list was obtained through the KEREA members' website [12]. Subsequent respondents were obtained via a snowballing sampling technique. Respondents' roles in the Kenya RE market are shown in Fig. 2. Findings from the survey were used to inform questions that were to be used in the key informant interviews.

Key informant interviews were carried out in March 2015. The study interviewed six respondents who had various roles in the Kenyan renewable energy market, namely, a government official, consultant, international lender, consumer/capacity builder, solar water heater consumer and a RE local investor. Findings from these interviews were used to triangulate the outcomes of the questionnaire survey and further the understanding of the responses.

After the key informant interviews, two site visits were carried out to Green Park Estate (Fig. 3a), namely, Athi River and Olkaria 1 Geothermal Plant [13]. Green Park Estate [14] is a gated community with 1550 house units along Mombasa Road, Nairobi, Kenya, which use solar water heating systems to supplement the grid power supplied by KPLC. Olkaria 1 Geothermal Power Station (Fig. 3b) was the first geothermal power plant in Africa, commissioned in June 1981. Since its commissioning, the plant has an availability factor of 95 % [15]. The visit to these two sites was important for the research as it provided some perspective on the responses obtained from the online survey and the key informant interviews.

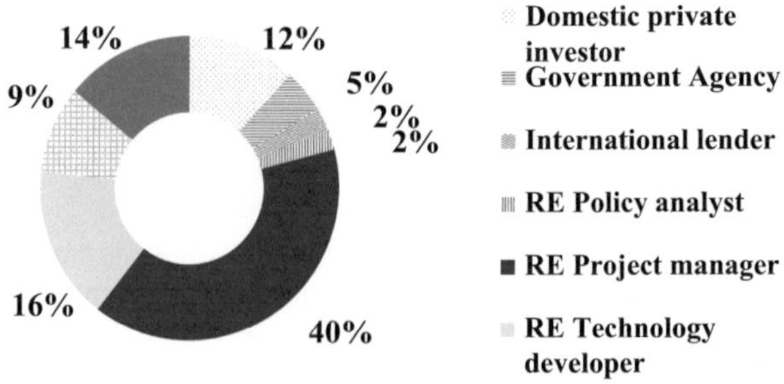

Fig. 2 Respondents renewable energy roles

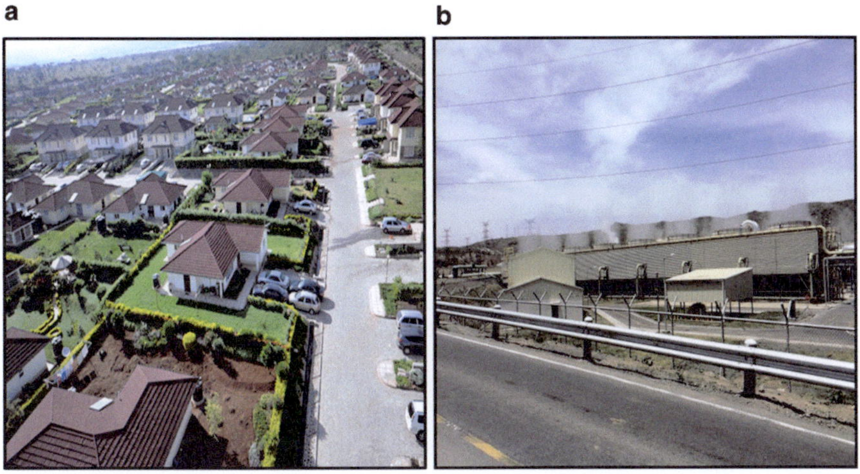

Fig. 3 (**a**) Green Park Estate with solar water heating systems on rooftops and (**b**) Olkaria Power Plant

2.2 Data Analysis

The study undertook two levels of data analysis, namely, summary statistics and regression analysis, both using Microsoft Excel analytical functions. The summary statistics established the main findings, after which a regression analysis was performed, though the authors will only discuss those results significant at the 99 % confidence level. Feedback from the key informant interviews was used to explain findings from the two levels of analysis.

3 Results

3.1 Key Findings from Summary Statistics

3.1.1 Preferred Technology

Out of all respondents ($n = 43$), 74 % stated that solar PV will be the favoured technology in sub-Saharan Africa over the next 5 years. Only 7 % of the respondents preferred geothermal, 5 % biomass, 5 % wind, 2 % hydro and 7 % other technologies. Of those who stated a preference for solar PV ($n = 32$), 56 % thought that it was suitable for rural, off-grid applications, and 28 % of them stated that the technology matches Africa's renewable resources (Fig. 4).

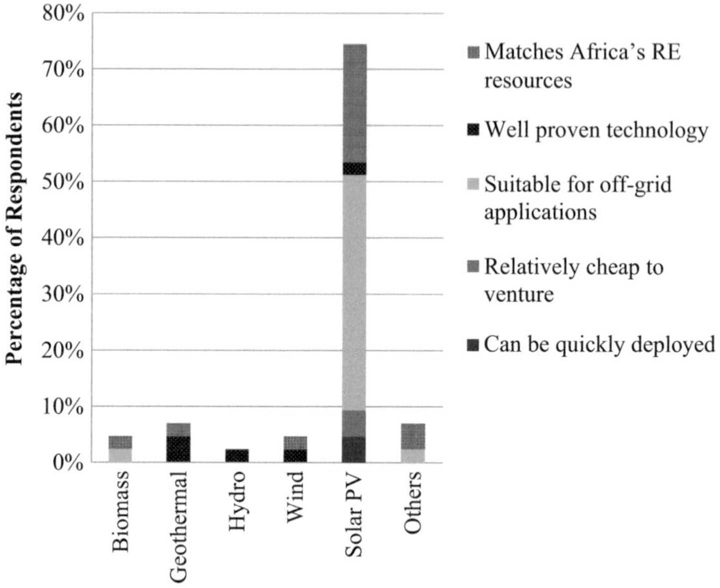

Fig. 4 Renewable energy technology and reasons for preference

3.1.2 Political Stability

All respondents stated that political stability was important for the viability of renewable energy investment in Kenya. Moreover, among a list of other possible risks experienced in the Kenyan renewable energy market, 42 % of the respondents stated that political risk is the most prominent risks that must be considered when financing sub-Saharan Africa renewable energy projects (Fig. 5). The second most important source of risk was regulatory risk, which is also partly tied with political risk, as the government has the capacity to quickly modify the regulatory regime, greatly changing the outlook of renewable energy.

For instance, the value-added tax (VAT) regimes of solar products have substantially changed over the past 5 years. In 2009, the solar products were zero rated (taxable, but at 0 % tax rate of tax on their input supplies). Subsequently in 2013 the incoming government made solar products taxable (at 16 % tax rate) in a bid to increase government revenue. However, the Kenyan government then once again decided to dismiss this tax on solar products in 2014 in a move to cut costs of renewable energy products by making them tax exempt [16].

3.1.3 Multilateral Lenders

All respondents stated that multilateral financial and development institutions (such as World Bank, Japan International Cooperation Agency, European Investment

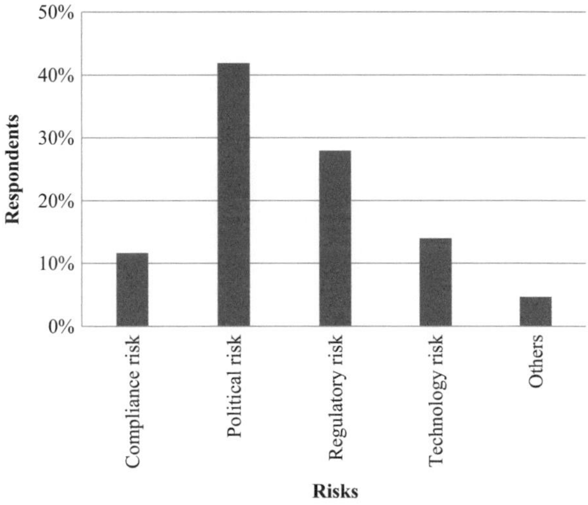

Fig. 5 Perceived risks in the Kenyan renewable energy market

Bank, etc.) will have a vital role to play in financing African renewable energy projects over the next 5 years. Moreover, 53 % of respondents stated that multilateral lenders will be the most important source of debt financing over the next 5 years (see Fig. 6). Nongovernment lenders (such as private banks, micro finance facilities, cooperative societies etc.) were selected as the second most important source of debt financing for renewable energy projects (26 % of respondents).

3.1.4 Feed-in Tariffs (FITs)

Forty-six percent of respondents ($n = 43$) stated that FITs are the most effective mechanism that should be used to attract private sector investment to RE technologies in Kenya (See Fig. 7). Of those who stated FITs were the most effective policy mechanism ($n = 20$), 85 % of them thought that solar PV is the most preferred renewable energy technology. This might imply that FITs could be the most suitable type of financing for solar PV technology in Kenya. Public-private partnerships were selected as the second most effective policy mechanism to attract private investment (35 % of respondents).

3.2 Key Findings from Regression Analysis

The key findings from summary statistics were subjected to a regression analysis by treating them as dependent variables and the rest of the responses as independent variables. A series of regressions were carried out between dependent variables (solar photovoltaic, multilateral lenders, political stability and feed-in tariffs) and

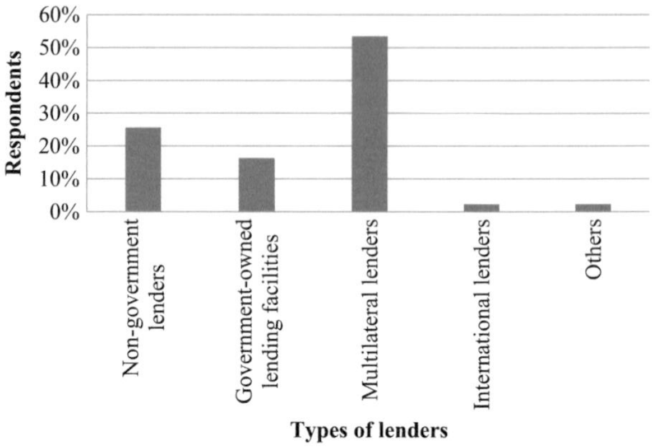

Fig. 6 Sources of debt financing for renewable energy projects

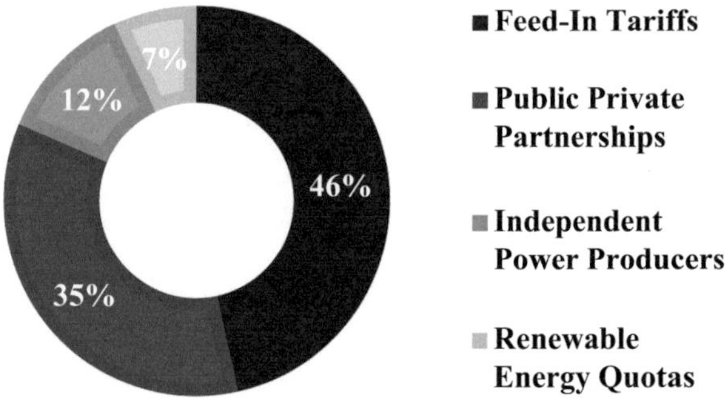

Fig. 7 Perceived as most effective policy tools available in the Kenyan RE market

each category of independent variables (RE market roles, reason for technology preference, sources of funds, risks, barriers and drivers of RE projects and mechanisms to promote RE technology). The only findings that will be discussed are those that were significant at the 99 % level of confidence, given the small number of respondents.

Two significant findings were obtained (Table 1). The first finding is that the preference for solar PV has a negative correlation with the technology being well proven. This implies that the reason solar PV is preferred is not necessarily because the technology is well proven but because it is suitable for rural, off-grid applications (as discussed in the summary statistics). The choice of feed-in tariffs as the most effective policy tool in attracting private investments in the Kenyan renewable energy market has a positive correlation with the source of equity for renewable energy projects being government investments and grants. This means that respondents who stated FITs as an effective mechanism to attract private sector investment also believe that an important source of equity is government investments and grants.

Table 1 P-values of significant regressions

Dependent variable	Independent variable	Coefficient	P-value
Solar PV	Technology is well proven	−0.80	1.93 %
Feed-in tariff	Government investments and grants	0.57	4.97 %

3.3 Key Findings from Key Informant Interviews

The results of the key informant interviews added a number of insights into the renewable energy market in Kenya, with the most significant of these outlined below.

3.3.1 Ignorance of Renewable Energy Policy Provisions

Most small-scale end users, especially those with solar PV panels installed in their houses, were not aware that the feed-in-tariff policy (2012) provides tariffs for both on-grid and off-grid solar PV systems. Moreover, informants who were aware of the energy policy provisions also stated that the transaction costs associated with the negotiation process when entering into power purchase agreement were high, especially for small-scale power producers. However, the feed-in-tariff policy (2012) provides standardised power purchase agreements (PPAs) templates to be used as a basis for negotiations as well as the various feed-in-tariff levels. Although the Kenyan government has attempted to increase the appeal of the renewable energy market, such feedback implies that many more improvements are still needed to ensure that all stakeholders—whether operating on a large or on small scale—are provided with equitable policy provisions.

3.3.2 Technology Challenges

The informants also highlighted how for small-scale solar PV, batteries have a much smaller lifespan (approximately 2 years) than solar panels. An option that could possibly counteract this would be feeding any excess power during the day directly the grid. However, some respondents claimed that KPLC is not always ready to sign a power purchase agreement with small-scale power producers.

3.3.3 Renewable Energy Financing Complexity

Much thinking is done at the level of international partners as well as African entrepreneurs and financiers. Moreover, there are many different market segments and business models currently being deployed, which greatly differ in characteristics and requirements. For instance, when comparing solar lighting (<10 MW) with large-scale solar utility scale power generation (>20 MW), the only thing they have in common is that electricity is produced and consumed and that the technology is based on solar PV, everything else is different. Hence to "establish" one viable financing mechanism might be too ambitious.

4 Results

4.1 Mismatch Between Ministry of Energy and Petroleum and Other Renewable Energy Stakeholders

Findings from the present study show that the renewable energy sector prefers solar PV technology mainly because it is suitable for off-grid, rural applications. The Kenyan government has largely taken a hands-off approach in the photovoltaic market. Moreover, it has also liberalised foreign exchange and import regimes, and this has allowed private entrepreneurship in the photovoltaic market to flourish. Hence, over the years, this market has grown gradually both technologically and commercially, making it accessible to lower-income users [17]. In terms of technology, the photovoltaic units have become cheaper and smaller in size. Commercially, innovative technologies such as MKOPA solar systems, which make use of mobile payments that allow people to buy these products on credit, have revolutionised asset financing of solar products in Kenya [18].

However, the Kenyan government is trying to meet its Vision 2030 development goals and increase the availability of electricity, with all efforts to reach the target of 5000 MW by 2030 mainly being done through the promotion of geothermal energy. Although geothermal resources in Kenya have an estimated potential of between 7000 and 10,000 MW [13], there is need to establish why the renewable energy practitioners are insisting that solar PV should also be promoted. In this sense it is important to note how the World Bank mainly funds geothermal projects in Kenya and that the power generated is fed into the national grid. This implies that the population not connected to the grid would not be able to benefit from the increased electricity generation and reduced charges (KPLC is the only corporate body mandated to distribute on-grid electricity in Kenya). According to KPLC's Annual Report as at June 2014, electricity is currently accessible to 35 % of the Kenyan population [19]. From an economic point of view, it would appear paradoxical how, given that less than 50 % of the population have access to electricity, having the national grid produce a surplus could mean huge wastage of electricity given the lack of connections of those in the more remote areas of the country.

4.2 The Issue of Storage and Smart Grids

The main components of solar home systems include solar cell modules, lead-acid batteries, charge controllers, low-voltage direct current (DC) appliances and other accessories such as switches, module mounts, etc. For the purpose of the urban and more developed parts of Kenya, it is clear that the issue of how to store electricity from PV panels will become crucial in the future. As the installed PV capacity increases, the life of batteries will become a fundamental problem for the financial sustainability of such a system. New technologies that produce batteries that can store energy in the presence of the sun and when utilities rates are low, as well as provide backup electricity supply, can go a long way in making solar home systems and even industries independent of the national grid (e.g. the proposed Powerwall by Tesla Motors [20]). However, cost implications should also be put into consideration when looking at such possibilities.

4.3 The Issue of Off-Grid Access to Electricity in Kenya

When talking about electricity in Kenya, it is important to remember that at present 65 % of the country is still not connected to the grid. Though the government is spending considerable amounts of money (for instance, KPLC government-funded projects comprising construction of substations and lines at various locations, amounting to US$ 23.38 million [16]) to increase electricity penetration, by 2020 it is expected that 70 % of the country would be connected, still leaving 30 % without access to grid electricity.

However, it is important to note that electricity is important for economic development and poverty alleviation, especially in rural areas. Without electricity, access to education is limited as students are not able to study at night, commerce would be crippled, businesses and hospitals dependent on electricity would not be able to operate, and people would end up using kerosene and charcoal to cook, which have serious associated health hazards. Moreover, mobile phone banking, by use of the M-PESA service, is a core source of small-scale financial services in Kenya, with money transfer and credit systems having 17 million active and registered users in the country as of 2013 (over 30 % of Kenyan population) [21]. M-PESA has led to local economic expansion, security, capital accumulation and increased levels of employment [22]. To have access to this service, the subscribers need mobile phones, which clearly need electricity to operate.

Thus, the issue of off-grid access to electricity in Kenya will be important for many years to come. In order to promote the sustainable development of the more remote parts of the country, the government should not only concentrate on increasing the on-grid installed capacity but also consider promoting off-grid projects to ensure a larger energy access in the country. Increasing on-grid capacity should also be coupled with an equal investment in transmission lines and increased number of electricity connections at affordable costs to various groups of consumers (industrial versus residential, urban versus rural, etc.). The feed-in-tariffs policy would go a long way in encouraging private investments in the sector. Initiatives by the government such as introducing a standardised power purchase agreement for small-scale power producers are a step in the right direction as it enables a less complicated and less costly negotiation process between the power producer and KPLC.

5 Summary

The authors conducted questionnaire surveys and interviews with key informants that showed how there is a discrepancy between the energy policy followed by the government and the views of many energy experts. Although the Kenyan government has exempted solar products from value-added tax and set specific feed-in tariffs for the various sources of renewables, a policy support regime has only been set for geothermal energy (+5000 MW programme) and not for other renewable energy sources of energy such as solar energy, which is readily available.

The solar market in Kenya has been able to thrive without direct government intervention. However, instead of the government taking a hands-off approach, it should facilitate solar energy market growth with more incentives and specific targets to ensure an increase in the solar energy uptake, especially in rural areas that are disconnected from the grid.

Finally, the FIT policy seems to be effective in the renewable energy market in Kenya. The policy also gives provision for revisions every 3 years. The government should use this opportunity to create a more robust market by putting in place provisions that promote the coexistence of both the public and private renewable energy markets for the benefit of the entire Kenyan population.

References

1. Brundtland G et al (1987) Our common future: report of the 1987 World Commission on Environment and Development. Oxford University Press, Oxford
2. U.S. EIA (2012) International energy statistics. Retrieved 20 May 2015 from U.S Energy Information Administration: http://www.eia.gov/cfapps/ipdbproject/iedindex3.cfm?tid=2&pid=2&aid=7&cid=regions,&syid=1998&eyid=2012&unit=MK
3. IEA (2012) World energy outlook. Retrieved 20 May 2015 from International Energy Agency: http://www.worldenergyoutlook.org/resources/energydevelopment/energyaccessdatabase/
4. United Nations Environment Programme (2012) Financing renewable energy in developing countries. United Nations, United Nations Environment Programme Finance Initiative. Instaprint, Geneva
5. Odero WO, Reeves WA, Kipgeyo N (2015) KENYA 2015. AfDB; OECD; UNDP, African Economic Outlook

6. Government of the Republic of Kenya: Energy Regulatory Commission (2012) Feed-in-Tariffs policy for wind, biomass, small hydros, geothermal, biogas and solar resources. Policy document. Government of the Republic of Kenya, Ministry of Energy and Petroleum, Nairobi
7. The World Bank (2015) World development indicators: electricity production, sources and access. Retrieved 25 June 2015 from The World Bank website: wdi.worldbank.org/table/3.7
8. Government of the Republic of Kenya: Ministry of State for Planning, National Development and Vision 2030 (2008) Kenya vision 2030. Retrieved 20 May 2015 from Kenya Vision 2030: www.vision2030.go.ke
9. Government of the Republic of Kenya: Ministry of Energy and Petroleum (2013) 5000+MW by 2016: power to transform Kenya. investment prospectus. Republic of Kenya, Ministry of Energy and Petroleum, Nairobi
10. The World Bank (2015) Kenya overview: results. Retrieved 25 May 2015 from The World Bank website: http://www.worldbank.org/en/country/kenya/overview#3
11. Energy Regulatory Commission (2014) Electricity supply industry in Kenya. Retrieved 21 Jun 2015 from ERC: Energy Regulatory Commission: www.erc.go.ke/index.php?option=com_content&view=article&id=107Itemid=625
12. Kenya Renewable Energy Association (2012) Our members: KEREA. Retrieved 4 Jun 2015 from KEREA website: http://kerea.org/membership/our-members/
13. Matek B (2015) 2015 annual U.S. & global geothermal power production report. Annual, Geothermal Energy Association
14. Superior Homes Kenya (nd) Green park: superior homes Kenya. Retrieved 22 June 2015 from Superior Homes Kenya website: http://www.superiorhomes.co.ke/
15. Kenya Electricity Generating Company: Geothermal Plants (2015) Retrieved 28 June 2015 from Kenya Electricity Generating Company Limited http://www.kengen.co.ke/index.php?page=business&subpage=geothermal&id=1
16. Climate Innovation Center, Kenya (2015) Tax regimes on solar products in Kenya. Fact sheet. Strathmore University, Strathmore Energy Research Center, Nairobi
17. Hankins M (2000). A case study on private provision of photovoltaic systems in Kenya. http://www.worldbank.org/html/fpd/esmap/energy_report2000/ch11.pdf
18. Mkopa Solar (2015) Products. Retrieved 28 Jun 2015 from Mkopa Solar: http://www.m-kopa.com/products/
19. Kenya Power and Lighting Company Limited (2014) Annual report and financial statements: financial year ended 30 June 2014. Annual, Kenya Power and Lighting Company Limited, Nairobi
20. Tesla (2015) Powerwall. Retrieved 28 Jun 2015 from Tesla Motors: http://www.teslamotors.com/powerwall
21. Safaricom (nd) MPESA statistics: Safaricom. Retrieved 26 Jun 2015 from Safaricom Limited website: www.safaricom.co.ke/personal/m-pesa/statistics
22. Plyler MG, Haas S, Nagarajan G (2010) Community-level economic effects of M-PESA in Kenya: initial findings. Financial Services Assessment, College Park, p 5

Oil and Gas Industry's Role on the Transition to a Low-Carbon Future in Thailand

Warathida Chaiyapa, Miguel Esteban, and Yasuko Kameyama

Abstract The present study examines role of oil and gas (O&G) industry in Thailand towards the transition to a low-carbon future or, in other words, the diversification of the energy mix to be less dependent on fossil fuels and increase the penetration of renewable and alternative energy. It proposes that although government policy is important to drive changes in the energy mix, the O&G industry, an established energy supplier, could arguably be a key player for sustainable energy development. After reviewing the energy policy of Thailand, and in particular the climate change mitigation and renewable and alternative energy development policy, the study distributed an online questionnaire survey and conducted interviews with six O&G companies in Thailand to gain first-hand data on their corporate response to climate change and low-carbon energy sources. Findings revealed proactive responses in climate change mitigation with various activities undertaken signifying a range of degree of commitment among the various companies. Finally, the authors outlined policy recommendations based on an analysis of incentives and rationales for companies to take part in low-carbon energy development in Thailand.

Keywords Climate change mitigation • Low-carbon energy development • Oil and gas industry • Thailand

W. Chaiyapa (✉) • M. Esteban
Graduate Program in Sustainability Science – Global Leadership Initiative (GPSS-GLI), Graduate School of Frontier Sciences, The University of Tokyo, 5-1-5 Kashiwanoha, Kashiwa, 277-8563 Chiba, Japan
e-mail: wwlnajah@gmail.com

Y. Kameyama
Center for Social and Environmental Systems Research (Sustainable Social Systems Section), National Institute for Environmental Studies, 16-2 Onogawa, Tsukuba, 305-8506 Ibaraki, Japan

M. Matsumoto et al. (eds.), *Sustainability Through Innovation in Product Life Cycle Design*, EcoProduction, DOI 10.1007/978-981-10-0471-1_13

1 Introduction

Thailand, the second largest economy in ASEAN (Association of Southeast Asian Nations), is endowed with limited petroleum resources found both onshore and offshore. The country is ranked fifth in Southeast Asia for oil and natural gas proven reserves (0.5 billion barrels and 0.3 trillion cubic meters, respectively) [1]. However, due to rapid urbanization and economic development, primary energy demand has been continuously increasing at 2.3 % on average per year between 2011 and 2035, making it the second highest in ASEAN after Indonesia [1]. The country has in the past been heavily dependent on petroleum products. In 2012 oil accounted for 37 %, and gas for 44 %, of total primary energy supply [2]. Yet, a significant share of the oil and gas supply is imported, rendering the country vulnerable to energy price volatility and insecurity. Crude oil has been mostly imported from the Middle East and natural gas through a pipeline from neighboring countries like Malaysia and Myanmar or in the form of liquefied natural gas (LNG). Together with a high dependence on petroleum, Thailand's greenhouse gas emissions, as well as air pollution, is projected to increase in the future. Thailand's Second National Communication submitted to the United Nations Framework Convention on Climate Change (UNFCCC) reported its total national emissions, including Land-Use, Land-Use Change and Forestry (LULUCF), in 2000 as 229.08 $MtCO_2$ equivalent, whereas the World Resources Institute (WRI) reported the total emissions with LULUCF in 2011 as 352.8642 $MtCO_2$ eq [3, 4]. Power generation, industry, and domestic transport are the top three emitting sectors.

To enhance energy security and maintain economic development, while addressing rising GHG emissions, has become a big challenge not only for Thailand but also for other developing countries. One promising solution to this challenge is to head towards an energy transition to low-carbon and renewable energy sources [5]. Daniel Yergin, a vice chairman of IHS Cambridge Energy Research Associates who served as an Energy Community Chair between 2011 and 2012 in the World Economic Forum, commented that "transition" is "not an abrupt change from one reality to another but rather a shift that unfolds generationally over considerable time, and one that may lead to greater diversity in the energy marketplace" [6].

Along the lines of such a remark, the seed to transition to low-carbon energy has been cultivated and growing in Thailand for some time. The Thai government, through the Ministry of Energy, has launched an energy policy which focuses on five strategies: "energy security, alternative energy, supervising energy prices and safety, energy conservation and efficiency, and environmental protection" [7]. Two concrete policy plans have been promoted: the Energy Efficiency Development Plan (EEDP 2011–2030) and the Renewable and Alternative Energy Development Plan for 25 % in 10 Years (AEDP 2012–2021). The former aims to reduce the country's energy intensity by 25 % by 2030, compared to that in 2010, while the latter sets a target to increase renewable and alternative energy supply to 25 % of total energy consumption by 2021 [7, 8]. Thanks to rich agricultural products – in

Table 1 Alternative and renewable energy consumption in three sectors: electricity generation, heat generation, and biofuels in 2011–2014 and target in 2021

Year	2011	2012	2013	2014	Target 2021
Energy for electricity generation (MW)					
Biomass	1790	1959.95	2320.78	2451.82	4800
Biogas	159.17	193.40	265.23	311.50	3600
Solar	78.69	376.72	823.46	1298.51	3000
Wind	7.28	111.73	222.71	224.47	1800
Waste	25.48	42.72	47.48	65.72	400
Small hydropower less than 12 MW	95.70	101.75	108.80	142.01	324
New form of energy	0	0	0	0	3
Energy for heat generation (ktoe)					
Biomass	4123	4346	4694	5144	8500
Biogas	402	458	495	528	1000
Waste	1.7	78.2	85	98.1	200
Solar	2	3.5	4.5	5.1	100
Energy for biofuels (million liter per day)					
Ethanol	1.2	1.4	2.6	3.2	9
Biodiesel	2.1	2.8	2.9	2.9	7.2

The table was produced using raw data from Department of Alternative Energy Development and Efficiency (DEDE), Ministry of Energy [10]

particular sugar-based crops – Thailand has great potential to develop a number of alternative energy resources such as biofuels, biomass, biogas, biodiesel, and ethanol. In addition, its geographical location along the equator offers Thailand a plentiful supply of renewable energy, such as solar PV power – with average radiation of 18.2 MJ/m2/day – and wind power [9]. Although fossil fuels remain dominant in energy mix, the country has seen gradually but consistently increasing its renewable and alternative energy consumption. Table 1 shows the energy consumption in various types of renewable and alternative energy in three main sectors, (1) electricity generation, (2) heat generation, and (3) biofuels for transportation in 2011–2014. The table also shows the 2021 targets set by "AEDP 2012–2021."

In addition to the two master plans, Thailand recently made a pledge at the COP 20 in Peru 2014 to cut at least 7 % of GHG emissions below the business-as-usual scenario in energy and transportation sectors by 2020 (announced by the Minister of Natural Resources and Environment in the Joint High-level segment of the 20th session of the Conference of the Parties to the UNFCCC [11]). The pledge was significant since it was the first official GHG reduction commitment of Thailand through the international climate mitigation agreement after ratifying Kyoto Protocol since 1999. All aforementioned policies are a clear signal that Thailand is steering its efforts towards a transition to a low-carbon future.

Nonetheless, the road to the low-carbon future, or in other words the diversification of energy supply, requires more than encouraging government policies or subsidies. Jacobsson and Bergek [12] constructed a model of inducement or blocking mechanism for some renewable energy technology, including three core elements: government policy, firm entry/activity, and feedback from market formation. According to them, the government can provide supports in forms of "investment subsidies, demonstration programs and legislative changes," the positive feedback of market formation such as increased sales helps in accumulating capitals for renewable energy development, yet "the ambiguous and/or opposing behavior of some established energy suppliers and capital goods suppliers has reduced legitimacy of renewable energy technology and has thus blocked the supply of resources and guided the direction of search away from these technologies" [12]. The behavior of energy suppliers has been considered as a key factor that can influence sustainable energy development [13].

The oil and gas (O&G) industry, a primary energy supplier, is among those directly affected by the renewable and alternative energy development and climate change mitigation efforts. Considering the fact that renewable energy is not the core business of O&G companies which may reduce profits and that climate change mitigation was viewed as threat to O&G companies [14, 15], it would appear obvious that the O&G industry in general is unlikely to positively respond to renewable energy promotion. However, in order to ascertain whether this is indeed the case, the present study will investigate O&G companies in Thailand under the circumstance described earlier. Interestingly, the study finds that some O&G companies are conducting various climate change mitigation activities, to different degrees of commitment. Notably, the Thailand state-owned company has been investing in renewable energy such as solar PV and conducted active R&D of alternative energies to fossil fuels.

However, corporate strategies and their incentives for actions vary from company to company. The present study thus aims to present and discuss how the O&G companies in Thailand response and prepare themselves for the transition to a low-carbon future. The main issues that will be discussed are (1) current O&G companies' climate change mitigation activities, (2) incentives or rationales of their actions, and (3) the implications of their corporate responses to climate change.

Understanding how O&G companies in Thailand are responding to climate change mitigation and low-carbon energy development policy is important for two reasons. First, the results contribute to the argument that the business strategies of established energy suppliers like the O&G industry are crucial for the transition to low-carbon future societies. While the external push from the government and market is necessary, the change within the energy sector is vital due to their long history and powerful financial and political influence. Also, the present research provides a case study of Thailand's O&G industry as a starting point to take into account this industry in other developing countries which are facing similar challenges in balancing economic development and environmental protection. Policy makers in Southeast Asian countries could thus benefit from the experience of Thailand, especially regarding how to foster positive responses towards a transition to a low-carbon future from the O&G industry in their countries.

2 Methodologies

2.1 Conceptual Framework

The transition to a low-carbon future in this present study aligns itself with the climate change mitigation effort which, according to 5th Assessment Report of the Intergovernmental Panel on Climate Change, or IPCC 5AR (2014), is defined as "a human intervention to reduce the sources or enhance the sinks of greenhouse gases (GHGs)" [16]. The O&G industry, which is part of energy supply sector, can contribute to global climate mitigation through various options. There are, for example, energy efficiency improvements, fugitive emission reduction in fuel extraction as well as in energy conversion, transmission and distribution systems, fossil fuel switching, carbon capture and storage (CCS) technology installation, alternative and renewable energy development, as well as planting forests to increase carbon sinks.

Although the aforementioned options could potentially reduce GHG emissions, some of them may have a side effect in terms of encouraging a lock-in on fossil fuel products. Considering the increasingly severe impact of global climate change, as well as the scarcity of petroleum resources, the long-term phase-out of fossil fuels and their substitution by low-GHG alternative energy resources would appear to be imperative [17]. In this regard, the present study gives more focus on renewable energy development in O&G industry over other climate change mitigation activities that the O&G companies are pursuing.

2.2 Research Approaches

Mixed qualitative research approaches were adopted. After extensively reviewing secondary documents, i.e., company websites and annual and corporate social responsibility (CSR) reports, the authors distributed an online questionnaire survey to obtain primary data on the climate change mitigation efforts that companies were conducting. Then, semi-structured interviews were carried out with companies who responded to the online survey and other stakeholders who were working in the issue of climate change policy in Thailand, namely, government authorities, NPOs, and academics. The research was pursued within a 2-year period of study from 2013 to 2015.

2.3 Targets of Study

The present study tackles the O&G companies that operate in Thailand, which are comprised of multinational corporations and a national (state-owned) petroleum company. Many of them are only operating in the upstream industry (petroleum exploration and production), whereas a few operate in the complete production cycle including midstream (refinery) and downstream (retail distribution) sectors. The study approached ten O&G companies, the largest companies in the sector. Six of them, from the UK, the US, Japan, Hong Kong, and Thailand, responded and agreed to be interviewed. They altogether accounted for a cumulative production volume of 89.62 % of crude oil, 91.82 % of natural gas, and 97.1 % of condensate [18]. It is worth noting that respondents were company personnel whose positions were related to climate strategies in the company, for instance, the Quality, Health, Safety and Environment (QSHE) manager, an environmental specialist, or an engineer. In the present study, the name of respondents and companies will be kept anonymous and company respondents will be referred to by the company's home country.

To triangulate answers from company respondents, the authors approached a group of non-company stakeholders comprising (1) government officials working in climate change mitigation policy-related government offices, (2) NGOs' staff whose work is related to the O&G industry, and (3) academics who have expertise in climate change mitigation and GHG emissions reduction. In total four interviewees were carried out: two government officials from the Department of Mineral Fuels (DMF), the Ministry of Energy and Thailand Greenhouse Gas Management Organization (TGO), one lecturer from the Faculty of Engineering, Chiang Mai University, and one NPOs staff from the Petroleum Institute of Thailand (PTIT). Once again their names and positions are kept anonymous.

3 Results and Discussion

3.1 Corporate Strategies Related to Climate Change Mitigation and Renewable Energy Development

Considering the fact that Thailand has not yet imposed any binding regulations on climate change, especially on the O&G industry, the study examined voluntary corporate responses to climate change by means of an online questionnaire survey. The results reflect to what extent the O&G companies internalize climate change mitigation into their business operation in Thailand. The study offered eight possible climate change activities and asked the company respondents to choose any that apply to their company. The activities offered were (1) measuring and reporting GHG emissions data; (2) setting a voluntary reduction target; (3) increasing energy efficiency; (4) reducing flared gas; (5) R&D to produce green products such as biofuel; (6) installing carbon capture and storage technology; (7) changing portfolio, such as investing in renewable energy; and (8) planting forests to increase carbon sinks. Table 2 presents the responses of the six O&G companies on climate

Table 2 Responses on climate change mitigation activities of O&G companies in Thailand

Climate change mitigation strategies		Company/country of origin					
		Thai	USA 1	USA 2	UK	Hong Kong	Japan
GHG reduction	Measure and report GHG data	√	√	√	√	√	√
	Set reduction target	√	√	X	X	X	X
	Increase energy efficiency	√	√	X	X	X	X
	Reduce flared gas	√	√	X	X	√	X
	R&D to produce green products	√	X	X	X	X	X
	Install carbon capture and storage (CCS) technology	X	X	X	X	X	X
	Invest in renewable or alternative energy	√	X	X	X	X	X
Increase carbon sink	Plant forests	√	√	√	√	√	√

change mitigation activities for reducing GHG emissions and for increasing carbon sinks.

From Table 2 it is clear that O&G companies have pursued climate change mitigation efforts to various degrees. First and foremost, all six companies have measured and reported GHG emission data. The Thailand-based company and US-based company (1) are the only two companies who have set up a reduction target as well as attempting to increase energy efficiency during their operations. The Thailand-based, US-based (1), and Hong Kong-based companies have reduced the flaring of gas. However, only the Thailand-based company has developed green products and invested in renewable energy. Interestingly, none of them have installed carbon capture and storage technology, while at the same time all have planted forests to increase carbon sinks.

3.2 Rational of the Companies' Response to Climate Change Mitigation and Renewable Energy Development

The questionnaire survey provides preliminary results on the climate change mitigation activities that the O&G companies are performing. The study approached the company respondents to further examine their rationales to undertake each mitigation option, with the following sections outlining the most important interview findings.

Due to government policy, every O&G company has measured and reported GHG emission data.

Since 2013 the Department of Mineral Fuels (DMF), Ministry of Energy, has implemented "the guideline for measuring and reporting GHG emissions for petroleum exploration and production sector." Prior to this guideline, some companies started measuring and reporting GHG emission data on a voluntary basis by applying different guidelines and measuring methodologies. The Thailand-based company informant explained that the company started this activity in 2008. It has formed its own guideline called "XXX (company name) GHG Accounting and Reporting," adopting some prominent GHG standards such as ISO 14064-1, GHG Protocol, IPIECA, API Compendium, and the IPCC. The guideline is employed by the headquarters and its subsidiaries. The US-based company (1) also applied its own GHG accounting guideline which its headquarter assigned to all local branches in different operating countries.

However, the key informants from government authorities (DMF and TGO) and academics seem to question companies' GHG emission data. Firstly, the company can choose to report GHG emissions by either operation control or equity share, which results in different final figures. Secondly, the companies apply diverse guideline, and thus it is difficult to compare the data between them. Third, only a small number of companies have voluntarily measured emissions. To address these concerns, the government authority (DMF) decided to implement a common guideline to harmonize the diverse procedures that companies have applied in the past and to establish an industrial GHG inventory for the first time. The guideline was assigned to every O&G company as a "Notification of the Director-General of Mineral Fuels Department" with penalty of a 50,000 baht fine (approximately US$ 1,524) for those who fail to comply.

Nevertheless, feedback on the guideline was mixed. The Hong Kong-based company informant noted that the government's guideline lessened their workload since they do not have to search for GHG standards from international organizations. On the other hand, some companies are concerned of the reliability and compatibility of the government's guideline with international standards. The US-based company (2) informant remarked that the DMF should follow up implementation of the guideline to see whether the GHG emission data is compatible with international standards or not.

Companies can reduce production costs and gain more income from mitigating GHG emissions.

Climate change mitigation activities, and in particular energy efficiency improvement, flared gas reduction, and green products development, are expected to bring about positive monetary benefits to companies. By increasing energy efficiency (using less amount of fuels), companies can reduce production costs. In addition, rather than flaring gas, companies found that they can obtain another product and gain additional income. Last but not least, green products such as biofuels help companies obtain additional profits since consumers are concerned of the environment and keen to choose environmentally friendly products.

The interview with the Hong Kong-based company informant revealed that the company's CEO is concerned with environmental issues and assigned staff to find ways to reduce gas flaring, especially the small amount of gas that is associated with crude oil, their main product. The engineers of the company then recommended that this should be used to produce liquefied natural gas (LNG) instead of just burning it. However, the project is waiting for the approval from the government authority since it may violate the concession right law, which allows companies to only extract petroleum but not produce new products. The Thailand-based company is the only one which has produced green products, i.e., biodiesel and gasohol, thanks to having operations in midstream and downstream (the other five companies contacted only have upstream operations). In terms of biodiesel, the company has cooperated with various organizations to carry out R&D for producing biodiesel from palm, *Jatropha curcas* (Barbados nut), and marine microalgae. Regarding palm oil, the company has invested since 2007 in palm plantations in Indonesia through its subsidiary "PTT Green Energy Pte., Ltd (PTTGE)." The PTTGE in 2010 acquired about 200,000 Ha land bank, mainly located in Kalimantan, and is expected to expand to 500,000 ha in 2020 [19]. Also, through another subsidiary in Thailand, "Bangchak Petroleum Public Company Limited," the company produces biofuels from palm oil, having a production capacity of 360,000 l per day in 2014, while Thailand's total production volume is 3.212 million liters in the same year [20, 21]. Moreover, the company established a partnership with the Commonwealth Scientific and Industrial Research Organization (CSIRO), Australia's national science agency, and invested around 90 million baht (approximately US$ 3 million) to produce algae-based biofuels. The project is now at pilot scale and expected to be commercialized in 2017 [22].

For gasohol, the company's subsidiary, Bangchak Petroleum, is producing ethanol at around 400,000 liter per day in Ayutthaya province [23]. In addition, the company has recently conducted a joint venture with Mitr Phol Biofuel Company, the largest ethanol plant in Thailand, with production capacity of 500,000 liter per day. The project aims at constructing production and distribution center for E85 in Chaiyaphum province. The center is expected to be the largest plant for gasohol E85 in the northeastern region of Thailand [24]. In 2013, there were in total 20 ethanol plants in the country, which have production capacity as 4.79 million liters per day [25]. With the production capacity of these two ethanol plants, the Thailand-based company is occupying a large share of the biofuel market.

It is worth noting that the company's active response is very much the result of government policy, namely, the cancelation of the sale of benzene octane 91 on January 1, 2013. This policy was part of the AEDP 2012–2021 to increase alternative fuels and reduce energy imports. As a result, only benzene octane 95 is still available to serve people who have cars produced before 1995 (which cannot use gasohol). However, as older cars are gradually being phased out and the retail price of benzene 95 is around 6.16 baht (US$ 0.18) per liter more expensive than gasohol 91 (E10) and 12.28 baht (US$ 0.36) than gasohol E85 (see Table 3 for a summary of current retail prices), there has been a gradual increase in gasohol consumption, in particular gasohol 91 (E10) which is a substitute for benzene octane 91. Figure 1 shows the consumption of benzene octane 95 compared with gasohol in 2014, 1 year after the government canceled the sale of benzene octane 91. Essentially, current government policy is to promote gasohol E20 and E85 which would reduce substantially the dependence on imported crude oil.

As a national company which has a mission to provide energy security, the Thailand-based company diversifies its core business by investing more in renewable energy.

Table 3 Retail price of fuels on June 26th, 2015 [26]

Fuels	Bath per liter	US$ per liter
Benzene 95	35.66	1.05
Gasohol 91 (E10)	28.68	0.85
Gasohol 95 (E10)	29.5	0.87
Gasohol 95 (E20)	27.28	0.81
Gasohol 95 (E85)	23.38	0.69

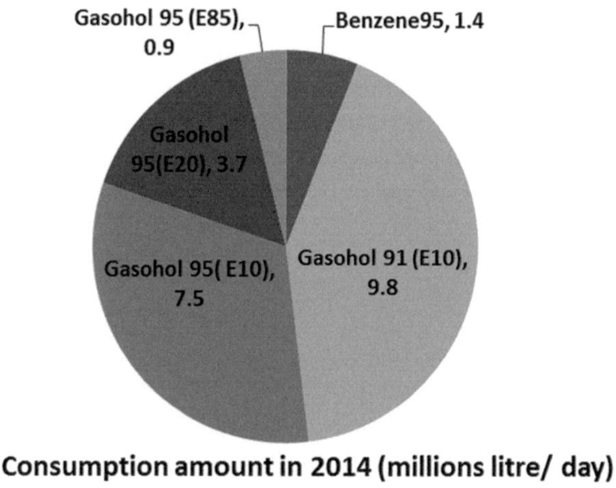

Fig. 1 Fuel consumption by types in 2014 [27]

Among the six company respondents, the Thailand-based company is the only company which has invested in renewable energy. The other five companies are multinational corporations that were granted a concession right to operate in Thailand. The interviews with the foreign company respondents revealed their prime concerns relate to the decreasing production volume, especially crude oil, and uncertainties regarding the new concession bidding round. As the Japan-based company informant pointed out, the daily production volume from its onshore oil wells has been decreasing and is not reliable. The situation is so worrying that the company may not able to obtain a return on its investment before the current concession right expires.

Unlike the foreign companies, whose main focus is on extracting petroleum over their concessions, the Thailand-based company has started to diversify portfolio from only petroleum to alternative and renewable energy. It even set the company vision to become "Thai Premier Multinational Energy Company" [28]. In 2013, the company invested 40 % of total share in Thailand Solar Renewable Company (1,450 million baht or approximately US$ 43 million) to construct electricity generating plants from solar PV. The project aims to build 10 power plants with a total 80 MW capacity [29].

When asked about the rationale for renewable energy development, the Thailand-based company informant referred directly to the company's mission as a national energy company to sustain Thailand's energy security. This goal aligns with its mission to the country, published in the company website as "ensuring long-term energy security by providing adequate high-quality energy supply at fair prices to support economic growth" [28].

There is no GHG reduction policy imposed on the O&G industry in Thailand; thus, CCS technology is not yet necessary.

According to key informants from the government (DMF) and academics (interviewed in 2014 before Thailand declared a reduction pledge), Thailand, a non-Annex I party to the Kyoto Protocol, does not have an obligation to cut down GHG emissions. As a result, there have been no GHG emission reduction regulations imposed on any sectors and in particular the O&G industry. Climate change mitigation measures in Thailand had so far have taken the shape of Nationally Appropriated Mitigation Actions (NAMAs), which have focused on either energy efficiency improvement or renewable energy development. Also, some sectors have voluntarily participated in CDM (Clean Development Mechanism) projects, which brought about economic benefits to participants. Although the Thai government would like to implement a GHG emission reduction policy in the future, the O&G industry may not be considered the prime target due to its relatively lower share of GHG emissions compared to other sectors such as transportation and electricity generation. According to a government informant (DMF), the O&G industry is an energy producer which does not release as much GHG emissions as sectors like electricity generating plants, which is considered to be the main energy user.

Due to the lack of binding GHG emission reduction policy, O&G companies considered it unnecessary to pursue CCS technology at present. Moreover, the technology itself requires a massive capital investment, thus taking some time to provide a return on investment. Apart from the Thailand-based company, which is state-owned, the other five foreign companies raised concerns regarding the approaching expiring date of their concession right in 2022 and the uncertainty of their future in Thailand due to the current suspension of the new concession bidding round. Thus, given present uncertainties, they were reluctant to invest the huge amounts of capital required for such technology.

Planting forests is part of the companies' corporate social responsibility (CSR) activities, which fosters good relationship with communities and local government.

All six companies showcased their forest planting as part of their efforts at environmental conservation, as well as community relation-building activity. However, the degree and scope varied in each company. The Thailand-based company launched "the one-million rai reforestation project in honor of the King" in 1997 and successfully "bestowed upon the forested area to H.M. the King on 16 November 2002" [30]. The countrywide reforested areas have claimed to "sequester about 18.17 million tons of carbon dioxide (worth about 6.052 billion baht) along with about 14.5 million tons of oxygen emission" [31].

Foreign companies like the Hong Kong-based, US-based, UK-based, and Japan-based companies have launched reforestation efforts with the communities and local governments nearby their onshore operation sites. However, their reforestation activity appeared to primarily serve a company-community relations building purpose rather than increasing carbon sinks for mitigating climate change. The Japan-based company informant mentioned that the company once encountered misleading criticism by local NGOs that tried to tempt local communities to protest against the company. However, thanks to its good CSR activities with communities, in particular forestation, the communities did not follow NGOs' thoughts but trusted that the company had followed good governance principles. Similarly, the UK-based company provided financial supports to local government projects to plant trees in degraded areas or alongside roads. This was done partly to build good relations with government authorities.

3.3 The Implications of O&G Industry's Responses to Climate Change Mitigation in Thailand

The O&G industry as a main primary energy supplier has an important role to play in the transition to a low-carbon future or, more specifically, sustainable energy development. Opposition from companies could slow down or inhibit the transition to a low-carbon future, despite government subsidies or supporting policies. Thus, understanding corporations' business strategies are crucial for policy makers. The present study examines both national and foreign O&G companies, providing an opportunity to compare between the behaviors of these two kinds of the corporations. First, the corporate responses to climate change mitigation of the national O&G company appear to be more proactive than those of foreign companies in both quality and quantity. This can be seen by the fact that the Thailand-based company is the only company that has invested in renewable and alternative energy. Such difference in business strategy could be the result of the nature of the national company, which has at its core its responsibility towards the citizens of its home country, compared to foreign companies who are accountable only to shareholders. Thus, investigating climate change mitigation activities of foreign companies in their home countries should be carried out to see how proactive in mitigating climate change they are, and whether their behavior depends on the context of country in which they operate.

Another important implication is that voluntary climate change mitigation activities will be pursued if the companies expect to obtain benefits from them, in particular economic benefits. This can be seen from the implementation of energy efficiency, flared gas reduction, R&D of green products, and reforestation. In this regard, the positive feedback from consumers is important for the O&G companies to take mitigation efforts. In the case of Thailand, the consumers are keen to buy alternative energy sources such as gasohol E85 and biodiesel B100. This is due to the prices of these alternative fuels being cheaper than normal gasoline. However, it is noted that the lower prices are a result of government's interference in energy price management. On the other hand, GHG emission measuring and reporting is fundamental for the companies to know how many emissions they are responsible for and start a reduction program but does not promise economic or social benefits to the companies. Publishing emission data could even be considered a sensitive issue since companies may undergo severe scrutiny from the government or NGOs if they release high levels of them. In the case that a voluntary effort maybe not possible, a firm government policy might be required, together with the implementation of laws or regulations that every company would have to comply with.

The petroleum concession law, which allows the concessionaire to only explore and produce crude oil, condensate, and natural gas, appear to limit the choices of O&G companies in commercializing the gas that they would otherwise flare. According to the Petroleum Act, B.E. 2514 (1971) and the latest amended version in B.E.2550 (2007), the concessionaires are to conduct "petroleum operations" which was defined as "exploration, production, storage, transport, sale, or disposal of petroleum" [32], in which "to produce" means "to undertake any operation in order to obtain Petroleum from its reservoirs; and also includes to employ any process other than refining and petrochemical operation in order to render petroleum in saleable or disposable forms" [32]. The concessionaires are thus limited at the upstream operation and prohibited to produce petroleum products from those extracted crude oil or natural gas. As a result, the Thai government should consider revising the Petroleum Act by providing rooms for the concessionaires to be able to use the extracted resource for producing products such as LNG. More importantly, the Act does not mention any climate change mitigation, as it was implemented decades ago. Thus, the authors recommend that climate change mitigation is added onto the Act, in order to make companies responsible for the mitigation of GHG emissions.

4 Summary

Thailand is blessed with abundant natural resources and has great potential for renewable energy development. Achieving its GHG emission reduction pledge submitted to the UNFCCC while maintaining economic prosperity is a challenge for the country. The present study examines one promising solution to this challenge which is to head towards an energy transition to low-carbon and renewable energy sources. However, the energy transformation requires not only experience and knowledge but also a tremendous, long-run investment [33]. Given their accumulated wealth and know-how, it seems appropriate that the O&G companies take on the main role in developing alternative or renewable energy. The present study examines the corporate responses to climate change of the O&G industry by reviewing secondary sources as well as employing online questionnaire survey and semi-structured interviews to obtain primary data. Findings suggested that companies generally appear to comply with the climate change mitigation policy and the alternative and renewable energy development plan of the Thai government. They have undertaken a number of mitigation activities, ranging from basic ones like measuring and reporting GHG emission data to more proactive projects such as flared gas reduction and green product development. However, only one company, from Thailand, has made active efforts to invest in renewable energy (solar PV) and set a target for itself to become an “energy company”. All findings suggested that there is clear difference between the climate change mitigation activities of the Thailand-based and foreign companies. While the former actively implements diverse mitigation efforts (including renewable energy investment), the latter seem concerned mainly with the petroleum exploration in the sites where they have been granted concession rights. Thus, it appears that most of incentives or rationales regarding the O&G industry’s mitigation efforts are very likely to be based around economic and social positive benefits. However, the Thailand-based company has the mission of being a national company, thus upholding energy security of the country as its prime purpose. In conclusion, the transition to a low-carbon future could be helped if the Thai national company takes a leading role that persuades other O&G companies to pursue mitigation action more proactively in the future.

Examining the O&G industry in Thailand can help understand the way that companies perceive incentives and their rationales, thus offering a more complex analysis on how O&G industry may help the transition to a low-carbon future. Yet the case of Thailand only fills a small gap in existing literature. The present study would like to encourage more study of O&G industry in other countries and regions. The authors believe that steering the enormous resources of this powerful industry to be part of the solution to global climate change is the only way to fundamentally tackle energy problems worldwide.

References

1. IEA (2013) Southeast Asia energy outlook: world energy outlook special report
2. Energy Statistic of Thailand (2013) Retrieved 15 May 2015, from http://www.eppo.go.th/info/cd-2013/pdf/info1.pdf
3. Thailand Second National Communication. Retrieved 15 May 2015, from https://unfccc.int/files/national_reports/non-annex_i_natcom/submitted_natcom/application/pdf/snc_thailand.pdf
4. CAIT 2.0 – WRI’s climate data explorer. Retrieved 15 May 2015, from http://www.wri.org/resources/websites/cait
5. WEF (2013) Energy vision 2013, energy transitions: past and future
6. WEF (2013) Energy vision 2013, energy transitions: past and future, p 2
7. Asia Pacific Energy Research Centre (APERC) (2013) APEC Energy overview. Retrieved 20 April 2015 from http://aperc.ieej.or.jp/file/2014/3/18/APEC_Energy_Overview_2013.pdf, p 223
8. DEDE (Department of Alternative Energy Development and Efficiency) (2011) The renewable and alternative energy development plan for 25 percent in 10 years (AEDP 2012–2021). Retrieved on 16 May 2015, from www.dede.go.th/dede/images/stories/dede_aedp_2012_2021.pdf
9. Department of Alternative Energy development and Efficiency (DEDE). Annual report 2012. Retrieved 16 May 2015, from http://weben.dede.go.th/webmax/content/annual-report-2012-0

10. http://www4.dede.go.th/dede/index.php?option=com_content&view=article&id=1841%3A2010-09-22-07-02-07&catid=128&lang=th. Retrieved 16 May 2015
11. NAMA Pledge in text received from an academic informant
12. Jacobsson S, Bergek A (2004) Transforming the energy sector: the evolution of technological systems in renewable energy technology. Ind Corp Change 13(5):815–849
13. Pinkse J, van den Buuse D (2012) The development and commercialization of solar PV technology in the oil industry. Energy Policy 40:11–20
14. Kolk A, Levy D (2003) Multinationals and global climate change: issues for the automotive and oil industries. Glob Strateg Manag 9:171–193
15. Kolk A (2008) Development in corporate responses to climate change in the past decade. In: Hansjurgens B, Antes R (eds) Climate change, sustainable development and risk: an economic and business view. Physica Publishers, Heidelberg
16. IPCC (2014) Mitigation of climate change fifth assessment report: working group 3. Retrieved 15 May 2015, from http://www.ipcc.ch/report/ar5/wg3/
17. IPCC (2014) Technical summary, fifth assessment report: working group 3, Retrieved 15 May 2015, from http://www.ipcc.ch/report/ar5/wg3/
18. Department of Mineral Fuels (DMF) Ministry of Energy of Thailand, 2013. Annual report. Bangkok
19. http://www.pttge.com/EN/SitePages/PTTGE.aspx, Retrieved 16 May 2015
20. Bangchak Petroleum (BCP) Company Annual Report (2014) Retrieved 20 May 2015, from http://www.bangchak.co.th/th/downloads-annual-report.aspx
21. Department of Energy Business, Ministry of Energy http://www.doeb.go.th/info/data/datadistribution/biodieselB100_Manufac.pdf. Retrieved on 20 May 2015
22. http://www.nationmultimedia.com/business/PTT-plans-to-commercialise-algae-based-fuel-by-201-30209903.html. Retrieved on 20 May 2015
23. BCP Annual Report (2014) Retrieved 16 May 2015, from http://www.bangchak.co.th/th/downloads-annual-report.aspx
24. http://www.mitrphol.com/index.php/th/news_and_activity/newsdetail/46. Retrieved on 20 May 2015
25. http://webkc.dede.go.th/testmax/node/159. Retrieved on 20 May 2015
26. http://www.eppo.go.th/retail_prices.php. Retrieved on 26 June 2015
27. http://www.eppo.go.th/info/Situation/monthly/2014_12.pdf. Retrieved on 26 Jun 2015
28. http://www.pttplc.com/EN/About/pages/Vision-Mission-Values.aspx. Retrieved on 20 May 2015
29. http://www.siambioenergy.com/bio65/index.php/78-news/145-tse-40. Retrieved on 20 May 2015
30. http://hq-srvtst-s08.pttplc.com/internet_test/en/social-activities-environment-reforestation.aspx. Retrieved on 20 May 2015
31. http://www.pttplc.com/en/media-center/news/social-environmental/pages/news-2014-11-28.aspx. Retrieved on 20 May 2015
32. http://thailaws.com/law/t_laws/tlaw0430.pdf. Retrieved on 26 June 2015
33. Lee ZH, Sethupathi S, Lee KT, Bhatia S, Mohamed AR (2013) An overview on global warming in Southeast Asia:CO2 emission status, efforts done and barriers. Renew Sust Energ Rev 28:71–81

Material Recovery and Environmental Impact by Informal E-Waste Recycling Site in the Philippines

Atsushi Terazono, Masahiro Oguchi, Aya Yoshida, Ruji P. Medina, and Florencio C. Ballesteros Jr.

Abstract Waste electrical and electronic equipment (WEEE or e-waste) contains both valuable and hazardous substances and there is great demand for metal scrap. To better understand both material recovery and the environmental impacts by informal recycling of e-waste, and to find potential ways to improve the process, we carried out a field survey at an informal recycling site in the Philippines. We identified the Au recovery process used at the site and evaluated the layout of the recycling site. We collected 31 soil samples at the recycling site and analyzed the metal contents of each sample to clarify the metal distribution within the site. We determined that valuable substances (such as Au), as well as hazardous substances (such as Pb), were scattered throughout the soil at the informal recycling site. The results of our cluster analysis indicated that Au, Ag, Pb, and Sb were categorized in the same group. Improvements are needed in the metal recovery process and in hazardous substance emission control in the informal recycling.

Keywords E-waste • Informal recycling • Hazardous substances • Valuable substances

1 Introduction

Waste electrical and electronic equipment (WEEE or e-waste) is of great concern among researchers and policymakers because of its potential as a source of serious environmental pollution. Both domestically generated and imported wastes are of concern. E-waste contains both valuable and hazardous substances, however, and there is great demand for metal scrap. Oguchi [1, 2] showed the contents of toxic metals, precious metals, and other common metals in the printed circuit boards of

A. Terazono (✉) • M. Oguchi • A. Yoshida
Center for Material Cycles and Waste Management, National Institute for Environmental Studies, Tsukuba, Japan
e-mail: terazono@nies.go.jp

R.P. Medina • F.C. Ballesteros Jr.
University of the Philippines, Diliman, Quezon, Philippines

M. Matsumoto et al. (eds.), *Sustainability Through Innovation in Product Life Cycle Design*, EcoProduction, DOI 10.1007/978-981-10-0471-1_14

21 and 24 types of WEEE, suggesting the importance of controlling these substances in recycling and waste management processes. Valuable substances such as precious and rare metals should be recovered effectively while at the same time the emissions of hazardous substances into the environment should be controlled.

Although the EU countries, as well as Japan, South Korea, and China, have introduced e-waste management systems [1, 3–6], China and other developing Asian countries still face environmental problems in informal recycling sectors. There is no clear definition of the concept of "informal," and the term "informal" has often been characterized as being beyond the reach of different levels and mechanisms of official governance; lacking in regulation, structure, and institutionalization; and as non-registered and illegal [7].

Many studies of the health and environmental impacts of informal recycling of e-waste have been reported and reviewed. Sepúlveda et al. [8] examined improper recycling techniques such as dumping, dismantling, and inappropriate shredding, burning, and acid leaching; they also reviewed the related environmental impacts. Fujimori et al. [9] investigated the metal content of surface matrices such as soil and dust at both formal and informal recycling sites in the Philippines and determined that the health risks posed by toxic metals in dust were high for workers—even those at formal sites.

Although there have been many reports about the environmental impacts of hazardous substances in e-waste, very few studies have examined the fate of valuable substances at informal recycling sites. Keller [10] carried out excellent work for assessment of Au recovery processes in India, including cyanide leaching, mercury amalgamation, and Au stripping, but the smelting is common in the Philippines and other Southeast Asian countries. To better understand both material recovery and the environmental impacts by informal recycling of e-waste, and to find potential ways to improve the process, we carried out a field survey at an informal recycling site in the Philippines. We specifically focused on a precious metal recovery process and investigated the fate of metals as well as metal distribution on the surface soil at the site.

2 Materials and Methods

2.1 *Surveyed Recycling Site*

Field visit was carried out in January 2012. The recycling facility is situated in the City of Meycauayan, Bulacan, Philippines, and has been in operation for more than 30 years. It is located in a mixed residential and commercial neighborhood, and a perimeter fence effectively isolates it from the neighboring business establishments, residences, and a vacant lot near the back of the facility. The facility primarily buys e-waste from industrial facilities for dismantling. The parts are dismantled in a segregation area and then sold to junkshop owners, who usually pick up the materials at the facility. The facility is also engaged in the transportation of hazardous wastes. Precious metals—specifically Au and Ag—are recovered in a smelting area. However, the precious metals recovery is usually done by outsiders who bring in raw materials sourced from elsewhere. The facility's owner occasionally decides to extract precious metals using terminals and integrated circuit (IC) chips of electronics gathered from within the facility, but this is done only when the supply of raw materials becomes abundant, because processing small amounts is normally not profitable. E-waste stock and various reagents and other input materials for precious metal recovery are stored in a stockyard area. The layout of the informal recycling site is shown in Fig. 1.

Eighty percent of the workers in the facility are men. They do the activities that require greater strength, such as lifting, hammering, and dismantling. Women generally do less strenuous activities such as cutting and stripping wires and dismantling and separating small metal parts. The facility employs an average of 20 workers, of whom 5 males are regular employees. The other workers are seasonal and work when there is a large amount of e-waste to be processed. Part-time workers are also employed to work in groups to process specific stocks of e-wastes and work anywhere from 1 to 6 months a year. Likewise, the smelters are called in to work only when needed.

The regular workers are paid US$47 a week, and the seasonal workers are paid US$35 a week for men and US$25 a week for women. The smelters receive special rates because they are directly exposed to more harmful working conditions. In addition to wages, all workers are provided with free meals and snacks within the 8-h daily work period. The average age of the workers is 33 years, with the youngest at the time of our visit being 15 and the oldest 56.

Most of the workers reside within the city, and a few come from neighboring areas. The majority of the workers are elementary school graduates. Although some have attended high school, none of them has attended college. Ninety percent of the workers are married and support families with four or five members.

2.2 *Gold Recovery Process*

2.2.1 Process Overview

We reviewed two Au recovery processes in our field survey: a chemical (acid leaching) process and an IC burning process (called *sunog* IC in Filipino, where *sunog* means burning). Both processes utilize smelting after preprocessing, so the differences are primarily in the preprocessing and the types of materials to be applied.

The chemical Au recovery process is used for terminals, printed circuit boards, and jewelry, whereas IC chips are processed by using the IC burning process.

At the preprocessing stage, the chemical process uses cyanide to extract Au, whereas open burning is used to remove organic substances as part of the IC

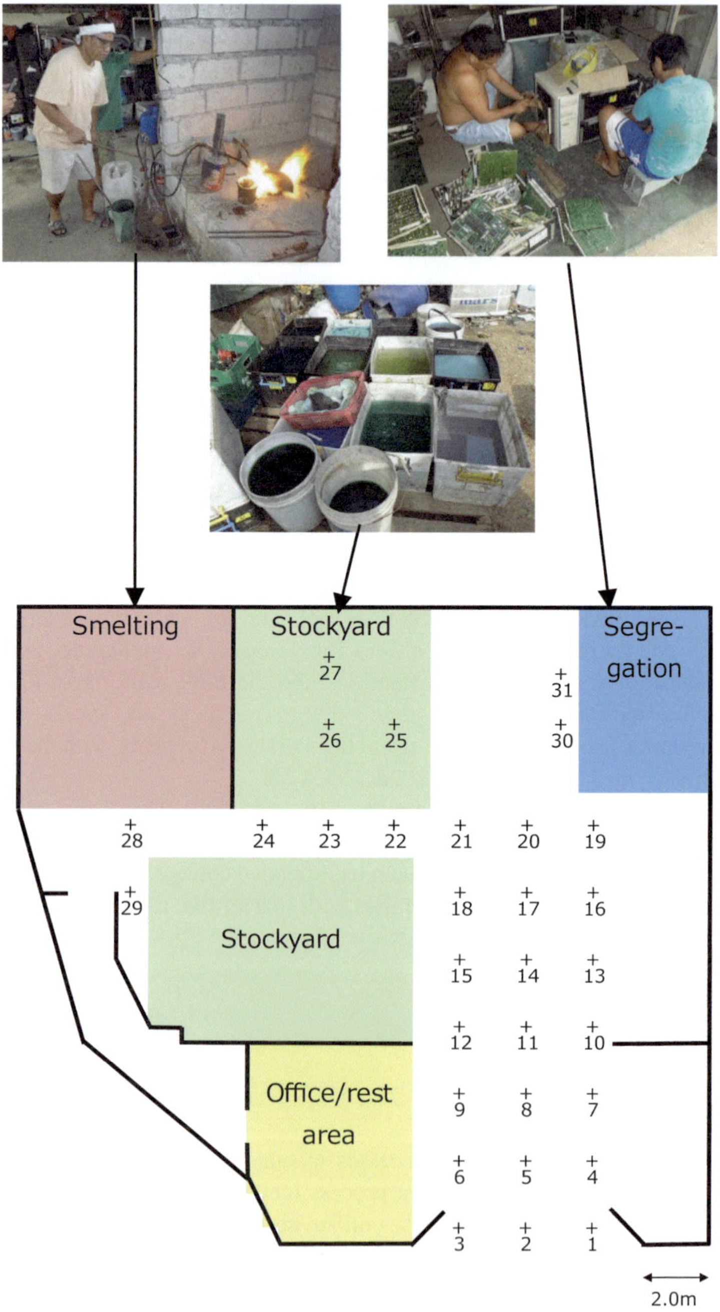

Fig. 1 Layout of the surveyed informal recycling site. Numbers with (+) refer to the location of soil samples

burning process. Open burning of ICs is strictly prohibited, however, because of environmental concerns. Therefore, the chemical process is almost exclusively used for Au recovery, and it was used in our test sample.

2.2.2 Mass Balance

The amounts of reagents used and the output (Au) obtained are largely affected by the quality and quantity of the material inputs. In this study, we used electrical conducting terminals containing Cu with Au coatings (also coated with Ni and Sn) as test samples to evaluate the mass balance for input and output of materials.

In order to confirm the mass balance of the process, we weighed each of the material used for and produced from the test run at the process: raw material (terminals before the cyanide treatment); borax; Pb nuggets; terminals after the cyanide treatment; two types of slag, one of which was borax-derived (slag A) and the other was lead-derived (slag B); and recovered Au nugget. We then sampled each of the input and output materials for the metal concentration analysis. Terminal and Au nugget samples were digested with aqua regia. Slag samples were pulverized and digested with aqua regia followed by alkali fusion with sodium carbonate and boric acid. Metal contents were determined by using inductively coupled plasma—optical emission spectroscopy (ICP–OES, SII NanoTechnology, SPS3500).

2.3 Metal Content of Surface Soil at the Informal Recycling Site

To better understand the distribution of metals at the study site, 31 grid-based surface soil samples were obtained for analysis for 63 metal elements (Ag, Al, As, Au, B, Ba, Be, Bi, Ca, Cd, Ce, Co, Cr, Cs, Cu, Dy, Er, Eu, Fe, Ga, Gd, Ge, Hf, Ho, In, Ir, K, La, Li, Lu, Mg, Mn, Mo, Na, Nb, Nd, Ni, Pb, Pd, Pr, Pt, Rb, Rh, Ru, Sb, Sc, Se, Si, Sm, Sn, Sr, Ta, Tb, Te, Ti, Tl, Tm, V, W, Y, Yb, Zn, Zr).

The soil samples were air dried at room temperature for at least 1 week. The dried soil samples were sieved with a 2-mm sieve. Approximately 1 g of the samples was digested with 20 mL of aqua regia on a hot plate. For silver analysis, the solvent was replaced with 6 M chloric acid after the aqua regia digestion to avoid silver chloride precipitation. Concentrations of metals were determined by using ICP–OES (Seiko Instruments, VISTA-PRO) and ICP – mass spectrometry (ICP–MS, Agilent Technologies, 7500cx).

Cluster analysis was then conducted to categorize the metals in the soil samples by using the Ward method.

3 Results

3.1 Gold Recovery Process

3.1.1 Process Overview

The chemical gold recovery process starts with dousing of the raw materials with a sodium cyanide solution prepared by dissolving 50 g of sodium cyanide powder and 75 g of a separating agent (85 % sodium m-nitrobenzene sulfonate and 15 % sodium carbonate) in 1 L of water (Fig. 2). Sodium cyanide is used to dissolve the Au, and the separating agent is used as a stripper to remove Ni from the base metal. The Au is plated over the Ni coating, so removing nickel therefore removes the Au from the base metal.

The raw material—cyanide mixture—is then heated and swirled continuously to facilitate the removal of Au (Reaction 1), as indicated by the loss of the gold color in the coating of the raw materials. The solid Cu materials, with their tin coatings exposed, are then separated from the solution and washed with water to ensure that all traces of cyanide are removed. The washings, however, are mixed with the Au–cyanide solution.

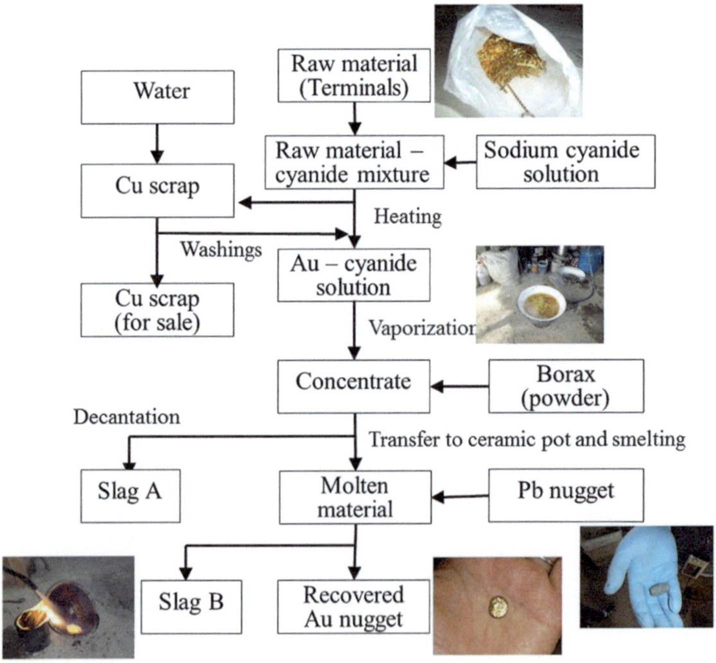

Fig. 2 Schematic of the chemical gold recovery process at the surveyed informal recycling site

$$4Au + 8NaCN + O_2 + 2H_2O \rightarrow 4NaAu(CN)_2 + 4NaOH \qquad (1)$$

The water in the Au–cyanide solution is then boiled off. When the final solution thickens, borax powder is added and mixed in. The resulting paste is then transferred to a ceramic pot or crucible and subjected to firing. Pb nugget (in the form of a Pb bar) is added to the molten material to obtain metallic Au (Reaction 2). Non-gold components are removed by decantation, and the recovered Au is transferred to a different container, where it cools and solidifies.

$$2NaAu(CN)_2 + Pb \rightarrow Na_2Pb(CN)_4 + 2Au \qquad (2)$$

3.1.2 Mass Balance

We identified 0.25 L of cyanide solution, 1.75 L of water, 274 g of borax powder, and 228 g of Pb nugget as input materials for the 989 g of terminals used as our test sample.

The results of the analysis are shown in Table 1. Au and Cu contents of the input terminals were 0.92 % and 89.1 %, respectively. The Au content of the recovered Au nugget was 85.2 %. Although the Pb content of the Pb nugget used as an input was not analyzed, the Pb content was determined to be 51.4 % in slag B, and the Pb concentration was 2.9 % in the recovered Au.

On the basis of the results for the output materials, 5.5 g of Au was recovered as Au nugget and 1.5 g of Au remained on the processed terminals, suggesting approximately 80 % of Au was recovered from the raw materials by using this process. Pb was mostly transferred to slag; however, a small portion (0.1 %) was contained in the recovered Au nugget.

Table 1 Metal contents of the input and output materials in our test sample (mg/kg-dry)

	Input	Output			
	Terminal (original input material)	Terminal (after sodium cyanide solution treatment)	Recovered Au nugget (output material)	Slag A (from borax addition)	Slag B (from Pb addition)
Au	9200	1500	852,000	20	40
Cu	891,000	867,000	93,800	2400	29,300
Sn	71,100	67,700	<10	4300	24,700
Ni	13,800	13,400	20	1200	74,200
Pb	<10	20	29,200	124,000	514,000
Si	90	90	80	136,000	36,300
Mass (g)	989	969	6.4	194	217

Table 2 Metal contents of the input and output materials in our test sample (mg/kg-dry)

Soil sample number	Metal contents									
	Ag	Al	As	Au	Ba	Ca	Cd	Ce	Co	Cr	Cu
1	2.0	37,049	2.3	0.15	157	31,882	5.0	19	27	27	120
2	1.8	37,573	1.7	0.22	114	33,303	0.7	12	30	36	91
3	0.5	44,549	1.6	0.05	64	40,523	0.5	12	36	13	110
4	1.9	35,031	2.0	0.27	121	32,139	1.2	14	33	27	121
5	2.9	38,114	1.8	0.40	85	36,604	2.2	12	30	22	128
6	3.5	40,082	1.7	0.17	112	33,908	1.1	13	33	36	109
7	17.6	37,730	2.2	4.59	153	39,620	2.7	17	31	29	1230
8	3.4	41,703	1.8	0.20	89	35,971	1.2	14	34	16	106
9	4.0	39,752	1.8	0.52	105	33,995	1.3	16	34	17	126
10	5.0	43,121	1.6	0.96	51	39,227	3.0	10	36	16	310
11	4.3	40,153	1.9	1.30	94	36,548	1.2	14	33	17	556
12	3.0	44,959	1.7	5.00	120	38,028	2.0	16	35	13	130
13	6.4	41,597	2.1	0.21	130	33,005	9.0	15	34	16	306
14	2.5	39,920	2.2	0.47	153	30,917	1.3	21	29	16	138
15	5.9	41,777	2.3	0.27	155	30,310	1.3	22	29	16	102
16	2.0	45,270	2.1	0.17	127	34,344	2.0	17	34	14	160
17	10.9	37,898	3.5	3.27	197	32,517	2.7	25	34	27	234
18	5.0	43,251	3.1	0.75	350	29,955	3.0	40	37	18	270
19	16.0	37,147	2.9	1.26	153	39,724	18.0	19	35	26	1100
20	17.8	38,423	2.7	1.95	124	37,355	9.6	17	30	33	336
21	20.0	26,813	30.6	5.00	329	82,988	40.0	46	43	160	1000
22	75.7	21,783	6.4	7.91	504	26,581	214.8	17	244	262	2444
23	160.0	48,463	3.8	2.00	174	14,620	35.0	72	150	49	1900
24	26.0	57,560	3.3	9.00	885	12,581	9.0	75	47	39	7300
25	3.1	85,114	8.6	1.95	523	7693	22.8	43	52	47	2513

26	2.0	65,115	3.1	0.63	260	9340	36.0	63	34	41	310
27	3.1	65,244	3.1	0.80	297	9595	2.9	64	44	42	516
28	240.0	18,166	14.8	21.00	159	32,147	96.0	13	78	111	3600
29	150.0	49,329	4.6	5.00	216	11,482	7.0	65	48	40	2100
30	14.0	64,889	3.9	1.00	1056	12,004	18.0	45	56	78	630
31	3.0	73,432	3.8	0.29	514	11,201	1.0	40	52	67	290
Geometric mean	7.1	42,484.6	3.0	0.9	181.6	26,234.6	4.6	23.2	41.1	31.5	390.9
	Dy	Fe	Mn	Nd	Ni	Pb	Sb	Si	Sn	Zn	
1	3.4	69,441	1470	9.5	37	120	3.5	705	20	255	
2	3.5	68,421	1299	7.5	23	55	0.25	423	12	156	
3	3.9	83,776	1522	7.4	18	28	1.3	553	4	153	
4	3.2	72,361	1313	7.5	32	122	1.1	285	38	243	
5	3.4	71,258	1387	7.2	23	111	1.4	284	23	207	
6	3.8	74,321	1437	7.8	21	105	0.8	302	23	178	
7	4.1	67,706	1701	9.7	96	164	1.4	333	86	281	
8	3.8	78,519	1480	7.7	24	131	1.3	455	157	189	
9	3.7	73,868	1492	8.6	43	183	1.0	580	33	162	
10	4.0	80,178	1408	7.4	30	140	2.2	927	94	244	
11	3.7	76,874	1463	8.0	29	386	2.8	285	34	171	
12	3.8	77,383	1554	8.4	41	130	2.3	524	48	147	
13	3.6	75,106	1544	8.1	46	55	0.25	532	155	150	
14	3.9	67,065	1410	10.9	18	166	0.6	334	242	143	
15	3.5	66,737	1439	9.9	25	57	0.25	235	75	139	
16	3.7	76,728	1537	8.9	3918	41	1.1	585	47	173	
17	3.5	73,973	1783	10.1	76	533	5.1	325	648	231	
18	3.9	70,826	2284	13.5	26	230	4.0	604	77	178	
19	3.3	74,858	1448	9.3	341	450	12.5	641	220	779	

(continued)

Table 2 (continued)

Soil sample number	Metal contents									
20	3.4	69,778	1335	8.6	59	412	3.7	262	71	307
21	3.2	67,331	1450	22.9	188	1400	83.2	554	122	1232
22	2.0	142,493	3465	15.2	500	4840	51.6	284	1150	3031
23	5.0	58,249	2268	27.8	125	2000	41.2	817	328	422
24	25.4	60,348	3082	29.1	336	1200	20.3	994	191	218
25	2.4	90,631	1040	12.2	88	5066	40.3	426	888	300
26	5.5	61,665	2213	29.7	33	91	2.1	711	47	169
27	5.5	74,052	2829	26.5	43	225	2.6	1093	239	326
28	2.1	56,027	724	12.4	534	9000	179.3	548	1098	1062
29	4.8	54,120	2497	26.9	158	4300	65.2	663	362	313
30	2.0	125,251	2315	10.4	229	800	16.2	716	1178	636
31	2.2	116,547	1924	11.4	128	300	0.5	386	361	654
Geometric mean	3.7	74825.2	1657.8	11.4	71.7	299.0	3.5	483.8	110.6	284.1

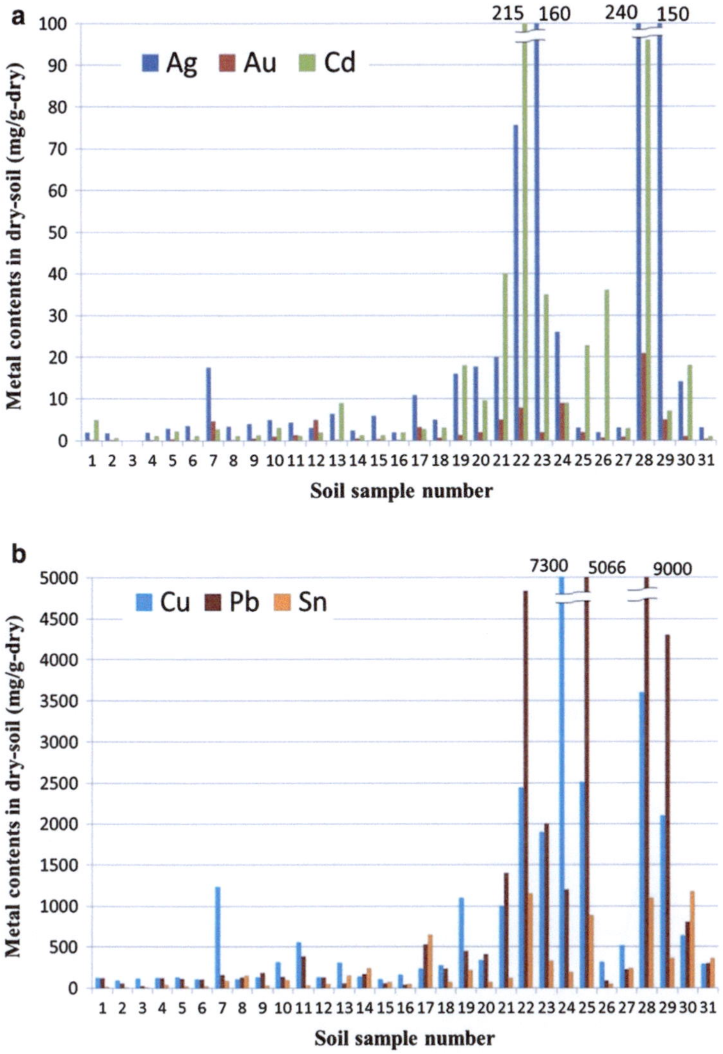

Fig. 3 Selected metal contents in surface soil samples at the surveyed recycling site (**a**) Ag, Au, Cd (**b**) Cu, Pb, Sn

3.2 Metal Contents of Soil Samples

The metal contents of soil sampled at the surveyed recycling site are shown in Table 2. Figure 3 shows the results for selected metals (Ag, Au, Cd, Cu, Pb, and Sn). The metal contents of Au, Pb, and Cu are shown by sample location in Fig. 4.

The highest concentration of Au—21 mg/kg-dry—was found in the soil closest to the smelting area; surprisingly, this concentration was higher than the level in gold ore (about 5 mg/kg) [11].

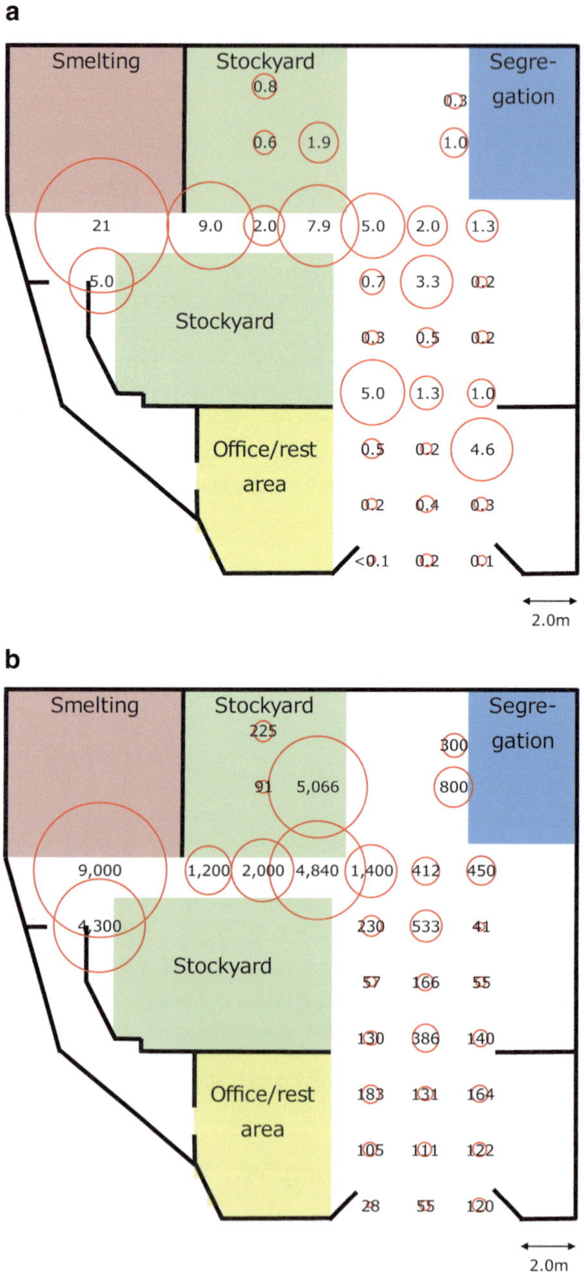

Fig. 4 Metal contents in soil samples at the surveyed informal recycling site (mg/kg-dry). *Red circles* show relative concentrations (**a**) Au (**b**) Pb (**c**) Cu

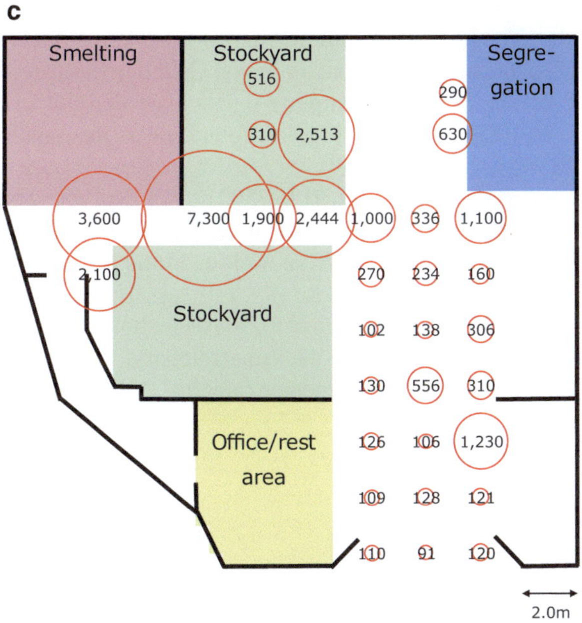

Fig. 4 (continued)

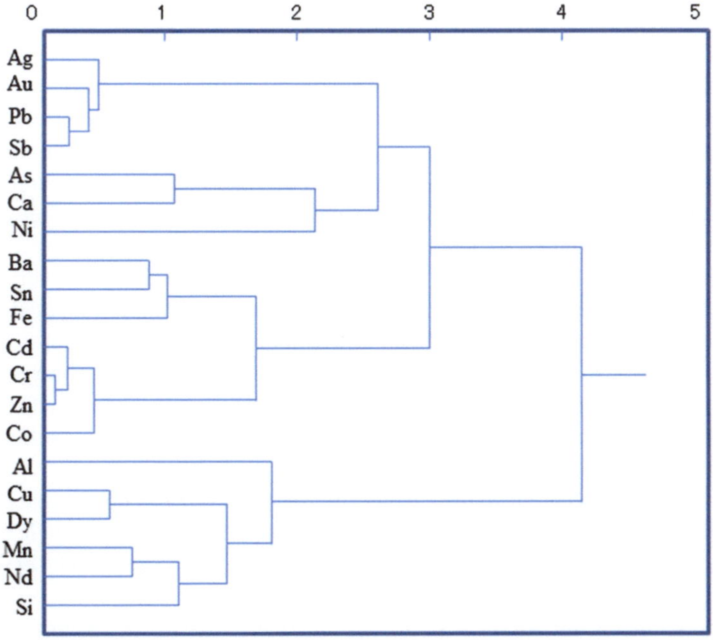

Fig. 5 Dendrogram of metal characterization based on the concentrations of metals in soil samples

The Pb concentration ranged from 28 to 9000 mg/kg-dry, with a geometric mean of 299 mg/kg-dry. The Pb content criterion in soil considering direct ingestion risk, as specified by the Japanese Soil Contamination Countermeasures Act, is 150 mg/kg-dry, and 18 of the 31 (58 %) sampling points exceeded this criterion. Higher concentrations (>1000 mg/kg-dry) were found near the smelting area and stockyard.

The Cu concentration ranged from 91 to 7300 mg/kg-dry, with a geometric mean of 391 mg/kg-dry. High concentrations of more than 1000 mg/kg-dry were found near the stockyard and smelting areas, as well as in the segregation area and the nearby entrance.

A dendrogram resulting from the cluster analysis is shown in Fig. 5. It shows that Ag, Au, Pb, and Sb are grouped into the same category, whereas common metals such as Cu and Al are grouped into another category.

4 Discussion

4.1 Gold Recovery Process

We investigated a chemical Au recovery process at an informal recycling facility in the Philippines. The process can be divided into a preprocessing stage and a smelting stage. In the preprocessing stage, sodium cyanide solution was used to extract Au. We have observed different reagents being used in other field visits—for example, nitric acid and hydrogen peroxide, nitric acid, and aqua regia at sites in Vietnam and China. It seemed that this was related with informal recyclers' attempt for a suitable combination of materials and reagents in each local community.

The results of the mass balance analysis showed that Pb from input additives was shifted into output slag and even the recovered Au nugget. Dust containing Pb was also most likely scattered to areas surrounding the smelting area.

The environmental impacts of discharges and residues from the recovery process need to be addressed. Gaseous discharges other than lead-containing dust are also produced. During the heating of the raw material—cyanide mixture—and the boiling of the resulting Au–cyanide solution, water vapor is discharged openly and directly to the atmosphere. The vapors also contain cyanide and NO2 and are invariably toxic. Notwithstanding this hazard, the facility is operated without safety equipment—not even exhaust fans.

Liquid discharges are also produced. The chemical process produces a cyanide-laced dilute muriatic acid solution used by the processors to wash their hands to remove cyanide solution. These liquids are discharged into canals and septic tanks or poured directly on the ground.

In the chemical process, Cu scraps are obtained after the Au plating has been removed. These materials are placed in sacks and sold to junkshop buyers who, in turn, sell them to traders for export to other countries. During the firing process, slags are produced from the solidification of the molten materials. These slags are normally left in the hearth or buried underground once a sufficiently large amount has been produced. The processes also generate used ceramic pots. Because these pots still contain minute quantities of precious metals, they are stored in the facility until there are enough of them to make Au extraction feasible. It may take several months to accumulate enough pots, and the recovery rate is normally very low.

4.2 Distribution of Metals Within the Recycling Site

The results of the soil analyses indicated that valuable substances (such as Au) as well as hazardous substances (such as Pb) were scattered within the informal recycling site. Taking into consideration that Pb is used as an input in the smelting process, the Pb and Au may have been emitted from the smelting process.

Fujimori et al. [12] studied the distribution of heavy metals (Pb, Cu, and Zn) in surface soil samples at an e-waste recycling site in the Philippines by using field-portable X-ray fluorescence. The results indicated that heavy metals had a relatively "small" pollution scale and largely remained inside the original workshop, as compared with toxic organohalogen compounds. Our results are supported by these findings.

Our cluster analysis showed that Ag, Au, Pb, and Sb were categorized into the same close group. As discussed in Sect. 3.1, Pb is an important input for Au recovery. Ag might be present in the printed circuit boards treated for Au recovery, or Ag recovery treatments might also be implemented in the same smelting area. Some kinds of the printed circuit boards may contain Sb as flame retardant substances, and Sb may have also been emitted from the Au recovery process.

In a cluster analysis by Fujimori et al. [9], As, Co, Fe, Mn, and Ni in soil samples from other informal recycling facilities were categorized as crust-derived metals. At those informal sites, physical activities such as dismantling and crushing were mainly conducted; the chemical Au recovery process was not. The crust-derived metals were categorized into groups other than the Au group in our study. For clarification, we need to better understand the contents of substances and the activities at stockyard and segregation areas.

4.3 Health Risks

The recycling activities expose the workers to several health risks. Not only are workers exposed to hazardous substances in the soil; they also inhale metallic dust from smelting activities and from activities in the stockyard and segregation areas. The handling of batteries and other equipment also exposes workers to corrosive liquids and vapors. Extraction of metals, which generates toxic vapors, also exposes them to health risks, especially from the smelters themselves. Although most workers wear long-sleeved shirts and long pants, they do not use any personal protective equipment (PPE) such as masks and gloves. It is particularly worrisome to see the workers extract the metal without using PPEs, despite the fact that they are handling highly toxic and hazardous cyanide and Pb.

According to worker interviews, the most common health problems, experienced by half of the workers, are the occasional coughs and colds. Cigarette smokers, who comprise 40 % of the worker population, and alcohol drinkers, who make up 70 % of the worker population, are more prone to these conditions. A few workers have skin allergies, and one worker doing metal recovery has visual impairment.

5 Conclusion

We carried out a field survey at an informal e-waste recycling site in the Philippines. We identified the Au recovery process used at the site and evaluated the layout of the recycling site. We evaluated the whole Au recovery process, including the metal content of all the input and output materials. Pb from inputs was recovered from output slag and even from recovered Au nugget.

We collected 31 soil samples at the recycling site and analyzed the metal contents of each sample to clarify the metal distribution within the site. Pb contents ranged from 28 to 9000 mg/kg, whereas Au ranged from <0.1 to 21 mg/kg. The highest contents for both were measured at the points closest to the smelting area; high contents were generally found around the smelting area and stockyard. We determined that valuable substances (such as Au), as well as hazardous substances (such as Pb), were scattered throughout the soil at the informal recycling site. The results of our cluster analysis indicated that Au, Ag, Pb, and Sb were categorized in the same group. Given that Pb is used as an input in the smelting process, it is likely the hazardous substance (Pb) and the valuable substance (Au) were emitted from the same smelting process.

Improvements are needed in the metal recovery process and in hazardous substance emission control. Establishing closed system of the smelting process and treatment of flue gases are recommended. Protective measures for workers are also needed. Many informal workers do not understand well the health impact by recycling activities or think lightly of that. Thus, workers' awareness raising of the need for protective measures is significant. If these types of measures are implemented, formalization of the informal recycling sector is expected to proceed.

Acknowledgments This study was funded by the Environment Research and Technology Development Fund from the Ministry of the Environment, Japan (K2347 and 3 K143010). We thank Melissa Cardenas of Environweave and Hidefumi Yokoo and Hidetaka Takigami of NIES for their sincere cooperation in the analyses.

References

1. Oguchi M, Sakanakura H, Terazono A (2013) Toxic metals in WEEE: characterization and substance flow analysis in waste treatment processes. Sci Total Environ 463–464:1124–1132
2. Oguchi M, Murakami S, Sakanakura H, Kida A, Kameya T (2011) A preliminary categorization of end-of-life electrical and electronic equipment as secondary metal resources. Waste Manag 31:2150–2160
3. StEP Initiative (2009) On the revision on EU's WEEE directive – COM(2008)810 final. http://www.step-initiative.org/files/step/_documents/StEP_TF1_WP_WEEERecast_rvsd_22March2010.pdf. Accessed on 17 July 2015
4. Aizawa H, Yoshida H, Sakai S (2008) Current results and future perspectives for Japanese recycling of home electrical appliances. Resour Conserv Recycl 52:1399–1410
5. Jang Y-C (2010) Waste electrical and electronic equipment (WEEE) management in Korea: generation, collection, and recycling systems. J Mater Cycles Waste Manag 12:283–294
6. Chung SS, Zhang C (2011) An evaluation of legislative measures on electrical and electronic waste in the People's Republic of China. Waste Manag 31:2638–2646
7. Chi X, Streicher-Porte M, Wang MYL, Reuter MA (2011) Informal electronic waste recycling: a sector review with special focus on China. Waste Manag 31:731–742
8. Sepúlveda A, Schuelp M, Renaud FG, Streicher M, Kuehr R, Hagelueken C, Gerecke AC (2010) A review of the environmental fate and effects of hazardous substances released from electrical and electronic equipments during recycling: examples from China and India. Environ Impact Assess Rev 30:28–41
9. Fujimori T, Takigami H, Agusa T, Eguchi A, Bekki K, Yoshida A, Terazono A, Ballesteros FC Jr (2012) Impact of metals in surface matrices from formal and informal electronic-waste recycling around Metro-Manila, the Philippines, and intra-Asian comparison. J Haz Mat 221–222:139–146
10. Keller M (2006) Diploma thesis at ETH. Assessment of gold recovery processes in Bangalore, India and evaluation of an alternative recycling path for printed wiring boards. http://www.empa.ch/plugin/template/empa/*/59244. Accessed on 17 July 2015
11. JOGMEC (2012) 4. Au. Mineral resources material flow 2011. 41 (in Japanese), http://mric.jogmec.go.jp/public/report/2012-05/4.Au_20120619.pdf. Accessed on 17 July 2015
12. Fujimori T, Takigami H (2014) Pollution distribution of heavy metals in surface soil at an informal electronic-waste recycling site. Environ Geochem Health 36:159–168

Part III
Business Design for Sustainability

Actors and System Maps: A Methodology for Developing Product/Service Systems

Avni Desai, Mattias Lindahl, and Maria Widgren

Abstract This paper presents a refined version of the Actors and System Map methodology. The refined method is more clearly defined and follows a structured process. This method can be used for describing actors and activities between a provider and a customer within a product/service system. The methodology helps in providing an overview of participating actors within a system, which is advantageous since there are often differences of opinion regarding the actors involved. An example of when the methodology has been used is presented.

Keywords Integrated product service offering • EcoDesign • Service design

1 Introduction

In traditional product sales, the focus is often on developing products that are cheap to manufacture and enable a profitable aftermarket consisting of, e.g., service, spare parts, and consumables.

However, increasing environmental requirements, in combination with increasing competition in the marketplace, sets higher demands on building relationships with the customer and fulfilling their needs in a better way [1]. As a result, some companies have started to offer services throughout the whole life cycle of the product or service as a source of competitive advantage. Incorporating additional services with what is provided has both financial benefits and lowers the environmental impacts associated with the product's life cycle [2]. Companies, e.g., telecom operators, have gone from being product providers to providing integrated services, where tangible products and intangible services are combined in an integrated system [3]. When a telecom operator purchases a service, it usually buys the equipment from its supplier and runs the equipment itself. This means that it needs to have more responsibility in service operations and that the equipment may remain in the ownership of the buyer [4].

A. Desai • M. Lindahl (✉) • M. Widgren
Environmental Technology and Management, Department of Management and Engineering, Linköping University, Linköping, Sweden
e-mail: mattias.lindahl@liu.se

M. Matsumoto et al. (eds.), *Sustainability Through Innovation in Product Life Cycle Design*, EcoProduction, DOI 10.1007/978-981-10-0471-1_15

This type of offering is often called a product/service system (PSS) [5, 6] or an integrated product service offering [7]. PSS offerings consist of combinations of physical products, services, and systems that have been integrated and optimized from a life cycle perspective in relation to the customer value (modified from Meier et al. [5]).

Services are intangible and therefore more complex when it comes to measuring the quality of the final outcome. This sets a higher demand on companies to deliver services which not only satisfy the customer's demand but also their experience. The complexity also makes it more difficult for the customer to make demands on the service that is going to be delivered [8]. The customer requirements and the increased competition make it advantageous for companies to identify the role of the inbound actors. In many cases, for the service to be able to evolve and improve, the actors from both the provider and the customer side must have access to important information, must know how information is transferred, and must have an idea of who will use the information they possess.

To conclude, information flow is key to realizing effective PSS [9]. In order to make the information useful, it is important to get a good understanding about transmitters and receivers so that information is utilized in a proper manner. Therefore, the main objective is to present a refined version of the Actors and System Map method developed by Lindahl et al. [10]. This method can be used for describing actors and activities between a provider and a customer within a PSS.

The structure of the paper is as follows. After this introduction, the Actors and System Map methodology is presented. This is followed by an example of when the method has been used: the "Orange and Ericsson case" by Desai and Widgren [11]. After the following discussion, some concluding remarks and future research are presented.

2 The Actors and System Map Methodology

This section describes the background behind the methodology as well as how the method works.

2.1 Background

Several mapping methods exist, e.g., "Activity Modeling Cycle" [12], "Actor Maps," "Map of Interaction" and "System Organization Map" [13], "Customer Value Chain Analysis" [14, 15], and "Service Ecology Map" [16]. However, after reviewing these, the conclusion drawn by Lindahl et al. [10] was that they lacked the features that were considered to be necessary and sufficient. Instead, they developed a methodology that covered the desired features: one that was easy for users to understand and use, one that was easy to communicate (e.g., high level of visualization), and, finally, from a clear PSS perspective, one that would cover actors, products, services, information, and activities. Their focus was on getting an overall perspective without too many details.

2.2 *Overall Methodology Description*

The overall aim of this methodology is to, on a relevant level of detail, visualize the network of actors involved in an existing or potential PSS or similar type of offering and show the flow of products, services, and information between them, as well as how the activities through these are managed (see also Lindahl et al. [10]). These visualizations are useful, for example, in identifying important flows of information, services, products, and activities. They can also be used for instance to identify the lack of information flows and activities or to show that existing ones not are optimal or necessary. This kind of knowledge is useful when developing a PSS or improving an already existing one, and this knowledge is often missing or is at a non-sufficient level of detail, especially concerning the overall understanding of the actors involved and the information flows and activities. This type of argumentation is, for example, in line with what Ulrich and Eppinger [17], Ullman [18], and Smith and Reinertsen [19] have argued as reasons for documenting and defining the product development process.

This type of knowledge can also be transformed into PSS requirements that can be used for designing a PSS or when evaluating different PSS options.

The methodology consists of six steps, as seen in Fig. 1. How these steps are performed is further described in the following six sections. The result from step 3 is an Actors Map, while the result from step 5 is a System Map.

The recommendation is that one facilitator/project manager is assigned to manage the work, e.g., to assemble all identified results (Actors and System Maps) as well as manage workshops and their outcomes.

2.3 *Step 1. Define the PSS That Will Be Analyzed*

The first step is to define the PSS that will be analyzed. This includes setting up the focus and ambition level for the analysis. Furthermore, define how to perform the analysis. As further presented in the coming sections, the data collection and analysis can be done in various ways depending on the situation and focus of analysis.

Fig. 1 Process steps

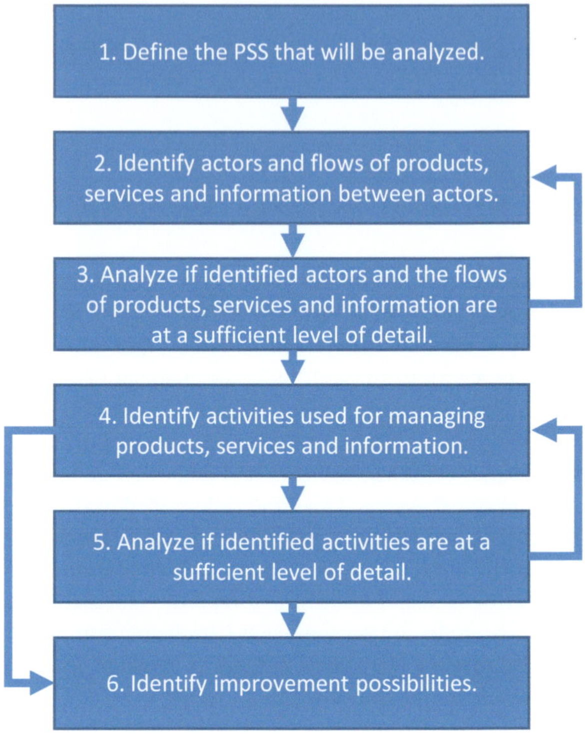

2.4 *Step 2. Identify Actors Involved and Flows of Products, Services, and Information Between Actors*

The second step is to identify the actors involved as well as the flows of products, services, and information between the PSS actors.

The first approach within the second step is to identify the actors. By asking actors within the PSS their view of how the PSS is provided, many actors can be identified and then further described. In general, a large number of actors are involved when providing a PSS [10], and it is common that they have different views on who is involved and what is happening during the execution process. Actors can be an individual, a group, a function, or a department; they can be, for example, designers, service staff, sales staff, users of the offering, and expendable providers. When identifying actors, it is important to not neglect actors that traditionally would be considered not to be involved in a PSS, e.g., NGOs and legislative functions that might influence the PSS [10].

In order to make it easier to overview the map, it is preferable to classify actors into different types and designate them using different colors, e.g., as shown in Fig. 2; this is a common method for classifying actor types (see, e.g., Lindahl et al. [10], Donaldson [15] and Tan [13]).

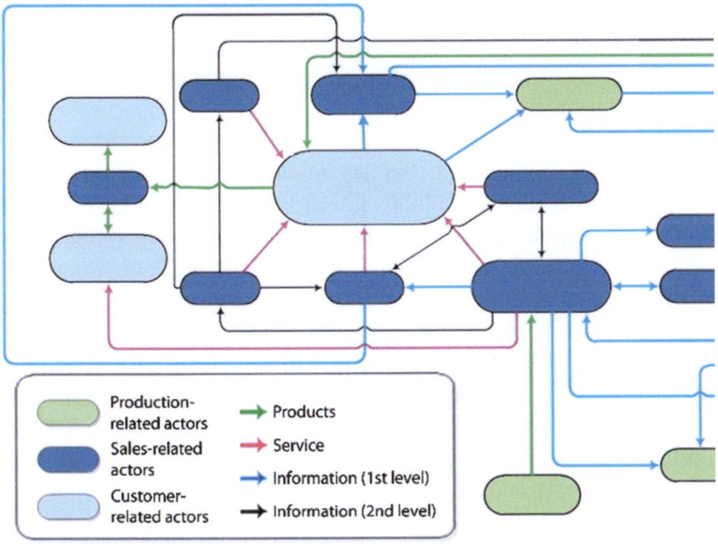

Fig. 2 Part of an example of an Actors Map [10]

The second approach within the second step is to identify important interactions (flows and directions) between the different identified actors, i.e., plot out flows of products, services, and information between actors involved in the PSS (minor flows should not be included). A product is a tangible object that is transferred from one actor to another, while a service is, e.g., education, support (such as how to use the PSS or performance calculations), or maintenance. Information is normally divided into two types, first- and second-level information:

- First-level information is directly related to the specific PSS and the ability to provide it. It could, e.g., be information about service needed, when service is needed, or information about payments.
- Second-level information is indirectly related to the specific PSS and is, e.g., about how to provide the PSS process or a future PSS. It could, e.g., be information about requirements that a future PSS ought to fulfill or approaches for improving a future PSS.

When plotting out the interactions between actors, if it is a clear two-way interaction, the line should have an arrow on both ends. Furthermore, in order to make it easier to overview the map, it is preferable to use different colors for products, services, and first- and second-level information, as seen in Fig. 2.

The identification process can be performed in many ways and combinations; this paper presents two principal approaches. One approach is to ask different actors related to the offering to plot out their view of which actors are involved in the analyzed PSS and, furthermore, to plot out flows of products, services, and information between those actors. When a sufficient number of actors have plotted their views (and when no additional important actors, products, service, and information have been identified), all maps are merged into one overall Actors Map. The other approach is to assemble different actors and then ask them to, together in workshop style, identify the actors involved and the flows of products, services, and information between those actors.

A pro with the first approach within the second step is that different actors will not influence other perspectives, but a con is that this method is very time-consuming and can require more loops in order to get an overall Actors Map that respondents agree upon. A pro with the second approach is that within one workshop, a good overall Actors Map can be achieved. Cons are, e.g., the risk that they influence each other and the difficulty in coordinating such a workshop that involves many different actors.

Independently of the approach, a whiteboard is a useful tool when plotting the actors and their interaction (products, services, and information), since actors quite often want to make changes and make their descriptions more detailed during the plotting process. It is also common that after a while they realize that they need to break down an overall actor into several actors, e.g., break down the service department into actors like service technicians, service planners, and service managers, and describe their interactions internally and with surrounding actors.

Finally, the actors' resolution level is normally quite detailed in order to be useful, but it depends on the situation, for example, different departments or functions/people involved in the PSS provided. At the same time, there is a risk of not making the Actors Map detailed enough, since important actors and interactions can get neglected and become less visible in a more aggregated, untransparent overview.

2.5 Step 3. Analyze if Identified Actors and the Flows of Products, Services, and Information Are at a Sufficient Level of Detail

Since it is common that different actors have different perspectives, it is preferable to have a workshop in order to analyze if identified actors and the flows of products, services, and information are at a sufficient level of detail. This is especially important when many different actors (first approach within the second step) have plotted their own perspectives, plots that then have been merged into an overall actor map.

During the workshop, it is determined if a sufficient level of detail has been reached. Sometimes during the workshop, changes are made or a decision is made that further actors need to be involved.

Finally, depending on the purpose of the analysis, the following step after this one is six. It could be, e.g., if the purpose is to identify nonoptimal distances between actors, such as the distance between the actors responsible for PSS requirement management and the main actor with the most important requirements

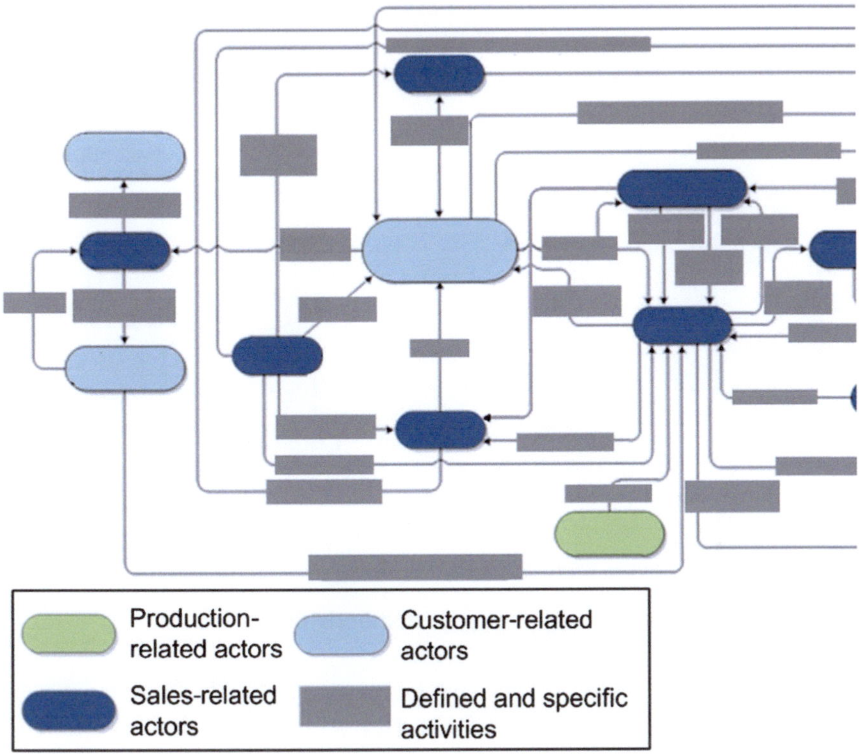

Fig. 3 Part of an example of a System Map

[10]. It could also be if the purpose is to support the PSS management to identify non-value giving links that can be omitted in the chain of actors involved [10].

2.6 *Step 4. Identify Activities Used to Manage Products, Services, and Information*

What distinguishes a System Map from an Actors Map is that it visualizes, in a more detailed way, the specific activities utilized to obtain interactions as well as the different types of interactions of product, service, and first- and second-level information (see Figs. 3, 4, 5, and 6; compare with Lindahl et al. [10]).

In the System Map, specific activities are spelled out and illustrated by boxes. An activity is characterized by the condition in which things are happening or being done, and it includes, for example, support systems, methods, and processes. As for the Actors Map, information is normally divided into two types, first- and second-level information (see Fig. 6 in chapter "Reducing Conflicts of Interest in

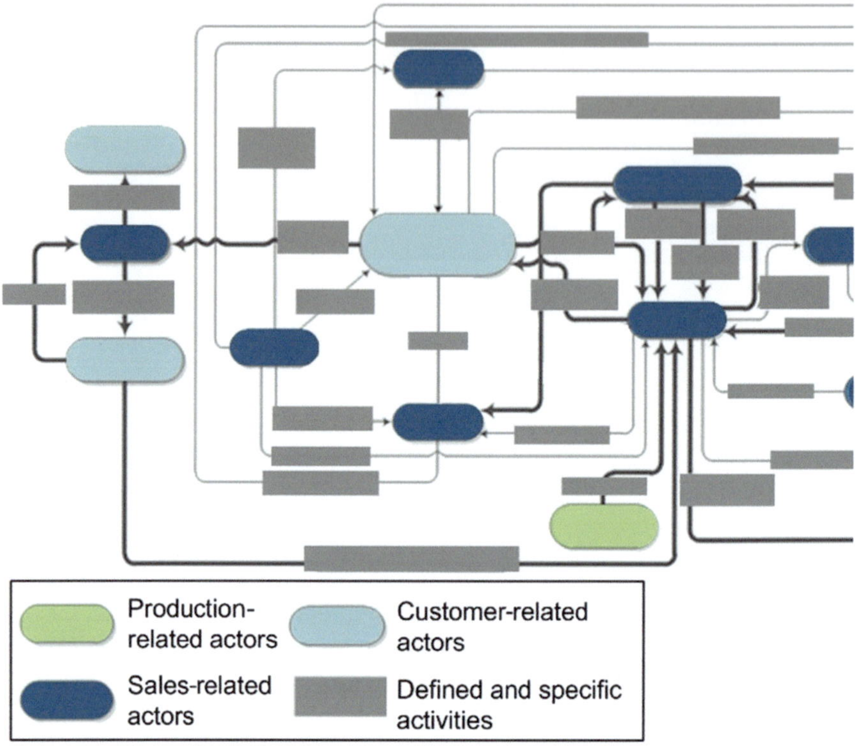

Fig. 4 Part of an example of a System Map that illustrates actors involved in the ongoing execution of a PSS and what activities they perform

Eco-design: The Relation of Innovation Management and Eco-design in the Automotive Sector").

An alternative for boxes is to designate numbers that are further explained in separate text (see Fig. 6 in chapter "Reducing Conflicts of Interest in Eco-design: The Relation of Innovation Management and Eco-design in the Automotive Sector"). This approach can be preferable when the System Map is very complex, and also since it enables a more detailed description of the activity and how products, service and information are managed within that activity.

Different life cycle phases of the PSS imply different activities and focus, something which can be illustrated in a System Map (see Fig. 4). This is why a System Map is useful when communicating and developing a PSS [10]. The given example illustrates actors involved in the ongoing execution of a PSS and what activities they perform. The flow in the system is illustrated by the darker, thicker lines.

The identification process of activities for the System Map can be done in the same manner as for step 2. Also for this step, a whiteboard is a useful tool since actors often want to make changes and make their descriptions more detailed during

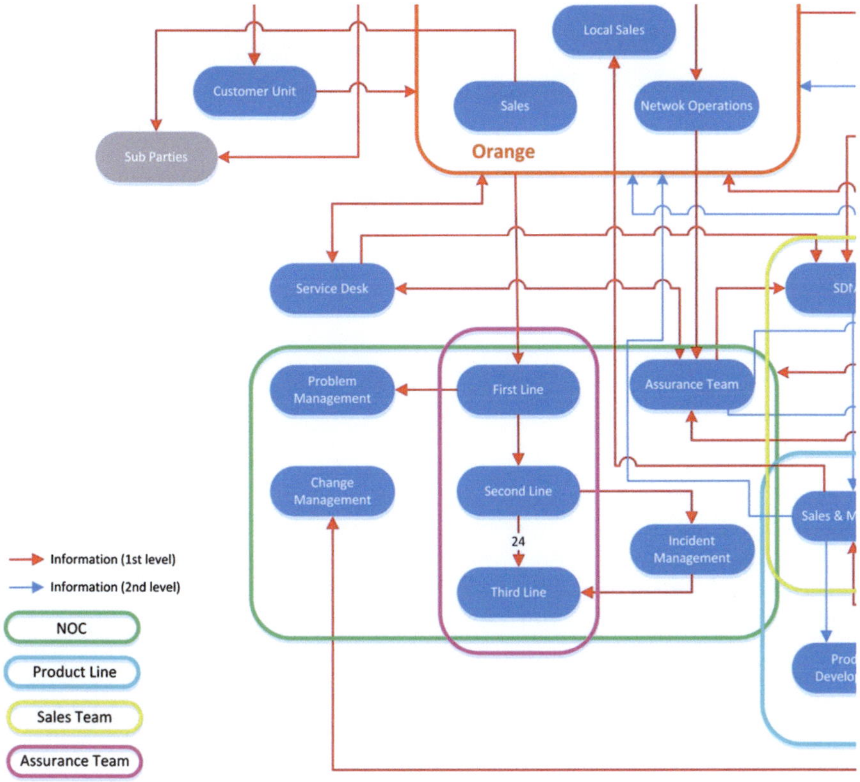

Fig. 5 Part of an Actors Map that illustrates the participating actors and their connections within Ericsson's Device Connection Platform

the plotting process. Finally, the resolution level is normally detailed in order to be useful, but the focus should always be to concentrate on the major and important activities, not cover them all.

2.7 Step 5. Analyze if Identified Activities Are at a Sufficient Level of Detail

The objective of this step is to analyze if the identified activities are at a sufficient level of detail. This is especially important if different actors (the first approach within the second step) have been plotting their own perspectives independently to other actors, plots that then have been merged into an overall System Map.

This step is often done during a workshop with different actors involved in the PSS. During the workshop, it is determined if a sufficient level of detail has been

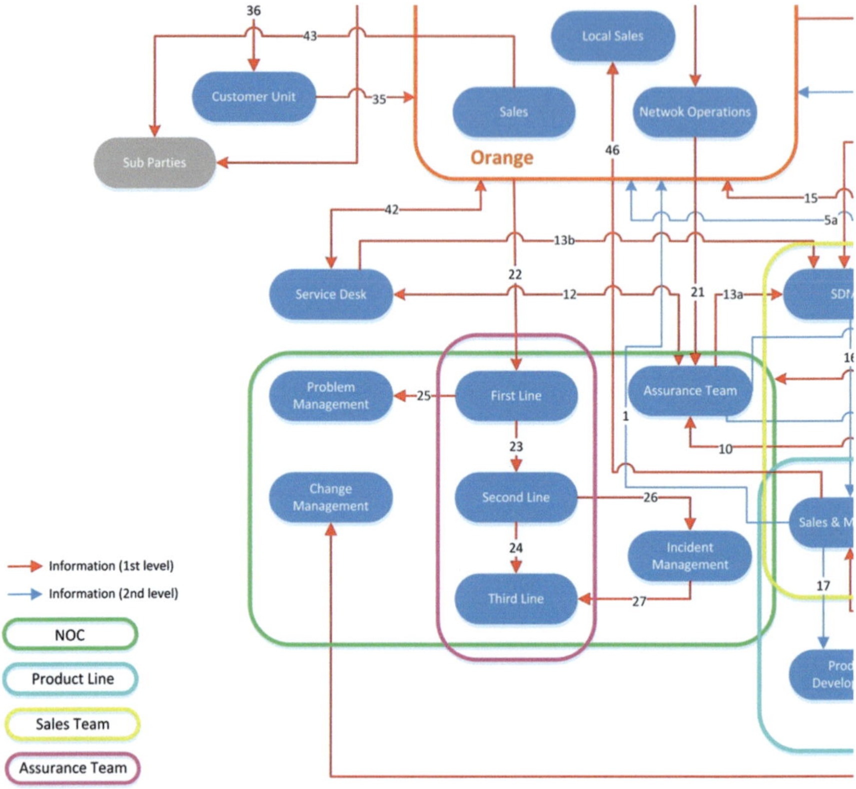

Fig. 6 Part of a System Map that illustrates what kind of information is transferred between actors within Ericsson's Device Connection Platform

reached or if further activities need to be included. During the workshop, changes in the System Map can be made.

2.8 *Step 6. Identify Improvement Possibilities*

This step is either done in, e.g., separate workshops or in conjunction with the workshop described in step 5.

Improvement possibilities can be of various kinds and levels, for example, to get better information flows between different actors; to reduce the risk that important information, e.g., about PSS requirements, is lost; and to avoid activities that provide low or negative values for the offering, e.g., unneeded steps.

However, the purpose of the Actors and System Map could also be just to provide a better overview of the PSS, e.g., in order to make it easier to follow up on a PSS, introduce new staff, or communicate with customers. This can be compared to arguments by, e.g., Ulrich and Eppinger [17], Ullman [18], and Smith and Reinertsen [19] for documenting and defining the product development process.

3 The Orange and Ericsson Case

An example of when the methodology has been used can be found in the study of the "Orange and Ericsson case," described below.

3.1 Background

Ericsson's Device Connection Platform (DCP) is one of the provided services that Ericsson offers to other telecom companies. The interaction between actors within Ericsson's DCP was investigated by using the six steps of the method in order to identify improvement opportunities in the interaction between the provider Ericsson and, in this case, the customer Orange.

The aim was to deliver a map with Ericsson's view of the interactions between actors in order to enhance the knowledge of how it can be improved.

In this case, two of the coauthors, Desai and Widgren, acted as facilitators/ project managers when performing the method. They visited and talked with different Ericsson actors involved in the DCP. They also assembled all data collected into the Actors and System Maps.

3.2 The Actors Map

The Actors Map of the service shows the different perspectives of the actors involved, how they are connected, and their main assignments from both Ericsson's and each respondent's point of view. The Actors Map from Ericsson's point of view will support the identification of the included actors and their assignments within the DCP.

When Ericsson provides the DCP for Orange, it is important for them to know what Orange expects, but also for them to explain what they expect from Orange as a customer.

> Once we have a project conference meeting, where the customer is included, we will have discussions in terms of describing the service offering of the DCP. Then we also describe what we need from the operator for us to be able to set up the platform. (Employee at Ericsson)

The Actors Map, presented below, illustrates the participating actors and their connections mentioned by different actors at Ericsson, as seen in Fig. 5.

Each actor plays a part in the DCP and it is important for Ericsson to be able to provide it. Also, Ericsson's perspective of the actors within Orange is included in the Actors Map. In some cases, the description of each actor is similar from different respondents, while in other cases it differs. As can be seen in the Actors Map below, those actors that are mentioned as a group of several actors are encircled in different colors.

3.3 The System Map

The System Map provides a visual overview of what type and what kind of information is transferred between the actors within Ericsson and between Orange and Ericsson. A System Map can also help in identifying non-value adding links in the information flow between actors. The System Map of the DCP is presented in Fig. 6.

In order to facilitate what each connection and information flow implies, all connections are numbered. As for the Actors Map, the actors including other actors are encircled in different colors in the System Map. Some of the actors are connected to the encircled actor, while others are connected to the actors within the encircled actor. There are two types of information flows, first and second levels. The first-level information flows are designated by red lines between different actors and describe the main information flows regarding a specific DCP offering. The second-level information flows are designated by blue lines between different actors to describe the main information flows aimed for developing future potential offerings. This is done to present the connections and information flows within and between both Ericsson and Orange more visibly.

4 Discussion

4.1 Application of the Methodology

The facilitator should consider that the respondents sometimes have different points of view and also different levels of knowledge about the participating actors in the service. Some respondents can have extensive knowledge regarding participating actors yet are unable to describe every actor's activities or responsibilities. Others do not have the extensive knowledge about all included actors, but can describe the ones they know further and in more detail. As Lindahl et al. [10] also describe, the respondents' view of the included actors and how they are connected differs. This results in the need to balance different descriptions and find similar descriptions of the actors.

One way to analyze the credibility of the connections mentioned by the respondents is for the facilitator to count the distance in number of connections from the respondent to the mentioned actor. The facilitator might conclude that the respondent who has fewer connections might have a better overview of the actors they are close to. On the other hand, the facilitator should bear in mind that counting the connections can be challenging. If it is a big actor such as Orange or R&D, there are almost certainly more connections within these actors.

The created Actors and System Maps can differ significantly depending on the respondent. When the respondent creates the Actors Map, the facilitator should not interpose regarding the way they chose to visualize the map. Instead, if needed, additional questions should be asked for more information about a specific actor or a connection. If the respondent on the other hand mentions information not relevant for the result, in those cases, the facilitator should not interrupt because of the risk of losing important information.

In both the first and second approaches within the second step, face-to-face meetings are more appropriate compared to meetings via videoconference. During face-to-face meetings, it is easier for the facilitator to understand the expressed thoughts of the respondents while they are creating the Actors Map. If anything is unclear, additional questions can be asked and the respondent can easily explain what they mean by pointing out the problem in the Actors Map. Therefore, more information can be gained from the face-to-face meetings when the facilitator is able to see the Actors Map at the same time as the respondent is creating it.

During the first step, the facilitator should avoid eliminating the actors that are not connected to any other actors. The risk is to create extra work later in the process, when important connections get lost when eliminations are made in such an early state. Actors that do not seem to be important can turn out to be so, even if accurate information regarding them is missing in the first step. Therefore, to avoid an extra workload later in the process, eliminations should not been made in step one until after the respondents are contacted again to confirm the importance, the activity, and the connections of those actors.

As Lindahl et al. [10] describe, many can be considered as customers within a PSS, so the concept of "the customer" can sometimes be misleading for the facilitator. As for the example of the "Orange and Ericsson case," the sub-parties are customers of the service as well but were not included.

The final result of the Actors and System Map can differ significantly from each respondent's map that is created during the meetings. This indicates that mapping out the actors and their connections is a good way to visualize how the interactions and the information flows within a company and between two companies actually looks, in comparison to what each employee seems to think it appears to be.

4.2 Pros and Cons with the Methodology

The refined and more structured Actors and System Map methodology presented in this paper (compare with Lindahl et al. [10]) helps to compile a great overview of all included actors within a PSS and all the connections between them. The maps help to identify which actor has the most connections, thereby aiding in analyzing the importance of that actor. This is line with Tan's conclusion [12]. The number of connections can signify if it is a big actor with several people included with different tasks or if it is an actor who may need assistance given that it has many connections to other actors. Since there are different points of view about the importance of each actor, but also regarding who the participating actors are, the method plays an important part in providing an overall picture, which is of importance for a company. As stated earlier, this is in line with arguments by, e.g., Ulrich and Eppinger [17], Ullman [18], and Smith and Reinertsen [19] for documenting and defining the product development process.

The implementation of the method requires that some assumptions are made, e.g., that some actors are described similarly and therefore are assumed to be the same actor. However, this may affect the result regarding the actors or the number of existing actors. Another identified difficulty that Lindahl et al. [10] also point out is that in many cases, the connections between actors can be quite long. This could indicate that an individual, a group, a function, or a department that possesses relevant information and the individual, group, function, or department that is in use of that information are far from each other and that important information can get lost along the way.

The information is not always taking the formal route, or at least not always taking the route it was intended, e.g., when the customer has been in contact with R&D in an early state and feels more comfortable contacting that department again regarding requests for enhancements or new features. This is often problematic since several steps are skipped in the information flow, which affects actors that are using the information and who hear about it much later. However, this is difficult to detect with the method since it is not apparent by studying the compiled map. To make it possible to detect these information flows, a much more detailed study must be done.

The respondents expressed that the method was easy to use without previous mapping method experience, which also was the expectation of the authors.

The Actors and System Map is an easy method to use since it does not require much preparation or any specific, costly tools. However, the author realized preparations are essential to state what result is expected. If the preparations are not made properly, creating the maps could take more effort than is gained. Since there are no limits on how detailed the maps can be constructed, delimitations of the study are of importance to ensure the desired result is obtained.

5 Concluding Remarks and Future Research

This paper presents a refined version of the Actors and System Map methodology that is clearly defined and follows a structured process, compared with Lindahl et al. [10]. The Actors and System Map helps providing an overview of participating actors within a system, which is advantageous since there are often differences of opinions regarding the actors involved. The method is advantageous since it can be used to study different cases, e.g., by studying only a service or a PSS. The method can also be applied on either a small scale or large scale, e.g., by studying the actors and information flows within a small department working on the PSS or applying the method to the whole PSS.

The Actors and System Map can result in being quite complex if it is done too detailed. Therefore, the preparations are essential to state what result is expected and to set up delimitations in an early state.

For future research, a means of controlling the credibility of the information gained regarding participating actors and their connections would be useful to eliminate the process of gathering information.

Acknowledgments Ericsson AB and Orange are acknowledged for providing valuable information for the presented case. The authors would like to express special gratitude for financial assistance from VINNOVA, and especially its research program called "Strategiska Innovationsområden." This research was also partly supported by the Mistra REES (Resource Efficient and Effective Solutions) program funded by Mistra (The Swedish Foundation for Strategic Environmental Research).

References

1. Sundin E, Ölundh Sandström G, Lindahl M, Öhrwall Rönnbäck A, Sakao T, Larsson T (2009) Challenges for industrial product/service systems: experiences from a learning network of large companies. In: CIRP industrial product-service systems (IPS^2) conference. Cranfield University, Cranfield
2. Lingegård S, Lindahl M, Sundin E (2010) Organizational changes in connection with Integrated Product Service Offerings. In: CIRP's 2nd IPS^2 conference. CIRP, Linköping
3. Elfving SW, Lindahl M, Sundin E (2015) Ericsson – the history from product to solution provider and challenges and opportunities in an evolving environment. Proc CIRP 30:239–244
4. Lindahl M, Sundin E, Sakao T (2014) Environmental and economic benefits of integrated product service offerings quantified with real business cases. J Clean Prod 64:288–296
5. Meier H, Roy R, Seliger G (2010) Industrial product-service systems – IPS^2. CIRP Ann Manuf Technol 59(2):607–627
6. Mont O (2004) Product-service systems: panacea or myth? In: The International Institute for Industrial Environmental Economics (IIIEE), Lund University, Lund, p 233
7. Lindahl M, Sakao T, Öhrwall Rönnbäck A (2009) Engineering implications of integrated product service offerings. In: 17th international conference on engineering design (ICED'09). Stanford University: Design Society
8. Åhlström P (2004) Lean service operations: translating lean production principles to service operations. Int J Serv Technol Manag 5(5/6):545–564

9. Sakao T, Öhrwall Rönnbäck A, Ölundh Sandström G (2013) Uncovering benefits and risks of integrated product service offerings – using a case of technology encapsulation. J Syst Sci Syst Eng 22(4):421–439
10. Lindahl M, Sakao T, Carlsson E (2014) Actor's and system maps for integrated product service offerings – practical experience from two companies. Proc CIRP 16:320–325
11. Desai A, Widgren M (2015) A collaboration in product service systems for telecom networks – "Orange and Ericsson case" study, Linköping University, LiU-IEI-TEK-A-15/02213-SE
12. Matzen D (2009) A systematic approach to service oriented product development. DTU Management, Kgs. Lyngby, p 186
13. Tan AR (2010) Service-oriented product development strategies. In: Danmarks Tekniske Universitet. DTU Management, DTU Management Engineering
14. Ishii K (2001) Customer value chain analysis (CVCA). In: Ishii K (ed) ME317 dfM: product definition coursebook. Stanford Bookstore, Stanford University, Stanford, pp 1.3.1–1.3.8
15. Donaldson K, Ishii K, Sheppard S (2006) Customer value chain analysis. Res Eng Des 16 (4):174–183
16. Moritz S (2005) Service design – practical access to an evolving field. International School of Design, University of Applied Sciences Cologne, p 245
17. Ulrich KT, Eppinger SD (2011) Product design and development, 5th edn. McGraw-Hill Education, Boston, p 432
18. Ullman DG (2002) The mechanical design process, 3rd edn. McGraw-Hill Higher Education, New York, p 415
19. Smith PG, Reinertsen DG (1997) Developing products in half the time: new rules, new tools, 2nd edn. Wiley, New York, p 320

PSS Without PSS Design: Possible Causes, Effects, and Solutions

Johannes Matschewsky

Abstract Product-service systems (PSS) are seen as a cornerstone of a future circular economy. However, in order to achieve the desired environmental benefits, the industrial implementation of PSS design is key. This chapter discusses the apparent lack of an adaptation of design processes to PSS or adoption of PSS design methods within companies, which are nevertheless successfully offering PSS. Based on experiences at two companies and under close regard of the relevant literature, possible causes of this lack of method adaptation/adoption are discussed, and the effects this may have are deliberated. Lastly, potential solutions to this issue and ways forward are introduced and reflected upon focusing on the companies.

Keywords Method adoption • Process adaptation • Business model • Integrated product service engineering • EcoDesign

1 Introduction

1.1 Background, Motivation, and Goals

Resource scarcity is becoming an increasing pressure, as, although faced with a finite planet, resource use on earth currently only knows one direction – up [1]. Confronted with this challenge, particularly the industries of developed countries have identified one critical component in coping with this: doing more with less, i.e., decoupling resource use and value creation, in one word, dematerialization. This has been a cornerstone of product-service systems (PSS) since an intensive phase of research on PSS commenced in the late 1990s [2, 3]. Through the efficiency aspects associated with them, PSS are hoped to become a key component of a still utopian circular economy [4].

However, the benefits of PSS as a business innovation, which can provide tremendous opportunities, do come at a cost: The implementation of good PSS

J. Matschewsky (✉)
Division of Environmental Technology and Management, Department of Management and Engineering, Linköping University, Linköping, Sweden
e-mail: Johannes.Matschewsky@LiU.se

M. Matsumoto et al. (eds.), *Sustainability Through Innovation in Product Life Cycle Design*, EcoProduction, DOI 10.1007/978-981-10-0471-1_16

design. Good PSS design takes the entire life cycle into account and provides a high level of integration of products and services, beginning at the earliest stages of design (see, e.g., [5–7]).

During on-site research activities with two industrial PSS providers, a broad understanding was created for the design processes and procedures employed at the companies. Due to reasons of anonymity, these companies shall be called "Levor" and "Navitas" within this chapter. Both are providers of industrial goods and PSS and are among the world leaders in their respective domain.

Under the course of the research, it became increasingly clear that the design processes applied were either not or only slightly adapted to the altered requirements brought about by the introduction of PSS. This is surprising, as both companies are successful in offering PSS, with one of them having a higher number of PSS sold than traditional product sales offerings in certain markets. The question arose, whether or not the academic presumption that design processes must be adjusted to PSS is simply wrong – or whether the lacking adoption has other causes. In the course of the research, it was found that the latter is the case. Consequently, implications of this as well as possible ways of remedying the situation were then taken into focus. An additional research motivation was the lack of literature on this matter in the area of PSS design.

Based on the knowledge gathered, this chapter aims to *examine* possible *causes* for the lack of process adaptation and method adoption with respect to PSS, to *clarify* the *effects* of this non-transition, and to point out viable *first* steps aimed at *improving* this situation.

1.2 Research Approach

The information collected at "Levor" and "Navitas" was evaluated with close focus on state-of-the-art literature in the field. Three areas were identified as possible causes for the described phenomenon. Probable (future) implications of the lacking adaptation of design processes to PSS are discussed with respect to business as well as the environmental considerations. Potential solutions to this issue are discussed within the same boundaries, and their implications at the case companies are briefly reported. Even though the presented threefold of possible causes, effects, and solutions can neither be considered universally applicable nor comprehensive, the strong integration of literature employed in the reasoning ensures that the new knowledge which is presented is relevant to both academics and practitioners alike. For academics, this chapter may be considered an impulse to examine own industrial experiences in order to contribute to its extension. For practitioners it may serve as a further clarification of PSS benefits, an incentive to more quickly move to a full integration of PSS and to embrace both challenges and opportunities PSS can provide.

1.3 Structure

The chapter commences by explaining the methodological approach of the research in Sect. 2. Subsequently in Sect. 3, the need for PSS design as such is discussed. Hereafter, in Sect. 4, possible causes for a lack of adaptation of design processes to PSS are evaluated. In Sect. 5, business- as well as environment-related effects and possible solutions are discussed. Sect. 6 then concludes the chapter, summarizing its content and giving an outlook on forthcoming research efforts on this topic.

2 Methodology

2.1 Literature Review

In order to relate the empirical findings and suggestions for process improvement to the state of the art in PSS design research, a thorough literature study was performed. A part of the literature used was gathered for a recent literature-focused paper on implementability of PSS design methods [8]. Further literature was collected through keyword searches, predominantly in the Scopus engine. Keywords or strings used included but were not limited to *PSS design benefits*, *PSS design efficiency*, *PSS design methodology*, *organizational challenges and PSS*, and *managerial support and PSS*. Due to space restrictions, only the most relevant publications were used.

2.2 Actor Maps

In an effort to fully understand the connections between the actors involved in designing a PSS offering as well as the information exchanged between them, interviews to derive "Actor Maps" were conducted with designers and project managers at "Levor." A full account of the "Actors and System Maps" method can be found in [9]. Each of the participants of the interviews was informed in advance through a one-sheet about the method that would be used during the interview. They were further informed that the focus would be on their immediate working tasks, the information they exchanged while fulfilling these tasks, and the methods used for fulfilling them. During the interview, the information given by the participant was mapped on a whiteboard and subsequently digitalized using Microsoft Visio.

This chapter utilizes the information and understanding gathered as a point of departure for a deeper examination of the possible causes of the identified low adaptation of design processes to the integrated offerings provided by "Levor."

2.3 *Understanding Product/Service Integration: The "Navitas" Case Study*

A PSS design methodology was tested and evaluated at "Navitas" for several years. During this process, a profound understanding for the inner workings of the company with respect to product and service design was derived, after many meetings and interviews with a substantial number of employees active in design of products, services, as well as management. In light of the data and knowledge gathered during this time, the results reported here were obtained in order to investigate the perceived contradiction between offering PSS and the challenge to adapt processes to the new business model. Further, an effort is made to present a convergent result (cf. [10]).

3 Why PSS Design?

Before discussing the issue of a lack of PSS design in companies offering PSS, first the need for new design approaches to be implemented by practitioners must be clarified. A significant portion of the research on PSS has been dedicated to the topic of PSS design (see, e.g., [5]). Implementing integrated offerings in a company previously active in the sale of products is a tremendous challenge, and researchers involved with PSS appear to largely take the relevance and necessity of a particular approach toward the design of these offerings as a given. Nonetheless, a few significant points made in literature are discussed in the following in order to support the reasoning in the coming sections of this chapter.

The design divisions of companies are faced with a substantial complexity due to the need to integrate products and services if a PSS implementation is to be successful. A PSS design framework helps cope with this, while traditional engineering methods are also considered useful if enhanced to fit the new challenges [11]. Particularly the possibility of clarifying the relationship between service and the value created as a benefit of PSS design is seen as an advantage [6]. This is extended further in [12] by stating that in addition to the traditional, product-focused paradigm of functional requirements, PSS design serves to fulfill customer value. This goes beyond functional requirements, and, therefore, an approach must be taken that goes beyond traditional product design. When clarifying the concept of PSS design, [13] describes PSS design as "design toward stakeholders [which] provides designers with new degrees of freedom and covers an earlier phase of design." This difference is particularly apparent when considering the sustainability aspects of PSS [13]. A further particularity that PSS design aims to tackle is the iterative nature of the development [4], going so far as to describe the process of PSS design as based on "trial and error." This can be attributed to the quite early stage that PSS are in with respect to their industrial deployment, and it all the more clarifies the need for fitting methods to be used. A lack of adaptation of the physical design when transitioning toward PSS is a further issue in case the broad and life cycle-oriented perspective of PSS design is omitted [14]. On a similar note, integrated design of products and services is seen as a cornerstone of providing environmentally beneficial PSS, particularly with respect to result-oriented PSS offerings [15]. Special focus is also put on balancing the issue of customer value with internal company capability [16], which traditional product design paradigms do not offer. Lastly, a lack of focus on service during the design process may serve to increase operational cost [17].

Overall, a number of benefits of and needs for implementing PSS design have been identified. Nonetheless, the implementation of PSS design methods in industrial practice is lacking [18]. After discussing the adaptation of PSS design methods and tools to support their implementation in the industry [8], this chapter aims to shed a light on the company side of this challenge in the coming sections.

4 Why Companies Are Not Utilizing PSS Design

In this section, three overall issues are discussed that may lead companies that offer PSS to not implement PSS design methods or adapt their design processes to the changing needs in light of this new offering. The understanding for the processes within two industrial companies offering PSS, which was created during two industry-focused research projects, serves as the basis for deriving the causes for companies not to implement PSS. Literature is used to interpret and support the findings and observations made.

4.1 Reduced Customer Pressure

PSS dilute the boundaries between customer and provider, particularly in the case of highly integrated (availability- or result-oriented) business models [6]. The traditional border between customer and provider, which also marked the border between one actor establishing requirements and the other actor fulfilling these requirements, is increasingly disappearing. An illustration of the product sales-oriented relationship is given in Fig. 1, visualized by the dark green color.

In product sales, customers have strong incentives to set requirements concentrated on both matters of efficiency (achieving the desired goal with minimal resource use) and effectiveness (achieving the desired goal in the required quality). The provider aims balance both factors in order to deliver value to the customer (see Fig. 1). In the case of "Levor," the company has been traditionally excellent in taking account of customer requirements through principles of lean [19]. The path of handling customer requirements as perceived by an engineering designer is illustrated in Fig. 2 (blue solid path). This process is similar to processes utilized by many product-oriented industrial companies (see, e.g., requirement specification

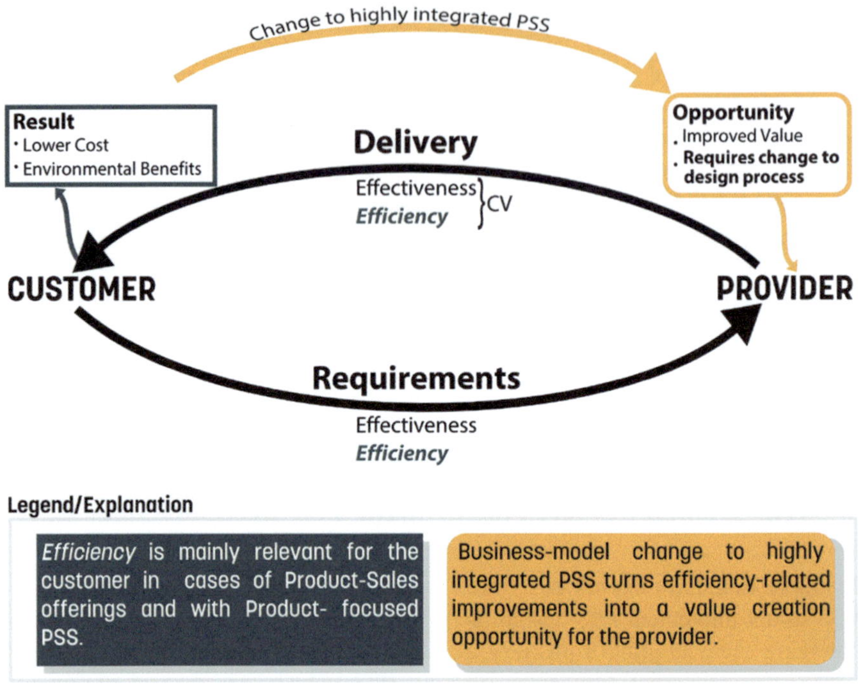

Fig. 1 New value creation opportunities for providers when transitioning to integrated PSS business models

in [20], pg. 146). Today, "Levor" offers, besides their traditional product sales tier, a model which can be described as a highly integrated PSS [6]. Through this, as further shown in Fig. 1, efficiency improvements are no longer prioritized by companies offering highly integrated PSS, especially with regard to customer-oriented processes. This is particularly notable, as many of the efficiency-related requirements have a direct impact on environmental performance and sustainability, an area on which many PSS design methods lay particular focus (see, e.g., [21]). Without the pressure of fulfilling increasingly demanding customer requirements, companies have a much lower incentive to adapt their processes to PSS design and to implement PSS design methods. A reduced use of the established process of addressing customer requirements and the lack of a new process to handle the internalized efficiency requirements may lead to a lower perceived need for process adaptation and, thus, the adoption of PSS design methods and tools. The high technological aptitude of "Levor" to fulfill the set requirements on effectiveness could further strengthen the impression that no change to current processes is needed. This may be even further enhanced by the fact that offerings for traditional sales and PSS are jointly developed at "Levor."

In summary, it can be stated that a lacking awareness of new value creation opportunities (cf. yellow parts of Fig. 1) and therefore an impeded adaptation of the

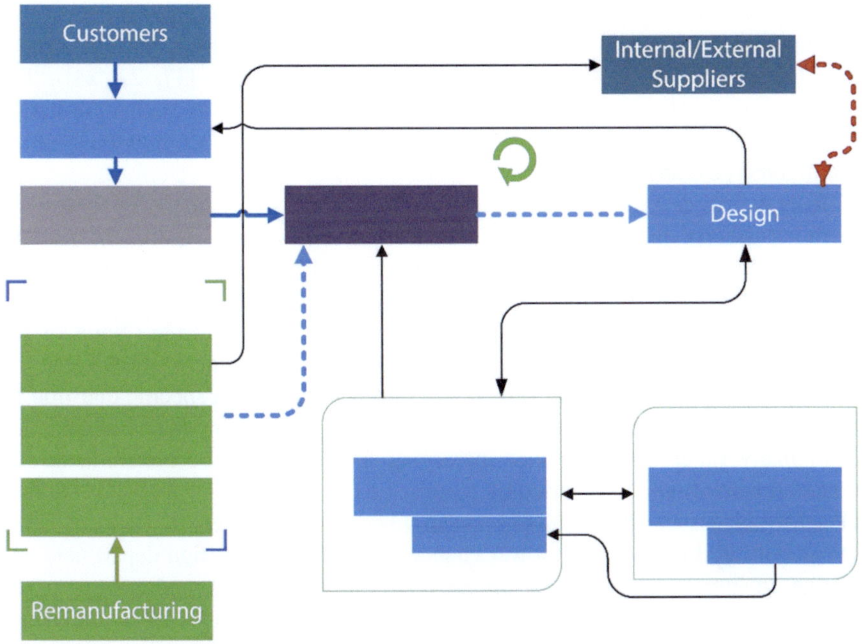

Fig. 2 Information between manufacturing and design during conceptual (*blue dotted*) and detailed design (*red dotted*), informal feedback stream from remanufacturing (*green solid*), and customer input (*blue solid*). Blurred due to confidentiality requirements

handling of internal requirements may play a role in an underdeveloped adaptation of design processes to PSS offerings.

4.2 *Lack of Involvement of Internal Stakeholders Who Are Key to PSS Design*

A significant portion of the impact of lean engineering on the manufacturing industry stems from the concept of cooperation during the design and development process, particularly between manufacturing and design [22]. This close relationship is reflected in the process employed at "Levor" as seen by a design engineer, signified by the blue and red dotted lines in Fig. 2. Again, this process is likely to be similar in other companies due to the lean design approach [19]. Because of that, the identified items bear general relevance beyond this case company.

PSS design has a strong impact on many actors and stakeholders throughout the life cycle of an offering. However, service/maintenance and remanufacturing are among the activities most strongly affected by this business model change [14]. Due to this, service/maintenance and remanufacturing are sensitive to well-executed PSS design, i.e., they benefit from positive developments, and lacking progress becomes noticeable to them earlier and more profoundly than to other stakeholders. Service/maintenance and remanufacturing are also in particular focus in literature on PSS design and PSS design methods (e.g., [17, 23, 24]). The importance of information being available and the quality of such information both for remanufacturing and design departments have also previously been highlighted [25, 26].

In "Levor," despite the fact that remanufacturing is part of the PSS offering, there is currently only little information fed back to the design department from the remanufacturing facility. This feedback is not formalized at this point. The pathway shown in Fig. 2 is therefore not nearly as established as the consideration for customer requirements and the coordination with manufacturing. None of the interviewees mentioned remanufacturing as an important input to the design process, and the feedback stream shown was only named after a specific question by the interviewers. Further, the discovered feedback track also collects information from many other actors, further increasing the likelihood of remanufacturing-related information being buried by feedback from more established stakeholders. Due to this, issues that regularly occur when products used in PSS are being remanufactured have a much lower chance of reaching the design department and having an impact on design. A similar observation has been made assessing the information feedback and use of information provided by service and maintenance staff in [9]. The feedback processes from these actors are less formalized than, e.g., those concerning manufacturing, which may lead to a lower prioritization of these issues. In effect, the historically strong product focus of companies transitioning from being a product to becoming a PSS provider continues to be strengthened, inhibiting a profound transition.

From the findings presented, it can be deduced that the lack of a direct information pathway from key internal stakeholders benefiting from good PSS design (such as remanufacturing or service/maintenance divisions) may lead to reduced incentives to adapt existing design processes.

4.3 Managerial Focus and Product/Service Integration

Organizational changes undergone by companies when bringing integrated offerings of products and services to market are a widely researched subject. Observations made during an extensive project aimed at testing and implementing a PSS design methodology [27] at "Navitas" have shown that not only bringing PSS to the market initially but also the continuous development toward a higher integration and better utilization of PSS benefits (see Sect. 3) is highly dependent on relationships between service and design divisions, as well as management.

In 2007, [28] stated that in order to succeed in offering integrated solutions, managers of previously product-focused companies should "Create independent business unit to get focus while still maintaining strong links to line organization – top management's support needed" (pg. 75). The experience of evaluating [27] at a machine manufacturer shows that this statement also holds when integrated solutions are already brought to market. Further, as early as 1999, [29] identified a need for a change in organization and corporate culture in order to successfully market sustainable offerings comprised of products and services. In addition, [3] has identified the close integration of product and service organization as a key aspect in allowing the economic benefits created by services to be transmitted to the manufacturing organization, in order to drive manufacturing and design changes with a PSS perspective.

The main challenge during the evaluation of the PSS design methodology at "Navitas" was to be found in the organizational separation of product and service development and the relationship between the different divisions. Historically, the product design division was less inclined to support the implementation of new design methods aimed at embracing the changes that offering PSS brings about. This may also be due to a lack of understanding of the possible benefits of a fully PSS-oriented design approach: This resounds the findings of [30], according to which manufacturing organizations often do not understand the benefits of higher reliability and maintainability, so long as competitors are not superior in these fields. As "Navitas" is successful in its business area, this may be an aspect leading to a lack of support for a full integration of PSS in design processes, particularly from a product design perspective.

The deployment of an integrated design method for PSS was also hampered by the substantially differing size of the design and service divisions. The employees involved with the evaluation of the PSS design method pointed out their satisfaction with the method in general, although they remarked that the complexity of such a method is a hindrance to implementation in an industrial environment. This points back to the fact that it is likely that the methods available are in part at fault for this issue, as discussed extensively in [8]. At the end of the evaluation process, the results were communicated to senior management staffs, and a modular implementation was suggested in order to cope with the complexity problem. Among management, a strong product focus could be perceived. It has previously been described that senior managers lack detailed knowledge about design and innovation processes in product design [31]. This may be a further aspect that can pose a hindrance to implementing a PSS design method.

Overall, insufficient managerial support and strong product focus as well as the operational and strategic separation of product and service design organizations led to no immediate changes in development processes. In addition, the benefits of the proposed method were not communicated to management in a way that would have warranted a stronger focus on the implementation of the method (e.g., cost-benefit calculations, new KPIs).

Table 1 Summary of key findings of possible causes of lacking adaptation of design processes to PSS

Possible causes for not implementing PSS design	Related literature
Customer pressure and incentives	[6, 21]
Reduced customer pressure toward process adaptation; this results from the responsibility shift between customer and PSS provider (Fig. 1)	
Previous customer requirements (efficiency oriented) have not been internalized by the provider side	
Stakeholder impact on design	[3, 9, 14, 25, 26, 30]
Lack of involvement of key internal stakeholders in PSS design process: remanufacturing and service	
Less formalized feedback processes	
Lack of time to develop a relationship as existing between manufacturing and design	
Management view and integration of product and service organizations	[6, 28, 29]
Persistence of a strong product focus	
Imbalance between product- and service-oriented divisions (financial, power)	
Lack of understanding of current processes and benefits of PSS design by management staff	

4.4 Summary

Within this section, a number of possible reasons for companies not to adapt their processes to PSS design were identified, and their general applicability was discussed with respect to literature. In Table 1, the information gathered is presented in condensed fashion in order to facilitate discussion and to allow industrial practitioners to get a concise overview of the challenges identified.

5 Discussing Effects of Not Adapting Design Processes to PSS and Potential Solutions to Bridge Existing Gaps

In this section, business- and environment-related effects of the identified hindrances for design process adaptation to PSS are discussed. Further, possible solutions, which may support the clarification of PSS design benefits and may drive practitioners to more readily adapt their processes, are presented. In addition, the impacts of these findings within the case companies are briefly reported.

5.1 Business-Related Matters

5.1.1 Capturing Lost Value Creation Opportunities

Due to the customer requirements shifting away from efficiency and concentrating largely on effectiveness, as described in Sect. 4.1 and Fig. 1 of this chapter, there is a chance of a significant part of the value creation opportunities to be lost. Traditional product-oriented design processes are not fit to capture the value contained in an increased effectiveness, if requirements are not set by the customer side. Thus, the opportunities brought about by the business model change, namely, to make use of the value of an offering optimized for efficiency, may go unnoticed. Although processes are in place to take in customer requirements for offerings to be designed, there are no processes that assist with internalizing previously external requirements and the associated value creation opportunities that arise from offering highly integrated PSS.

In order to increase the understanding of practitioners for these transfers of value creation opportunities, the concept of provider value may be viable [32]. Through this, additional possibilities to gain value beyond monetary value alone can be identified in direct relation to product and service components and different stakeholders. Further, provider-oriented quality function deployment (QFD), focusing particularly on processes taken over that were previously handled by customers, may be a viable approach to internalize previously external value-related aspects (see, e.g., [33]). Within large industrial companies, internal divisions may have customer/provider relationships to one another [9], suggesting the viability of an internal QFD process. Further, methods such as PSS-focused LCC [34] may help with grasping the lifetime cost of included components, leading to a broader understanding of the entire product life cycle.

Overall, it should not be forgotten that also the implementation of customer requirement-oriented processes in the past decades was neither simple nor painless for most organizations [35], indicating that time is also a key factor when transitioning to PSS-focused design processes.

5.1.2 Supporting Product and Service Design for PSS

Particularly the low impact of stakeholders that are dependent on good PSS design can lead to design decisions that may be adversarial to the performance of products and services within the offering. As mentioned in Sect. 4.2, the integration of customer requirements and the increasing collaboration between design and manufacturing departments have led to notable advancements in product design and customer value. Similarly, a close collaboration between the beneficiaries of PSS-oriented design processes (here identified as service/maintenance as well as remanufacturing departments) would likely lead to a substantial improvement in the performance of offerings in the use and end-of-(first)-life stages of the life cycle of a

PSS. This may be achieved, e.g., through direct delegates from these divisions who are involved in decision-making processes in the design stage of PSS development.

In order to be able to implement such a cooperation, creating a deep understanding of the relations among the actors and systems involved in the provision of offerings may be a suitable first step, as this often differs from the perception official documents may give [9].

A strong focus on life cycle considerations that spans the entire development process is naturally a very important aspect when seeking to improve the utilization of the benefits PSS can offer [36]. In integrating Service CAD [37] and life cycle simulation, [38] presented a tool which could be of great value in this regard. In a similar approach, [39] presented an illustrative tool developed to visualize life cycle value in computer-aided design (CAD) systems. [8] state that particularly tools like these, which do not substantially alter existing processes and are fit to bring about gradual changes, may be successful in steadily shifting processes toward a better integration of PSS design principles.

5.2 The Environment and Its Relevance for Future PSS Provider Success

As [4] states, "PSS is not the sustainability panacea," mainly the issue of dematerialization and reduced resource use is mentioned in literature as a cornerstone of PSS [2, 3]. Quantifiable data verifying the environmental benefits of PSS has long been lacking [4]; however, results have recently been published underlining efficiency benefits of PSS over traditional product sales offerings [15]. Nonetheless, thoughtful and integrated design of products and services from the earliest stages of development is essential in order to have the ability to obtain these benefits [7]. In line with findings in PSS-oriented literature discussed above, fully integrated design of products and services, aimed at efficiency and high customer value, will benefit providers, customers, and the environment. When discussing the matters found from a perspective of moving toward a circular economy, the environmental aspect may rightfully be the one given a maximum weight and importance. Nevertheless, in the reality of a growth-oriented economy, the topic of efficiency and dematerialization carries a profound implication, particularly for the provider side: Even though some countries are beginning to succeed in reducing the resources used within their economies, overall, resource use is still increasing substantially with every passing year [1]. As PSS are hoped and expected to play a significant role in moving toward a circular economy [4], the coming resource scarcity may very likely shift optimized efficiency from currently being a strong incentive for *additional* provider value to becoming a *matter of surviving* as a provider in an increasingly competitive market environment. For a more and more dematerialized economy, which creates sufficient value to allow all to pursue a life with no poverty and without deteriorating the base of our existence, highly efficient and integrated offerings are needed. A strong focus on matters of resource efficiency, dematerialization, and the environment is becoming increasingly crucial for producing companies. How quick they are on the uptake on these matters may have a noticeable impact on their medium-and long-term performance.

5.3 Outcomes Within Companies

Together with the observations and deductions presented here, next steps have been planned at the participating companies to further examine and possibly alleviate the presented causes for an underdeveloped adaptation of design processes to PSS design paradigms.

Within the ongoing project at "Levor," ProVa (Provider Value Evaluation [32]) and Actors and System Maps [9] will be evaluated for their feasibility and usability as learning tools to enhance the general understanding of current processes and possible adaptations needed. Further, a possible adoption of the methods through the company will be assessed. These efforts will focus particularly on design for remanufacturing, which may likely lead to improvements for the PSS offerings in general.

At "Navitas," the evaluation of the PSS design method with support from researchers was concluded without management deciding upon an immediate implementation plan. However, the need for adaptation of processes was understood – as a result, a new position for evaluating and further pursuing the integration between product and service as well as the development of integrated offerings has been filled.

6 Concluding Discussions and Further Research

The cases of the companies "Levor" and "Navitas" on first sight appear as curious anomalies. Seemingly without substantial adjustments to design processes, which, according to literature [5–7, 13], are seen as critical for the success of PSS offerings, they have managed to become successful in offering PSS that deliver high customer value. Based on experience and information gathered during projects with the two companies, this chapter has discussed this paradox: Companies start offering PSS without a substantial adjustment of their design processes – and nonetheless they are successful. Can we therefore cease all research on PSS design and continue with "business as usual"? Based on the results of the research presented, the answer is a firm *no*.

Departing from an understanding of two companies' design processes and internal relations when developing products and PSS, possible causes for the absence of an adaptation of existing design processes to PSS offerings were identified. Subsequently, potential results of the non-adaptation of design processes were discussed and viable solutions to the respective challenges examined.

Although the causes for non-adaptation of processes/non-adoption of PSS design methods may be individual and dependent on the circumstances within the respective company, generalizations were possible based on the literature identified and introduced throughout the chapter. Further, the solutions suggested can be useful even if the particular causes are not found within a company or if the identification stage is simply skipped due to a lack of an external audit (e.g., through researchers, consultants). It must be assumed that this outsider's perspective and neutral position toward the case companies have been helpful in identifying the challenges presented.

Overall, particular focus should be brought to the environment- and efficiency-related issues identified. Currently, these may serve as added value, as the companies and their success appear to rely on their technological aptitude and superiority in the marketplace. However, as resources are becoming scarcer, maximum efficiency will likely transition from being an *added benefit* or provider value to becoming a *necessity* to remain on the market. Taking steps toward attaining this goal by developing and implementing design processes that integrate products and services, have a true life cycle perspective, and are concentrated on effectiveness and efficiency *now*, out of a position of strength and success, appears much more promising than a forced transition out of pure economic necessity *tomorrow*, when the room for error is much diminished.

As indicated in Sect. 5.3 in the coming months, the usefulness of the proposed methods to identify, communicate, and address the detected challenges in adapting existing design processes to PSS will be evaluated in industrial practice. Further, their usefulness in conveying hitherto unutilized PSS benefits within companies' offerings will be assessed and presented in a future publication.

Acknowledgments I wish to extend my gratitude to the many practitioners who participated in the projects and contributed to this research with their knowledge, time, and dedication. I further wish to thank VINNOVA (particularly its research program "Strategic Innovation") as well the anonymous company "Navitas" for funding parts of this research.

References

1. OECD (2015) OECD green growth studies: material resources, productivity and the environment. OECD Publishing, Paris
2. Goedkoop MJ, Van Halen CJG, Te Riele HRM, Rommens PJM (1999) Product service systems, ecological and economic basics. Econ Aff 36:132
3. Mont O (2002) Clarifying the concept of product–service system. J Clean Prod 10:237–245
4. Tukker A (2015) Product services for a resource-efficient and circular economy – a review. J Clean Prod 97:76–91
5. Vasantha GVA, Roy R, Lelah A, Brissaud D (2012) A review of product-service systems design methodologies. J Eng Des 23:635–659

6. Meier H, Roy R, Seliger G (2010) Industrial product-service systems-IPS2. CIRP Ann Manuf Technol 59:607–627
7. Lindahl M, Sundin E, Rönnbäck A et al (2006) Integrated product and service engineering–the IPSE project. Perspect Radic Chang Sustain Consum Prod, pp 315–324. http://citeseerx.ist.psu.edu/viewdoc/download?doi=10.1.1.124.7387&rep=rep1&type=pdf#page=315
8. Matschewsky J, Lindahl M, Sakao T (2015) Facilitating industrial adoption of design methods for product-service systems. In: ICED15 20th international conference on engineering design, Milan, pp 301–310
9. Lindahl M, Sakao T, Carlsson E (2014) Actors and system maps for integrated product service offerings – practical experience from two companies. Procedia CIRP 16:320–325
10. Yin RK (2013) Case study research: design and methods. SAGE Publications, Thousand Oaks
11. Kimita K, Shimomura Y, Arai T (2009) Evaluation of customer satisfaction for PSS design. J Manuf Technol Manag 20:654–673
12. Geng X, Chu X (2012) A new importance–performance analysis approach for customer satisfaction evaluation supporting PSS design. Expert Syst Appl 39:1492–1502
13. Sakao T (2011) What is PSS design? – Explained with two industrial cases. Procedia Soc Behav Sci 25:403–407
14. Sundin E, Lindahl M, Ijomah W (2009) Product design for product/service systems. J Manuf Technol Manag 20:723–753
15. Lindahl M, Sundin E, Sakao T (2014) Environmental and economic benefits of integrated product service offerings quantified with real business cases. J Clean Prod 64:288–296
16. Pezzotta G, Pinto R, Pirola F, Ouertani M-Z (2014) Balancing product-service provider's performance and customer's value: the SErvice Engineering Methodology (SEEM). Procedia CIRP 16:50–55
17. Aurich JC, Fuchs C, Wagenknecht C (2006) Life cycle oriented design of technical product-service systems. J Clean Prod 14:1480–1494
18. Baines TS, Lightfoot HW, Evans S et al (2007) State-of-the-art in product-service systems. Proc Inst Mech Eng B J Eng Manuf 221:1543–1552
19. Womack J, Jones D (2010) Lean thinking: banish waste and create wealth in your corporation. Simon and Schuster, New York
20. Pahl G, Beitz W, Feldhusen J, Grote K-H (2007) Engineering design: a systematic approach. Springer, London
21. Vezzoli C, Ceschin F, Diehl JC, Kohtala C (2015) New design challenges to widely implement "Sustainable Product-Service Systems". J Clean Prod 97:1–12
22. Dankbaar B (1997) Lean production: denial, confirmation or extension of sociotechnical systems design? Hum Relat 50:567–583. doi:10.1177/001872679705000505
23. Sakao T, Mizuyama H (2014) Understanding of a product/service system design: a holistic approach to support design for remanufacturing. J Remanufacturing 4:1
24. Sundin E, Ostlin J, Rönnbäck A et al (2008) Remanufacturing of products used in product service system offerings. In: Mitsuishi M, Ueda K, Kimura F (eds) Manuf. Syst. Technol. New Front. SE – 110. Springer, London, pp 537–542
25. Pigosso DCA, Zanette ET, Filho AG et al (2010) Ecodesign methods focused on remanufacturing. J Clean Prod 18:21–31
26. Lindkvist L (2014) Exploring product life-cycle information flows with a focus on remanufacturing. Linköping University Electronic Press, Linköping
27. Sakao T, Birkhofer H, Panshef V, Dörsam E (2009) An effective and efficient method to design services: empirical study for services by an investment-machine manufacturer. Int J Int Manuf Serv 2:95–110
28. Windahl C (2007) Integrated solutions in the capital goods sector: exploring innovation, service and network perspectives. Department of Mangement and Engineering, Linköping university, Linköping

29. Manzini E (1999) Strategic design for sustainability: towards a new mix of products and services. In: Proceedings first international symposium environmentally conscious design inverse manufacturing IEEE, Tokyo, pp 434–437
30. Markeset T, Kumar U (2005) Product support strategy: conventional versus functional products. J Qual Maint Eng 11:53–67
31. González FJM, Palacios TMB (2002) The effect of new product development techniques on new product success in Spanish firms. Ind Mark Manag 31:261–271
32. Matschewsky J, Sakao T, Lindahl M (2015) ProVa – provider value evaluation for integrated product service offerings. Procedia CIRP 30:305–310
33. Akao Y (2004) Quality function deployment: integrating customer requirements into product design. Productivity Press, New York
34. Sakao T, Lindahl M (2015) A method to improve integrated product service offerings based on life cycle costing. CIRP Ann - Manuf Technol 64:33–36.
35. Beskow C, Johansson J, Norell M (1998) Implementation of QFD: identifying success factors. In: IEMC'98 Proceedings. int. conf. eng. technol. manag. Pioneer. new technol manag issues challenges third millenn. (Cat. No.98CH36266). IEEE, pp 179–184
36. Umeda Y, Takata S, Kimura F et al (2012) Toward integrated product and process life cycle planning—an environmental perspective. CIRP Ann Manuf Technol 61:681–702
37. Arai T, Shimomura Y (2004) Proposal of service CAD system – a tool for service engineering. CIRP Ann Manuf Technol 53:397–400
38. Komoto H, Tomiyama T (2008) Integration of a service CAD and a life cycle simulator. CIRP Ann Manuf Technol 57:9–12
39. Bertoni A, Bertoni M, Isaksson O (2013) Value visualization in product service systems preliminary design. J Clean Prod 53:103–117

A Method of Selecting Customer-Oriented Service and Delivery Modes in Designing Environmentally Benign Product Service Systems

Yutaka Dairokuno, Juri Matsumura, and Shozo Takata

Abstract The increasing seriousness of environmental problems is forcing manufacturing companies to make their businesses sustainable. For improving environmentally friendliness of their business, therefore, manufacturing companies pay attention to product-service system (PSS), because it can increase the effectiveness of life cycle options, which are means of reducing the environmental load. We propose a design method for environmentally benign PSS business, in which the provider offers products through a combination of services, delivery modes, and life cycle options. Such business can meet the different customer needs while enabling the implementation of delivery modes and life cycle options requiring a group of customers. The method is verified via application to managed document services.

Keywords Environmentally benign business • Design method • PSS • Delivery mode • Life cycle option

1 Introduction

The increasing seriousness of environmental problems is forcing manufacturing companies to make their businesses sustainable operations that minimize environmental impacts and resource consumption while providing necessary functions to customers. This is why environmentally benign businesses based on the product-service system (PSS) have gained attention recently. Manufacturing companies find it difficult to employ the life cycle options that could reduce the environmental load during product life cycles when they simply sell to customers [1]. However, they can maintain controllability of the products not only in the beginning of the life cycle (BoL) but also in the middle of the life cycle (MoL) and end of the life cycle (EoL) phases by, for example, changing their delivery mode from selling to leasing

Y. Dairokuno • J. Matsumura (✉) • S. Takata
Department of Business Design and Management, Waseda University, Tokyo, Japan
e-mail: dairokuno@ruri.waseda.jp

M. Matsumoto et al. (eds.), *Sustainability Through Innovation in Product Life Cycle Design*, EcoProduction, DOI 10.1007/978-981-10-0471-1_17

or adding a maintenance service during the product's usage phase. This would enable companies to implement various life cycle options.

This study defines an environmentally benign PSS business (ePSS business) as any business that provides products and the services designed to reduce the environmental load through their life cycle options and delivery modes. In such businesses, it is very important to offer the customer the proper combination of product and services along with the life cycle options and delivery modes best suited to the individual customer needs. The provider must also have the group of customers required to implement life cycle options and delivery modes such as sharing and batch processing. This requires a method of selecting the appropriate combination of product and services as well as the life cycle options and the delivery mode suited to customer needs. In addition, a method is needed for grouping the similar customers to enable the implementation of certain delivery modes and life cycle options.

A number of studies have been conducted on PSS design methods [2]. Although environmental friendliness is regarded as inherent to PSS, not many works have explicitly discussed how to improve it. Moreover, most works deal with PSS evaluation methods rather than PSS design procedures [3, 4]. As a result, there are a few works on design procedure of ePSS [5, 6]. Nakamura et al. proposed 18 rules for designing eco-businesses, which include eight rules for the promotion of eco-consciousness, based on their investigation of 28 eco-businesses. Maxwell et al. proposed a checklist for improving the sustainability of product and services.

However, few works (except for our previous work [7]) have explicitly discussed the relationships among the components of ePSS (i.e., services, life cycle options, and delivery mode) and their selection according to their relationships. We analyzed how services and delivery modes affect the effectiveness of life cycle options and proposed a method of composing ePSS by combining services, life cycle options, and delivery mode so as to minimize the environmental load and costs to customers. However, we have not examined the differences among individual customer needs. Customers' willingness to buy a product and service depends on the consistency between the customers' needs and the functions provided to them. Therefore, we need to design ePSS in a way that considers customer needs and the constraints on the number of customers needed to implement specific delivery modes.

We propose a design method for the ePSS business. We assume that the provider offers a certain product to a predefined community or company, each with different needs. The provider tries to integrate the appropriate services and life cycle options and provide them via the appropriate delivery mode in order to satisfy customer needs and reduce the environmental load. We define an ePSS type as any combination of services, life cycle options, and the delivery mode. The provider conducts an ePSS business by adopting a set of ePSS types to meet the different customer needs while enabling implementation of delivery modes and life cycle options to a group of customers that will accept this ePSS type. We apply the proposed method to managed document services (MDS) in order to demonstrate its effectiveness.

A managed document service is a PSS business that provides document management functions to customers using copiers and related services.

The rest of this paper is organized as follows. Section 2 explains the framework of our design method. Section 3 explains the design procedure. In Sect. 4, the proposed method is applied to the MDS to demonstrate its effectiveness. Section 5 concludes the paper.

2 Relationship Between ePSS Type and Customer

Our aim is to develop a method of designing an ePSS business consisting of the optimum combination of ePSS types to minimize the environmental load while maximizing customer value.

Before discussing the framework for designing the ePSS business, we provide below the definitions of "services," "delivery mode," and "life cycle options" used in this study.

Services are classified into action services, information services, and financial services. Because financial services do not have a direct relationship with environmental load, we do not consider them in this study. Services influence both functions and costs. For example, when a maintenance service is provided, the quality of printed material (a function of printers) will be maintained; however, this service increases costs, which the customers must pay.

Delivery modes are ways of providing functions to customers. Table 1 lists the items characterizing delivery mode proposed by Matsumoto [7]. A delivery mode can be defined through the selection of one of the options for each item. Typical examples of delivery mode are shown in Table 2.

Life cycle options are ways of reducing the environmental load [8]. They are classified in terms of the life cycle phase in which they are applied (i.e., BoL, MoL, and EoL). Table 3 shows the life cycle options considered in this study and the intended reductions for reducing the environmental load.

Because the effects of the life cycle options in the BoL phase are not directly related to services or delivery modes, these options (such as design change) are not considered in this study. In addition to the life cycle options in the EoL phase (such as reuse and recycling), which are widely recognized as major life cycle options, we deal with those in the MoL phase (such as proper use, increasing operation rates, and batch processes). Proper use reduces the consumption of energy and consumables by operating under certain conditions and installation environments. Increasing operation rates reduces the number of products needed for operations by processing multiple needs through time sharing. Batch processing reduces the required operational time by consolidating several small batches into single ones to lower usage frequencies.

Figure 1 shows the relationship between ePSS types and customers. The provider offers a set of functions to the customers through the product and service. The utility the customer obtains from these functions depends on the customer's

Table 1 Items characterizing delivery mode

Items of delivery mode	Product delivery mode						
	Manage			Operation			
	Owner	Contract period	Charging	Operator	Installation location	Usage mode	Customer
Possible options	Customer	Unlimited	Bullet payment	Specific customer	Installation location	Exclusive use	Specific customer
	Provider	Specific	Specific payment	Specific group of customer	Customer	Time sharing	Specific group of customer
			Installment payment	Nonspecific customer	Provider	Pooling	Nonspecific customers
			Pay per	Provider	Third party		

Table 2 Typical examples of delivery mode

Delivery mode type	Service type	Owner	Contract period	Charging	Operator	Installation location	Usage mode	Customer
Selling product	–	Customer	Unlimited	Bullet payment	Specific	Customer	Exclusive	Specific
Sharing	–	Provider	Specific	Specific payment	Specific group	Third party	Time sharing	Specific group
Rental	–	Provider	Unlimited	Pay per	Nonspecific group	Customer	Exclusive	Nonspecific group
Performance service	Action	Provider	Specific	Specific payment	provider	Provider	Exclusive	Specific

Table 3 Life cycle options considered in this study [9]

Life cycle stages	Life cycle options			Intended reduction
Middle of life cycle	Reduce	Proper use	Operate under proper conditions and environment	Energy and consumable
		Maintenance	Apply proper maintenance at the right time	Virgin material
		Increasing operation rates	Process multiple needs through time sharing	Products needed for operation
		Batch processing	Consolidate small batches into single ones to lower usage frequencies	Required operational time
End of life cycle	Reuse	Product reuse	Collect and treat properly	Virgin material
		Parts reuse	Collect and treat properly	Virgin material
	Recycling	Recycling	Collect and treat properly	Virgin material

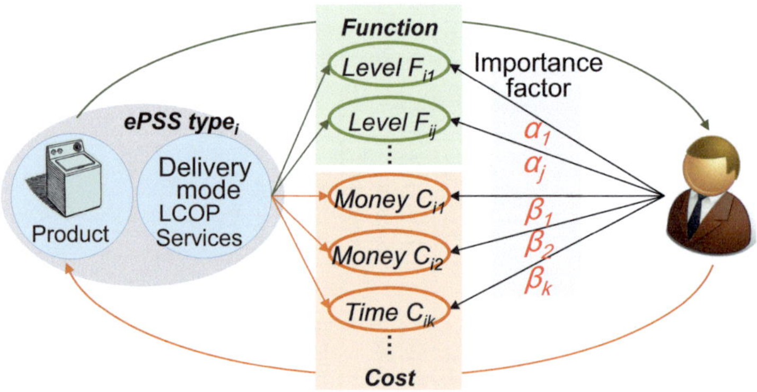

Fig. 1 Relationship between ePSS type and customer

preferences among them. In exchange for gaining the utility, the customer bears the burden produced by a set of costs the customer must pay to receive the functions. The term "cost" is used here in a broad sense, involving not only monetary costs but also costs in terms of time, effort, and environmental load. The customer's burden depends on the customers' willingness to pay. We represent the difference between the customers' preferences and willingness to pay in terms of the importance factors α_j and β_k, as indicated in Fig. 1.

The levels of the functions and costs are determined not only by the product and services but also by the delivery mode and the life cycle options. For example, a sharing delivery mode can reduce the cost per use, while a proper use life cycle option can enhance the quality of printed material.

We assume that the customer value V_i obtained by a specific ePSS type is determined by the ratio of the weighted sum of the function F_{ij} to the cost C_{ik}, as expressed in Eq. 1, where i, j, and k denote the index of the ePSS types, the function items, and cost items, respectively, and where m and n are the number of function items and cost items.

$$V_i = \sum_{j=1}^{n} F_{ij}\alpha_j / \sum_{k=1}^{m} C_{ik}\beta_k \tag{1}$$

In the first step of our proposed method, the best ePSS type, which maximizes each customer's value, is identified based on Eq. 1. Then, the customers in the target market (the given community or company) are segmented into groups according to ePSS types in order to construct the best ePSS business in terms of cost and environmental load.

3 Proposed Design Procedure

3.1 Outline of the Design Procedure

Figure 2 shows the outline of the proposed design procedure, consisting of two stages. In the first, the ePSS types are prioritized for each customer according to his or her needs. In the second, the best combination of ePSS types, configuring the ePSS business, is selected for the target community or company.

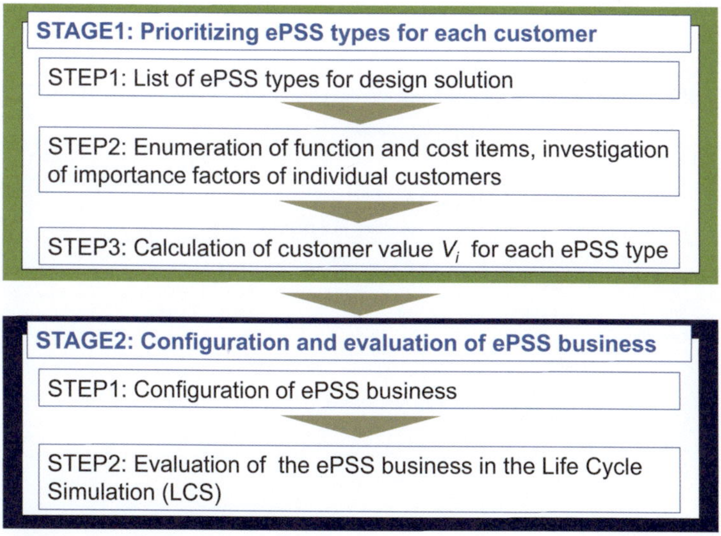

Fig. 2 Outline of ePSS design procedure

3.2 *Prioritizing ePSS Types for Each Customer*

The first stage of the design procedure consists of three steps. In the first step, the effective ePSS types are listed as candidates for the ePSS type to be provided to each customer. To create this list, effective life cycle options are first determined based on the result of the product's life cycle assessment (LCA). Then, we select the services and delivery modes most effective in increasing the effects of the selected life cycle options. For this purpose, we use the correspondence tables shown in Tables 4 and 5, which represent how each service or delivery mode could increase the effects of the life cycle options. The symbol "++" in the *i*, *j* cell of the tables indicates that the adoption of the *i*-th service or delivery mode guarantees a 100 % execution of the *j*-th life cycle option. The symbol "+" indicates that the *i*-th service or delivery mode could increase the execution rate of the *j*-th life cycle option to a certain extent. The exact figures of the execution rate increment for the life cycle options should be investigated via a customer questionnaire survey concerning each product.

Table 4 Effects of services on execution rate of life cycle options [7]

		Life cycle options						
Service type	Life cycle stages	Proper use	Batch process	Maintenance	Product reuse	Increasing operation rates	Product and Parts reuse	Recycling
Action	Usage	++	++		++	++		
	Collection						++	++
	Maintenance			++				
Information	Usage	+	+		+	+		

Table 5 Effects of changing delivery mode on execution rate of life cycle options [7]

				Life cycle options						
Delivery mode	Possible options			Proper use	Batch process	Maintenance	Product reuse	Increasing operation rates	Product and parts reuse	Recycling
Owner	Customer	→	Provider			++			++	++
Period	Unlimited	→	Specific						+	
Operator	Specific customer	→	Provider	++	++		++	++		
		→	Specific group	–			+	+		
		→	Nonspecific	–	–		+			
Installation location	Customer	→	Third party	–						
	Provider	→								
Usage mode	Exclusive	→	Time sharing					+		
	Exclusive	→	Pooling		+			+		

After the effective life cycle options, services, and delivery modes for the given product are selected, the candidates of the ePSS type to be provided to the customer are configured by combining them.

In the second step, the functions and cost items are enumerated according to the product features. Then, the customer's importance factors for the functions and cost items are determined via the pair comparison method [9].

In the third step, the customer value V_i of i-th ePSS type is calculated by Equation (1), and the ePSS types are prioritized in order of the customer values.

3.3 Configuration of ePSS Business

The second stage of the design procedure consists of two steps. In the first, the potential combinations of ePSS types, configuring the ePSS businesses, are identified considering the constraints on the number of customers per ePSS type required to implement the life cycle option and delivery mode. In the second step, the potential ePSS businesses are evaluated through a life cycle simulation, and the best ePSS business is selected.

In the first step, the candidates of the ePSS businesses are configured by selecting a certain number of ePSS types and allocating each customer to them. In the second step, we select the best ePSS business—which minimizes the object function represented in Eq. 2—where P_i and E_i denote the provider's cost and the environmental load of i-th ePSS type. In the optimization, we consider two constraints. First, the value of each customer V_i must be greater than that in the case of the delivery mode of "selling product" as expressed by Eq. 3. Second, the number of customers allocated to each ePSS type, sizec_i, must be greater than or equal to that of the bare minimum L_i, which is required to implement the ePSS type, as expressed by Eq. 4.

$$\min S = \sum\nolimits_{n=1}^{n} P_i E_i \tag{2}$$

$$V_i \geq V_{\mathrm{selling}} \tag{3}$$

$$\mathrm{sizec}_i \geq L_i \tag{4}$$

To evaluate the ePSS businesses in terms of environmental load and life cycle cost, we use the life cycle simulation (LCS), which we developed based on the system proposed by Komoto [10]. The LCS is executed by generating the events that occur during the product life cycle and evaluating the results induced by them. The events comprise "product use," "maintenance," "failure," and "end of use," in which such events as "collection," "reuse," "recycling," "illegal dumping," and "new product installation" can occur.

Product use is triggered by customer demands. It increases a product's cumulative operating time and then increases the failure rate according to the product's failure distribution. In addition, "product use" also induces the consumption of energy and consumables, influencing the monetary costs and environmental loads involved. The life cycle options related to this event are "proper use," "increasing operation rates," and "batch processing." These options can reduce resource and energy consumption.

Failure occurs according to the failure distribution function of each product and the cumulative operating time at each time point. In this system, the failure distribution function is represented by a Weibull distribution.

The effect of maintenance is represented as the reduction of the cumulative operating time. Whether proper maintenance is applied depends on the product owner and the adoption of maintenance service. For example, when the product owner is the provider, the recovery rate is higher than when the owner is the customer because the provider has enough knowledge to conduct the proper maintenance.

End-of-use is triggered by failure or obsolescence. End-of-use types are divided into collection or illegal dumping. The ratio of the two is determined by the collection rate, which depends on the owner, collector, and collection location, which are in turn determined by the delivery mode and the adoption of a collection service.

After collection, the products are checked to determine if reuse or recycling is possible according to the remaining life. If the remaining lives of all product parts are greater than the guaranteed life, the product is judged to be reusable. Even if the product as a whole is not reusable, a part is reusable if its remaining life is greater than the guaranteed life. An item's reuse rate is influenced by its owner. Reuse helps reduce the number of newly manufactured products and parts.

4 Application Example

4.1 Case Study Scenario

We applied the proposed method to MDS. We assume that the provider has a blanket contract with a company to execute an ePSS business. We also assume that the company has 30 offices, each with 30 employees. We regard each office as one customer with individual needs and assume that every employee in the same office has the same needs. As mentioned, we assume that one type of product is provided to the customers with different combination of services. The features of the product are shown in Table 6. The parameters used in the LCS are set based on these features. We also refer to data in [11] and to company websites [12, 13] to estimate life cycle costs (LCC) and to execute LCA during the LCS.

Table 6 Product feature, LCC, and LCA data

Product feature	Value
Copy speed (/sheet)($\times 10^{-3}$)[b]	0.83
Electricity consumed in operation (KW)[b]	1.50
Electricity consumed in waiting (KW)[b]	0.11
Electricity consumed in mode of electric power saving (KW)[b]	0.08
Cost	Rate
Retail price (/product)[a]	1
Manufacture cost (/product)[a]	0.16
Payment of print sheet in specific payment (/sheet)[a]	0.64×10^{-5}
Payment of print sheet in bullet payment (/sheet)[a]	0.19×10^{-5}
Electricity rate (work)(/hour)[a]	0.022
Electricity rate (wait)(/hour)[a]	0.0016
Recycle cost (/product)[a]	0.05
Action service for usage (/sheet)[b]	0.21×10^{-4}
Action service for collection (/frequency)[b]	0.0091
Action service for maintenance (/frequency)[b]	0.18
Information service for usage (/month $\times$ product)[b]	0.0091
Exchanging expendables (/frequency)[b]	0.068
CO_2 coefficients (kg-CO_2)	
Manufacture (/product)[c]	435.33
Product delivery (/product)[c]	27.96
Electricity (/KWh)[c]	0.41
EOL (/product)[c]	0.96

[a]Based on [11]
[b]Based on [12]
[c]Based on [13]

4.2 *Listing of ePSS Types*

We first select the life cycle options, services, and delivery modes that could be used to compose the ePSS types applicable to this example. Figure 3 shows the relative proportions of CO_2 emissions for copiers at each life cycle phase [13]. The figure shows that the environmental load for manufacturing and usage are dominant. We thus select the life cycle options "proper use," "increasing operation rates," and "maintenance" in the usage phase and "product reuse" and "parts reuse" in the manufacturing phase as effective options for reducing the environmental load. Since copiers do not have the features that would allow us to consolidate small batches into one, the "batch processing" life cycle option cannot be taken.

Then, the services that can facilitate the effectiveness of the life cycle options are identified on the basis of Table 4: collection service for product and parts reuse, maintenance service and service for exchanging expendables for executing proper maintenance, operation service for proper use, and advising service for proper

Fig. 3 Composition ratio of CO2 emissions in each life cycle phase [13]

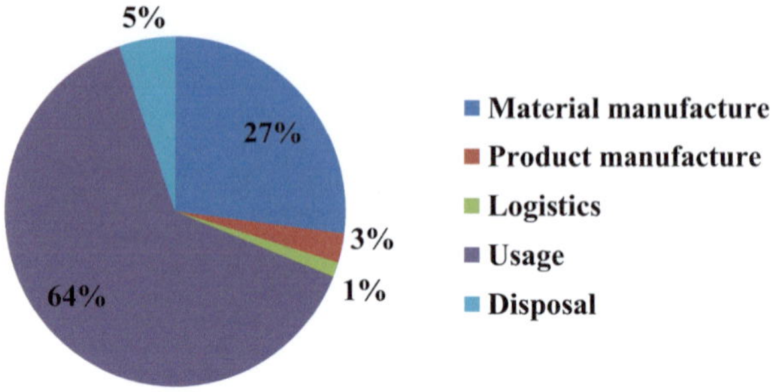

operation. Finally, the delivery modes that could facilitate the selected life cycle options are chosen on the basis of the correspondence table (Table 5).

In listing the possible ePSS types, not all combinations of the selected life cycle options, services, and delivery modes are implementable. We need to select the feasible combinations. For example, if "proper use" is selected as the life cycle option, operation service or advising service should be selected as added services. In addition, the operator in delivery mode should be the "provider" in the former case and the "customer" in the latter case. We listed the 96 ePSS types applicable to this example.

4.3 *Estimation of Customers' Importance Factors*

In this example, we focused on print quality as the product function and listed three function items: "reproducibility," "alignment," and "clarity." We also listed ten cost items, which are divided into monetary cost, time, effort, and environmental load. Table 7 shows the effects of changes in ePSS components (i.e., services, life cycle options, and delivery modes) on the levels of functions and costs. The symbols "+" and "−"indicate whether the employment of the services, life cycle options, and delivery modes increase functionality and costs relative to the simple sales of the products. For example, employing the operation service reduces the time spent on copying but increases the customers' cost.

The methods of estimating the levels of the functions and cost items are shown in Table 8. We normalize each item level by its average value when calculating the customer values. To estimate the importance factors, we conduct a questionnaire survey on 30 customers via the pair comparison method. We assume that each respondent expresses the customer needs in each office

Table 7 Effects of changes in ePSS components on the level of functions and costs

Items of function and cost			Monetary			Time			Effort			Environment	Print quality		
Changing items of ePSS			Buying expenses	Operational cost	Repair cost	Printing time	Replacing consumable time	Repair time	Printing effort	Replacing consumable effort	Repair effort	Environmental load	Reproducibility	Alignment	Clarity
Product delivery mode	Owner	Provider	–									–	+		
	Contract period	Specific	–												
	Charging	Specific payment	–	–	–										
	Customer	Specific group	–												
		Nonspecific	–									+			
	Operator	Provider		+		–	–		–	–		–		+	+
		Specific group		+								–		–	–
		Nonspecific		+								–		–	–
	Installation location	Provider				+			+						
		Third party		+		+			+			+			
	Usage mode	Time sharing		–								–			
		Pooling		–								–			
Life cycle options	MoL	Proper use		–		–	–		–	–		–		+	+
		Maintenance			+			+		–	–	–	+		
		Increasing operation rates	–	–								–			
		Batch processing	–	–								–			
	EoL	Product reuse	–									–			
		Part reuse			–							–			
		Recycling										–			

(continued)

Table 7 (continued)

Items of function and cost			Monetary			Time			Effort			Environment	Print quality		
Changing items of ePSS			Buying expenses	Operational cost	Repair cost	Printing time	Replacing consumable time	Repair time	Printing effort	Replacing consumable effort	Repair effort	Environmental load	Reproducibility	Alignment	Clarity
Services	Action	Operation		+		−			−			−		+	+
		Collecting										−			
		Maintenance			+		−	−		−	−	−	+		
	Information	Advising		+		−			−			−		+	+

Table 8 Estimation method of the levels of function and cost items

The items of function and cost	Measuring method
Buying expenses	Number of manufactured product × manufacturing cost
Operational cost	Payment of print sheet + electricity expense + service fee
Repair cost	Number of maintenance × labor cost + parts cost
Printing time	Copy speed × number of print
Replacing consumable time	Replacing time × frequency
Repair time	Repair time × frequency
Printing effort	Weight of print sheet × number of print × distance for copier × frequency
Replacing consumable effort	Weight of print sheet × number of print × distance for copier × frequency
Environmental load	LCA
Reproducibility	Number of print sheet/replacement frequency of module
Alignment	Number of misprint × 0.6 × execution of proper rate
Clarity	Number of misprint × 0.4 × execution of proper rate

4.4 Calculation of Customer Value V_i

We calculate the individual customer value V_i to identify the ranking of the ePSS types. Table 9 shows examples of the top three ePSS types in terms of the individual customer values. We assume that each customer is at least somewhat satisfied with any of the top three ePSS types because all customer values for these ePSS types are greater than those when the delivery mode is "selling product" for every customer. In this example, there are 15 ePSS types included in the top three lists of the 30 customers. Table 10 offers detail on the 15 ePSS types, with No. 0 representing "selling product." These 15 ePSS types are used to configure the ePSS business in stage 2.

4.5 Configuring Candidates for ePSS Businesses

Using the 15 ePSS types selected in the previous section, we configure the candidates for the ePSS businesses. The number of ePSS types used in one ePSS business is limited to three to avoid combinatorial explosion. This method generates eight candidates. These ePSS businesses satisfy the constraint in terms of the necessary number of customers allocated to each ePSS type.

Table 9 The ranking of ePSS types for each customer

		First		Second		Third		Selling products
		ePSS type	V_i	ePSS type	V_i	ePSS type	V_i	V_i
Customers	A	85	1.96	89	1.86	87	1.86	0.77
	B	85	3.36	63	3.00	7	2.52	0.80
	C	85	1.92	89	1.76	65	1.73	0.83
	D	48	3.70	25	3.58	71	3.53	0.80
	E	86	1.95	25	1.93	66	1.93	0.96
	F	90	2.35	66	2.35	86	2.31	1.05
	G	85	3.15	90	3.10	66	3.10	0.97
	H	84	4.52	86	4.41	85	4.40	1.52
	I	90	1.94	66	1.94	86	1.89	1.00
	J	25	4.77	71	4.71	95	4.71	0.87
	...	...	...	...	...	...	...	...

4.6 Evaluation of the Number of ePSS Businesses in the LCS

We conduct an LCS for each candidate for the ePSS business for a period of 10 years with a simulation cycle of 1 week. We assume that the number of prints is 880 per employee for a period of 1 week [14]. The "proper use" option can reduce printing by 25 % by using two-sided printing. The execution rate of "proper use" is 0.464 for the "selling product" delivery mode [7]. Based on a survey on human error [15], we assume an 11.3 % misprint rate. The "proper use" option can reduce misprinting and printing time because of the reduction in the number of prints.

We compare the results of the simulation for the eight ePSS businesses in Table 11. Figure 4 shows the total CO_2 emissions and the LCC for these eight ePSS businesses, along with the result of the "selling product" delivery mode case, shown in the leftmost bar chart in the figure.

4.7 Discussion

The best ePSS business, which minimizes the object function, is the combination of the No. 66, 71, and 85 ePSS types. The No. 66 type is "time-sharing use" of the copier in a print room without installing copiers in their offices (which are close to a print room). The No. 71 type is also time-sharing use, but the provider is the one providing the operation service. The necessary number of copiers is installed in a document center, which processes all the printing for the customers. The No. 85 type provides copiers for exclusive use in individual customers' offices at specific periods. This ePSS type is similar to what has been widely adopted in

Table 10 15 ePSS types for ePSS business

	Product delivery mode							Action services				Information service
ePSS type	Owner	Contract	Charging	Customer	Operator	Installation location	Usage mode					
0	Customer	Unlimited	Bullet	Specific customer			Exclusive					
7	Customer	Unlimited	Installment	Specific customer		Provider	Exclusive					
48	Provider	Specific	Bullet	Specific customer	Provider		Exclusive	Operation	Maintenance	Collection		
95	Provider	Specific	Bullet	Specific group	Provider	Customer	Time sharing				Exchanging expendables	
82	Provider	Specific	Installment	Specific customer		Third party	Exclusive					
25	Customer	Unlimited	Bullet	Specific group	Provider		Time sharing	Operation				
71	Customer	Unlimited	Bullet	Specific group	Provider	Customer	Time sharing					
84	Provider	Specific	Bullet	Specific customer			Exclusive		Maintenance	Collection		Advising
85	Provider	Specific	Installment	Specific customer			Exclusive					
86	Provider	Specific	Bullet	Specific group		Customer	Time sharing					
87	Provider	Specific	Installment	Specific group		Customer	Time sharing					
89	Provider	Specific	Installment	Specific customer		Provider	Exclusive					
90	Provider	Specific	Bullet	Specific group		Provider	Time sharing					
63	Customer	Unlimited	Installment	Specific group		Customer	Time sharing					
65	Customer	Unlimited	Installment	Specific customer		Provider	Exclusive					
66	Customer	Unlimited	Bullet	Specific group		Provider	Time sharing					

Table 11 Eight ePSS business

	ePSS business								
	0	1	2	3	4	5	6	7	8
Combination of ePSS types	0	25	66	66	25	71	85	25	71
		66	71	85	85	85	86	85	85
		85	85	95	86	86	95	90	90

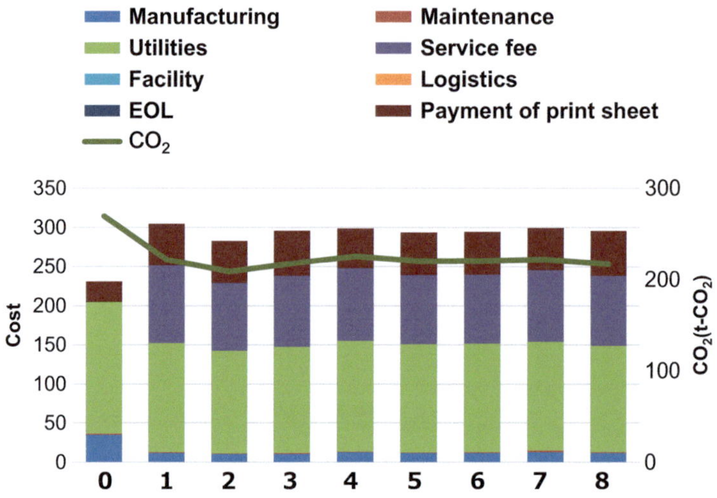

Fig. 4 Total CO2 emissions and cost

offices. The breakdown of the number of customers for each ePSS type is eight customers for No. 66, eight for No. 71, and 13 for No. 85.

In all eight ePSS businesses evaluated in this case study, the monetary cost is greater than it is in the "selling product" delivery mode case because the customers are predisposed to consider costs as less important. Thus, the services that reduce time, effort, and environmental load, such as operation service, are selected to increase customer value.

5 Summary

The increasing seriousness of environmental problems is forcing manufacturing companies to make their businesses sustainable operations that minimize environmental impacts and resource consumption while providing socially necessary functions. We propose a design method for ePSS business in which the provider offers a product using the appropriate combination of services, delivery mode, and life cycle options to a community or company containing customers with different needs. We propose adopting a set of ePSS types to meet the different customers' needs while enabling the implementation of delivery modes and life cycle options that require a group of customers to accept the ePSS type being used. The proposed method is applied to MDS to demonstrate its effectiveness.

References

1. Mont O (2002) Clarifying the concept of product-service system. J Clean Prod 10(3):237–245
2. Tukker A (2015) Product services for a resource-efficient and circular economy – a review. J Clean Prod 97:76–91
3. Tran T, Park JY (2015) Development of a strategic prototyping framework for product service systems using co-creation approach. In: 7th CIRP industrial product-service systems conference, Saint-Etienne, vol 30, 1–6
4. Kimita K, Shimomura Y, Arai T (2009) Evaluation of customer satisfaction for PSS design. J Manuf Technol Manag 20(5):654–673
5. Nakamura N, Mandai K, Fukushige S, Umeda Y (2012) Proposal of a methodology for supporting generation of new eco-business ideas. Des Innov Value Sustain Soc 78(2):136–142
6. Maxwell D, Van der Vorst R (2003) Developing sustainable products and services. J Clean Prod 11(8):883–895
7. Matsumoto E, Ohtake J, Takata S (2012) A method for selecting delivery modes in environmentally benign product service system design. In: Proceedings of the 4th CIRP international conference on industrial product-service systems, Tokyo, 375–380
8. Kimura F, Umeda Y, Takahashi S, Tanaka N, Nagata K, Fujimoto J, Matsuhashi R, Mitsuhashi T (2004) Handbook of inverse manufacturing. Maruzen Company, Tokyo (In Japanese)
9. Saaty T (1980) The analytic process. McGraw-Hill, New York
10. Komoto H, Tomiyama T, Nagel M, Sivester S, Brezet H (2005) Life cycle simulation for analyzing product service systems. In: Proceedings of EcoDesign, Tokyo, 2B-1-2F
11. Sakai T, Takata S (2012) Reconfiguration management of remanufactured products for responding to varied user needs. CIRP Ann Manuf Technol 61(1):21–24
12. Fuji Xerox Corporation (2012) DocuCentre-IV C2263 brochure
13. Life Cycle Assessment Society of Japan (JLCA) (2014) LCA Japan forum news. vol 65. LCA Society of Japan, Tokyo (In Japanese)
14. Nikkei Business Publications (2014) NIKKEI BP environment report. Nikkei Business Publications, Inc., Tokyo (In Japanese)
15. Moriyama Professional Engineer Office (2008) Human error and risk assessment. Moriyama professional Engineer office, Kanagawa (In Japanese)

Design for Remanufacturing and Circular Business Models

Sharon Prendeville and Nancy Bocken

Abstract Remanufacturing has significant market potential, and the purported benefits of remanufacturing are compelling. Remanufacturing offers a means to retain control of products and materials throughout the product life cycle, and, therefore, through remanufacturing, businesses can insulate against material price shocks and future material scarcity issues. Remanufacturing has therefore been identified as a 'sleeping giant' whose potential, once tapped, can fast track companies to increased profits, while, in parallel, realising circular practices within industrial systems. However, it is widely accepted that the majority of products that are currently remanufactured have not been designed for remanufacturing and business models to support remanufacturing are complex. In light of these combined issues, this paper presents the story of a business in transition. The core aim of this paper is to build understanding of design for remanufacturing and remanufacturing-oriented business models. The paper reports on the linkages between design and business model strategies by presenting a case of a business-led pilot study conducted to explore the commercial viability of remanufacturing. The results show how life cycle considerations and a combination of design and business strategies can accelerate transition to resource-efficient business models. The paper illuminates the topic of remanufacturing by showcasing a dynamic real-world business case, from which other companies can learn.

Keywords Design for remanufacturing • Ecodesign • Eco-innovation • Sustainable business models • Circular economy

1 Introduction

Remanufacturing presents a significant market opportunity, and the purported benefits of remanufacturing are compelling [1, 2]. In the UK alone, market value of remanufacturing activities is estimated to be worth up to £5.4 billion, excluding

S. Prendeville (✉) • N. Bocken
Product Innovation Management, Industrial Design Engineering, TU Delft, Delft, Landbergstraat 15, 2628 CE, The Netherlands
e-mail: S.M.Prendeville@tudelft.nl

M. Matsumoto et al. (eds.), *Sustainability Through Innovation in Product Life Cycle Design*, EcoProduction, DOI 10.1007/978-981-10-0471-1_18

the economic benefits of jobs created [2]. Remanufacturing offers a means to retain control of products and materials throughout the product life cycle, and, therein, through remanufacturing, businesses can insulate against material price shocks and future material scarcity issues. Remanufacturing has therefore been identified as a ‘sleeping giant’ whose potential, once tapped, can fast track companies to increased profits, while, in parallel, realising circular practices within industrial systems.

The concept of the circular economy has gained traction in industry and policy as a pathway to deliver resource efficiency. While not a new concept, its widespread appeal has never been greater. One driver for this interest in circularity is due to increasing awareness of resource scarcity as a key risk to business competitiveness and the potential impact this may have on businesses in the long term. A 2010 study reported that demand for metals is likely to increase tenfold, putting unimaginable stresses on our natural environment [3]. For example, only a quarter of gold and palladium is currently recovered through waste electrical and electronic equipment (WEEE) recycling centres.

Ecodesign is a strategic approach to designing products to reduce environmental impacts across the whole product life cycle. In doing so, it can identify layers of waste and also layers of value. Ecodesign practices are fundamental to remanufacturing, and, through ecodesign, businesses can deliver products suited to servicing, leasing and hiring options to support remanufacturing. When coupled with innovations in product design, remanufacturing is perceived as a high-value strategy commensurate with circular economy thinking.

To date the topic of design for remanufacturing has been integrated within an overarching ecodesign research agenda [4]. Remanufacturing can be differentiated from other similar approaches (refurbishment/reuse) by focusing on the reprocessing of products to a like new or better than new state.

Barriers to integration of ecodesign in companies have been written about extensively, and the findings of these studies converge [see 5, 6]. For example, cost, lack of resources and knowledge are typical barriers. Some authors propose that cultural aspects (poor management commitment, lack of supply chain integration, supply chain relationships, employee attitudes, company’s moral standing) as well as limited demand-side pull are key factors in determining why companies fail to act. More recently, these issues have been framed from the perspective of ‘barriers to moving to circular economy business models’ [7]. Similarly, while the business benefits to remanufacturing are evident, remanufacturing is challenging for companies to adopt. The reasons for this are diverse and wide-ranging, from legislative quandaries (such as perverse incentives and barriers to market entry) to barriers stemming from limited business capabilities (such as lack of reverse logistics infrastructure and lack of design know-how).

In addition, it is widely accepted that the majority of products that are currently remanufactured have not been designed for remanufacturing. This could be attributed to the lack of coordinated design for remanufacturing research, the deficit of companies actively engaging with design for remanufacturing, as well as the lack of integration of design for remanufacturing processes within companies. It is therefore unsurprising that other authors report a shortfall of rich case studies exploring remanufacturing-oriented business models [4].

In light of these issues, this paper tells the story of a business in transition. The paper discusses the linkages between design and business model strategies by presenting a business-led pilot study, conducted to explore the commercial viability of remanufacturing for an original equipment manufacturer (OEM).

The business case was identified through recent research undertaken by the European Remanufacturing Network (ERN). The ERN project seeks to map the range of remanufacturing businesses operating throughout Europe from a design, processes and business models perspective. Through ERN a number of companies 'in transition' to remanufacturing have been identified, an evidence to suggest that industry increasingly sees remanufacturing as a potential route to offer new value propositions to customers.

2 Literature Review

2.1 Design for Remanufacturing: The State of Play

Well-coordinated product development processes require skill sets of two distinct areas within design: those expert in strategic design and those skilled in detailed product design. This is universal and extends to product development processes centred on remanufacturing. Design for remanufacturing involves 'considering the product strategy (marketing, reverse logistics) and the detail engineering of the product in terms of remanufacture' [8].

The design for remanufacturing literature has almost exclusively focused on the latter of these two areas [see, e.g. 9, 10]. This literature discusses elements such as a product's 'suitability for remanufacturing', and from this good practice, guidelines that determine a product's suitability for remanufacturing have been developed. For example, automatic transmissions are identified, being high-value products, with a steady pace of technology evolution. Despite this area of the research having the most focus, Hatcher et al. [4] state that 'Design for X' research areas (discussed further in section 'Design Guidelines for Remanufacturing') must take more of a life cycle perspective.

In recent years, other authors are beginning to explore the benefits of service design tools in enabling remanufacturing within industry [11, 12]. This begins to inform a more strategic design approach, which relates closely to wider business strategy. This is because strategic design begins with strategy definition by distilling market insights to inform the direction of future product portfolios, understanding client value and needs through user insights and service design principles and utilising communication design and branding to position products and companies.

Table 1 summarises a selection of key papers on design for remanufacturing. Yang et al. [9] and Sundin et al. [10] integrate product design properties and design

Table 1 Literature summary of design strategies suited to remanufacturing [9, 10, 13–16]

Design for technology integration	Design for reverse logistics	Detailed design and materials' selection	Design for standardisation and reducing complexity
Ensure technology exists to restore product [14]	Design for reverse logistics [9]	Connectors: number of different components in each group, number of different components, number of connectors and tools [15]	Product is made up of standard interchangeable parts [14]
Product technology is stable over more than one life cycle [14]	Define reverse logistics [14]	Consider fastening methods [16]	Modular design [14]
Technology must be able to extract a component without damage [14]	Establish logistics for collection and distribution of goods and analysis of uncertainty characteristics, including probabilistic returns [13]	Use non-corrosive materials to prevent rust [16]	Identify design standards [13]
Identify opportunities for upgrade [13]	Collection of core parts [14]	Increased part fragility [16]	Reduce complexity [16]
Platform design [14]	Define how product will be collected [14]	Wear resistance [10]	
Upgrade: functional and interfaced decoupling [15]			
Architecture structure [15]			

strategies with the process of remanufacturing. This includes 'design for disassembly and reassembly' as well as 'design for cleaning, handling and maintenance'. Aside from these, the following four key areas have been identified from the literature (summarised in Table 1): design for technology integration, design for reverse logistics, detailed design and material selection and designing for standards and reducing complexity.

2.2 *Design Guidelines for Remanufacturing*

Ecodesign strategies are heuristics to operationalise ecodesign, during new product development (NPD) activities, by targeting specific areas across the product life cycle. Unlike ecodesign tools, ecodesign strategies offer easy integration of good design practices without the burden of formal tools, which, it has been shown,

Table 2 Business models suited to remanufacturing and the circular economy more broadly [23–25]

Business model	Description	Link to remanufacturing and circular economy	Examples
Access and performance model	Providing the capability or services to satisfy user needs without needing to own physical products	The manufacturer or service provider retains ownership of the product and is therefore incentivised to make the product last as long as possible	Launderettes, car club models, bicycle-sharing schemes, aerospace service contracts and printer/copier service/lease contracts
Extending product value	Exploiting residual value of products – from manufacture to consumers and then back to manufacturing – or collection of products between distinct business entities	This model is focused on exploiting the residual value of products. In an ideal case, companies themselves refurbish or remanufacture, but often *other* companies see the opportunity first and start these practices ('gap exploiters', [25])	Companies such as LEAPP in the Netherlands and Gazelle in the USA who collect and sell refurbished electronics. OEMs such as Apple and Dell are selling their own certified refurbished laptops
Classic long life	Business models focused on delivering long product life, supporting design for durability and repair	To support product long life, design for repair, refurbishment and remanufacturing are essential. This would fit in the company's normal design ethos	Companies who create products that last 'beyond a lifetime' (e.g. durable washing machines, caravans, high-end watches)
Encourage sufficiency	Solutions that seek to reduce consumption through durability, upgradability, service, warrantees, reparability and a non-consumerist approach to marketing and sales	This model follows the classic long-life model and encourages companies to take a non-consumerist approach to sales. Design for repair, refurbishment and remanufacturing can support slow consumption	Patagonia Don't Buy this Jacket campaign (http://www.patagonia.com/email/11/112811.html), Vitsœ Against planned obsolescence approach (https://vimeo.com/18996295)

companies fail to internalise. Examples of these strategies include design for remanufacturing, function integration, light weighting or right weighting and material substitution. The literature in this area is thematically similar and is commonly presented in the form of unordered lists of ecodesign strategies [17]. Similarly, much of the literature on design for remanufacturing centres on guideline approaches (see Table 1).

Importantly, distinctions between design for recycling, design for remanufacturing and other DfX methodologies are beginning to emerge. For example, other authors state that certain strategies typically associated with remanufacturing (ease of access, handling, disassembly and reassembly) are simply not important when considering recycling [18].

In contrast, one study on design for recycling found that considering design for easy separation of parts is a valid design strategy and can generate important solutions to improve automated recycling processes [19]. Some other conflicting findings have been identified. It has also been suggested that certain product characteristics (the high value of certain parts, reverse logistics), which make some products suited to remanufacturing, are outside the remit of control of the designer [18]. A more recent study stipulates that design for reverse logistics is in fact within the product designer's responsibility and reverse logistics needs to be considered during product development processes [9].

2.3 Remanufacturing Business Models

A business model is about the way you do business [20] and typically consists of a value proposition (what value is provided and to whom), value creation and delivery (how value is provided) and value capture mechanisms (how the company makes money from providing value) [21].

In general, remanufacturing is offered through two distinct approaches: remanufacturing as a separate business activity or remanufacturing as an original equipment manufacturer (OEM) service. The literature on remanufacturing-oriented business models sits between the field of product-service systems research [11, 12] and wider business model research focused on sustainable consumption and production [22–24].

Few empirical studies have been undertaken on remanufacturing business models, and other authors state that there is insufficient knowledge of the mechanisms of remanufacturing business models at present. However, work is emerging in the field of circular economy. One study states that in a circular economy product design and business model innovation go hand in hand and this process starts with a visionary goal [22] (see Table 2).

2.4 Selected Case Studies from the Literature

Other authors have illustrated the role of design in remanufacturing, through standalone case studies on remanufacturing practices. These case studies serve to evidence what is widely acknowledged by the remanufacturing community. For example, through case studies of three Chinese companies, Hatcher et al. [18] find that current design of electronic and electrical components are not well suited to remanufacturing. Critically, other authors have found that remanufacturing of energy-using products is not consistently beneficial for the environment, particularly when energy efficiency can be improved through new technological innovations [26]. This is also reflected in a separate paper which discusses the shifting of life cycle hotspots on account of (1) efficiency improvements and (2) consumption patterns [27]. These studies illustrate the need to integrate life cycle and environmental assessment approaches within remanufacturing studies.

A recent case study, commissioned by the Scottish Government, shows first-hand the impacts of the lack of a system-wide remanufacturing approach in the automotive sector. Here light weighting, to foster fuel efficiency, takes priority as a design strategy. Light weighting makes it increasingly difficult to remanufacture parts and components down the value chain. This is because material substitution activities compromise the durability of the parts. Thus, material selection needs to consider trade-offs along product life cycles for a broader perspective when applying what appear to be straightforward design guidelines. To address this, Prendeville et al. [28] developed a qualitative eco-innovation model to probe decision-making to identify trade-offs. Similar trade-offs have also been discussed by other authors [29, 30].

Mont et al. [31] discuss a theoretical case study of a pram, illustrating how a potential remanufacturing business model might be implemented, combined with an evaluation of the cost benefits. The study shows that the economic case is clear for the company, whereas the environmental assessment is less so. This supports recommendations by Hatcher et al. [4] who state in their review article that more evidence is required to prove the environmental benefits of remanufacturing activities. This review also identifies that there are very few case studies of design for remanufacturing and real-world examples of business models suited to remanufacturing.

2.5 Summary: Research and Practice Gap

While a number of seminal papers and key authors have propelled the research forward, overall, remanufacturing research is unsystematic and lacks strategic direction. The current state of design for remanufacturing research is limited to detailed product design engineering, with a lack of knowledge on strategic design elements. A number of other research gaps have been identified, including the need for integration of life cycle thinking within design for remanufacturing, more knowledge on design for remanufacturing within the product development process (key performance indicators, design metrics) and more case studies showcasing best practices in both design and business models for remanufacturing.

3 Method

In light of the research gaps identified in Sect. 2.5, the core aim of this paper is to build understanding of design for remanufacturing activities through a real-world example of a business seeking to transition to remanufacturing.

The research takes an action research approach to develop a remanufacturing case study undertaken at a business-to-business design and manufacturing company. The case study was conducted at a critical time during which the company undertook a remanufacturing pilot study. The company's intention for the pilot study was to explore the commercial viability of a remanufacturing-oriented business. The pilot study focuses on a single product, an office task chair, typically sold to corporate clients on the European market since 1999.

3.1 Case Study Selection

The product, a high-end office task chair, was chosen as it has high residual value, a resilient design, easily replaceable parts, good supply (of returned product) and good demand. The product was not designed for remanufacturing but was 'built to last' and was identified (by the company) as having good potential for remanufacturing. This product was also chosen due to the availability of 1000 chairs returned from a client, which provided a unique opportunity to explore the viability of remanufacturing.

3.2 Case Study Activities

The pilot study involved the following core activities: five company meetings to discuss and review the development of the pilot activities, three on-site visits to facilitate understanding through observation in situ during product development, streamlined environmental assessment, desk research to review project review reports and presentations generated through the course of the pilot study. The environment assessment was undertaken by the research team. Two types of assessment were undertaken and these are described briefly in Table 3.

3.3 Case Study Analysis

The data was analysed iteratively and in light of (1) interviews with remanufacturing experts undertaken as part of the ERN project and (2) the literature review.

Table 3 Environmental assessment methods

Overview of LCA and statistical methods used
Method 1: Life cycle assessment
Life cycle assessment (LCA) comparing the 'pilot product' with similar products in the company's own range. LCA generated using streamlined software as part of the action research activities. Streamlined LCA was chosen for quick and easy results, to support decision-making during NPD. The modelling adopted a best-fit approach, when data wasn't available. Scenarios for remanufacturing involved modelling additional parts as well as extending the life of the product over a given functional unit
Method 2: Statistical inference
Inference assessment approach to assess typical environmental impacts of the product (across weight and carbon indicators) by calculating the mean and standard deviation of a sample of products from publicly available data ($n = 30$)

4 Results

This section presents the results of the pilot study. The results are shown according to the four key areas of the pilot study: product evaluation, environmental assessment, commercial research and product take-back activity. Table 4 presents an overview of the questions addressed during this pilot study, its key stakeholders and activities undertaken.

4.1 *Product Evaluation*

Up to 1000 products, received from a client, were assessed using a bespoke evaluation framework. This involved a visual and mechanical inspection of products, to assess typical wear and to identify parts needing to be replaced. The results are presented in Fig. 1 and shows that of the 12 major sub-assemblies three parts typically need to be replaced (Part B, Part J and Part K).

4.2 *Environmental Assessment*

The company set out to achieve a fourfold reduction of environmental impacts through remanufacturing this product. The findings of the assessment indicated that:

- Minimising the material cost of remanufacturing is imperative, i.e. avoiding the replacement of as many parts as possible and particularly any large components.
- Using polypropylene in place of polyamide in key parts lowers the overall impact of the product.

Table 4 Key barriers to remanufacturing faced by the case company: coupled with design and business model solution strategies

Barrier to remanufacturing (faced by the case company)	Key collaborative stakeholder	Business solution strategy	Design solution strategy
New product sold through a dealership structure: risk of damaging relationships if dealers already offer a reman./refurb. service	Dealer	Develop collaborative partnerships (through incentives) with existing dealership network to exploit existing sales channels	High-end product design, combined with well-developed OEM brand, harnessed to enter new business-to-consumer market and incite 'desirability'
Limited access to client and end-of-first-life product. Supply of product potentially erratic	Client	Lower cost remanufactured alternative attractive to many clients but low uptake of service	Service design tools explored to uncover value and increase client uptake of product return service
Future legal and financial costs: risks of longer warranties need to be accounted for	End user	Costs of warranties incorporated in financial forecasts	More resilient product design through the development of durable products and evaluation of returned products
Existing dealer network barrier to resale into existing market	Market	New business-to-consumer market identified to have good potential for second life product	Development and design of communication and visual online presence to target a new domestic market through Web-based sales
High labour costs of remanufacturing process	Third parties	Collaboration with existing reuse charities may provide opportunities potential cost saving	Sales outlets that have less demanding product quality requirements (minor, visual imperfections on parts)

- Maximising recycled content in all parts, particularly any steel parts, would be beneficial.

4.3 Commercial Research

The (secondary) market value of this product was identified through an analysis of resale of similar product through channels such as eBay and second-hand dealers. The findings were positive, and the company proposed that the price could be increased due to brand value associated with being sold by the OEM. The company was not seeking to restructure its means of revenue generation, through leasing, for example, but rather needed to retain control of products as well as identify new

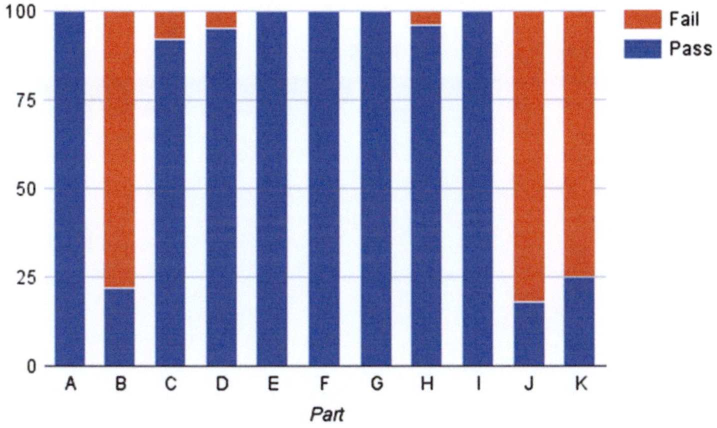

Fig. 1 Product evaluation: parts and sub-assemblies

sales and distribution channels for remanufactured products. Hence, rather than a radical business model innovation, it was seeking to modify its existing business model. For this reason a rebate system was used as the basis for calculations. Finally, through the commercial research activities, an emerging business-to-consumer target market was identified.

4.4 Product Take-Back Pilot

The pilot coincided with a factory relocation due to significant company growth in the previous years. This, serendipitously, provided a physical space for the remanufacturing trial, i.e. at the old factory. This allowed the company to test a process for collection of cores, reprocessing them through a test remanufacturing cell. In particular this pilot activity identified the following three needs:

1. To pre-assess products (suitability for remanufacturing) at the customer's site to control the quantity of product returning to the facility
2. For a product evaluation framework to support the remanufacturing team with on-site evaluations
3. For multiple end-of-life strategies for products (worn parts to be reworked, degraded parts to be recycled)

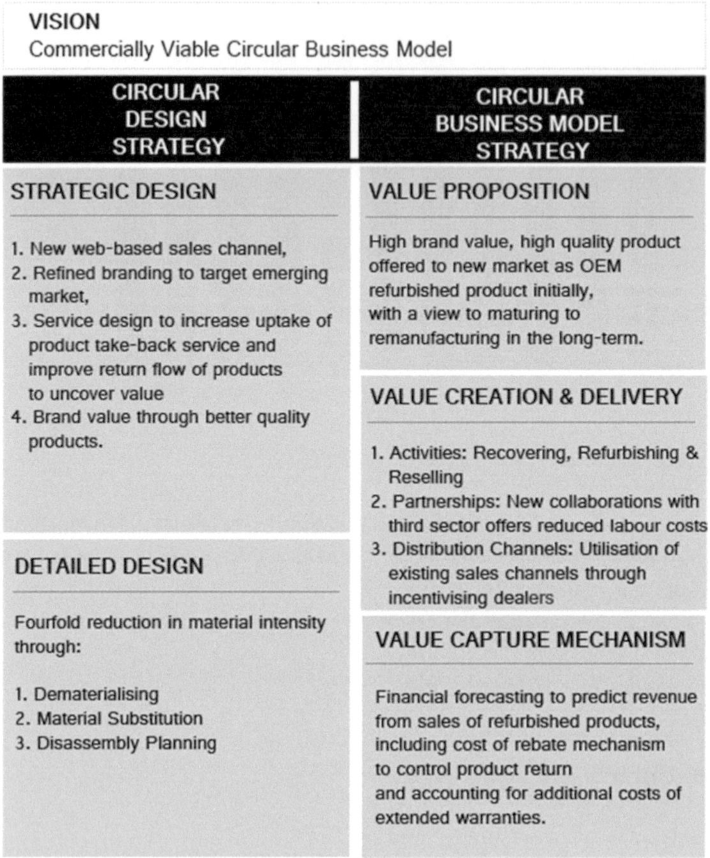

Fig. 2 Circular design and business model strategies (Adapted from [7])

5 Analysis and Discussion

This paper shows the interplay between the business model and design strategy in the context of remanufacturing, illustrating how these are interdependent and complementary. A number of barriers to remanufacturing were identified through the case study, and the paper discusses a combination of design and business model solution strategies adopted to overcome these. Figure 2 summarises this combined approach (specific to this case).

The design and business model approaches described in this case study can be mapped onto generic frameworks published by other authors. For example, here, the business approach is very close to an 'extended product value' model described in Table 2. In addition, Fig. 2 shows how a company's circular vision is comprised of both a design strategy and a business innovation strategy. This design strategy can be broken into strategic design actions (brand, design of service delivery) and detailed design actions (product evaluation framework, environmental assessments, ecodesign strategies) to inform future research.

While some questions remain open for the company, the pilot provides a starting point to explore the viability of remanufacturing. In doing so, this study shows key areas of uncertainty which are relevant to other companies. Four areas have been identified that require actions by companies seeking to transition to remanufacturing, including development of a product evaluation framework, undertaking an environmental assessment, commercial research and undertaking a product take-back pilot to test this process.

The study shows how mitigating the risks of business model adaption, through testing and piloting, can act as positive internal catalysts. For example, it was decided that a more robust financial assessment would be required. This shows how a simple financial forecast can act as an internal lever in the company, incentivising senior team members to undertake a more thorough financial assessment.

In this study the company prioritised a design for remanufacturing approach, where in reality, sold the product as a refurbished item into a new market. This illustrates the need for research to take a broad interpretation of design for remanufacturing, in order to identify the best design insights. In this sense the pilot is seen (by the company) as a stepping stone to a more mature remanufacturing capability in the future. Similar approaches have been identified at other companies participating in the European Remanufacturing Network. This is important as it illustrates how business model innovation is a dynamic and transitional process. In light of this, future design and business model research on remanufacturing should consider 'company maturity' as a key aspect of the research.

6 Conclusion

The literature review discussed the lack of detailed design and business cases on remanufacturing and the need to integrate environmental assessments within remanufacturing research studies. This case study advances existing knowledge by showcasing a multi-strategy approach which includes a supportive environmental analysis of the product, in combination with an exploration of design and business model strategies.

The paper provides a simple, but structured, framework of questions to investigate remanufacturing from a multi-strategy perspective. These are outlined in Table 4 and Fig. 2, respectively, and can be generalised to other studies and adapted for future research activities. In doing so, the paper provides insights to support other companies who seek to undertake similar activities.

Acknowledgments This research is supported by the European Union's H2020 programme through the European Remanufacturing Network (ERN).

References

1. Guitini R, Gaudette K (2003) Bus Horiz 46(960):41–48
2. All-Party Sustainable Resource Group (2014) Triple win: the economic, social and environmental case for remanufacturing. Policy Connect
3. UNEP (2013) Metal recycling: opportunities, limits, infrastructure. A Report of the Working Group on the Global Metal Flows to the International Resource Panel (Reuter MA, Hudson C, van Schaik A, Heiskanen K, Meskers C, Hagelüken C (eds)). UNEP, Nairobi
4. Hatcher G, Ijomah W, Windmill J (2011) Design for remanufacture: a literature review and future research needs. J Clean Prod 19(17–18):2004–2014
5. O' Rafferty S, O' Connor F (2010) Regional perspectives on capacity building for ecodesign. In: Facilitating sustainable innovation through collaboration. Springer, Dordrecht, pp 159–183
6. Prendeville S et al (2013) Motivations and barriers to ecodesign in industry. (ENEC) Available at: http://www.ecodesign-centres.org/projects.htm
7. Bocken N, Holgado M, Evans S (2015) Challenges and opportunities in circular business model innovation for clothing from a retailer perspective. ISIE conference, 7–10 July 2015, Surrey, UK
8. Nasr N, Thurston M (2006) Remanufacturing: a key enabler to sustainable product systems. Rochester Institute of Technology, New York
9. Yang SS, Ong SK, Nee AYC (2015) Towards implementation of DfRem into the product development process. Procedia CIRP 26:565–570
10. Sundin E, Östlin J, Rönnbäck AÖ, Lindahl M, Sandström GÖ (2008) Remanufacturing of products used in product service system offerings. In: Manufacturing systems and technologies for the new frontier. Springer, London, pp 537–542
11. Costa F, Prendeville S, Beverley K, Teso G, Brooker C (2015) Sustainable product-service systems for an office furniture manufacturer: how insights from a pilot study can inform PSS design. Procedia CIRP 30:66–71
12. Sakao T, Mizuyama H (2014) Understanding of a product/service system design: a holistic approach to support design for remanufacturing. J Remanufacturing 4(1):1–24
13. Pigosso DC, Zanette ET, Guelere Filho A, Ometto AR, Rozenfeld H (2010) Ecodesign methods focused on remanufacturing. J Clean Prod 18(1):21–31
14. Gray C, Charter M (2008) Remanufacturing and product design. Int J Prod Dev 6 (3–4):375–392
15. Willems B, Dewulf W, Duflou JR (2008) A method to assess the lifetime prolongation capabilities of products. Int J Sustain Manuf 1(1–2):122–144
16. Hammond R, Amezquita T, Bras B (1998) Issues in the automotive parts remanufacturing industry: a discussion of results from surveys performed among remanufacturers. Eng Des Autom 4:27–46
17. Fiksel J (1996) Design for environment: creating eco-efficient products and processes. McGraw-Hill Professional Publishing, New York
18. Hatcher GD, Ijomah WL, Windmill JF (2013) Design for remanufacturing in China: a case study of electrical and electronic equipment. J Remanufacturing 3(1):1–11
19. Fakhredin F, Bakker CA, Geraedts JMP Designer focused quickscan recyclability assessment method. In: Proceedings of CARE innovation "Going Green" conference, Vienna, 17–20 November 2014. Austrian Society for Systems Engineering and Automation
20. Magretta J (2002) Why business models matter. Harv Bus Rev 80(5):86–92

21. Short S, Bocken N, Barlow C, Chertow M (2014) From refining sugar to growing tomatoes. J Ind Ecol 18(5):p603–p618
22. Bocken NMP, Bakker C, De Pauw I (2015) Product design and business model strategies for a circular economy, Sustainable design & manufacturing conference, Seville, 12–14 April 2015
23. Bocken N, Short S, Rana P, Evans S (2014) A literature and practice review to develop sustainable business model archetypes. J Clean Prod 65:42–56
24. Bocken NMP, Rana P, Short SW (2015) Value mapping for sustainable business thinking. J Ind Prod Eng 32(1):p67–p81
25. Bakker C, van Hollander M, van Hinte E, Zijlstra Y (2014) Products that last: product design for circular business models. TU Delft Library, Delft
26. Bakker C, Ingenegeren R, Devoldere T, Tempelman E, Huisman J, Peck D (2012) Rethinking eco-design priorities; the case of the Econova television. In: Electronics goes green 2012+ (EGG), 2012. IEEE 1–7
27. Gutowski T, Sahni S, Boustani A, Graves S (2011) Remanufacturing and energy savings. Environ Sci Technol 45(10):4540–4547
28. Prendeville S, O'Connor F, Palmer L (2014) Material selection for eco-innovation: SPICE model. J Cleaner Prod (Special Issue on Lean and Green Manufacturing) 85:31–40
29. Brennan G, Tennant M, Blomsma F (2015) Business and production solutions 10: sustainability, key issues. Earthscan Routledge, New York, p 219
30. Prendeville S, O'Connor F, Sherry J, Palmer L (2013) Ecodesign trade-offs in new product development. In: LCA Avnir conference. Lille
31. Mont O, Dalhammar C, Jacobsson N (2006) A new business model for baby prams based on leasing and product remanufacturing. J Clean Prod 14(17):1509–1518

Development of Low-Carbon Society Businesses in Japan

Takashi Iwamoto and Hidetaka Aoki

Abstract Activities to develop low-carbon society businesses in Japan started in 2009 and have been executed until now. Governmental funds were invested intensively in four areas – City of Yokohama, Toyota City, Keihanna (Kyoto), and City of Kitakyushu – and the verification projects were executed. The research was conducted in order that low-carbon society businesses will be smoothly developed after the verification projects. It was found through the projects that the following points were important to develop low-carbon society businesses. The first point was that the governmental support should be based not on element technologies but on future business models. The second point was that technologies should be selected based on the future business models. The third point was prioritization and the fourth point was the management function to develop business models.

Keywords Low-carbon society business • Smart city • Business model

1 Introduction

After the economic crisis in 2008, Japan's most important issue was to reform the industrial structure. This means that Japan should create new industries urgently. The authors with their colleagues who worked for Dream Incubator (DI) Inc. executed new industry creation and new business development in various new industry fields such as energy and environment, agriculture, forestry, fisheries and livestock, town development, infrastructure, logistics and transportation, medical equipment and pharmaceuticals, etc. [1, 2].

Figure 1 shows DI's activities to create new industries. As the first step, DI creates an umbrella of new industries. This means that DI designs concepts of new industries, makes scenarios to grow new industries, and coordinates new businesses development with governmental organizations. After making the umbrella, DI

T. Iwamoto (✉)
Graduate School of Business Administration, Keio University, Yokohama, Japan
e-mail: iwamoto@a8.keio.jp

H. Aoki
Global Brain Corporation, Tokyo, Japan

M. Matsumoto et al. (eds.), *Sustainability Through Innovation in Product Life Cycle Design*, EcoProduction, DOI 10.1007/978-981-10-0471-1_19

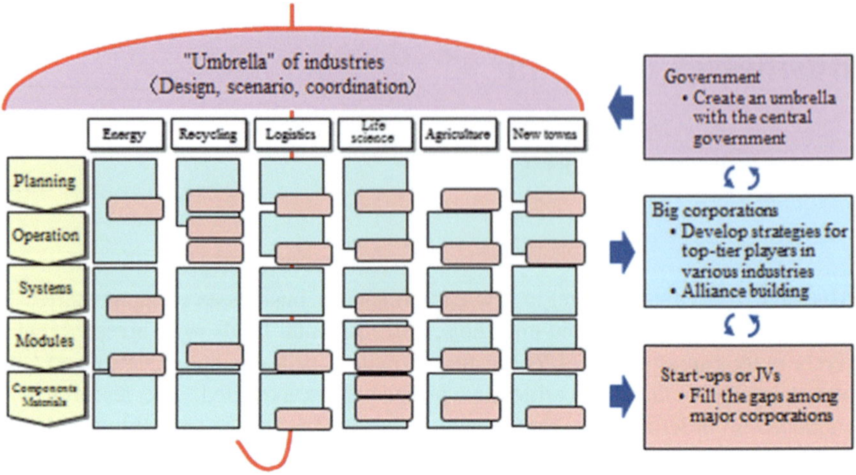

Fig. 1 Activities to create new industries

develops strategies for top-tier players in various industries and builds alliances when necessary. When there are lacking functions to develop new industries, DI arranges to make joint ventures or start-ups.

In this chapter, the activities to develop low-carbon society businesses in Japan were summarized, and the important points to be successful for the business development were extracted in order that low-carbon society businesses will be smoothly developed.

2 Research Methods

Figure 2 shows the process to develop new industries through public-private collaboration. At the beginning, the public sector develops public policies. After the public policies are effective, the private sector executes verification projects to develop business models and the public sector supports the verification projects. Then, after the verification projects finish, the private sector starts to deploy businesses based on the verified business models. The authors have been working on development of low-carbon society businesses since the beginning.

Figure 3 shows the approach for this research. At first, all the information related to low-carbon society businesses in Japan was collected. From all the information, key outcomes achieved so far were extracted. Then, interviews with main players were made to clarify the important points to develop low-carbon society businesses. Finally, the important points to develop low-carbon society businesses were summarized. The interviewed persons include governmental officials, management people who joined smart city projects, and experts in energy and environment industries.

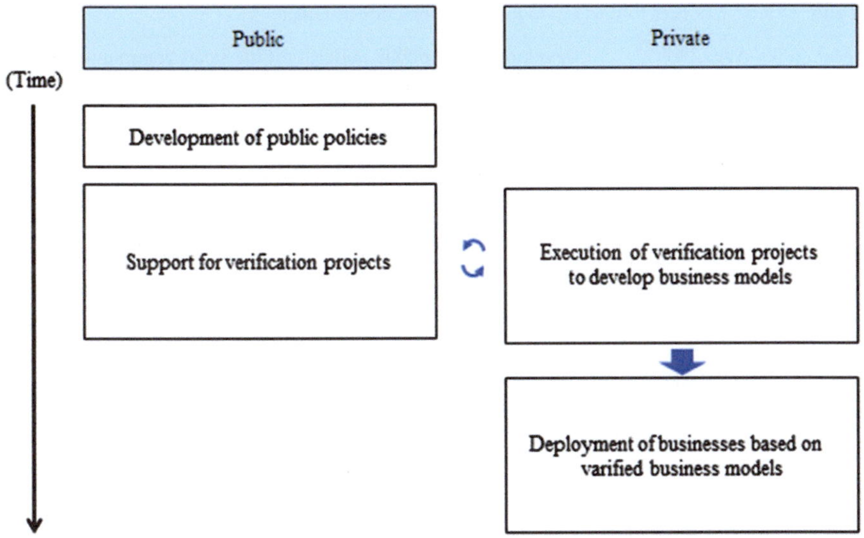

Fig. 2 Process to develop new industries through public-private collaboration

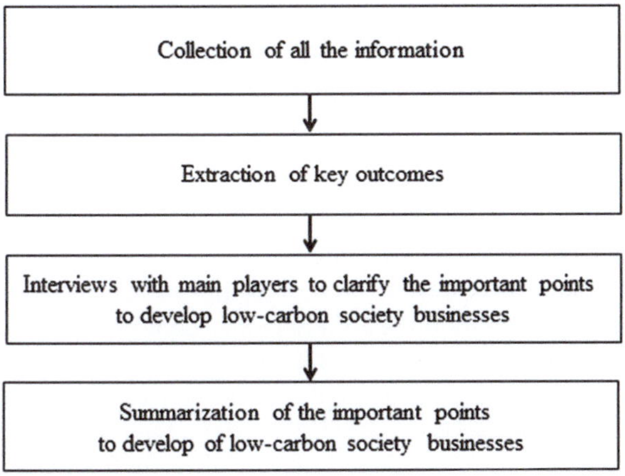

Fig. 3 Approach for this research

3 Smart City Projects

3.1 Overall Flow

The projects to develop low-carbon societies in Japan started with the proposal to the Ministry of Economy, Trade and Industry (METI), which is one of Japanese governmental organizations, by the authors in January 2009. Table 1 shows the key activities to develop low-carbon societies in Japan. After the proposal, the authors

Table 1 Key activities to develop low-carbon societies in Japan

Mar. 2009–Aug. 2009	Development of public policies for the smart community project
Apr. 2009–Mar. 2010	Planning for the smart community project by each area
Apr. 2010–Mar. 2015	Execution of the smart city projects by the selected 4 areas

developed public policies and the public policies became public in August 2009 [3]. From April 2009 to March 2010, about 20 areas in Japan worked on planning to apply for the smart community verification projects and 4 areas out of about 20 areas were selected as next-generation energy and social system demonstration areas. The four areas comprise City of Yokohama, Toyota City, Keihanna (Kyoto), and City of Kitakyushu. The name of the smart community projects was changed to the smart city projects after the projects started, and four areas executed the projects from April 2010 to March 2015 [4].

3.2 *Public Policies*

Public policies to execute the smart community projects were developed from March to August 2009 and were chosen as METI's most important public policies for FY2010 [1–3]. FY stands for fiscal year and the fiscal year in Japanese governments is April to March.

The key points of these public policies are shown in Fig. 4. The first point was to draw the vision which was called smart city concept at that time. In this vision, what kind of communities Japan should have in 2,050 were clarified. The second point was to grow promising technology modules. The essence of this point was to prioritize the technologies. The third point was smart city management and the party to manage the smart city should be responsible for developing business models.

Figure 5 is the figure to show the essence of the public policies. The high-priority technologies were categorized based on future business models and the government asked the areas which would like to apply for the projects to make plans based on the future business models.

The public policies made this time were very different from the previous public policies and the differences are shown in Fig. 6. There were main differences in three aspects – selection of technologies, scale of experiments, and continuation.

3.3 *Verification Projects*

As described above, 4 areas out of about 20 areas in Japan were selected for smart city projects. The definition of smart city was a new style of city providing sustainable growth and designed to encourage economic activities that reduce the burden on the environment while improving QoL (Quality of Life) [4].

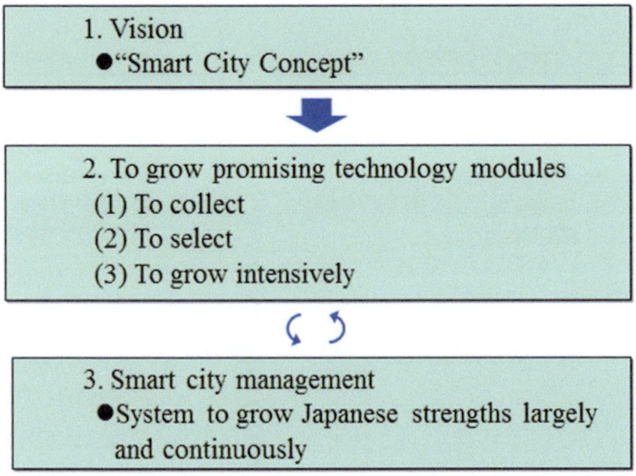

Fig. 4 Key points of the public policies for the smart community projects

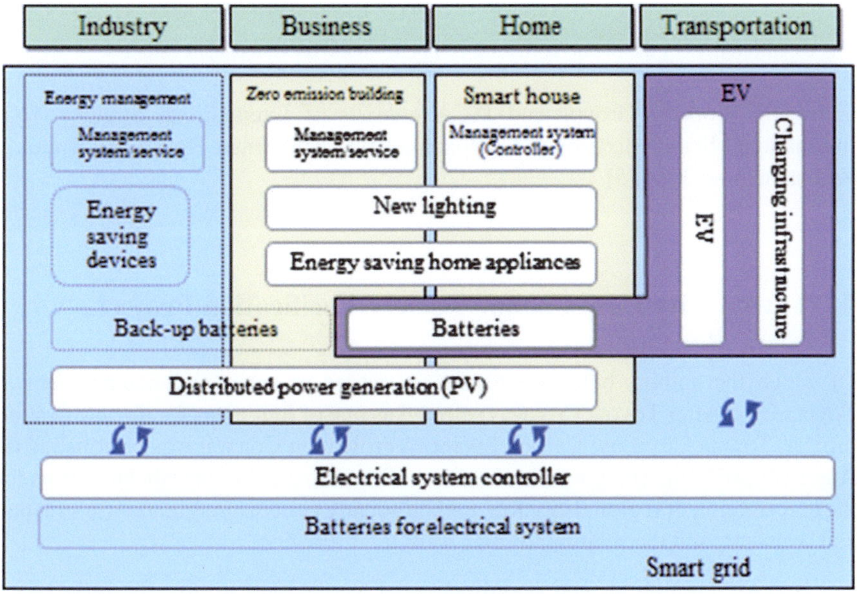

Fig. 5 The figure to show the essence of the public policies for the smart community projects

The smart city projects continued for 5 years since April 2010 until March 2015 and the major outcomes are shown in Table 2. WG, SWG, EMS, and JSCA in Table 2 stand for Working Group, Subworking Group, energy management system and Japan Smart Community Alliance, respectively.

JSCA was established in April 2010 with the aim of resolving and overcoming the obstacles of individual organizations through collaboration of the public and private sectors. The number of the members was 283 as of June 12, 2015 and the organizations of JSCA were shown in Fig. 7. JSCA's activities have been made

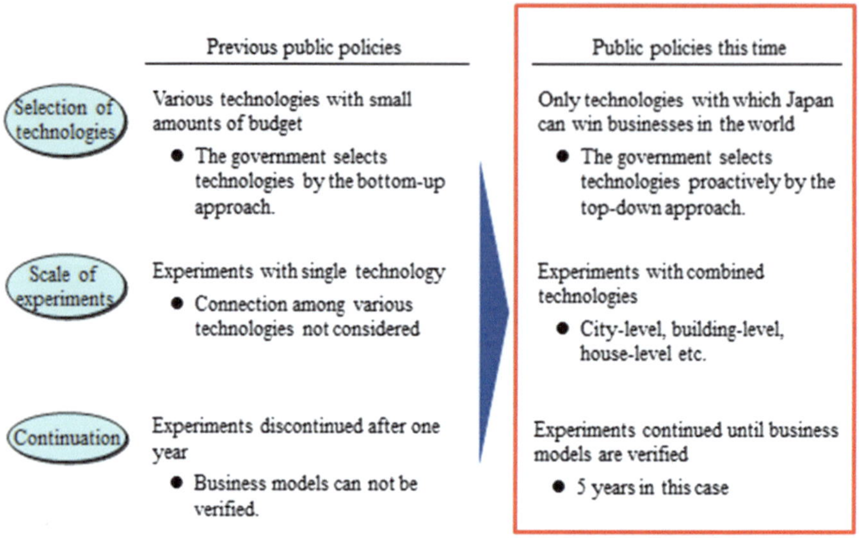

Fig. 6 Differences of the public policies this time from the previous public policies

under the Working Groups and JSCA consists of International Strategy WG, International Standardization WG, Roadmap WG, and Smart House and Building WG as of June 2014 [5].

4 Toyota City Low-Carbon Society Verification Project

DI which the authors belonged to was in charge of smart city management in Toyota City. After Toyota City was selected as one of four areas for the verification project, Toyota City Low-Carbon Society Verification Council was established on August 5, 2010 [6]. The purpose of the council was to carry out verification tests for the construction of regional city low-carbon society systems for greater use in Japan and overseas, and the major activities were the following:

1. Planning, promotion, and coordination of verification testing for low-carbon society systems
2. Coordination with various related agencies and organizations
3. Provision of information and PR agencies outside the council
4. Other activities that are required to achieve the purpose of this council

Figure 8 shows the organization of Toyota City Low-Carbon Society Verification Council. Forty-eight organizations joined the council and four groups were organized under the management group based on the public policies. The management group consisted of Toyota City, Toyota Motor, Chubu Electric Power, Denso, and DI. Four groups consisted as follows:

Table 2 Major outcomes of the smart city projects

Year	Day	Title
2010	Aug. 8	Selection of Next-Generation Energy and Social System Demonstration Areas
	Aug. 11	Announcement of master plans for the Demonstration of Next-Generation Energy and Social systems
2011	Oct. 11	Adoption Results of Feasibility Studies for the Overseas Development of Smart Communities and Other Projects
	Dec. 16	Results of the Second Meeting of the Smart House Standardization Study Group, JSCA International Standardization WG EMS-SWG
2012	Feb. 24	Summary by the smart House Standardization Study Group, JSCA International Standardization WG
	Mar. 12	Launching International Feasibility Studies on the "Smart Community" Concept
	Mar. 23	Signature of first content for smart community project under Delhi-Mumbai Industrial Corridor
	Mar. 30	Announcement of successful applicants for the Smart House International Standardization Research Program
	Apr. 17	Selected Areas in creating Master Plans under the Project for Promoting Introduction of Smart Communities
	Jun. 22	The First Meeting of JSCA Smart House/Building Standardization and Business Promotion Study Group
	Nov. 5	Launching international feasibility studies on "Smart Community"
2013	Aug. 2	FY2013 International Feasibility Studies on "Smart Community
	Aug. 10	METI to Launch a Smart Mansion Assessment System
	Oct. 25	FY2013 International Feasibility Studies on "Smart Community" (Second Invitation)
	Nov. 25	Demonstration Project for the Incentive-based Demand-Response Program will Start
2014	Jun. 26	FY2014 International Feasibility Studies on "Smart Community"

1. Optimizing energy use in homes (12 organizations)
   Leader: Denso
   Member: Chubu Electric Power, Toyota Motor, Toyota Housing, Toho Gas, Sharp, KDDI, Toyota Smile Life, Mitsubishi Corporation, Yazaki, Secom, and Aisin Seiki
2. Construction of a low-carbon transportation system

   2.1 Promoting introductions of next-generation automobiles (two organizations)
   Leader: Toyota City, Toyota Motor

   2.2 Installing and expanding the charging infrastructure and hydrogen stations (six organizations)

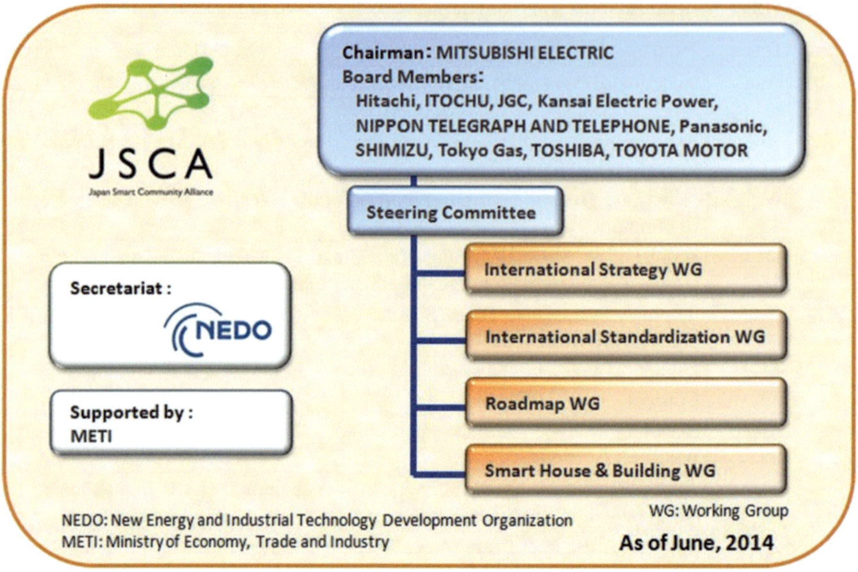

Fig. 7 Organization of JSCA

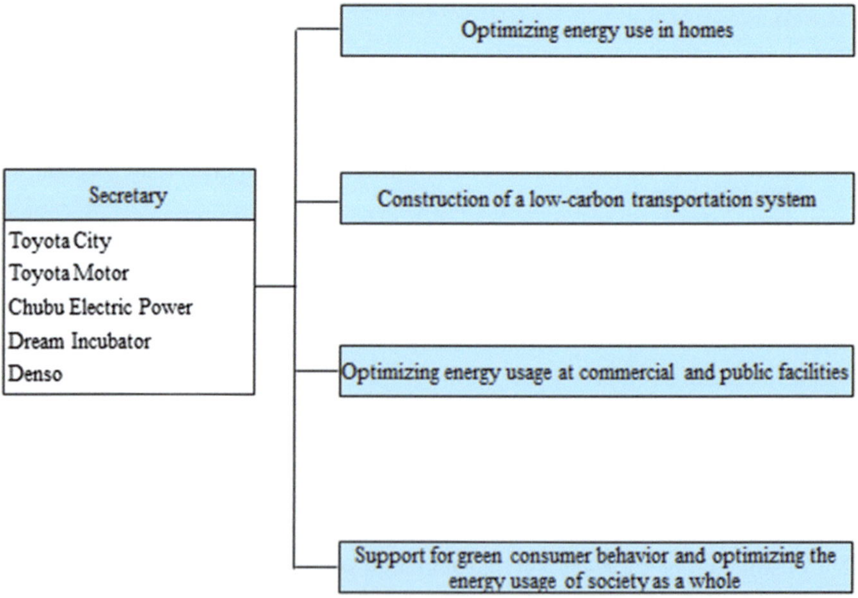

Fig. 8 Organization of Toyota City low-carbon society verification promotion council

Leader: Toyota City, Toyota Motor
Member: Toho Gas, Toyota Tsusho, and Central Nippon Expressway

2.3 Relieving traffic congestion and promoting low-environmental-impact driving through the use of ITS (Intelligent Transport Systems) (two organizations)
Leader: Toyota City, Toyota Motor

2.4 Promoting the use of public transit (three organizations)
Leader: Toyota City, Toyota Motor
Member: Nagoya Railroad

2.5 Sharing of personal mobility and next-generation automobiles (five organizations)
Leader: Toyota City, Toyota Motor
Member: KDDI, Toyota Industries, and Yamaha Motor

3. Optimizing energy usage at commercial and public facilities

3.1 V (Vehicle) to CVS (Computer-controlled Vehicle System)/School (four organizations)
Leader: Toyota Motor
Member: Toyota City, Circle K Sunkus, and Toyota Industries

3.2 Introduction of energy storage equipment at commercial and other facilities (five organizations)
Leader: Denso, Toyota Tsusho
Member: Toyota Motor, Systems Engineering Consultants, and Yamato Transport

3.3 Hybrid EMS (energy management system) (two organizations)
Leader: Toho Gas
Member: Toyota City

4. Support for green consumer behavior and optimizing the energy usage of society as a whole

4.1 EMS/EDMS (Energy Data Management System) (eight organizations)
Leader: Toyota Motor
Member: Fujitsu, Chubu Electric Power, ENERES, DI, Toshiba, Hewlett-Packard Japan, and Sumitomo Electric Industries

4.2 Providing incentives, such as the use of Eco-Points (five organizations)
Leader: Toyota City, DI
Member: Nagoya University, Toyota Motor, and Nagoya Railroad

4.3 Implementation of environmental learning and awareness activities (two organizations)
Leader: Toyota City
Member: Aichi Prefecture

4.4 Preparation of low-carbon society model district (five organizations)
Leader: Toyota City
Member: Aichi Prefecture, Toyota Motor, Chubu Electric Power, and Toho Gas

Table 3 Major outcomes of Toyota smart city project

Year	Day	Title
2012	Oct. 1	Car batteries proving a source of power in times of peace and in times of disaster
	Oct. 10	Intelligent HEMS Geared Toward "Locally Produced, Locally Consumed" Energy
	Oct. 23	Optimizing the control of the energy supply and demand throughout the entire community
	Nov. 20	Instant Searches on Connections between Automobiles and Public Transport
	Dec. 12	Urban car sharing for short trips
	Dec. 28	Using solar energy to its maximum benefit in commercial facilities
2013	Feb. 8	Restraining electrical power speaks when simultaneously recharging multiple PHVs and EVs
	Feb. 25	Verifying the effects of Demand Response in 160 households
	Apr. 4	Verification experiment into ideal societies that use hydrogen to attain in the wide-spread distribution of FCVs
	Apr. 5	Toyota Ecoful Town, where you can experience the future of low-carbon societies today
	Dec. 16	Electrical Power Supplied from FC (Fuel Cell) Buses to Evacuation centers during Disasters
	Dec. 18	Improvements and Actual Cutover to the Ha:mo Urban Transportation System
2014	Jan. 24	Making Use of Verification Experiments to Support Business Opportunities for Small- to Medium-Sized Companies
	Mar. 19	Using All Electrically Generated by Solar Power for Self-Consumption with HEMS
	Mar. 20	Building Systems to Reduce Carbon and Equalize Electricity Usage Based on Regional Electric Demand Forecasts
	Apr. 30	Optimizing PHV and EV Recharging in Accordance with Power Demand in the Community
	May 1	"Top Up" Recharges for Batteries Mounted on Collection and Delivery Vehicles
	May 14	Verification Experiment into Communications Technology that Supports Smart Grids
	Jun. 9	"Mirai no Futsu" (Normal in the Future) Toyota Ecoful Town Hands-on Experience Facility now Completed
	Jun. 11	Cut Backs on Electricity Consumption Confirmed Through DR Verification Experiments Targeting 160 households
2015	Mar. 5	Environmentally Progressive City Conference Held with UN as Co-Host
	Mar. 6	Initiative Picks up Pace to Build EV Car Sharing Business
	Mar. 6	Suggestions for Energy Conservation and How to Use PHV, Tailored to Each Household's Lifestyle
	Apr. 10	As a Future Challenge City, We Welcome Corporate Proving Experiments – An Interview with Toyota City Mayor Toshihiko Ota
	May 28	Launching Sales of a New HEMS Model Using Demonstration Results

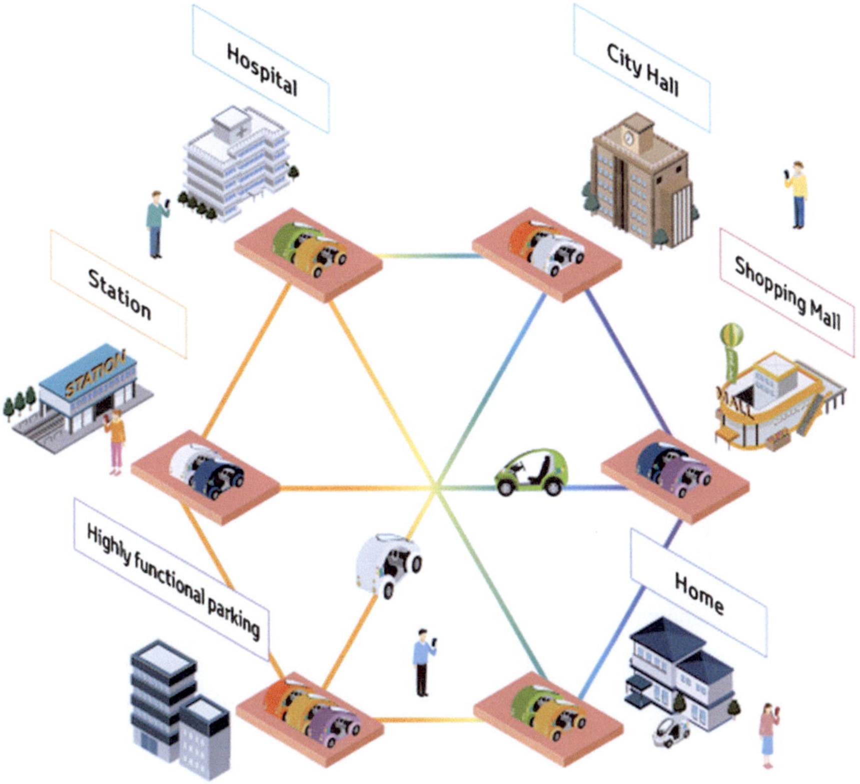

Fig. 9 The figure to show Ha:mo RIDE

Various outcomes were obtained in this 5 years' project and the major outcomes are shown in Table 3.

HEMS, PHV, EV, FCV, and DR in Table 3 stand for Home Energy Management System, Plug-in Hybrid Vehicle, Electric Vehicle, Fuel Cell Vehicle, and Demand Response, respectively. Ha:mo in Table 3 is the next-generation transportation system, Harmonious Mobility Network, Toyota Motor has been developing. Ha: mo is for optimal use of cars and other personal vehicles in combination with public transportation [7].

Ha:mo is one of the major outcomes and real-life testing was deployed to Tokyo and Grenoble, France, in addition to Toyota City. Figure 9 shows the mobility service called Ha:mo RIDE. Ha:mo RIDE is the service for mobility sharing for short trips, such as from the train station to a store, from the office to visit a client, etc., when necessary.

In addition to transportation system, house-level energy management system, commercial- and facility-level energy management system, and city-level energy management system were developed, and the business models were considered through verification activities as expected when public policies for smart cities were developed.

5 Conclusions

Activities to develop low-carbon society businesses have been made from 2009 to now. Public policies were developed in 2009, the verification projects had been executed from 2010 to 2015, and business deployment started in 2015. Considering the outcomes of smart city projects, it seems that the outcomes were produced as expected when the public policies were developed.

Through the experiences of these activities, it can be concluded that the important points to develop low-carbon society businesses were the following:

1. Public policies should be developed considering future business models.
2. Technologies should be selected based on future business models. In the case of low-carbon society, technologies were selected based on the categories defined by the future business models (city level, commercial and public facility level, house level, and transportation system level).
3. Prioritization is important to be successful. In the case of the smart city projects, only 4 areas out of about 20 areas were selected.
4. Management function should be implemented. In the case of Toyota City project, the management group was organized and four groups were managed under the management group.

The government support ended in March 2015 and private corporations should deployed businesses without the government support from April 2015. The authors expect that the verification projects from 2010 to 2015 will be of use when private corporations will establish business models.

References

1. Takashi Iwamoto (2012) Industry Producing Integrating Technology, Strategy and Public Policy. In: The Japan Society for Science Policy and Research Management 27th annual meeting proceedings, The Japan Society for Science Policy and Research Management, pp 709–712
2. Takashi Iwamoto (2013) Low-carbon society industry producing. In: Proceedings of EcoDesign 2013, 4–6 December 2013
3. Ministry of Economy, Trade and Industry (2010–2015) Smart Community. In: Website of the Ministry of Economy, Trade and Industry
4. New Energy Promotion Council (2010–2015) Japan Smart City Portal. In: Website of Japan Smart City Portal
5. Japan Smart Community Alliance (2010–2015) About JSCA. In: Website of the Japan Smart City Portal
6. Toyota City Low-Carbon Society Verification Promotion Council (2010–2015) About the Council. In: Website of Smart merit TOYOTA CITY
7. Toyota Motor (2013–2015) Ha:mo. In: Website of Toyota Motor

What Is 'Value' and How Can We Capture It from the Product Value Chain?

Jacquetta Lee, James R. Suckling, Debra Lilley, and Garrath T. Wilson

Abstract The mobile phone industry is based upon the rapid development of handsets and the high turnover of devices in order to drive sales. Phones are often used for shorter periods of time than their designed life, and when discarded it is often through channels that result in lost resource. This unsustainable business model places strain on resources and creates adverse environmental and social impacts. Through interrogation of a stock and flow model, a product-service system (PSS) for a small consumer electronic device, a mobile telephone, is proposed. The points at which value may be extracted from the PSS are identified. A quantitative measure of value is proposed in order to allow the evaluation of the most appropriate time to extract it. This value is not solely monetary, but is derived from the combination of indicators which encompass environmental, economic, and technological factors. A worked example is presented, in which it is found that the precious metals within the phone are the main determinants for value extraction. These metals are found in the printed circuit board, leading to a requirement to design phones for ease of extraction of these components in order to access the value within.

Keywords Product-service system • Smartphones • Value

1 Introduction

The mobile phone has become a ubiquitous technology which the consumer has embraced as part of their everyday lives with enthusiasm. From their introduction in the early 1980s, the mobile phone has undergone a constant and rapid development, increasing in processing power and device capability whilst reducing in physical size. However, their swift advancement has led to a legacy of waste and lost resource in which consumers are incentivised to replace their phones before an

J. Lee (✉) • J.R. Suckling
Centre for Environmental Strategy, University of Surrey, Guildford, UK
e-mail: j.lee@surrey.ac.uk

D. Lilley • G.T. Wilson
Loughborough Design School, University of Loughborough, Loughborough, UK

M. Matsumoto et al. (eds.), *Sustainability Through Innovation in Product Life Cycle Design*, EcoProduction, DOI 10.1007/978-981-10-0471-1_20

end of life driven by product failure and absolute obsolescence [1]. Furthermore, the high turnover of phones is exacerbated by the lack of drivers to return phones through appropriate channels for recycling [2]. Instead, the old, replaced phones are often kept by the consumer and left unused in storage or hibernation [3]. At the start of hibernation, the mobile phones have a technological and monetary value which could be extracted through reuse in secondary markets, but at the end of hibernation, they have often decreased in value so much as to be worthless to the consumer and business [4]. Due to their small size, many often end up in landfill, their value essentially lost. The situation may be further exacerbated when phones are sold for reuse: they may be sold or donated to lower value markets in developing countries where the recycling infrastructure is not in place to allow for effective collection at end of life [5].

All of the above factors combine to create a leakage of precious resource and a loss of value during the products' potential useful lifetime as well as a negative environmental and social impact, from pollution from landfill and the constant need to replace the materials from virgin sources, when recycling could be more environmentally efficient [6].

This paper will explore the value contained in phones and what 'value' means in the context of developing a sustainable product-service system business model. Further discussion will be given on the methods which may be used to maximise the capture of value and at what stages of the product's life cycle it can be best extracted.

2 Stocks and Flows

2.1 A System-Based Approach

The current mobile phones' business model is built around the concept of maximising flows, producing and selling as many handsets as possible within a given period of time. The more phones a business can sell, the more successful it is. This flow can broadly be defined as the movement of handsets from the manufacturer, through the consumer(s), to the eventual disposal route, e.g. landfill or recycling. With consumer demand typically satiated by replacement with a new handset that has been manufactured by mining evermore virgin material and with relatively few phones returned for recycling, resource is commonly lost at the end of the life cycle [2, 7]. This is not a resource-efficient business model, and in a world with finite size and mineral reserve, this is clearly not a sustainable solution. In Europe the Waste Electrical and Electronic Equipment (WEEE) Directive [8] has gone some way to alleviate the flow of electronic waste to landfill, but it is not perfect, with loopholes being exploited to enable export of waste to markets with informal recycling [9], creating deleterious implications for environmental and human toxicity [10].

An alternative approach to improving resource efficiency is to consider a system based upon maintaining the quality and quantity of *stocks* [11]. In this approach the phones in use are considered a stock which must be maintained through the careful management of flows around the system in order that they are utilised most effectively for an *appropriate* length of time. The stock is maintained at sufficient quality by multiple levels of intervention, for example, reusing the stock as is, without modification, if it is still able to perform its function adequately. Reuse of whole phones can be supplemented by reuse of components and remanufacture of phones to optimise the effective time that each part may be used. Finally, recycling the materials to ensure closure of the economy can reduce the burden on virgin stock by minimising extraction.

Figure 1 shows a possible mobile phone stock and flow diagram. The stock of phones, S, is maintained for a time period, T, by channelling the phones returned to the manufacturer through the reuse or remanufacture loops. Those that cannot be reused are sent for component value assessment and disassembly. Parts of the phone that may be reused are directed to a stockpile of components, whereas those that cannot are reprocessed and recycled. The recycled material is fed back into the manufacturing process and used to reduce, but not eliminate, the quantity of virgin material required: the mobile phone market is growing, plus the recycling loop is dissipative, so there will always be a requirement for additional 'new' flows into this system. However, this new systemic approach should minimise this resource burden. Given the technological advance of mobile phones, the stockpile of components will only have a given shelf life before becoming obsolete. Once obsolete, they too are returned for reprocessing and recycling or possibly exported to a lower value market in which they still have worth. At all steps there will be inevitable waste which exits the system (dashed lines). The waste is seen as an imbalance in the flows into and out of the stock. An example of this is the waste flow from remanufacture; the waste is the components which are replaced. A flow of equal magnitude is required from the stockpile of components to balance this out.

If a strategy of prioritising virgin stock use is pursued, with reprocessed material making up the shortfall in supply, the effort in reusing the components becomes disproportionate. This leads to a simplified version of Fig. 1 in which the obsolete stock feeds directly into the reprocessing.

Additionally, each pool of stock requires an intervention, denoted by an i (blue text), in order to be maintained. These interventions take the form of input into the stock, such as labour or energy, and emissions, for example, the greenhouse gases associated with energy production. At each point the appropriate level of intervention must be calculated to ensure an efficient use of the stock.

The flow of stock is divided at certain points into fractional flows. These fractions are indicated by the parameters, f_1, f_2, etc. Each of the fractions is dictated by the stock into which the flow is feeding. For example, the flow of material into the manufacturing process from mining virgin material, $(1\text{-}r_2)(1\text{-}r_1\text{-}u_1)p$, is dictated by the difference between the flow required to maintain the stock, $(1\text{-}r_1\text{-}u_1)p$, and the supply of reprocessed and recycled stock available, $r_2(1\text{-}r_1\text{-}u_1)p$. The assessment of the fractional flows between the stocks leads to an understanding of the

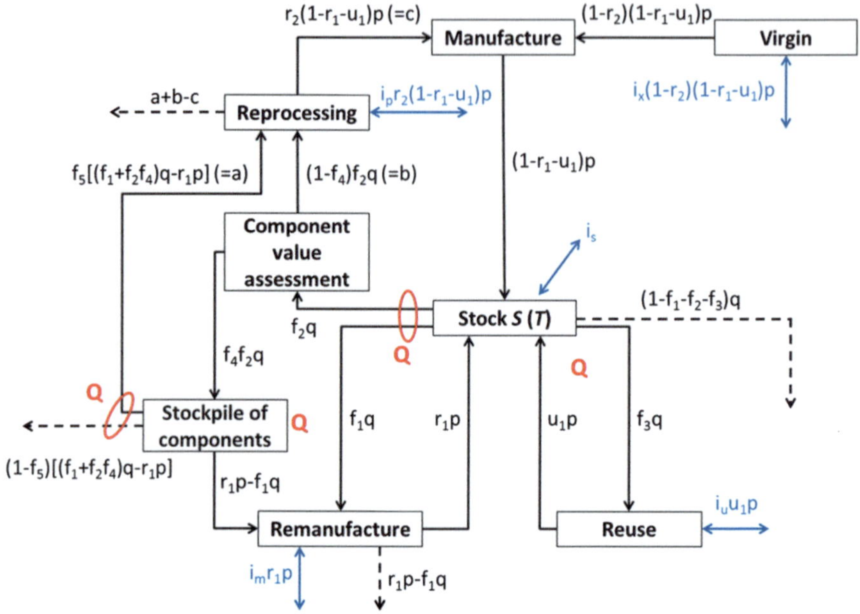

Fig. 1 Stock and flow system of a mobile phone

sustainability of the entire system. To achieve a targeted sustainable level may itself lead to the definition of desired values for f_i, r_i and u_i.

2.2 *Product-Service Systems and Stocks and Flows*

PSSs provide a business model whereby the *function* or *value* that satisfies consumer *needs* becomes the focus of business moving away from '*designing* (*and selling*) *physical products only*, *to designing* (*and selling*) *a system of products and services which are jointly capable of fulfilling specific client demands*' [12]. As illustrated in the PSS proposed by Wilson et al. [13], it is not necessary that the hardware of an electronic device be owned by the consumer; indeed, the hardware, including circuit boards (PCBs), chips (ICs) and other electronic components, can be leased as part of a service to the consumer due to their low intrinsic emotional value. The emphasis of the PSS and the system, therefore, can shift from producing more hardware towards maintaining the hardware in use, which may be regarded as one method of retaining the quality and quantity of the stock. Logically, the most effective businesses operating PSSs are the ones that can provide the service with the most efficient maintenance possible; therefore, PSSs are suited to the stock and flow model through their use of effective design to ensure reliability, ease of maintenance, reuse and remanufacture of devices through the carefully planned application of labour.

The analysis of an effective stock and flow model can inform the development of a PSS by identifying the points in the process where a decision must be made and hence the behaviour of the participants in the PSS. A decision is distinct from an intervention: the latter is an impact incurred through maintaining the stock; the former dictates through which channel the flow is directed. The decision points are indicated in Fig. 1 by red Q's. The decision points will not only inform and shape the consumer interaction with the service but will also inform and direct the flow of components through the loop which will allow the best extraction of value.

In order to make the best decision possible, the sustainable value of the phone and constituent components must be understood at points in both the stock and flow model and also time. The decisions become the comparison of the sustainable value against predefined thresholds. This is addressed later in this article.

2.3 Hierarchy of Sustainability

There is a hierarchy of preferred routes within those shown in Fig. 1. The closer that the phones and components can remain in form to the stock, S, the more sustainable the system. Reusing the phones without modification is preferable to remanufacture, which is itself preferable to complete dismantling in order to recover the components and materials. This leads to a hierarchy which will enable keeping the flow as close to the stock as possible. As such prioritisation should be to:

1. Increase T: time spent where the phone fulfils its designed functions to a level required by the market, in order to reduce p.
2. Increase reuse, f_3.
3. Increase remanufacture, f_1.
4. Increase stockpile, f_4.
5. Increase reprocessing after stockpiling, f_5.
6. Reduce inputs into manufacture, i_m.

If desired the list may be extended further to include ever more incremental advantages.

2.4 Value and Thresholds

In order to control the flow of phones and components around the system, the decision points are used to affect the routing of a particular flow. The decisions relate to the ability of the complete phone or the components within it to perform a function within the stock and flow system as a whole. Therefore, it is proposed that at each decision point, a threshold must be set against which the items are assessed.

If they meet this threshold, they are directed along one particular route, or along the other if they do not. In order to set the threshold and assess the worth of an item, it must be assigned a value and that value assessed at that point in the flow at a given time. The assessment of value at given points in the flow system will inform the creation of a PSS and its interaction points with the consumer.

3 Understanding Value

In order to extract most value from the stocks and flows at the most appropriate juncture, it is necessary to have a thorough understanding of the value embodied within the phones and components at any given moment in time. The assessment of value will be used at each of the identified decision points in the stock and flow system as well as to help in defining the consumers' touchpoints.

The term value has been used in this article much already, but it may mean many different things to different people; thus it is necessary to provide a definition of value for this work. The main consideration of value in typical product-oriented business models at present is given towards the economic value that can be extracted. This gives no consideration to the indefinite sustainability of the business into the future. To that extent other key values are ignored. For example, the value of retaining a certain metal through recapture and recycling so that it may be available again in the future must not be overlooked or the value of allowing a person to live a healthy and productive life. These are values which are harder to measure in current terms, but are no less important to creating a truly sustainable business and society.

3.1 Indicators

Sustainability can be discussed in terms of three pillars: environment, society and economy [14]. Any consideration of value operating within a sustainable PSS should also incorporate these. At the initial stage of framework development, it is proposed that value within the PSS is judged on a streamlined set of indicators to reduce the information capture that is required. They are proposed to act as a proxy for other impacts within the mobile phone sector. For the purposes of this article, the focus is given to economic and environmental indicators and a new, proposed technological indicator required for the fast moving mobile phone sector.

3.1.1 Environmental

Global warming potential (GWP) is a well-understood environmental impact [15], widely reported in, for example, carbon footprinting [16]. It is proposed as a rough proxy for intensity of resources in the manufacture of components and as part of the consideration of recycling metals within those components. It is recognised that there will be situations where this relationship does not hold.

Water use is also proposed, as this stock cannot be substituted by other materials. Semiconductor manufacture uses significant quantities of water [17], and it is proposed as one of the key planetary boundaries [18]. It is possible to extract data on both the GWP and the quantity of water used from life cycle assessment (LCA) databases, such as ecoinvent 3.1 [19].

An indicator of remaining reserve and consumption is proposed. The use of materials and their remaining availability can be used as a proxy for future sustainability of the metal in question. Further discussion of this is offered later in this article.

3.1.2 Economic

The price of metals and components is the main driver for recycling, aside from possible regulation of toxic substances [8], in the current business climate. Therefore, price of the metals per given weight and price of components is proposed as an economic indicator. Access to resources is a combination of many factors, but price (albeit perhaps slightly lagging from events) does reflect aspects of availability.

3.1.3 Technological

A technological indicator must capture the ability of a component to perform its function within the context of a smartphone. These devices are typically comprised of many different components performing different functions, each of which may be measured using multiple different metrics. For example, a CPU may be measured by clock speed, bus speed, number of cores, algorithm optimisation, and so on. Collecting all of the data for a good cross comparison between multiple processors becomes an intensive process especially when the factors may carry different weighting for importance. Therefore, it is proposed that price may act as a proxy of the technical capability of the component.

A component such as a CPU will reduce in price as it is superseded by more capable versions, indicating its ability to perform against the backdrop of the average of the products available in that product category. A similar trend may be observed across many other technologies as their specification sheet becomes poorer compared to new products. Furthermore, it is combined with the economic indicator, reducing the need for data collection, and price is easily accessible to any consumer or manufacturer, increasing ease of use. For this reason it is proposed for use in this value assessment.

3.2 Value Calculations

The calculation of value must allow a threshold to be set so that when it is met, an action is triggered. This can be achieved by simply comparing the value at time t against the value at the point of creation of the phone (t_0):

$$Value(t) = \frac{\text{Value of phone}(t)}{\text{Value of phone}(t_0)} \tag{1}$$

Equation 2 shows the value of a phone being the product of the components under consideration.

$$\text{Value of phone} = \prod_i (\text{Value of part}_i) \tag{2}$$

The equation is given as a product, as opposed to a sum as the value of any one component must have the ability to control the overall value calculation. For example, a component with relatively little value will not contribute significantly to a sum calculation. If the component becomes unable to perform its function and its technological value drops to zero, it will have little impact on the value of the phone even though the phone can no longer perform *its* function. Therefore, it is necessary to allow the component to control the overall value; in a product, if component value falls to zero, so does that of the phone. The value of the components is then calculated as

$$\text{Value of part}_\text{i}(t) = (GWP_i(t_0))(H_2O_i(t_0))(\text{rcy}_i(t))(\text{res}_i(t))(\text{price}_i(t))x_i \tag{3}$$

where GWP_i is the global warming potential required to create component i, H_2O_i is the water required to create component i, rcy_i is the recycling indicator for the metals used in component i, res_i is the resource indicator the metals used in component i and price_i is the monetary value of component i. The factor x_i may be allocated as considered appropriate, but it is proposed that component mass is the best metric. The time of assessment of the different factors within Eq. 3 is different. GWP_i and H_2O_i are assessed at the time of manufacture of the component. These two factors are essentially an indicator of what has gone into making the component. Avoiding replacing the component for as long as possible allows a spread of impact within these categories over a longer period of time and therefore a net reduction in impact per year of use. The factors rcy_i, res_i and price_i should be assessed at the given point in time of value calculation. These in effect represent the value that can be extracted from the components at time t.

Equation 4 shows the calculation to derive the rcy_i indicator. This acts to show the importance of recycling a metal from an environmental and economic perspective. Again simplicity is the goal of this indicator. Other methods of assessing recyclability tend to focus on the environmental responsibility of recycling, but less the economic advantages [20–22]. Furthermore, other indicators often include efficiency of the recycling system in recovering the metal. For the purposes of a PSS, it can be assumed that the collection rate is near 100 %; therefore the focus is on the value of recycling the metal itself:

$$rcy_i = \sum_l (\$/\mathrm{kg}_l)(GWP_l)y_l \tag{4}$$

where $\$/\mathrm{kg}_l$ is the price per kg of metal l, GWP_l is the global warming potential indicator for metal l and y_l is the fraction of metal l by weight in the component. GWP_l is expanded as

$$GWP_l = (GWP_{\mathrm{raw}} - GWP_{\mathrm{rec}}) \tag{5}$$

where GWP_{raw} is the GWP associated with producing one kg of metal l from virgin ore and GWP_{rec} is the GWP associated with creating one kg of metal l from recyclate. The form of the equation prioritises metals that have a large saving in GWP from recycling and cause a large GWP by being produced. If no recycling information is found, GWP_{rec} may be set to 0. This simultaneously gives priority to metals which are not recycled and also avoids penalising metals which are recycled but for which the information is not available.

Equation 6 shows the res_i indicator. This acts as a measure of the pressure on known reserves by current consumption:

$$\mathrm{res}_i = \sum_l \frac{U_l}{R_l} y_l \tag{6}$$

where U_l is the annual production of metal l and R_l is the known reserve of metal l in tons and y_l is the fraction of metal l by weight within the component. The indicator uses production as a proxy for consumption. Production data can be readily found within the same publications as reserve [23] and will often reflect consumption of the same metal over a similar period.

3.3 Incomplete Bill of Materials

It is apparent that the consideration of the value of a phone or components is only in terms of weighted fractions of the materials of interest. This presents two options: ascertain the benefit of every material present in a mobile phone or focus on those which represent the best potential for sustainability, i.e. the metals. These may be recycled indefinitely and cannot be synthesised chemically in the same manner as organic materials.

3.4 Costs

Each time a component within the phone is replaced, a cost is incurred. This is defined as a negative impact against each of the indicators, where appropriate. For example, emissions against GWP_i and H_2O_i will occur, and a monetary expense is created, e.g. through wages. There are not anticipated to be costs associated with the res_i, rcy_i or price_i indicators. Therefore, the cost can be calculated as

$$\text{cost}_i = (GWP_m)(H_20_m)(\text{expense}_m) \tag{7}$$

where GWP_m is the associated emissions, H_2O_m is the water use and expense_m is the monetary costs of *replacing* the component. It should be noted that the impacts of creating the components themselves are not included in the cost: those are factored into the calculation of value, as outlined in Eq. 2, and double accounting should be avoided where possible.

Upon replacement of a part, the value of the phone becomes the subtraction of the value calculated in Eq. 7 from that of Eq. 2:

$$\begin{aligned}&\text{Value of phone at point of replacement} =\\&\prod_i \{(GWP_i)(H_2O_i)(\text{rec}_i)(\text{res}_i)(\text{price}_i) - (\text{cost}_i)\}\end{aligned} \tag{8}$$

It is clear that upon replacement of the component, the cost may keep the value of the phone as a whole below the threshold for replacement. This would indicate that the gain from replacement is not worth the cost of the intervention. In these cases the value of the individual components must be made separately to ascertain their worth in stockpiling for use in other phones. It is entirely possible that a component will never be worth the cost of replacing, and it is here that the value calculation can inform the best development of the PSS for sustainability.

When a component is replaced within the phone, care must be taken to maintain the relative time factors, t. All of the components must have a t_0 set to the moment that they are incorporated into the phone. Therefore, the original components will have a t_0 set to the time of manufacture, and the replacement components will have a different t_0 set to the time of replacement.

3.5 Calculations at the Decision Points

Value is calculated at each of the decision points within the stock and flow system, Fig. 1. The factors used to calculate the value at each point do not change, but the form of the calculation will change depending on the particular decision point.

3.5.1 Reuse

If a phone is returned, for whatever reason the first value assessment may determine if it has sufficient residual value to be reused without modification. Those found to be above the threshold may be routed through f_3q.

3.5.2 Remanufacture

If found not be to sufficiently valuable for reuse, it can be considered for remanufacture. If the value can be restored sufficiently with replacement of one or more parts, including associated costs, it may be sent round route f_1q. If not, the phone is sent for disassembly and value assessment, f_2q. This calculation is the same as that for reuse, except the cost is applied on a component by component basis; therefore, it is a level of disaggregation compared to the reuse assessment.

3.5.3 Stockpile

Once disassembled the components are analysed for their ability to return the existing stock of phones to a sufficient value to re-enter the stock pool. Again, the equations used are the same as those already presented, only with a different time point of assessment. If the components are deemed insufficient, they are sent for reprocessing, $(1\text{-}f_4)f_2q$, to recapture the raw materials, otherwise they are sent for stockpiling. Whilst being stockpiled they may be periodically tested for retained value. If components are found to not have enough value, they are removed from the stockpile for reprocessing.

3.5.4 Post-stockpile

Upon exiting the stockpile, the components may be analysed for residual value in potential reprocessing or for export to lower value markets. It is possible that some components will never realise sufficient value via reprocessing to overcome the associated costs and that best extraction of value is through export to lower value markets, even with the associated uncertain disposal routes. The assessment of value at this juncture may be split into two separate equations which can be compared directly against one another. These are shown in Eqs. 9 and 10:

$$\text{Value in UK} = (\text{rec}_i)(\text{res}_i) - (\text{cost}_i) \tag{9}$$

$$\text{Value export} = (GWP_i)(H_2O_i)(\text{price}_i) - (\text{landfill}_i) \tag{10}$$

where landfill_i is the product of the impact from landfill of a component arising against the indicators as outlined previously in this article, much like cost, and the rest of the terms are defined the same as before. The larger value arising from the calculation determines whether the component are exported, $(1 - f_5)[(f_1 + f_2 f_4)q - r_1 p]$, or routed for reprocessing, $f_5[(f_1 + f_2 f_4)q - r_1 p]$.

It is at this decision point that social indicators would be most useful to consider. The loss of components to a low value market with poor recycling may result in a negative social impact. This is either through leaching of toxic substances to landfill [24], informal recycling or recycling with lower environmental standards than the EU [10]. An appropriate proxy for health may be human toxicity potential [15] but even this does not capture the possibilities for benefit that mobile telephony may bring to those markets [5]. These factors must be understood on a case-by-case basis and must overcome the current problem of relatively uncontrolled export of electronic waste which occurs even in Europe's highly regulated sector [9].

3.6 Calculation of Initial Value

The calculation of value has been presented as a method of determining the direction of flows between stocks within the PSS. This is a time-sensitive calculation, but before it can be made, the value within the phone itself must be understood to know which components or assemblies may be targeted for value extraction. The result of a component value assessment is presented here.

A sample set of components within a phone are modelled to demonstrate the application of value. The phone is modelled as consisting of a subset of components including the main PCB, mounted with processor and memory ICs, the display (LCD), the battery, antenna and the metal case. The component weights and compositions are based upon a Fairphone [25] with supplementary data for PCB [22], battery [26] and LCD [27] composition used where appropriate. The antenna and LCD are assumed to contain a PCB and ICs as part of their construction. Information for GWP and H_2O indicators is taken from the ecoinvent 3.1 database accessed via SimaPro LCA software, PRé Consultants. Data on resources is sourced from Minerals and Commodity Summaries 2015 [23]. Due to volatility of price of materials, these were accessed at time of writing from a variety of online sources. The value is calculated as a snapshot at the point of manufacture of the phone; therefore, it does not consider how price of components may vary with time.

All of the metals found in the literature relating to the components are considered: copper, gold, lithium, cobalt, silver, indium, stainless steel, iron, chromium, nickel, manganese, aluminium, lead, tin, zinc, molybdenum, antimony and palladium.

Figure 2 shows the phone by component mass. The mass includes both metals and non-metals. It is shown that the housing contains the majority of phone mass (38.4 %) followed by the battery (26.4 %), with the PCB the lightest at 4 %.

Figure 3 shows the value of the components. It is clear that the PCB dominates the value, taking over 84 % of the total. Most of the remainder is taken by the

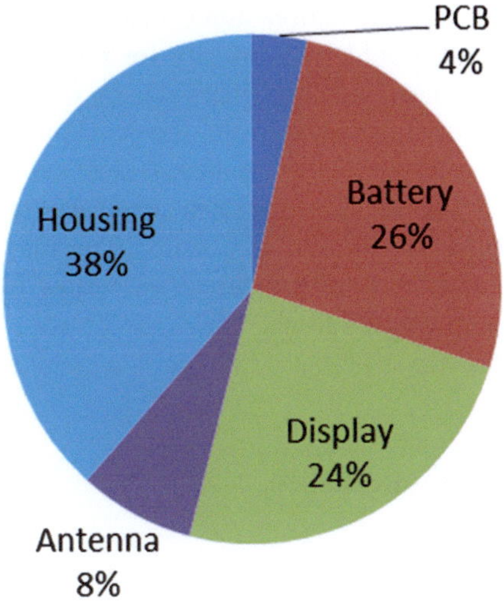

Fig. 2 Component mass fraction within the example phone

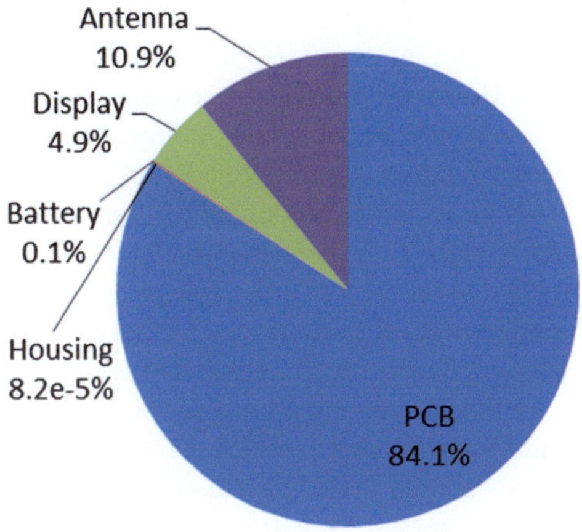

Fig. 3 Percentage value of the components

antenna and display. The housing no longer contributes any significant value to the phone at only 8.2×10^{-5} %.

The value of the phone is dominated by the precious metals it contains: gold, silver and palladium. Despite the tiny combined content of those metals within the phone, a total of 0.032 %, their high price per kg (Au = $42279, Ag = $584, Pd = $8096 at time of writing) and large GWP for extracting virgin material (Au = 17100, Ag = 485, Pd = 8613 kgCO_2-eq/kg [19]), as opposed to recycling, makes their contribution to the net value the largest. Any PCB within the phone has all three of these metals; therefore, any component containing a PCB has an increased total value both economically and environmentally. By contrast, the housing becomes negligible due to the low price and ready availability of the steel from which it is made.

The calculation shows that in the design of the PSS, PCB-containing components would be a priority for reuse wherever possible, and the flows should be directed accordingly between the stocks to achieve this. This is especially true at the time of removal from the stockpile, when the PCB would always be better reprocessed for the materials than exported to a lower value market.

The design of the phone itself should be to maximise the opportunity for reuse of the PCB where at all possible. This may be achieved, for example, by design of the architecture of the circuitry to allow for the efficient replacement of components with long lifetimes.

4 Conclusion

The stock and flow model for a circular economy has been discussed for its ability to identify decision points which must be addressed in order to direct flows through the most appropriate loops for sustainability. These decision points require a metric which can be applied to the value contained within a mobile phone so that thresholds may be set in order to ensure that the value is extracted at an appropriate time. However, mobile phones are made up of multiple components each with a different potential or actual lifetime. In the current business model, phones are often replaced before the components are truly obsolete [28] or can no longer perform a function. In consideration of the components, they may have a lifetime which is dictated by loss of performance, for example, the battery which will degrade over a predictable period of time [29], or by technological advance, or even through user interference, for example, accidental damage to a display. Each of the lifetimes will have a different predictability of the moment when replacement is needed.

Building upon the prior work of the authors [13, 28], we suggest that such a PSS could be developed in which components with low emotional value and that require regular upgrade, the internal electronic components, could be owned by the business and leased as part of the service to the consumer. By the handset being provided by the PSS supplier as part of the service, this allows for control and retention of the valuable metals within the phone and the best opportunity for extraction of the value from the phone at each point within the PSS.

Within this suggested PSS, the consumer has two touchpoints: during purchase (e.g. e-commerce shopping, traditional bricks-and-mortar retail, local franchises or independent community upgrade shops) and during servicing. A servicing operation is triggered when a consumer requires or desires a hardware upgrade (due to one or a combination of reasons for obsolescence; absolute, functional, technological, societal, etc. [30–32]) or if there is a breaching of a given operational threshold for the phone as a whole or a component within it (as defined by the service contract). By returning the phones for regular upgrade, the service will satiate consumer demand for the latest hardware whilst also enabling businesses to retain and assess components for value individually, thereby defining the treatment that the components will undergo within the backroom activities of the PSS and facilitating efficient metal use and recovery.

Therefore, the components must be assigned a time-varying value which can be measured against a threshold, the reduction below which will trigger an interaction between the service provider and the consumer to bring their service level back to required specification. This creates an interaction of the consumer with the PSS in which the consumer returns the phone to a suitable servicing platform, and the phone is examined with internal hardware (such as the deteriorated battery) and software upgrades installed as requested or as necessitated.

The refurbished phone is returned to the customer, thereby maintaining its optimum performance whilst also ensuring the return of handsets and/or components to the business for further treatment, stockpiling or reprocessing as appropriate. In order to understand when to extract the best value possible at each point within the PSS, this article has explored the concept of quantitative value incorporating the economic and environmental pillars of sustainability. The addition of a new technological pillar is also proposed in order to assess the value of a mobile phone in terms of its ability to operate at a required performance level.

The value of a simplified phone is calculated. It is shown that precious metals, gold, silver and palladium, dominate the value despite being only a tiny fraction of the total mass. In this example, they are predominantly found within PCBs and components containing them. The dominance of value is due to their high price per kilogram and the large GWP associated with extracting them from the ground compared to recycling. The high value of the PCB highlights the need to ensure that these components are used for as long as possible and that when considered for disassembly, they may be removed from the phone with minimal intervention.

Acknowledgments The authors wish to thank the Engineering and Physical Sciences Research Council (EPSRC) for their financial support of this work as part of the CLEVER (Closed Loop Emotionally Valuable E-waste Recovery) project, funded under grant number ep/k026380/1. The authors confirm that data underlying the findings are available without restriction. Details of the data and how to request access are available from the University of Surrey repository: http://epubs.surrey.ac.uk/id/eprint/808005.

References

1. Oguchi M, Murakami S, Sakanakura H, Kida A, Kameya T (2011) A preliminary categorization of end-of-life electrical and electronic equipment as secondary metal resources. Waste Manag 31:2150–2160
2. Tanskanen P (2013) Management and recycling of electronic waste. Acta Mater 61:1001–1011
3. Ongondo FO, Williams ID (2011) Greening academia: use and disposal of mobile phones among university students. Waste Manag 31:1617–1634
4. Geyer R, Blass VD (2010) The economics of cell phones reuse and recycling. Int J Adv Manuf Technol 47:515–525

5. Osibanjo O, Nnorom IC (2008) Material flows of mobile phones and accessories in Nigeria: environmental implications and sound end-of-life management options. Environ Impact Assess Rev 28:198–213
6. Navazo JMV, Méndez GV, Peiró LT (2014) Material flow analysis and energy requirements of mobile phone material recovery processes. Int J Life Cycle Assess 19:567–579
7. Chancerel P, Rotter VS (2009) Assessing the management of small waste electrical and electronic equipment through substance flow analysis – the example of gold in Germany and the USA. IEEE International Symposium on Sustainable Systems and Technology 2009, Tempe. IEEE, New York, pp 380–385
8. Waste Electrical and Electronic Equipment Directive 2012/19/EU.
9. Huisman J et al (2007) Waste electrical and electronic equipment (WEEE) final report. United Nations University, Bonn
10. Wong MH, Wu SC, Deng WJ, Yu XZ, Luo Q, Leung AOW, Wong CSC, Luksemburg WJ, Wong AS (2007) Export of toxic chemicals – a review of the case of uncontrolled electronic-waste recycling. Environ Pollut 149:131–140
11. Stahel WR (2010) The performance economy, 2nd edn. Palgrave MacMillan, Basingstoke
12. Manzini E, Vezzoli C (2003) A strategic design approach to develop sustainable product service systems: examples taken from the 'Environmentally Friendly Innovation' Italian prize. J Clean Prod 11(8):851–857
13. Wilson GT, Bridgens B, Hobson K, Lee J, Lilley D, Scott JL, Suckling J (2015) Single product, multi-lifetime components: challenges for product-service system development. PLATE conference 2015. Nottingham Trent University, College of Art and Design and Built Environment (CADBE), Nottingham
14. Bhamra TA, Lofthouse VA (2007) Design for sustainability a practical approach. Gower Publishing Limited, Aldershot
15. Goedkoop MJ, Heijungs R, Huijbregts M, Schryver, A De, Struijs J, VanZelm R (2009) ReCiPe 2008, a life cycle impact assessment method which comprises harmonised category indicators at the midpoint and endpoint level. 1st edn. Report 1: Characterisation
16. PAS 2050:2008 Specification for the assessment of the life cycle greenhouse gas emission of goods and services
17. Williams ED, Ayres RU, Heller M (2002) The 1.7 kilogram microchip: energy and material use in the production of semiconductor devices. Environ Sci Technol 36:5504–5510
18. Rockström J et al (2009) A safe operating space for humanity. Nature 461:472–475
19. ecoinvent v3.1, ecoinvent, Zurich, Switzerland. http://www.ecoinvent.ch/
20. Legarth JB, Alting L, Danzer B, Tartler D, Brodersen K, Scheller H, Feldmann K (1995) A new strategy in recycling of printed circuit boards. Circ World 21:10–15
21. Huisman J, Boks CB, Stevels ALN (2003) Quotes for environmentally weighted recyclability (QWERTY): concept of describing product recyclability in terms of environmental value. Int J Prod Res 41:3649–3665
22. Le H-L, Yamasue E, Okumura H, Ishihara KN (2013) MEMRECS – a sustainable view for metal recycling from waste printed circuit boards. J Environ Prot 4:803–810
23. United States Geological Survey (2015) Mineral commodity summaries 2015
24. Kiddee P, Naidu R, Wong MH (2013) Metals and polybrominated diphenyl ethers leaching from electronic waste in simulated landfills. J Hazard Mater 252–253:243–249
25. Guvendik M (2014) From smartphone to futurephone: assessing the environmental impacts of different circular economy scenarios of a smartphone using LCA. MSc thesis Industrial Ecology, Delft University of Technology and Leiden University
26. Castillo S, Ansart F, Laberty-Robert C, Portal J (2002) Advances in the recovering of spent lithium battery compounds. J Power Sources 113:247–254
27. Savvilotidou V, Hahladakis JN, Gidarakos E (2014) Determination of toxic metals in discarded liquid crystal displays (LCD). Resour Conserv Recycl 92:108–115
28. Suckling J, Lee J (2015) Redefining scope: the true environmental impact of smartphones? Int J Life Cycle Assess. doi:10.1007/s11367-015-0909-4

29. Takeno K, Ichimura M, Takano K, Kamaki J (2005) Influence of cycle capacity deterioration and storage capacity deterioration on li-ion batteries used in mobile phones. J Power Sources 142:298–305
30. Cooper T (2004) Inadequate life? Evidence of consumer attitudes to product obsolescence. J Consum Policy 27:421–449
31. Packard V (1967) The waste makers. Penguin, London
32. van Nes N, Cramer J (2006) Product lifetime optimization: a challenging strategy towards more sustainable consumption patterns. J Clean Prod 14:1307–1318

How Japanese Companies Can Contribute to Water Sustainability

Yoshihiko Sakamoto and Takashi Iwamoto

Abstract Technologies and qualities of Japan's water infrastructures are of the highest levels in the world. However, the presence of Japanese water businesses in the world is very low because the businesses have been deployed mostly in Japan. In this research, case studies were done to reveal elements necessary to deploy water businesses in the world smoothly. Japanese companies have the strengths in equipment and EPC (engineering, procurement, and construction) but the weakness in operation because most of Japanese companies have relied on municipalities for years. As a result, Japanese companies cannot deploy their businesses as systems. Japanese companies should increase the involvement in operations. In other words, Japanese companies should reject to the concept of "All-Japan" and find global partners to operate water infrastructures for years.

Keywords Water businesses • Equipment • EPC • Operation • Global partner

1 Introduction

Table 1 shows water resources in the world and Japan [1]. The annual amount of precipitation is on average 1668 mm per year, double the amount of the world average. On the other hand, the amount of water supply is only 3401 m3 per person in Japan, half the world average. Japan's total land area is small and has many steep slopes making it very difficult to store rainwater, and Japan has limited water supply as a result despite of the ample rainfall. Despite of the limited water supply, Japanese people do not feel the shortage in their daily lives, and this may be the result from the extremely high levels of technologies and services of infrastructure.

The Japanese water businesses are limited to the Japanese domestic market and have minimal presence in the global market. If high-level technologies and services of water infrastructures in Japan expand into the world markets, it will likely contribute widely to consistent water supply.

Y. Sakamoto (✉) • T. Iwamoto
Graduate School of Business Administration, Keio University, Yokohama, Japan
e-mail: yoshihiko.skmt@keio.jp

M. Matsumoto et al. (eds.), *Sustainability Through Innovation in Product Life Cycle Design*, EcoProduction, DOI 10.1007/978-981-10-0471-1_21

Table 1 Water resources in the world and Japan

	World	Japan
Total area (1000 km2)	133,769	378
Total population (mio inhab)	7049	126
Average precipitation in depth (mm/year)	807	1668
Precipitation in depth (m3/year · inhab)	15,320	4986
Total water resources (km3/year)	54,813	430
Water resources per capita (m3/year · inhab)	7634	3401
Total water withdrawal (km3/year)	3885	90
Water withdrawal per capita (m3/year · inhab)	551	712

The Japanese government has implemented the support for Japanese companies expanding globally with the "strategic export of infrastructure." The author et al. published a paper to discuss global deployment of Japan's infrastructure systems including water infrastructure system, and the key points for global deployment of Japan's infrastructure systems were the following [2]:

- Scale
  Any global infrastructure system business is "scale" business, and all the firms which occupy the market share have capabilities of technologies related to the systems and global sales and support capabilities.
- Influence with operation
  It is important to have influences with operations because operation firms will control the specifications of infrastructure systems.
- Collaboration with government
  Government support is necessary in two aspects, sales and finance. Infrastructure is controlled by government in any country, and government-to-government negotiation is important to receive orders. And R&D expense, CAPEX, and OPEX are basically high in any infrastructure system and financial support until business models are established is important.

In this research, study specific to water infrastructure businesses was deepened, and the key points for Japanese companies to expand their businesses into the global water markets were extracted.

2 Research Approach

The approach of this research is shown in Fig. 1. Based on gathered public information, the best practices were selected. The best practices were benchmarked, and points for success and barriers overcome to succeed were discussed. At the same time, hypothesis was created and verified through interviews with experts in this field. Finally, the key points are summarized, and the strategic options that Japanese options should focus on are proposed.

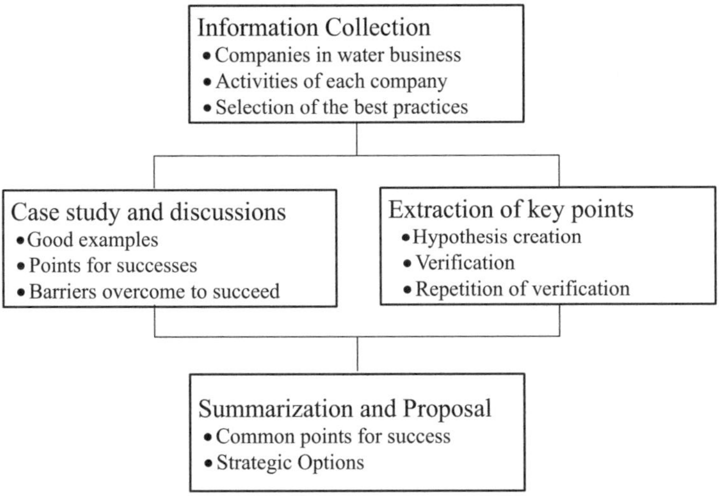

Fig. 1 Research approach

Water businesses are a generic name encompassing all businesses related to water. The scope of this research includes supplying water for waterworks, industrial and agricultural water, desalination (such as seawater desalination), wastewater treatment, recycled water, and environmental conservation for river basins and more.

Participant companies' business lineups in the industry span widely from the examination and planning of water supply development and equipment installation to plant engineering and the operation, maintenance, and management of facilities.

Among the expansive water businesses, this research defines water businesses as the businesses by private companies focused on waterworks and sewage (including waste water management) as well as industrial water supply and desalination.

3 Water Businesses in Japan

According to the trial calculation of the Ministry of Economy, Trade, and Industry (METI), the water business industry was a 36 trillion yen industry in 2007 and is expected to rise to 87 trillion yen by 2025. Figure 2 shows the forecast of water business market. It is expected that the markets of equipment and EPC will grow from 17 trillion yen to 49 trillion yen, and the markets of O&M (operation and maintenance) will grow from 19 trillion yen to 38 trillion yen [3].

Figure 3 shows the domains of the water businesses, and water businesses consist of the following three domains:

- Equipment which includes materials, parts, and equipment
- EPC (engineering, procurement, and construction)

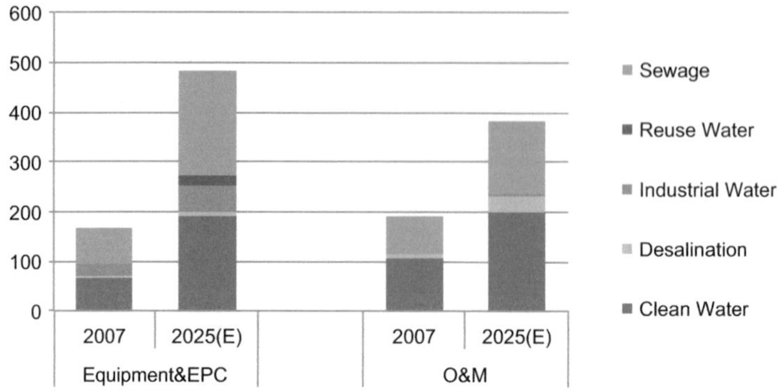

Fig. 2 Forecast of the water business market

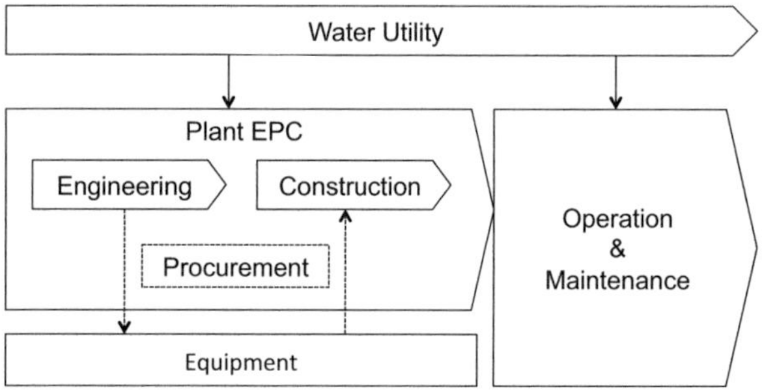

Fig. 3 Domains of water businesses

- O&M (operation and maintenance)

Figure 4 shows the LTVs (lifetime values) in water businesses. O&M occupies the largest portion in terms of LTV in water businesses, while Equipment and EPC occupy smaller portions, because O&M contributes over a long period of time, although the single-year contribution is small.

Figure 5 shows companies in water businesses. Most of Japanese companies focus on one domain, while overseas companies cover multiple or all domains.

Table 2 shows the requirements to enter water businesses. Japanese companies have high global market shares in the technology fields such as the reverse osmosis membrane. Moreover, Japanese companies are highly regarded for their project management skills and energy reduction technology in the EPC domain and have proven performance in areas such as the Middle East. However, municipalities have a history of bearing burden in the O&M domain, and entrants of private firms have been scarce as a result. While companies such as general trading companies have begun to establish industrial companies in the specific areas, domestic privatization has lagged. Looking abroad, Japanese general trading companies have entered

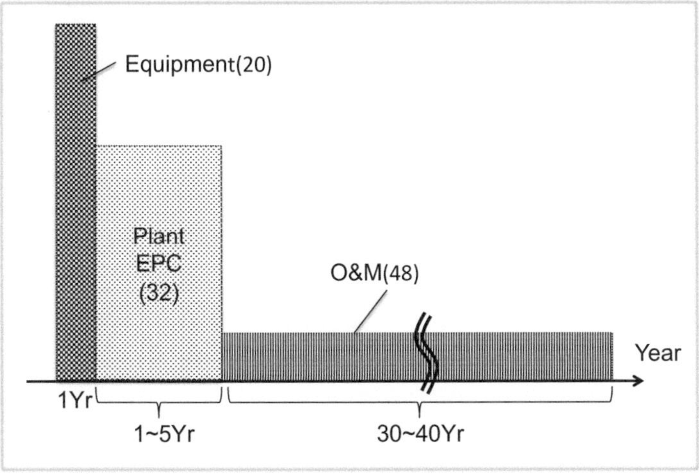

Fig. 4 LTVs in water businesses

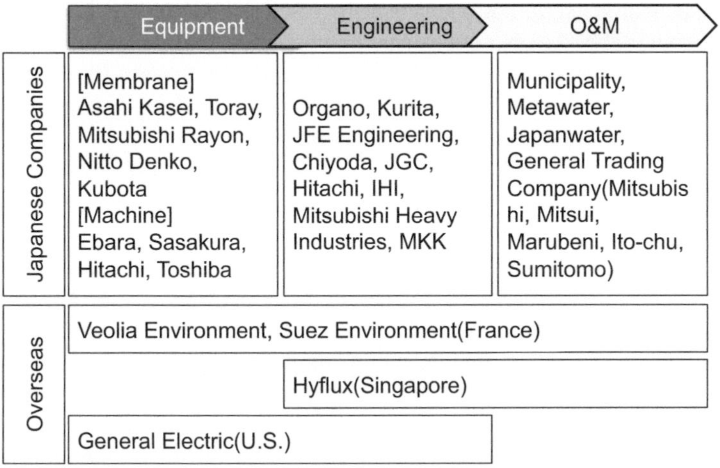

Fig. 5 Firms in water businesses

Table 2 Requirements to enter water businesses

	Requirement
Equipment	High technology
	Advanced safety for clean water
	Removal of most materials
EPC	Adaptation to regulations
	Project management
O&M	Long-term commitment
	Experiences of daily operations
	Proper maintenance

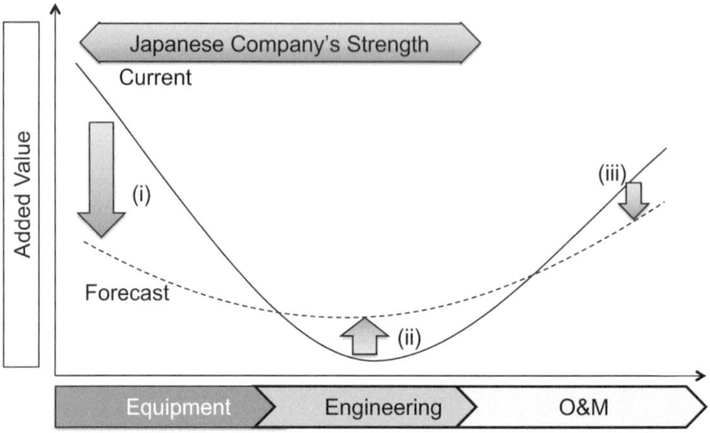

Fig. 6 Added value curve

limited markets through means of acquisitions, and some municipalities have licensed technologies globally; however the results have been dismal.

Compared to the top 2 water business companies – Veolia and Suez – that cover all business domains and emerging forces of Hyflux and General Electric (GE) that are putting up a challenge, Japanese companies are only able to supply machinery or partial engineering partner or as a subcontractor.

Figure 6 shows the added value curve. The current strengths of Japanese companies in equipment such as membranes and machinery are expected to weaken as Chinese and Korean manufacturers adapt technologically (i). The added value of their other strength in EPC will increase as the trends strengthen for energy reduction and efficiency (ii). Due to increased competitiveness, the added value is expected to decrease slightly in O&M (iii). Thus O&M players will expand their business domain to EPC in order to avoid competition by integrated management. In addition, recent EPC tenders occasionally require commitment to long-term O&M.

Based on these expectations, accelerated entry into the O&M domain needs to go forward. With this understanding, the Japanese government implemented a support program to help All-Japan export of infrastructure.

But there have been no specific results worth mentioning.

4 Benchmarking of Global Companies

Three companies – Veolia, Hyflux, and General Electric (GE) – were selected for benchmarking.

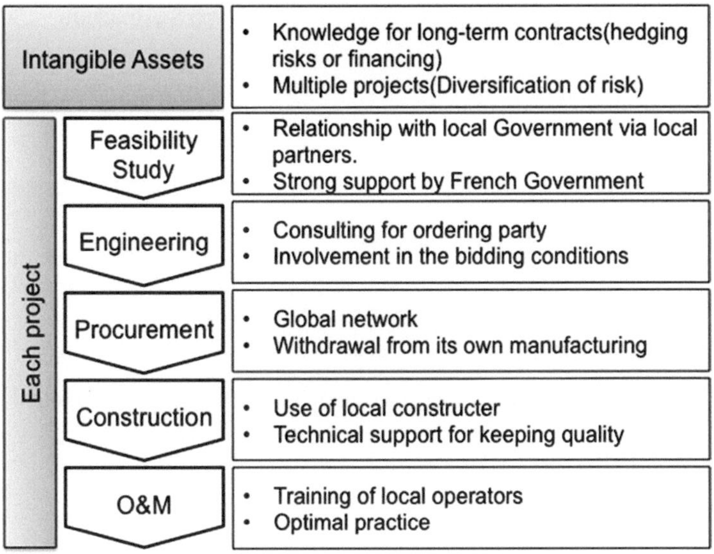

Fig. 7 Veolia's strategy

4.1 Veolia Environment [4, 5]

Veolia was founded in 1853 in Leon, France, to do businesses in waterwork maintenance, hygiene improvement, and infection prevention. Veolia has the world's top water-supplied population and a thorough differentiation strategy with its project composition, plant construction, and business management aspects as shown in Fig. 7.

In terms of project composition, Veolia enters target markets through acquisitions or joint venturing to obtain project information from local governments quickly and becomes involved in the setting of the bidding conditions to leverage the company's strengths. In addition, Veolia sometimes receives support from French government for important projects, appeals to organizations such as the World Water Forum, and focuses on building government relationships. Veolia has plentiful past achievements, know-hows to hedge business risks, the creditworthiness of the company, and the know-hows to benefit from public funds.

For plant construction, Veolia looks for the best materials globally and is able to realize low costs by helping and supporting local companies.

After the management of a new plant begins, Veolia is able to realize efficient management based on the accumulated know-hows by training the local staff and creating a stable O&M management structure.

Moreover, their businesses consist of water, waste, and energy, and thus Veolia is able to propose a customized system based on the customer's needs for any water business domain.

From Veolia's case, the following success factors were extracted:

1. Information gathering
   - Acquisition of local companies or through joint ventures
   - Government relationship-building activities
2. Building up know-how
   - Understanding and classifying risks
   - Obtaining financing
   - Efficient operation
3. One-stop proposal
   - Customized proposals based customer needs including compound electric systems

4.2 Hyflux [6]

Since Singapore was founded as a country, over 50 % of the country's water supply has been imported from Malaysia. However, as Malaysia started to negotiate to increase water prices, Singapore government decided to look for ways to increase the water self-sufficiency rate.

Hyflux was founded in 1989 and initially focused on chemical products for water treatment and membrane treatment. To respond to the Singapore government's policy, Hyflux expanded their business domains to planning, manufacturing, and management of plants. This is especially evident post-2000 as Hyflux received orders and therefore expanded their business domains for sewage-reclaimed waters and large-scale seawater desalination. With the cooperation of a number of external partners, Hyflux built up performance domestically, and now it is expanding aggressively to the global market.

The characteristics of Hyflux are in the capabilities of providing integrated services from desalination businesses including seawater desalination and reclaimed water to plant planning and manufacturing and even O&M. By participating in O&M and taking into account LTV, it is able to have a strategic bid price for EPC giving them a favorable competitive advantage to achieve orders.

In terms of financing, it is able to construct project financing not only internally but also externally as well.

From the Hyflux's case, the following success factors were extracted:

1. EPC bidding based on LTV
   - Competition based on EPC alone is difficult.
   - Obtaining O&M know-how.

2. Involving flexible partners
   - Finding partners for different aspects including but not limited to plant construction, management, and financing

4.3 General Electric [7]

Well known as a comprehensive manufacturer, GE entered the water business with the acquisition of Canada's Glegg Water in 1999. Since, GE has increased their water business domain presence from chemical products, membranes production for water treatment to plant EPC. GE has top market share for chemical products and membrane treatment and has a strong competitive advantage as an equipment supplier.

GE's distinctive characteristic is that they have grown by packaging water treatment as a solution with their other divisions such as the power plant and medical equipment users. In other words, by cross-selling with other group companies as a total solution and not water treatment as a stand-alone product, GE has been able to maximize their added value. As a matter of fact, the company's water business is named "Power and Water" suggesting that the two are inseparable.

From the GE's case, the following success factors were extracted:

1. Using other internal businesses as resources
   - Strengthening the businesses not by selling the water treatment products as stand-alone products but by cross-selling products of other divisions.

5 Conclusions

Japanese companies have strengths in equipment and EPC and weakness in O&M because municipalities were depended on in O&M. As a result, EPC once seen as strength is also now being said to have fallen behind global competitors. To break through this situation, the Japanese government has proposed the All-Japan movement for both public and private companies to work together to expand globally.

The success factors in the water business extracted in this research were summarized in the following three points:

- The level of coordination of the entire project
- Accepting and involving a flexible partner
- Proactive government relationship building

In terms of the coordination level of the entire project, the project valuation that Veolia and Hyflux utilize to evaluate the O&M and LTV is a key to expand water businesses globally. Furthermore, the know-how of the entire project that is developed and accumulated over a number of years will become an intangible asset and therefore the foundation for a new competitive advantage for Japanese companies.

With the performance of Japanese companies falling behind, finding and involving partners for operations and other similar business are essential. This is especially true in areas that require water treatment technology yet have insufficient electricity. In these cases, much like how Veolia has the biomass power generation and component selling as well as GE's cross-selling in a broader sense, Japanese companies need to consider such opportunities as well.

With the third factor of proactive government relationship building, government support is a necessity with the current limited know-how of the Japanese companies and the fact that entering the O&M market is an extremely long-term obstacle. Moreover, to maintain the current strength in the Equipment domain, the development of a new standard and as a result a proactive lobbying will be required. The key is to involve governments of markets with a small GDP/capita in order for Japanese companies to accumulate and build up business performance in these areas.

From these findings three strategic options will be proposed as follows:

(a) Public-private partnership

The first strategic option deals with public and private partnership as shown in Fig. 8. This is similar to the current All-Japan infrastructure export initiative of the Japanese government. The merit of this strategy is being able to propose a packaged based on the proven high level of Japanese water infrastructure technology. However, there are countless players in Japan, and as there is no leading firm, and therefore not having a successful example, it is a major problem.

In addition, it is difficult to determine if the know-how of the municipalities O&M has sufficient competitiveness and efficiency making it a potential obstacle for competitiveness.

(b) Global collaboration

The second strategic option is to have Japanese companies participate as a total coordinator managing projects from all standpoints as shown in Fig. 9. This is a method that general trading companies have already taken, and this method does not restrict to All-Japan or the collaboration of Japanese firms only but covers to find global competitive firms that have the core elements for the overall strategy. The merit of this method is that best practices can consistently be utilized, and the know-how of global partners can be absorbed and accumulated. The difficulty of this strategy is that there needs capability of total coordination and that financial risks including investment and financial arrangements should be taken.

(c) Acquisition

The third option is expanding and growing business domains through acquisitions as shown in Fig. 10. This is just as how Veolia has acquired local firms to expand their business domains. The merit of this is that one firm will be able to cover all business domains, while the demerit is that the financial risks will significantly increase with this strategy.

Fig. 8 Public-private partnership

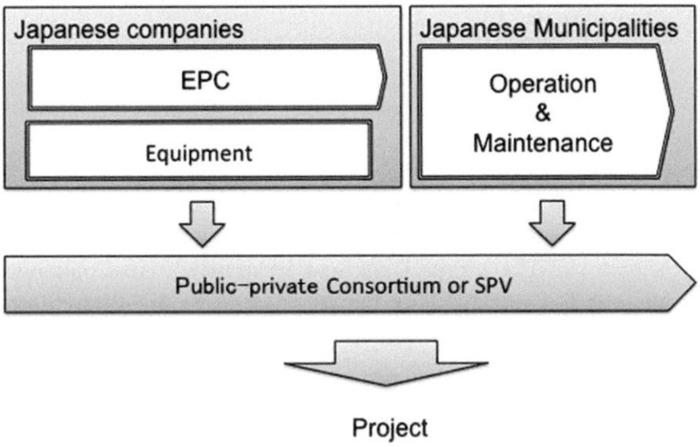

Fig. 9 Global collaboration

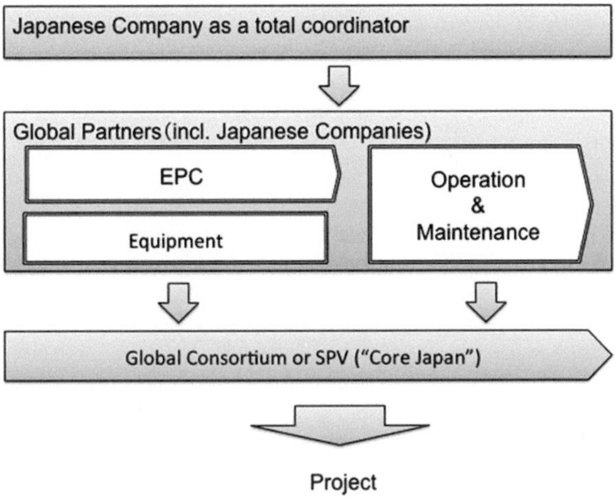

Fig. 10 Acquisition

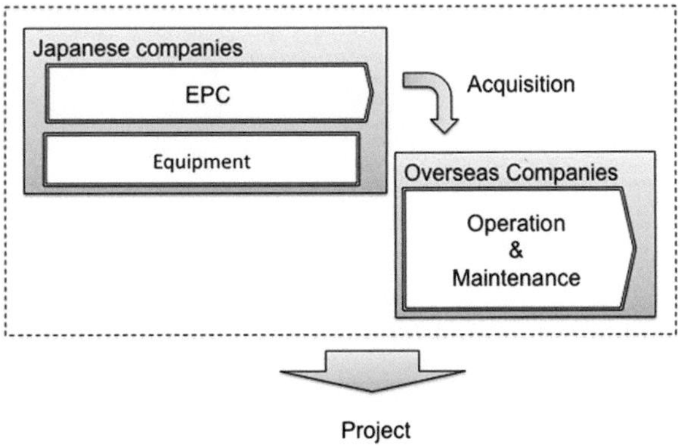

References

1. Food and Agriculture Organization of the United Nations (2015) Aquastat. In: Website of FAO Water
2. Iwamoto T, Aoki H, Hanao E, Yin F, Sakamoto Y (2015) Global deployment of Japan's infrastructure systems. In: 2015 Proceedings on PICMET'15: management of the technology age, Portland, Keio in Japan
3. Ministry of Economy, Trade and Industry (2010) Challenges and concrete measures for the international expansion of the water business. In: Website of the Ministry of Economy, Trade and Industry
4. Veolia (~2015) Information about Veolia. In: Website of Veolia
5. Yoshinobu Konomi (2010) Veolia. Keio Business School Case Library
6. Hyflux (~2015) Information about Hyflux. In: Website of Hyflux
7. General Electric (~2015) Information about General Electric. In: Website of General Electric

Analysis of Disassembly Characteristics and PSS Proposal by Component Reuse of Mobile Phones

Hideyuki Sawanishi, Yuji Sasaki, and Nozomu Mishima

Abstract In Japan, the annual number of used smartphones is rapidly increasing. Such used smartphones often outflow to overseas, and domestic material recycling system will be endangered, if smartphones become dominant in the recycling market. Component reuse (remanufacturing) is a hopeful way to extract larger value from used products. In focusing on component reuse, the paper carries out disassembly experiments to know the bottlenecks of disassembly. The paper proposed design improvement plans and also proposed conceptual PSS (product service system) ideas. It concluded that component reuse of used mobile phones can be a good way to bridge the gap between product reuse and material recycling.

Keywords Component reuse • Manual disassembly • Used mobile phones • Easy to disassemble • PSS

1 Introduction

Management of used electronics is an emerging problem in all over the world. In Japan, nearly 40 million mobile phones face end-of-life stage per year [1]. Circular use of such used products is an urgent problem. Although material recycling is a basic way to utilize them, total material value included in one used mobile phone only amounts 100 yen or so [2]. Therefore, it might be difficult to cover recycling cost such as transportation cost, labor cost, facility cost, and so on, only by recovered materials. In addition, the annual number of used smartphones that often have high retail value as second-hand product is rapidly increasing. Such used smartphones often outflow about 6,700,000 to overseas [3]. This fact means that domestic material recycling system will be endangered, if smartphones become dominant in the recycling market. Of course, one solution is to reuse such products as second-hand products. But, since product lives of mobile phone is rather short, the next step should be considered too. Actually, product reuse cannot be a major choice for used mobile phones to be treated. Although the numbers of mobile

H. Sawanishi • Y. Sasaki • N. Mishima (✉)
Graduate School of Engineering and Resource Science, Akita University, Akita, Japan
e-mail: nmishima@gipc.akita0u.ac.jp

M. Matsumoto et al. (eds.), *Sustainability Through Innovation in Product Life Cycle Design*, EcoProduction, DOI 10.1007/978-981-10-0471-1_22

phones sold in the market are steadily increasing, the product is becoming more and more common as it is shown in Fig. 1 [4]. The number of mobile phones plus other personal information devices in all the households in Japan has reached more than 150 million units at the end of fiscal year 2014. Recently, smartphones occupy more than half of the market [5]. However, the collected number of mobile phones is gradually decreasing (Fig. 2 [6]), and the number of reused mobile phones per year is still about 2 % (0.14 million units) of the number of collected mobile phones as of 2012. It means that even the ratio of smartphones which are possible to reuse as second-hand products is getting higher and higher, "product reuse" has not become common as it is expected. To increase collection rate and establish better circular use of mobile phones and valuable materials included in the products, many countermeasures should be considered and combined suitably. Component reuse (remanufacturing) in between product reuse and material recycling is a hopeful way

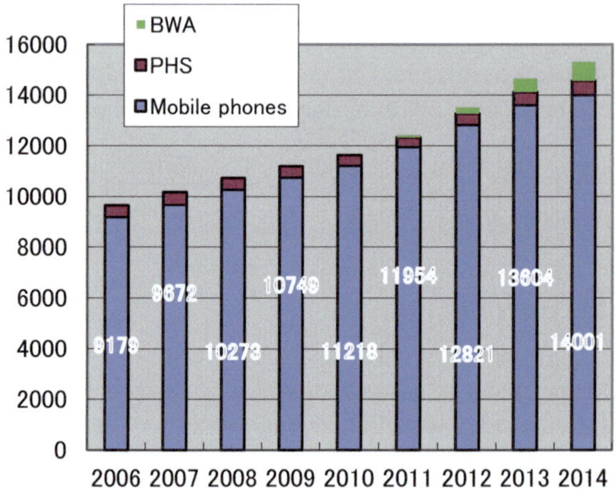

Fig. 1 Increasing commodity of mobile phone

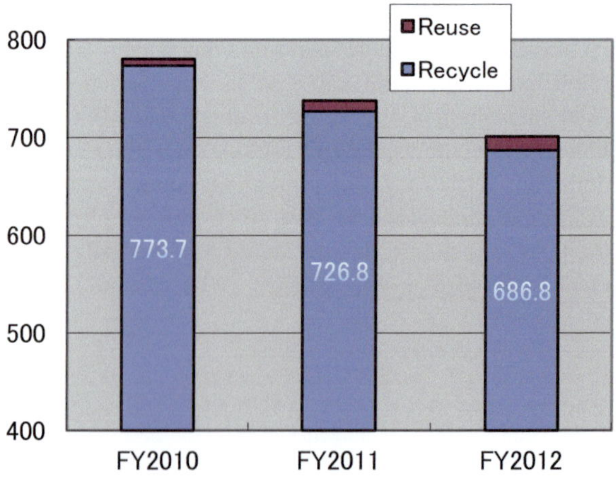

Fig. 2 Situation of reuse and recycle of used mobile phones

to extract larger value from used products. This paper focuses on component reuse of mobile phones. In order to reuse components of used mobile phones, manual disassembly process is irreplaceable. Time of disassembly directly affects the cost of component reuse, and sometimes it might affect the quality of recovered components. Thus, the paper carries out disassembly experiments to clarify the bottlenecks of manual disassembly of used mobile phones and tries to propose some design improvement plans as well. The paper proposes a business model to utilize recovered components.

2 Disassembly of Used Mobile Phones

2.1 Disassembly Experiments

Since manual disassembly and detachment of components are necessary procedures for component reuse, the paper carried out some disassembly experiments. The purpose of the experiments was to know the difficulties in disassembly for both feature phones and smartphones. And then, disassembly characteristics of both types were compared in order to clarify what will be the situation when smartphones become dominant in end-of-life market, in the future. (The ratio of smartphones in annual production has already exceeded 50 % [5].) Since LCD is the most promising target of component reuse, disassembly time of LCD and total disassembly time were separately measured. In the experiments, measured time of disassembly was compared by standard disassembly time calculated by WFA (work flow analysis) [6]. Table 1 shows the standard disassembly time of each step of disassembly work, corresponding to types of joints and other features of detachment work. In the analysis, the total process of disassembly is considered to be a repetition of the eight steps, "grasp," "detach joint," "pill out," "adjust direction," "move," "regrasp," "put," and "change attitude." Basically, raw data (measured time) was considered as the disassembly time for each component. However, sometimes the disassembly operator took a long time in finding the right way to disassemble, or because of damaged parts such as worn out screws, certain disassembly step took very long time. When the measured disassembly time exceeds three times of the estimated disassembly time based on the standard time shown in Table 1, it is recognized that the disassembly step was not normal. In such case, Eq. 1 was used in determining the expected disassembly time. To decide the standard time and eliminate the difference of skill of workers, six people carried out 12 disassembly experiments. The average time for total disassembly and to detach LCD is shown in Table 2.

Table 1 Standard disassembly time estimated by WFA

Disassembly motion	Variations	Standard disassembly time [min]
Grasp	–	0.01
Detach joint	Snap fit	0.02–0.08
	Screw	0.04/unit (0.1 min for the first one)
	Ring joint	0.08
	Press fit	0.12
	Calking	0.12
	Solder	0.08
	Glue	0.12
	Welding	0.12
Pullout	Not necessary	0
	Fitting is loose	0.01
	Fitting is tight	0.02
Adjust direction	Upper surface	0
	Inclined upper surface	0.005
	Side surface	0.01
	Inclined lower surface	0.015
	Lower surface	0.02
Move	Light and small	0.01
	Light but large	0.015
	Heavy and small	0.015
	Heavy and large	0.02
Regrasp	Easy recognizable shape	0
	Not easy recognizable shape	0.02
Put	–	0.01
Change attitude	Same surface	0
	Other surface	0.02

Table 2 Average time of 12 disassembly experiments

	Total disassembly time	Minimum disassembly time of LCD
Average	12 min 18 s	5 min 20 s

$$Te_k = Ts_k \frac{\sum_{i=1}^{i=k-1} Tp_i + \sum_{i=k+1}^{i=n} Tp_i}{\sum_{i=1}^{i=k-1} Ts_i + \sum_{i=k+1}^{i=n} Ts_i} \quad (1)$$

where Te_k is the expected disassembly time of step k, Ts_k the standard disassembly time for step k based on WFA, Tp_i the measured disassembly time of step i, and Ts_i the standard disassembly time of step i.

Table 3 Disassembly time of feature phones

	Total disassembly time	Minimum disassembly time of LCD
Provider S, type 814	7 min 48 s	6 min 20 s
Provider A, type I	13 min 4 s	6 min 26 s

2.2 *Disassembly of Feature Phones*

Finally, based on WFA shown in the former section, disassembly time of two types of feature phones were measured as in Table 3. In the disassembly experiments, total disassembly time means the time to detach all the detachable parts separately. Beyond this point, some physical breakage methods such as jaw crusher, cutting mill, etc. will be necessary. Plus, since it is well known [7] that LCD (liquid crystal display) units are the hopeful targets of component reuse, minimum times to separate LCD units were analyzed from the total disassembly flow.

2.3 *Disassembly Characteristics of Smartphones*

Disassembly experiments of two types of smartphones were also carried out. The data are basically measured time of actual disassembly work. But, as well as the feature phones, expected disassembly time calculated by Eq. 1 was used when there was any trouble in the disassembly process. Table 4 shows the results of the disassembly. The table shows that there was a big difference both in the total disassembly time and minimum disassembly time of LCD.

2.3.1 Comparison of Disassembly Characteristics

Although it is unsure whether the abovementioned results can be generalized in all the feature phone and smartphone designs, based on our previous experiences to disassemble used feature phones, we tried to extract some information regarding design for disassemble. Table 5 is the comparison of the numbers and types of joints of four models.

From this investigation, it is shown that there is no big difference in jointing methods of two feature phones. About the two smartphones, the big differences are the number and area of glues. This fact suggests these points. These will be the base in considering design improvement plans for easy to disassemble.

- The design of feature phones is almost standardized and there are not so many variations.
- On the other hand, the design of smartphones is still in trial-and-error stage and there can be many variations.

Table 4 Disassembly time of smartphones

	Total disassembly time	Minimum disassembly time of LCD
Provider D, type Xp	18 min 20 s	13 min 45 s
Provider D, type L	6 min 41 s	4 min 48 s

Table 5 Types and numbers of joints

Model	Make S, type 814	Make A, type I	Make D, type Xp	Make D, type L
Number of PCBs	3	3	1	2
Snap fits	19	15	17	15
Screws	4	3	3	2
Glues	7	7	10	2
Number of attitude changes	5	5	3	5

- The number (and area) of glue is the largest factor to determine disassemble time.
- Some components require large forces to detach because of wide glued area.
- The existence of hidden screws, worn screws, and rusted screws is the bottlenecks of disassembly.

Especially, glues will be the important points in designing smartphones. Although it can be effective in reducing assembly cost and time, it will be bottlenecks for remanufacturing, repair, and recycling [8]. Thus, for establishing a better circular use of smartphones in any measures, some design improvements will be necessary [9]. Concrete plans for design improvement regarding disassembly characteristics will be discussed in the next section.

3 Design Improvement Plans

3.1 General Trends in Disassembly

From this investigation, it is shown that there is no big difference in jointing methods of two feature phones. About the two smartphones, the big difference is the number of glue points. From our previous paper [10] regarding disassembly characteristics of used mobile phones, general guidelines of easy-to-disassemble design shown in Table 6 are suggested. Of course, theoretically, if the number of snap fits increases, total disassembly time will increase. However, in practical designs, components have to be attached together anyway. So, when the number of snap fits increases, numbers of other joints such as screws, glues, etc. will decrease. That will make disassembly easier. To propose practical ideas to improve disassemble characteristics, these information can be the hints.

Table 6 Factors to determine disassembly characteristics

Factors	Easy to disassemble	Difficult to disassemble
Glued area	Narrow	Wide
Number of glued points	Less	More
Number of screws	Less	More
Color of screws (screw covers)	Recognizable	Difficult to recognize
Hidden screws	Not exist	Exist
Number of snap fits	More	Less
Visibility of joints	High	Low
Modularity	High	Low
Number of attitude changes	Less	More

3.2 *Design Improvement Plan for Feature Phones*

As for feature phones, because of the long history of design considerations in the products, there are not so many ideas to make them easy to disassemble. However, still some bottlenecks in disassembly can be found. Figure 3 shown next is a schematic view of the design improvement proposals. Points of improvement are also shown below.

- Since the locations of screws are often difficult to recognize, marks to show the locations of screws in different colors are added.
- To understand the direction of detaching LCD easily, slide-in structure is used.
- To make LCD easy to separate, the area of glue to joint top cover and housing is decreased.
- PCBs are concentrated to one and is located in the under housing.

3.3 *Design Improvement Plans for Smartphones*

Since designs of smartphones are rather different from those of feature phones regarding disassembly characteristics, different strategies are necessary. Basically, the structures of smartphones are simpler than feature phones. But, often because of wide area of glues, disassembly is difficult or takes a longer time. To improve this point, following improvement suggestions were considered. Figure 4 is the schematic view in visualizing the improvement ideas.

- Modular design of the components should be enhanced. This feature will also decrease manufacturing costs.
- Instead of glues to attach LCD units and housing, sandwich structure to ensure easy-to-assemble and easy-to-disassemble feature at the same time is introduced.

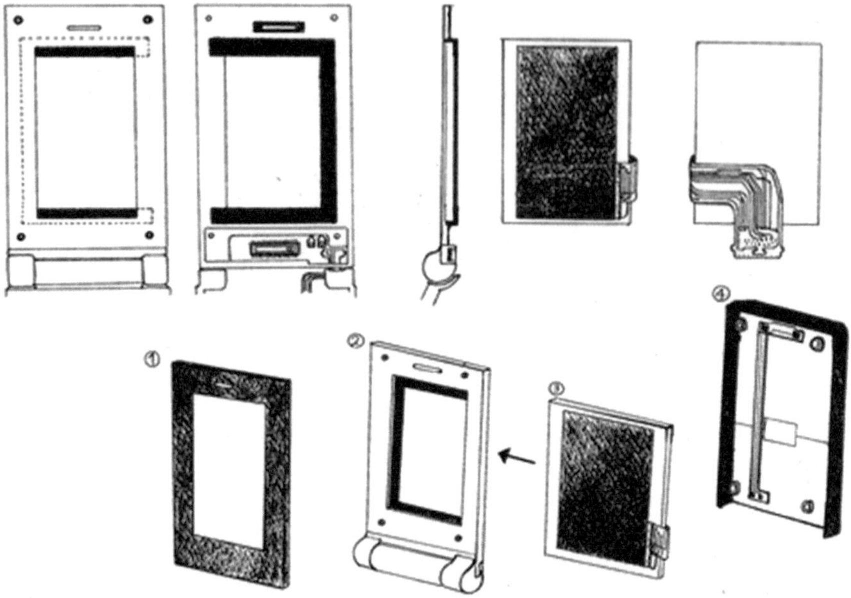

Fig. 3 Design improvement plans of feature phones

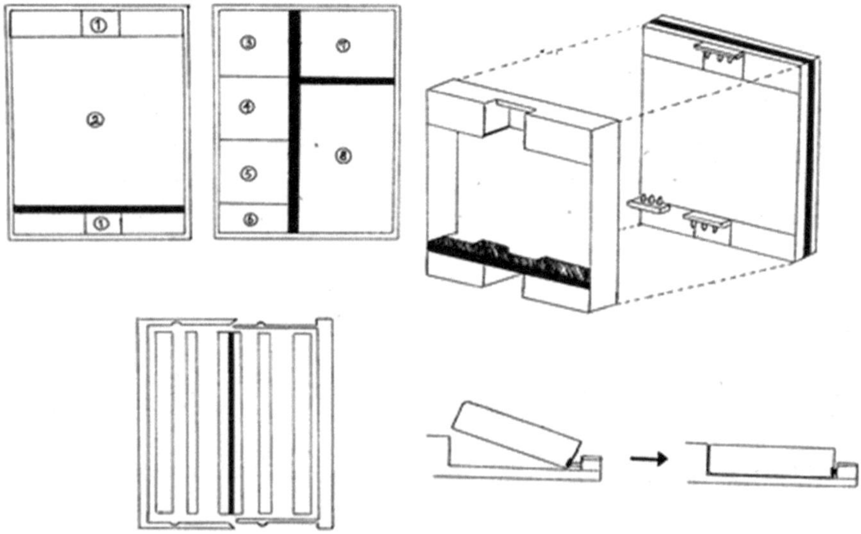

Fig. 4 Design improvement plans of smartphones

Table 7 Estimated time to detach LCD in improved design

	Estimated time
Feature phone	45.6 s
Smartphone	12.6 s

3.4 Estimated Disassembly Time

Based on the abovementioned design improvement plans both for feature phones and smartphones, we tried to estimate the disassembly time after design improvements. The estimation based on the predicted work flow is shown in Table 7. As shown in the table, time to detach LCD can be greatly shortened.

4 Proposal of PSS Featuring Remanufacturing of Components

Design improvement plans for smartphones were suggested in the previous section. In this section, business models in combining with the design improvement plans will be discussed. Actually, a new business model [11] regarding modular design of smartphones has been proposed. In the proposal, all the components of smartphones are designed in modulus. The concept has a large similarity with the design improvement plans on this paper. However, there are some advantages in the proposal of this paper, as shown below:

- By introducing sandwich structure which place components between top cover and bottom cover, falling of components can be prevented.
- By introducing snap fits in attaching modular components, assembly and disassembly will be easier.

Since there are some advantages in the design improvement plans in this paper, we hereby propose a PSS (product service model) based on remanufacturing of used LCD units.

- By using second-hand LCD for the display module, it is possible to reduce the cost and environmental impact at the same time.
- Basically, the top cover seen in Fig. 4 is proposed to protect the main components and to prevent the components from falling. But, it is also possible to apply attractive colors or illustrations.
- It is also possible to provide decorated covers as spare parts.
- By choosing only necessary modules, it is possible to reduce the price. For example, if the consumer does not need a camera or loud speaker, such functions can be reduced.
- When there is a malfunction, repair cost can be decreased by using used components. (Breakage of LCD unit is one of the most common reasons of malfunction of smartphones [12].)

- If it is possible to build stockyards of used components, it is possible to shorten the repair time drastically. This point is pointed as a big advantage of ongoing remanufacturing business [13].
- As an additional business model, it will be good to discount the price of new products, when the consumers ring in their used smartphones.

Anyway, by providing this business model, it might be possible to control the decrease in number of collection of used mobile phones. This can be an important step in establishing a circular use of mobile phones.

5 Summary

This paper explained that for used mobile phones which are still getting more and more popular, component reuse will be one of the keys to reduce environmental burden and to establish better circular economy.

Since manual disassembly is the necessary procedures to reuse components, disassembly experiments to know the characteristics of disassembly were carried out. The time and procedure for manual disassembly process to detach the LCD was precisely analyzed based on WFA which is a powerful tool in "design for disassembly." In the experiments, two feature phone models and two smartphone models were disassembled, and disassembly characteristics were compared. According to the analysis, designs of feature phones are rather consistent, while smartphone designs are rather different. Although smartphone outlooks are rather similar, some features such as area of glue were different. This feature makes disassembly process of smartphones inconsistent.

The paper also proposed some design improvement plans to achieve easy-to-disassemble features, based on the result of the disassembly experiments. It was estimated that by carrying out such design improvements, disassembly time of used mobile phones will be greatly reduced.

Finally, the paper tried to propose a business idea to utilize component reuse. Since most of the practical examples of component reuse of mobile phones are reuse of LCD (liquid crystal display), the paper also considered ideas to reuse used LCD. Some ideas to utilize LCD for inexpensive spare parts or original LCD units of low-end model of smartphones were proposed.

In future, more precise examination of disassembly characteristics of used mobile phones along with the feasibility assessment based on practical proposal of a product in which the recovered LCD is used will be necessary. Plus, design and prototyping of such products will be useful too.

References

1. http://www.soumu.go.jp/main_content/000299758.pdf. Accessed, 27/05/15
2. Takahashi K, Nakamura J, Otabe K, Tsuruoka M, Matsuno Y, Adachi Y (2009) Resource recovery from mobile phone and the economic and environmental impact. J Jpn Inst Met 73(9): 747–751
3. http://www.meti.go.jp/committee/summary/0003198/020_03_00.pdf. Accessed 27 June 2015
4. http://www.mobilerecycle.net/index.html. Accessed 20 Apr 2014
5. http://businesstech.co.za/news/mobile/33217/smartphones-to-outship-feature-phones/. Accessed 4 Apr 2015
6. Yamagiwa Y (2008) Design for assembly disassembly. J Jpn Soc Precis Eng 74(8):805–808
7. http://www.ekisho-ichiba.com/. Accessed 20 Apr 2014 (in Japanese)
8. Suga T, Hosoda N (2000) Active disassembly and reversible interconnection. Publication Year, pp 330–334
9. Kroll E, Hanft TA (1998) Quantitative evaluation of product disassembly for recycling. Res Eng Des 10(1):1–14
10. Sawanishi H, Torihara K, Mishima N (2015) A study on disassemblability and feasibility of component reuse of mobile phone. Proc CIRP 2015:740–745
11. Ministry of Environment, Ministry of Economy, Trade and Industry (2010) Report of the study group about recovery of the rare metal and proper processing of used small household appliances (In Japanese)
12. http://matome.naver.jp/odai/2134015968780366101. Accessed 4 June 2015
13. Matsumoto M (2009) Business frameworks for sustainable society: a case study on reuse industries in Japan. J Clean Prod 17(17):1547–1555

Seller-Buyer Matching for Promoting Product Reuse Using Distance-Based User Grouping

Yuki Yamamori, Yumihito Yokoki, and Hiroyuki Hiraoka

Abstract Transforming today's mass consumption society into one that circulates products is one of the key concepts of sustainability. Reuse of furniture and home appliances within neighborhoods already occurs in many countries. This paper describes a system that finds an appropriate path between households in a locality for increasing the degree of reuse of some products. Based on a predefined map that has information on houses, the proposed system determines a pair of houses that are separated by the shortest distance, among the combinations of houses, with the appropriate conditions for reusing a product. An agent-based technology is used to determine grouping based on distance between houses. A simulator visualizes the grouping for performing the matching operations.

Keywords Reuse • Agent • Grouping • Matching

1 Introduction

Realization of a circulating society is important in order to change today's mass consumption society to a sustainable one [1]. Promotion of the 3 Rs (reuse, recycle, and reduce) increases circulation. In order to promote circulation, we are developing a system based on network agents for management of individual parts required for effective reuse [2]. The network agent associated with a part in this system is called a "part agent." A part agent moves across the network and interacts with other agents of various users. This system can realize an efficient product life cycle.

Considering the reuse among houses in a neighborhood, the distances required to transfer the used products as well as their prices and performances influence users' satisfaction. To avoid reuse of products between distant users, the agents created groups in which the distance among houses are limited. In each group, a matching server, played by a part agent, creates appropriate pairs of agents using a matching algorithm that satisfies the users.

Y. Yamamori (✉) • Y. Yokoki • H. Hiraoka
Department of Precision Mechanics, Chuo University, Tokyo, Japan
e-mail: yamamori@lcps.mech.chuo-u.ac.jp

M. Matsumoto et al. (eds.), *Sustainability Through Innovation in Product Life Cycle Design*, EcoProduction, DOI 10.1007/978-981-10-0471-1_23

This paper examines the validity of the agents' groups. Simulations of part agents allow visualization of the groups in several conditions.

Section 2 describes the concept of part agent. Section 3 explains a matching scheme of agents in each group. Section 4 describes a method to create groups of agents. Simulation of grouping users and its results are shown in Sects. 5 and 6. Section 7 offers a conclusion.

2 Part Agent System

The proposed part agent system is based on the following usage scenario. The system uses a part agent to manage all information about an individual part throughout its life cycle. The proposal assumes the spread of networks and high-precision radio frequency identification (RFID) technology [3].

The part agent is generated at the manufacturing phase of the main parts, when an RFID tag is attached to its corresponding part. The part agent identifies the RFID tag throughout the part's life cycle and tracks the part's transfer through the network. RFID tags were chosen for identification because they have a higher resistance to environmental stress than printed codes such as bar codes, which may deteriorate or become dirty over a part's life cycle. Moreover, one can read, write, and store data in RFID. These functions are not feasible by other print-based identification methods.

For related research, product embedded identifier (PEID) [4] has been developed which involves a small computing chip, an RFID tag, and sensors to support the middle and end of life of the products. In contrast to the PEID system, our system aims to promote multiple reuse of individual parts that may go beyond the manufacturer's management. This requires a "lightweight" system that can be used repeatedly without maintenance of sophisticated hardware.

Figure 1 shows the conceptual scheme of the part agent. The part agent collects the information needed to manage its corresponding part by communicating with various functions within the network. These functions may involve a product database that provides product design information, an application that predicts the deterioration of parts, one that provides logistic information, or one that provides market information. Furthermore, the part agent communicates with local functions on-site, such as sensory functions that detect the state of the part, storage functions for individual part data, and management and control functions of the product. Communication is established using information agents that are subordinate network agents generated by the part agents.

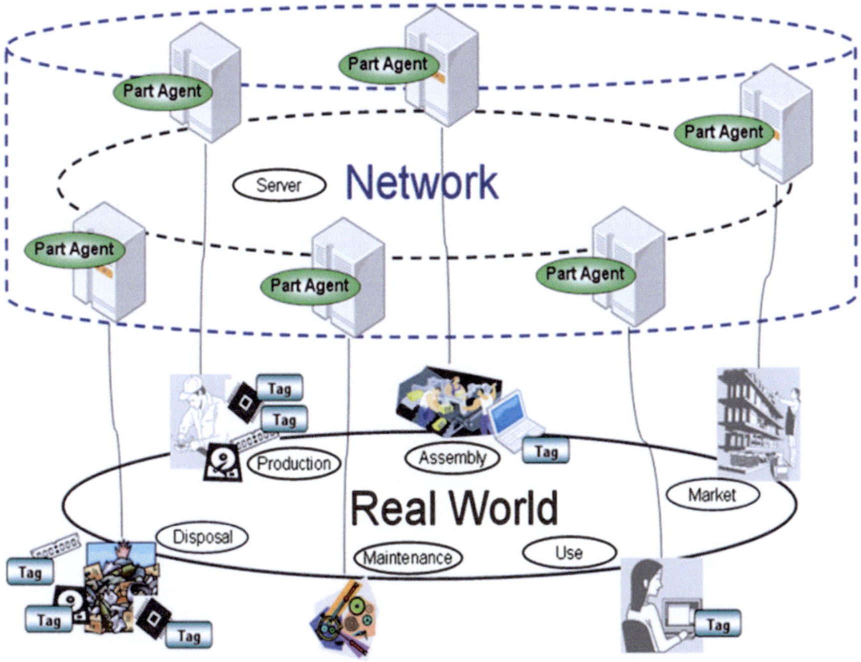

Fig. 1 Conceptual scheme of the part agent

3 Framework of Part Agent for Local Reuse of Products

We focus on reuse of home electric appliances among neighboring homes in a local area.

Figure 2 shows the concept of the reuse of products using matching function based on a user's satisfaction [8].

A part agent advises the user on replacement of a product when its product deteriorates. The part agent proposes to replace the user's product with a used product by the following procedure. First, agents make a group. Next, a matching server matches agents in a group.

For the reuse of products among neighboring homes, the distance between users is important. We call "the limit of the distance that a user can move" the user's maximum distance. We perform the grouping using this user's maximum distance.

An agent's grouping progresses in the following procedure. First, an agent checks whether a. If a group exists, an agent investigates using the user's maximum distance to determine whether the agent can join the group. The existing group member evaluates the agent who is going to join the group. If all the agents are closer than the user's maximum distance, the agent can join the group. The details of this grouping procedure are described in Sect. 4.2.

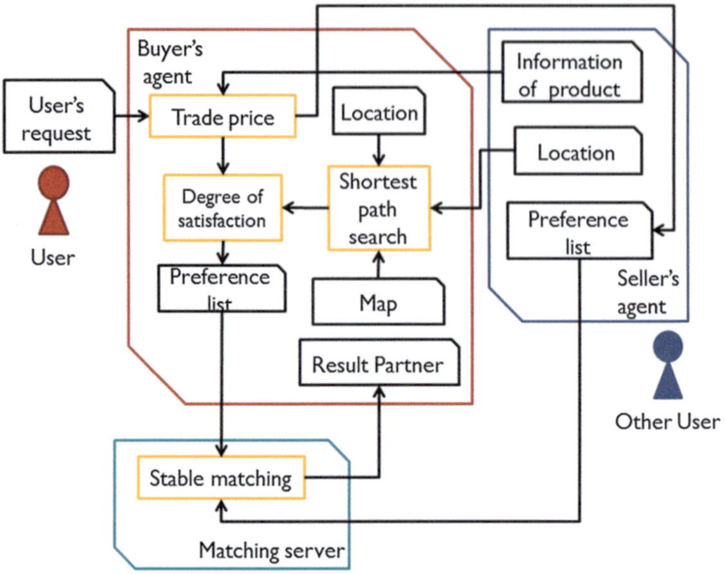

Fig. 2 Concept of the reuse of products using a matching function based on user's satisfaction

The next step is matching users in the group. A part agent collects information on the same kind of products that are on sale by sellers as candidates to be purchased. The agent evaluates every candidate product using an index of degree of satisfaction described later.

Each agent transmits the evaluated results to the matching server in the group as a preference list of its user. The matching server matches a buyer and a seller based on this preference list using the Gale-Shapley method [5] and notifies every matched user via its agent. Thereby, a buyer can purchase a used product from an appropriate seller.

Degree of satisfaction A buyer's agent calculates the degree of satisfaction DoS using a formula (1) on a home appliance owned by a seller. It is assumed that a buyer's satisfaction is determined by the following four elements:

1. The distance to the seller (shown as D in the formula). An agent calculates this distance by Dijkstra method [6].
2. The difference ratio between the price of candidate product and the price the buyer wishes (shown as P in the formula). We obtain this value P by calculating the difference of the price of the candidate product (shown as in P_P in the formula) and the price that the buyer wishes (shown as P_U in the formula) and then dividing it by P_U.
3. The difference ratio between the specification of a candidate product and the buyer's specification (shown as S in the formula). We obtain this value S by calculating the difference in the specification of the candidate product (shown as

S_P in the formula) and the buyer's specification (shown as S_U in the formula) and then dividing by S_U.

4. The residual life of candidate product (shown in L in the formula) [8]

$$
\begin{aligned}
\mathrm{DoS} &\equiv \frac{1}{a_1 P + a_2 S + a_3 D} + a_4 L \\
P &\equiv \frac{|P_P - P_U|}{P_U} \\
S &\equiv \frac{|S_P - S_U|}{S_U}
\end{aligned}
\tag{1}
$$

4 Multiple Groups for the Matching

4.1 Grouping Based on the Distance Between Users

People have to travel distances to exchange products such as home appliances to perform reuse in a local area. A user will not go out of his reasonable range to exchange a product even if it fits well with his demand. To avoid mismatching due to the distance between participants, we introduced grouping of participants based on distance in the matching procedure, assuming that all participants have the same maximum traveling range.

Instead of K-mean method [7] that is known for dividing a group of members into a set of suitable groups, we employ the following method where agents create groups by themselves.

Part agents create groups in which distances among users are less than a predefined maximum distance. Reuse of products is performed as shown in the next section only between a pair of users in a group, that is, within the maximum distance.

A group must satisfy the following conditions:

- A group must include at least one member.
- A member can belong to only one group.
- Each group has a leader part agent that will play a role of a matching server described in the next section.

Finally, part agents are divided into groups. An example consisting of 100 part agents is shown in a Fig. 3.

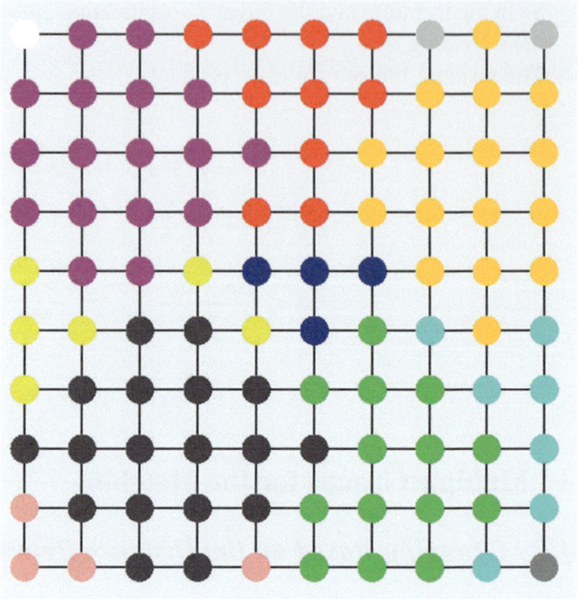

Fig. 3 The agents divided into groups

4.2 *Grouping*

4.2.1 Entry in a Group

Part agents create groups. Figure 4 shows the procedure and Fig. 5 exemplifies the process.

First, an agent enters a group using the following procedure:

1. Part agent with user request: A part agent whose user requests the reuse of its product, i.e., release of the existing product, and/or purchase of a product, searches for other part agents that have the same requests.
2. Other agents within a user's maximum distance: If the part agent finds such agents, it computes distance to the user based on a map it contains. If the distance is within the maximum distance, the part agent records its group as a candidate group.
3. Other groups within a user's maximum distance: The part agent calculates distances to every agent in a candidate group. If distance to an agent in the group is longer than the maximum distance, the group is removed from the candidate list.
4. Selection of a group: When two or more groups exist as candidate groups, the part agent selects the group with the largest number of agents as the group it enters.

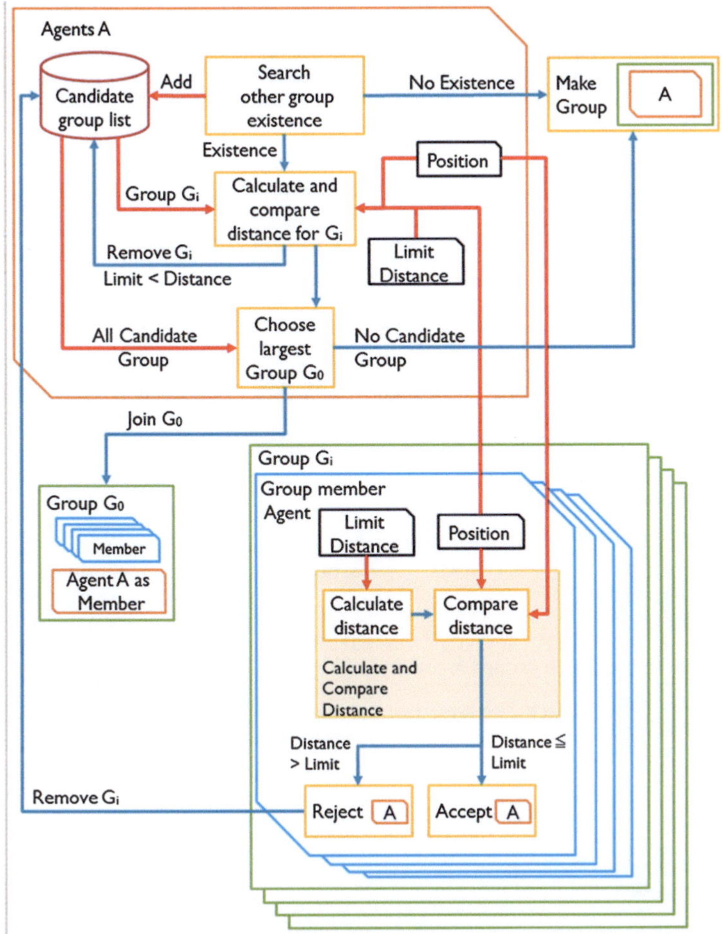

Fig. 4 Group division using maximum distance

4.2.2 Entry in a Group

A part agent transfers from one group to another when the number of agents in the second group is larger than that in the first:

- When the leader of a group leaves the group, it nominates a new group leader at random from the remaining group members.
- If the last agent leaves a group, the group is removed.

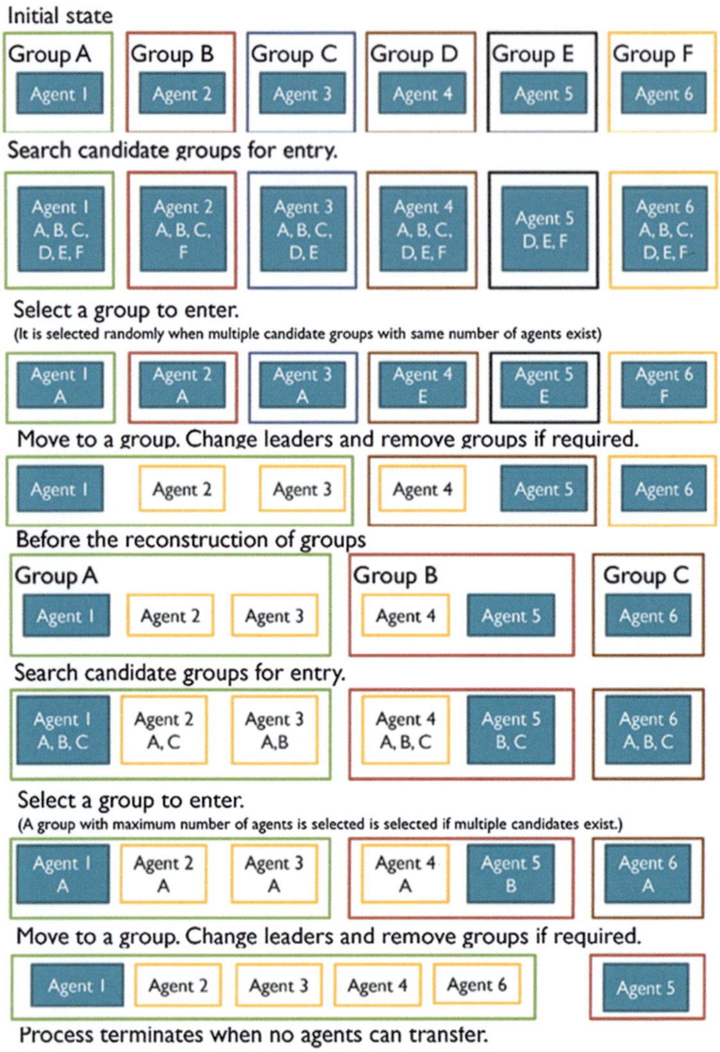

Fig. 5 Group division using maximum distance

4.2.3 Reconstruction of Groups

As part agents can move from one group to another, groups may be reconstructed. Reconstruction is performed in the following procedure:

1. Every part agent searches candidate groups in the same way as entry procedure described in Sect. 4.2.1.
2. If a selected group for a part agent is not the current group to which the agent belongs, the agent expresses the request to transfer.

3. For every part agent requesting to transfer, transfer to its selected group is made as described in Sect. 4.2.2.
   This procedure is repeated until no agents request to transfer.

5 Simulation of Grouping

5.1 *Simulator*

We compare the number of the groups created and the number of agents in a group against a user's maximum distance in order to evaluate the grouping.

A simulation is performed in a flow as shown in Fig. 6. A simulator performs simulation on each agent, after setting up its conditions. For each agent, operation progresses in the following order: a selection process, a grouping process, a matching process, and a reuse process. This flow of processes is performed in a single unit of time. The simulation is carried out for multiple units of time.

The reason for dividing operations of the simulation into five processes is that some tasks should be performed only after all the agents finish preceding tasks. For example, since matching of agents can only be performed after all the groups are created, we have divided matching and grouping into separate processes.

Each process is described below. See the reference [9] for the details of experimental elements and simulator's process.

We performed the simulation eight times for one condition. Each experimental result is shown with the average value on each condition.

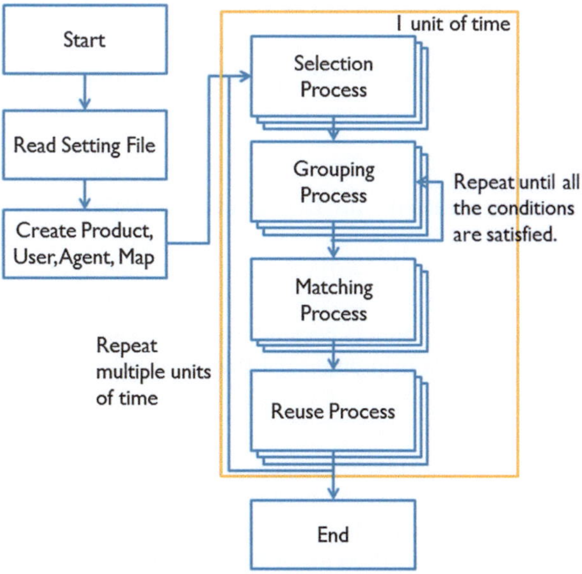

Fig. 6 Simulation flow

5.2 *Experimental Condition*

5.2.1 Map

Streets are arranged in a grid structure with 10 × 10 intersections. We installed a house in each intersection. We set the distance of each road connecting the intersections of the grid as 300 m

5.2.2 Product

As an initial condition, we distributed one refrigerator at each home. We assigned its residual life in a uniform distribution with the value between 1 % and 100 %.

We made the capacity of the product with the average of 400 L and distributed with a standard deviation of 30 L.

The price of a new product is assigned with the average of 100,000 yen and distributed with a standard deviation of 20,000 yen.

All the refrigerators deteriorate 4 % in one unit time.

5.2.3 Product's Conditions on Sales

In this simulation, in order to evaluate the group of agents, we assume that an agent proposes reuse in every unit of time. We made the following conditions on sales of a product:

- When a user uses a product in one unit of time, a user wishes for its purchase or sale by the probability of 100 %.
- When a user does not possess the product, the user wishes to purchase one.
- When a user has two products, the user wishes to sell one with shorter residual life.

5.2.4 Coefficients for Degree of Satisfaction

We set the value of coefficients a_1, a_2, a_3, and a_4 in the formula (1) of the degree of satisfaction as 1. These values of coefficients equalize the effects of P, S, and distance in the formula because maximum value of P and S is 1.0 and the distance in the experiment is in order of 1.0 when it is measured in kilometer.

5.2.5 User's Maximum Distance

We experimented with grouping by setting the user's maximum distance as 900 m, 1200 m, 1500 m, 1800 m, 2100 m, 2400 m, 2700 m, and 3000 m.

6 Simulation Result

The groups obtained by this simulation are shown in Figs. 7, 8, and 9. A circle in the figures denotes an agent and a line denotes a path on a map. The circles with same color show the agents who belong to the same group.

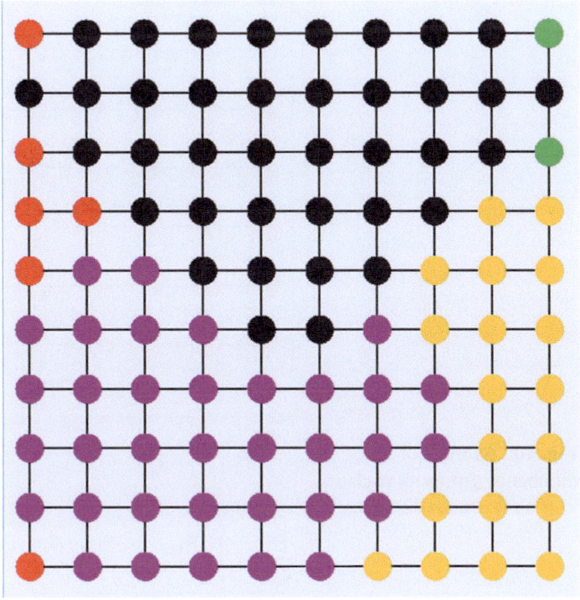

Fig. 7 Grouping in maximum distance = 2700 m

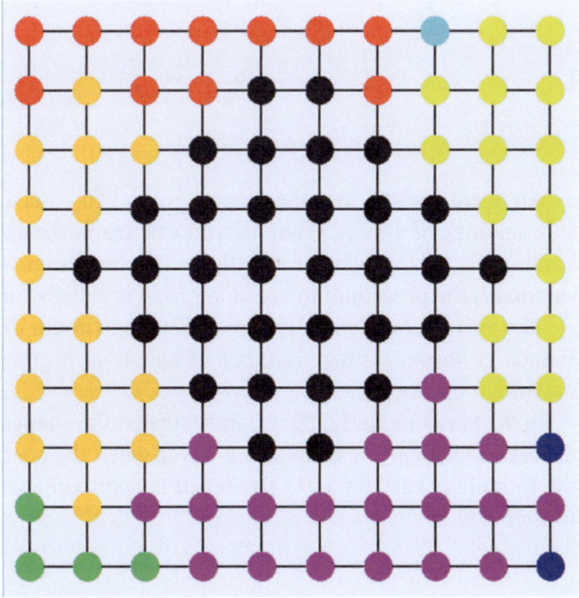

Fig. 8 Grouping in maximum distance = 2100 m

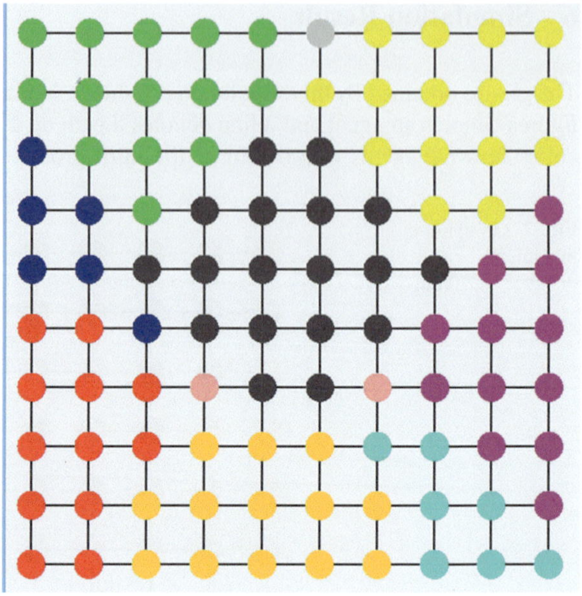

Fig. 9 Grouping in maximum distance = 1500 m

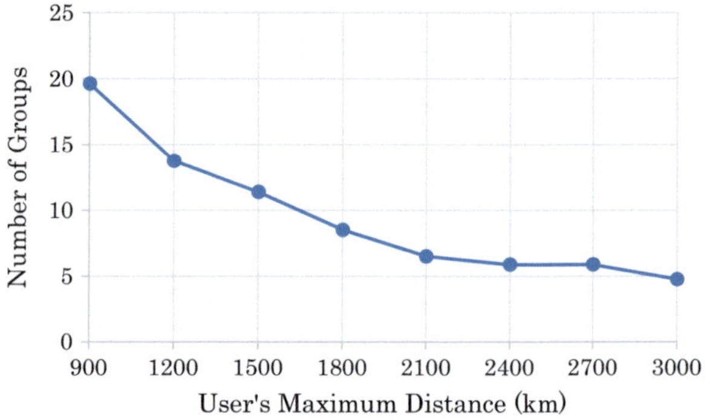

Fig. 10 Average of the number of groups in each user's maximum distance

An agent tries to join the largest group. This makes small groups exist in the surroundings of a large group as you can see in the experimental results. We also confirm that all the agents are settled within maximum distance in all the groups. In summary, the procedure to create a group is satisfactory.

The average number of groups against maximum distances is shown in Fig. 10. Figure 11 shows average number of agents in a group and its variations against maximum distance.

In the previous study [9], we investigated the change of satisfaction *DoS* against distance *D* between a seller and a buyer when the coefficients a_1, a_2, a_3, and a_4 in the formula (1) are set as 1. The result is approximated by the following quadratic formula (2):

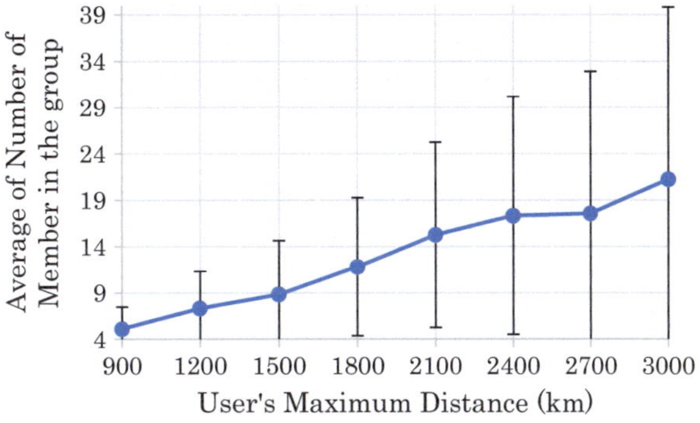

Fig. 11 Average number of agents in a group and its variations against maximum distance

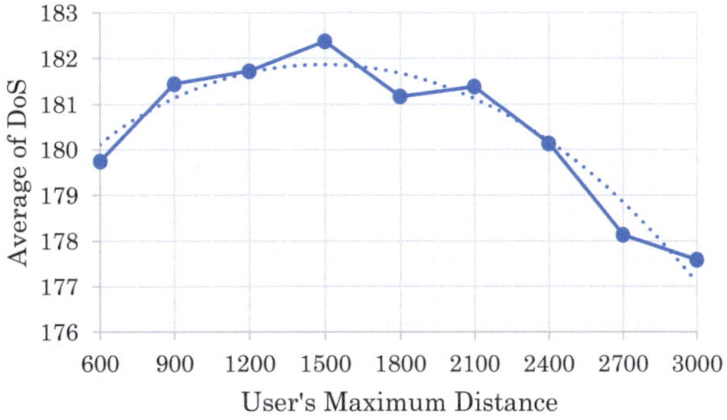

Fig. 12 Average of total value of the degree of satisfaction resulting from all the pairs of agents in group

$$DoS = 6 \times 10^{-8} D^2 - 3 \times 10^{-3} D + 2.0214 \quad (2)$$

We estimate user's satisfaction using this formula (2) based on the simulated distance between two agents in a group. Figure 12 shows the average of total value of the degree of satisfaction resulting from distance between all the pairs of agents in every group.

The degree of user satisfaction was highest when the user's maximum distance was about 1500 m. In other words, when the coefficient of a degree of satisfaction is set as one, the user's maximum distance to satisfy many users is about 1500 m. On the other hand, relevance of the variation in the number of groups and the number of the members was not observed.

7 Conclusion

We developed a system that finds an appropriate path between houses in a locality to increase the degree of reuse of products. We also developed a function to create groups by considering the distance between two houses. We evaluated the effectiveness of the function in visualizing different groups for a given set of conditions for reuse. Further research is needed to reduce the computational complexity of the proposed method. We will conduct an experiment in which users have various maximum distances.

Acknowledgments This work was supported by JSPS KAKENHI Grant Number 24560165.

References

1. Hauschild M, Jeswiet J, Alting L (2005) From life cycle assessment to sustainable production: status and perspectives. Annals of CIRP 54(2):535–555
2. Hiraoka H, Ueno, Kato K, Ookawa H, Arita M, Nanjo K, Kawaharada H (2013) Part agent advice for promoting reuse of the part based on life cycle information. 20th CIRP international conference on life cycle engineering, Singapore, pp 335–340
3. Borriello G (2001) RFID: tagging the world, communications of the ACM; 2005, 44/9:34–37own B, Aaron M
4. Jun HB, Kiritsis D, Xirouchakis P (2007) Research issues on closed-loop PLM. Comput Ind 58:855–868
5. Miyazaki S (2013) Stable matching problem. Inf Process 54(10):1064–1071 (in Japanese)
6. Dijkstra EW (1959) A note on two problems in connexion with graphs. Numer Math 1:269–271
7. Miyamoto S (1999) A guide to cluster analysis - the theory of fuzzy clustering and application, Morikita Publishing, Tokyo (in Japanese)
8. Yamamori Y, Nanjo K, Hiraoka H (2014) Reuse of products considering distance among users and their satisfaction, EcoDe P&S, p.131–134, (in Japanese)
9. Yamaori Y, Nanjo K, Hiraoka H (2014) Matching reused products with users considering their satisfaction, Eco Balance, 0915

User Model in the Life Cycle Simulation of Mechanical Parts Based on Prospect Theory

Yumihito Yokoki, Yuki Yamamori, and Hiroyuki Hiraoka

Abstract To realize the effective reuse of mechanical parts for developing a sustainable society, it is essential to manage the individual parts over their entire life cycle. Therefore, we are developing a part agent system using network agents. In this paper, we assume that a user acts based on the expected value obtained by current and future actions such as use, repair, replacement, and disposal. We employ prospect theory to calculate these expected values, and user behavior is discussed in the life cycle simulation using a part agent system based on prospect theory.

Keywords Part agent • Prospect theory • Life cycle

1 Introduction

In recent years, it has become necessary to develop a method for realizing a sustainable society. To realize the effective reuse of mechanical parts for developing a sustainable society [1], it is essential to manage the individual parts over their entire life cycle. Manufacturers, if they perform reuse-based production, need to estimate the quality and quantity of the parts that are returned for reuse. However, it is difficult for them to predict the quality and quantity of the used parts, owing to the uncontrollable and unpredictable diversity of user behavior. In addition, product users also have difficulties carrying out appropriate maintenance on the many various parts in their products [2]. Based on these considerations, we propose a scheme whereby a part "manages" itself and supports user maintenance activities. Network agents that are programmed to follow their real-life counterpart parts throughout their life cycle are being developed. These network agents are referred to as "part agents" [3]. These network agents provide users with appropriate advice on the reuse of its part and promote the circulation of reused parts.

Previous works have proposed methods of user support when using a part agent [2–4]; however, the user may choose not to accept the proposed actions

Y. Yokoki (✉) • Y. Yamamori • H. Hiraoka
Department of Precision Mechanics, Chuo University, Tokyo, Japan
e-mail: yokoki@lcps.mech.chuo-u.ac.jp

M. Matsumoto et al. (eds.), *Sustainability Through Innovation in Product Life Cycle Design*, EcoProduction, DOI 10.1007/978-981-10-0471-1_24

recommended by the part agent because of his or her preferences. For example, even when the failure probability of a product is low, the user might want to repair or replace it with a new product considering the risk of failure seriously. Even if a product exhibits good performance, the user might want to replace it with another one with lower environmental load. Therefore, it is necessary to develop a user model and method that conform to the behaviors of real users.

In this paper, we propose a user behavior model based on "prospect theory" [8]. Prospect theory is a behavioral economic theory that describes the way people choose between probabilistic alternatives that involve risk where the probabilities of outcomes are known. The theory states that people make decisions based on the potential value of losses and gains rather than the final outcome and that people evaluate these losses and gains using certain heuristics. This study observes the simulated life cycle of a part by using a part agent and describes how the part agent provides advice for the user based on this model.

First, the concept of a part agent is described. Then, the mechanism of a part agent that provides advice on the reuse of a part is explained in Sect. 3. In Sect. 4, Prospect theory is described. Next, the expanded life cycle used in this work and the application of Prospect theory are described in Sect. 5. In Sect. 6 a life cycle simulation is conducted to evaluate the mechanism and described with some initial results. Finally, the paper is concluded in Sect. 7.

2 Part Agent System

A part agent manages all information regarding its corresponding part throughout its life cycle. The proposal assumes the spread of networks and high-precision radio frequency identifier (RFID) technology [5].

A part agent is generated during the manufacturing phase of core parts where an RFID tag is attached to its corresponding part. The part agent identifies the ID of the RFID tag during the life cycle of the part, tracking the part through the network. We chose an RFID tag for identification because RFIDs have a higher resistance to smudge or discoloration than printed bar codes and will last for the life of a part.

Figure 1 shows the conceptual scheme of the part agent. The part agent communicates with various functions within the network and collects the information needed to manage its corresponding part, such as product design information, predicted deterioration of parts, logistic information, or market information. It also communicates with local functions on-site, such as sensory functions that detect the state of the part, storage functions for individual part data, as well as management and control functions of the product. Communication is established using information agents that are subordinate network agents generated by the part agents [2].

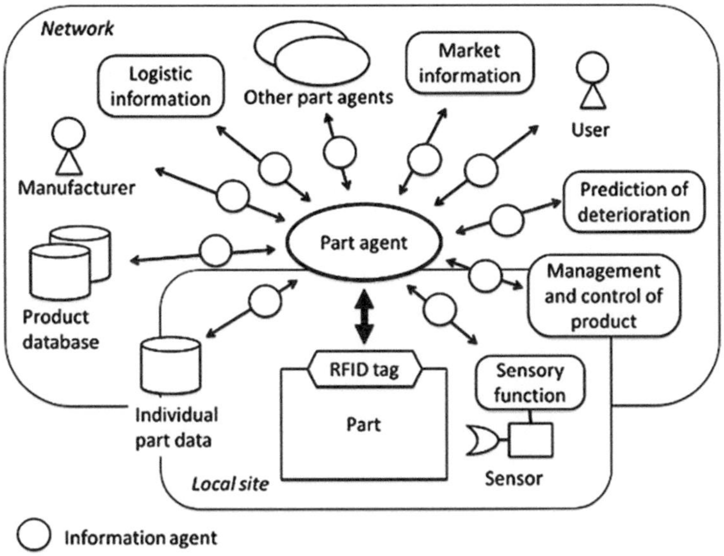

Fig. 1 Conceptual scheme of the part agent

3 Creation of Advice by Part Agent

3.1 Framework of Part Agent for Advice Generation

Figure 2 shows a framework for a part agent to advise its life cycle model of a part. At each time step, the part agent predicts the possible states of the part in the near future and evaluates those options in order to provide advice to the user [5].

A part agent generates advice in the following order. First, the part agent expands the life cycle of the part by using the current status of the part and its evaluated values based on deterioration model. All possible candidate paths in the life cycle are estimated using a user model based on prospect theory. Based on the estimation, an appropriate course of action for the part is selected and recommended to the user as an advice. In this study, we have simplified the mechanism of failure prediction in order to verify the effect of prospect theory. Details on the expansion and estimation of the life cycle of a part are described in Sect. 5.

3.2 Evaluation of Profit, Cost, and Environmental Load Based on Deterioration

A variety of information is required for part agents to evaluate appropriate maintenance actions. This includes the profit acquired from the part, the required cost for

Fig. 2 Framework of part agent for advice generation

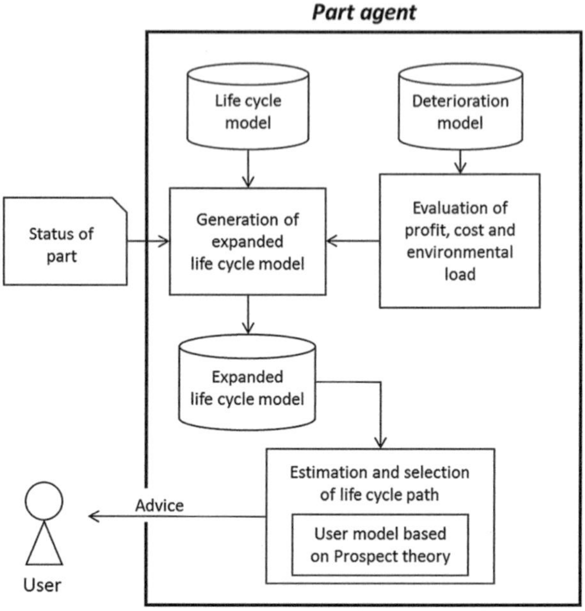

the part, and the environmental load placed on the part. As these values change primarily due to deterioration, a deterioration model of the part is required, which shows the manner in which the mechanical performance of the part deteriorates.

Based on this information, the total performance index (TPI) [6] is derived for evaluating the life cycle. The TPI provides a proper evaluation of the performance of a part throughout its life cycle by balancing its profit, environmental load, and costs. The TPI is found using following Eq. 1.

$$TPI = \frac{\text{profit}}{\sqrt{\text{cost} \times \text{environmental load}}} \quad (1)$$

Note that the numerator of the TPI is defined as value in the original paper [6]. But this paper defines the numerator as profit in order to capture the direct effect.

The TPI is an index showing the ratio profit to cost and environmental load. It does not have a unit. A large TPI indicates that profit is high compared to the environmental load and cost. The expected TPI of the stage (described in Sect. 5) is calculated using Eq. 1 based on the expectations of profit, cost, and environmental load. The part agent then proposes to the user a choice of life cycle with the highest expected TPI.

4 Prospect Theory

Prospect theory is a behavioral economics theory published by D. Kahneman and A. Tversky in 1979.

In expected utility theory [7], risk aversion and risk seeking are determined solely by the utility function. However, choices among risky prospects exhibit several pervasive effects that are inconsistent with the basic tenets of utility theory [8].

Prospect theory has two characteristic equations. They are explained simply below.

4.1 Value Function

The value function represents the mental value for gain or loss. In prospect theory, if we define v as the mental value that a human being feels they will gain or lose x, v is concave above the reference point $x = 0$ $(v''(x) \leq 0,\ x \geq 0)$ and convex below the reference point $(v''(x) > 0,\ x < 0)$, where $v''(x)$ denotes second-order differential of $v(x)$. A positive or negative value of $v''(x)$ indicates whether the value function $v(x)$ is convex or concave. Furthermore, v as the mental value is steeper for losses than for gains $(v'(x|x \geq 0) > v'(x|x < 0)$, where v' is the derivative of $v(x)$). The first two conditions reflect the principle of diminishing sensitivity: the impact of a change diminishes with distance from the reference point. The last condition implies the principle of loss aversion [9].

From these features, the value function $v(x)$ is denoted by a graph, as shown in Fig. 3. The axis of the abscissa of a graph represents the gain or loss x. This graph is drawn using Eq. 2.

$$v(x) = \begin{cases} x^{\alpha}\ (x \geq 0) \\ |-\lambda(-x)^{\beta}\ (x < 0) \end{cases} \tag{2}$$

where the coefficients $\alpha = 0.88$, $\beta = 0.88$, and $\lambda = 2.25$. These coefficients were calculated through an experiment by A. Tverski and others [9].

This equation was used to evaluate the property values for stages in the life cycle.

4.2 Weighting Function

The weighting function is obtained from the experiment to represent the observed nonlinearity of preferences [9].

Fig. 3 Value function

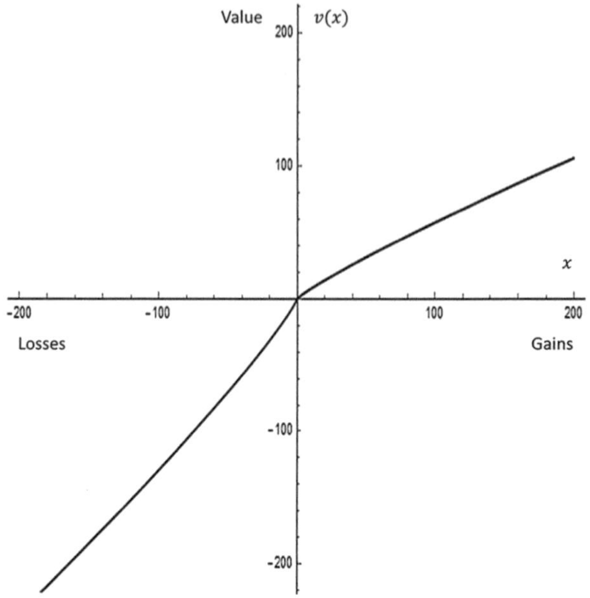

The weighting function $w^{+}(p)$ and $w^{-}(p)$ are given in the following Eqs. 3a and 3b for gain and loss, respectively:

$$w^{+}(p) = \frac{p^{\gamma}}{\{p^{\gamma} + (1-p)^{\gamma}\}^{1/\gamma}} \tag{3a}$$

$$w^{-}(p) = \frac{p^{\delta}}{\left\{p^{\delta} + (1-p)^{\delta}\right\}^{1/\delta}} \tag{3b}$$

where p is the probability of an event and $\gamma = 0.61$ and $\delta = 0.69$ are coefficients obtained in the experiment [8].

The graph of $w(p)$ against the probability p is drawn based on these equations, as shown in Fig. 4. The solid line represents the original probability of an event.

This graph shows that the shape of the curve changes from concave to convex at approximately 35 % of a probability of an event. It means that if the probability of events increases to more than 35 %, a human feels that the event occurs in lower probability than the original probability, and if it is less than 35 %, a human would feel that the event occurs in higher probability than the original probability.

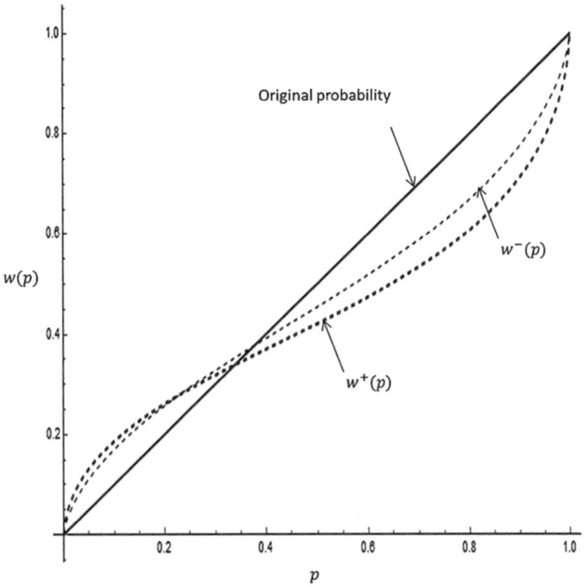

Fig. 4 Weighting functions for gains and for losses

5 Expansion of Life Cycle of Part

5.1 Life Cycle Model

Figure 5 shows a simple example of the life cycle model. The life cycle of a part consists of life cycle stages and life cycle paths that connect them. The circles represent the life cycle stages, which include produce, sell, use, repair, and dispose. The arrows indicate the life cycle paths.

The part agent expands the life cycle of the part. It represents possible changes in the life cycle of the part over time. Figure 6 shows an example of an expanded life cycle of a part starting from the "use" stage. The life cycle of each stage depends on properties such as profit, cost, and environmental load. Each expanded life cycle path relies on the probability that the part will take that path. The part agent guides the path to the stage with the best expectation.

5.2 Application of Prospect Theory to the Estimation of the Life Cycle of the Part

Figure 7 shows how a part agent selects the next stage using the expanded life cycle of the part. The figure shows a situation in which the current life cycle stage is a use stage and the possible next stages are Stage1, Stage1′, and Stage1″. A circle denotes an expanded life cycle stage with its property value—*V*1, *V*2, *V*3, etc. A line

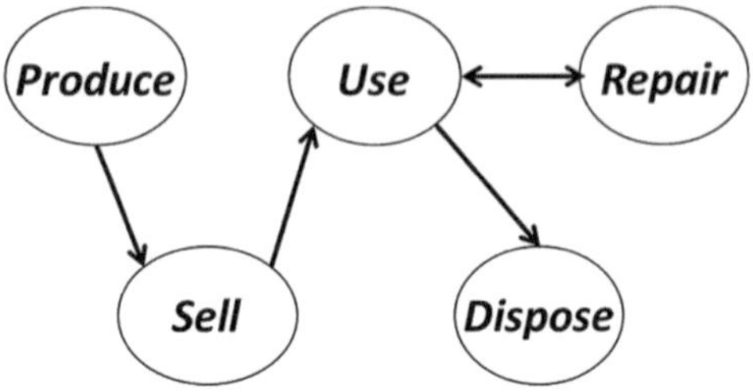

Fig. 5 Simple life cycle model

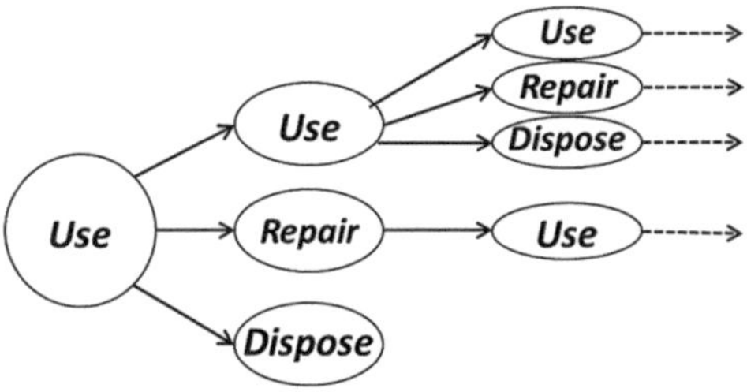

Fig. 6 Expanded life cycle model

denotes an expanded life cycle path with its probability—*p*12, *p*13, *p*14, etc. To evaluate each possible stage, its expected value of properties is calculated, taking into consideration the series of paths in the future. A series of stages connected with the paths is defined as a "route." The property values for the next stages and their probabilities are collected for all possible routes that could occur in the future. The expectation is then calculated for each by multiplying the sum of the property values and product of probabilities, as shown in Eq. 4.

$$EV = \sum_{\text{Route}} \left(\sum_{\text{Stage in route}} V * \prod_{\text{Path in route}} P \right) \tag{4}$$

where *EV* is the expected value of a candidate next stage with the following future stages. *V* is the sum of the property values for the stages in the route, and *P* is the accumulated probability of the paths in the route.

For example, in Fig. 7, the expected property value of Stage1 is calculated via Eq. 5.

$$\begin{aligned} &\text{expected value(Stage1)} = \\ &(V1 + V2 + V5)*(p12*p25) + (V1 + V2 + V6)*(p12*p26) \\ &+(V1 + V2 + V7)*(p12*p27) + (V1 + V3 + \ldots)*(p13*\ldots) + \ldots \end{aligned} \tag{5}$$

Prospect theory has been applied to the calculation of expected values as shown in Eq. 6.

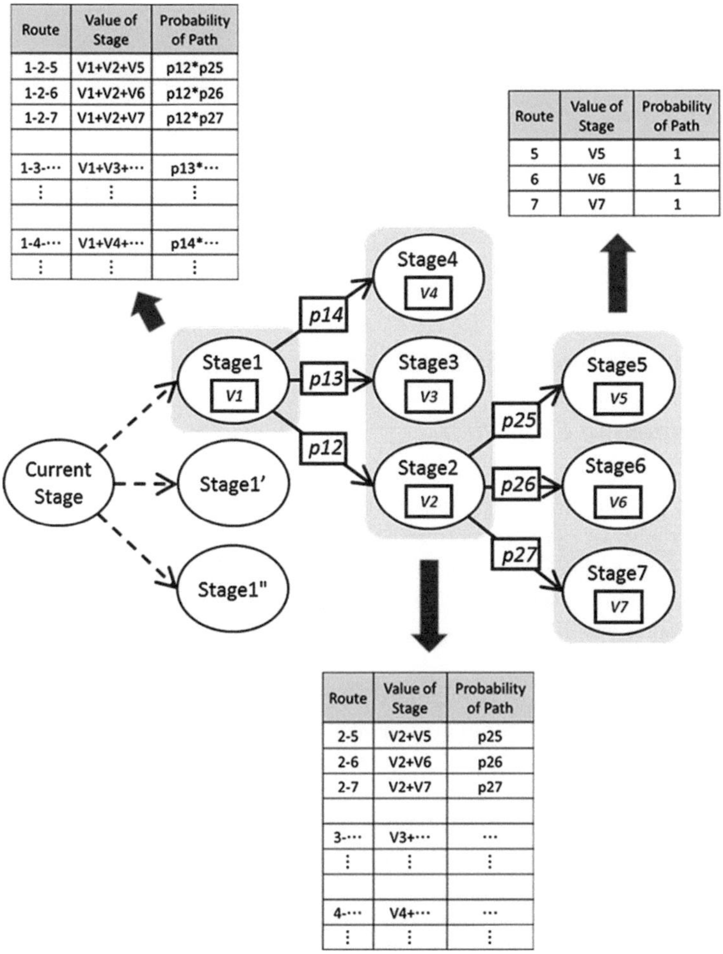

Fig. 7 Process of accumulating the values of the life cycle

$$EV = \sum_{\text{Route}} \left(v\left(\sum_{\text{Stage in route}} V \right) \times w\left(\prod_{\text{Path in route}} P \right) \right) \tag{6}$$

where $v()$ is the value function and $w()$ is the probability weighting function. In this example, the value function is applied to accumulated property values such as $(V1 + V2 + V5)$, and the weighting function is applied to accumulated probability such as $(p12 * p25)$.

The profit, cost, and environmental load properties of the part are predicted based on the deterioration of the part. In this case, profit is defined as gain, and cost and environmental load are defined as losses.

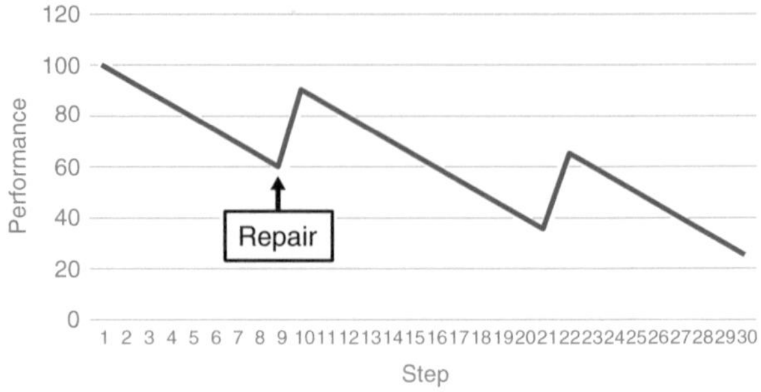

Fig. 8 Deterioration model

6 Life Cycle Simulation

6.1 Simulation Condition

To perform a simulation to estimate the future state of a part, we assume that a part deteriorates with its operational time. Our deterioration model of the part is shown in Fig. 8. The mechanical performance of the part decreases with its operational time and is recovered when the part is repaired. In this simulation, it decreases by 5 % of initial performance for each step of use and recovers 40 % within initial performance on repair.

The properties of parts are updated in the stages for each step as shown in Table 1. P denotes the current performance of the part and P_{max} denotes its initial performance as a new product. It has been assumed that the cost and environmental load depend on P and P_{max} [2].

Table 2 shows the probabilities of the life cycle paths. Pr and Pu represent the probabilities that the part is either repaired or is not repaired. These are calculated using the following Eqs. 7 and 8:

$$Pu = 1.0 - Pr \tag{7}$$

$$Pr = 0.05 \times \frac{P_{max} - \text{performance}}{P_{max}} \tag{8}$$

6.2 Simulation Result

To evaluate this method of applying prospect theory to the estimation of the life cycle of a part, a simulation of a part agent was performed for 30 steps. The part agent chose the stage with the highest expected value of TPI as the appropriate next stage.

Figure 9 shows the TPIs from the life cycle simulation for each step. The red line represents the results with prospect theory and the blue line represents the results

Table 1 Properties of stages

	Performance	Profit	Cost	Environmental load
Produce	P_{max}	0	0	0
Sell	P	0	P	$0.1 \times P$
Use	$P-0.05 \times P_{max}$	P	$0.2 \times P_{max}-0.1 \times P$	$0.2 \times P_{max}-0.1 \times P$
Repair	$P+0.4 \times P_{max}$	0	$0.5 \times P_{max}$	$0.2 \times P_{max}$
Dispose	0	0	$0.2 \times P_{max}$	P_{max}

Table 2 Probabilities of paths

To / From	Produce	Sell	Use	Repair	Dispose
Produce		1.0			
Sell			1.0		
Use			Pu	Pr	0.05
Repair			1.0		
Dispose		1.0			

without prospect theory. Repair is performed when the TPI reaches zero. The disposal stage was not identified in this simulation.

Simulations for both cases result in the same TPI. This result indicates that the numerical changes that occurred while utilizing prospect theory did not influence the selection by the agent.

Figure 10 shows the expected TPIs for each step. The dotted lines represent the results without prospect theory, and the solid lines represent the results with prospect theory. Compared with the case where prospect theory is not applied, lower TPIs are observed for the case with prospect theory. This result shows that, although the values of the TPI change when prospect theory is applied, the priority of stages selected does not change. Therefore, the application of prospect theory was found to be indifferent in this particular case.

6.3 Discussion

The disposal stage was not included because there is no merit expected over the other stages based on our definition of property values shown in Table 1. This needs to be improved.

In this study, the coefficient obtained in the experiment by Tversky and others was used for the functions of prospect theory. These coefficients should be experimentally calculated according to the conditions of this study. It is necessary to evaluate the influence of the difference in a coefficient.

In the simulation, the stages were selected based on the TPI. We believe that this is the reason for any faults in our results. The effect of prospect theory will likely be clearer if the simulation were to address failures of parts and their accompanying

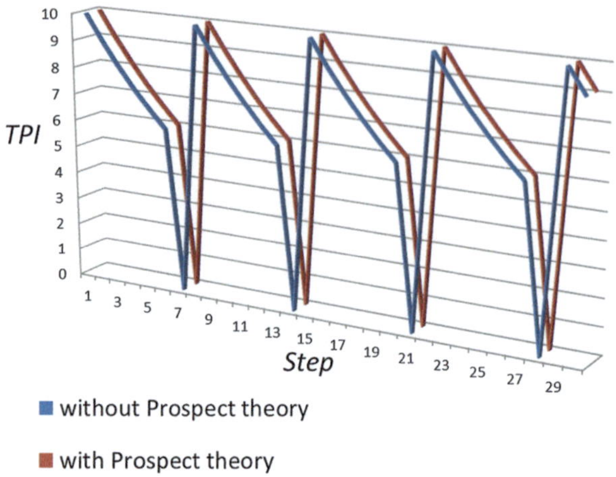

Fig. 9 TPI in life cycle simulation for each step

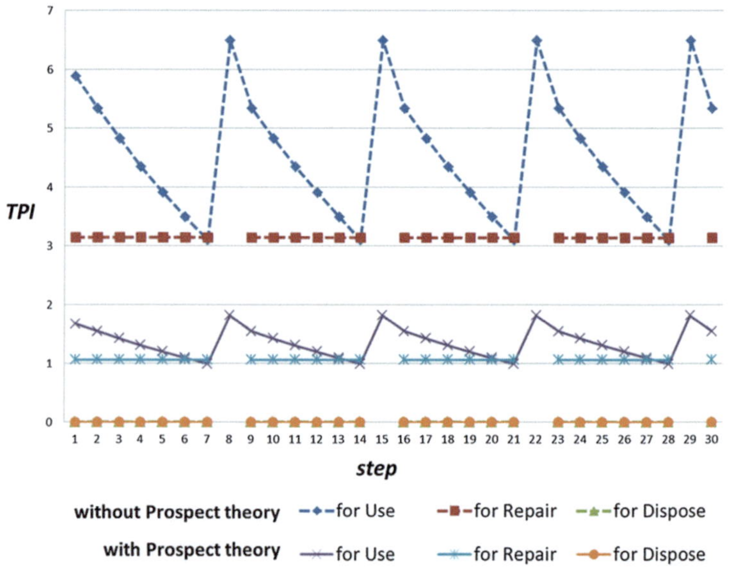

Fig. 10 Expected TPI for each step

losses. By applying this consideration to the model, a clearer result will be obtained, showing the choice under risk condition as failure.

The application of prospect theory to different life cycle models expressed in this paper should also be examined.

7 Conclusion

In this paper, we proposed a model of user behaviors based on "prospect theory." This paper describes how the part agent creates advice for the user based on this model.

We observed the simulated life cycle of a part by using a part agent which provides advice based on this method. We also discussed the effects of the part agent on its decisions with this method, as well as the problems that occurred.

We have some remaining issues and future prospects. First, the consideration of failure and the phenomenon accompanying it should be applied and examined. The coefficient of prospect theory as applied to the life cycle model should also be examined. This method could be applied to other models. Finally, more realistic models on failure and the degradation of parts and more realistic user behavior are required.

Acknowledgments This work was supported by JSPS KAKENHI Grant Number 24560165.

References

1. Hauschild M, Jeswiet J, Alting L (2005) Life cycle assessment to sustainable production: status and perspectives. Ann CIRP 54(2):535–555
2. Nanjo K, Yamamori Y, Yokoki Y, Sakamoto Y, Hiraoka H (2015) Maintenance decisions of part agent based on failure probability of a part using Bayesian estimation. In: The 22nd CIRP conference on life cycle engineering, Sydney, 7–9 April 2015
3. Hiraoka H, Ueno T, Kato K, Ookawa H, Arita M, Nanjo K, Kawaharada H (2013) Part agent advice for promoting reuse of the part based on life cycle information, 20th CIRP international conference on life cycle engineering, Singapore, 2013, pp 335–340
4. Nanjo K, Yamamori Y, Kato K, Ookawa H, Kawaharada H, Hiraoka H (2013) Part agent that proposes maintenance actions for a part considering its life cycle. The 11th global conference on sustainable manufacturing, Berlin, pp 235–240
5. Borriello G (2005) RFID, tagging the world. Commun ACM 44(9):34–37
6. Kondo S, Masui K, Mishima N, Matsumoto M (2008) Total performance performance analysis of product life cycle considering the uncertainties in product-use stage. Advances in life cycle engineering for sustainable manufacturing businesses, Tokyo, pp 371–376
7. Mongin P (1997) Expected utility theory. Handbook of economic methodology, Edward Elgar, London, pp 342–350
8. Kahneman D, Tversky A (1979) Prospect theory: an analysis of decision under risk. Econ J Econ Soc 47:263–291
9. Tversky A, Kahneman D (1992) Advances in prospect theory: cumulative representation of uncertainty. J Risk Uncertain 5(4):297–323

Research on Corporate Social Responsibility Advertising Design

Tsai-Feng Kao and Jui-Che Tu

Abstract To carry out green marketing, enterprises should strive to reach a balance between commercial interests and environmental protection, identify the factors that influence consumers' purchase decisions on green products, and develop appropriate marketing strategies, in order to be more competitive. This paper discusses the advertising design of cause-related marketing (CRM) and green processes (GP) of corporate social responsibility (CSR), and control group (CG) advertisings, and analyzes the influence of CSR advertising design and film information of consumers. This study adopted the quasi-experiment mode and conducted empirical research according to 3×2 between-subjects factorial designs. The results showed that consumers have a relatively positive response to CSR advertising designs and film information. After viewing the CRM film, the subjects changed their attitude regarding CRM advertising, but no effect was found in the GP advertising. The findings of this study can provide an understanding to the effects of the two CSR advertising designs and film information for reference to future green marketing planning.

Keywords Green design • Cause-related marketing • Green processes • Film information

1 Introduction

It is important for companies to possess the capability to achieve establish sustainable development. Stakeholders today are well educated, possess professional knowledge, and have more demands. A good company reputation is the best competitive advantage for enterprises and the key asset in market competition.

T.-F. Kao (✉)
School of Information Technology and Design, Department of Multimedia and Game Design, Overseas Chinese University, Taichung, Taiwan
e-mail: g9730808@gmail.com

J.-C. Tu
Graduate School & Department of Creative Design, National Yunlin University of Science & Technology, Yunlin, Taiwan

M. Matsumoto et al. (eds.), *Sustainability Through Innovation in Product Life Cycle Design*, EcoProduction, DOI 10.1007/978-981-10-0471-1_25

Company reputation or image is determined by how a company carries out its business. To earn a good company reputation or image, a company needs to have sustainable development capacity, actively plan and adapt to the changing environment, and increase their ability to face the problems related to sustainable development [24]. In recent years, more and more companies have taken CSR as their long-term business strategy [14, 25], and research supports that corporate social responsibility (CSR) influences marketing effects [14, 23, 29, 30]. Generally, the main reason for companies to adopt CSR strategy is to promote a better society and a cleaner environment. In an emergency, CSR is the key for companies to succeed [11, 12].

CSR includes many corporate marketing activities, which are mostly focused on the stakeholders, such as charity organizations, community organizations, employees, suppliers, customers, and descendants' benefits [29]. CSR activities mainly integrate social needs into a responsible business pattern to promote good moral development and generate long-term benefits [10]. In terms of marketing methods, one important project is monetary or product donations to charity organizations or community organizations, and such an activity can be called cause-related marketing (CRM) [22, 26]. CRM can be used as a strategic tool to promote brand image and increase the sales volume, which is particularly effective on selected consumers [1, 8, 22, 28]. CRM has an effect on consumption behavior and is especially important to increase brand and business fitness [1, 14, 26]. Another green marketing mode is that companies emphasize the green features of their products by using certification and labeling. According to Rahbar and Wahid [27], the green marketing tools of eco-labeling and eco-branding have certain influence on consumers. Therefore, this study aims to explore whether the advertising designs of CSR marketing activities have the expected effect on consumers.

The visual stimulation of advertising influences customers' advertising and brand attitudes [13, 18], as advertising images help consumers to understand product information. Visual information includes stories, pictures, and sounds, which form a convincing website, so that consumers would feel more confident about the products and have increased consumption attitudes [18]. Film can bring good advertising effect [5, 19]; however, such image simulation effect is not always positive. Jr. et al. [15] suggested that, when the advertiser is a charity organization, such advertising is created in order to solve social problems or seek aid. However, the information receivers may be disturbed by other information or depicted scenes and have no response to the advertising appeal.

This paper discusses the influence of CSR advertising designs and film information on advertising effect. The method adopted is cited from the effect of CSR marketing in social network sites (SNSs) by Jeong et al. [14]. Regarding the influence of film information on attitude, the concepts proposed by Cialdini [5] and Litvine and Wüstenhagen [19] are adopted. The film intervention is regarded as an impact factor. This research can help to understand the factors influencing consumers in judgments of green advertising, as well as the relationship of various factors in green advertising effect.

2 Theoretical Background

2.1 CSR Activities

CSR is defined as organizational responsibility and aims to respect personal rights and promote human welfare [20]. In other words, corporate consideration and response are beyond the problems of economy, technology, and legal norms in the narrow sense and aim to realize social and environmental benefits [6]. In marketing, the contents of CSR activities influence consumer's attitude and behavioral intention. Through CSR activities, consumers may have positive emotion. It is therefore very important to develop proper CSR advertising to actively impact consumer's attitude and behavior [16]. CSR activities have different purposes and methods. It is worth discussing what kinds of activities have significant influence on consumers and which kinds of CSR activities have better marketing effects.

According to Garriga and Melé [10], as based on interactions between enterprises and society, CSR is classified into the following four types. The first is instrumental theories, which are mainly to realize economic goals and achieve marketing purposes through social activities. It is supposed that CSR is the tool to create corporate wealth, and only economic consideration is involved in the relationship between enterprises and society. The second is political theories, which focus on the company's social power in the political arena, in particular, the power to obtain related political forces. The third is integrative theories, which hold that enterprises should integrate social needs. It is generally thought that the corporate business can gain sustainable development and growth by depending on society, and even the business itself depends on society. The last one is ethical theories, which suggest that moral value is added to the relationship between enterprises and society. CSR should be regarded from the perspective of morals, and enterprises should take it as a responsibility in moral obligation to accept society. Jeong et al. [14] classified CSR into six types. The two most commonly used types are CRM and cause sponsorship. According to Sprinkle and Maines [29], CSR activities not only include participation in related organizational activities (e.g., charity organizations, community organizations, employees, suppliers, customers, and later generations) but also donation activities that provide monetary or product donations to charity organizations or community organizations. Some companies encourage their employees to participate in charity or public benefit activities in order to seek social acceptance; some focus on employee safety and benefits, provide safe workplaces, education welfare, and health care; some take consumer wishes and future generation protection as their mission, such as reducing excessive packing to attract customers to pay attention to issues of environmental protection, water sources, and energy usage issues; and some concern about animal protection and the influence on endangered animals' habitats, as it is forbidden to conduct animal testing and endanger animal survival in product R&D, production, and selection of ingredients and materials. "Green" production is also emphasized in corporate activities, such as energy saving, carbon reduction, materials recycling, reducing packaging materials, and using local raw materials.

2.2 CRM and GP

CRM is a kind of marketing activity in which the companies promise to contribute a certain amount of money to society or business when customers buy their products or services [16]. It is taken by companies as a way to differentiate from competitors and increase corporate brand value [17]. Such marketing effect can be taken as a method to improve corporate brand image and increase the sales volume [17, 22, 26]. CRM activities mostly show dual purposes: on the one hand, to improve public awareness to support public service units (e.g., Welthungerhilfe), to receive donations, and other social factors and, on the other hand, to improve the corporate reputation and customer loyalty and enable companies to gain financial benefit [3]. Through CRM activities, companies can successfully establish reputation, image, and identity, which allow consumers to know their brands, thus achieving the purpose of CSR.

A very important key of CRM development is to resonate with consumers [2]. Although CRM strategies have been effective increasing the sales volume, CRM-related disputes still exist. Thus, companies need to determine when and how to correctly use CRM [2]. Most CRM activities involve donations by the companies when consumers buy their products. Some creative CRM approaches include involving consumer participation, which is irrelevant to the purchase of products [7]. Therefore, in CRM activities, companies should seek the support of consumers in order to succeed.

Another green marketing approach is certification and labeling for the emphasis of the green features of the products, and especially, the influence of eco-seals on consumers is worthy of notice. Rahbar and Wahid [27] adopted the Persuasion Knowledge Model to research the influence of eco-labeling and eco-branding in green marketing tools on consumers. The results showed that reliable eco-labeling and eco-branding affect the consumer's purchase behavior. Advertisers have highlighted their green certification through guarantee or certification methods. Consumers' characteristics (environmental concern and brand familiarity) and advertisers' control characteristics (sealed source and advertising appeal) are the factors influencing the effect of eco-seals. Consumer's thoughts on environmental problems and eco-seals are determined by brand familiarity, ecological sealed source, and advertising attraction [4]. In addition, it is a method to attract consumers by properly manifesting CSR action, but not excessively seeking interest. For example, in its corporate responsibility report, IHG points out that it educates and stimulates affiliated subsidiaries to create innovative environmental protection measures every year and introduces their various methods to customers. Through online tools, companies can emphasize the strategies adopted for sustainable development and establish green signs [9].

This paper discusses CSR marketing advertising, as based on the four types of CSR proposed by Garriga and Melé [10], and the CRM and social responsibility business practices in six types of CSR, as proposed by Jeong et al. [14]. This paper integrates the key points of CSR activities, as proposed by Sprinkle and Maines [29], selects the charity organization activities in CSR activities, and emphasizes the green production mode. It takes the CSR marketing advertising type as a variable influencing consumer attitude and regards it as the key thinking of marketing strategy. As consumers may have positive emotions through CSR activities [16], the following hypotheses are given:

H1: CSR marketing advertising has positive influence on the advertising attitude of consumers.

H1a: The subjects contacting CRM advertising have more positive advertising attitude than those contacting general advertisings.

H1b: The subjects contacting GP advertising have more positive advertising attitude than those contacting general advertisings.

2.3 *Film Stimulation*

Visual stimulation is effective communication. When applied in advertising, consumers may have positive responses [13]. With the increased quality and amount of information provided on the internet, there is an increasing quantity of information options provided to consumers. Advertising images may help consumers to understand the product information and form a strong consumption attitude. Taking the tourism industry, for example, mental image processing is regarded as convincing communication. With increased amount of personal stories, pictures, and sounds on tourism websites, along with consumer-specific contents, a convincing website can be constructed to strengthen the confidence of consumers and help them to form a long-term consumption attitude [18]. Matthes et al. [21] treated American consumers as the subjects and examined the effects of four kinds of visual advertisings (functional type, emotional type, mixed type, and control type) on the subjects. When products are promoted with a mix of functional and emotional green advertising, pleasant natural scenery is adopted in order to express the environmental advantage of green advertising. The results showed that emotional and mixed advertisings have a significant influence on brand attitude. These actions do not rely on the green participation of consumers. Functional advertising is used when it is necessary to measure the green purchase behavior or green product attitude. In addition to image-based advertising, the information effect of film integrating image, sound, and plot is also worthy of notice. For example, the film "An Inconvenient Truth" by Al Gore caused carbon offsets to be increased by 50 % [19]; American TV advertising of the environmental problem of "Iron Eyes Cody spot" resulted in a great echo throughout the USA and won countless awards and millions of dollars in donations. Therefore, PSA (public service advertising) is regarded as a mass medium able to obtain success and benefit [5]. The following hypotheses are hereby put forward:

H2: Film information influences consumers' advertising attitude.
H2a: The subjects watching such films have a more positive advertising attitude than those not watching such films.

Commercial advertising often stimulates consumers to imagine that they are using the products or services. However, when an advertiser is a charity organization, the advertising is aimed to correct a social problem and seeks people related or unrelated to the event to lend a hand, such as soliciting monetary donations for victims of tsunami, hurricane, or famine, to help fight against the persecution of wild animals, or scholarships to assist financially disadvantaged students. In many cases, these appeals try to stimulate watchers to resonate and imagine that they are the potential beneficiaries of aid, in order to provide aid; however, not all cases have this result. Sometimes the information receivers may imagine that they are in two different roles (sponsor and donation receiver), which may disturb information processing and reduce the effectiveness of the appeal. Therefore, such advertising information is detrimental to the consumers' responses [15]. CSR activities are aimed to persuade consumers, and a positive result may appear, or certain emotions, such as hope, may be enhanced under negative situations. If consumers feel suspicious of such activities, they may think that the company involved is attempting to benefit from controlling consumer emotion. Therefore, it is important for enterprises to understand how to develop strategic CSR advertising to actively influence the attitude and behavior intention of consumers [16].

Research has verified that visual stimulation in advertising influences the advertising attitude and brand attitude of consumers [13, 18]; however, whether the effect of film information is positive or negative is yet to be explored. Tu et al. [31] used the "positive and negative 2Cs," developed by Wen-chien Chen, as a stimulant of green information. The result verified that green film information has an effect on subjects, and such advertising effect is caused under the effect of green information. However, the stimulation effects of pictures and images are not fully positive. According to Jr. et al. [15], in many cases, when an advertiser is a charity organization, the advertising involved aims to solve social problems or seek aid. Such kind of appeal attempts to stimulate the public to resonate and provide aid, but information receivers may be disturbed by other information and fail to respond to the appeal. Therefore, this study holds that consumers may have different responses to CSR marketing advertising under the effect of film information than under the condition without film. The following hypotheses are proposed:

H3: CSR marketing advertising and film information are interactive.
H3a: With film information, the subjects watching CRM advertising design have a more positive advertising attitude than those watching ordinary advertising design.

H3b: Without film information, the subjects watching GP advertising design have a more positive advertising attitude than those watching ordinary advertising design.

H3c: Without film information, the subjects watching CRM advertising design have the same advertising attitude as those watching ordinary advertising design.

H3d: Without film information, the subjects watching GP advertising design have the same advertising attitude as those watching ordinary advertising design.

3 Method

3.1 Research Design and Description of Subjects

This study adopted the experimental method to verify the hypotheses. The subjects were asked to watch green product advertising and provide their response. In the experimental design, 3 (advertising design: CRM advertising vs. GP advertising vs. CG advertising) $\times$ 2 (film: yes vs. no) between subject factorial design was adopted. Advertising attitude was taken as the dependent variable. By convenient cluster sampling, this study treated the students from a university in central Taiwan as the subjects, whose age ranged from 18 to 22 years old. They were randomly distributed into six experimental groups, with 18–30 subjects in each scenario, for a total of 210 subjects. After eliminating the invalid questionnaires, there were 167 effective questionnaires. The analysis results showed that among the consumer environmental products currently available in Taiwan, green footwear advertisings adopt CRM as the advertising strategy, and some even highlight the environmental protection features of the products. Therefore, green footwear was selected as the targeted product of this study. The green footwear advertising design was used after removing the brand logo. Three advertisings were presented, which are similar except for different slogans and advertising subtitles.

3.2 Independent Variable

3.2.1 CSR Advertising Design

Regarding the advertising text selection, pretest was conducted. The green footwear advertising commentaries were collected through networks, and three CRM and GP advertising titles were sorted with seven advertising subtitles to design the pretest. To avoid the subjects' biases on the same type of appeals, the advertising texts were arranged in a staggered manner. The subjects of the pretest were 37 students from a university in southern Taiwan. CRM advertising commentaries were described according to the definition of CRM by Kim et al. [16]. In the questionnaire, CRM advertising design is a kind of marketing activity, defined by the donation of a company to the society or charity organizations when customers make a purchase. The GP description is based on the concept of Bickart and Ruth [4] and on the CSR activities and sustainable development strategy of Gao and Mattila [9]. The "advertising design" stresses on the least damage and minimal impact (side effect on) to the environment, as well as the maximum resource utilization efficiency, in part or the entire life cycle, from design, production, packaging, and transportation to usage and discarding of products. Based on the features of the two kinds of advertisings, the subjects were asked to measure the fitness of the advertising texts in the experiment and select the response.

A Likert 7-point scale was used for scoring, ranging from 7 (CRM) to 1 (GP). The scores were averaged. There were 35 valid samples. In the CRM advertising, the main caption was "Buy one and give one," and the subtitle was "You buy a pair of shoes, we will give a new pair to a needy child for you. Through your purchase, children may avoid death by catching diseases due to no shoes to wear." In the GP advertising, the caption was "Love the earth for sustainable development," and the subtitle was "From shoelaces to shoe soles, more than 75 % of shoes are made from recycled materials; and through our utilization of recycled PET products (a kind of plastic made from recovered PET bottles), our company has consumed 78 million PET bottles for the benefit of the earth." The two advertisings significantly differed in the average number of titles and subtitles (title: $M_c = 6.54$, SD = 0.611, $M_g = 1.57$, SD = 1.065, t_{c-g} (35) =21.963, $p < 0.001$; subtitle: $M_c = 6.71$, SD = 0.519, $M_g = 1.37$, SD = 0.942, t_{c-g} (34) =26.095, $p < 0.001$; $c_=$ CRM advertising; $g_=$ GP advertising; SD =standard deviation; t=T-Test; p= P-value).

3.2.2 Film Control

In this study, information manipulation was done by the provision of film information. The experimental material of film was obtained from pretest, and the film information was based on the contents of four green footwear advertisings, including two advertising films of People's Walk and one advertising film of TOMS Company. The subjects watched the green films and then answered the questionnaire, which contains two items, including "This film attracts my attention and concern" and "After watching this film, I became more concerned about environmental protection." A Likert 7-point scale was used for scoring. The results showed that the film of TOMS has the highest score and has a significant difference from other films ($M_1 = 6.89$, SD = 0.323; $M_2 = 6.63$, SD = 0.547; $M_3 = 2.37$, SD = 1.767, t_{1-2} (34) = 2.491, $p < 0.05$; t_{1-3} (34) = 1.467, $p < 0.001$). Thus, the advertising film of TOMS was taken as the experimental material of the film information.

3.3 Dependant Variable

The communication effect of advertising representation, namely, the measurement to produce the effect of consumers after accepting the advertising, includes advertising representation, brand attitude, and purchase intention survey [32]. As advertising is aimed to persuade consumers to buy products, and eventually increase product sales volume, the change of consumer attitude is related to their purchase behavior. Hence, the advertising effect is indeed the communication effect of advertising representation. Therefore, this study observed the advertising effect from the aspect of advertising attitude.

Referring to Cheng and Wu [33] and Priester et al. [34], this study modified their questionnaires on advertising attitude and included the advertising attitudes of "intelligible," "vivid," "pleasant," "exhilarating," and "satisfying the requirements." A Likert 5-point scale was used for scoring, ranging from 7 (strongly agree) to 1 (strongly disagree).

The items on "advertising attitude" were designed with reference to related literature, and the scale has good content validity. The internal consistency reliability was tested, and the advertising attitude was $\alpha = 0.869$, which indicates a high reliability as it is over 0.70.

3.4 Experimental Procedure

The formal experiment was repeated six times. In each experiment, the researcher first introduced the research purposes and then distributed the questionnaire. The first part of the questionnaire collects personal background information; the second part focuses on the manipulations of film information; the third part measures the attitude of the subjects. For all the groups, the first part and third part were exactly the same. For the second part, the film viewing group was shown with selected films. According to the film information effects proposed by Litvine and Wüstenhagen [19] and Cialdini [5], and concerning the emotional responses of the subjects, two items were presented regarding the film, in order to measure the film information. Only the answers that identify with the film information effect were included as valid samples. The group that did not view a film only needed to fill out the first part and third part of the questionnaire. The entire experiment lasted about 10–15 min.

Table 1 Results of ANOVAs (CSR Ad and film information)

	Source of variation	SS	DF	MS	F	η^2
Ad attitude	CSR Ad (A)	125.94	2	62.97	4.516*	0.046
	Film information (B)	96.266	1	96.266	6.904**	0.035
	A×B	8.206	2	4.103	0.294	0.003
	Error	2635.177	189	13,943		
	Total	64382	195			

*$p < 0.05$, **$p < 0.01$

4 Research Result and Analysis

4.1 Verifying the Advertising Effect of Individual Factors

This study mainly aimed to discuss the influence of two independent variables (CSR advertising and film information) on advertising attitude. Two-way ANOVA (Analysis of Variance, ANOVA) was conducted for statistical analysis. The results are shown in Table 1. As seen, there is no interactive effect between CSR advertising and film information, while CSR advertising is significantly different from the film information in individual effect.

The verification of the research hypotheses is as follows. H1 (CSR marketing advertising has positive influence on the advertising attitude of consumers) suggests that the subjects contacting CRM and GP advertisings have more positive advertising attitude than those contacting general advertisings. The results showed that, regarding CRM and CG advertisings, the subjects displayed significantly different advertising attitudes. The advertising attitude in CRM advertising was significantly higher than the advertising attitude in the CG ($M = 18.627$ vs. 16.526, $F_{(2,194)}=8.904$, $p < 0.001$). The advertising attitude in GP advertising was significantly higher than the advertising attitude in the CG ($M = 18.03$ vs. 16.526, $F_{(2,194)}=8.904$, $p < 0.01$). This indicates that, regarding CRM and GP advertisings, consumers have a relatively positive advertising attitude. Therefore, H1 is supported. The subjects watching CRM and GP advertisings displayed more positive advertising attitude than watching general advertisings. Thus, H1a and H1b are both supported.

H2 (film information influences consumers' advertising attitude) suggests that those who watched the film have a better response than those who did not watch the film. As shown in Table 1, the individual factors of film information are significantly different in advertising attitude, thus, H2 is supported. A comparison of means revealed that the group that watched the film had significantly higher scores than the group that did not watch the film ($M = 18.477$ vs. 16.978, $t\,(193) = 3.741$, $p < 0.001$). In other words, the subjects that watched the film have more positive advertising attitudes, thus, H2a is supported.

Table 2 Verification to simple main effect of CSR advertising and film information on advertising attitude

Source of variation	SS	df	MS	F	p	Post hoc comparison
Ad (A) in B_1	30.416	2	15.208	1.172	0.314	
Ad (A) in B_2	121.485	2	60.743	4.047	0.021^*	CR > CG
Film (B) in A_1	13.762	1	13.762	1.691	0.198	
Film (B) in A_2	37.145	1	37.145	1.792	0.183	
Film (B) in A_3	49.378	1	49.378	3.778	0.057	

A_1 CRM advertising, A_2 GP advertising, A_3 CG advertising
B_1 with film, B_2 without film
CR CRM advertising, *CG* CG advertising
$^*p < .05$

4.2 *Verifying the Advertising Effect of CSR Advertising and Film Information*

H3 discusses the relationship between CSR advertising and film information. As the interactive effect between CSR advertising and film information is not obvious, H3 is not supported. H3a–d concerns the subjects' views regarding different CSR advertising designs after watching the film. Thus, the "simple main effect" of two independent variables was further verified to discuss the influence on advertising effect in different CSR advertising and film information situations. The verification results are as shown in Table 2.

As seen in Table 2, for the groups with film information, the subjects did not show significant difference in advertising attitude after viewing the CRM advertising, the GP advertising, or the CG advertising. For the groups without film information, the subjects who viewed CRM advertising have significantly higher advertising attitude than those who viewed the CG advertising ($M = 18.16$ vs.15.52, $F = 4.047$, $p < 0.05$), and there was no significant difference in other aspects. Therefore, H3a is supported. In other words, without the effect of film information, the subjects have more positive advertising attitude to CRM advertising design than to CG advertising design. H3b is supported. The subjects who did not watch the film have no significant difference in advertising attitude toward GP advertising design and CG advertising design. In other words, without watching the film, GP advertising design does not trigger more responses from the subjects. H3c and H3d discuss the film effect. As shown in Table 2, H3c and H3d are supported. This suggests that with film information, the subjects showed no differences in advertising attitude regarding CRM advertising design or CG advertising design. On the other hand, with film information, there is no difference in advertising attitude regarding GP advertising design or CG advertising design. In other words, with film information, the phenomenon of close advertising attitude is formed between CRM advertising, GP advertising, and CG advertising.

Although H3c and H3d are both supported, the film effect of H3d remains unclear. As indicated by H3b, without film information, GP advertising design has a result similar to the CG advertising design, indicating that the film has no clear effect on the subjects' attitude toward GP advertising. Hence, although H3d (with film information, the subjects have no differences in advertising attitude to GP advertising design and CG advertising design) is supported, the effect of film information still cannot be applied into GP advertising.

This study further discussed the role of film information in CSR advertising. According to the differences of the means of the subjects with and without film information, in the CRM advertising group, the subjects with film information have more positive advertising attitude than those not without film information ($M = 19.09$ vs.$18.16, F = 1.691, p > 0.05$). In the GP advertising group, the subjects with film information have better advertising attitude than those without film information. In the CG, those with film information have better advertising attitude than those without film information ($M = 17.53$ vs.$15.52, F = 3.778, p > 0.05$). For the three advertising designs, the film effect did not result in significantly different responses; however, the means of the three designs revealed that the subjects watching CRM advertising and GP advertising are most unlikely to be influenced by the film, while the subjects in the CG may have a more positive advertising attitude under the effect of film information. Although no significant difference was found, it has the most significant response among the three.

5 Discussion

This study discussed the influence of CSR advertising and film information on advertising effect. According to H1, consumers prefer CRM and GP advertisings over CG advertisings, that is, CSR marketing advertising is more popular than ordinary advertisings. Many researchers have found that CSR may have marketing effects [14, 23, 29, 30], which is the key for enterprises to succeed [11, 12]. This study reached the same conclusion and further confirmed the advertising effect of CRM and GP advertisings for CSR marketing. This finding can serve as reference for marketing advertising design.

As for H2, the advertising effect of individual factors showed that the group with film information has better advertising attitude than that without film information. That is, under the film effect, consumers form a better advertising attitude. This finding is consistent with Hartmann and Apaolaza-Iba'n˜ez [13], who suggested that consumers may have positive response under visual stimulation, as well as Lee and Gretzel [18], who indicated that mental image processing is a kind of persuasive communication.

H3 concerns the relationship between CSR marketing advertising and film information, as well as the effect of film information on advertising attitude under different CSR advertising designs. It is found that, without film information, there is a significant difference in CG advertising and CRM advertising. On the other hand, with film information, the subjects' advertising attitude changes, and the subjects in the **CG** have a good impression of the advertising and form a relatively positive advertising attitude. This is close to the advertising attitude in the CRM advertising group. This effect is not found in GP advertising. Moreover, the comparison of different responses under the film effect in three advertising designs revealed that the subjects in the CG are likely to have a more positive advertising attitude under the effect of film information, while the subjects watching CRM advertising and GP advertising are most unlikely to be influenced by film information. Although the above results have no significant differences and a specific theory is unable to be formed, it still indicates possible influence of film information on different advertising designs. According to Litvine and Wüstenhagen [19], the film information effect should be noted. Cialdini [5] regarded that film was the method to obtain the benefits of marketing success. The results of this study indicated that as the film actually plays a role in CRM advertising, no clear results are pointed out in GP advertising. This is possibly because, as Jr. et al. [15] mentioned, in many cases, the information receivers may be disturbed by other information and fail to respond to the advertising appeal. Therefore, as the film attributes to partial effect, it is unable to respond to all advertising types. Possibly, different films may have different corresponding relationship with CRM advertising or GP advertising. In the future, it is necessary to further explore different film types and advertising design types and concern the individual factors of consumers, in order to obtain more complete results. Finally, this study provides an understanding of the factors influencing consumers in judgments of green consumption, as well as the relationship of different CSR advertising designs and film information in green advertising effect. The importance of CSR advertising design in a green marketing system is verified.

Acknowledgments We are grateful to all the participants in our studies. We gratefully acknowledged the Overseas Chinese University Taiwan and the National Yunlin University of Science and Technology Taiwan for assistance in collecting the data. This research was supposed under the Ministry of Science and Technology of the Republic of China funding scheme (project number 104-2410-H-240-007).

References

1. Barone MJ, Miyazaki AD, Taylor KA et al (2000) The influence of cause related marketing on consumer choice: does one good turn deserve another. J Acad Mark Sci 28(2):248–262
2. Barone MJ, Norman AT, Miyazaki AD et al (2007) Consumer response to retailer use of cause-related marketing: is more fit better? J Retail 83(4):437–445
3. Berglind M, Nakata C (2005) Cause-related marketing: more buck than bang? Bus Horiz 48:443–453
4. Bickart BA, Ruth JA (2012) Green eco-seals and advertising persuasion. J Advert 41(4):51–67
5. Cialdini RB (2003) Crafting normative messages to protect the environment. Curr Dir Psychol Sci 12(4):105–109
6. Davis K (1973) The case for and against business assumption of social responsibilities. Acad Manag J 16(2):312–322
7. Ferle CL, Kuber G, Edwards SM et al (2013) Factors impacting responses to cause-related marketing in India and the United States: novelty, altruistic motives, and company origin. JBR 66(3):364–373

8. Folse JAG, Niedrich RW, Grau SL et al (2010) Cause-relating marketing: the effects of purchase quantity and firm donation amount on consumer inferences and participation intentions. J Retail 86(4):295–309
9. Gao Y, Mattila AS (2014) Improving consumer satisfaction in green hotels: the roles of perceived warmth, perceived competence, and CSR motive. Int J Hosp Manag 42:20–31
10. Garriga E, Melé D (2004) Corporate social responsibility theories: mapping the territory. JBE 53(1–2):51–71
11. Govindan K, Kannan D, Shankar KM et al (2014) Evaluating the drivers of corporate social responsibility in the mining industry with multi-criteria approach: a multi-stakeholder perspective. J Clean Prod 84:1–19
12. Hamann R, Kapelus P (2004) Corporate social responsibility in mining in Southern Africa: fair accountability or just green wash? Development 47(3):85–92
13. Hartmann P, Apaolaza-Ibáñez V (2010) Beyond savanna: an evolutionary and environmental psychology approach to behavioral effects of nature scenery in green advertising. J Environ Psychol 30:119–128
14. Jeong HJ, Paek HJ, Lee M et al (2013) Corporate social responsibility effects on social network sites. JBR 66(10):1889–1895
15. Jr RSW, Hung IW, Jiang Y et al (2008) Visual and verbal processing strategies in comprehension and judgment. J Consum Psychol 18(4):244–257
16. Kim EEK, Kang J, Mattila AS et al (2012) The impact of prevention versus promotion hope on CSR activities. Int J Hosp Manag 31(1):43–51
17. Lafferty BA, Goldsmith RE (2005) Cause–brand alliances: does the cause help the brand or does the brand help the cause? JBR 58(4):423–429
18. Lee W, Gretzel U (2012) Designing persuasive destination websites: a mental imagery processing perspective. Tour Manag 33(5):1270–1280
19. Litvine D, Wüstenhagen R (2011) Helping "light green" consumers walk the talk: results of a behavioural intervention survey in the Swiss electricity market. Ecol Econ 70(3):462–474
20. Manakkalathil J, Rudolf E (1995) Corporate social responsibility in a globalizing market. SAM Adv Manag J 47:29–32
21. Matthes J, Wonneberger A, Schmuck D et al (2014) Consumers' green involvement and the persuasive effects of emotional versus functional ads. JBR 67(9):1885–1893
22. Müller SS, Fries AJ, Gedenk K et al (2014) How much to give? – the effect of donation size on tactical and strategic success in cause-related marketing. IJRM 31(2):178–191
23. Oppewal H, Alexander A, Sullivan P et al (2006) Consumer perceptions of corporate social responsibility in town shopping centres and their influence on shopping evaluations. J Retail Consum Serv 13:261–274
24. Papasolomou I, Demetriou M (2005) Cause-related marketing (CRM): branding financial service organisations in cyprus by aligning CSR with business motivations. Int Rev Publ Non Profit Mark 2(2):70–86
25. Pirsch J, Gupta S, Grau SL et al (2007) A framework for understanding corporate social responsibility programs as a continuum: An exploratory study. JBE 70:125–140
26. Pracejus JW, Olsen GD (2004) The role of brand/cause fit in the effectiveness of cause-related marketing campaigns. JBR 57(6):635–640
27. Rahbar E, Wahid NA (2011) Investigation of green marketing tools' effect on consumers' purchase behavior. Bus Strat Ser 12(2):73–83
28. Ross JK, Patterson LT, Stutts MA et al (1992) Consumer perceptions of organizations that use cause-related marketing. J Acad Mark Sci 20(1):93–97
29. Sprinkle GB, Maines LA (2010) The benefits and costs of corporate social responsibility. Bus Horiz 53:445–453
30. Swindley D (1990) UK retailers and global responsibility. Serv Ind J 10:589–598
31. Tu JC, Kao TF, Tu YC et al (2013) Influences of framing effect and green message on advertising effect. Soc Behav Personal 41(7):1083–1098

32. Wang KC, Hsieh AT, Chen WY et al (2002) Is the tour leader an effective endorser for group package tour brochures? Tour Manag 23:489–498
33. Cheng FF, Wu CS (2010) Debiasing the framing effect: the effect of warning and involvement. Decis Support Syst 49:328–334
34. Priester JR, Godek J, Nayakankuppum DJ, Park K et al (2004) Brand congruity and comparative advertising: when and why comparative advertisements lead to greater elaboration. J Consum Psychol 14(1–2):115–123

Part IV
Sustainable Production and Material Recovery

Systems Approach to Resource Efficient and Cleaner Production Solutions: Method and Implementation

Jurgis Kazimieras Staniskis and Egle Katiliute

Abstract Resource efficient and cleaner production (RECP) means a continuous application of an integrated preventive environmental strategy to processes, products and services to increase overall efficiency. This leads to improved environmental performance, cost savings and reduction of risks to humans and the environment. Additionally, RECP strategy includes a clearly defined RECP management system, which ensures continuous improvement of environmental and economic performance. Each action to reduce consumption of raw materials and energy, and to prevent or to reduce generation of waste, increases productivity and brings financial benefits to an enterprise. A preventive approach means that environmental problems are addressed before they arise at the time when choices are made concerning processes, raw materials, design, transportation, services, etc. The paper describes a method that comprises logical steps of RECP assessment that are embedded in a special procedure. For implementation of the RECP innovations, the Revolving Facility SPIN (the System for Preventive Innovations) has been developed. SPIN addresses the needs to enhance the financing opportunities of the RECP innovations that otherwise would not have been implemented. The Facility constitutes a follow-up of various RECP programmes and projects supported by the UN worldwide. The facility had considerable catalytic effects by demonstrating to other financiers and enterprises that the financing of priority cleaner production investments yields significant environmental and economic benefits. The paper presents an overview of more than 200 successful cases of RECP innovation implementation. In each case, realised savings have been verified and compared with the expected savings.

Keywords Resource efficient and cleaner production • Assessment methodology • Financing and implementation • Industrial companies

---

J.K. Staniskis (✉)
Institute of Environmental Engineering, Kaunas University of Technology, Kaunas, Lithuania
e-mail: justa@ktu.lt

E. Katiliute
School of Economics and Business, Kaunas University of Technology, Kaunas, Lithuania

M. Matsumoto et al. (eds.), *Sustainability Through Innovation in Product Life Cycle Design*, EcoProduction, DOI 10.1007/978-981-10-0471-1_26

1 Introduction

We cannot measure the future. What we know, because we can prove it with simple math and physics, is that global economy is now bigger than the planet, which means that the economy at some point will stop growing. Whether that was in 2008 or is still to come in 2012 or 2015 is of historical interest only. It is certainly going to come. That has led to endless scientific, philosophical, ideological and political debates. Many argued that the warnings were wrong or at least exaggerated. Others argued that the warnings were right but humans are smart and always come up with new technologies and behaviour through the market. The minority have argued that consumerism is a bad culture for humanity because it leads to bad social outcomes and lives without meaning, and that is why we should try to develop a better macroeconomic model for society in order to enhance our quality of life [1].

The linear 'take, make and dispose' model, the dominant economic model of our time, relies on large quantities of easily accessible resources and energy and as such is increasingly unfit for the reality in which it operates. Working towards efficiency – a reduction of resources and fossil energy consumed per unit of economic output – will not alter the finite nature of their stocks but can only delay the inevitable. A deeper change of the operating system is necessary. The notion of the circular economy has attracted attention in recent years. The concept is characterised, more than defined, as an economy that is restorative and regenerative by design and aims to keep products, components and materials at their highest utility and value at all times, distinguishing between technical and biological cycles. It is conceived as a continuous positive development cycle that preserves and enhances natural capital, optimises resource yields and minimises system risks by managing finite stocks and renewable flows. It works effectively at every scale [2].

Sustainable industrial development could be defined as a process of adopting business strategies and activities to meet the needs of an enterprise and its stakeholders today while protecting, sustaining and enhancing the human and natural resources that will be needed in the future. To ensure sustainable industrial development, systematic application of the following measures is needed:

- Resource efficient and cleaner production
- Environmental and integrated management systems
- Product-related measures of sustainable industrial development (e.g. eco-design, life-cycle approach)
- Sustainability reporting

Experience shows that companies having identified cost-effective and technically feasible RECP options may not be able to make necessary RECP investments to realise the financial benefits and environmental advantages. Financing of RECP innovations varies among countries and types of the projects. Domestic and international efforts to strengthen environmental financing still face a number of serious obstacles, many of which are related to profound economic, political and social problems.

Resource efficient and cleaner production is a recognised and proven strategy for more efficient use of natural resources and minimising waste, pollution and risks at the source where they are generated. RECP should be an essential part of any comprehensive pollution management system at an enterprise or national level. In many cases, the adoption of RECP improvements can reduce or even eliminate the need for end-of-pipe investments and can have both environmental and economic benefits [3].

Sustainable industrial development is defined as a pattern or patterns of development that balance a country's concerns for competitiveness, for social development and for environmental soundness. Either absolutely or comparatively, such development accomplishes three tasks:

- It encourages a competitive economy, with industry producing for export as well as for the domestic market.
- It creates productive employment, with industry bringing long-term employment and increased prosperity.
- It protects the environment, with industry efficiently utilising non-renewable resources, conserving renewable resources and remaining within the functional limits of the ecosystem.

Businesses and societies can find approaches that will help to move towards all three goals at the same time: environmental protection, social well-being and economic development. To achieve circular economy in today's context of market and private sector-driven development, the developing countries, the least developed countries in particular, require methodological and financial support from the industrialised countries to build up their basic capacities.

The Institute of Environmental Engineering (APINI) (EU Centre of Excellence in Sustainable Industrial Development (APINI – SID)) initiated implementation of RECP concept in Lithuania and is the main institution in the country working in this area. In 1992–2007, in co-operation with different foreign and local partners, APINI experts have developed the RECP methodology based on the systems approach, financing and implementation scheme and have supported more than 150 Lithuanian companies to implement RECP innovations.

2 Methodology

Resource efficient and cleaner production is a recognised and proven strategy for improving the efficient use of natural resources and minimising waste, pollution and risks at the source where they are generated. Significant reductions in pollution loads can often be obtained at a little cost, and more efficient use of resources and reduction of pollution in industrial production are clearly preferable to reliance on end-of-pipe treatment. In many cases, adoption of RECP improvements can reduce or even eliminate the need for end-of-pipe investments. Therefore, RECP is associated with both environmental and economic benefits. Despite its conceptual simplicity, it took until the late 1990s for RECP to become recognised as a valuable approach for achieving the dual objectives of environmental improvement and industrial development. Resource efficient and cleaner production is being increasingly recognised as an essential route towards sustainable development.

RECP should not be considered only as an environmental strategy, because it also relates to economic considerations. In this context, waste is considered as a 'product' with negative economic value. Each action to reduce consumption of raw materials and energy, and to prevent or to reduce generation of waste, increases productivity and brings financial benefits to an enterprise.

Sustainable development is a good business in itself. It creates opportunities for suppliers of 'green consumers', developers of environmentally safer materials and processes, firms that invest in eco-efficiency and those that engage themselves in social well-being. These enterprises will generally have a competitive advantage. They will earn their local community's goodwill and see their efforts reflected in the bottom line.

A preventive approach means that environmental problems are addressed before they arise when choices are made concerning processes, raw materials, design, transportation, services and more. Such an approach effectively addresses the wasting of natural resources since pollution not only leads to environmental degradation but is also a sign of inefficient production processes or management. In practice, resource efficient and cleaner production means the following:

- Avoiding or reducing the amount of waste produced
- Using energy and resources efficiently
- Producing environmentally sounder products and services
- Generating less waste, reducing costs and increasing profits

Resource efficient and cleaner production is achieved by applying know-how, by improving technology, and/or by changing attitudes. RECP strategy includes the following prevention practices:

1. *Good housekeeping*: taking appropriate managerial and operational provisions to prevent leaks and spills (such as preventive maintenance schedules and frequent equipment inspections) and to enforce the existing working instructions (through proper supervision, training etc.)
2. *Input substitution*: substitution of input materials by less toxic or by renewable materials or by adjunct materials (for instance, lubricants, coolants, cleansing agents, etc.), which have a longer service life-time in production
3. *Better process control*: modification of the working procedures, machine instructions and process record keeping in order to run the processes at a higher efficiency and lower waste and emission generation rates
4. *Equipment modification*: modification of the (existing) productive equipment and utilities, for instance, through addition of measuring and controlling devices, in order to run the processes at a higher efficiency and lower waste and emission generation rates

5. *Technology change*: replacement of the technology, processing sequence and/or synthesis pathway in order to minimise waste and emission generation during production
6. *Product modification*: modification of the product characteristics in order to minimise the environmental impacts of the product during or after its use (disposal) or to minimise the environmental impacts of its production
7. *Using energy efficiently*: energy is a very significant source of environmental impact. Energy use may result in effects on land, water, air and biodiversity, as well as in the production of large quantities of solid wastes. The environmental impacts resulting from energy use can be decreased by improved energy efficiency as well as by using energy from renewable sources, such as the sun and wind
8. *On-site recovery/reuse*: reuse of waste materials in the same process or for another useful application within the company [5]

   The below described logical steps of RECP assessment are embedded in an organised procedure. This procedure is instrumental in organising the RECP efforts, informing the necessary stakeholders within the company and bringing together the persons that can develop, evaluate and implement RECP innovations. RECP assessment, which basically was developed by authors and colleagues from Norway, is divided into five phases (see Fig. 1).

1. *Planning and organisation*. RECP planning is a systematic method for identifying options to reduce or avoid the generation of waste. The process:

   - Ensures the selection and implementation of the most cost-effective RECP innovations.
   - Consistency of RECP objectives and activities with those identified in the organisation's broader planning process.
   - Facilitates broader investment analysis and decision-making (such as capital budgeting and purchasing).
   - A documented RECP plan may be a condition for receiving financing or insurance at more attractive rates.

2. *Pre-assessment*. The prime objective of the phase is the selection of one or a few assessment focuses. It requires a preliminary identification and evaluation of the RECP potential at the plant level. While doing so, first inventories of obvious options as well as a preliminary estimate of the waste generation costs are made.
3. *Assessment*. The phase focuses on the production processes that involve waste streams and is based on examination and re-evaluation of the production processes. The re-evaluation consists of 'source identification' followed by 'cause diagnosis' and 'option generation'.

   - For the *source identification*, an inventory is made of the material flows entering and leaving the company with the associated costs. This results in a process flow diagram allowing the identification of all sources of waste and emission generation.

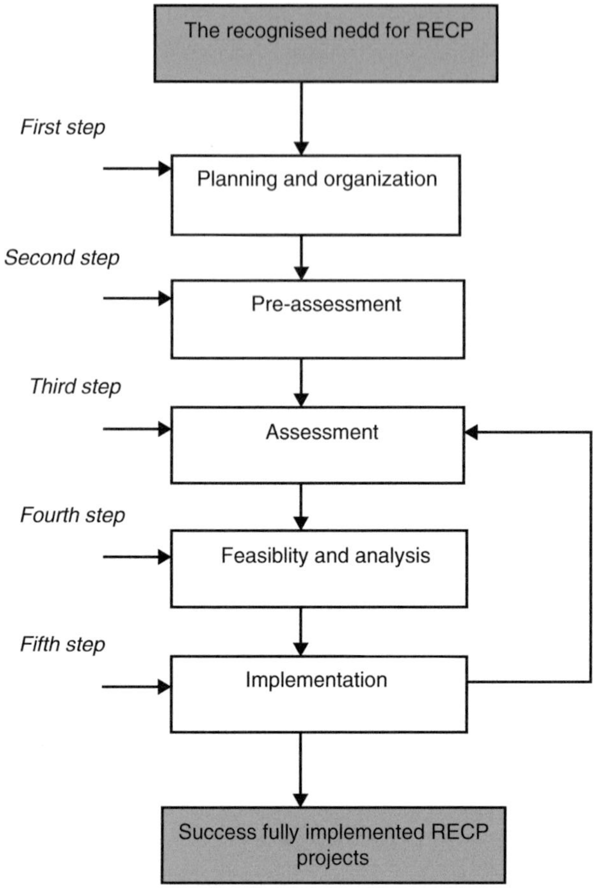

Fig. 1 Scheme of simplified RECP methodology

Material flow analysis (MFA) is a systematic assessment of the flows and stocks of materials within a system defined in space and time. It connects the sources, pathways and intermediate and final sinks of a material.

- The next step (*cause diagnosis*) is based on the material and energy balance and investigates the factors that influence the volume and composition of the waste and emissions generated. Evaluation of the relative importance of each of the possible waste generation causes is performed.
- The purpose of the *option generation* is to create a vision of how to eliminate or control each of the causes of waste and emission generation.

4. *Feasibility studies*. These studies have to prove whether each of the options is technically and economically feasible and whether each option contributes to the environmental improvement. The level of detail in the feasibility studies needs to be tailored to the nature of the option, since options may vary from simple

operational improvements and the use of alternative materials to installation of advanced equipment.

- *Technical evaluation* consists of two interrelated parts. First, it should be evaluated whether the option can be put into practice, i.e. a check of the availability and reliability of equipment, the effect on product quality and productivity, the expected maintenance and utility requirements and the necessary operating and supervising skills. Second, the changes in the technical specifications can be converted into a projected material balance, reflecting the input and output material flows and energy requirements after implementation of the RECP innovations.
- *Economic evaluation* consists at least of data collection (regarding investments and operational costs, as well as benefits), choice between evaluation criteria (payback period, net present value (NPV) or internal rate of return (IRR)) and feasibility calculations. The economic data collection is based upon the results of the technical evaluation. In order to properly incorporate the long-term economic advantages of RECP, it is highly recommendable to apply total cost assessment (TCA) principles to the economic evaluation (especially for high cost options).
  It is suggested that all significant environmental costs (and impacts) should be traced to the responsible product (as opposed to the use of overhead accounts). That is, material tracking could be used for material costs and other related product costs (such as energy, water use and so on), as well as for waste streams. Where tracking does identify costs, these should be clearly identified with the tasks or processes generating the costs.
- The objective of *environmental evaluation* is to determine the positive and negative impacts of the option to the environment. It is of little value if an option reduces one problem and creates another, potentially a worse one. In order to be comprehensive, an environmental evaluation must take into account the whole life-cycle of a product or service.

5. *Implementation and continuation.* The feasible prevention measures are implemented and provisions taken to assure the ongoing application of RECP. The development of such ongoing programme requires the monitoring and evaluation of the results achieved by the implementation of the first batches of prevention measures. The expected result of this phase is threefold: (i) implementation of the feasible prevention measures, (ii) monitoring and evaluation of the progress achieved by the implementation of the feasible options and (iii) initiation of ongoing RECP activities [4].

The integration of environmental management accounting (EMA), which provides for the transition of data from financial and cost accounting to increase material efficiency, reduce environmental impact and risk and reduce environmental costs, is the most valuable part of the assessment methodology. The approach introduced in the methodology has the underlying assumption that all purchased materials must by physical necessity leave the company either as product or waste and emission. Therefore, when calculating environmental costs, not only disposal fees are regarded, but also the wasted material purchase value and the production costs of waste and emissions are added [5].

Investments in RECP can have attractive economic benefits due to reduction of input costs for materials, energy and water and reduced expenditures related to waste treatment and disposal. Several international organisations, banks and donors have initiated and implemented projects to facilitate introduction of RECP investments. However, these domestic and international efforts to strengthen environmental financing still face a number of serious obstacles, many of them being related to profound economic, political and social problems.

Furthermore, an effective financing system requires environmental strategies with clear goals and priorities. There is a need for training and education in environmental management and financing. In many countries, capacities for preparing financially and environmentally sound projects should be increased.

3 Implementation

Companies that have identified cost-effective and technically feasible innovations may be unable to make necessary RECP investments allowing realising the financial benefits and environmental advantages. RECP investments may not generate as much economic value for the company in comparison to other types of investments, or the company may not succeed in accessing finances at all due to a poor balance sheet or low profitability in the past.

The starting point is that the RECP assessment according to the APINI methodology results in feasible preventive innovations, followed by the process of input to capital budgeting, in which the profitability of innovations is compared with the profitability of alternative investment options (e.g. market expansion, new product development, etc.). This may result in RECP investment proposal, which may be regarded as the company's intention to invest in a particular innovation, provided that finances can be made available. Whether the planned investment will eventually take place depends on the company's ability to collect capital either from its own resources or by attracting external funds.

The Revolving Facility SPIN (the **S**ystem for **P**reventive **In**novations) addresses the needs to enhance the financing of the RECP innovations that otherwise would not have been implemented with a high leverage for donors and an effective use of scarce resources. The Facility enables the financing of smaller investment projects in the target countries. It constitutes a follow-up of various RECP programmes and projects supported by donors in Central and Eastern European countries. A scheme of SPIN system for preventive innovation development and implementation in industry is presented in Fig. 2.

In terms of RECP investment financing, the BDSP experts play a crucial role in RECP innovation identification, evaluation, implementation and reporting, because they:

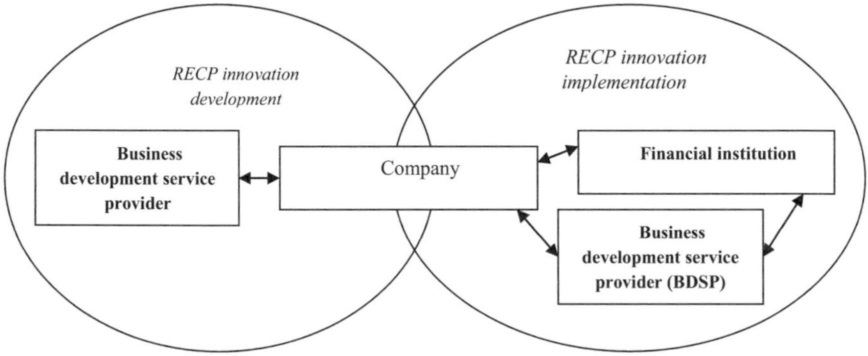

Fig. 2 SPIN system for RECP innovation development and implementation in industry

- Prepare a loan application on behalf of the applicant, including assistance in calculation of costs savings and environmental benefits
- Assist the financing institution in communication with the applicant and preparation of loan documentation, project description and reporting requirements
- Prepare the project progress and completion reports to be presented as part of the borrower's disbursement request
- Assist in the project monitoring and supervision, including the supervision of procurement and project implementation progress as compared to the budgets and implementation plan as well as the project objectives

The facility had considerable catalytic effects in Lithuania by demonstrating to other financiers and enterprises that the financing of priority cleaner production investments yields environmental and economic benefits. The main objective of the facility is to finance, on favourable terms, implementation of high-priority RECP innovations with a rapid payback, which yield environmental and economic benefits ('win-win projects'). The innovations should be commercially viable with an identifiable and secure stream of earnings to be used to repay the loan. The basis for providing a loan is a cash flow of RECP investment and ability of the enterprise to repay the loan over the agreed period [6].

4 Results

In Lithuania, business development service provider's role was fulfilled by KTU Institute of Environmental Engineering (APINI) and the role of financing institution – by Nordic Environment Financing Corporation (NEFCO). RECP assessment is one of core elements for RECP investment project development according to the methodology presented above, which covers:

- RECP pre-assessment (RECP audit preparation, division of processes into unit operations, construction of process flow diagrams linking the unit operations)

- Material balance (determination of process inputs and outputs), derivation of a material balance (gathering the input and output information, deriving a preliminary material balance, evaluating and refining the material balance)
- Synthesis (identification of RECP innovations; environmental, technical and economic evaluation; design of RECP action plan)
- The company's 'economic health' analysis
- The development of RECP investment project

For each RECP investment project, supervision and monitoring plan is prepared as a part of the loan application. The company has to provide APINI experts a possibility to inspect goods, sites, factories, installations and construction sites included in the investment project, any related documentation, and to supervise the implementation of the project. Financial and physical performance indicators are specified in the loan agreements. Environmental performance indicators (EPIs) monitor the company's effectiveness and efficiency of resource management. This applies mainly to physical resources (materials, energy).

Based on this methodology, environmental issues in the RECP innovation development focus on the following:

- Location of the object with respect to population centres, sensitive local land issues and existing levels and sources of pollution
- Pollution category (air, surface water, groundwater, hazardous waste)
- The scale of pollution impact
- The effect associated with the pollution, including possible toxicity to human health, a possible impact to climate change and damage to the natural ecosystems and habitat

For each project, the realised savings are verified and compared with the expected savings. A standardised reporting format is provided, with a focus on savings in energy use, water use, use of chemicals, etc. The environmental effects of each project are verified. This verification should document the reduction in emissions and wastes and the reduction in inputs (water, energy, chemicals, etc.). The mentioned standardised format ensures the management of all RECP innovation data in a special database at APINI. The data in the database are continuously updated.

In conventional cost accounting, the aggregation of environmental and non-environmental costs in overhead accounts results in their being 'hidden' from the management. There is substantial evidence, based on experience of Lithuanian industrial companies, that the management tends to underestimate the extent and growth of such costs. By identifying, assessing and allocating environmental costs, environmental accounting allows the management to identify opportunities for costs savings. For example, it can be said that more than 29 % of environment-related measures in the Lithuanian companies that had developed RECP investment projects were re-estimated using the proposed approach, allowing to relocate the investments and to get significant environmental and economic benefits. The

Table 1 Results from implemented CP innovations

Number of companies	126
Number of implemented CP innovations	211
Environmental results (yearly):	
El. energy consumption reduced	27,584,000 kWh
Heat energy consumption reduced	60,518,000 kWh
Waste amount reduced	86,700 t
Chemicals consumption reduced	850 t
Air emission reduced	79,500 t
Drinking water consumption reduced	297,500 m^3
Diesel consumption reduced	387,000 l
Environmental results (yearly):	
Natural gas consumption reduced	5,883,000 m^3
Fuel consumption reduced	656,800 t
Wastewater amount reduced	622,500 m^3
Industrial water consumption reduced	468,900 m^3
Economic profit:	
Total investment in RECP innovations	16,529,000 EUR
Yearly savings from RECP innovations	9,605,000 EUR

success of RECP in Lithuania can be measured by environmental and economic benefits. The real data from the APINI RECP database are presented in Table 1 [7].

It is evident that RECP investment project was developed and implemented in accordance to the proposed methodology:

- Increases profitability and lowers production costs
- Provides a rapid return on any capital or operating investments required
- Leads to a more efficient use of energy, natural resources and raw materials
- Increases staff motivation through reduced worker risks
- Reduces the risk of environmental accidents
- Reduces/eliminates regulatory compliance costs

In the framework of UNEP and UNIDO projects, Lithuanian experience was transferred to several countries in Africa, Central America, South-East Asia, China and Russia.

5 Conclusions and Discussion

Some key learning points concerning cleaner production are the following:

- RECP approach reduces pollution generation at every stage of the production process and thus minimises or eliminates waste that needs to be treated at the end of the process.
- RECP is not a one-time event; it is a continuous process.

- RECP can be achieved through good operating practices, process modification, technology changes, raw material substitution and redesign and/ or reformulation of product.
- Effluent treatment, incineration and even waste recycling outside the production process are not regarded as RECP, although they remain necessary activities to achieve a low environmental impact.
- The economic advantages of RECP are based on the fact that RECP is more cost effective than pollution control. The systematic avoidance of waste and pollutants increases process efficiency and improves product quality. Through pollution prevention at the source, the cost of final treatment and disposal is minimised.
- The environmental advantage of RECP is based on the fact that it solves the waste problem at its source. Conventional end-of-pipe treatment often only moves the pollutants from one environmental medium to another.
- The barriers to RECP are mostly related to human rather than technical factors. The end-of-pipe approach is well known and accepted by industry and engineers. Existing government policies and regulations often favour end-of-pipe solutions. There is a lack of communication between those in charge of production processes and those who manage the wastes that are generated. Managers and workers, who know that the factory is inefficient and wasteful, are not rewarded for suggesting improvements.
- Taking into account that RECP addresses the problem at several levels at the same time, introduction of an industry/ plant level programme requires the commitment of top management and a systematic approach to cleaner production in all aspects of the production process.

In spite of the results presented above, industry rarely considers cleaner production initiatives as an option for improving productivity in parallel to a reduction of the impact on the environment. End-of-pipe solutions are still commonly applied if environmental demands are to be met.

Experience from RECP projects implemented in Lithuania has shown that it is difficult to ensure the follow-up of RECP activities on a company level after the project end. Therefore, a decision has been made to integrate RECP and environmental management systems (EMS) and to start the implementation of preventive environmental management systems:

- Firstly, cleaner production is not a continual and normal practice for industry.
- Secondly, environmental management system (EMS) does not guarantee that RECP will be applied or even that an environmental performance above regulatory requirements will be obtained, i.e. EMS focuses on the quality and strengths of procedures and not the actual outcome of these procedures regarding pollution abatement and improved resource productivity.

Many companies recognised the importance of the environmental impacts of their products and began to incorporate significant environmental aspects into their product design and development processes. This requires the identification of key environmental issues related to the product throughout its entire life-cycle. The key issues include problematic activities, processes and materials associated with the product from raw materials acquisition, manufacturing, distribution, use and disposal, in other words, the entire life-cycle.

Product design and development relating to improved environmental performance has many definitions, e.g. design for environment, ecological design, environmental design, environmentally conscious design, environmentally responsible design, socially responsible design, sustainable product design, sustainable product development, green design and life-cycle design. In this article, we use eco-design as a term of choice [8].

Product eco-innovations based on life-cycle assessment were successfully introduced to 15 industrial companies for products like electricity metre, refrigerator, firewood stove, buckle switch, corner sofa, etc. For instance, implementation of eco-design tools for an electricity metre has reduced environmental impact 2.73 times in Ecoindicators' 99 mPt (from 1400 to 423) [7].

It could be also mentioned that a number of EU environmental directives and regulations (e.g. IPPC Directive, REACH) having direct impact for SMEs have been introduced, but many enterprises have problems in their practical application. Therefore, there is a need for a thorough analysis of the SME needs and challenges they are facing to ensure enforcement of the directives. One of the obvious reasons why many SMEs struggle or fail to respond to EU directives and regulations intended to enforce eco-efficient production processes as they don't have sufficient capacity in RECP innovation development and implementation. In this context, the following specific problems could be highlighted:

- Existing eco-innovations aimed at sustainable production are not systematically demonstrated and used.
- Incentives for SMEs to develop and to apply RECP innovations are insufficient.
- Links between demand and supply of technological and organisational/managerial innovations are weak.
- Sustainability aspects are not sufficiently integrated in the innovation processes.

Finally, it could be stressed that development and implementation of RECP innovations mainly depends on co-operation level between business and research institutions. The best result could be achieved when knowledge of research institutions and technical skills of enterprises is merged. The most significant problem is often not financing availability for innovation implementation, but lack of thorough analysis of existing situation in enterprises, understanding of the problem reasons and development of innovations that correspond to the principles of the circular economy.

References

1. Gilding P (2011) The great disruption. How the climate crisis will transform the global economy. Bloomsbury Publishing, London, 292 p
2. Delivering the circular economy. A toolkit for policymakers (2015) Ellen Macarthur foundation. 176p
3. Huhtala A, Ciccozzi E (2003) Financing cleaner production investments – UNEP experience/ clean technologies and environmental policy, 5/2, 87–91
4. Staniskis JK, Arbaciauskas V (2004) The sustainable industrial development:reality and vision. Technological choices for sustainability. Springer, New York, 375p
5. Staniskis JK, Stasiskiene Z, Chr J (2008) Environmental management accounting, investment appraisal and financing. Nova Science Publishers, New York, 346p
6. Staniskis JK, Jayaraman C (2010) Cleaner production and energy conservation for sustainable development. Daya Publishing House, Delhi, 288p
7. Staniskis JK ed (2010) Sustainable innovation in lithuanian industry: development & implementation. Technologija, Kaunas, 458 p. (in Lithuanian)
8. (2010) Sustainable consumption and production in the Asia_Pacific Region. Effective responses in a resource constrained world. IGES white paper III. Institute for Global Environmental Strategies. 295 p

Integrated Production and Transportation Scheduling for Low-Carbon Supply Chains

Yoshitaka Tanimizu, Hiromasa Ito, and Kenta Matsui

Abstract This paper presents a model for integrated production and transportation scheduling in order to develop a low-carbon supply chain. A genetic algorithm and heuristic rules are applied to production scheduling problems, and beam search technique is used for transportation scheduling problems in consideration of both the tardiness of products and the amount of carbon dioxide emitted in manufacturing and transportation processes. A simulation system is developed to verify the effectiveness of the model.

Keywords Carbon dioxide emission • Supply chain • Genetic algorithm • Beam search • Heuristic rule

1 Introduction

Sustainability issues, such as greenhouse gas emissions, hazardous industrial waste, and environmental pollution, have become serious problems facing the international community. Greenhouse gas emissions have been blamed for raising the Earth's temperature. The Kyoto Protocol, negotiated in December 1997 in the city of Kyoto, is an international agreement that aims to reduce carbon dioxide emissions and the presence of greenhouse gases [1]. About 40 industrial countries have promised to cut emissions of greenhouse gases by 5 % on average by 2012 to lower the level to that of 1990. According to a government report, total greenhouse gas emissions for FY 2011 were 1380 million tons [2]. Compared to the total emissions in the base years stipulated in the Kyoto Protocol, the total emission of carbon dioxide arising from energy increased by 4.0 %.

A green supply chain has been attracting attention as a new approach which seeks to minimize a product or service ecological footprint. Logistics and transportation activities attract the most interest among companies which claim to have established a green supply chain. Most of the companies have made changes to their transport and logistics operations, and half have made cuts in transportation. The

Y. Tanimizu (✉) • H. Ito • K. Matsui
Graduate School of Engineering, Osaka Prefecture University, Osaka, Japan
e-mail: tanimizu@me.osakafu-u.ac.jp

M. Matsumoto et al. (eds.), *Sustainability Through Innovation in Product Life Cycle Design*, EcoProduction, DOI 10.1007/978-981-10-0471-1_27

report reveals that 83 % of the companies claim to factor environmental concerns into their strategic decisions, but fewer have implemented green supply chains [3].

In general, a supply chain consists of different organizations, such as raw material suppliers, part suppliers, subassembly manufacturers, and assembly manufacturers, as well as among customers in order to improve the efficiency of material and information flows among the organizations. As information technology has progressed, the trend has been toward even more flexible supply chains or dynamic supply chains [4, 5]. Our previous studies proposed a supply chain model for a group of organizations having a make-to-order production system that has no inventories of raw materials, assembled parts, and completed products. A two-layered supply chain model consists of two kinds of organization: clients and suppliers [6, 7]. The model provides a negotiation protocol to determine suitable prices and delivery times for ordered products through the iteration of the negotiation processes between the clients and suppliers and through the modification processes of supplier production schedules. The proposed model was extended in consideration of transportation functions between the suppliers and clients [8, 9]. The extended model provided a method that suppliers could use to estimate suitable delivery times and prices for products by optimizing production and transportation schedules in terms of both the profits of suppliers and the amount of carbon dioxide emitted in transportation processes [10].

This paper presents a new model for integrated production and transportation scheduling problems in order to develop a low-carbon supply chain. The model consists of two optimization approaches for the production scheduling and the transportation scheduling. In the production scheduling process, the loading sequences of products are determined on manufacturing resources, such as machine tools, by using a genetic algorithm (GA) [11–14], and suitable processing times in manufacturing resources are alternatively selected for products by using heuristic rules in consideration of both the tardiness of products and the amount of carbon dioxide emitted in manufacturing processes. In the transportation scheduling process, the allocations of products to transportation vehicles moving between suppliers and clients are determined by using a heuristic search technique, such as beam search technique, which has the advantage of potentially reducing the candidates for searching. It can improve the loading ratio of transportation vehicles and decrease in the number of transportation vehicles for reducing carbon dioxide emissions. A prototype of multi-agent simulation system is developed for a low-carbon supply chain in consideration of production and transportation scheduling problems. Experiments are carried out on a two-layered supply chain model consisting of a supplier and a client in order to verify the effectiveness of the proposed model from the viewpoint of both the total profit of supplier and the amount of carbon dioxide emitted.

The reminder of this paper is organized as follows. Section "Background" presents a background. Section "Basic Supply Chain Model" explains the previous basic supply chain model. Section "Production and Transportation Scheduling Considering Carbon Dioxide Emmisions" formulates the extended supply chain model considering carbon dioxide emission reductions. Section "Experiments" assesses the effectiveness of the proposed method by using a developed supply chain simulation system and demonstrates results of experiments. Finally, section "Conclusions" concludes this study.

2 Background

Environmental issues are gaining justifiable popularity among society, governments, and industry due to negative environmental developments [15]. A green supply chain is expected to reduce waste, minimize pollution, save energy, conserve natural resources, and reduce carbon emissions. Bloemhof-Ruwaard et al. [16] introduced some OR models and techniques to deal with the environmental issues. Cholette and Venkat [17] calculate the energy and carbon emissions associated with the transportation links and warehousing activities in food and beverage supply chains, particularly in the wine industry. A carbon labeling system of products has been emphasized to help consumers' environmental decisions in the supply chain. The implication is that an effective mechanism to encourage green cooperation along the supply chains still needs to be developed [18]. Many businesses that sell environmentally friendly, green or sustainable, low-carbon emission products tend to cost more [19]. This cost is usually passed onto the consumers in the form of higher prices. Sundarakani et al. [20] propose that companies that do not measure and manage carbon emissions along their supply chains and in collaboration with their supply chain partners will place themselves at a disadvantage. As only papers dealing with supply chain management are taken into account, it is logical that economic issues are addressed. The green and sustainable supply chain management is the management of material, information, and capital flows as well as cooperation among companies along the supply chain while integrating goals from all three dimensions of sustainable development, i.e., economic, environmental, and social, which are derived from customer and stakeholder requirements [21]. Therefore, a method for green supply chain management should be superior from the viewpoint of both the economical aspect and environmental aspect.

3 Basic Supply Chain Model

3.1 Two-Layered Dynamic Supply Chain Model

3.1.1 Model Components

In this paper, we consider a two-layered dynamic supply chain model as a basic model of dynamic supply chains, as shown in Fig. 1. The model consists of two stages or levels including two kinds of components respectively, they are a client and a supplier. The clients require products and send orders for them to all the

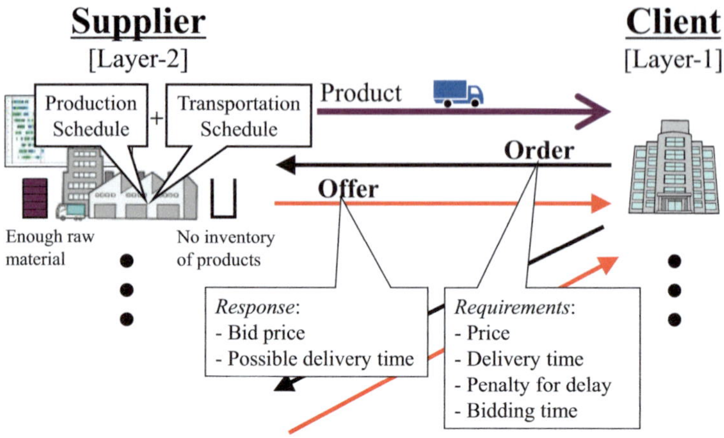

Fig. 1 Two-layered dynamic supply chain model

suppliers. The suppliers generate offers and send them to the client. The offers include information about bid prices and possible delivery times for the products, which have been determined by the modification processes of the production and transportation schedules and by the negotiation processes between the clients and the suppliers.

3.1.2 Negotiation Protocol

Steps in the negotiation process between the client and suppliers are as follows.

1. A client generates a new order and sends it to all suppliers when a product is needed. The orders include information on the required price, the required delivery time, the penalty charge factors for delays, and the bidding time. Instead of just tolerating them, the clients can receive penalty charges for any delays in deliveries.
2. The suppliers have production schedules including the production of pre-ordered products already under contract. After adding the newly ordered product to the existing production schedules, the suppliers improve the production schedules by using genetic algorithms (GA) and a heuristic rule and determine transportation schedules by using a search method. Then, the suppliers generate offers for the client. The offers include information on the bid price and the possible delivery time of the newly ordered product.
3. The client evaluates all the offers sent from the suppliers. The client accepts the offer with the lowest bid price if the offer satisfies the requirements in terms of both delivery time and the price. However, if no offers satisfy the requirements of the order, the client relaxes the requirements and sends the modified order to the suppliers.

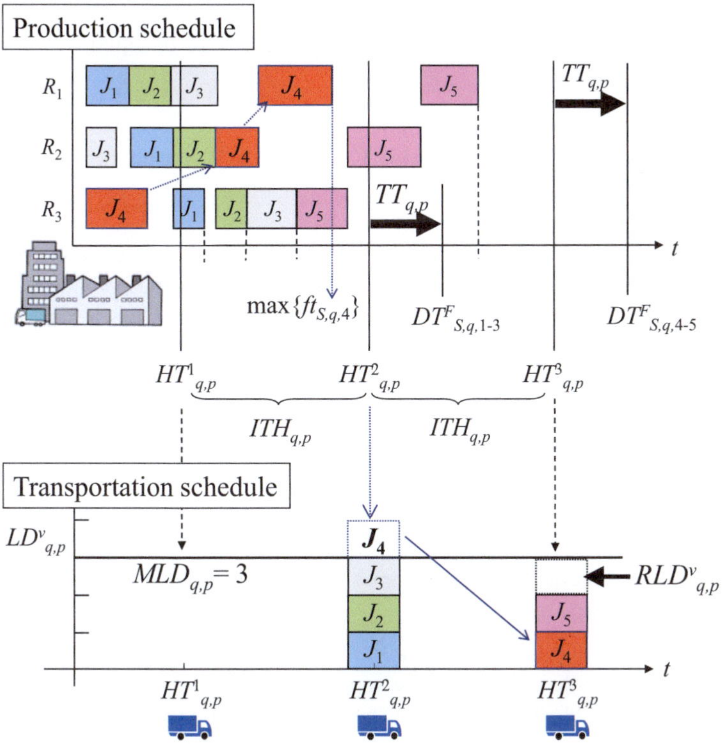

Fig. 2 Combination between a production schedule and a transportation schedule

The negotiation processes are repeated until an offer satisfies the order and one of the suppliers enters into a contract with the client. Suitable price and delivery time of products can be determined through the negotiation processes.

3.2 Transportation Model

3.2.1 Transportation Conditions

Figure 2 shows a combination between a production schedule and a transportation schedule. The following conditions are assumed as constraints in the transportation model.

1. A product is transported form a supplier to a client by a transportation vehicle. The transportation vehicle does not travel to two or more clients in one delivery.
2. Each transportation vehicle leaves a supplier at a constant time interval and transports products in a given time.

3. The capacity for loading products is limited on a transportation vehicle. The quantity of products loaded on a vehicle is less than or equal to the limit of loading capacity.

3.2.2 Estimation of Carbon Dioxide Emissions in Transportation Process

The improved ton-kilometer method [22] is used for the calculation of carbon dioxide emissions in truck transportation. This calculation method has been stipulated in the Manual for Calculation and Report of Greenhouse Gas Emissions by the Ministry of Environment and the Ministry of Economy, Trade, and Industry. The following equation represents the carbon dioxide emissions in truck transportation.

$$CDT^{v}_{q,p} = LD^{v}_{q,p} \times WU_{q,p} \times TD_{q,p} \times FU^{v}_{q,p} \times \frac{1}{1000} \times HU_{q,p} \times EF_{q,p} \times \frac{11}{3} \quad (1)$$

where,

$CDT^{v}_{q,p}$ Carbon dioxide emissions from a transportation vehicle on the v-th shipment from supplier S_q to client C_p.
$LD^{v}_{q,p}$ Loading amount. Quantity of products that are loaded on the transportation vehicle on the v-th shipment from supplier S_q to client C_p. The transportation vehicles are not able to load products beyond maximum loading capacity $MLD_{q,p}$. If the load $LD^{v}_{q,p}$ is 0, the transportation vehicle does not leave the supplier.
$WU_{q,p}$ Weight of a product.
$TD_{q,p}$ Transportation distance between supplier S_q and client C_p.
$FU^{v}_{q,p}$ Fuel consumption rate per unit ton-kilometer in truck transportation.
$HU_{q,p}$ Heating value per unit volume.
$EF_{q,p}$ Emission factor.

The value of fuel consumption rate $FU^{v}_{q,p}$ is determined based on the types of transportation vehicles, their maximum loading capacity, and the loading ratio. The heating value per unit volume $HU_{q,p}$ and the emission factor $EF_{q,p}$ are constant values determined by the types of fuel, such as heavy fuel oil, light oil, and gasoline. For truck transportation using light oil, the fuel consumption rate is given by the following equation.

$$\ln\left(FU^{v}_{q,p}\right) = 2.71 - 0.812\ln\left(\frac{LR^{v}_{q,p}}{100}\right) - 0.654\ln\left(MLD_{q,p} \times WU_{q,p} \times 1000\right) \quad (2)$$

$$LR^{v}_{q,p} = \frac{LD^{v}_{q,p}}{MLD_{q,p}} \times 100 \quad (3)$$

where,

$LR^{v}_{q,p}$ Loading ratio of a transportation vehicle on the v-th shipment

$MLD_{q,p}$ Maximum loading capacity. Upper limit of quantity of products that can be loaded onto transportation vehicle for shipping from supplier S_q to client C_p

According to the equations provided in the improved ton-kilometer method, it is recognized that the carbon dioxide emissions can be reduced under the conditions of truck transportation with higher loading ratios.

3.3 Profit of Supplier

The suppliers can estimate their profits when they enter into contracts with the client. The total actual profit TPF_q is estimated using the following equation.

$$\begin{aligned} TPF_q = & \sum_{h=1}^{p}\sum_{j=1}^{n}\left(RW_{q,p,N(h,j)} - PN_{q,p,N(h,j)} + \sum_{g=1}^{G}\Delta PN_{q,p,g}^{Rv(h,j,r)}\right) \\ & -\sum_{h=1}^{p}\sum_{j=1}^{n}\left(k_{h,j} \times max\left\{DT_{q,p,N(h,j)}^{A} - DT_{q,p,N(h,j)}^{C}, 0\right\}\right) \\ & -\sum_{h=1}^{p}\sum_{f=1}^{v}\left(RLD_{q,h}^{f} \times HC_{q,h}\right) \end{aligned} \tag{4}$$

where

$RW_{q,p,n}$ Reward required for product.
$PN_{q,p,n}$ Penalty charge due to delay in actual delivery time as compared to contracted delivery time.
$\Delta PN_{q,p,g}^{Rv}$ Penalty charge due to delays in delivery times of contracted orders g ($g = 1,2,\ldots,G$), if the addition of new product causes delays.
$k_{p,n}$ Penalty charge factor, representing penalty charge per unit time. It is included in order $O_{p,n}$.
$DT_{q,p,n}^{A}$ Actual delivery time of product.
$DT_{q,p,n}^{C}$ Contracted delivery time of product.
$RLD^{v}_{q,p}$ Amount of room for loading on a transportation vehicle.
$HC_{q,h}$ Shipping cost per product.

This equation shows how the suppliers obtain profits. First, the suppliers should enter into contracts to obtain rewards for products. Second, the suppliers should reduce the penalty charges that are paid to clients due to delays in product delivery times. Third, the suppliers should eliminate wasteful shipping costs by increasing the load ratios of their transportation vehicles, since suppliers pay the shipping cost per vehicle. Unused loading room on transportation vehicles decreases suppliers' total profits if the vehicles are loaded to less than their maximum capacity.

3.4 Optimization Problem

Two objective functions are simultaneously considered in the supply chain model, as shown in the following equations.

$$f_1 = W_{q,p,n} \times TR_{q,p,n} + \sum_{g=1}^{G} \left(W_{q,p,g} \times TR_{q,p,g} \right) \rightarrow \min \tag{5}$$

$$f_2 = \sum_{p=1}^{P} \sum_{v=1}^{V} CDT_{q,p}^{v} + \sum_{p=1}^{P} \sum_{n=1}^{N} \sum_{j=1}^{J} CDP_{q,p}^{j,n} \rightarrow \min \tag{6}$$

where

$W_{q,p,g}$ Weight
$TR_{q,p,g}$ Tardiness of product
$CDP_{q,p}^{j,n}$ Carbon dioxide emitted from manufacturing process of machine tool R_j for product $N_{p,n}$ in supplier S_q.

The first objective function represents the minimization of total weighted tardiness. This equation improves the production schedules in order to enhance suppliers' chances of entering into contracts and to reduce the penalty charges. The second objective function represents the minimization of carbon dioxide emissions from transportation vehicles. This equation improves the transportation schedules and reduces the amount of carbon dioxide emitted in truck transportation.

4 Production and Transportation Scheduling Considering Carbon Dioxide Emissions

This paper presents a new model for integrated production and transportation scheduling problems in consideration of carbon dioxide emissions in not only transportation processes but also machining processes in order to develop a low-carbon supply chain model. This section firstly explains an estimation method of carbon dioxide emitted in machining processes. The following subsections describe two optimization approaches for the production scheduling and the transportation scheduling.

4.1 Estimation of Carbon Dioxide Emissions in Machining Processes

Narita [23] proposed an estimation method of environmental burden, such as carbon dioxide emissions, in machining processes of a machine tool by using a machining simulation and NC program. A calculation formula for total environmental burden was derived as shown in the following equation.

$$Pe = Ee + Ce + LOe + \sum_{i=1}^{N} Te_i + CHe \tag{7}$$

where

Pe Environmental burden of machining operation [kg-CO2]
Ee Environmental burden due to electric consumption of a machine tool [kg-CO2]
Ce Environmental burden of coolant [kg-CO2]
LOe Environmental burden of lubricant oil [kg-CO2]
Te_i Environmental burden of cutting tool [kg-CO2]
CHe Environmental burden of metal chip [kg-CO2]

Electric consumption of a machine tool, environmental burden of coolant, and environmental burden of lubricant oil are decreased in proportion to the increase of machining time and operating time. On the other hand, environmental burden of cutting tool is increased in proportion to the machining speed. Environmental burden of metal chip has a constant value under the condition of machining features and materials.

The relationship between the cutting speed and the carbon dioxide emissions was represented by a parabolic equation as shown in Eq. 8 which was obtained by using Eq. 7 and simulation results of a vertical machining center with a carbide ball end mill for machining on PX5 material.

$$CDP_{q,p}^{j,n} = 3.38 \times 10^{-2} \times \left(vc_{q,p}^{j,n}\right)^2 - 27.0 \times vc_{q,p}^{j,n} + 6.02 \times 10^3 \tag{8}$$

where

$vc_{q,p}^{j,n}$ Cutting speed of machine tool R_j for product $N_{p,n}$ in supplier S_q [m/min]

In case where the cutting speed is faster than the optimal speed with the lowest carbon dioxide emissions, there is a trade-off relation between the cutting speed and the carbon dioxide emissions in machining processes as shown in Fig. 3. High-speed cutting process causes to emit a large amount of carbon dioxide. If the faster processing time of machining operation is required in the scheduling process, the machine tool emits a larger amount of carbon dioxide.

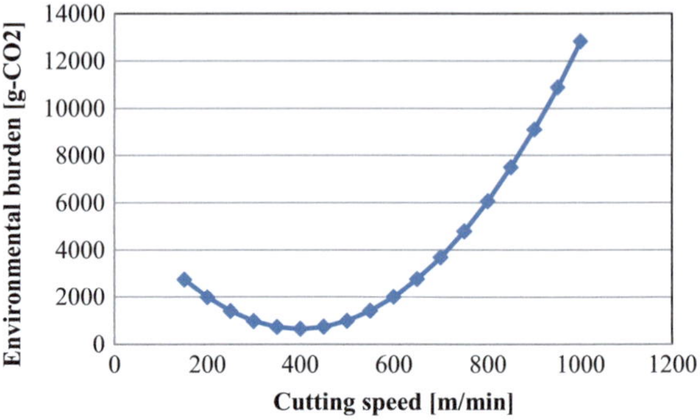

Fig. 3 Relationship between environmental burden and cutting speed estimated by machining simulation

4.2 *Integrated Production Scheduling and Transportation Scheduling*

Suitable production schedules and transportation schedules are determined through the iteration of production scheduling processes and transportation scheduling processes. Figure 4 shows the processes which include the evolutionary algorithm and heuristic search method. Three decision variables should be determined in this study as follows.

1. Loading sequences of jobs on each machine tools
2. Cutting speeds in machine tools
3. Allocation of jobs to transportation vehicles

4.2.1 Determination of Loading Sequence of Jobs by GA

Our previous researches proposed a GA-based scheduling method to determine suitable loading sequences on machine tools. After adding the newly ordered job in the existing production schedule, the first individual in the initial population is created from the production schedule. The chromosome of the individual consists of job names which are allocated to an array in the order of the execution sequence of the manufacturing operations. Other individuals in the initial population are randomly created by changing the positions of the genes of the first individual. In the case of a decoding process of an individual, a production schedule is generated by allocating jobs to machine tools in the order of job names described in a chromosome in consideration of a given process plan.

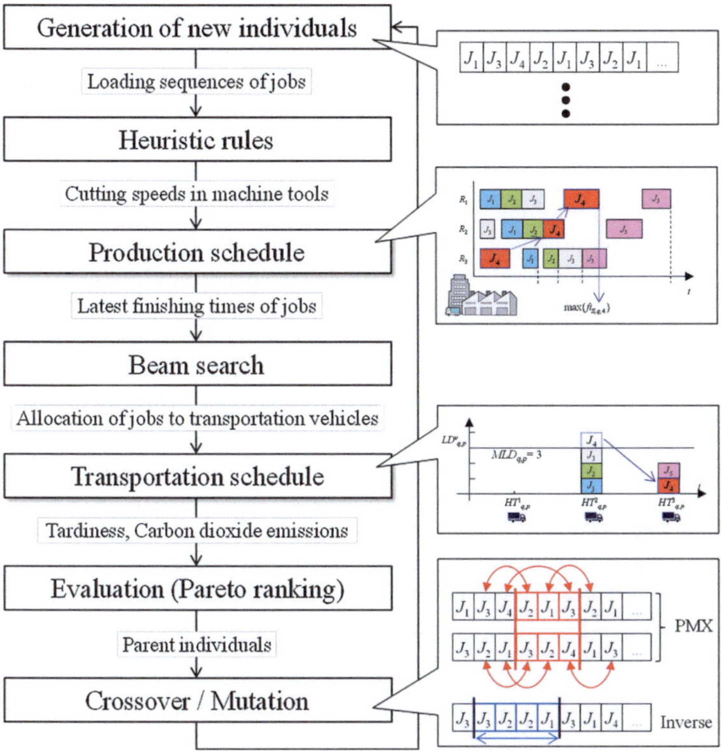

Fig. 4 Integrated production and transportation scheduling process

4.2.2 Selection of Cutting Speed by Heuristic Rules

Cutting speeds are determined for machining processes of jobs on machine tools. This research proposes the following four heuristic rules in order to select a cutting speed from several alternatives of cutting speeds.

1. RANDOM: Randomly selecting a cutting speed from candidates.
2. SPT (shortest processing time): Selecting the fastest cutting speed from candidates.
3. LCT (lowest carbon dioxide emitted per unit time): Selecting a cutting speed with the lowest carbon dioxide emissions from candidates.
4. DC (due date and low-carbon dioxide emissions): Firstly, selecting a cutting speed with the lowest carbon dioxide emissions. In case where the delay of due date occurred in transportation scheduling process, the fastest cutting speed is reselected for the delay job from candidates.

Then, all the processing times are estimated for the jobs on the machine tools. Based on the loading sequences and the processing times, a production schedule is generated and the latest finishing times of the jobs are estimated in the production scheduling process.

4.2.3 Estimation of Job Allocation to Vehicles by Beam Search

Based on the latest finishing time of the jobs, the jobs can be allocated to transportation vehicles. We proposed beam search method in the previous study to improve the loading ratio of transportation vehicles and decrease in the number of transportation vehicles for reducing carbon dioxide emissions.

Beam search [24] is like breadth-first search in that it progresses level by level. Unlike breadth-first search, however, beam search moves downward only through the best w nodes at each level; the other nodes are ignored. Therefore, there are only w nodes under consideration at any depth. Beam search has the advantage of potentially reducing the candidates for searching, since non-promising candidates are pruned at each step in the searching processes. Beam search is a heuristic search technique, since heuristic rules are used to prune non-promising candidates. The following heuristic function was proposed in the previous study [25] to find the best partial solution representing suitable allocation of jobs to transportation vehicles.

$$h = \sum_{v=1}^{V} RLD_{q,p}^{v} + \frac{1}{ND_{q,p}^{v'} + 1} \tag{9}$$

where,

$ND_{q,p}^{v'}$ Total number of products which can be loaded on the transportation vehicle on the v'-th shipment from supplier S_q to client C_p and still not be allocated

This heuristic function finds the promising transportation schedule which has not only the smaller total amount of room for loading on transportation vehicles but also the larger total number of jobs not being allocated. The products can be allocated to a small number of transportation vehicles with high load ratio and can be delivered before the required delivery time.

4.2.4 Evaluation by Pareto Ranking and Genetic Operations

A Pareto ranking method is applied for evaluating the individuals to the production and transportation schedules from the viewpoint of the minimization of both the total weighted tardiness and the carbon dioxide emissions as described in Eqs. 5 and 6. The evaluation values are used to calculate the fitness values of the individuals and determine the ranks of the individuals in the populations. The most preferable individual is selected from the Pareto-optimal set in the generation. The genetic operations, such as a crossover and a mutation, are applied to the other individuals for creating and updating the Pareto-optimal set of individuals in order to improve production and transportation schedules. The partially mapped crossover (PMX) and the inverse are used as a crossover operation and a mutation operation, respectively.

4.2.5 Iteration

These optimization processes are repeated until a bidding time. The most preferable individual in the last generation is selected from the Pareto-optimal set just before the bidding time. The supplier generates an offer based on the production and transportation schedules which are generated from the selected individual.

5 Experiments

5.1 Simulation System

A prototype of multi-agent real-time simulation system has been developed for a low-carbon supply chain in consideration of production and transportation scheduling problems in order to verify the effectiveness of the proposed model from the viewpoint of both the total profit of supplier and the amount of carbon dioxide emitted. The prototype system was implemented using Windows-based networked personal computers (Intel Core i7-3770 3.40 GHz CPU with 8.00 GB of RAM) for ease of applicability to real organizations.

Experiments were carried out on a two-layered supply chain model consisting of one supplier and one client by using two networked computers. The supplier has ten machine tools in the manufacturing system. Before starting the simulations, the supplier is provided with a job-shop-type production schedule consisting of ten products. Each machine tool can change three types of cutting speeds, such as a low speed, a medium speed, and a high speed. The machine tool emits a larger amount of carbon dioxide as the cutting speed is higher. Transportation conditions are summarized in Table 1. Population size, crossover rate, and mutation rate of the GA were 30, 0.8, and 0.2, respectively.

The client continuously generated 30 new orders and negotiated with the supplier for about 6 h in one simulation. The prototype system created a suitable production schedule and a transportation schedule in consideration of both the carbon dioxide emissions and the profit of supplier.

Table 1 Transportation conditions

Parameters	Values
Transportation time $TT_{1,1}$ [s]	6000
Interval of time between shipping $IHT_{1,1}$ [s]	1500
Maximum loading capacity $MLD_{1,1}$	3
Transportation distance $TD_{1,1}$ [km]	100
Shipping cost $HC_{1,1}$ [JPY]	3000
Fuel type of vehicle	Light oil
Heating value per unit volume $HU_{1,1}$ [GJ/k]	38.2
Emission factor $EF_{1,1}$ [kg-C/GJ]	18.7

5.2 Performance Evaluation

This section evaluates the optimization performance of the four heuristic rules for determining cutting speeds of machine tools from the viewpoint of both the minimization of carbon dioxide emissions and the maximization of the profit of supplier. Computational experiments were carried out ten times for each of the four heuristic rules under the same experimental conditions. The results of the experiments were compared with each other. The results are summarized in Fig. 5 and Table 2. Figure 5 shows the carbon dioxide emitted from manufacturing processes in the supplier and from transportation processes between the supplier and the client. Table 2 summarizes not only the total carbon dioxide emitted from the supply chain but also the number of contracts with the supplier and the total profit of the supplier. Both the LCT rule and the DC rule are more effective in reducing the amount of carbon dioxide emitted from manufacturing processes than the RANDOM rule and the SPT rule. According to the profit of supplier, the DC rule is superior to the LCD rule.

To assess the profitability and the effectiveness on environmental burden, the return on carbon (ROC) is usually used as a management index. The ROC evaluates the amount of carbon dioxide emission reduction under the consideration of economy as shown in the following equation.

$$ROC_q = \frac{TPF_q}{\sum_{p=1}^{P}\sum_{v=1}^{V} CDT_{q,p}^{v} + \sum_{p=1}^{P}\sum_{n=1}^{N}\sum_{j=1}^{J} CDP_{q,p}^{j,n}} \tag{10}$$

Figure 6 summarizes the average values of ROC on different heuristic rules. As shown in this graph, the DC rule is superior to the other rules from the viewpoint of both economical aspect and environmental aspect.

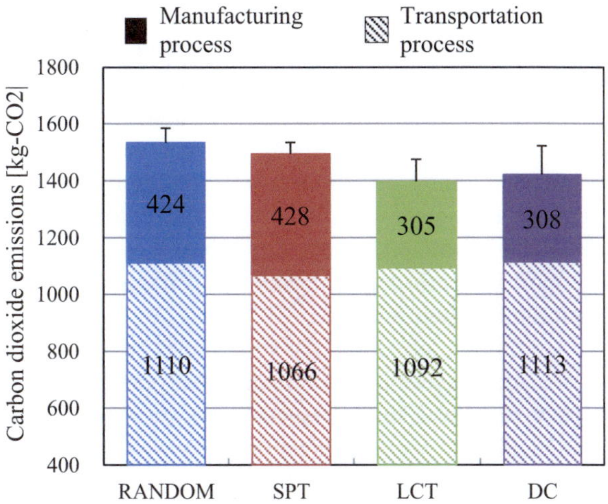

Fig. 5 Carbon dioxide emitted in supply chain

Table 2 Experimental results of profits and carbon dioxide emissions

	RANDOM	SPT	LCT	DC
CO_2 (av.) [kg-CO2]	1534	1494	1397	1421
Number of contracts	39.5	39.8	38.8	39.1
Profit (av.) [$*10^3$ JPY]	170.6	169.4	164.7	168.6

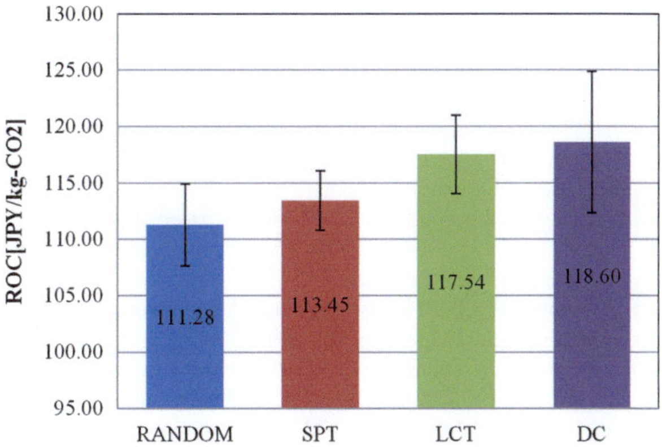

Fig. 6 Return on carbon

6 Conclusions

This paper presents a new model for integrated production and transportation scheduling problems in order to develop a low-carbon supply chain. The model consists of two optimization approaches for production scheduling and transportation scheduling. In the production scheduling process, the loading sequences of jobs are determined by the individuals in the GA-based scheduling method, and the suitable processing times of machining operations on machine tools are alternatively selected for products by using heuristic rules in consideration of both the tardiness of products and the amount of carbon dioxide emitted in manufacturing processes. In the transportation scheduling process, the allocations of products to transportation vehicles moving between suppliers and clients are determined by using beam search technique which can improve the loading ratio of transportation vehicles and decrease in the number of transportation vehicles for reducing carbon dioxide emissions. The suppliers repeat these optimization processes by modifying the individuals through the genetic operations, such as a crossover and a mutation, until the bidding time.

A prototype of simulation system was developed for a low-carbon supply chain in consideration of production and transportation scheduling problems. The prototype system was implemented using networked personal computers for ease of applicability to real organizations. Experiments were carried out on a two-layered supply chain model consisting of a supplier and a client by using two networked personal computers. The optimization performance of the four heuristic rules for determining cutting speeds of machine tools is evaluated from the viewpoint of both the minimization of carbon dioxide emissions and the maximization of the profit of supplier. Experimental results show that the DC rule is superior to the other rules from the viewpoint of both economical aspect and environmental aspect.

References

1. Ministry of Economy (2008) Kyoto protocol target achievement plan, https://www.env.go.jp/en/earth/cc/kptap.pdf
2. http://www.meti.go.jp/english/policy/energy_environment/global_warming/kyoto_protocol.html
3. BearingPoint Inc (2008) 2008 supply chain monitor "How mature is the green supply chain?" http://www.bearingpoint.com
4. Emerson D, Piramuthu S (2004) Agent-based framework for dynamic supply chain configuration. In: Proceedings of 37th Hawaii international conference on system science, 70168a, CD-ROM, Big Island, Hawaii
5. Piramuthu S (2005) Knowledge-based framework for automated dynamic supply chain configuration. Eur J Oper Res 165:219–230
6. Tanimizu Y, Yamanaka M et al (2006) Multi-agent based dynamic supply chain configuration considering production schedules. In: Proceedings of 2006 international symposium on flexible automation, Osaka, pp 572–578
7. Tanimizu Y, Ozawa C et al (2007) Credibility of supplier in dynamic supply chain. In: Proceedings of 40th CIRP international seminar on manufacturing systems, CD-ROM, Liverpool
8. Tanimizu Y, Harada K et al (2010) A two-layered model for dynamic supply chain management considering transportation constraint. J Adv Mech Des Syst Manuf 4(5):913–923
9. Tanimizu Y, Harada K, et al (2010) Lean production and transportation scheduling for dynamic supply chain management. In: Proceedings of the 8th IEEE international conference on industrial informatics, Osaka, pp 869–874
10. Tanimizu Y, Amano K et al (2012) Multi-objective production and transportation scheduling considering carbon dioxide emissions reductions in dynamic supply chains. Int J Autom Technol 6(3):322–330

11. Hollanad JH (1975) Adaptation in natural and artificial systems. University of Michigan Press, Ann Arbor
12. Goldberg DE (1989) Genetic algorithm in search, optimization and machine learning. Addison Wesley, Reading
13. Tanimizu Y, Sugimura N (2002) A study on reactive scheduling based on genetic algorithm. In: Proceedings of the 35th CIRP-ISMS, Soul, pp 219–224
14. Tanimizu Y, Sakaguchi T, Iwamura K, Sugimura N (2006) Evolutional reactive scheduling for agile manufacturing systems. Int J Prod Res 44(18–19):3727–3742
15. Gungor A, Gupta SM (1999) Issues in environmentally conscious manufacturing and product recovery: a survey. Comput Ind Eng 36:811–853
16. Bloemhof-Ruwaard JM, Beek P, Hordijka L, Wassenhove LN (1995) Interactions between operational research and environmental management. Eur J Oper Res 85(2):229–243
17. Cholette S, Venkat K (2009) The energy and carbon intensity of wine distribution: a study of logistical options for delivering wine to consumers. J Clean Prod 17:1401–1413
18. Sarkis J, Zhu Q, Lai K (2011) An organizational theoretic review of green supply chain. Int J Prod Econ 130:1–15
19. Hassini E, Surti C, Searcy C (2012) A literature review and a case study of sustainable supply chains with a focus on metrics. Int J Prod Econ 140:69–82
20. Sundarakani B, Souza R, Goh M, Wagner SM, Manikandan S (2010) Modeling carbon footprints across the supply chain. Int J Prod Econ 128:43–50
21. Seuring S, Muller M (2008) From a literature review to a conceptual framework for sustainable supply chain management. J Clean Prod 16:1699–1710
22. The Ministry of Environment and the Ministry of Economy (2006) Trade and industry, guideline of calculation methods for carbon dioxide emissions in logistics (in Japanese). http://www.greenpartnership.jp/pdf/co2/guidelinev3.0.pdf
23. Narita H (2012) Environmental burden analyzer for machine tool operations and its application. Manufacturing system. InTesh 247–260
24. Winston PH (1984) Artificial intelligence, 2nd edn. Addison-Wesley, Reading
25. Ito H, Tanimizu Y et al (2014) Work allocation in transportation scheduling considering carbon dioxide emissions reductions in dynamic supply chains. In: Proceedings of ISCIE/ASME 2014 international symposium on flexible automation, 105L, Hyogo

Usage of a Digital Eco-factory for a Printed Circuit Assembly Line

Yasuhiro Sudo, Michiko Matsuda, and Fumihiko Kimura

Abstract A concept called digital eco-factory (a collection of software agents that virtually represent the hardware and software facilities involved in a manufacturing system) has been developed to examine the environmental performance, productivity, and manufacturability, simultaneously, of a manufacturing system. This study reports three applications of the digital eco-factory in evaluating a printed circuit assembly (PCA) line. The first application deals with the determination of environmental performance and productivity of a PCA line for various production plans without changing the configuration of the line. The second application deals with the evaluation of the adequacy of the production plans for various PCA lines. The last application deals with the effects of an accident (e.g., machine failures) while running a PCA line. The successful implementation of the proposed digital eco-factory in the above three cases helps identify the general configuration of the digital eco-factory that can be used for other production lines.

Keywords Digital factory • Computer-aided planning • Environmental assessment

1 Introduction

A digital factory has been used to examine a production scenario by simulating the manufacturing operation, production process, and manufacturing system configuration at the design and planning stage. A digital factory provides various data for consideration of productivity and manufacturability, and for the determination of a process plan [1, 2]. Multiagent technology is applied to construct a digital factory [3, 4]. All factory elements such as machine tools, assembly machines, robots, AGVs, and workers are configured as software agents.

Y. Sudo (✉) • M. Matsuda
Department of Information and Computer Sciences, Kanagawa Institute of Technology, Kanagawa, Japan
e-mail: sudo@ic.kanagawa-it.ac.jp; matsuda@ic.kanagawa-it.ac.jp

F. Kimura
Department of Mechanical Engineering, Hosei University, Tokyo, Japan
e-mail: fumihiko.kimura@hosei.ac.jp

M. Matsumoto et al. (eds.), *Sustainability Through Innovation in Product Life Cycle Design*, EcoProduction, DOI 10.1007/978-981-10-0471-1_28

On the other hand, manufacturing industries have to consider not only productivity but also the sustainability of the global environment and the whole life cycle of a product. For example, the ISO 20140 series are concerned with evaluation of energy efficiency of the production system. This is provided as methodologies for environmental sustainability [5]. Therefore, a digital factory is required to simulate environmental performance in addition to productivity and manufacturability [6]. As a solution for the requirements, a digital eco-factory has been proposed [7–9]. A digital eco-factory is a virtual factory on which production line configuration and the production scenario are examined from an environmental viewpoint.

To proceed with the practical usage of the digital eco-factory, a multiagent-based construction of the digital eco-factory is applied and a trial digital eco-factory has been implemented for a PCA (printed circuit assembly) line [10]. This trial shows how to construct the digital eco-factory with high accuracy and usability and, especially, how to make software agents of components such as machines on the PCA production line and manufactured PCA (printed circuit assembly). In this paper, a more practical implementation is enacted and the system is used experimentally for various use cases. Experimental usages show the availability and utility of the digital eco-factory.

2 Application of a Digital Eco-factory to a PCA Production Line

2.1 Proposed Digital Eco-factory

A digital eco-factory has been proposed for a simultaneous examination of environment assessment, productivity, and manufacturability [7–9]. When the proposed digital eco-factory is used, the environmental performance of the planned production scenario is examined in addition to productivity and manufacturability at the same time with various granularities such as machine level, product level, and factory level. The digital eco-factory is constructed on the virtual production line modeling an actual production line and its components. Multiagent technologies can be applied to modeling them.

The conceptual structure of the proposed digital eco-factory is shown in Fig. 1. The virtual production line mirrors the structure of the shop floor in the actual factory. In these lines, all components such as machine tools, assembly machines, robots, and workers are configured as software agents. These agents are called "machine agents." Manufactured products such as machined workpieces and assembled parts are also configured as software agents. These agents are called "product agents." Product agents lead and control execution of virtual production to finish themselves as a completed product. In the digital eco-factory, there are two panels. One is the plant panel and the other is the product panel. The operator of the digital factory can input configuration of the production line and production scenarios including control policy for the production line, an energy saving policy, granularity of environmental indexes, etc. through panels. The operator can also

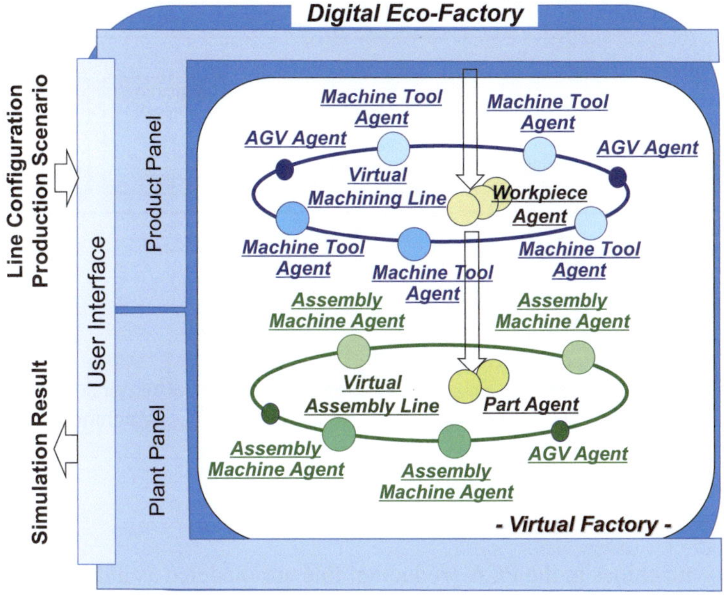

Fig. 1 Concept of the proposed digital eco-factory

observe progress and results of the virtual production through panels. The product panel monitors the condition of virtual production progress from the product view. The plant panel also monitors the status of virtual production progress from the equipment/device view and the line view. Environmental performance indexes such as carbon dioxide emissions and energy consumption are monitored through panels.

2.2 *Construction of Digital Eco-factory for PCA Line*

The concept of a digital eco-factory applies to the PCA production line [10]. A PCA line consists of a solder paste printer, several electronic part mounters, a reflow soldering oven, and a visual inspection machine. These machines are connected in sequence from the printer to the inspection machine. When a blank printed circuit board (PCB) is input to a solder paste printing machine, the production process is started. The second process is to mount parts on the PCB. The number of electronic part mounters is changed depending on the capacity of the line. In the part-mounting process, a number of mounters and processing speeds used differ, depending on the board type such as board size or a number of mounted parts. The third process is to solder. A reflow soldering oven consumes the most energy in a PCA line. A reflow soldering oven is placed, one per PCA line. At the final process, the completed PCA is inspected by the visual inspection machine.

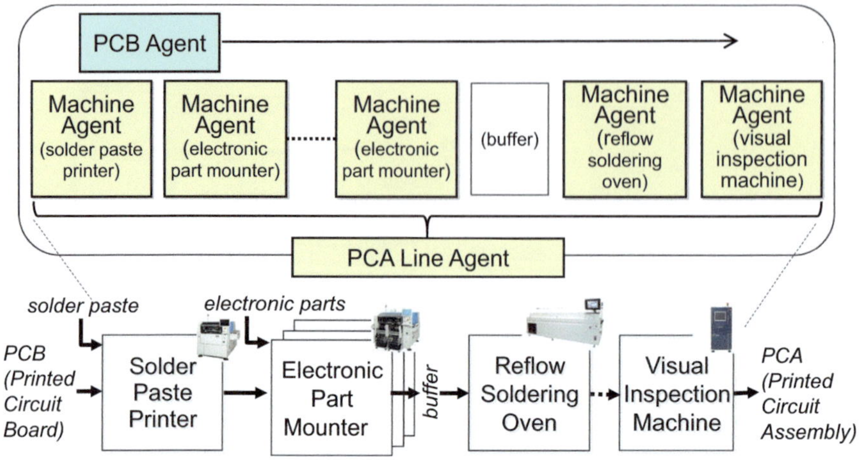

Fig. 2 Construction of digital eco-factory for PCA

These machines in the PCA production line are modeled as individual machine agents and connected in sequence as shown in Fig. 2. Usually, there are several PCA lines in the factory. A PCA line agent monitors the progress of virtual production on the line. A PCB is input to the production line and a PCA is output from the line. A PCB is also modeled as a part agent which is one of the product agents.

2.3 *Usage of Digital Eco-factory for PCA Line*

Three use scenarios of a digital eco-factory for a PCA line are assumed as shown in Table 1. In use scenario 1, the configuration of the PCA line is determined through repeating the simulation by changing component machines, and the performances of newly introduced machines are examined by virtual execution according to the testing scenario from the energy efficiency view and productivity view. This procedure starts the step of selecting templates of machine models in the machine catalogue e-library, fulfilling the templates, and describing the connecting relationship between machines, control policy, etc. In use scenario 2, after process planning, which is supported by the product design tool such as a CAD/CAM system, the assembly plan for the product is evaluated by virtual production at a virtual PCA line. This evaluation is repeated by changing parameters and plans until satisfactory environmental efficiency and productivity data are obtained. In use scenario 3, reference data such as prediction data of the PCA line status for machine trouble are obtained by simulation to assist PCA line operation. Monitored data from the actual production line is compared with reference data.

Table 1

Use case name	Contents
Use scenario 1 PCA line configuration	Preliminary reviews in PCA line design for the energy consumption minimization
	Prior assessment of energy consumption and productivity in PCA line design for new products
	Preliminary evaluation of the increase or decrease of energy consumption by installing new machine
Use scenario 2 Assembly plan determination	Evaluation of energy consumption of PCA line in consideration of the standby of equipment
	Evaluation of energy consumption for optimal PCA line selection by production items and production quantity
	Forecast of energy consumption and productivity for optimizing operation plan
Use scenario 3 PCA line operation assist	Prediction of PCA line status during machine trouble
	The discovery of abnormal state by comparison of the predicted and measured values of energy consumption
	Progress prediction of production plan

3 Implementation Details of the Virtual PCA Production Line

3.1 System Structure

The digital eco-factory for a PCA is implemented as a trial. The virtual PCA production line is constructed on the commercially available multiagent simulator "Artisoc" [11]. The implemented system structure is shown in Fig. 3. Machine agents which construct a virtual PCA line are set by filling up agent forms which are a schema of agents according to the configuration data of PCA line. A PCA line agent is also set. PCB agents for each PCA product are generated according to the PCA production plan. When the PCB agent is generated, virtual production is started. Conditions of virtual production such as productivity and power consumption are monitored.

3.2 Configuration of the PCA Line

A virtual PCA production line agent is configured as a multiagent system. The PCA line and each machine on the PCA line are modeled as a software agent according to the provided configuration data. By modeling as a software agent, not only the static characters of the machine but also the dynamic behavior of the machine can be modeled.

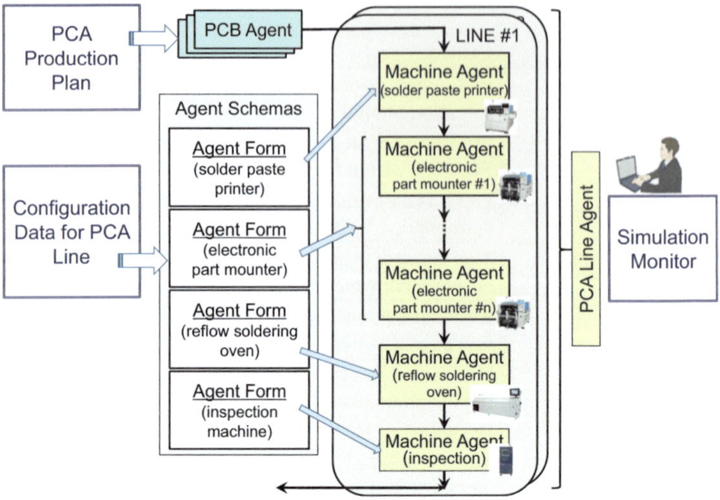

Fig. 3 Implementation of digital eco-factory for PCA

3.2.1 Configuration Data

Configuration data of PCA production lines such as the composition of a production line and specifications of component machines are provided to the system to construct a virtual production line in the digital eco-factory. Table 2 shows the detail of a line configuration for PCA production. A factory usually has several PCA lines. A PCA line consists of a solder paste printer, several electronic part mounters, a reflow soldering oven, and a visual inspection machine. Specification data for each of the machines are also provided. Contents of specification data differ, depending on the machine type. Each machine agent is generated according to these configuration data and specification data. Once a target PCB is put into the assigned line, the PCB goes in sequence in the same PCA line. Configuration data is provided through an MS-Excel file in the implemented system.

3.2.2 Machine Agents

There are four types of machine agents on the PCA line. They are an agent for a solder paste printer, agents for several electronic part mounters, an agent for a reflow soldering oven, and an agent for a visual inspection machine. A solder paste printer is a screen printing machine used for painting solder paste on PCBs, using metal masks. A solder paste printer takes the first position on the PCA production line. To set up a solder paste printer agent, needed parameters such as standby power consumption, printing power consumption, metal mask size, and printing time are provided with the configuration data. The behavior of a solder paste printer agent is shown in Fig. 4 using a UML activity diagram. After power on, a printer is

Table 2 Contents of configuration data for PCA line

Items	Data contents			
Machine type	Solder paste printer	Electronic part mounter	Reflow soldering oven	Visual inspection machine
Machine ID				
Connection to	Mounter ID	Reflow oven ID/mounter ID/buffer ID	Inspection machine ID	
Belonging line ID				
Machine data	Power consumption [kwh] <standby>	Power consumption [kwh] <standby>	Power consumption [kwh] <standby>	Power consumption [kwh] <standby>
	Power consumption [kwh] <metal mask change>	Power consumption [kwh] <part setting>	Power consumption [kwh] <start-up>	Power consumption [kwh] <checking>
	Power consumption [kwh] <paste replenishment>	Power consumption [kwh] <board set>	Power consumption [kwh] <shutdown>	Power consumption [kwh] <paste replenishment>
	Power consumption [kwh] <board set>	Power consumption [kwh] <board recognition position correction>	Power consumption [kwh] <temperature profile change>	Power consumption [kwh] <board set>
	Power consumption [kwh] <metal mask set>	Power consumption [kwh] <nozzle selection>	Power consumption [kwh] <board set>	Power consumption [kwh] <board output>
	Power consumption [kwh] <printing>	Power consumption [kwh] <part picking>	Power consumption [kwh] <residual heating>	Checking time [sec]
	Power consumption [kwh] <metal mask release>	Power consumption [kwh] <part mounting>	Power consumption [kwh] <heating>	Changeover time [min]
	Power consumption [kwh] <board output>	Power consumption [kwh] <board output>	Power consumption [kwh] <cooling>	
	Printing time [min]	Part set time [min] <large>	Power consumption [kwh] <board output>	
	Metal mask size [mm]	Part set time [min] <medium>	Operation time [min]	
	Air compression [MPa]	Part set time [min] <small>	N2 flow [L/min]	
	Airflow [L/min]	Number of part type		

(continued)

Table 2 (continued)

Items	Data contents			
Machine type	Solder paste printer	Electronic part mounter	Reflow soldering oven	Visual inspection machine
			Changeover time [min] <low –> high>	
	Changeover time [min]	Air compression [MPa]	Changeover time [min] <high –> low>	
		Airflow [L/min]		
		Changeover time [min]		

in the standby state. In the standby state, the printer consumes standby electric power. When a PCB with produced PCA data is input to the printer, the printer agent starts the simulation of the operation sequences and calculations of printing power consumption and production time. As a preparation before printing, a metal mask is changed if needed. The first operation is setting up the PCB. The second is setting of the mask metal. The third is stage moving. The fourth is printing. The fifth is releasing the metal mask. The sixth and last is the output of the PCB. Then the printer reverts back to the standby state [8].

An electronic part mounter is used for picking and placing surface-mount electronic components such as capacitors, resistors, and integrated circuits onto the PCBs. An electronic part mounter takes the second position on the PCA production line. Usually, several part mounters are in the production line. Some of the part mounters are not used depending on the types and number of mounted parts. An electronic part mounter is also modeled as a software agent.

A reflow soldering oven is used for reflow soldering of surface-mount electronic components to PCBs. A reflow soldering oven is modeled as a software agent. A visual inspection machine is also modeled as a software agent.

3.2.3 PCA Line Agent

The PCA line agent is set corresponding to each PCA production line. A PCA line agent monitors the condition of machines on the PCA line and reports results of the virtual PCA production. Figure 5 is an activity diagram for a PCA line agent. In the implemented system, when a PCA line agent accepts the job reports from machine agents, a PCA line agent displays the whole line conditions. A PCA agent communicates with other agents when some trouble happens on the line for troubleshooting.

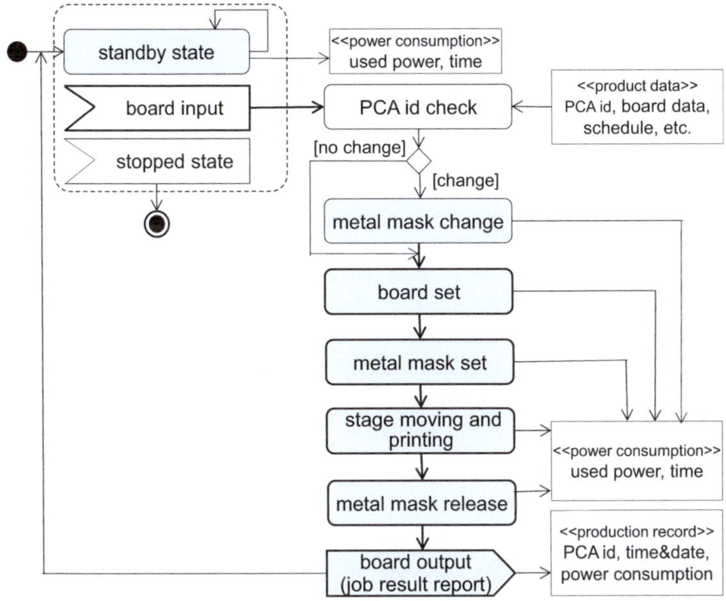

Fig. 4 Activity diagram for agent of a solder paste printer

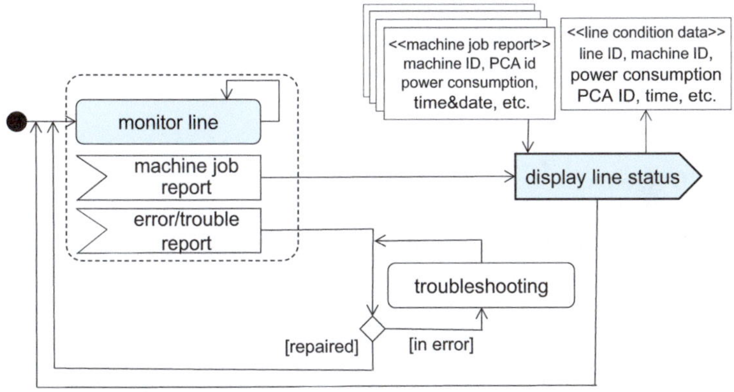

Fig. 5 Activity diagram for agent of a PCA line

3.3 Simulated Production Scenario and PCB Agent

An examined production scenario is provided as a production plan. A PCB agent for each PCA product is generated according to the production plan and input to the assigned PCA line in assigned order. PCB agents carry out production procedures in the virtual PCA production line and provide monitoring data from the product view.

Table 3 Example of PCA production plan data

Production order	1
Product ID	PCA#00101
Production line	Line #02
Input time [epoch sec]	1430739000
Number of mounted parts	1900
Board size W [mm]	450
Board size D [mm]	500
Board size H [mm]	5
Temperature profile	High
Assembly drawing file	PCB_0001.brd
Deadline [epoch sec]	1430813000
Production order	2
Product ID	PCA#00101
Production line	Line #02

3.3.1 Production Plan Data

A production plan for PCAs consists of produced PCA data and production management data. Assigned production line and input order to the line are provided for each PCA. An example of a production plan data is shown in Table 3. The board size, types and amount of mounted parts, kind of soldering paste, type of temperature profile, etc. have an effect on the operations of the PCA line, reflecting this in the energy consumption of machines as a result. Production plan data is provided through an MS-Excel file in the implemented system.

3.3.2 PCB Agents

The behavior of the PCB agent is shown in Fig. 6 using a UML activity diagram. When a production is ordered according to the production plan, a PCB agent is generated. Parameters such as board size, types and amount of mounted parts, kind of soldering paste, and type of temperature profile are set, based on the production plan data when the agent is generated. The PCB agent asks the solder paste printer agent in the assigned production line whether it is vacant. When the printer is occupied, the PCB agent waits until the machine becomes vacant. When the printer is vacant, the PCB agent asks the job and receives the report about process condition such as power consumption, processed time, and result. After finishing the process, the PCB agent asks the machine for the next process, whether it is vacant, and repeats the same procedure until the final process of the production line is reached. A PCB agent can access the whole production condition data about itself [10].

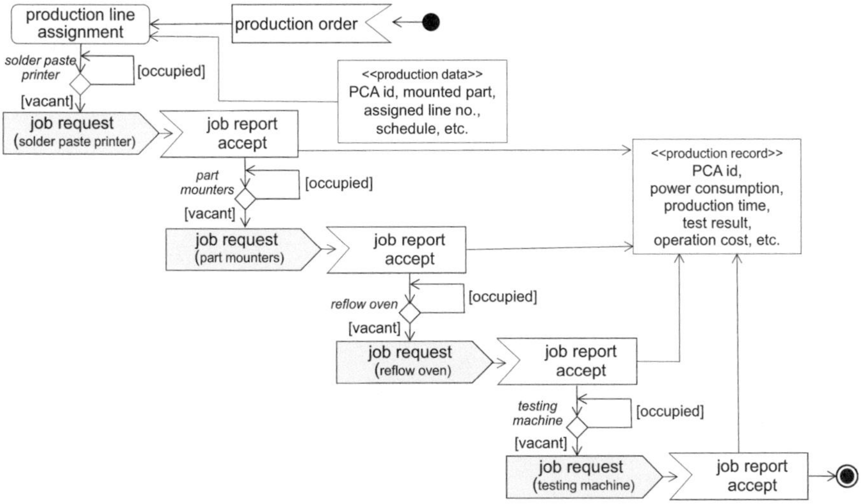

Fig. 6 Activity diagram for agent of a PCB

4 Experimental Use of Virtual PCA Production Line

4.1 Overview of Experimental Usages

The virtual PCA production line is used experimentally. Experimental use cases are set corresponding to the three use scenarios in Table 1.

Figure 7 is a screenshot of monitoring windows for the execution example. In Fig. 7, the agents' behaviors are visually displayed in the window in area A in real time. Area B displays time-series data of power consumptions of each of the machines on a virtual PCA line. The horizontal axis represents the "step." In this experimentation, one step corresponds to 6 s of temporal granularity. Area C shows production volume and progress on each PCA line. Area D displays accumulated data of power consumption for each of the machines.

4.2 Experimental Example for Use Scenario 1

The first use case is experimentation for a PCA line configuration. This use case assumes the situation for selecting machines when a new PCA line is constructed. The production plan used has 12 PCA products, then 6 PCB agents are put into each production line. All PCA products are divided into three types depending on the size of the PCB and the temperature of the solder. Two cases were examined. Two PCA production lines are constructed on the virtual factory floor. Case (a) uses machines that have the general efficiencies assumed in the real factory. Another

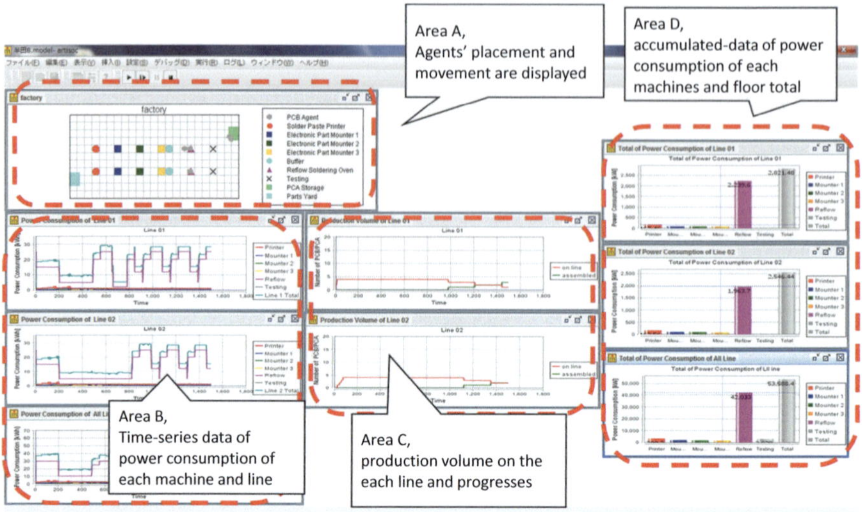

Fig. 7 Monitoring windows of virtual PCA production lines using "Artisoc"

case (b) uses machines providing with slightly higher efficiency but with higher power consumption. The production plan used is completely the same in both cases.

Figure 8 shows the simulation results. The time to complete all productions for case (a) is longer than for case (b). The peak of power consumption on the factory floor for case (b) is higher than for case (a). However, the total power consumption in both cases is very close to each other. In designing a new PCA production line, if a machine's catalogue and e-library is obtainable, users can construct virtual PCA production lines. By comparing the performance of machines, users determine the line configuration and the floor layout at the design stage of the manufacturing system.

4.3 Experimental Example for Use Scenario 2

The second use case is an experimentation for the assembly plan determination. If the same production line configurations are used, the production efficiency differs for each production plan. This means that the virtual PCA production line can be used to refine a production plan. Input products are the same as in example 4.2, but the input timing and order are different.

Figure 9 shows simulation results. In case (a), peak power consumption is higher than in case (b). The major cause of this is the synchronous working of reflow soldering oven machines in case (a). In the PCA line, a reflow soldering oven is using the highest electric energy. In case (b), reflow soldering ovens are working alternately in PCA line 01 and 02. It enables to cut the peak of power consumption by controlling the input timing of products. However, the time to complete all

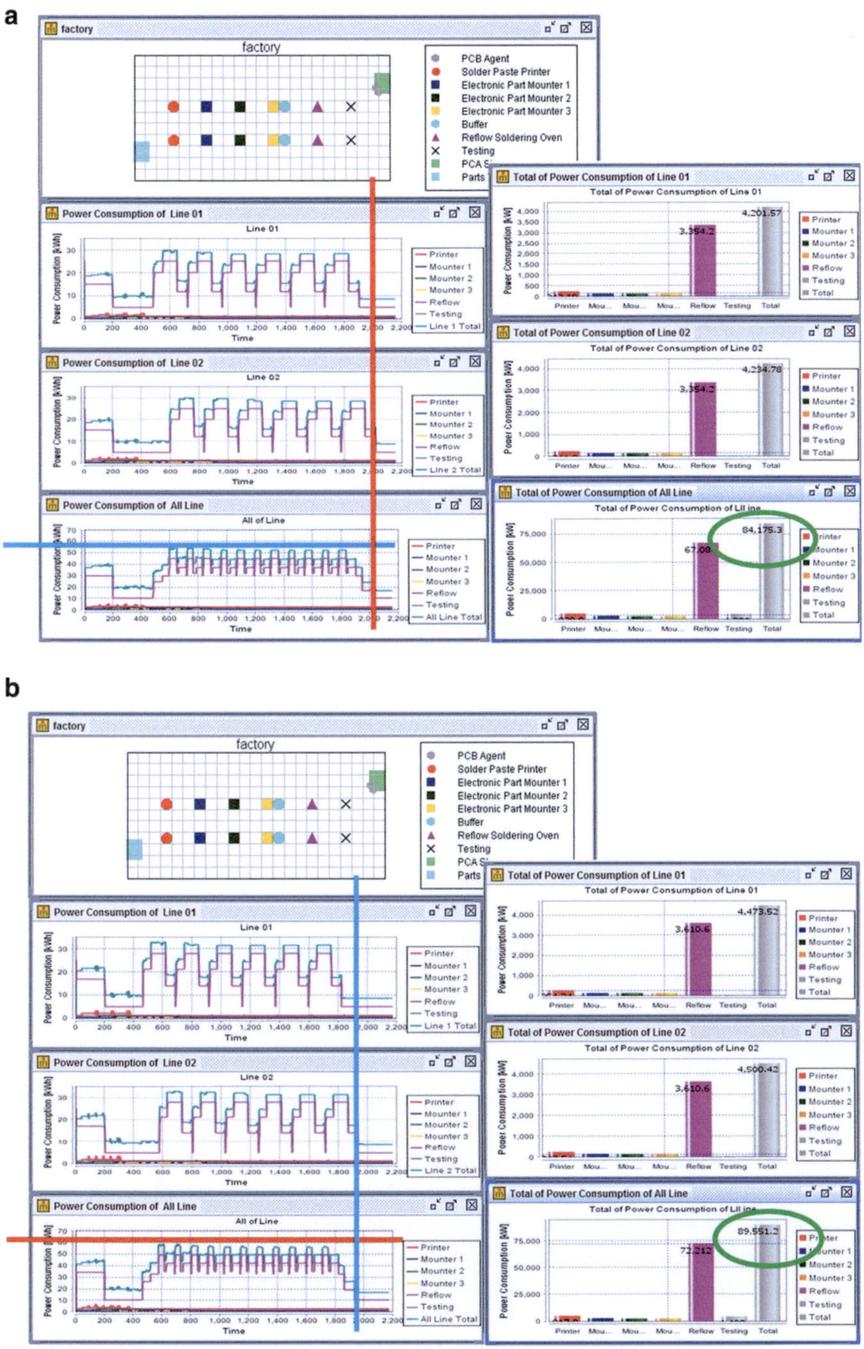

Fig. 8 Experiment for comparing performance in different machine effects (**a**) Machines have general efficiencies assuming real factory (**b**) Better efficiency just a little than case (*a*)

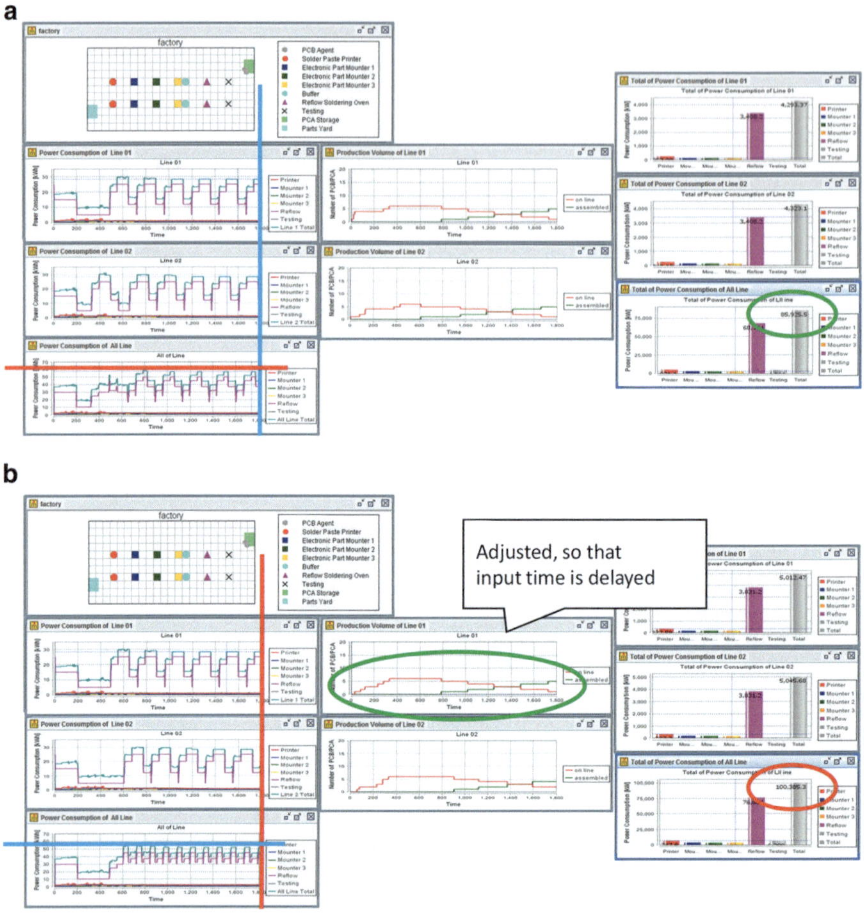

Fig. 9 Experiment for comparing performance in different production plans (**a**) Both of PCA lines are working synchronously (**b**) Reflow soldering ovens are working alternately

productions in case (b) is longer than in case (a). And the used total power consumption is also higher than in case (a). For effective utilization of the existing production line, the tuning of the production plan is important.

Figure 10 shows simulation results from using different numbers of PCA lines. In case (b), there are three PCA lines on the floor. The production plan has 12 PCA products of the same type. Four PCB agents are put into each production line. As a corollary, quick completion of products is confirmed in case (b). But the totally used power consumption and peak point are higher than in case (a). These simulations are assuming production planning PCA line(s) "go to sleep," for example. Especially, a reflow soldering oven needs start-up heating time and energy. Therefore, decreasing energy consumption is a complex problem.

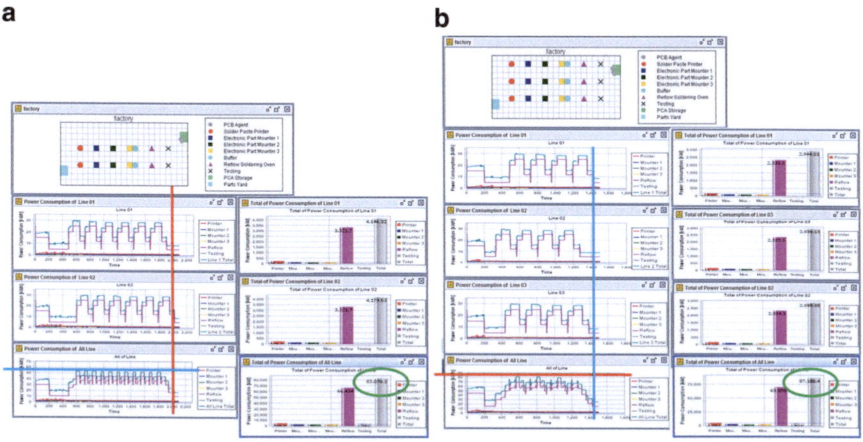

Fig. 10 Experiment for comparing performance in different production line configurations (**a**) Two PCA lines on the factory floor (**b**) Three PCA lines on the factory floor

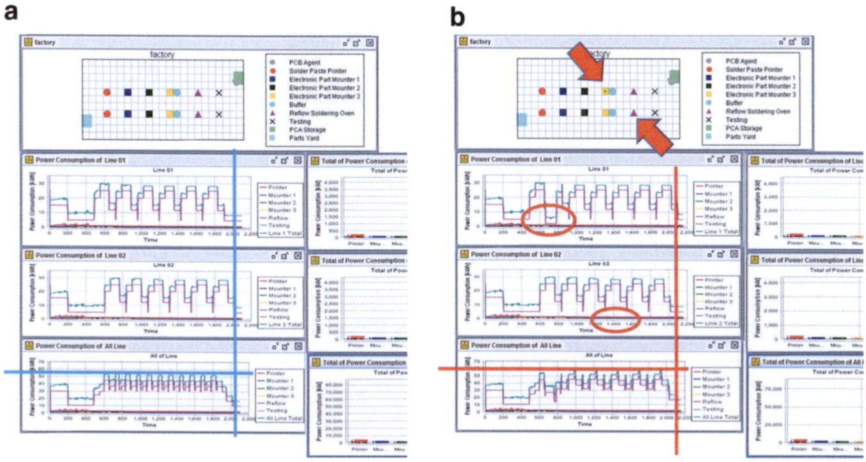

Fig. 11 Experiment to observe the effect of sudden accidents (**a**) PCB agents are working on the schedule (**b**) The production plan was delayed by intentionally–machine troubles

4.4 Experimental Example for Use Scenario 3

The third use case is an experimentation for assisting the PCA line operation. This use case includes minimizing unexpected losses. More specifically, to find the effects caused by introducing intentional machine troubles in the real production plan.

Figure 11 shows simulation results with and without machine troubles. The input product plans are completely the same, but in case (b) an emergency machine stop has occurred. The red circle on line 01 shows that the reflow soldering oven stopped for 12 min. The red circle on line 02 shows that electric parts mounter three stopped for 20 min. In case (a), PCB agents are working on the schedule; reflow soldering ovens are working alternately in PCA lines 01 and 02. However, in case (b), proceeding of production plan was delayed, and the reflow soldering oven is working synchronously. As a result, the peak of power consumption is increased. These simulations show that the rescheduling for a peak power cut is needed. Thus users are able to prepare robust production plans.

5 Summary

A collection of software agents called "the digital eco-factory" that virtually represent the hardware and software facilities involved in PCA lines has been developed to examine its (PCA line) environmental performance, productivity, and manufacturability in a concurrent manner. The digital eco-factory has been used in various real-life situations, and performance evaluation results have been obtained for PCA lines under various production plans and unexpected situations (accidents). As demonstrated in this study, when a digital eco-factory is available, it is possible to know the implications of the production plans beforehand. This way, the digital eco-factory helps identify the most appropriate production plan and helps maintain a balance between the environmental performance and productivity/ manufacturability.

Acknowledgments This work is supported by JKA and its promotion funds from KEIRIN RACE. The authors thank members of Technical Committee "DEcoF (Digital Eco-factory) (Oct.2012–) by FAOP (FA Open Systems Promotion Forum) in MSTC (Manufacturing Science and Technology Center), Japan, for fruitful discussions and their supports [12]. The authors are grateful to Dr. Udo Graefe, retired from the National Research Council of Canada for his helpful assistance with the writing of this paper in English.

References

1. Freedman S (1999) An overview of fully integrated digital manufacturing technology. In: Proceedings of the 1999 winter simulation conference, pp 281–285
2. Bley H, Franke C (2004) Integration of product design and assembly planning in the digital factory. CIRP Ann Manuf Technol 53(1):25–30
3. Monostori L, Váncza J, Kumara S (2006) Agent-based systems for manufacturing. CIRP Ann Manuf Technol 55(2):697–720
4. Matsuda M, Kasiwase K, Sudo Y (2012) Agent construction of a digital factory for validation of a production scenario. In: Procedia CIRP vol 3 (special issue on) 45th conference on manufacturing systems, Elsevier, pp 115–120
5. ISO 20140-1: 2013 Evaluating energy efficiency and other factors of manufacturing systems that influence the environment — part 1: overview and fundamental principles
6. Stefa ID, Draghicia D, Draghici A (2013) Product design process model in the digital factory context. Procedia Technol 9:451–462

7. Matsuda M, Kimura F (2012) Configuration of the digital eco-factory for green production. Int J Autom Technol (Spec Issue Green Manuf Supply Chain) 6(3):289–295
8. Matsuda M, Kimura F (2013) Digital eco-factory as an IT support tool for sustainable manufacturing, digital product and process development systems. IFIP Adv Inf Commun Technol 411:330–342
9. Matsuda M, Kimura F (2015) Usage of a digital eco-factory for sustainable manufacturing. CIRP J Manuf Sci Technol 9:97–106
10. Matsuda M, Sudo Y, Kimura F (2015) A Multi-agent based construction of the digital eco-factory for a printed-circuit assembly line. In: Procedia CIRP (special issue on) 48th conference on manufacturing systems, Elsevier (in print)
11. Kozo Keikaku Engineering Inc., Users manual of artisoc, http://www.mas.kke.co.jp/cabinet/manual-en.pdf
12. MSTC (Manufacturing Science and Technology Center), http://www.mstc.or.jp/development/faop/

A Negotiation Model for Closed-Loop Supply Chains with Consideration for Economically Collecting Reusable Products

Kenta Matsui and Yoshitaka Tanimizu

Abstract A remanufacturer generates an order for a used product considering economic profits through a disassembly schedule. This requires a negotiation model embedded in a closed-loop supply chain where the stakeholders (such as client, manufacturer, supplier, and remanufacturer) interact for achieving their respective objectives. This study proposes a new negotiation model for increasing the degree of reuse of products and, thereby, reducing the waste. A simulation system based on closed-loop supply chain is also developed to evaluate the effectiveness of the proposed negotiation model.

Keywords Closed-loop supply chain • Reverse supply chain • Reuse • Negotiation • Genetic algorithm

1 Introduction

Environmental problems are recognized as one of the most serious issues in these last few decades. Many manufacturing enterprises focus on incorporating not only economical but also environmental concerns into their strategic decisions [1]. Green supply chain management [2, 3] and sustainable supply chain management [4–7] have gained increasing attention within both academia and industry. The concept of the green supply chain management (GrSCM) covers all the phases of a product's life cycle, from the extraction of raw materials through the design, production, and distribution phases, to the use of products by consumers and their disposal at the end of the product's life cycle including reconditioning, reuse, and recycling of products.

Products and materials are returned from customers to suppliers or manufacturers through reverse supply chains in order to be recycled, reused, or reconditioned. Gungor and Gupta [8] indicated that the effort must be made for environmentally conscious manufacturing and product recovery systems to be profitable so that the incentive for development and planning of these systems

K. Matsui • Y. Tanimizu (✉)
Graduate School of Engineering, Osaka Prefecture University, Osaka, Japan
e-mail: tanimizu@me.osakafu-u.ac.jp

M. Matsumoto et al. (eds.), *Sustainability Through Innovation in Product Life Cycle Design*, EcoProduction, DOI 10.1007/978-981-10-0471-1_29

continues. Then, it is required to establish a method for reconditioning, reuse, and recycling of used products in consideration of the feasibility of realizing a balance between environmental and economic concerns.

Previous studies have proposed a basic model of closed-loop supply chains and a negotiation protocol for reusing products in consideration of economic efficiency [9]. The negotiation protocol provides a method for synchronizing the demand of reusable parts and the supply of used products among the organizations in the closed-loop supply chains. However some used products may become waste products by inconsistency between the demand and the supply of the used products from the viewpoint of product value. This study improves the negotiation protocol for further increasing the reused products and reducing the waste. A customer usually discards a product by stochastic means in consideration of its life cycle. New protocol provides a customer with a method for discarding a product systematically. A remanufacturer can prospectively enter into a contract with a customer through negotiation with the customer in consideration of both the required prices from the remanufacturer and the possible time to dispose of the used product from the customer.

The reminder of this paper is organized as follows. Section 2 shows literature review. Section 3 briefly describes the previous supply chain model. Section 4 explains new negotiation protocol. Finally, Section 5 demonstrates experimental results.

2 Literature Review

A great deal of literature has been produced on the closed-loop supply chain where the authors have studied numerous problems such as network design problems, product acquisition management problems, marketing problems, and alike [10]. Guide et al. [11] have taken a contingency approach to exploring the factors that affect the production planning and control of the closed-loop supply chains incorporating the product recovery. Nielsen and Bruno [12] have analyzed the possibility of utilizing the closed-loop supply chains from the viewpoint of mass customization setting and elicited the requirements for designing the product, manufacturing, and supply chain. Hassini et al. [6] have indicated that quite a few papers have addressed the pricing issue from the context of sustainable supply chain. Some products that are highly significant in terms of sustainability (e.g., electric vehicles and wind turbines) require rare earth metals. Rare earth metals are subjected to high supply risk (critical ones (dysprosium, terbium, europium, neodymium, and yttrium) and near-critical ones (cerium, indium, lanthanum, and tellurium)) [13]. Therefore, proactively addressing the material supply risks and preventing supply chain disruptions are important issues of sustainable economy. In other words, how to achieve the globally diverse supplies, identify the appropriate substitutes, improve the capacity of recycle/reuse, and improve the yield of critical materials have made the supply chain analysis even more significant from the context of sustainable economy. Takata et al. [14] have analyzed the actual life-cycle data of photocopy machines. Based on the data analysis, they have proposed several models for identifying the collection rate and discard rate of the product (photocopy machine). Umeda et al. [15] have analyzed the life cycles of several products (such as one-time-use camera, photocopy machine, and automatic teller machine (ATM)) from the viewpoint of reusability. They point out that a product can be reused only when a manufacturer manufactures and sells same types of products in the market. In consideration of the reusability of product in the market, the used product should be returned from customers to suppliers or manufacturers as soon as possible, since the product life cycle appears to becoming shorter in recent years.

A lot of used products may become waste products in traditional recovery approaches, since customers discard products without consideration for reuse of the products whenever they want. A strategy for economical reuse of products is required to collect suitable products from customers and to provide manufacturers with reusable parts disassembled from the used products. Our study proposes a negotiation protocol which synchronizes the demand of reusable parts and the supply of used products among the organizations in the closed-loop supply chains. In the case where a lot of usable parts are required for generating new products, customers are stimulated to discard products for reuse. On the other hand, when few parts are required, customers are discouraged from discarding products. As if the remanufacturer deals with customers like virtual warehouses, the remanufacturer can economically collect suitable used products without various kinds of additional cost for warehouse and safety stock.

3 Previous Supply Chain Model

3.1 Modeling of Forward Supply Chains

This section briefly describes a model of forward and reverse supply chains which we have proposed in previous studies [8, 16, 17], in order to help understand our closed-loop supply chain strategy. The previous studies have represented a framework for dynamically forming and reconfiguring a supply chain as a dynamic supply chain [16, 17]. Each organization in the supply chains can change business partners for every order to find suitable business partners and enter into profitable contracts.

A three-layered supply chain model consisting of a client, a manufacturer, and a supplier has been proposed as a basic model of forward supply chain which involves make-to-order (MTO) companies with no inventories [17]. It is assumed that only the supplier has sufficient raw materials but no inventory of final products. The model has provided with both the negotiation process to determine the suitable prices and delivery times of products among the three-layered organizations and the

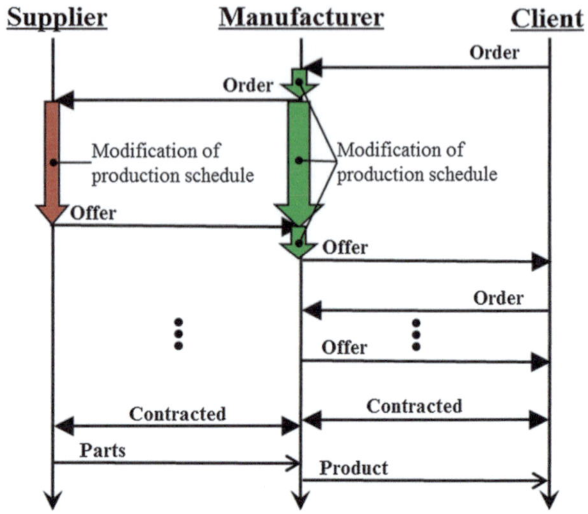

Fig. 1 Negotiation process among a supplier, a manufacturer, and a client

modification process of the production schedules in the supplier and the manufacturer. A client and a manufacturer sequentially send orders which include the requirements of delivery times and prices of the ordered products. A supplier and a manufacturer modify their existing production schedules by using a genetic algorithm (GA) and send offers for the orders, respectively. The offers include information about possible delivery times and bid prices of the ordered products. The actual delivery times and prices are determined by the negotiation processes among the organizations. Figure 1 represents the negotiation process among a client, a manufacturer, and a supplier. The vertical axes show the time.

3.2 *Modeling of Closed-Loop Supply Chains*

The forward supply chain model extended to a closed-loop supply chain model consisting of both MTO companies in forward supply chains and remanufacturing-to-order companies in reverse supply chains [8]. Figure 2 shows a basic model of a closed-loop supply chain. A model component in the reverse supply chain receives a used product from a client and provides assembly manufacturers with usable parts. The component is referred to as a remanufacturer in the study.

A remanufacturer has neither stock of usable parts nor used products. A remanufacturer generates an order for used product and sends it to clients when the remanufacturer receives an order for a usable part from a manufacturer. After receiving an offer for a used product from a client, a remanufacturer generates an offer for a usable part and sends it to a manufacturer. A remanufacturer modifies its disassembly schedule and estimates the possible delivery time $dtr^F_{s,h,n}$ and the bid price $pcr^F_{s,h,n}$ of the usable part by using the following equations:

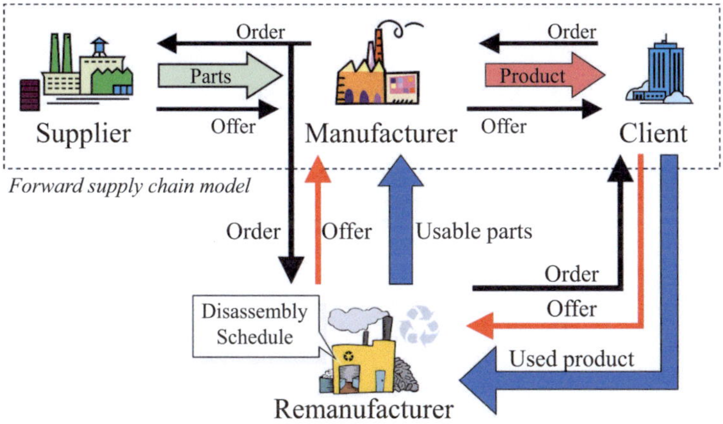

Fig. 2 Closed-loop supply chain model

$$dtr^F_{s,h,n} = ctr_{s,h,n} \tag{1}$$

$$pcr^F_{s,h,n} = tcr_{s,h,n} + rwr_{s,h,n} - pnr_{s,h,n} + \sum_{g=1}^{G} \Delta pnr^{Rv(s,n,r)}_{s,h,g} \tag{2}$$

where

$ctr_{s,h,n}$ completion time of a usable part recovered by remanufacturer R_s
$tcr_{s,h,n}$ total cost for buying, disassembling, and repairing of a used product
$rwr_{s,h,n}$ reward for a usable part required by remanufacturer R_s
$pnr_{s,h,n}$ penalty charge due to delay in delivery time of a usable part

The estimated profit $epf_{s,h,n}$ is calculated by the following equation:

$$epf_{s,h,n} = rwr_{s,h,n} - pnr_{s,h,n} + \sum_{g=1}^{G} \Delta pnr^{Rv(s,n,r)}_{s,h,g} \tag{3}$$

Figure 3 represents the negotiation process among a manufacturer, a remanufacturer, and a client. A remanufacturer considers a balance between supply and demand of reusable parts in the negotiation protocol. When a lot of usable parts are required by manufacturers, a remanufacturer stimulates clients to discard products for reuse by indicating high required prices for the used products to the clients. The clients discard products even if the products don't satisfy the condition of their life cycles estimated based on the Weibull distribution which is commonly used to model life data [18]. On the other hand, when few usable parts are required, remanufacturers indicate low required prices to discourage clients from discarding products. Then, the remanufacturer can increase the amount of reused products whenever they want and reduce waste products.

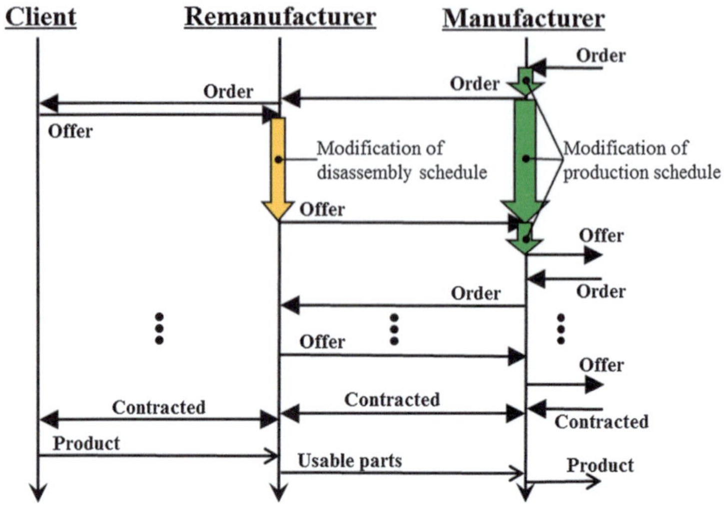

Fig. 3 Negotiation process among a manufacturer, a remanufacturer, and a client

4 Negotiation Protocol

4.1 Previous Negotiation Protocols

Previous study improves the negotiation protocol to increase the reusability of products. There are three ways for a remanufacturer to obtain used products in the following:

1. Push-type discarding strategy: This is a conventional approach. A client discards a product by stochastic means in consideration of its life cycle. Failure rate of product is usually used to estimate when a client discards a product. The Weibull distribution is used in the study to estimate product's life cycle. A client discards a product, when it has been used beyond its life cycle. The discarded products should be stored temporarily, since a client does not consider the condition of reuse of the product. A remanufacture can obtain only the stored product for reuse. In the case where no remanufacturers need the stored product for a certain period, it must be disposed as a waste product.
2. Pull-type discarding strategy: This is proposed in the previous research. A remanufacturer stimulates the intention of a client to discard a product, when a used product is required for reuse. A remanufacturer indicates high required prices for a used product to a client. The product's value is estimated based on the Weibull distribution in the study. It decreases with the time of usage of the product. The client evaluates a motivation for discarding a product by using the following equations and determines which product is discarded. If a much lower price is indicated for a used product, it is stochastically discarded and may become a waste product.

$$mv_{p',n'} \geq rn \tag{4}$$

$$mv_{p',n'} = \frac{pcr^{O}_{s,n} - \left(PV(t) - dcc_{p',n'}\right)}{pcr^{O}_{s,n}} + cif \tag{5}$$

$$PV(t) = PV(0) \times rr \times \exp\left\{-\left(\frac{t}{\eta}\right)^{\beta}\right\} \tag{6}$$

$$PV(t) = PV(0) \times \exp\left\{-\left(\frac{t}{\eta}\right)^{\beta}\right\} \tag{7}$$

where

$mv_{p',n'}$ motivation for providing a remanufacturers with a product for reuse.
rn random numbers.
$pcr^{O}_{s,n}$ required price for a product. It is determined by a remanufacturer.
$PV(t)$ product's value. It is estimated based on the Weibull distribution.
$dcc_{p',n'}$ cost for discarding a product by client C_p'.
cif client's incentive factor.
$PV(0)$ initial value of product. It equals to the purchase price of product.
β shape parameter which determines the mode of failure.
η scale parameter which defines the life of product.
t time of usage of product.
rr rate of reusable part to used product.
dcf decommissioning cost factor.

3. Negotiation-type discarding strategy: In the case where a client does not accept an order from a remanufacturer for discarding a product from the viewpoint of the required price, the client provides the remanufacturer with a possible time to dispose of the product as an offer. The product's value decreases with the time of usage of product according to the Weibull distribution, as shown in Fig. 4. The client estimates the time when the product's value equals to the required price sent from the remanufacturer by using the following equation:

$$\ln t = \ln\eta - \frac{1}{\beta}\ln\left\{\ln\left(\frac{PV(t)}{PV(0)}\right)\right\} \tag{8}$$

The remanufacturer evaluates the offer from the client by modifying a disassembly schedule and estimates a profit by using Eq. 3. The remanufacturer prospectively enters into a contract with the client by accepting the offer from the client. Then, the remanufacturer can obtain a lot of used products by negotiating with the client in consideration of both the required prices from the remanufacturer and the possible time to dispose of the used products from the client.

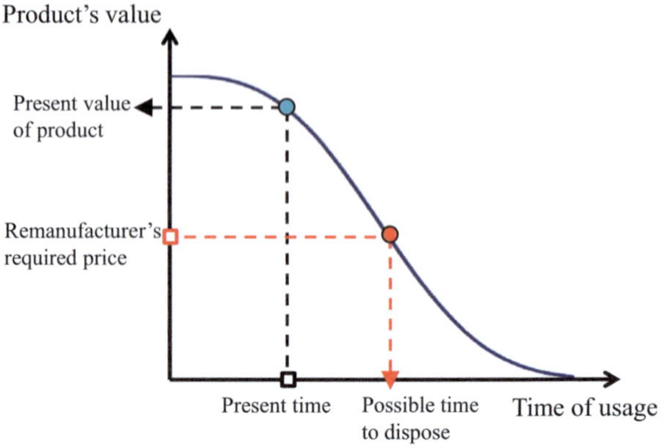

Fig. 4 Estimation of time to dispose of product based on the Weibull distribution

4.2 New Negotiation Protocol

This study improves the negotiation-type discarding strategy to increase the reusability of products and to collect reusable products economically. New negotiation-type discarding strategy is summarized in the following.

A remanufacturer generates new order information for used products. The order information includes not only the required price but also the required delivery time. It can increase the amount of reused parts and a profit of a remanufacturer by providing clients with the appropriate price and time for collecting the used products. When a client gets the order from a remanufacturer, the client evaluates a product's value by using the following equations and determines which product is discarded, as shown in Fig. 5. In the case where the client had no products satisfying with Eq. 9, the client sends the remanufacturer the information about a possible time to dispose of the product as an offer.

$$pcr_{s,n}^{O} \geq pcr_{p',n'} \tag{9}$$

$$pcr_{p',n'} = PV(0) \times rr \times \exp\left\{ -\left(\frac{dtr_{s,n}^{O}}{\eta} \right)^{\beta} \right\} \tag{10}$$

where

$pcr_{p',n'}$ acceptable product's value to dispose.
$dtr_{s,n}^{O}$ required delivery time price for a product. It is determined by a remanufacturer.

Figure 6 represents the improved negotiation process among a manufacturer, a remanufacturer, and a client. The negotiation process can be summarized as follows.

Step 1. A remanufacturer receives orders from all manufacturers.

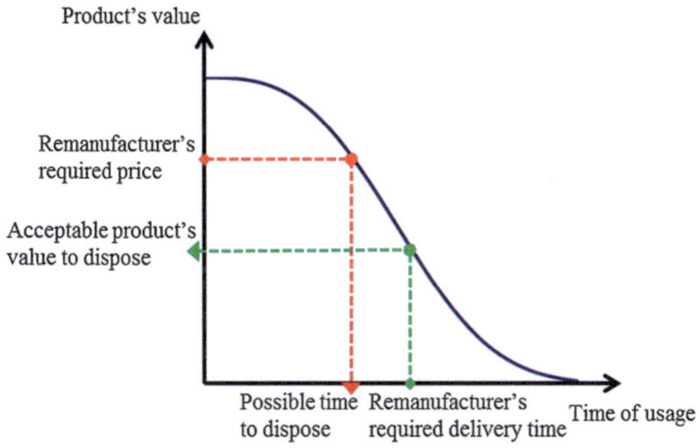

Fig. 5 Estimation of time and price to dispose of product based on the Weibull distribution for new negotiation type

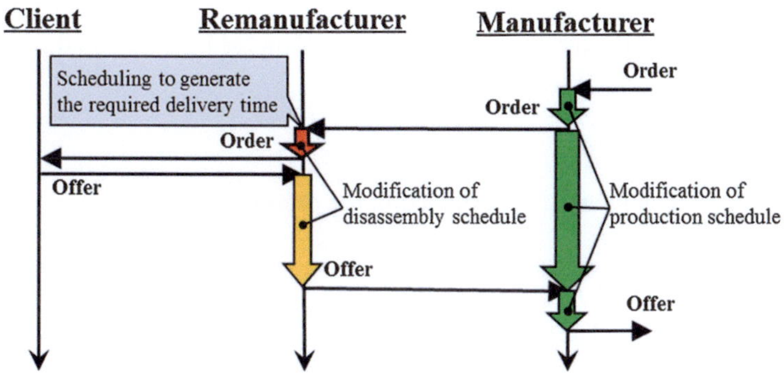

Fig. 6 Negotiation process among a manufacturer, a remanufacturer, and a client

Step 2. The remanufacturer uses the GA to improve the disassembly schedule for a short time. The remanufacturer generates and sends an order to a client. The order includes the required price and the required delivery time given by the following equations:

$$dtr^{O}_{s,n} = \min_{1 \leq j \leq J} \{st_{s,n,j}\} \tag{11}$$

$$pcr^{O}_{s,n} = pcm^{O}_{h,s,n} - tcr_{s,h,n} - rwr_{s,h,n} \tag{12}$$

$$tcr_{s,h,n} = dc_{s,h,n} + mc_{s,h,n} \tag{13}$$

$$mc_{s,h,n} = bv_{s,h,n} \times (1 - avd) \tag{14}$$

where

$dtr^{O}_{s,n}$ required delivery time included in the order sent to a client
j number of remanufacturing operations
$j = 1, 2, \ldots, J$
$st_{s,n,j}$ starting time of jth remanufacturing operation in the sth remanufacturer

$pcr^{O}_{s,n}$ required price included in the order sent to a client

$pcm^{O}_{h,s,n}$ required price included in the order sent from a manufacturer to a remanufacturer

$tcr_{s,h,n}$ total repair cost for parts

$dc_{s,h,n}$ disassembly cost for parts

$mc_{s,h,n}$ mending cost for parts

$bv_{s,h,n}$ basic values for mending

avd minimum allowable value of the deterioration degree according to the Weibull distribution

$rwr_{s,h,n}$ reward of remanufacturer

Step 3. When a client receives the order, the client generates an offer by using new negotiation-type discarding strategy.

Step 4. After the remanufacturer receives the offer, the remanufacturer modifies the disassembly schedule in consideration of the constraint given by Eq. 15. The remanufacturer generates and sends the manufacturer an offer which includes the bit price and the possible delivery time estimated by Eqs. 1 and 2.

$$st_{s,n,j} \geq dtc^{F}_{s,n} \tag{15}$$

$dtc^{F}_{s,n}$ possible delivery time included in the offer sent from a client

5 Computational Experiments

5.1 Experimental Conditions

A prototype of a simulation system for closed-loop supply chains has been developed in order to evaluate the effectiveness of the proposed model and negotiation protocol. The prototype system was developed using Windows-based networked computers (Intel Core 2 Duo E8500 3.16 GHz CPU with 1.99 GB of RAM) for ease of applicability to real organizations. Two suppliers, two manufacturers, a remanufacturer, and a client were implemented as agents on different six computers, respectively, as shown in Fig. 7.

The client continuously generated 100 new orders and negotiated with two manufacturers during the experiments for about 6 h. The manufacturers generated orders and negotiated among the two suppliers and one remanufacturer every 30 s of bidding time in order to generate offers for the client. Population size, crossover rate, and mutation rate of the GA were 30, 0.8, and 0.2, respectively. The client has already contracted 100 products with other manufacturers. In the initial condition, 40 products have been already delivered to the client and used by the client in the market. The purchase prices of the products are determined based on the contracted prices of products. The parameters of the client are summarized in Table 1. These

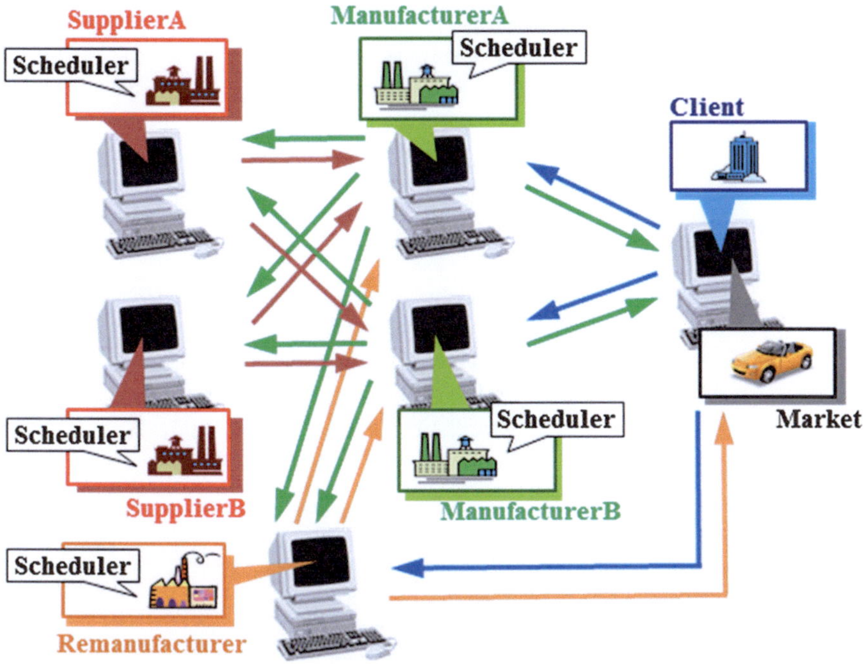

Fig. 7 Simulation system of closed-loop supply chains

parameters are used in Eqs. 5, 6, 7, 10, and 14. On the same condition of the used products, the new model is compared with the previous model. The parameters of the manufacturers, suppliers, and remanufacturer are summarized in Tables 2 and 3. In the initial conditions, the manufacturers had job-shop-type production schedules consisting of ten resources and 20 contracted products. The suppliers and remanufacturer had job-shop-type production schedules consisting of five resources and 20 contracted products, respectively. The delivery time factor and the penalty factor are randomly determined from the uniform distributions.

5.2 *Comparisons of Experimental Results*

Experimental results of the model with new negotiation protocol including new negotiation-type discarding strategy were compared with ones of the model with the previous protocol including negotiation-type discarding strategies. Ten experiments were carried out for the new model and the previous model, respectively. The experimental results are summarized in Table 4. The remanufacturer in the new model can increase the amount of reused products more than the one in the previous model. The new model can reuse more products than the previous one. It is recognized through analysis of the experimental results that the new negotiation

Table 1 Value of each parameter of the client

Parameters	Values
Client's incentive factor: *cif*	0.5
Decommissioning cost factor: *dcf*	0.1
Rate of reusable part to used product: *rr*	0.4
Allowable value of the deterioration degree: *avd*	0.3

Table 2 Value of each parameter of the manufacturers

Parameters	Values
Number of resources	10
Number of initial jobs	20
Number of initial operations	200
Processing time of each operation(s)	*U*[80, 600]
Delivery time factor	*U*[3.0, 5.0]
Penalty factor ($/s)	[1.0, 1.5]

U [*a*, *b*]: uniform distribution between *a* and *b*

Table 3 Value of each parameter of the suppliers and remanufacturer

Parameters	Values
Number of resources	5
Number of initial jobs	20
Number of initial operations	100
Processing time of each operation (s)	*U*[8, 60]
Delivery time factor	*U*[2.2, 5.0]
Penalty factor ($/s)	[1.0, 1.5]

U [*a*, *b*]: uniform distribution between *a* and *b*

Table 4 Comparison of experimental results

[Average]	Previous model	New model
Number of reused products	31.7	33.7
Profit of remanufacturer ($\times 10^3$$)	64.9	73.4

protocol can reuse a large amount of used products and provide more profit to the remanufacturer by the appropriate order price and time for collecting used products.

6 Conclusion

Previous studies proposed a basic closed-loop supply chain model and a negotiation protocol for increasing the amount of reused products and reducing waste products. This study improves the negotiation protocol between a remanufacturer and a client in order to increase the reusability of products and economically collect reusable products. A remanufacturer can enter into a contract with a client by the appropriate order price and time for collecting used products. Then, a lot of used products are economically collected by a remanufacturer through negotiation with a client in consideration of both the required delivery time from the remanufacturer and the acceptable product's value to dispose of the used products from the client. Some experiments were carried out by using the developed simulation system for a closed-loop supply chain. Experimental results of the model with new negotiation protocol were compared with the one of the previous protocol. The results show that the new protocol can reuse a larger amount of used products and give more profits to the remanufacturer.

References

1. BearingPoint: 2008 Supply chain monitor "How mature is the green supply chain?", Survey report (online). Available from http://www.supplychainstandard.com
2. Srivastava SK (2007) Green Supply -Chain management: a state-of-the-art literature review. Int J Manag Rev 9(1):53–80
3. Sarkis J, Zhu Q, Lai K (2011) An organizational theoretic review of green supply chain. Int J Prod Econ 130:1–15
4. Seuring S, Muller M (2008) From a literature review to a conceptual framework for sustainable supply chain management. J Clean Prod 16:1699–1710
5. Gupta S, Palsule-Desai OD (2011) Sustainable supply chain management: review and research opportunities. IIMB Manag Rev 23:234–245
6. Hassini E, Surti C, Searcy C (2012) A literature review and a case study of sustainable supply chains with a focus on metrics. Int J Prod Econ 140:69–82
7. Seuring S (2013) A review of modeling approaches for sustainable supply chain management. Decis Support Syst 54:1513–1520
8. Gungor A, Gupta SM (1999) Issues in environmentally conscious manufacturing and product recovery: a survey. Comput Indust Eng 36:811–853
9. Tanimizu Y, Shimizu Y, Iwamura K, Sugimura N (2013) Modeling and simulation of closed-loop supply chains considering economic efficiency. In: Advances in production management system, part I, Springer, Heidelberg, pp 461–468
10. Llgin MA, Gupta SM (2010) Environmentally conscious manufacturing and Product Re-covery (ECMPRO): a review of the state of the art. J Environ Manage 91:563–591
11. Guide VDR, Jayaraman V, Linton JD (2003) Building contingency planning for closed-loop supply chains with product recovery. J Oper Manag 21:259–279
12. Nielsen K, Bruno TD (2013) Closed Loop supply chains for sustainable mass customization. In: Advances in production management system, part I, Springer, Heidelberg, pp. 425–432.
13. The U.S. Department of Energy (2011) Critical materials strategy
14. Takata S, Watanabe M, Ohbayashi Y (2006) Collection rate estimation model in closed-loop manufacturing. In: Proc. of CIRP international conference on life cycle engineering, Leuven, pp. 601–606
15. Umeda Y, Kondoh S, Sugino T, Yoshikawa H (2006) Analysis of reusability using 'Marginal Reuse Rate'. CIRP Ann Manuf Technol 55(1):41–44
16. Tanimizu Y, Yamanaka M, Iwamura K, Sugimura N (2006) Multi-agent based dynamic supply chain configuration considering production schedules. In: Proc. of international symposium on flexible automation, Osaka, pp. 572–578
17. Tanimizu Y, Ozawa C, Shimizu Y, Orita B, Iwamura K, Sugimura N (2013) Flexible multi-layered dynamic supply chain models with cooperative negotiation. Int J Autom Technol 7 (1):128–135
18. Mazhar MI, Kara S, Kaebernick H (2007) Remaining life estimation of used components in consumer products: life cycle data analysis by Weibull and artificial neural networks. J Oper Manag 25:1184–1193

Concept Proposal and Feasibility Study of Remote Recycling: Separation Characteristics and Cost-Profit Analysis

Kenta Torihara, Yuta Kadowaki, Jun Ooki, and Nozomu Mishima

Abstract Management of used electronics is an emerging problem in all over the world. In order to utilize the valuable metals contained in used electronic equipment, efficient system for recycling is necessary. The authors have proposed a concept of "remote recycling" which is to carry out separation of valuable metals by teleoperation. As the first step of the feasibility study, rough crushing and visual separation of particles were carried out. By measuring compositions of the particles, it was shown that only by rough crushing and handpicking, the percentages of metals were much higher than printed circuit board (PCB) itself. The experimental results showed that remote recycling is technically feasible. The study also tried some cost-profit analysis based on the prices of metals contained in the selected particles and showed economical feasibility of the concept. Finally, the study concluded that remote recycling will be a totally new and promising concept which is suitable for handling used small-sized electronics.

Keywords Mobile phones • Teleoperation • PCB • Recycling • Material composition

1 Introduction

Management of used electronics is an emerging problem in all over the world. In Japan, so-called urban mine [1], which means unutilized materials contained in used products, is under discussion. Unlike large-sized products, newly enforced law for small-sized products does not require recycling fee to cover recycling cost, from consumers. Thus, in order to utilize the urban mine, efficient social system for recycling is necessary. In the previous papers [2, 3], the authors have proposed a concept of "remote recycling" which is to carry out some of the manual operations of the recycling process by teleoperation. In the concept, roughly crushed particles are carried by conveyer and will be separated by some sorting devices into two groups. By implementing remote recycling at locations where labor cost is

---

K. Torihara • Y. Kadowaki • J. Ooki • N. Mishima (✉)
Graduate School of Engineering and Resource Science, Akita University, Akita, Japan
e-mail: nmishima@gipc.akita-u.ac.jp

M. Matsumoto et al. (eds.), *Sustainability Through Innovation in Product Life Cycle Design*, EcoProduction, DOI 10.1007/978-981-10-0471-1_30

relatively inexpensive or by utilizing potential labor power, the authors think it is possible to reduce the recycling cost.

As the first step of the feasibility study, the study takes used mobile phone as an example and carries out experiments to clarify the metal separation characteristics. Based on the previous survey [4], high percentages of valuable metals contained in small- and medium-sized home appliances are contained in used mobile phones (Table 1 and 2), since mobile phone is one of the most common electronic products in modern society and some precious elements are used to realize high performances of mobile phones. Plus, it is well known [5] that most of the valuable metals such as gold (Au), silver (Ag), copper (Cu), palladium (Pd), and so on are contained in PCB. It means that if we can collect PCBs of used mobile phones efficiently and send to recycling facilities, the recycling system of small-sized electronic equipment can be efficient, since transportation cost of the used products is a cost driver. In addition, there are only a few metal smelters in the world that can extract various metals from PCB. One of the few [6] exists in Akita Prefecture that locates rather far from the metropolises in Japan. It might be good that we can disassemble used mobile phones in the urban area and send only PCBs to the smelter. Manual disassembly of PCB can be a good way to increase the metal concentration in the recycling process. It is possible to decrease unnecessary transportation of

Table 1 Estimation of valuable elements contained in small-sized electronic equipment [4]

Product	Pd	Au	Ag	Cu
Mobile phone	0.55	2.1	12.2	486
Portable game player	0.05	0.4	1.3	265
Non-portable game player	0.03	0.1	2.1	67
Portable CD/MD player	0.01	0	0.1	9
Digital audio player	0	0	0	0
Digital camera	0.07	0.3	2.4	87
Driving navigation system	0.09	0.1	1.4	109
Video camera	0.21	0.1	2.2	49
DVD player	0.09	0.4	5.4	507
Others	0.08	2.5	24.9	2262

Table 2 Estimation of valuable elements contained in small-sized electronic equipment (contd.) [4]

Product	Ta	W	Nd	Dy
Mobile phone	4.12	3.44	3.93	0.08
Portable game player	0.94	0.14	0.73	0.02
Non-portable game player	0.24	0.13	0.12	0.01
Portable CD/MD player	0.18	0	0.01	0
Digital audio player	0	0	–	0
Digital camera	2.85	0.24	0.12	0.02
Driving navigation system	0.99	0.14	0.28	0.07
Video camera	2.19	0.15	0.21	0.02
DVD player	2.51	0.47	0.44	0.11
Others	3.20	1.31	0.08	0.14

non-valuable parts. However, since manual disassembly is a big cost driver, the idea is to replace it by a combination of rough crushing and visual separation from a remote site. In order to implement remote recycling, at least, it is necessary to be able to achieve higher concentration of metals by visual separation than manually disassembled PCBs.

Therefore, this study carries out separation experiments as the first step to know the technical feasibility of the concept. In the experiments, rough crushing and visual separation of particles that seem to contain metals will be carried out. The study estimates whether the separation of valuable metals only by visual information is efficient enough or not. The study also tries some cost-profit analysis based on the prices of metals contained in the selected particles. The goal of the study is to show that remote recycling is technically and economically feasible and can be a promising way to handle used small-sized electronics.

2 The Concept of Remote Recycling

The previous paper [7] regarding cost-profit analysis of recycling process of used mobile phones indicates that the total of material values recovered from used mobile phones is equivalent or smaller than the total of labor cost, transportation cost, facility cost, and other costs. Technological development to reduce cost is ongoing. But most of the development focuses on facility costs. The concept of the remote recycling proposed in this study can drastically decrease labor cost and transportation cost. The study carries out basic experiments to show technical feasibility of the concept. A schematic view of the concept is shown in Fig. 1. As it is shown in the figure, roughly crushed particles are carried by conveyer and are separated by some sorting devices into two groups. The visual information is sent by web camera via internet. Operators stay in remote sites where labor cost is inexpensive. Or it might be possible to use voluntary labor power by providing an attractive software to operate the remote operation as if it is an online game. In this case, the labor cost will be zero.

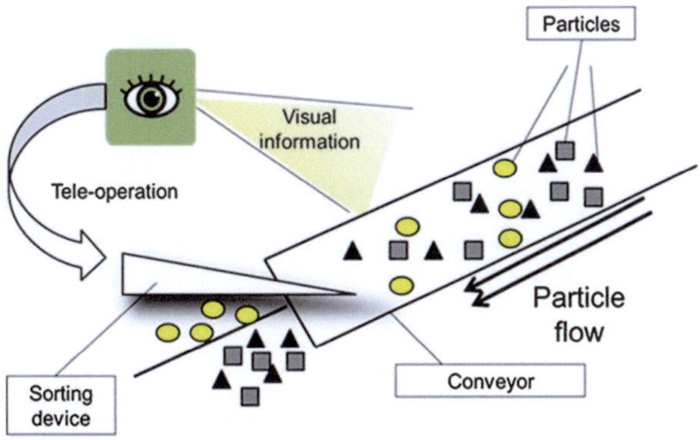

Fig. 1 Conceptual sketch of remote recycling

The advantages of this system are shown as below.

- Since no chemical process is necessary, environmental impact of the process will be small.
- By utilizing inexpensive or voluntary labor, it is possible to reduce labor cost of the recycling process.
- It is also possible to reduce transportation cost, by screening unnecessary parts near users' locations.
- When there is a requirement to recover additional element, it is possible to change screening strategy flexibly.

3 Separation Experiments

3.1 Experiment Process

One of the reasons of high labor cost in current recycling process is manual disassembly process. The process is rather common [8]. On the other hand, manual disassembly can enhance quality of the recycling process. Since PCB contains most of the valuable material, manual disassembly of PCB is effective in increasing the metal concentration. Thus, if it is possible to achieve higher or at least equivalent percentages of metals by a less expensive method, it will be helpful in reducing the recycling cost. In our proposal, remote separation can replace this manual disassembly process. In order to clarify the technical feasibility of the concept, separation characteristics are investigated through experiments. The procedures of the experiments are shown in Fig. 2. In the left-hand process, used mobile phones are

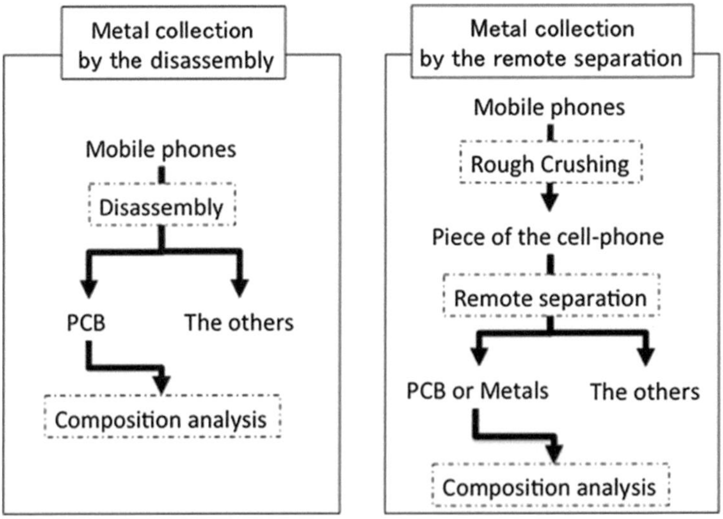

Fig. 2 Procedures of the experiments

manually disassembled and PCBs are detached. Then, the PCBs are crushed by a rotary mill. In the right-hand process, whole mobile phones are crushed first, and then, the operator separate particles seem to be valuable from other particles. In both processes, batteries and LCD (liquid crystal display) unit which might emit toxic materials by crushing are detached in advance.

By comparing the results of the left-hand process and the right-hand process of the above figure, it will be possible to know the suitability of the remote separation. Basically, if the metal concentration of the separated particles in the right-hand process is higher than the left-hand one, we can say remote separation is efficient enough. In this study, since teleoperation system has not been prototyped, manual separation was carried out. However, possibility of the separation by visual information can be the basic fact to support the feasibility of the concept. After the separation, material composition of both samples are measured by X-ray fluorescence (XRF) [9] and inductively coupled plasma (ICP) [10]. ICP is used to know the concentration of gold and XRF is used to know compositions of other elements. Comparing the results of both particles is the last process of the experiments.

3.2 *Example of the Experiments*

In a practical experiment, PCB (Fig. 3) and other parts of a mobile phone are roughly crushed using a rotary cutting mill. Since the screen size of the cutting mill is 8 mm, particles will be smaller than 8 mm square. Average size of the crushed

Fig. 3 An example of disassembled PCB

Fig. 4 Separated particles

particles is 4–5 mm square. Next, Figure 4 shows the separated particles from the mixed particles. This means that the operator thought these particles were originally from PCB or other metal parts, if metal concentrations of Fig. 4 are better or equivalent to that of Fig. 3.

4 Experimental Results

4.1 Measurement of Material Composition by XRF

After separating the target particles, material compositions were measured by X-ray fluorescence (XRF) analysis. Figures 5 and 6 show material composition of PCB particles and visually separated particles correspondingly.

4.2 Measurement of Gold Density by ICP

Although XRF is a good way in order to know material compositions of various kinds of elements, very small density of element cannot be detected or is inaccurate. On the other hand, since the value of a used mobile phone heavily depends on gold contained in IC (Integrated Circuit), recovery of gold is one of the key issues in recycling. Since the measurement by XRF cannot detect the existence of gold, a

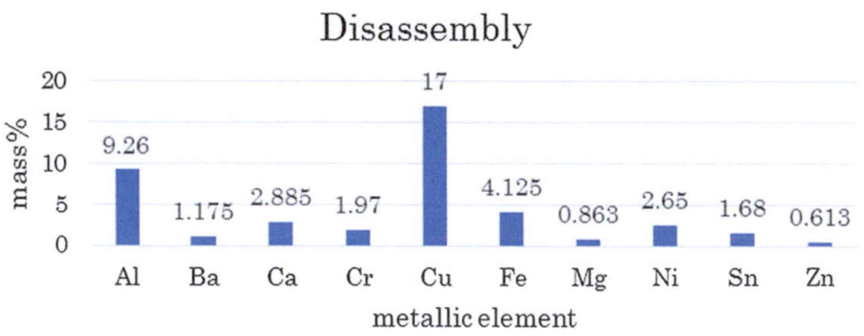

Fig. 5 Material compositions of PCB-origin particles

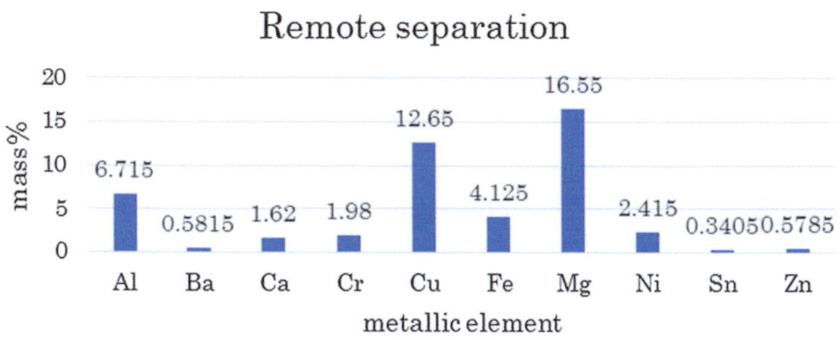

Fig. 6 Material compositions of visually separated particles

Table 3 Concentration of gold

	Au (ppm)
PCB-origin particles	459
Separated particles	463

more precise measurement of material composition based on ICP [10] has been carried out. By using ICP, it is possible to measure material densities up to a few parts per million. Table 3 shows the measured data of concentration of gold in PCB and separated particles. Basically, the table shows there is no difference of concentration between PCB and separated particles.

4.3 Interpretation of the Results

Based on the measurement of XRF shown in Fig. 5, material composition of both samples is not very different except for magnesium and copper. The reason of the differences in these two elements is assumed as below.

- Magnesium is contained in structure parts and can be recovered only by visual separation because of metallic gloss.

- Copper is contained both in PCB and other parts. Some of the copper attached to plastic parts cannot be recognized as valuable parts.
- There are no big differences in densities of nickel, chromium, and zinc. Since these elements are contained in resistors, capacitors, and terminals of PCBs, the fact supports PCB-origin particles can be recognized by vision.
- Density of tin decreased greatly in visually separated particles. Since tin is mainly contained in solder of PCBs, it is assumed it was not able to recognize solder as valuable parts.
- Densities of iron are equivalent. Iron contained in terminals of PCBs can be completely recovered.

Based on the ICP analysis shown in Table 3, the fact is rather simpler. Since most of the gold is contained in ICs of PCBs, the same density means that the PCB-origin particles were completely recognized and selected as valuable parts. Again, this is a rather strong supporting fact for separation based on visual information. Even for the particles after rough crushing, operators can recognize PCBs as PCBs. It means that visual separation is possible and the very first step of the feasibility study of remote separation has been cleared.

5 Cost-Profit Analysis

- Aforementioned analysis on material compositions basically suggested remote separation is feasible. However, it is also necessary to consider economical aspects of the concept. The author has explained that one of the advantages of the idea is reduction of the recycling cost. However, if the profit of the recycling process decreases a lot, such process is useless. To check this point, a simple cost-profit analysis was carried out. For the cost-profit analysis, again the previous paper [7] is helpful. The study provide a calculation of labor cost (145 JPY), transportation cost (50–100 JPY), and recoverable material price (112 JPY) of mobile phone recycling. As the experiments suggested, it is basically possible to recognize PCB-origin parts by vision. But of course, some of the elements cannot be well recognized. The assumptions in cost-profit analysis are shown as follows.
- Gold, nickel, tin, zinc, iron, and chromium can be 100 % recovered.
- Two-thirds of aluminum and copper contained in a whole mobile phone can be recovered.
- Under this assumption, recoverable material price of mobile phone will be about 90 JPY, while labor cost will be greatly decreased. If we assume that transportation cost is proportional to the weight of transported parts, the cost will be 40 % of the cost of transporting the whole mobile phones. Average weight of PCB is about 40 % of the total weight. Since the reduced transportation cost is larger than the decreased material price, remote separation can be helpful in reducing the overall cost-profit ratio. In addition, an important advantage of remote

Table 4 Cost-profit estimation

Items	Conventional recycling (JPY)	Remote recycling (JPY)
Labor cost	145	0?
Transportation cost	50–100	20–40
Extra facility cost	–	50
Recoverable material price	112	90
Total profit	−133 to −83	0–20?

recycling concept is that it has a potential to decrease labor cost. Therefore, cost-profit ratio can be much more improved. Contrarily, remote recycling requires some extra setups at users' sites. Based on the cost of the experimental setup and assumed capability to recycle mobile phones per year, facility cost can be calculated. It will be about 50 JPY per phone. Finally, rough cost-profit analysis of current system and remote recycling system is shown in Table 4.

- Of course, the analysis shown above is too abstract. It heavily depends on whether it is really possible to use voluntary labor power in remote separation process. But still the authors think that the total recycling process of used mobile phones can be self-profitable by realizing remote recycling concept.

6 Conclusions

This study proposed a totally new concept in the recycling process of small electronic equipment such as mobile phones and named remote separation. As the first step of the feasibility study, the study carried out separation experiments to check whether it is possible to recognize valuable parts only by visual information. Since it is said that valuable elements are mostly contained in PCBs of mobile phones, the point of the experiments was to compare manually disassembled PCB with particles through visual separation.

In the experiment, material compositions of crushed PCB and visually separated particles were measured by XRF and ICP and compared. For the experimental results and some information regarding the elements contained in each part of a mobile phone, the study concluded that it was possible to recognize PCB-origin particles as PCB-origin particles. For example, for gold, it is known that most of the gold in a mobile phone is contained in ICs that are mounted in PCB. So the experimental result showing the same density of gold in PCB-origin particles and separated particles means that the operator could recognize PCB-origin particles almost perfectly. As for the elements aforementioned, gold, aluminum, copper, magnesium, nickel, iron, and zinc were recoverable by remote separation.

In addition, an abstract cost-profit analysis based on the cost estimations and recoverable material prices was carried out. The result also suggested that it would be possible to reduce the labor cost and transportation cost greatly by remote recycling, and recycling process of small electronic equipment can be self-profitable. This technical feasibility and the economical feasibility will be a supporting fact to say remote recycling is a promising way to go.

Of course, there is some future work to be solved as well. It is necessary to measure some more low-density and high-value elements such as silver, tantalum, palladium, and so on to know the efficiency of remote separation. And since the samples were from one mobile phone and number of experiments was not enough, much more separation experiments will be necessary. In addition, remote separation is a method for initial screening to reduce weight and volume to be treated in the recycling process. A suitable combination with other physical separation method such as magnetic separation, electrostatic separation, etc. should be studied.

References

1. Nanjyo M (1988) Bulletin of the Research Institute of mineral dressing and metallurgy. Tohoku Univ 43(2):239–251
2. Matsumoto M, Mishima N, Masui K, Kondoh S (2009) Proposal and feasibility assessment of tele-inverse manufacturing. Int J Autom Technol 3(1):11–18
3. Torihara K, Kitajima T, Mishima N (2015) Design of a proper recycling process for small-sized E-waste. Procedia CIRP 26:746–751
4. (2010) Ministry of Environment, Ministry of Economy, Trade and Industry: Report of the study group about recovery of the rare metal and proper processing of used small household appliances. (In Japanese)
5. Shiratori T, Nakamura T (2007) Concept of "Artificial Deposit" 2 – transition of the metal potential of spent electric and electronic appliances. J MMIJ 4(5):171–178
6. http://www.dowa.co.jp/en/jigyo/metalmine_summary.html. Accessed 29 June 2015
7. Takahashi K, Nakamura J, Otabe K, Tsuruoka M, Matsuno Y, Adachi Y (2009) Resource recovery from mobile phone and the economic and environmental impact. J Jpn Inst Metals 73 (9):747–751
8. http://www.comgeo.net/archives/1183. Accessed 06 Apr 2014
9. Beckhopff B, Kanngießer B, Lnghoff N, Weddell R, Wolff H (2007) Handbook of practical x-ray fluorescence analysis, Springer Science & Business Media, Berlin, Germany
10. Hou X, Jones BT (2000) Encyclopedia of analytical chemistry – inductively coupled plasma – optical emission spectrometry. Wiley, Chichester, pp 9468–9485

Simulation-Based Uncertainty Quantification in End-of-Life Operations for Strategic Development of Urban Mines

Hitoshi Komoto, Shinsuke Kondoh, and Keijiro Masui

Abstract Strategic development of the life cycle of electric and electrical equipment such as mobile phones and laptop computers is critical for the recycling of precious and rare metals as well as the reuse and remanufacturing of components. The quality of collected products at the end-of-life (EoL) stage is uncertain due to variations in these products regarding the material composition and reusability. Quantification and reduction of such uncertainty is crucial for the appropriate assignment of EoL options such as reuse, remanufacturing, and recycling. The paper models and analyzes the inspection process as well as the process to assign specific EoL options to collected products. An entropy-based indicator is proposed for the quantification of the uncertainty derived from variations in collected products regarding their age. The impact of the sampling rate and accuracy of the inspection process on the number of the occurrences of EoL options is analyzed with life cycle simulation. It found that the simulation result identifies specific time periods, in which the inspection process is meaningful, and that the sampling rate positively correlates the number of the occurrences of reuse and remanufacturing processes.

Keywords End-of-life design • Life cycle simulation • Uncertainty

1 Introduction

1.1 Background

The Japanese manufacturing industry plays a crucial role in boosting the domestic economy and cultivating the employment. Among a wide range of products made in the country, electric and electrical equipment (such as laptop computers, smart phones, and other industrial devices) is known for its functionality and performance criteria. As these products become more advanced regarding these criteria, the

H. Komoto (✉) • S. Kondoh • K. Masui
Advanced Manufacturing Research Institute, National Institute of Advanced Industrial Science and Technology, Tsukuba, Japan
e-mail: h.komoto@aist.go.jp

M. Matsumoto et al. (eds.), *Sustainability Through Innovation in Product Life Cycle Design*, EcoProduction, DOI 10.1007/978-981-10-0471-1_31

amount of precious metals and rare metals in these products increases [1–3]. Thus, if significant amount of these products are collected the end of life (EoL), they are regarded as alternatives to natural metal mines or so-called urban mines [4].

Strategic development of urban mines of the above mentioned products is a challenge due to uncertainty derived from the variations in the quantity and quality of collected products [5, 6]. Such uncertainty is a cause of the difficulty in selecting appropriate EoL options such as reuse, remanufacturing, and recycling.

In order to manage such uncertainty regarding the collected products, these EoL processes should be investigated under rigorous scrutiny.

- Inspection of collected products for the classification with respect to their states (e.g., model year, use condition, material composition)
- Assignment of specific EoL options (e.g., reuse, remanufacturing, recycle) to the collected products with defined classification

In order to analyze the impact of these processes from a life cycle perspective, a life cycle model consisting of these processes is useful. Figure 1 shows such a life cycle model. The model expresses several crucial concepts in detailing the performance of these processes (other details are shown in the rest of the paper). First, some collected products are inspected directly, while the state of other collected products is estimated based on the inspection results. Second, assigned EoL operations are not always executable due to bad estimation (of the state of products) and inappropriate assignment (of an EoL option).

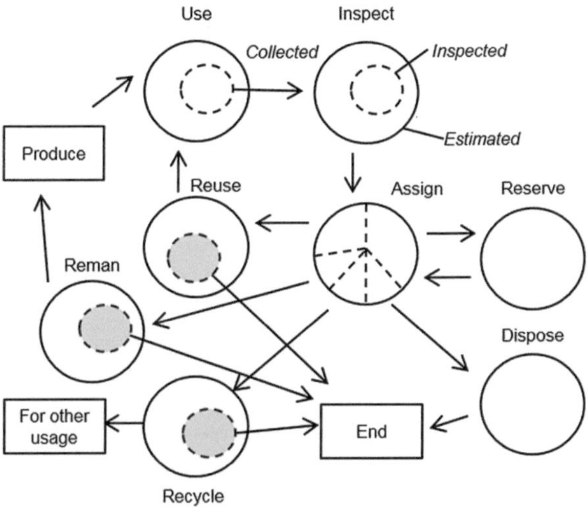

Fig. 1 A life cycle model with focus on inspection and assignment processes

1.2 Research Objective

The research objective of the paper is to quantify the uncertainty dealt in the inspection and assignment processes and investigate the impacts of these processes on the execution of EoL options from a life cycle perspective. Our approach to the quantification of such uncertainty is the introduction of entropy-based indicator, which deals with the variations in EoL options assigned to individual products. Life cycle simulation (LCS) [7, 8] is used for the simulation of the generation and state transition of individual products by evaluating the execution of processes shown in Fig. 1.

1.3 Structure of the Paper

The paper is structured as follows. Section 2 formalizes inspection and assignment processes, which are the main focus of the study. The concept of the uncertainty in these processes is introduced. Section 3 presents a life cycle model, which is developed for the analysis of these processes from a life cycle perspective. Section 4 shows how the life cycle model is implemented and simulated for this analysis. In Sect. 5, the impact of the uncertainty in these processes on the number of executed EoL options is analyzed based on the simulation results. Section 6 summarizes and concludes the paper.

2 Formalization

This section attempts to formalize the EoL decision process as a strategy π, which assigns specific EoL process to each collected product i.

$$\pi(i_1, \ldots, i_n) = (eol_1, \ldots, eol_n) \tag{1}$$

where n is the number of the products and eol is one of the EoL options currently available. With reference to Fig. 1, the EoL decision process consists of subprocesses called *inspect* and *assign*. In *inspect*, the state of collected products is estimated. While some of the products are directly inspected, the state of other products is estimated based on the state of the inspected products. In *assign*, EoL options (e.g., *reuse* and *recycle* in the case of the model in Fig. 1) are assigned to the collected products considering their estimated state. The rest of this section describes the detail of the subprocesses.

2.1 Inspection

By *inspect*, the state of products is estimated. Each *inspect* process applied to product i is formulated as follows:

$$\text{inspect}(i) = s_i^* \tag{2}$$

where s_i^* is the estimated state of product *i*.

The following two phenomena are considered as sources of the uncertainty regarding the state of collected products relevant to the inspection process:

- *Accuracy*: since some estimation techniques (e.g., machine learning) are used in the process, the estimated state s_i^* may not be the same as the actual state of the product s_i.
- *Sampling rate*: considering the process costs, not all collected products are inspected. In this case, the state of uninspected products is estimated.

Considering both the accuracy and sampling rate of the process, the *inspect* process is formulated as follows:

$$(i_1, \ldots, i_n) \rightarrow \left(s_1^*, \ldots, s_m^*, s_R^*, \ldots, s_R^*\right) \; [m \le n] \tag{3}$$

where *m* is the number of inspected products and s_R^* is the estimated state of uninspected products (from the $m+1$ th to *n* the products). The total (100 %) inspection indicates $m = n$. The estimated state of uninspected products s_R^* does not appear in the formulation in the case of the total inspection.

2.2 Assignment

The *assign* process is to assign a specific EoL option to each collected product, whose state is estimated directly or indirectly. When the assign process is applied to a product i, the process is formulated as follows:

$$\text{assign}(i) \rightarrow eol_i \in \text{EOL}_{\text{available}} \tag{4}$$

where eol_i is one of the EoL options currently available $\text{EOL}_{\text{available}}$. In the case of the model in Fig. 1, it includes some (or none or all) of *reuse*, *reman*, *recycle*, *dispose*, and *reserve*.

2.2.1 Availability of EoL Options

The EoL options currently available ($EOL_{available}$)) shown above may change with the progress of the life cycle of products. For instance, recycling of products may become possible after recycling plants for the products are built. Remanufacturing of products may drop out from the available EoL options, when manufacturing plants are closed at the end of the life cycle of products.

2.2.2 Applicability of EoL Options

Whether available EoL options are applicable to a product or not depends on the state of the product. This formulation indicates the refinement of the candidates of the selected EoL option eol_i:

$$eol_i \in EOL\left(s_i^*\right) \cap EOL_{available} \tag{5}$$

where $EOL(s_i^*)$ is EoL options applicable to product *i* with the inspected state s_i^*.

2.2.3 Evaluation of the Assigned EoL Options

EoL options assigned based on inaccurate inspection may be in appropriate (or unexecutable) considering the deviation between the estimated state of a product s_i^* and the actual state s_i. For instance, some components of old products are not compatible with new products. If old products are regarded as new products by mistake and the remanufacturing EoL option is assigned to them, they are not actually used for remanufacturing.

In such a case, $EOL(s_i^*)$ is not always the same as $EOL(s_i)$. This formulation states that the selected EoL option eol_i is not actually executable.

$$eol_i \in EOL\left(s_i^*\right) \cap EOL(s_i)^C \cap EOL_{available} \tag{6}$$

where $EOL(s_i)^C$ is the complement of $EOL(s_i)$. In Fig. 2, the evaluation regarding the executability is made at each EoL process, i.e., reuse, reman, and recycle. The colored elements with a broken circumstance in these EoL processes indicate a member of products with invalid assignment of EoL options.

2.3 *Quantification of Uncertainty*

Uncertainty of the state of urban mines is derived from the diversity of the state of collected products. To quantify this uncertainty, the paper adopts the concept of

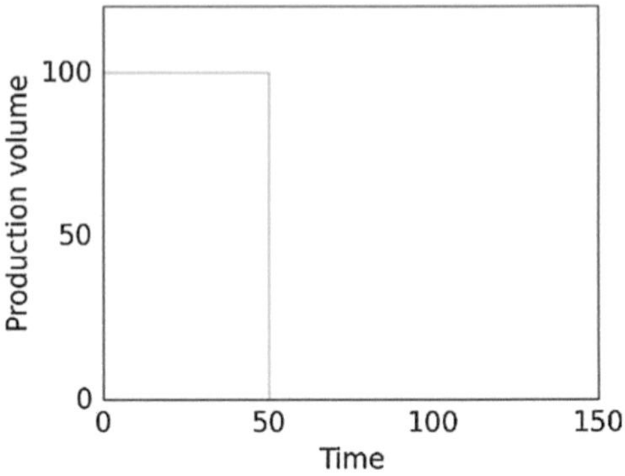

Fig. 2 Production volume as input

entropy in information science. In general, the entropy of a system I with N states is formulated in terms of the probability distribution of the system's states.

$$I = \sum_{i=1}^{N} p_i \log_2 p_i \tag{7}$$

where a distinct state of the system is denoted by i and the probability for the system to be in the state i is denoted by p_i.

The entropy obtained at *assign* process at time t regarding the chosen EoL options is defined similarly. In this case, i denotes one of the available EoL options, and p_i denotes the number of collected products, to which the EoL option i is assigned, over the number of all collected products.

3 Life Cycle Modeling

This section introduces the life cycle model, which is illustrated in Fig. 1. The model consists of the EoL decision process presented in Sect. 2. For the explanation of the model, the model is decided into four stages: production, use, inspection, and end of life (EoL).

In the life cycle model, the state of a product is defined by two parameters: its elapsed lifetime (i.e., age) t_l and the time of production (i.e., the time of birth) t_g. Both variables are related by the actual time in the life cycle t.

$$t_g + t_l = t \tag{8}$$

The elapsed lifetime of products increments until the products are recycled or disposed. When products are used for remanufacturing, the elapsed lifetime of the products is set to 0, and the time of remanufacturing is set to the time of production.

When products are used for reuse, the value of these parameters is not updated. The explanation of these EoL options is shown in Sect. 3.4.

3.1 Production

Figure 2 shows the production volume of products used in the model. Products are placed in the market (i.e., use stage). Although the given production plan terminates at $t = 50$, the reuse and remanufacturing processes supply products to the same market.

3.2 Use

A small portion of products is collected from the use stage. Each product has a parameter called *reliability*, which indicates the probability of the product to be collected during a given time interval (e.g., 0.05 % per month). Although the value of the parameter can be dependent on the state of the product, the study assumes that the parameter is a constant. Furthermore, the study assumes that all products have the same value so that it does not increase the complexity of the model to be analyzed regarding the aspects outside of the study. (This study focuses on the processes *inspect* and *assign*.)

3.3 Inspection

The accuracy of the inspection process is modeled as the deviation between the actual value and inspected value of the time of production (t_g and t_g^*). The inspected value t_g^* is defined as a probabilistic function of the actual value t_g as follows:

$$t_g^* = \max\left(0,\ N\left(t_g,\ \sigma_{\text{ins}}\right)\right) \tag{9}$$

where N is the normal distribution with the average and standard deviation. Furthermore, r is denoted by the sampling rate of the process used in Sect. 5.

3.4 End of Life

As shown in Fig. 1, the EoL options considered in the life cycle model are as follows:

- Reuse: Collected products are used in the market again.
- Reman (remanufacturing): Collected products are disassembled, and parts of the products are reused in the production process.
- Recycle: Collected products are recycled and obtained materials are reused for other usage.
- Dispose: Collected products are disposed.

Furthermore, the EoL option *reserve* is used when the selection of the EoL options is postponed. This option is chosen in practice when the number of collected products is not large enough to exceed the breakeven point of the specific EoL option, or more cost-effective EoL options are going to become available soon.

In the life cycle model, the listed EoL options are assigned to individual products and evaluated periodically with a given interval. The EoL option *reserve* is applied to all collected products between two periods.

The map shown in Fig. 3 defines the desired choice of the EoL options available and applicable as a function of the state of a product, which is formulated below.

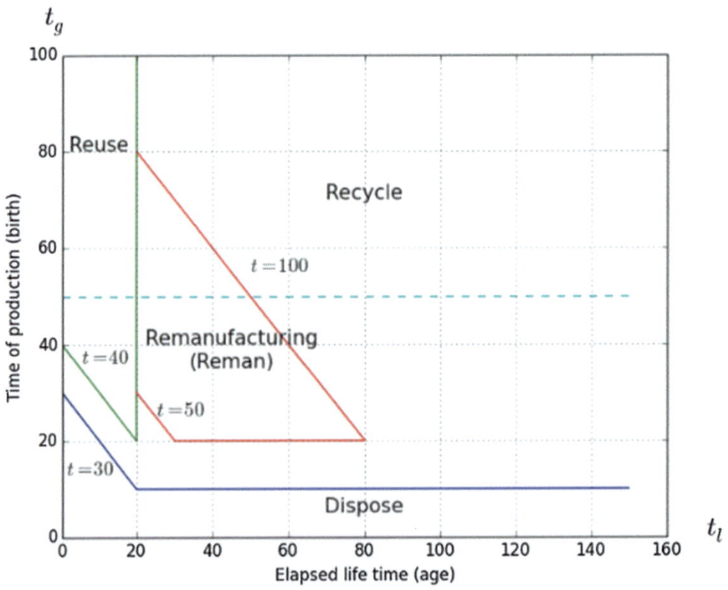

Fig. 3 Mapping from the state of products to EoL options

$$\text{map}(s) = eol \tag{10}$$

Since two variables (t_l and t_g) represent the state of products in the life cycle model, the map is defined by a two-dimensional map, which is divided into four regions, to which an EoL option is assigned.

The map is interpreted as a scenario of the introduction and termination of EoL options with consideration of the state of products. The scenario is described below:

- At $t = 0$: Dispose is the sole EoL option.
- At $t = 30$: Recycle becomes a member of the available EoL options. Products with $t_g^* > 10$ are considered recyclable (i.e., products produced at the earlier time are not recyclable).
- At $t = 40$: Reuse becomes a member of the available EoL options. Products with $t_l^* = t_l - t_g^* < 20$ are considered reusable (i.e., younger products are reusable).
- Between $t = 50$ and $t = 100$: Remanufacturing becomes a member of the available EoL options (i.e., the remanufacturing plant operates within the interval). Products with $t_g^* > 50$ are considered as those suitable to remanufacturing.
- The order of the priority of EoL options is reuse, remanufacturing, recycling, and dispose.

4 Implementation

In order to quantify the uncertainty of the state of an urban mine, whose constituent products have complex life cycles, life cycle simulation (LCS) is useful [7, 8]. With LCS, the behavior of individual products (e.g., physical deterioration and functional obsolescence) manufactured at different timings is computed based on discrete event simulation. In this section, an implementation of LCS is explained along with the construction of the life cycle model shown in Sect. 3.

4.1 Program Architecture

The architecture of LCS is shown in Fig. 4. An LCS program is made of three programs, the main program, a program describing the entities instantiated in the life cycle (entity model), and a program describing the processes executed in the life cycle (process model).

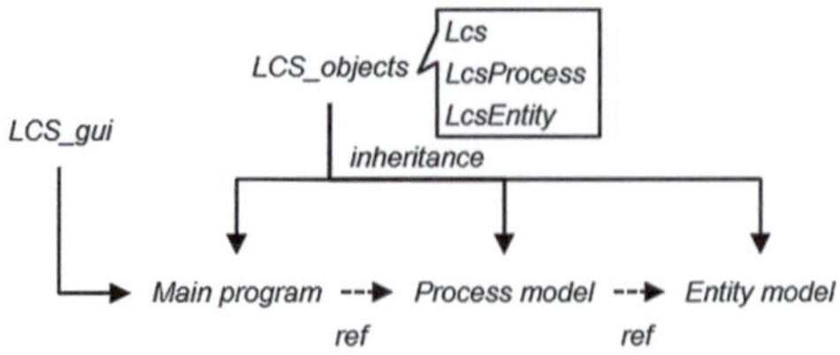

Fig. 4 LCS program architecture

```
def main_program():
    lcs=Lcs('entity.py','process.py')
    lcs.process('Initialization')
    lcs.process('Production')
    lcs.process('Use')
    lcs.process('Inspect')
    lcs.process('Assign')
    lcs.process('CalcEntropy')
    lcs.process('Reserve')
    lcs.process('Reuse')
    lcs.process('Reman')
    lcs.process('Recycle')
    lcs.process('Dispose')
    lcs.process('LifePassing')
    lcs.set_global_parameters({'volume':100,'lambda':0.03,'step':2,'sigma':1,'sample':0.9})
    lcs.run(step=250)
    save_results_as_csvfile(lcs,'ed2015a_res_0_%d_%d.csv'%(1,0.9))
    app=wx.App()
    viewer=LcsResultViewer()
    viewer.show_all(lcs)
    app.MainLoop()
```

Fig. 5 LCS main program

4.2 Main Program

The main program shown in Fig. 5 describes the procedure of a simulation experiment. It first declares the entity model and process model used for LCS (a). It then declares the processes used in LCS (b). The names in the process declaration (i.e., initialization) correspond to the name of (subclass of) LcsProcess class defined in the process model. These processes are executed iteratively following the order of declaration. Then, the global parameters of LCS are defined (c). In (d), LCS is executed with the given number of the iteration (i.e., step). The rest of the procedure (e) includes the visualization and export of the result of LCS.

4.3 Process Model

In the process model, processes are defined as subclasses of LcsProcess. They perform instantiation, state transition, and deletion of entities. Figure 6 shows the definition of the process "Use," which concerns with random selection of products to be collected from the market. The method *execute* specifies the concrete behavior of the process with reference to the entity (Product) and group (cond_Use).

```
class Use(LcsProcess):
    def __init__(self,name,timing):
        LcsProcess.__init__(self,name,timing)
    def execute(self):
        for i in range(self.calc_volume('cond_Use')):
            entity=self.get_entity(i,'cond_Use')
            if random.uniform(0,1)<self.model.get_global_parameter('lambda'):
                entity.set_state('phase',"inspect")
```

Fig. 6 The definition of use process in the process model

```
class Product(LcsEntity):
    def init_states(self):
        self.init_state('t_l',0,True)                    # incrementation in use stage
        self.init_state('t_g',0,True)                    # the year of production (actual)
        self.init_state('t_g_est','unknown',False)       # the year of production (inspected)
        self.init_state('phase','use',False)             # the phase in the life cycle
    def def_scenes(self):
        self.def_scene('cond_Use','phase=="use"')
        self.def_scene('cond_Inspect','phase=="inspect"')
        self.def_scene('cond_Assign','phase=="assign"')
        self.def_scene('cond_Reserve','phase=="reserve"')
        self.def_scene('cond_Reuse','phase=="reuse"')
        self.def_scene('cond_Reman','phase=="reman"')
        self.def_scene('cond_Recycle','phase=="recycle"')
        self.def_scene('cond_Dispose','phase=="dispose"')
        self.def_scene('cond_End','phase=="end"')
        self.def_scene('cond_End_Dispose','phase=="end" and eol=="dispose"')
```

Fig. 7 The definition of use process in the process model

As object-oriented programs support, the LCS allows the inheritance of the description of method and attributes. Thus, the size and complexity of the definition of each process is manageable. For instance, the process model prepared in the paper consists of 12 processes. The length of each process definition is as long as 20 lines in average.

4.4 *Entity Model*

Figure 7 shows a part of the definition of entity P1, which is the class description of all products in the life cycle. It defines the attribute and its initial value in the method *init_state*. As explained above, t_l, t_g, and t_g^* are the major parameters whose value characterizes the (actual and estimated) state of products. The parameters store temporal information or the history of state transition. The method *def_scenes* consists of a list of def_scene methods, which define intrinsic description of the state of products for grouping them. For instance, the location of individual products, which is within one of the circles in Fig. 1, is defined by the value of attribute *phase*.

5 Analysis

5.1 The Behavior of the Life Cycle Model

The behavior of the above life cycle model was computed with LCS. The sampling rate *r* and the standard deviation of the accuracy of the inspect process σ_{ins} are set to 90 % and one time step, respectively.

Figure 8 shows the number of executed EoL options in the life cycle model (with reference to the cumulative number of production, which is the integration of the production volume shown in Fig. 2). The horizontal axis is the progress of simulation time step. The vertical axis is the number of the occurrences of the execution of EoL options.

The validity of the life cycle model is confirmed with the simulation result. EoL options like *recycle*, *reman*, and *reuse* are introduced as indicated in the scenario shown in Fig. 3. Figure 9 shows the entropy of *assign* process with respect to the simulation. The entropy is 0 until the member of $\mathrm{EOL}_{\mathrm{available}}$ is *dispose*, when the inspection process is not meaningful. It increases step-by-step at $t = 30$, 40, and 50, when *recycle*, *reuse*, and *reman* are introduced, respectively. The drop of the entropy occurred around $t = 70$ is explained by the fact that the number of the occurrences of *reuse* of collected products has suddenly decreased.

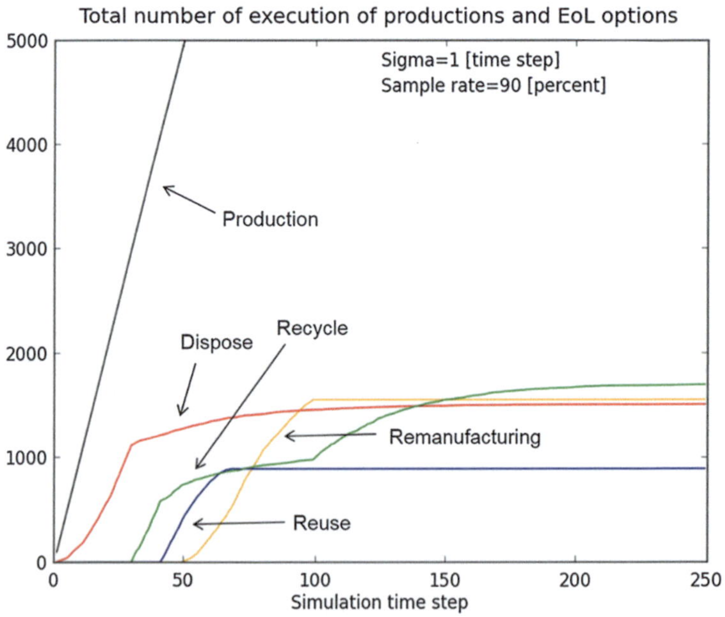

Fig. 8 The number of executed EoL options

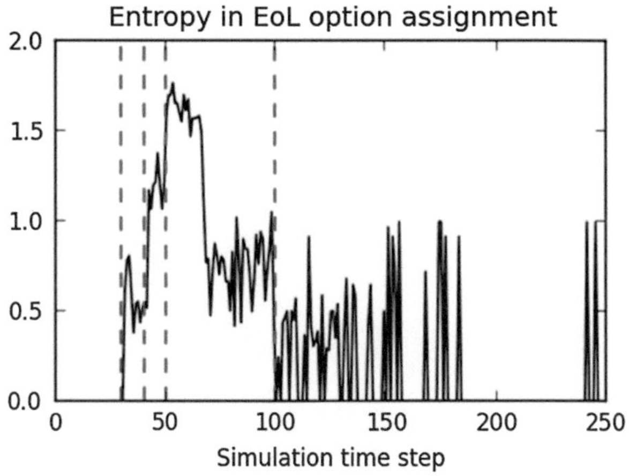

Fig. 9 Entropy of the process to select EoL options

5.2 *Comparative Analysis*

In order to analyze the relation between the variation of uncertainty in the inspection process and the number of the occurrences of EoL options, 1,000 simulations with a variety of input parameters were carried out. Due to limitation of computational resource, the production volume is 10 % of the original life cycle model. Input of each simulation is the standard deviation of the accuracy of the inspect process σ_{ins}, and the sampling rate r follows the normal distribution.

$$\sigma_{\text{ins}} = N(3, 1) \tag{11}$$

$$r = N(50, 15) \tag{12}$$

The output is the number of occurrences of EoL options *reuse*, *reman*, *recycle*, and *dispose* with a variety on the value of input parameters. Figure 10 summarizes the relation between two parameters out of the input and output parameters (i.e., six parameters in total).

In Fig. 10, the correlation coefficient c and a line indicating the linear regression are shown. The plots with a circle show that the input parameters are independent to each other (i.e., $c = 0$). The sampling rate of the inspection process and the number of occurrences of *reuse* and *reman* are positively correlated ($c = 0.78$ and 0.81, respectively). However, the standard deviation of the accuracy of the inspection process and the number of occurrences of all EoL options are not correlated ($-0.17 < c < 0.01$). The number of the occurrences of *reuse* and *reman* is not strongly correlated with that of the occurrences of *recycling* and *dispose*, because *reuse* and *reman* simultaneously decrease the occurrence of *recycling* and *dispose*, but reused or remanufactured products are either recycled or disposed in the long term.

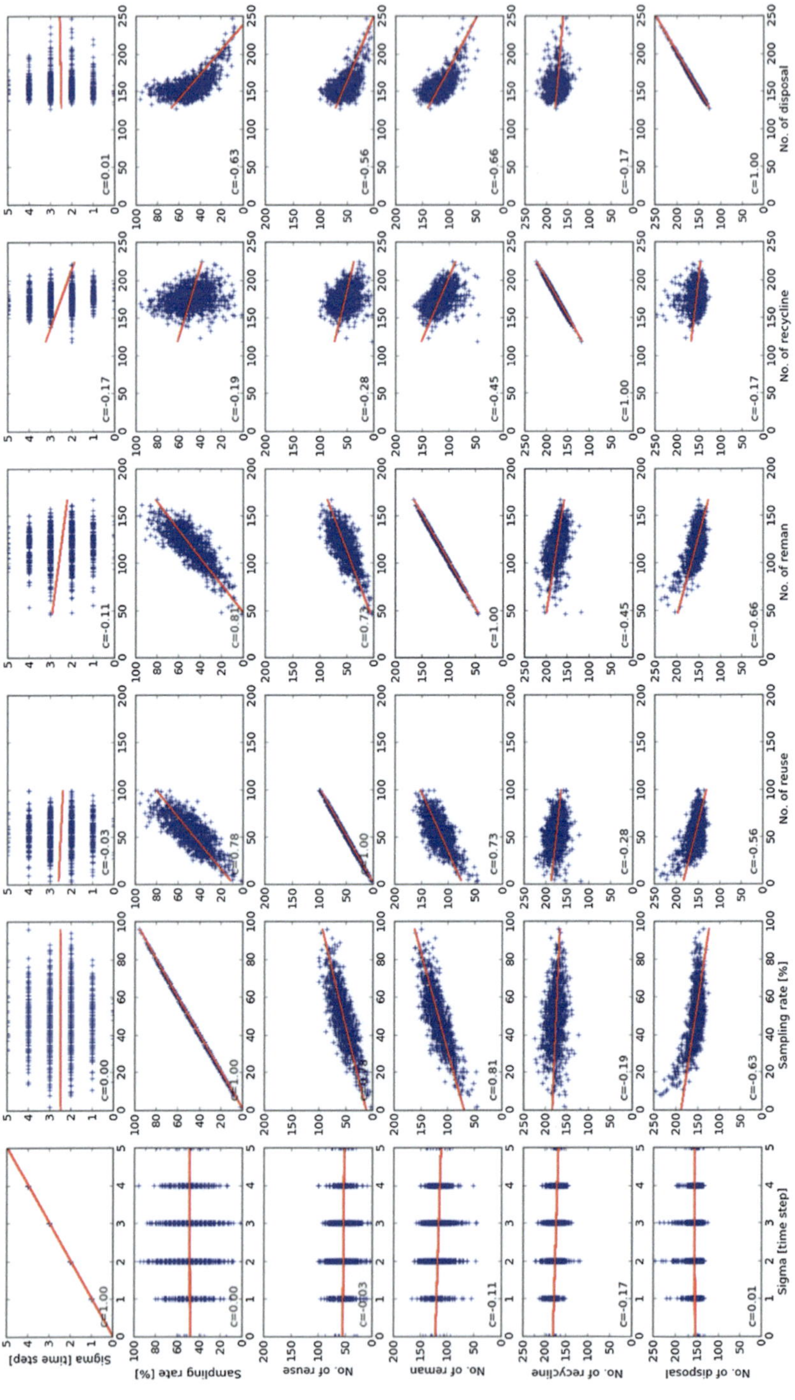

Fig. 10 Entropy of the process to select EoL options

6 Summary

Strategic development of the life cycle of electric and electrical products is critical for the recycling of precious and rare metals as well as the reuse and remanufacturing of components. In order to quantify and decrease the uncertainty in the EoL stage of these products, the paper has modeled and analyzed a life cycle model, which focuses on the inspection process and the assignment process, with life cycle simulation. It found that the simulation result identifies specific time periods, in which the inspection process is meaningful, and that the sampling rate positively correlates the number of the occurrences of reuse and remanufacturing processes.

References

1. Ministry of the Environment (2013) Act on promotion of recycling of small waste electrical and electronic equipment, enforced on April 2013
2. Chancerel P, Meskers CEM, Hagelueken C, Rotter VS (2009) Assessment of precious metal flows during preprocessing of waste electrical and electric equipment. J Ind Ecol 13(5):791–810
3. Binnemans K, Jones PT, Blanpain B, Gerven TV, Yang Y, Walton A, Buchert M (2013) Recycling of rare earths: a critical review. J Clean Prod 51:1–22
4. Nanjo M (1988) Urban mine, new resources for the year 2000 and beyond. Bulletin of the research institute of mineral dressing and metallurgy, Tohoku University 43(2):239–251
5. Wang F, Huisman J, Stevels A, Balde CP (2013) Enhancing e-waste estimates: improving data quality by multivariate. Waste Manag 33:2397–2407
6. Komoto H, Kondo S, Masui K (2013) Life cycle modeling and simulation of consumer products in context of the development of urban mines, In: Proceedings of EcoDesign 2013: 8th international symposium on environmentally conscious design and inverse manufacturing, Jeju, Korea, 4–6 Dec 2013
7. Umeda Y, Nonomura A, Tomiyama T (2000) A study on life-cycle design for the post mass production paradigm. Artif Intell Eng Des Anal Manuf 14(2):149–161
8. Komoto H, Tomiyama T (2008) Integration of a service CAD and a life cycle simulator. CIRP Ann Manuf Technol 57(1):9–12

The Potential of Additive Manufacturing Technology for Realizing a Sustainable Society

Shinsuke Kondoh, Toshitake Tateno, Yusuke Kishita, Hitoshi Komoto, and Shinichi Fukushige

Abstract Today, additive manufacturing (AM), which refers to a process by which digital design data is used to build up artifacts by decomposing material, is gaining growing interest from industry. The AM's capability for producing complex structure in extremely small lot size can enable more optimal design for today's manufacturing products. Through such optimal design of each product, energy and material consumption of society can be significantly reduced. As AM can produce a wide variety of components in one-by-one production, the total number of the products (and components) can be significantly reduced. In addition, the products made by AM can be optimally designed and manufactured for each particular purpose. This implies these products have no unused functions that may consume additional energy and materials. The objective of the paper is to propose the method for evaluating AM's potential for reducing environmental impact of society considering these factors caused by introducing AM technology into industry.

Keywords Additive manufacturing • Environmental load • Life cycle value • Life cycle simulation • Scenario analysis

1 Introduction

Today, additive manufacturing (AM), which refers to a process by which digital design data is used to build up artifacts by decomposing material, is gaining growing interest from industry. The AM's capability for producing complex

S. Kondoh (✉) • Y. Kishita • H. Komoto
National Institute of Advanced Industrial Science and Technology (AIST), Tsukuba, Japan
e-mail: kondou-shinsuke@aist.go.jp

T. Tateno
Meiji University, Ikuta, Japan

S. Fukushige
Osaka University, Osaka, Japan

M. Matsumoto et al. (eds.), *Sustainability Through Innovation in Product Life Cycle Design*, EcoProduction, DOI 10.1007/978-981-10-0471-1_32

structure in extremely small lot size can enable more optimal design for today's manufacturing products. Through such optimal design of each product, energy and material consumption of society can be significantly reduced. As AM can produce a wide variety of components in one-by-one production, the total number of the products (and components) can be significantly reduced. In addition, the products made by AM can be optimally designed and manufactured for each particular purpose. This implies these products have no unused functions that may consume additional energy and materials. Furthermore, AM also enables lightweight (but complex) design of components which may contribute significant reduction in energy consumption of products.

Although some researchers evaluated the environmental load of additive manufacturing process itself and compared it with that of conventional manufacturing processes [e.g., 1–4], little research focused on the possibility of environmental load reduction and increased product value through such optimal design of products. The objective of the paper is to propose the method for evaluating AM's potential for reducing environmental load of society considering these factors caused by introducing AM technology into industry. This paper is organized as follows. Section 2 describes the potential environmental load reduction through introduction of AM technologies. Section 3 proposes our approach to evaluate AM's potentials for realizing sustainable society taking into account its positive effect on product and service values as well as its impact for reducing environmental load. A simplified calculation example is also provided to explain our method. Section 4 summarizes the paper.

2 Impact of Additive Manufacturing

According to the Society of Manufacturing Engineers, AM is defined as the process of manufacturing a physical object through layer-by-layer selective fusion, sintering, or polymerization of a material [5]. There exist several types of AM, which differ mainly in terms of the materials they use and the processes used for creating slices of materials. Some AM, such as electron beam melting and selective laser sintering, uses higher-power laser to melt and sinter powder-based materials, and other AM, such as multi-jet modeling, use ink-jet-type nozzle to deposit molten materials and then to harden. Although AM processes vary like this way, they share the same concept and the same characteristics as discussed in following sections.

2.1 Impact on Production Stage

As AM adds material particle by particle until a final product is achieved, it theoretically consumes lesser raw materials than traditional subtracting manufacturing processes, where material is removed from an oversized block (or sheet). In addition, AM uses less complementally materials than conventional ones such as injection molding, die casting, and CNC cutting which require molds, fixtures, and tools. Thus, the production process itself can become cleaner by introducing AM. AM also produces minimal waste stream while reaching satisfactory shape.

As AM requires no molds nor tools that are specific to a given design of product, it enables customized components without incurring additional cost. Every component can be designed and made completely different from the others and thus the product can be customized to each individual user by using AM. Such customized products may require lesser raw materials and energy than mass-produced products because the latter must implement more functions (i.e., rarely used functions for most users that someone might use) than the former. The ability of customization may also contribute to the desirability and thus the longevity of these products, which also leads less consumption of raw materials. Such ability of customization and decreased lead time caused by AM may also enable order-to-made production, which results in significant reduction in production volume of society.

2.2 Impact on Usage Stage

Because most of the design constraints, which are generally issued by design for manufacturing and design for assembly (DFM/DFA) principles, are eliminated by introducing AM, it can make very complicated structure which cannot be achieved by conventional manufacturing processes. Thus, AM enables optimal design for products. For example, AM is used to improve the functionality of certain components such as those related to liquid and gas flow, many of which are manufactured far from optimal today. They are constructed with a combination of simple shapes such as planes and circular holes which are easy to produce by conventional manufacturing processes. For some transport-related products (e.g., vehicles, airplanes, etc.) and fluid-related equipment (e.g., gas turbines of power plants, pumps, etc.), the optimized design as a result of AM could lead to significant improvement in their energy efficiency. Lightweight design of components with high strength to weight ratios (e.g., honeycomb structure), which will also become possible by introducing AM, might result in significant reduction in fuel consumption during product usage as well.

AM also enhances upgradability of products, which is effective for prolonging product useful life. Since AM can produce customized components one by one in very short lead time, manufactures can easily redesign and provide components when they are worn out or become obsolete if the products are adequately designed taking into account AM's potential for product upgrade.

2.3 Impact on Supply Chain

AM also enables sustainable supply chain through compression of manufacturing processes that are necessary for obtaining final products. Conventional manufacturing chains consist of different types of technologies such as die casting, CNC cutting, heating, and so forth. The facility corresponding to each technology is generally distributed in wide area. As a result, total traveling distance over whole supply chains tends to be long. The number of work-in-process (WIP) products is also very large in general. As AM can create a variety of shapes by itself, many processes that are necessary in conventional manufacturing chains can be eliminated to shorten total traveling distance over whole supply chain of products. As a result of compression of supply chain, the number of WIP products can be significantly reduced as well.

By using AM, manufacturers can produce a component at anytime and anywhere they want without incurring considerable cost; there is no need for stockpiling components that will be required for repairing their products in future. The reduction in part stock also contributes to material saving.

2.4 Impact from Whole life cycle Perspective

AM is also effective for remanufacturing of worn out components by depositing material onto a surface of worn out component (e.g., turbine blade). The useful life of components can also be extended by using AM.

Figure 1 depicts schematic flow of material over whole product life cycle. Products, which are produced from raw materials, are used and then collected from users after their useful life. Some are repaired and directly reused in the market again and others are disassembled into components for reuse or recycling. Some components that have sufficient quality for building products are reused as components and others with insufficient quality are recycled to recover material or energy. Note that product reuse is generally better than component reuse and component reuse is generally better than material recycling from an environmental viewpoint. In short, the smaller the loop materials go through, the lesser their associated environmental load become in Fig. 1.

As described above, AM contributes to the reduction of environmental load at each life cycle stage in Fig. 1. Moreover, AM can reduce environmental load through whole product life cycle by making smaller loops of material circulation, because it improves the capability for product and component reuse as well as extending product useful life. AM's potential for reducing environmental load is thus much greater than aggregation of those at each life cycle stage.

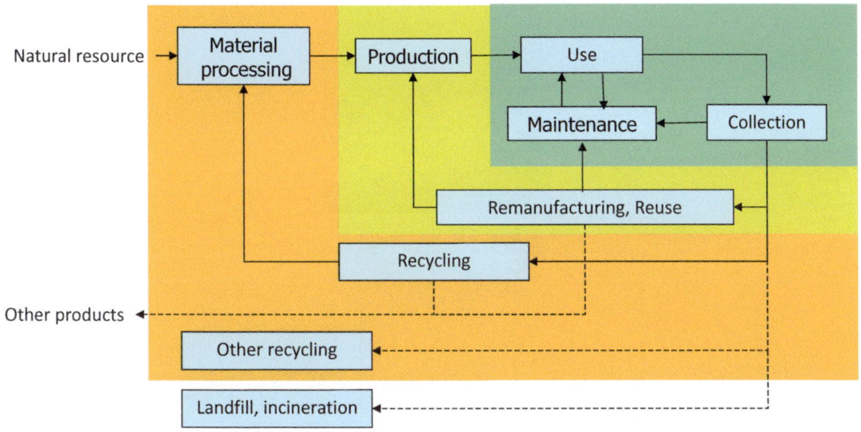

Fig. 1 Material circulation via remanufacturing, reuse, and recycling

3 Method for Evaluating the Impact of AM

3.1 Approach

3.1.1 Life Cycle Value

Since AM changes the way of product provision, it is not sufficient to evaluate its environmental load alone. In order to compare the effectiveness and efficiency of product and service provision before and after introduction of AM technology, the value improved by AM should also be compared with the one without AM.

As AM may improve product useful life, production lead time, and so forth, the value should be evaluated considering time factor. To solve this problem, this paper employs the concept of life cycle value, which is proposed by [6]. Detail formulation of life cycle value will be given in Sect. 3.2.

3.1.2 Scenario Analysis

AM's impact for environmental load reduction should be evaluated considering the transition in material flow: from *larger material circulation loops to smaller ones* as described in Sect. 2.4. Such change in material flow cannot be wholly understood by focusing on environmental load alone, because the behavior of each stakeholder (e.g., manufacturer, retailer, user, recycler, second-hand user, and so forth), which determines the amount of materials associated with each flow in Fig. 1, is driven by economic, societal, and ethical factors as well as environmental ones. The flow of each item in Fig. 1 is thus regarded as a complex function of cost, benefit, interests, and environmental consciousness of each stakeholder. However, it is not feasible to consider every factor at the same time to estimate material flow invoked by

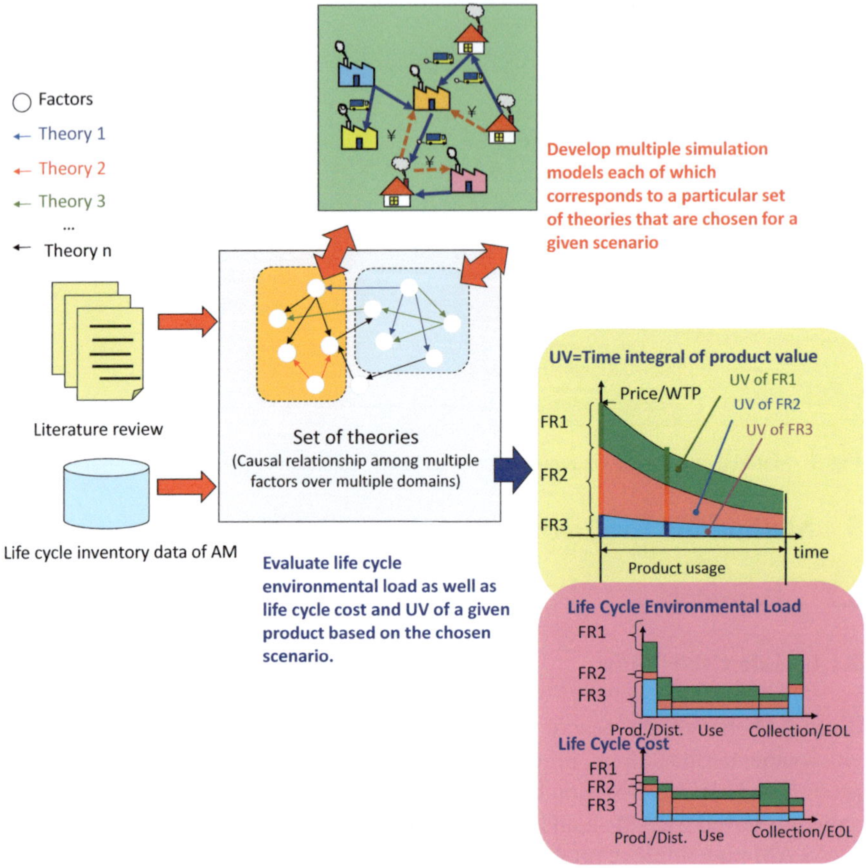

Fig. 2 Approach for evaluating the AM's impact on society

introduction of AM. A systematic and step-by-step refinement of calculation model is indispensable for solving this problem. To this end, scenario analysis [7] and discrete event simulation method [8] are employed in this study.

Figure 2 shows the approach for evaluating the AM's impact on society. The authors first (i) list up every factor that may affect the behavior of each stakeholder in addition with possible causal interrelationships among them and (ii) identify a set of theories, against which a component of simulation model is implemented. Then, the authors (iii) develop a set of scenarios that are robust against prediction uncertainties in each simulation model and construct corresponding simulation models to calculate life cycle value as well as life cycle environmental load and cost over whole society.

The procedure is summarized as follows:

Step 1: Identification of every environmental, economic, and societal factor that is affected by the introduction of AM

Step 2: Explicit representation of interrelationship among all the factors identified in step 1
Through literature review, interrelationships among all factors are comprehensively thought out to form a set of theories, each of which explicitly represents causal relationship among particular factors among them.

Step 3: Formulation of a set of scenarios that is robust against prediction uncertainties
A set of theory that is used for evaluating environmental impact of AM is selected to form a scenario to be analyzed in step 4. Based on the reliability of each theory, a set of scenarios is formed so that the result is robust against uncertainties in each theory.

Step 4: Development of calculation model for evaluating the impact of AM
Based on a set of theories that is included in each scenario, the calculation model for evaluating AM's impact on society is developed by using life cycle simulation methods [8]. Life cycle simulation model represents each individual stakeholder's behavior as well as that of each individual component of a given product based on discrete event simulation model. Life cycle simulation model first interprets each interrelationship among factors chosen as a basis for each scenario as that among parameters of each stakeholder and each component. Then, it simulates behaviors of every stakeholder as well as those of every component to calculate total environmental load, cost, and life cycle value produced in each scenario.

3.2 *Life Cycle Value*

Life cycle value (LCV) is defined as time integral of product value during its useful life. Figure 3 shows the concept of LCV.

In general, the LCV of a product increases as the product's functional performance and the length of its continued use increase. Thus, the LCV of a product is defined as time integral of product value, assuming that product value is strongly correlated with its functional performance.

LCV is given as follows:

$$\mathrm{LCV} = \sum_{k} LCV_k \tag{1}$$

$$\mathrm{LCV}_k = \int_{st_k}^{tt_k} \mathrm{V(t)dt} \tag{2}$$

where k, st_k, tt_k, $V(t)$, LCV, and LCV_k denote the index for life cycle stage, starting time, and termination time at life cycle stage k, product value at time t, LCV over all life cycle stages, and LCV at life cycle stage k, respectively.

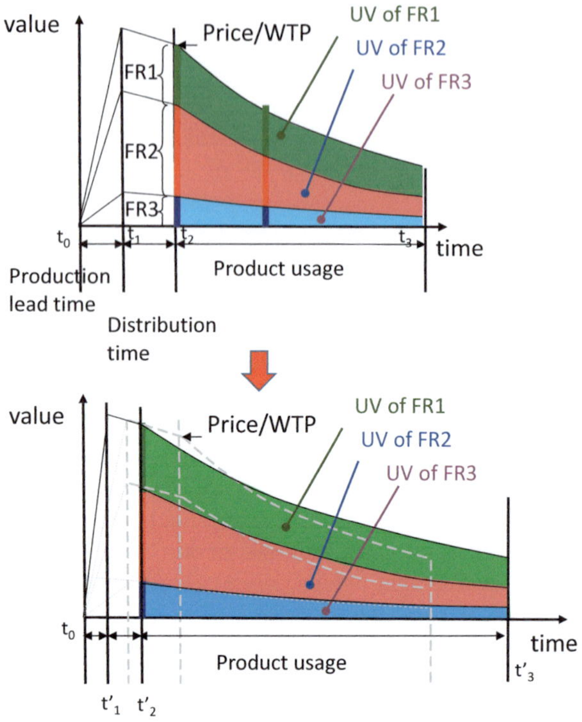

Fig. 3 AM's impact on life cycle value

The use value (UV) of a product, which is given by LCVs over all product use stages (the product life cycle can include multiple use stages), is especially introduced as follows to evaluate LCV during its useful life:

$$UV = \sum_{k \in S_u} LCV_k \tag{3}$$

where S_u denotes a set of all product use stages.

Based on multi-attribute utility theory (MAUT) [9], product value at time t can be allocated to its dominant functional requirements (FRs) given as follows:

$$V(t) = \sum_i V_i(t) \tag{4}$$

$$V_i(t) = w_i(t) FR_i(t) \tag{5}$$

where i, $V_i(t)$, $w_i(t)$, and $FR_i(t)$ denote index of FRs, product value allocated to FRi, weighted factor for FRi, and functional performance of FRi at time t, respectively.

Weighted factor for each FR represents its importance to the customers. In this study, we assume that a product value is measured by its market price or customer's willingness to pay (WTP). Therefore, importance of each FR can be estimated by conjoint analysis [10] of various products with different specification.

A product value deteriorates by following two causes: namely, (i) *physical causes* and (ii) *value causes* [11]. Physical causes include failure and degradation of product due to aging and wear. Value causes include obsolescence of FRs (including aesthetic quality) of a product. Since the value of a product is given as weighed sum of its functional performance, value deterioration along time is given by decreases in $FR_i(t)$ and $w_i(t)$.

For the sake of simplicity, we express deterioration of $FR_i(t)$ and $w_i(t)$ as linear equations as follows:

$$FR_i(t) = \begin{cases} c_i t + d_i, \ (st_k \leq t \leq tt_k, k \in S_u) \\ FR(st_k), \ (st_k \leq t \leq tt_k, k \in S_p \cup S_d) \\ 0, \ (st_k \leq t \leq tt_k, k \in S_c \cup S_e) \end{cases} \tag{6}$$

$$w_i(t) = a_i t + b_i \tag{7}$$

where c_i, d_i, a_i, b_i, S_u, S_p, S_d, S_c, and S_e denote deterioration rate, initial performance, obsolescence rate, initial importance of FRi, and the set of all life cycle stages of product usage, production, distribution, collection, and end-of-life treatment stages, respectively.

c_i and d_i are estimated by empirical data of deterioration of similar products at their use stage by applying reliability theory. a_i and b_i can be estimated by regression analysis on importance of each FR at various time t.

As described in previous section, AM can decrease production lead time, shorten supply chain, and improve product durability by encouraging repair and remanufacturing. Thus, AM's impact on UV is schematically illustrated as shown in Fig. 3. The upper part and the lower part of the figure depict LCVs of a product before and after introducing AM, respectively. Time t_0, t_1, t_2, and t_3 in the figure denote the start of production, distribution, usage, and termination of usage stage, respectively. By introducing AM, t_1 and t_2 changes to t_1' and t_2' as shown in the lower part of the figure. Shorter production and distribution time contribute to the improvement of product value as well as extension of product useful life because the value of a product deteriorates as it becomes obsolete. The possible improvement in product deterioration rate (and obsolescence rate) by AM technologies also leads the greater UV than in the case without these technologies.

Let's consider a product with one functionality as an example for evaluating the impact of AM on product life cycle value. Assuming the values of a_i, b_i, c_i, d_i, t_1, and t_2 are given as first row in Table 1, the product useful life is calculated as 50 [months] assuming that the product is thrown away when its associated value becomes zero. Using the value for useful life, UV of the product is calculated as 183,333 [yen*months]. Let us assume that AM changes t_1, t_2, and c_i as given in the second row in Table 1, and the useful life and UV of the product is calculated as 76 [months] and 313,344 [yen*months], respectively. Note that the product value at the beginning of the product usage and the product useful time becomes greater by

Table 1 Calculation example of UV before and after introduction of AM

		Start of production	Start of distribution	Start of usage	End of usage								
		t0	t1	t2	t3	ai	bi	ci	di	Product value at t0	Product value at t2	Useful life	UV
Row no.	Unit	[Month]	[Month]	[Month]	[Month]	[yen/month]	[Yen]			[Yen]	[Yen]	[Months]	[yen*month]
1	Without AM	0	5	10	60	−1	100	−2	100	10,000	9000	50	183,333
2	With AM	0	2	4	100	−1	100	−1	100	10,000	9600	96	313,344

introducing AM. The result shows that the slight improvement in deterioration rate and production and distribution lead time may significantly improve UV of the product.

4 Summary

The objective of the paper is to propose the method for evaluating AM's potential for reducing environmental load of society. The paper first discusses the possible outcome of introduction of AM from the environmental aspect. Then, the paper proposes the method for evaluating AM's impact on society considering the multiple interrelationships among factors that represents every stakeholder's behavior driven by societal, economic, and ethical reason. To this end, the concept of life cycle value and scenario analysis are employed in the paper. The paper also illustrates how AM affects the life cycle value of products with a simplified calculation example. Future work includes collection of life cycle inventory data of AM technologies as well as collection of theories that are represented as causal relationships among multiple factors. Development of standardized model building process for a given set of theories, which is chosen as a basis for a scenario, is also tackled in future work with a practical case study.

Acknowledgments This research is partially financially supported by Grant-in-Aid for Scientific Research (No.15K00715), JSPS, Japan.

References

1. Diegel O, Singamneni S, Reay S, Withell A (2010) Tools for sustainable product design: additive manufacturing. J Sustain Dev 2(3):68–75
2. Huang SH, Liu P, Mokasdar A (2013) Additive manufacturing and its societal impact: a literature review. Int J Adv Manuf Technol 67:1191–1203
3. Bourhis FL, Kerbrat O, Dembinski L, Hascoet JY, Mognol P (2014) Predictive model for environmental assessment in additive manufacturing process. Procedia CIRP 15:16–31
4. Faludi J, Bayley C, Bhogal S, Iribarne M (2014) Comparing environmental impacts of additive manufacturing vs. traditional machining via life-cycle assessment. Rapid Prototyp J. Laboratory for Manufacturing and Sustainability, UC Berkeley. Retrieved from: http://escholarship.org/uc/item/0gv882qk
5. SME (2010) Available from http://www.sme.org/cgi-bin/communities.pl?/communities/techgroups/ddm/what_is_ddm.htm&&&SME
6. Kondoh S, Masui K, Hattori M, Mishima N (2008) Total performance analysis of product life cycle considering the deterioration and obsolescence of product value. Int J Prod Dev 6 (3/4):334–352
7. Umeda Y et al (2009) Proposal of sustainable society scenario simulator. CIRP J Manuf Sci Technol 1(4):272–278

8. Komoto H, Tomiyama T, Silvester S, Brezer H (2006) Life cycle simulation of products in a competitive market. In: Proceedings of the 13th CIRP international conference on life cycle engineering, Leuven, pp 233–237
9. Keeney RL, Raiffa H (1976) Decisions with multiple objectives: preferences and value tradeoffs. Wiley, New York
10. Green PE, Srinivasan V (1978) Conjoint analysis in consumer research: issues and outlook. J Mark Res XV:132–136
11. Daimon T, Kondoh S, Umeda Y (2004) Proposal of decision support method for life cycle strategy by estimating value and physical lifetimes. In: Proceedings of the 11th international CIRP life cycle engineering seminar, Belgrade

Biodegradable Mechatronic Products by Additive Manufacturing

Toshitake Tateno, Yuta Yaguchi, and Shinsuke Kondoh

Abstract Mechatronic products are necessary for our life and industrial factories. Recently, most mechatronic products have self-monitoring function, which is based on the sensor data. This function is useful not only for efficient maintenance but also efficient product recovery. However, additional sensing and communication devices lead to increase of environmental costs by themselves. Therefore, eco-design of mechatronic products including these devices is expected. In this paper, the use of additive manufacturing (AM) with biodegradable materials is proposed. First, the realization type of the product is defined as the proposition of this paper. Then, characteristics of AM and biodegradable materials for contributing to reduce environmental costs are analyzed. Additive manufacturing (AM), which is usually called as 3D printing, has naturally advantage in eco-design. Biodegradable material is also known to reduce energy for product disposal processes. However, it is shown that the combinational use of them promotes the effectiveness. Next, the comparison between the conventional product and the present product of the production and disposal processes is discussed by using an electronic print board. The result shows that the production process becomes remarkably simple and the recovery process becomes efficient. Finally, a realization example is introduced. An RFID antenna module fabricated with PLA and an original biodegradable electric conductive gel is shown. This module is confirmed to work certainly by connecting with an IC chip as a mass-production module. The realization of the proposed product is discussed.

Keywords Biodegradable materials • Additive manufacturing • Printed electronics • Eco-mechatronic products

T. Tateno (✉)
Department of Mechanical Engineering Informatics, Meiji University, Kanagawa, Japan
e-mail: tateno@meiji.ac.jp

Y. Yaguchi
Department of Engineering Design, Tamagawa University, Kanagawa, Japan

S. Kondoh
National Institute of Advanced Industrial Science and Technology (AIST), Tsukuba, Japan

M. Matsumoto et al. (eds.), *Sustainability Through Innovation in Product Life Cycle Design*, EcoProduction, DOI 10.1007/978-981-10-0471-1_33

1 Introduction

Mechatronic products are increasing their complexity for adding new services especially on Internet applications. IoT (Internet of Things) [1] is one of recent topics of the Internet applications. It aims at user services based on a large number of sensor data, which are gathered by sensors attached to target objects. It has been developed with small sensors and small communication devices through the Internet. In the field of eco-design, a lot of applications can be considered, e.g. health monitoring of products, remote maintenance, reuse matching, transportation planning, and so on. Especially, the recovery supports of mechatronic systems are good targets of these applications. The authors [2] have indicated the importance of these applications and have shown the system architecture and a realized prototype system.

While these applications have a lot of merits for reduction of environmental costs, there are some demerits. One is the fabrication and disposal processes of the devices generate environmental costs. This is a trade-off between the effect of applications and the additional costs for production of the devices. Therefore, environmental conscious design of the devices should be considered.

As one innovative technology, additive manufacturing (AM) is focused. AM has much possibilities to change product design and design process because it uses different principles of material processing from subtraction processing, which has been mainly used in current production. For example, AM enables to fabricate customized parts without molds and dies. This direct manufacturing of real parts from 3D models naturally has advantage to make products with less energy. Moreover, some biodegradable materials, such as polylactic acid (PLA), can be used. Biodegradable electric conductive materials are under development, but some of them are available to use.

In this paper, the sensing and communication modules for mechatronic products, which are fabricated by AM with biodegradable materials, are proposed. After the previous works are introduced, the expected effects and the importance of additive manufacturing on reduction of environmental costs are explained. As a realization example, a radio frequency identification (RFID) tag fabricated by AM with biodegradable materials was introduced.

2 Proposition of Eco-Mechatronic Products

2.1 Eco-Mechatronic Products

We assume mechatronic products, which satisfy the following three definitions. We name the products eco-mechatronic products. The products having one of the definitions separately may be realized easily. However, the satisfaction of all definitions promotes the effectiveness of environmental cost reduction by the device.

Definition 1
The product has a part or a module, which has functions to sense the condition of the product and to transmit the data through the Internet.

Definition 2
The part or module, which has a part of functions written in definition 1, is made with biodegradable materials.

Definition 3
The part or module, which is made with biodegradable materials, is fabricated by AM at a time.

2.2 *Type of the Eco-Mechatronic Products*

Practically, the eco-mechatronic products will consist of some printed modules and some mass-production modules. The printed module is assumed to be fabricated by AM. The mass-production module is assumed to be manufactured with conventional processing. Two structure types are considerable as shown in Fig. 1.

Type 1
All functions on definition 1 are realized in one printed module. This is the ideal type. Unfortunately, this type module cannot be realized with current technology because the IC chips could not be realized with biodegradable materials. However, it is important to recognize that this type is ideally perfect state.

Type 2
A part of functions on definition 1 is realized in the printed modules, and other modules are realized in the mass-production modules. This type is virtually the proposition of this paper.

2.3 *Considerable Use of Eco-Mechatronic Products*

In this section, two considerable uses and their expected effects are introduced.

2.3.1 Expendable Parts with Sensors

One considerable use of the eco-mechatronic products is expendable parts attached with sensors. A printer toner cartridge, which has a sensor for measuring the amount of toner, is an example.

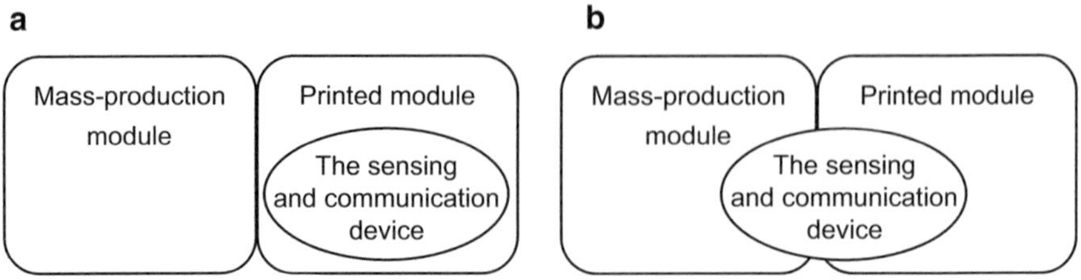

Fig. 1 Relation between the sensing and communication device and the mass-production and printed module on the definition of eco-mechatronic product types (**a**) Type 1 (**b**) Type 2

Most parts required for condition monitoring are expendable parts, which are replaced or maintained periodically. If we consider to add the sensing and communication functions to these parts, it is a novel way to build them as disposable parts, which are designed to be very low environmental costs in disposal process. Using biodegradable material for the expendable parts is reasonable because the entire part can be processed together without disassembling at the end of life.

As well as the disposal process, in the production process, AM has advantage to fabricate complex parts in one process without assembling [3]. If the expendable parts can be fabricated with the AM materials, the sensing and communication devices can be embedded in the parts. It can also make the parts small and simple.

From the logistic point of view, AM can reduce the amount of stock for spare parts [4]. Usually, a large number of them are prepared and stocked in distributed places because parts replacing operations are happened randomly in time and place. Since AM can fabricate directly from 3D model data, the parts can be provided anytime and anywhere.

From these points, it is reasonable to fabricate sensing and communication device in or on the expendable parts by AM with biodegradable materials.

2.3.2 Personal Fabrication of Mechatronic Systems

As one topic of AM, product fabrication by persons is focused as a production revolution [5], which is a change from the production by large companies to the production by persons. While the personal fabrication is currently limited in simple products such as accessories, it is said that the market will increase and the design will be extended to complex products in the near future. Especially, mechatronic products by AM are expected to bring a lot of values because complex mechanical products work with mechatronic systems. For example, if mechatronic products can be built directly with only an AM device without special assembling work, it enables users to get a preferable mechatronic product soon. However, there are some problems in the personal fabrication. One is the lack of design knowledge, which includes environment-conscious design. While personal AM machines enable persons to design and fabricate their original products, it is not sure that the design is well structured for disassembling and disposal processes. The appropriate disposing process cannot be guaranteed in the personal fabrication. In such case, the biodegradable material provides fail-safe function for environmental emission in disposing process.

3 Technologies and Their Contribution to the Environment

3.1 Sensing and Communication Devices

The small sensing and communication technology can contribute some services for reduction of environmental costs of products. The authors [2] have indicated the importance of using these technologies in maintenance, reuse, and recycle processes. Figure 2 arranges the effective use of the data. In maintenance field, the sensing data can be used for condition-based maintenance and maintenance planning. It can make a product life longer by reducing malfunction. In reuse process, the data are useful for searching a reuse target, in which a part extracted from an old product is used as a reuse part in a new product, because the current conditions and historical data of part are important to determine the remaining lifetime.

As well as the sensing and communication devices, the communication protocol will be standardized. The standardization promotes the use of the system and reduces redundancy of the system. The efficient system leads to low energy consumption especially in logistics.

3.2 Biodegradable Materials

Biodegradable materials naturally have advantage in reducing environmental costs because they don't need additional artificial processing without landfilling. This advantage will be enlarged when the entire product or module is made with biodegradable materials perfectly because disassembly processes and special treatment for landfilling are reduced in disposal process. Additionally, the transportation

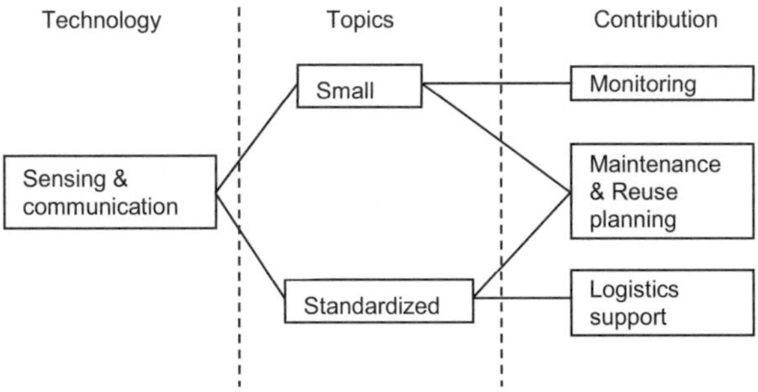

Fig. 2 Contribution to reduction of environmental costs by sensing and communication technologies

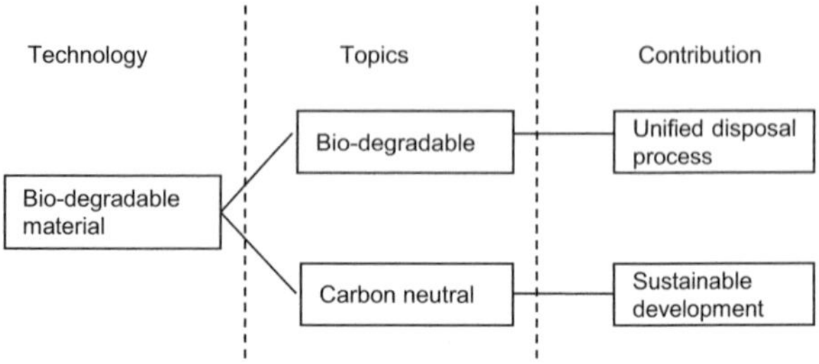

Fig. 3 Contribution to reduction of environmental costs by biodegradable materials

for disposition can be also reduced. In the relation with AM, some types of AM machines can use biodegradable materials such as PLA (polylactic acid). These materials have been studied to have additional properties such as electrical conductivity [6].

Most biodegradable materials are made with carbon neutral materials grown by sunlight. They contribute to sustainable development. Figure 3 arranges the effects.

3.3 Additive Manufacturing

AM has many possibilities in reduction of environmental costs because the fabrication process in AM realizes short cutting from conventional production process. Figure 4 arranges the effects.

Integrating Fabrication
Integrating fabrication is to make an assembled product as one process, which avoid the assembly process. AM has a possibility to realize it and has ability of multi-material fabrication. It reduces not only energy for assembling but also preparation for assembling facilities.

Mold and Die Less Fabrication
Usual parts are shaped by using molds or dies, which realize to make a large number of high quality parts at a low cost. Although the energy for making the molds and dies is large, its rate for each part becomes small if a lot of parts are fabricated with one mold or die. On the other hand, AM can fabricate parts without molds and dies. It reduces not only time and costs but also material and energy for making them. If the number of fabrication is small, the reduction effect for each part becomes large. It also saves the storage costs and energy of them for supplement of spare parts.

Direct Fabrication
AM can fabricate parts directly from 3D model data to a real object. It enables to fabricate parts at near place to be used. It can reduce environmental costs for transportation if the materials can be accommodated at the place.

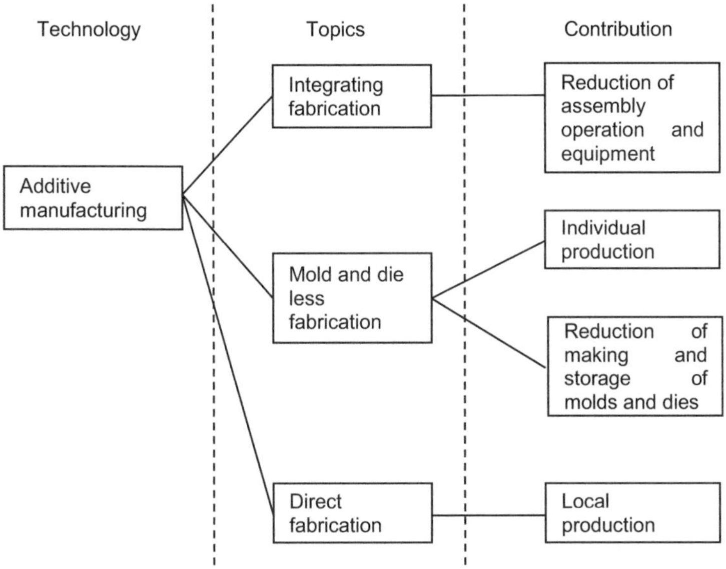

Fig. 4 Contribution to reduction of environmental costs by additive manufacturing

3.4 The Effects by Combinational Use and the Realization Possibilities

As mentioned above, three technologies separately have good points in reducing environmental costs. However, the important point is that their advantages are enhanced by combinational use. Figure 5 shows examples of the synergy effects.

It was mentioned that the sensing and communication devices were useful in condition monitoring for maintenance planning. This advantage is enlarged by the direct fabrication, which is a characteristic of AM because the repairing operations can be executed soon after the operation is planned without transporting spare parts.

The technical point for realizing the combinational use of the monitoring device and AM is to expand AM ability to fabricate both mechanical and electronic parts. The technology will be realized by two steps.

The first step is the realization of direct fabrication without IC chips. The technology has been developed in the printable electronic field [7]. Although the decisive method has not been found, the wires and body parts can be fabricated as one object by printing processes. If the economic costs are not considered, it can be used practically.

The second step is the realization of direct fabrication of the entire product including electronic elements. Recently, there are some reports to develop transistors [8] and capacitors [9] by using carbon nanotube as a printable material. The printing also reduces the manufacturing energy because it can be done at low temperature. The simple circuit can be made by printing processes. Since the

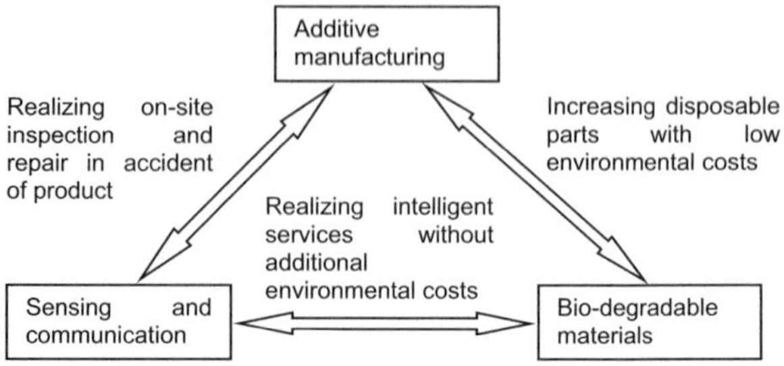

Fig. 5 Synergy effects by combinational use of three technologies

printing process has limitation in making integration circuits, it is a novel design to make multilayer circuits in the body by AM. These designs should be developed.

As the synergy effect between AM and biodegradable materials, the advantage of low-cost disposition of biodegradable product is enlarged by the integrating fabrication, which is a characteristic of AM, because the product can be landfilled without energy and equipment for disassembling. Some materials already exist. PLA is one example used widely in AM. However, if we consider the life cycle environmental cost of PLA, it is not so lower than conventional plastics because the making process of PLA needs a lot of energy.

Then, the full combinational use is desired. If the parts including sensing and communication function can be fabricated by AM with biodegradable materials, the use will have a large synergy effect and generate many kinds of services to use big data collected by ubiquitous sensors. Therefore, the total design of large systems is important. Since there are many studies on the biodegradable electronics [10], the development of the total design method over multidisciplinal fields is required to realize the proposed eco-electric products.

4 Comparison Between Conventional Mechatronic Products and Proposed Products

4.1 Electronic Device Production Process

Mechatronic products have mechanical parts and electronic parts. First, the life cycle process of typical electronic modules is discussed. Figure 6 compares between conventional and present products in a production process.

In fabrication process of conventional product, the product has hierarchical structure. First, the parts such as semiconductor parts are fabricated with lithography processing, which need a large amount of chemical liquid and water for processing and cleaning. Then, the electronic parts are mounted on the electronic board using solder, which usually need heating process for reflowing solder paste. Finally, the electronic devices are assembled with mechanical parts into the final products.

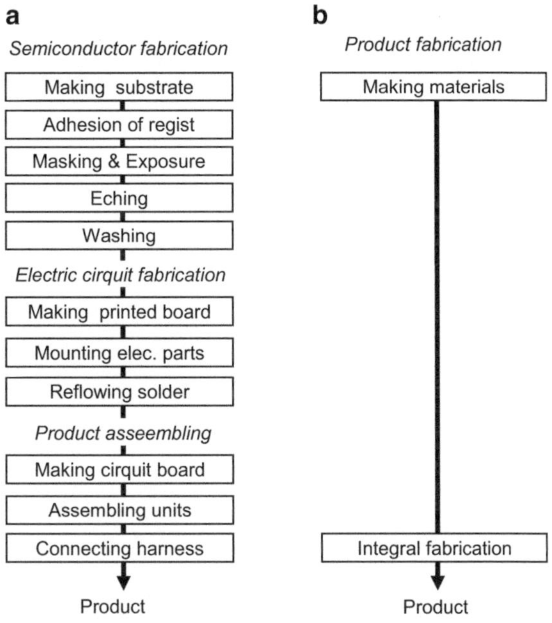

Fig. 6 Comparison between the conventional process and the proposed process in a production process (**a**) Conventional process (**b**) Proposed process

On the other hand, the proposed product has flat structure, which makes fabrication process simple. The process has only two steps, which are preparation of materials and printing. We can see that a lot of energy and required equipment for fabrication are drastically reduced.

4.2 Electronic Device Disposal Process

Figure 7 compares between conventional and present products in a disposal processes.

In disposal process of conventional products, after the product is transported to a disassembling factory, and separated into reusable or disposal parts, they are transported to different factories to add appropriate process, respectively.

In the case of the proposed product, the process looks similar with the conventional one, but it is different in quantity. First of all, the reuse and recycle plan is calculated by analysis of sensor data and historical data of the product. This process makes following process efficient. The transportation will be short because the nearest factory can be chosen by the analysis. After the parts, which are not biodegradable, are removed if it has, all parts are processed by biodegradation all together. We can see the process becomes simple and the environmental cost becomes lower than the conventional one.

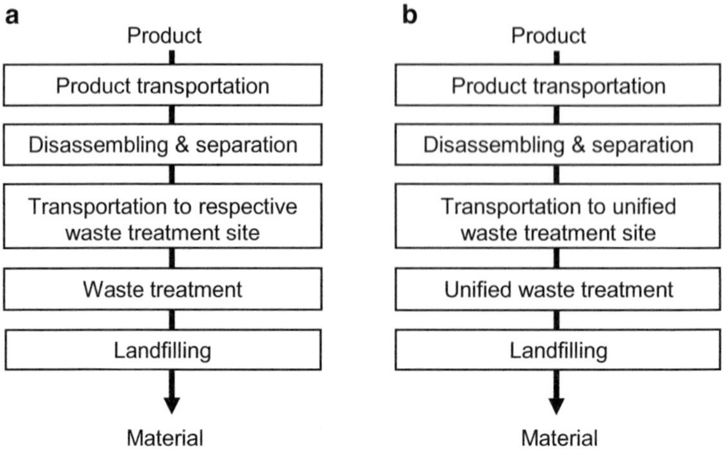

Fig. 7 Comparison between the conventional process and the proposed process in a disposal process (**a**) Conventional process (**b**) Proposed process

5 Practical Example

5.1 Product Example

As an example of the eco-mechatronic product, RFID tag for identifying the mechatronic part was developed [11]. Figures 8 and 9 show the developed module. This is a type 2 of the eco-mechatronic product described in Sect. 2.2, and the first step realization by direct fabrication described in Sect. 3.4.

The base and antenna belong to the printed module, and the IC chip (Japan Information System Co., Ltd.) belongs to the mass-production module. The antenna size is relatively large, but it may not be a problem because this printed module can be fabricated as one body with other mechatronic systems.

The base was fabricated with PLA by using a commercial material extrusion-type AM machine. The antenna was fabricated by an original material extrusion-type AM machine. The material was also produced with Ketjen black (Lion Corp.), Exeparl IPP (Kao Inc.), and carnauba wax. They are biodegradable materials usually used for cosmetics or commodities. By detaching the IC chip, the printed module can be disposed in one operation.

Experiment of detection by an RFID reader was conducted. As the result, the longest distance, in which the tag was detected by the reader, was 300 mm. Since the longest distance in the case of bare IC chip without antenna was 75 mm, the effectiveness of antenna was shown.

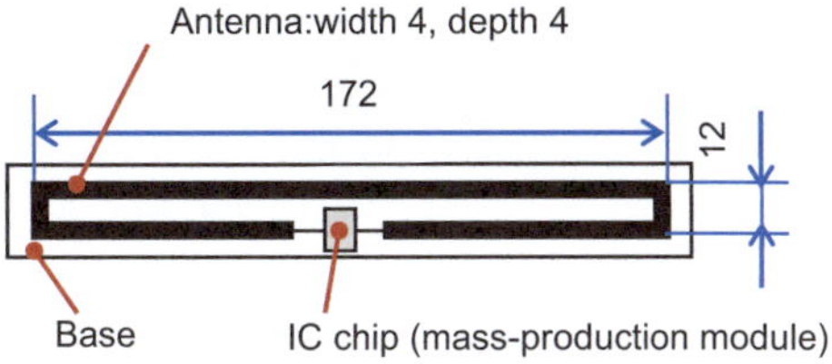

Fig. 8 Design of RFID as a realization example of eco-mechatronic products

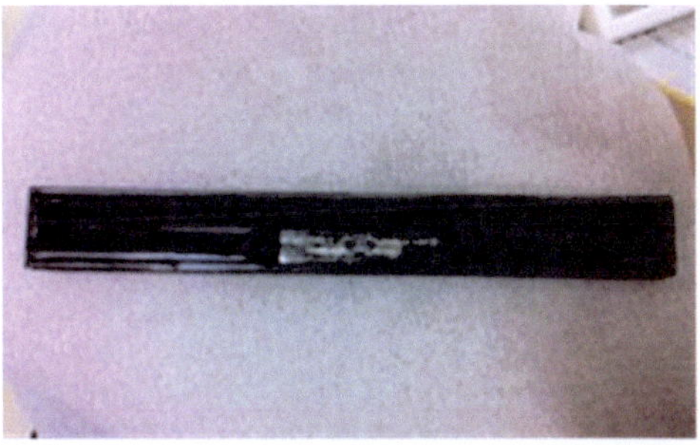

Fig. 9 Photo of the RFID prototype

6 Conclusions

(1) While sensors and communication devices attached to electric products are useful for creating services to reduce environmental costs, the problem was indicated such that the additional device generates environmental costs. In order to reduce the additional environmental costs, the use of additive manufacturing with biodegradable materials was proposed.
(2) After two considerable uses of the proposed products are introduced, the possible contribution of these technologies to reduction of environmental costs was analyzed and categorized.
(3) As a general evaluation of mechatronic products, comparison between the conventional process and the proposed process is discussed on production or disposal process.
(4) Realization of the proposed eco-mechatronic products was introduced with an RFID tag prototype, which was fabricated by AM with biodegradable materials.

References

1. Ashton K (2009) That 'Internet of Things' thing. RFID J 22:2009
2. Tateno T, Kondoh S (2009) Architecture of sensor network system for promoting condition-based maintenance and reuse. In: Proceeding of ASME international design engineering conference & computers and information in engineering conference, San Diego, California, Sept 2009, Paper no. DETC2009-87197

3. Gibson I, Rosen DW, Stucker B (2010) Additive manufacturing technologies: rapid prototyping to direct digital manufacturing. Springer, New York
4. Khajavi SH, Partanen J, Holmstrom J (2014) Additive manufacturing in the spare parts supply chain. Comput Ind 65:50–63
5. Anderson C (2012) Makers: the new industrial revolution. Crown Publishing Group, New York
6. Rupprecht L (2000) Conductive polymers and plastics: in industrial application. William Andrew, New York
7. Hon KKB, Li L, Hutchings IM (2008) Direct writing technology – advances and development. CIRP Ann Manuf Technol 57:601–620
8. Jones CS, Lu X, Renn M, Stroder M, Shih W (2010) Aerosol-jet-printed, high-speed, flexible thin-film transistor made using single-walled carbon nanotube solution. Microelectron Eng 87:434–437
9. Lawes S, Riese A, Sun Q, Cheng N, Sun X (2015) Printing nanostructured carbon for energy storage and conversion applications. Carbon 92:150–176
10. Irimia-Vladu M, Glowacki ED, Voss G, Bauer S, Sariciftci NS (2012) Green and biodegradable electronics. Mater Today 15(7):340–346
11. Yaguchi Y, Tateno T (2012) Printing of electrically-conductive gels for embedding electric circuits in elastic materials. In: Proceeding of international conference on positioning technology, Kaoshiung, Taiwan, Nov 2012, pp 259–262

The Monitoring of Three-Dimensional Printer Filament Feeding Process Using an Acoustic Emission Sensor

Pitchapa Lotrakul, Wimol San-Um, and Masaaki Takahashi

Abstract 3D printing or an additive manufacturing (AM) is an up-and-coming technology which receives a lot of attentions in the recent years. This technique features fabrication of a 3D object by layering material one layer over another. Nonetheless, there are still challenges in monitoring such technique, in particular, fused deposition modeling (FDM) type. Therefore, Acoustic Emission (AE) sensor which is one of nondestructive testing (NDT) methods that has shown its ability in monitoring rotating machine was utilized. Low-cost power supply and 3D-printed magnetic mounting were also equipped in the system to acquire efficient system in lower cost than commercial system. The signals were process through LabView using fast Fourier transform (FFT) to perform frequency-domain analysis to distinguish properly working machine from faulty machine. The simulated faulty conditions include the problem of filament stop feeding through extruder and no filament supply for the extruder. The result showed an ability of AE sensor and FFT in detecting machine with faulty condition.

Keywords 3D printing • Additive manufacturing • Filament feeding • Monitoring • Acoustic emission

1 Introduction

Additive manufacturing is a process which fabricates 3D object from CAD file by adding layers of material over another layer instead of subtracting material from material stock as in traditional process. This technique is becoming more popular in the present due to its numerous advantages such as increasing freedom in designing complex part, lightweight product, and reducing material wastage. There are many sub-techniques of this manufacturing procedure which are applied to manufacturing

P. Lotrakul • W. San-Um (✉)
Intelligent Electronic System Research Laboratory (IES), Faculty of Engineering, Thai-Nichi Institute of Technology (TNI), Bangkok, Thailand
e-mail: wimol@tni.ac.th

M. Takahashi
Department of Manufacturing, Institute of Technologists (IOT), Saitama, Japan

M. Matsumoto et al. (eds.), *Sustainability Through Innovation in Product Life Cycle Design*, EcoProduction, DOI 10.1007/978-981-10-0471-1_34

different type of materials [1]. However, there are still challenges such as improving materials and increasing the reliability and accuracy of equipment and processes [2] which are obstructing these techniques from being more widely used.

In particular, quality of the process has been the challenges of such technology for a long time. In 1998, T. Fang et al. used image analyzing in attempt to monitor layered manufacturing of ceramics. The results showed an ability to detect defects in linear region. Nonetheless, there was some struggle with the end region which was not linear [3]. Recently in 2014, H. Rieder et al. applied ultrasonic which is one type of nondestructive testing (NDT) to selective laser melting technique which was used to manufacture products from metal powder. The fusion of a single layer and a temporal formation of material defects can be infer from such analysis [4].

However, in the field of fused filament deposited manufacturing of plastic, the fault detection is not as widely studied. J. Yoon et al. used heterodyne technique to lower sampling rate of Acoustic Emission sensor in order to compare the result with piezoelectric strain sensor. The aim was to detect driving belt seeded looseness fault where narrow band filter around high peak in frequency domain was used to extract condition indicators to detect the fault [5].

Nonetheless, there are still other problems with such type of process to be studied. One of the problems that could occur to the machine during the printing process is filament stop feeding through extruder [6]. This could be due to many causes such as level of build plate or clog in extruder. This problem could cause energy wastage or even lead to machine damage. Although it can be spotted by a cautious observer, in a long period work, the process is usually run without an observer. A user could end up spending hours of processing time but receiving nothing but a piece of scrap and wastage of time and energy.

Acoustic Emission is one of nondestructive test (NDT) methods which allow specimen monitoring during the process without having to destroy it. AE is utilized in many conditions monitoring including many rotating and reciprocating machinery applications [7]. It has advantage of early detection due to its sensitivity [8]. It is also nondirectional which gives the ability to detect the whole machine with only one sensor instead of multiple sensors for multiple directions. In addition, fast Fourier transform (FFT) and frequency-domain analysis are commonly used in combining with AE to analyze and interpret such signal. FFT is a tool to convert time-domain signal to frequency-domain signal. It is useful in condition monitoring and fault prognosis which fault usually occurs in different frequency band from usual working frequency band.

This paper presents an alternative to monitor filament feeding process in fused deposition additive manufacturing using an Acoustic Emission sensor. Low-cost power supply and 3D-printed magnetic mounting are also equipped in the system for cost reduction and fine signal transmission. The signals are processed using LabView program in which fast Fourier transform and frequency-domain analysis are used as tools for the analysis. Finally, the comparison between normal and faulty machine is presented as a result. The proposed monitoring system makes it possible to build a system to inform the operator about the operating status and possible to reduce the loss of time and energy which is considered to lead to EcoDesign.

2 Component Preparations

Prior to the experiment, the AE system was assembled by combining the essential instruments. Despite preparation of commercial parts for the study, some parts were fabricated to suit the requirement of this particular experiment in low cost. These parts are power supply and sensor mounting.

2.1 *Low-Cost Power Supply*

In order to complete the system, a preamplifier was required to be connected to a power supply which also serves as a terminal to transfer signal to analog-to-digital (A/D) converter. Therefore, a power supply was fabricated using voltage reduction circuit to utilize 100V-AC household electricity. The objective is to fabricate low-cost power supply that could perform as a commercial power supply for AE sensor.

The power supply circuit consists of two main parts which are the voltage reduction circuit and the phantom power circuit which is the circuit to connect power supply with preamplifier and to acquire output signal. The schematic diagram of both parts of the circuit is shown in Fig. 1. The first part or the voltage reduction circuit used transformer to reduce 100V-AC household electricity to 50V-AC. Then, diode bridge was utilized to convert AC electricity to DC electricity. After that, LM317 adjustable voltage regulator was employed to adjust the output voltage to the desired voltage. This was achieved by potentiometer usage. The output voltage can be calculated from the following formula [9]:

$$V_o = V_{\text{ref}} \times \left(1 + \frac{R_2}{R_1}\right) + I_{\text{adj}} R_2 \tag{1}$$

where V_o is an output voltage, V_{ref} is an input voltage to LM317, R_2 is a resistance of potentiometer, R_1 is a resistance of 240 Ω resistor, and I_{adj} is the current of the

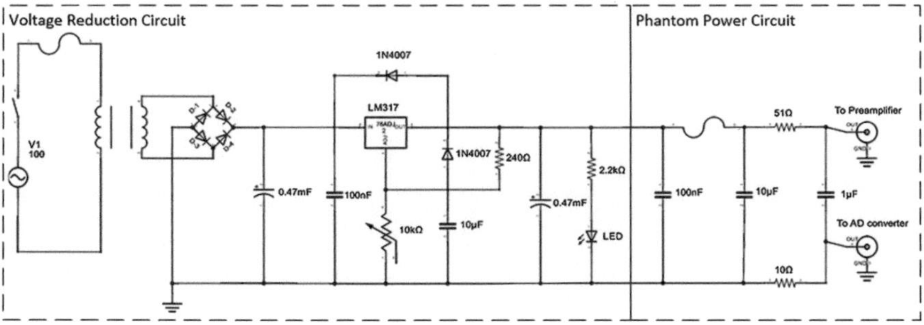

Fig. 1 Schematic diagram of low-cost power supply circuit

adjust pin. However, the device, LM317, was designed to minimize the current of the adjust pin. Therefore, the term $I_{\text{adj}} \times R_2$ can be neglected and the formula can be rewritten as

$$V_o = V_{\text{ref}} \times \left(1 + \frac{R}{240}\right) \tag{2}$$

In addition, the performance and the stability of the circuit are enhanced by utilizing several diodes and capacitors. Diodes are utilized in order to prevent misdirection electrical, while capacitors are utilized in order to stabilize electrical level in the circuit. The second part of the circuit or the phantom power circuit was adopted from the manufacturer's website. The objective of this part of circuit is for the sensor and the preamplifier to share an AE output signal and a preamplifier power within a single cable [10]. BNC connectors were employed as a low-noise connection platform. An LED lamp and switch were also utilized to fulfill the power supply as an indicator and a control of power supply status, respectively. The completely fabricated power supply, both inside circuit and outside apparatus, is shown in Fig. 2a,b, respectively.

2.2 D-Printed Magnetic Mounting

Attaching AE sensor to the testing machine was another consideration point. Well attachment would assure fine signal transmission which prevents detail loss in the process. Sensor coupling is an agent to substitute air between surface of the sensor and surface of the machine. Among different types of coupling, grease-based coupling has high viscosity which is suitable for rough surface and long-term stability. It also has the ability to withstand medium temperature [11]. This is appropriate for the application with 3D printer which has rough surface and

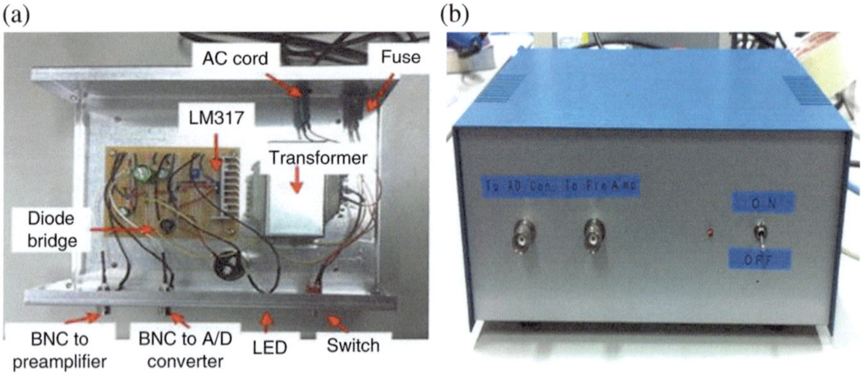

Fig. 2 (**a**) Inside circuit and (**b**) outside apparatus of the fabricated power supply

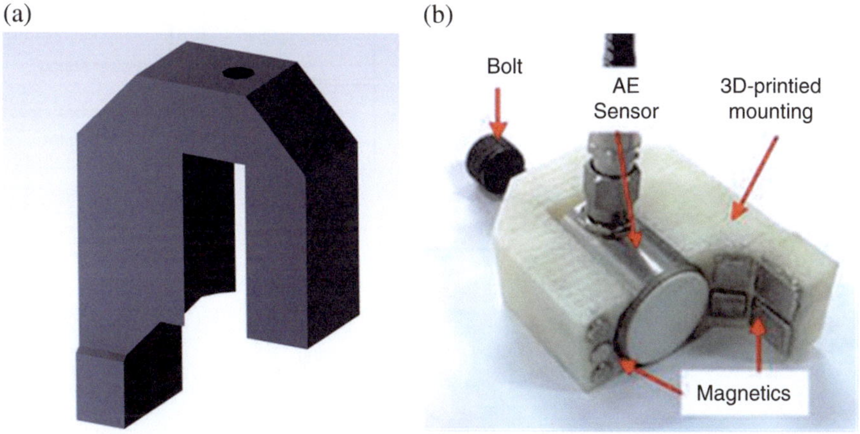

Fig. 3 (**a**) CAD prototype and (**b**) final assembly of AE sensor and magnetic mounting

vibration during the working process. Therefore, vacuum silicone grease was chosen as a sensor coupling in this study.

In order to use grease as sensor coupling, clamp was required. However, the dimension of the 3D printer motor block caused commercial mountings to be not so applicable. Therefore, AE sensor mounting was design using CAD software to perfectly match the dimension of the motor block. The rendering of CAD file of the mounting is shown in Fig. 3a and can be downloaded at [12]. Then, the mounting was fabricated using 3D printer which gave freedom in designing and also low-cost product. After that magnets and a bolt were assembled to the printed object to complete the fabrication as shown in Fig. 3b. The screw bolt was utilized in the design to help secure the sensor and benefits in removing trapped air between surfaces.

In addition, AE sensor is attach to motor block using double-sided tape as a simple connecting agent and collect data for normal working condition to test the effectiveness of magnetic mounting and vacuum grease as preconditioned experiment. This is to use as a reference data to compare with tested data in the main experiment.

3 Experimental Methodology

3.1 Experimental Setup

An overview of the test is shown in block diagram in Fig. 4 with arrows which show the flow of the signal through the test system. The test rig consisted of a 3D printer, an AE sensor, a preamplifier, a fabricated power supply, an A/D converter, and a computer. An AE sensor was attached to the extruder's stepper motor of the 3D

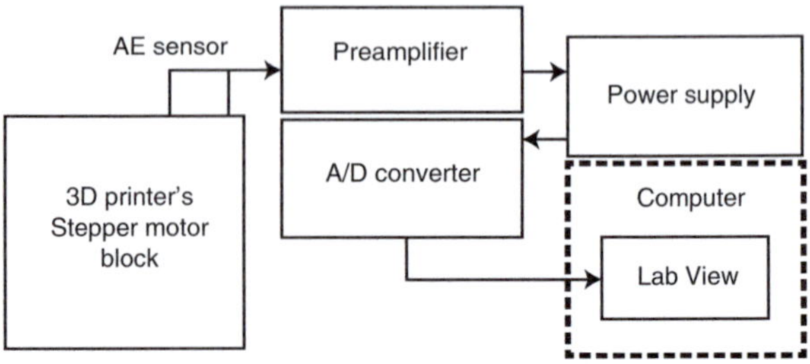

Fig. 4 Signal flow diagram of the test system

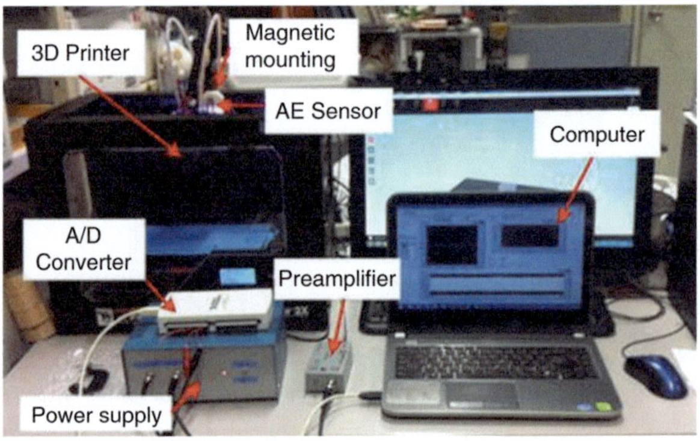

Fig. 5 Apparatus of the test rig

printer using a 3D-printed magnetic mounting which was fabricated in section 2A with silicone vacuum grease applied between AE sensor and stepper motor surfaces. The signal was collected from the stepper motor using AE sensor. Then, the signal was transmitted to the preamplifier with 60-dB gain to amplify the amplitude and the detail of the signal. After that, the signal was collected through an A/D converter and then analyzed using LabView program using FFT procedure. The experimental apparatus is shown in Fig. 5.

3.2 Attachment Comparison Experiments

Prior to the main experiment, preconditioned experiment was performed to discover the appropriate attachment methodology to be applied for the main experiment of this study. Two attachment methodologies which would be tested in this study were a double-sided tape and a fabricated magnetic mounting with vacuum silicone grease. Double-sided tape attachment is a simple attachment which allow strong bond between AE sensor and the specimen. It can be easily applied without any other components and can be attached to many different surfaces both vertical and horizontal. On the other hand, 3D-printed magnetic mounting and vacuum silicone grease were designed to suit this particular dimension of the motor block.

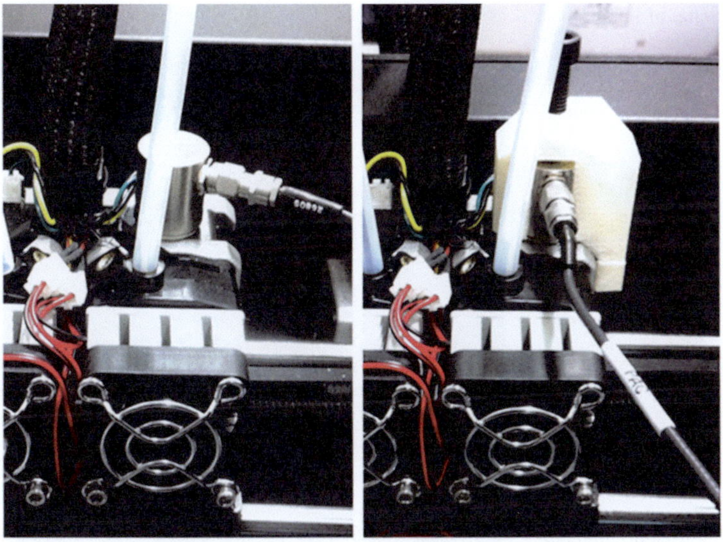

(a) (b)

Fig. 6 (**a**) Double-sided tape and (**b**) fabricated magnetic mounting experiment setup for an attachment comparison

Both methods were applied to attach an AE sensor to the stepper motor of 3D printer and then signals were acquired from normally working conditioned 3D printer. After that, these signals would be compared by mainly focusing on the amplitude and details of the signal. Effectiveness of both attachment methodologies would be compared in order to discover more suitable methodologies for the main experiment to assure well signal transmission for further analysis. Experiment setup for both attachment methodologies is shown in Fig. 6, (a) for double-sided tape and (b) for fabricated magnetic mounting.

3.3 Filament Feeding Condition Experiment

The main experiment of this study was the study of AE signals under different conditions of 3D printer. The experiment consisted of three conditions which were filament normally feeding condition, filament stop feeding through the extruder condition, and no filament supply for the extruder condition.

For the first condition, an AE sensor was attached to a properly working 3D printer to acquire signals as a reference signal. For the second condition which is filament stop feeding condition, in order to simulate filament stop feeding through the extruder, a C-clamp was attached to the filament, which was feeding to the extruder, in order to block the filament from feeding into the extruder. The components and setup of such experimental condition is shown in Fig. 7. The

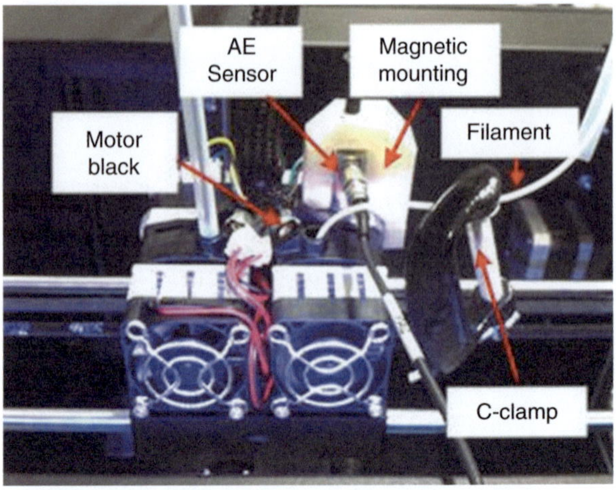

Fig. 7 Setup of filament feeding condition experiment

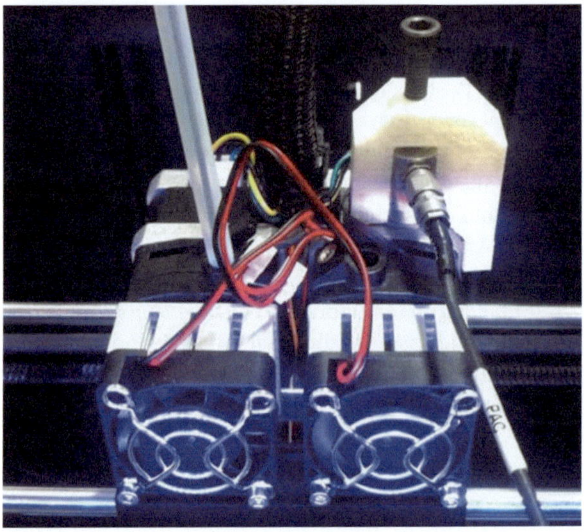

Fig. 8 Setup of no filament condition experiment

signal would be collected after C-clamp has settled on the top of the extruder and the filament has been perfectly stopped.

For the last condition, the no filament condition, filament was unloaded from the extruder using unload function of the printer. The printer was tested to assure that there is no filament going through the extruder before the signal would be collected. The experimental setup of such condition is shown in Fig. 8. The signals of each condition were collected separately and analyzed using LabView program. The analyzing program consisted of time-domain signal, frequency-domain signal acquiring from fast Fourier transform, and frequency and its value of peak amplitudes. The data from each condition was then compared in order to discover the distinguished parameter in the last step.

4 Results and Discussion

4.1 Result of Attachment Comparison Experiment

The time-domain signal of a normally working 3D printer using double-sided tape as an attachment of an AE sensor is shown in Fig. 9, while the time-domain signal of a normally working 3D printer using a magnetic mounting and vacuum silicone grease as an AE sensor coupling is shown in Fig. 10. Comparing both figures, the difference in amplitudes can be noticed that amplitude of the signal in Fig. 10 is about three times larger than in Fig. 9. This shows the ability of a magnetic mounting and vacuum grease as a coupling to attach AE sensor to the motor block. As a result of this preconditioned experiment, the magnetic mounting and vacuum grease was continually utilized as sensor coupling in the main experiment.

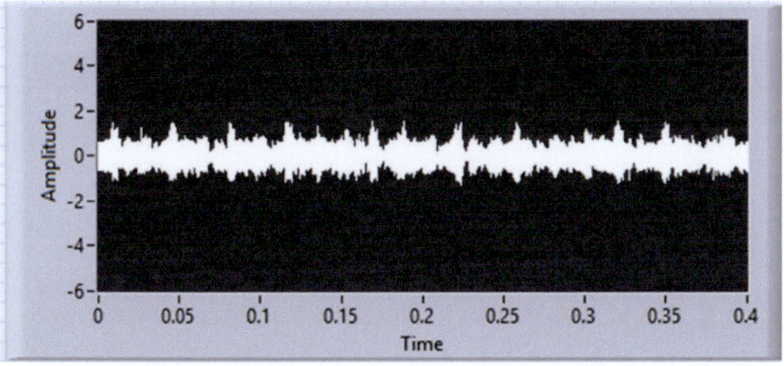

Fig. 9 Time-domain signal of normal condition using double-sided tape as an attachment

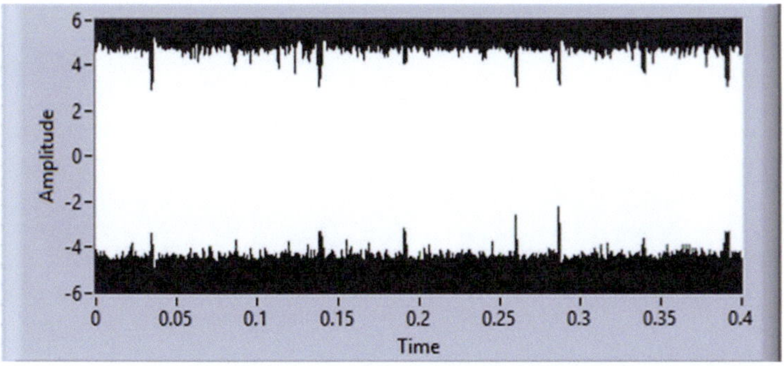

Fig. 10 Time-domain signal of normal condition using magnetic mounting and vacuum grease

4.2 Result of Filament Feeding Experiment

The time-domain signal for filament stop feeding condition is shown in Fig. 11. Comparing to the signal in normal condition in Fig. 10, these signals show very similar characteristics which are difficult to distinguish in faulty condition of the machine. Therefore, the fast Fourier transform was utilized and frequency-domain signal was acquired as shown in Figs. 12 and 13 for normal condition and filament stop feeding condition, respectively. Difference in frequency of maximum amplitude of the signal can be clearly noticed. Although high amplitude around 20 kHz and 50 kHz can be noticed from both conditions, normal condition has its maximum amplitude in 20 kHz frequency band, while filament stop feeding condition has its maximum amplitude in 50 kHz frequency band. In the other word, for normal condition, amplitude of 20 kHz frequency band is higher than in filament stop feeding condition, while for 50 kHz band, signal from filament stop feeding condition shows higher amplitude than normal condition. Consequently, it can be assumed that 50 kHz frequency band is the frequency of working motor, while the peak at 20 kHz band is caused by the process of filament feeding through the extruder.

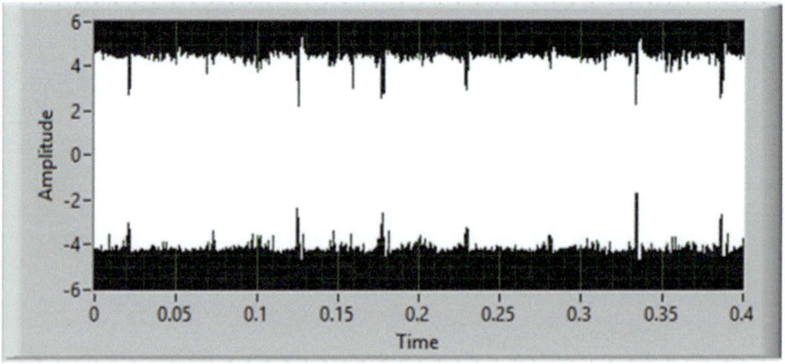

Fig. 11 Time-domain signal of filament stop feeding condition

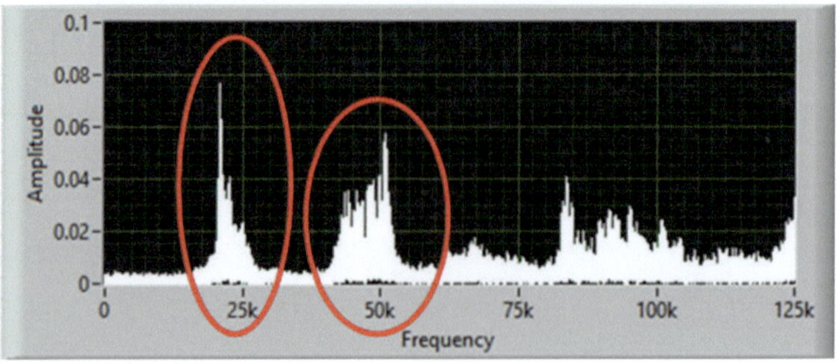

Fig. 12 Frequency-domain signal of normal condition

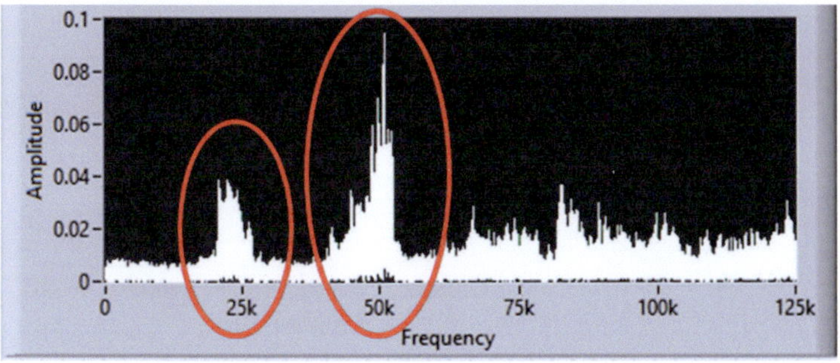

Fig. 13 Frequency-domain signal of filament stop feeding condition

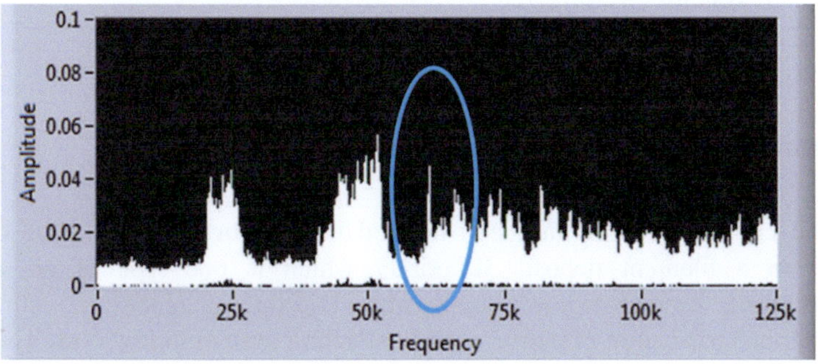

Fig. 14 Frequency-domain signal of no filament feeding condition

In addition, frequency-domain signal of no filament feeding condition is shown in Fig. 14. Comparing filament stop feeding and no filament feeding condition, both conditions show maximum amplitude at 50 kHz frequency band. This is consistent with the previous conclusion since no filament feeding condition can be considered in the same category as filament stop feeding because both conditions result in no filament through the extruder. However, the causes were different, blockage for filament stop feeding condition and lack of filament for no filament condition. Methodologies to resolve the problems were consequently different. Realizing such information, further comparison of both conditions was performed by a scatter plotted in Fig. 15. This plot supports the observation in Fig. 14 that when there is no filament supplying for the extruder, there would be an additional peak around 60 kHz frequency band. Therefore, no filament feeding condition can be distinguished from filament stop feeding by using this frequency band.

4.3 *Future Work Discussion*

For the further work, the result from this study should be applied to stop the printing process and shut down the printer when filament stops feeding, both filament blockage and no filament, is detected. Therefore, time and energy wastage can be

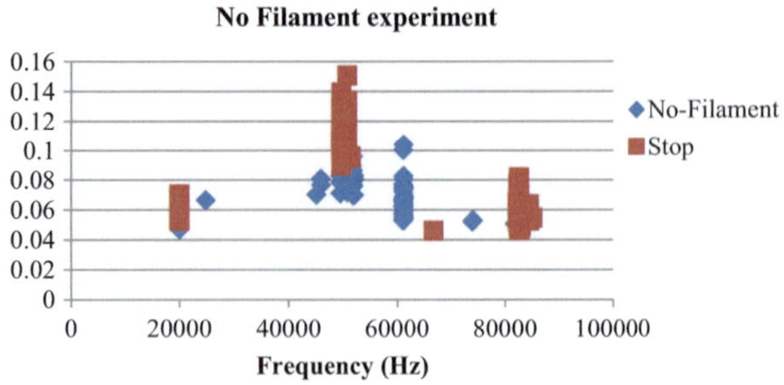

Fig. 15 Frequency-domain signal of no filament feeding condition

prevented. Band-pass filter can be applied in order to generate such feedback to the closed-loop system.

5 Summary

Acoustic emission sensor has been utilized to monitor the process of filament feeding in a filament deposit 3D-printer which is one type of an additive manufacturing. Low-cost power supply and 3D-printed magnetic mounting were also equipped to acquire effective signal collecting system in low cost. The signal was processed using fast Fourier transform and frequency-domain analysis through LabView program. The result shows noticeable difference between normally operating machine and machine with filament feeding problem including filament stop feeding condition and no filament feeding condition using frequency of maximum amplitude and amplitude of peaks of the signal as a distinguish tool. An ability of AE sensor in monitoring 3D printer is demonstrated.

Acknowledgments The authors are grateful to Research and Academic Services Division of Institute of Technologists (IOT) and Thai-Nichi Institute of Technology (TNI) for financial supports.

References

1. Wong KV, Hemandez A (2012) Review article: a review of additive manufacturing, ISRN Mechanical Engineering, vol. 2012, Article ID 208760, 10 pages
2. Sharon F (1998) Additive manufacturing technology: potential implications for U.S. manufacturing competitiveness. J Int Commer Econ 1:1–35, 2014
3. Fang T, Bakhadyrov I, Jafari' MA, Alpan G (1998) On-line detection of defects in layered manufacturing. IEEE international conference on Robotics & Automation Leuven, Belgium
4. Rieder H, Dillhofer A, Spies M, Bamberg J, Hess T (2014) Online monitoring of additive manufacturing processes using ultrasound". In: 11th European conference on non-destructive testing (ECNDT 2014)

5. Yoon J, He D, Hecke BV (2014) A PHM approach to additive manufacturing equipment health monitoring, fault diagnosis, and quality control. Annual conference of the prognostics and health management society
6. Replicator2 Desktop 3D printer troubleshooting and maintenance, MakerBot Industries, Brooklyn, New York
7. Al-Obaidi S, Leong MS, Hamzah RIR, Abdelrhman AM (2012) A review of acoustic emission technique for machinery condition monitoring: defects detection & diagnostic. Appl Mech Mater 229–231:1476–1480
8. Tana CK, Irvinga P, Mbab D (2007) A comparative experimental study on the diagnostic and prognostic capabilities of acoustics emission, vibration and spectrometric oil analysis for spur gears. Mech Syst Signal Process 21:208–233
9. LM317 data sheet, STMicroelectronics
10. MISTRAS Group, Inc., Phantom power. [Online]. Available: http://www.pacndt.com/index.aspx?go=products&focus=/sensors/phantompower/index.htm
11. National Physical Laboratory (2012) Guide on acoustic emission sensor couplants. [Online]. Available: http://www.npl.co.uk/acoustics/ultrasonics/research/guide-on-acoustic-emission-sensor-couplants
12. http://www.thingiverse.com/thing:1000086

Selective Volume Fusing Method for Cellular Structure Integration

Supachai Vongbunyong and Sami Kara

Abstract Cellular structure is one of the solutions for producing parts in a more resource efficient way by reduction of material and time used. These complex structures can be customised and built by using additive manufacturing technology. Cellular structures can be designed and modelled according to specific properties required. However, the modelling can be time and resource consuming due to the complexity and the size of the structure. This article presents a concept for integrating the cellular structure to the base 3D model that will be used in development of a design software tool. Hybrid B-Rep and polygon approach which is implemented with open-source libraries are used.

Keywords Cellular structure • Hybrid B-Rep • Geometry modelling • Additive manufacturing • Sustainable manufacturing • Rapid prototyping

1 Introduction

Additive manufacturing (AM) has been broadly used not only for rapid prototyping but also for producing functional parts. One of the outstanding advantages of AM is an ability to produce parts with highly complex geometry that is sometimes impossible to produce by using subtractive manufacturing. By adding material layer by layer, the parts can be produced regardless of complexity. As a result, the design can be done with less limitation due to manufacturability. Cellular structure is one of the possible approaches that can achieve material and weight reduction of the parts. As a result, manufacturing process requires less energy and time consumption.

S. Vongbunyong (✉)
Institute of Field Robotics, King Mongkut's University of Technology Thonburi, Bangkok, Thailand
e-mail: supachai_von@fibo.kmutt.ac.th

S. Kara
School of Mechanical and Manufacturing Engineering, The University of New South Wales, Sydney, Australia

M. Matsumoto et al. (eds.), *Sustainability Through Innovation in Product Life Cycle Design*, EcoProduction, DOI 10.1007/978-981-10-0471-1_35

1.1 Cellular Structures

Cellular structures are complex geometry formulated by connecting a number of *unit cells* together and aligned in a repetitive pattern. The unit cell is formulated with a geometry designed for particular tasks. They have been brought into attention due to manufacturability by using AM. Research regarding the properties and applications of cellular structures has been conducted in various fields, e.g. chemical, medical, mechanical, etc. [1]. They are used in two ways: as a support structure or a part of the model.

First, the support structures are necessary in a process for building models that include some features, e.g. overhanging parts. They are temporary platforms in which the material layer will be added on the top. Therefore, these structures are light yet strong enough to support the weight of the model [2, 3]. A number of strategies are developed to improve this property, for instance, using lattice structure with hollow truss that provides high peak strength [4], using Schoen gyroid and Schwartz diamonds as the unit cells that improve thermal resistance and significantly reduce the material required [5].

Second, with respect to design and optimisation, cellular structures are designed as a part of the model to satisfy particular requirements. The properties are classified as *stiffness*, *strength*, *compliance*, *thermal*, *dynamic*, and *visual* properties which are achievable with different geometry, dimension, and arrangement of the cell [1].

1.2 Problem in Modelling and Design

Problems in the design process of the cellular structures consist of four perspectives: size, parameterisation, validity and precision, and manufacturability [6]. Boundary Representation (B-Rep) and Stereo Lithography (STL) are compared regarding these perspectives.

The (a) *size* of the file representing cellular structure is usually large due to the complexity of the part. As a result the modelling process consumes a lot of computing resources and time accordingly. (b) *Parameterisation* is necessary in optimisation process of the structure to achieve required properties. Geometric features of the structure – e.g. beam profile of unit cell, dimension, etc. – should be variable in order to find an optimal solution. (c) *Validity and precision* involve the quality of the surface. These models can be manufactured if they are watertight, meaning that they are free from these artefacts: invert normal, crack, polygon self-intersection, etc. Due to the (d) *manufacturability*, non-watertight models need to be repaired a priori. The repair process is a highly resource consuming process that can be difficult, or sometimes impossible, depending on the complexity of the models.

Overall, B-Rep is more suitable for parameterisation and solid modelling but inefficient for representing complex geometry that contains a large number of triangle meshes. STL format is opposite in both perspectives. In practice, designers model the structure with solid modelling method in B-Rep format by using commercial CAD software and then convert the B-Rep model to an STL for being manufactured. The problems in the conversion process occur regarding validity and precision issues and lead to the problem in manufacturability.

Alternative modelling methods are developed to overcome these problems. By modelling polygon model directly, [6] used Function Representation method (F-Rep) and [2] used implicit method. The issue regarding validity of the model due to connectivity among cells can be resolved by using these methods; however, the design of geometry of unit cells can be much complicated. [7] proposed a hybrid geometric modelling (HGM) combining between design flexibility of B-Rep and manufacturability in polygon approach. Processing time and computing resources are improved from B-Rep method with format conversion. In this article, we propose a prefabrication hybrid geometric modelling (P-HGM) where the performance in generating structures with a large number of cells can be improved.

1.3 Scope and Overview

This article presents development of a software tool used to generate and integrate the cellular structure to the base geometry. However, analysis of physical properties of the resultant parts is not in the scope of this research.

This article is organised as follows. The framework of the software tool is presented in section "System framework", the methodology in section "Methodology", case study and experiment in section "Experiments", and discussion and conclusion in section "Conclusions".

2 System Framework

2.1 Software Tool: Free Form Design

For novices, working in 3D environment can be one of the considerable obstacles for CAD learners. People who have great ideas and want to express them, e.g. high school students, may refuse to learn CAD due to this difficulty. *Free Form Design*™ (FFD) is a software tool designed for non-experienced CAD users to be able to design 3D models intuitively and easily [8]. It is used as a platform where the module for cellular structure module (CSM) is built in. FFD is developed based on C/C++ and *Visualization Toolkits* (VTK) [9], while CSM is based on *Open Cascade Technology* (OCCT) [10] and VTK.

2.2 *Libraries*

2.2.1 Visualization Toolkits (VTK)

VTK 6.0 is a computer graphic library focusing on visualisation. This library handles a rendering pipeline and provides functions and algorithms for visualisation and modification of surface and volumetric models. In this case, only surface models are used. Similar to STL format which is a standard format for using in AM, data of the surface model is in the form of polygons, *vtkPolyData*. The models can be exported to STL file for production directly without further resource-intensive conversion.

2.2.2 Open Cascade Technology (OCCT)

OCCT 6.7.1 is a software development kit (SDK) based on Boundary Representation (B-Rep) as used in most of CAD software. It provides functions for geometric modelling, including solid and surface modelling.

2.3 *Framework*

B-Rep is well established and suitable for solid modelling, while STL is standard for AM. In FFD, both formats are interchangeable in order to integrate OCCT with VTK. An overview of the software framework is shown in Fig. 1e. CSM generates a cellular structure with particular unit cell. The structure is generated rapidly with P-HGM [11] (see section "Prefabrication hybrid geometric modelling"). The unit cell is modelled in B-Rep with OCCT. The entire structure is created by populating the unit cell by using VTK with polygonal surface model.

3 Methodology

3.1 *Prefabrication Hybrid Geometric Modelling*

P-HGM is a 3D modelling method designed specifically for generating cellular structure with uniform patterns [11]. This method is developed based on the concept of HGM [7] where hybrid modelling is a technique that is a conversion between B-Rep model and polygonal surface model. For HGM, each unit cell needs to be processed individually while placing on each location in the output structure. This makes HGM more flexible but consumes massive resource when the output structure contains a large number of cells. In comparison to P-HGM, by considering

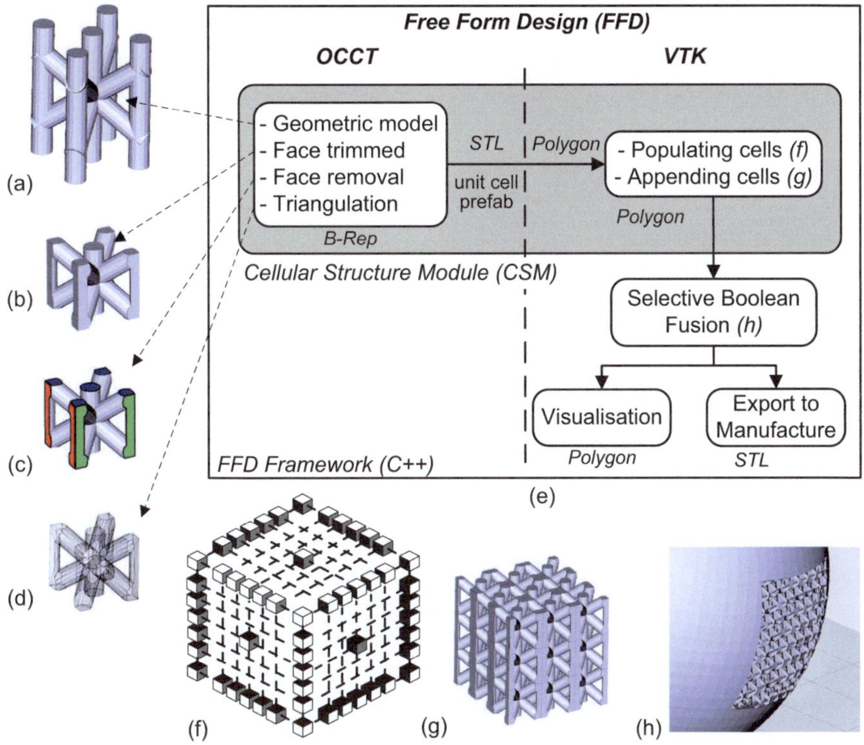

Fig. 1 Framework of CSM in free form design

only uniform structure as an outcome, time and resource consumption of the modelling process can be considerably improved. A set of *prefabricated unit cells* (prefab-cell) can be placed in a specific location to build the final structure rapidly regardless of the number of cell in the output structure.

P-HGM process can be divided into three phases: B-Rep modelling, prefabrication of unit cell, and creating output structure. An overview is illustrated in see Fig. 1.

Phase1: Modelling

Initially, the unit cells are modelled with B-rep due to the flexibility in the modelling process. In comparison to polygonal model, the geometry of the B-Rep model can be modified easily and represented by control parameters, e.g. dimension of the cell and beam profile. In addition to using the built-in modelling feature based on cell template (see example in Fig. 2), CAD software can be used to produce the model (e.g. Fig. 1a) which will be fully compatible if the cell is symmetrical.

Phase2: Prefabrication of Unit Cells

The expected outcome from this phase is a set of prefab-cells that will be used to construct the complete structure in *Phase3*. Prefab-cells will be connected with each other by appending instead of surface Boolean merging which is a resource-

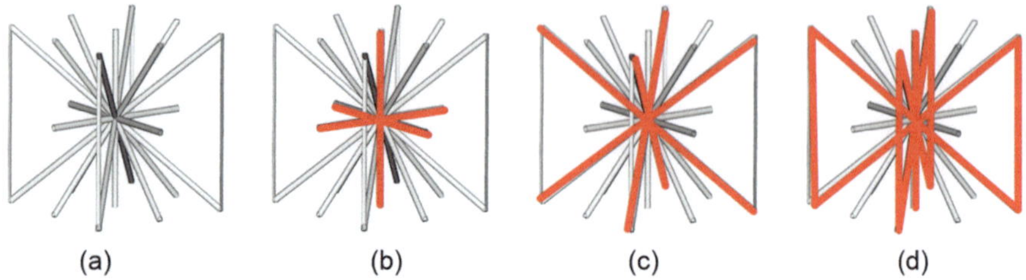

Fig. 2 Unit cell (**a**) template (**b**) Cross (**c**) X (**d**) XI

intensive process. Each unit cell will need to have certain faces removed, so that the edges can be connected seamlessly without producing bad edges.

After a symmetrical unit cell has been constructed, a set of prefab-cells (Fig. 3a) will be generated by considering the corresponding location in the output structure where the location can be known a priori (Fig. 3b). The location of the prefab-cell is considered with a cube with 3×3 unit cells. The prefab-cells can be classified into four characteristics according to the locations: *corner*, *edge of cube*, *centre of cube's face*, and *centre of cube*.

As a result, prefab-cells are generated in 27 types where the number and position of removed faces are different. In Fig. 3, the faces that will be visible in the complete structure will be maintained (white faces), while the others will be removed (gray faces). In summary, prefabrication process consists of three steps:

Step1: *Trimming with bounding box* (Fig. 1b): the excess parts of the B-Rep model will be trimmed with the bounding box of the unit cell.

Step2: *Remove faces* (Fig. 1c): particular faces of unit cells are removed according to the corresponding location.

Step3: *Triangulation* (Fig. 1d): the B-Rep model will be triangulated to have prefab-cells in polygonal model.

Phase3: Creating Output Structure

The prefab-cells will be populated over the entire volume of the bounding box of the output structure. Each prefab-cell will be placed at the equivalent location as the aforementioned four characteristics (Fig. 1f). Appending the cells is simple and not resource intensive as the data of the polygonal cell is appended from the rest without further calculation (Fig. 1g). Eventually, the output cellular structure with polygonal surface has been created.

3.2 Selective Volume Fusing

SVF is a 3D modelling method that modifies the base part by substituting a specified cutting volume – in this case, primitive geometry – with a cellular structure created by CSM (see Fig. 1h). The substitution is done with a sequence of surface Boolean operations.

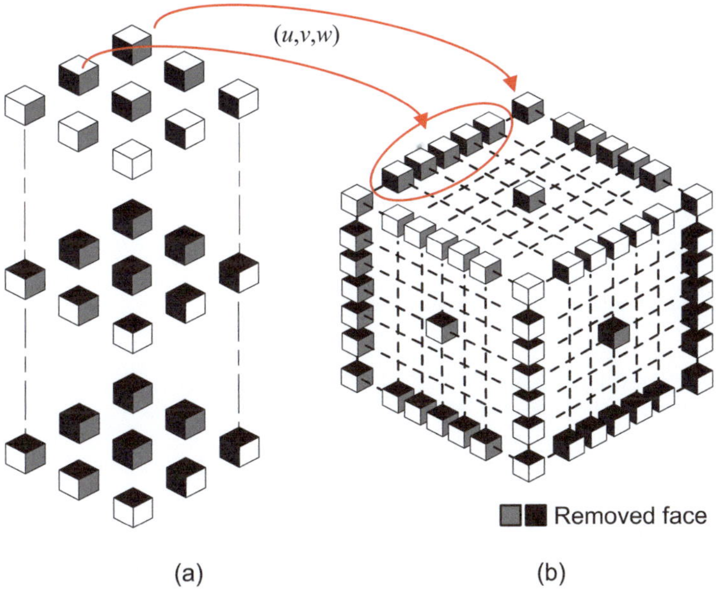

Fig. 3 Corresponding location (**a**) prefab-cell (**b**) output

As an overview, the surface Boolean operation is applied in three steps (see Fig. 4). First, (*Op-1*): the specified cutting volume is subtracted from the base part. Concurrently, (*Op-2*): the cellular structure is intersected with the cutting volume. The cavity resulted from *Op-1* will be exactly the same size as the structure produced from *Op-2*. Finally, (*Op-3*): the parts from *Op-1* and *Op-2* will be merged by adding them together.

In order to prove this concept, the process integration process in Fig. 4 is done manually in FFD with the following steps (see Figs. 5 and 6):

Step1: Specify Cutting Volume

The cutting volume can be any geometry (in this case, a rectangular box). The size of the bounding box of the cellular structure equals to the size of the bounding box of this cutting volume (see Fig. 6a).

Step2: Generate Cellular Structure

The cellular structure is generated by using CSM (see Fig. 5) according to the specified parameters: kernel types, profile thickness, polygon profile, cell size, and bounding box dimension. *Kernel types* are the configuration of the unit cell which is specified based on the templates shown in Fig. 2. The profile of the beams used in the unit cell is characterised with *polygonal profile* (n-side) and *thickness*. The size of the unit cell is set with *cell size*. The size of the entire cellular structure is determined by *bounding box*.

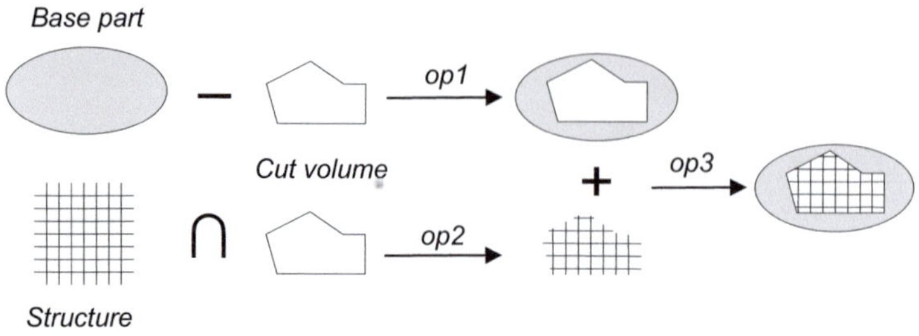

Fig. 4 Surface Boolean operation for part integration

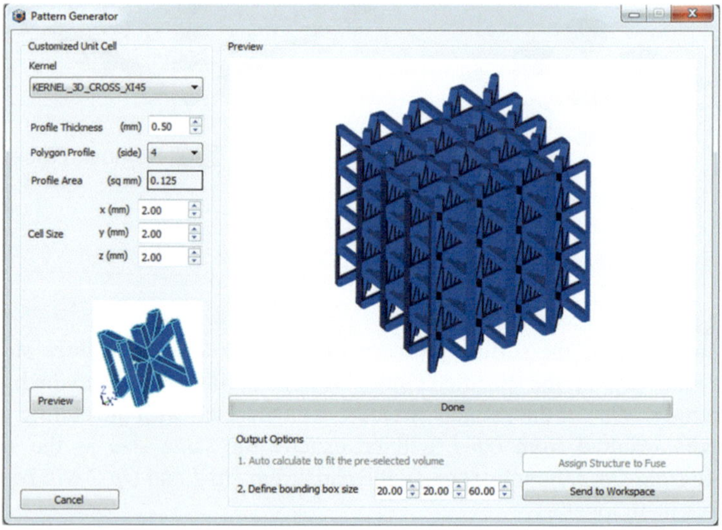

Fig. 5 Cellular structure generation module

Step3: Intersect Volumes

The intersection between the base part and the cutting volume (see Fig. 6b) will be used to obtain the exact intersected part of cellular structure (see Fig. 6c). The enclosure of the part will be aligned with the surface of the base part. However, the problem due to coplanar in the surface Boolean operation can occur. To avoid this issue, the cellular structure can be enlarged in every direction with minimal value (e.g. 0.001 mm) that will not affect the built quality in the manufacturing process.

Step4: Merge Volumes

The outcome cellular structure will substitute the cutting volume and then merge with the base part that has been cut (see Fig. 6c). The outcome is shown in (Fig. 6d, e). For VTK functions, *vtkSurfaceBooleanOperationFilter* are used for all surface Boolean operations, i.e. *Intersect*, *Add*, and *Subtract*. The parts are in the form of *vtkPolyData*.

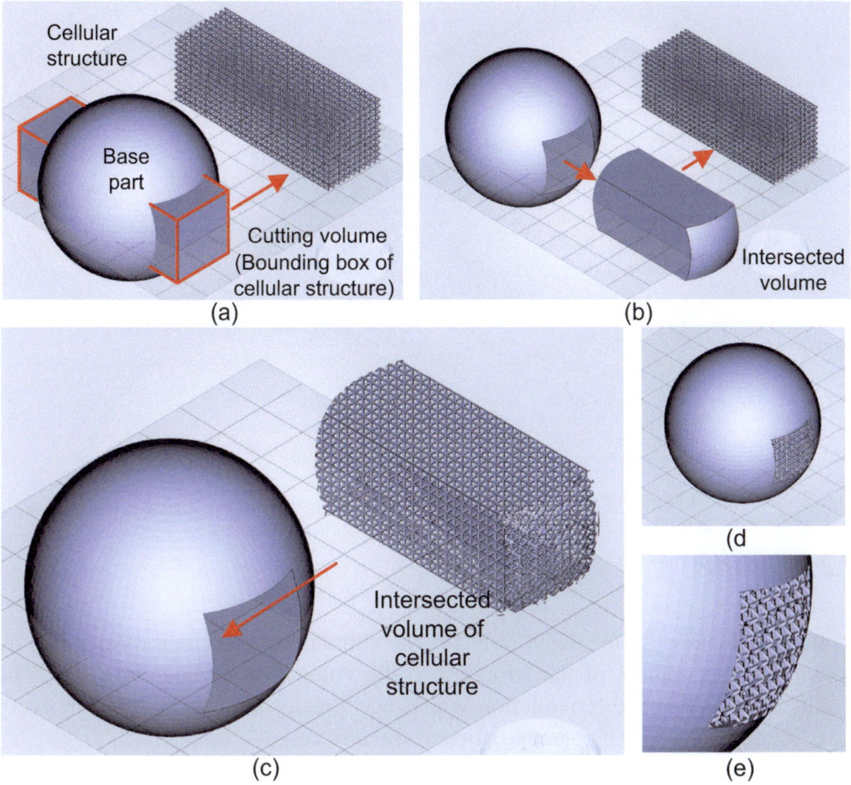

Fig. 6 Sequence of Boolean operations

4 Experiments

4.1 Experiment Method

The experiments are designed to measure the performance of this method, including the cellular structure generation phase and SVF phase. The cellular structures are generated with various set of and merged to a sphere (50 mm diameter with 9024 triangles) as shown in Fig. 5. The size of the *cutting volume*, cell *width, and profile thickness* is fixed. The cutting volume bounding box size = $20 \times 60 \times 60$ mm, cell width = 2 mm, and profile thickness = 0.5 mm. The characteristics of unit cell vary with two kernel types (Cross and X) and three polygonal profiles (n-side = 4, 8, and 16) as the samples shown in Fig. 7.

The experiment was done on a personal computer with Windows 7 64-bit, Intel Core-i7 230 GHz, 8GB RAM, and NVIDIA GeForce GT 630M Graphic Card.

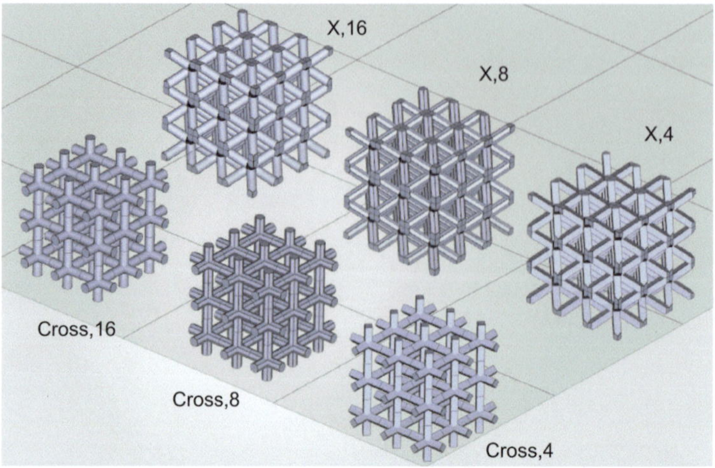

Fig. 7 Samples of structure with various parameters

4.2 Results

Volume reduction is calculated only for the cellular structure before merging by comparing the volume of the structure to the volume of its bounding box. The results show considerable reduction which was greater than about 75 % in every case due to the specified characteristics of the unit cells.

Regarding the complexity of the cellular structure, it can be indirectly determined by the number of triangles that vary according to the number of beams in the kernel (six beams for "Cross" and eight beams for "X") and the profile n-side. The processing time for generating the structure ranges 1–4 s due to the complexity of the unit cells. However, the variations are independent from the number of cells populated, thanks to the P-HGM method.

For the results due to the merging process (see Tables 1 and 2), processing time varied according to the complexity of the structures. For both kernel types, the merging process took longer as the number of triangles increased.

Little time variations among the cases and artefacts, e.g. bad edges and nonexisting cells, occurred due to the reliability issues of the surface Boolean operation of VTK. However, from the results, the proposed method was able to produce the outcomes in 50–500 s (see results in Fig. 8). The existing minor artefacts can later be repaired easily by using other commercial software.

Table 1 Experiment of unit cell kernel "Cross"

Properties		Profile n-side 4	8	16
Cellular structure	Generation time (s)	1.18	1.74	1.82
	Number of triangles	2.18E+5	2.96E+5	5.96E+5
	Volume reduction	91.8 %	88.6 %	87.8 %
Result	Merging time (s)	58	96	145
	Number of triangles	1.51E+5	4.64E+5	5.05E+5

Table 2 Experiment of unit cell kernel "X"

Properties		Profile n-side 4	8	16
Cellular structure	Generation time (s)	2.47	3.27	3.65
	Number of triangles	3.72E+5	4.98E+5	8.98E+5
	Volume reduction	74.2 %	75.9 %	82.3 %
Result	Merging time (s)	139	202	507
	Number of triangles	3.53E+5	4.64E+5	8.04E+5

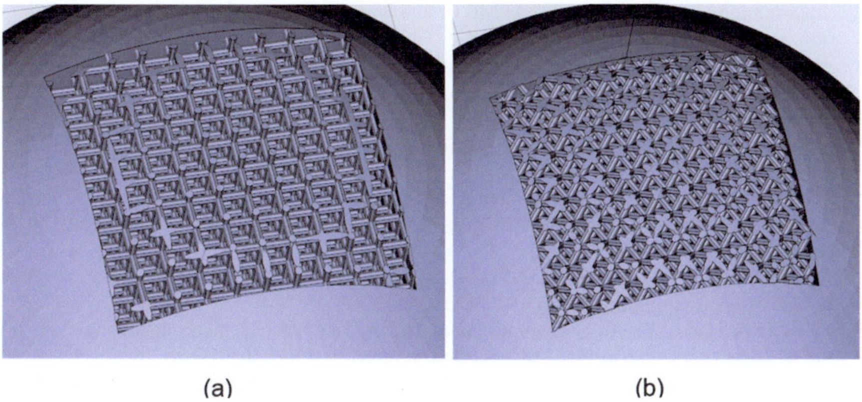

Fig. 8 Merging results with n-side = 8 (**a**) Cross (**b**) X

5 Conclusions

This article presents development of a modelling software tool for generating cellular structures and the concept of SVF for integrating the structure to the base models. The generation process is based on hybrid geometry modelling with prefabricated unit cell, P-HGM, which can generate the structure at minimal computing resource and processing time. The integration method, SVF, is able to merge the structure to any model at selected areas. The SVF is based on a sequence of surface Boolean operation.

In addition, the implementation of these methods is done in FFD which are based on VTK and OCCT. As a result, the experiments show the successful demonstration of these methods. The result also showed the parts with volume reduction due to the substitution of cellular structure. This does not only serve specific applications, e.g. aesthetics, but also reduce the material usage and manufacturing time.

For the future work, SVM method can be improved in many ways. The user interface for applying SVM should be more intuitive and more flexible which is not limited to primitive geometry only. In addition, the functionality of the outcome parts must be further analysed.

Acknowledgement This research is part of a project Free Form Design which is mainly funded by Advanced Manufacturing Service, Breseight Pty, Ltd, Australia. The research regarding the design and manufacturing cellular is partially supported by Tech Voucher programme, New South Wales Government, Australia, under grant number 01182228.

References

1. Doubrovski Z, Verlinden JC, Geraedts JMP (2011) Optimal design for additive manufacturing: opportunities and challenges. In: ASME design engineering technical conference, Washington, DC, USA, 28–31 August 2011, pp 635–646. doi:10.1115/DETC2011-48131. ISBN: 978-0-7918-5486-0. http://proceedings.asmedigitalcollection.asme.org/proceeding.aspx?articleid=1641854&resultClick=3
2. Strano G, Hao L, Everson RM, Evans KE (2013) A new approach to the design and optimisation of support structures in additive manufacturing. Int J Adv Manuf Technol 66 (9–12):1247–1254
3. Yan C, Hao L, Hussein A, Raymont D (2012) Evaluations of cellular lattice structures manufactured using selective laser melting. Int J Mach Tools Manuf 62:32–38. doi:10.1016/j.ijmachtools.2012.06.002
4. Queheillalt DT, Wadley HNG (2005) Cellular metal lattices with hollow trusses. Acta Mater 53(2):303–313, http://dx.doi.org/10.1016/j.actamat.2004.09.024
5. Hussein A, Hao L, Yan C, Everson R, Young P (2013) Advanced lattice support structures for metal additive manufacturing. J Mater Process Technol 213(7):1019–1026
6. Pasko A, Vilbrandt T, Fryazinov O, Adzhiev V (2010) Procedural function-based spatial microstructures. In: SMI 2010 – International conference on shape modeling and applications, proceedings, Aix en Provence, France, 21–23 June 2010, pp 47–56. doi:10.1109/SMI.2010.19. ISBN: 978-0-7695-4072-6. https://www.computer.org/csdl/proceedings/smi/2010/4072/00/4072a047-abs.html
7. Wang H, Chen Y, Rosen DW (2005) A hybrid geometric modeling method for large scale conformal cellular structures. In: Proceedings of the ASME international design engineering technical conferences and computers and information in engineering conference – DETC2005, Long Beach, CA, USA, 24–28 September 2005, pp 421–427. doi:10.1115/DETC2005-85366. ISBN: 0-7918-4740-3; eISBN: 0-7918-3766-1. http://proceedings.asmedigitalcollection.asme.org/proceeding.aspx?articleid=1588169&resultClick=3
8. Pty.Ltd. B (2014) Free form design™ – endless creativity with effortless modelling. http://www.free-form-design.com/
9. Kitware VTK (2013) Visualization toolkits. http://www.vtk.org/
10. OpenCascadeS.A.S (2014) Open cascade technology. http://www.opencascade.org/
11. Vongbunyong S, Kara S (2015) Rapid generation of uniform cellular structure by using prefabricated unit cells. Int J Comput Integr Manuf. http://dx.doi.org/10.1080/0951192X.2016.1187303 (In-review)

Recovery of Metals from E-Waste Mediated by Molten CRT Lead Glass

Hiroyuki Inano, Keiichi Tomita, Tatsumi Tada, and Naoki Hiroyoshi

Abstract Different metal recovery from unsorted e-waste by smelting was investigated. Metallic Pb (lead) can be recovered from cathode ray tube (CRT) glass containing PbO by reduction melting. During the reduction melting, molten Pb has the property of mixing with other metals, and basic experiments simultaneously recovering different metals in different kinds of e-waste were investigated. Reagents of major metal components contained in printed circuit boards (PCB) and liquid crystal display (LCD) panels were added to lead silicate glass. The mixture was melted in a reductive atmosphere, and the obtained glass and metal were analyzed. Better than 90 % of Au, Ag, Cu, and Ni were recovered together in metallic form, with the recovery rate of indium 60 %. Chemical thermodynamics calculations suggested that the recovered metals are more difficult to oxidize than CO. Further, the chemical thermodynamics calculations suggested that recovery of metals that were not tested would also be possible.

Keywords CRT glass • Lead • E-waste recycling • Precious metals • Reduction melting

1 Introduction

1.1 Treatment of Waste CRT Glass

Cathode ray tube (CRT) production has decreased steeply due to the replacement of CRT by liquid crystal displays (LCD), and it may be assumed that worldwide CRT production will terminate in the near future. As a result, it will not be possible to recycle CRT glass cullet in new CRT, the "closed loop recycling" that is presently going on.

H. Inano (✉) • K. Tomita • T. Tada
Industrial Research Institute, Hokkaido Research Organization, Sapporo, Japan
e-mail: inano-hiroyuki@hro.or.jp

N. Hiroyoshi
Graduate School of Engineering, Hokkaido University, Sapporo, Japan

M. Matsumoto et al. (eds.), *Sustainability Through Innovation in Product Life Cycle Design*, EcoProduction, DOI 10.1007/978-981-10-0471-1_36

In Japan, an annual 600–800 kt of CRT glass was produced at the peak, but since 2005 there has been no production of new CRT glass [1].

At present 2 million CRT TVs are recovered annually, generating 20 kt of glass cullet [2]. The CRT glass in the TVs contains heavy metals to shield against the X-ray radiation generated in the CRT.

The front panel glass of the CRT contains BaO and SrO as heavy metal oxides, and the glass in the funnel part contains up to 25 mass% of harmful PbO. This makes the chemical composition of the glass very different from that of mass-produced glass, such as that used in sheet glass and bottles [3], and CRT glass cullet cannot be directly utilized as raw material for other glass.

However, demand for Pb raw material for battery electrodes in automobiles and industry is increasing worldwide. In 2014, 11 million tons of Pb was produced worldwide [4], making the PbO containing CRT funnel glass a promising Pb resource.

The Pb contained in glass can be recovered, with Pb recovery methods that are categorized into two types, dry and wet processes. The dry process is enabled by the application of pyrometallurgy, melting at high temperatures, and is achieved through reduction melting [1, 5–8], chloride volatilization [9–11], and pyrovacuum [12] processes. In reduction melting, the glass is melted with reducing and viscosity-reducing agents and involves recovering of precipitated Pb. The chloride volatilization method melts glass with chloride and recovers evaporated $PbCl_2$. In the pyrovacuum method, the glass is melted with a reducing agent at low pressure and the recovered product is evaporated Pb. Commonly employed wet processes are the autoclave treatment [13] and mechanochemical extraction [14]. The wet processes involve fine grinding, very long reaction times, and wastewater treatment.

In Japan, several lead smelters accept CRT funnel glass at present. At first, dismantled CRT is separated into panels and funnels. The funnel glass is crushed locally and transported to a lead smelter. The valuable PbO content in funnel glass is 25 mass% of the crushed product, and even the low value glassy parts that comprise 75 % of funnel glass must be transported long distances.

1.2 Waste Containing Valuable Metals

Waste from electric household appliances contains valuable metals: liquid crystal displays (LCD) contain indium (In) which is included in transparent conducting oxide thin film as indium tin oxide (ITO); printed circuit boards (PCB) in PCs and mobile phones include gold (Au), silver (Ag), copper (Cu), and nickel (Ni) in the circuits, as well as tantalum (Ta), titanium (Ti), and barium (Ba) in the electronics parts and tin (Sn) in the solder. This has made the e-waste (electronic waste) become considered as an "urban mine". These kinds of e-waste and the included metals are listed in Table 1.

In Japan, collected PCB is incinerated to concentrate metal as a primary treatment in many locations, but the copper contents in the residue are still less than half.

Table 1 Kinds of e-waste and the included metals

Waste	Included metals
LCD panel	In as In_2O_3 in ITO thin film
PCB	Au, Ag, Cu, and Ni of circuits; Ta, Ba, and Ti in electronic components (capacitor); Sn in solder

To recover copper from PCB residue, even low value waste must be transported long distances.

1.3 Multi-Metal Recovery

When lead glass is melted at reductive conditions, the generated molten lead collects the coexisting metals in the glass, and when CRT glass is melted with metal containing e-waste, Pb would be precipitated with other valuable metals contained in the melt. The Pb alloy would separate from the metal-less glass residue, and the molten Pb can be employed as a collector of several metals. Among these, the Pb is an indispensable base metal, but the price is not very high at about 2000 US$/ton recently, much lower than Au and Cu. If a mixture of Pb including valuable metals is recovered, the product price would increase and CRT recycling would become economical. The various metals recovered together with the Pb can be separated in the Pb refining process.

If PCB and CRT funnel glass can be treated together, Pb and the valuable metals in the PCB can be extracted locally, and this would decrease transportation costs and CO_2 emissions, as well as the residue can be utilized at the widely dispersed treatment locations.

Basic experiments investigated the recovery of several metals simultaneous with Pb recovery from lead glass.

2 Experimental

2.1 Materials

A model glass was prepared to simulate actual CRT glass. Actual CRT funnel glass is composed of more than ten kinds of oxides and is a complicated component where it is difficult to know which of the components are involved in the reaction. For the experiments, a model glass composed of only four oxides was used. The model glass contained 25 mass% of PbO and has properties similar to actual funnel glass. The chemical composition of the model glass is shown in Table 2, and it was used with grain sizes smaller than 1 mm in the melting tests. The tests used Na_2CO_3

Table 2 Chemical composition of the model glass [15]

Component	Mass%
SiO_2	53.9
PbO	25.5
Al_2O_3	4.2
Na_2O	16.3

reagent as a viscosity-reducing agent and activated carbon powder as a reducing reagent. For the contained metals, Au, Ag, Cu, and Ni as present in PCB and In as in LCD were selected as the target metals to be recovered with the Pb, all in the form of metal powder reagents rather than actual waste.

2.2 Melting

Volumes of 20 g of model glass, 10 g of Na_2CO_3, 3 g of activated carbon (C), and each of 0.2 g of metal powders of Au, Ag, Cu, Ni, and In were mixed and placed in a 99.5 % alumina crucible. The crucible was set in an electric furnace and heated to 1473 K at a heating rate 300 K/H. After being kept at 1473 K for 1 h, the melted sample was poured into an iron mold, and after cooling to room temperature, the glassy and metallic phases were separated.

2.3 Characterization

The glassy phase was characterized as follows. The microstructure was observed by a digital optical microscope (KH-1300, Hirox, Japan). The crystalline phase was identified by powder X-ray diffraction (XRD, Ultima-IV, Rigaku, Japan). The chemical composition was analyzed by X-ray fluorescent analysis (XRF, ZSX Primus II, Rigaku, Japan). The thermal properties were analyzed by differential scanning calorimetry (DSC, Exstar DSC6200, Seiko Instrument, Japan), for 20 mg of sample heated at 10 K/min.

The metallic phase was observed by scanning electron microscopy (SEM, JSM-6610LA, JEOL, Japan), and the elementary analysis and mapping were carried out by energy-dispersive X-ray spectroscopy (EDS, JED-2300, JEOL, Japan).

The chemical states of the glassy and metallic phases were evaluated by X-ray photoelectron spectroscopy (XPS, Axis-HSi, Shimadzu/Kratos), with the binding energy calibrated using the contaminated hydrocarbon C1s line (284.6 eV). The standard peak positions of metal and oxide states for each element were obtained by measurement of the chemical reagents of the added metals and these oxides.

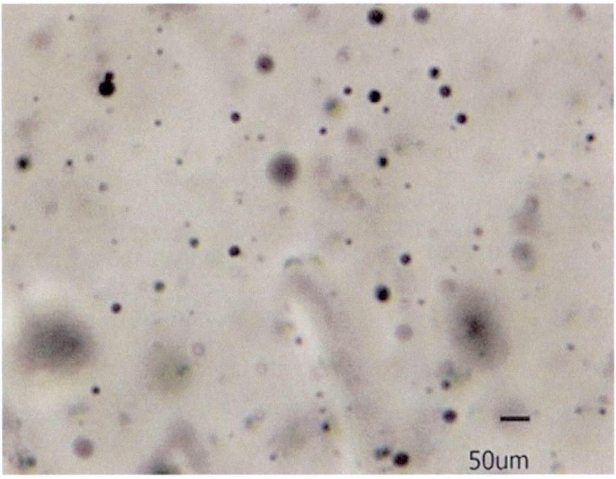

Fig. 1 Optical microscopic image of the obtained glassy phase

3 Results and Discussion

3.1 Characterization of the Glassy Phase

The obtained glassy phase looked opaque gray. A transmitted light microscopic image is shown in Fig. 1, with some micrometer to 10 s of micrometers of black particles in a transparent colorless glass matrix. An X-ray diffraction pattern of the obtained glassy phase is shown in Fig. 2. There are only Pb peaks on a halo indicating that it is an amorphous phase.

The result of the chemical analysis by XRF is shown in Table 3. Based on this result, the distribution ratios of the added metals between the glassy and metallic phases are shown in Fig. 3. Among the added metals, over 90 % of the Au, Ag, Cu, and Ni were in the metallic phase. However, for In only 61 % was in the metallic phase with the remaining 39 % in the glassy phase.

The DSC result is shown in Fig. 4. The horizontal axis indicates the temperature (K) and the vertical axis the DSC value. There is an endothermic peak starting from 573 K. The melting point of pure Pb is 600 K, and this melting point depression suggests that the Pb particles are not purely metallic but are present as an alloy. It is noteworthy that Pb particles in glass considered here are in a similar alloyed state.

3.2 Characterization of the Metallic Phase

A cross section of the obtained metal phase was observed by SEM and analyzed by EDS. The Pb from glass and the Au, Ag, Cu, and Ni of the added metals were detected by EDS. Elemental mapping for the detected metals was carried out, and a SEM image (a) and results of the mapping (b–f) are shown in Fig. 5. The scale at the

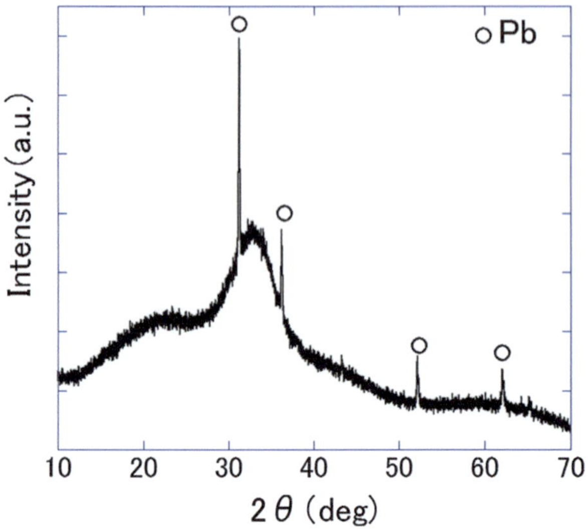

Fig. 2 X-ray diffraction pattern of the obtained glassy phase [15]

Table 3 Results of the chemical analysis of the glassy phase by XRF [15]

Component	Mass%
Na_2O	43.2
SiO_2	50.0
Al_2O_3	4.20
NiO	0.11
CuO	0.11
Ag_2O	0.06
In_2O_3	0.47
Au	0.01
PbO	1.74

Fig. 3 Distribution ratios of the added metals between the glassy and metallic phases

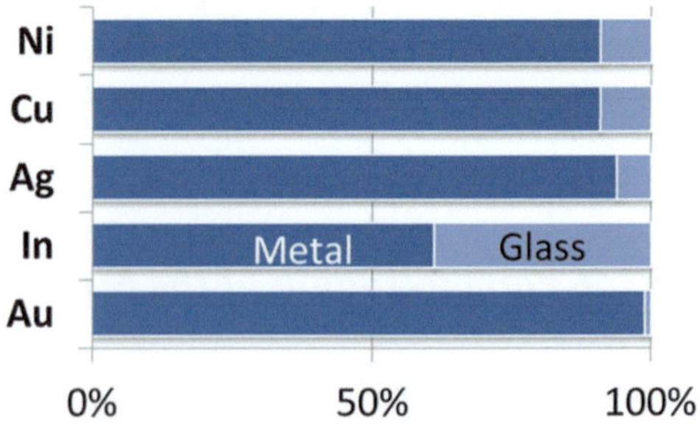

Fig. 4 DSC curve for the glassy phase

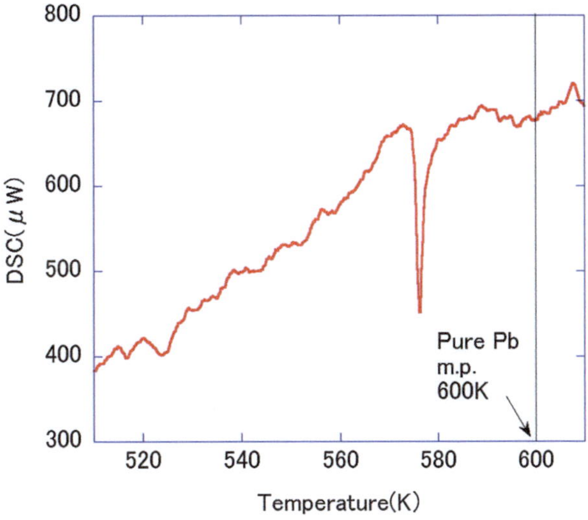

left of the map shows the gradient from high (white) to low (black) element concentrations. As shown in Fig. 5f, there is Pb throughout the whole area of the image. At the central area where Pb is not present, there are signals for Ni and Cu. At the lower right, there is Au and Ag. The results show that the major component in the metal phase is Pb and that there are several micrometer diameters Au, Ag, Cu, and Ni alloy particles present.

3.3 Details of the Chemical States of Each Element by XPS

The chemical states of each element contained in the glassy and metallic phases were compared and will be described here. The chemical states can be analyzed by X-ray photoelectron spectroscopy (XPS), and a chemical shift in the inner shell electron binding energy indicates the chemical states of an element.

The results of the metal phase analysis are described, first with a survey scan and qualitative analysis of the binding energy at 1100–0 eV. There were strong Pb peaks, and Au, Ag, Cu, and Ni peaks were identified. Using narrow scans for the inner shell photoelectrons and Auger electrons of these elements, comparing the peak positions, showed that the peak positions of each element agreed with or was close to the metal states, not the oxide states.

In the results of the glassy phase analysis, only indium among the added metals was detected by the survey scan. The photoelectron spectrum of the $In3d_{5/2}$ narrow scan is shown in Fig. 6, showing the standard peak positions of the chemical reagents indium and In_2O_3. The peak position of indium in the glassy phase is different from that of the metallic phase, shifted toward a higher binding energy

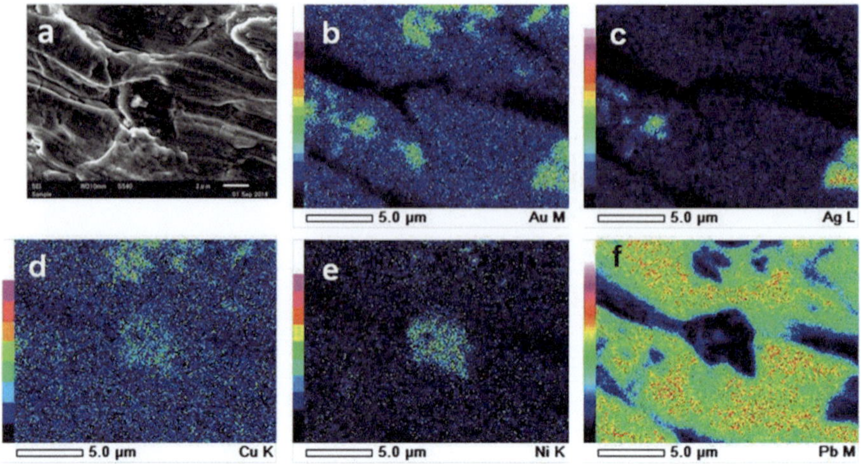

Fig. 5 SEM image (**a**) of the metallic phase and EDS maps of the same area for (**b**) Au, (**c**) Ag, (**d**) Cu, (**e**) Ni, and (**f**) Pb

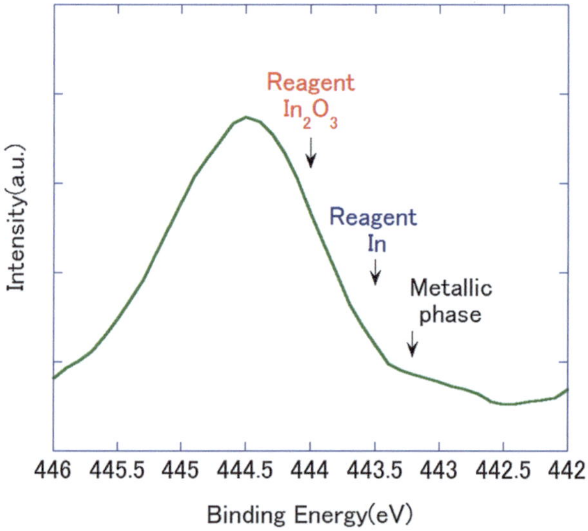

Fig. 6 XPS spectrum of the In $3d_{5/2}$ of the glassy phase [15]

than that of In_2O_3. This result suggests that the indium contained in the glassy phase is present as an oxide in the glass structure.

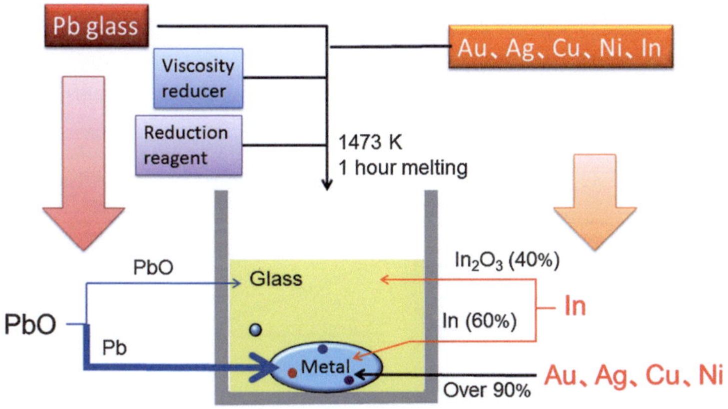

Fig. 7 Material flow of the reduction melting of the model glass (lead glass) and added metals (Au, Ag, Cu, Ni, and In) reported here

3.4 *Distribution of Elements*

The experiments here melted lead glass at 1473 K for 1 h with a reducing reagent, viscosity reducer, and selected metals (Au, Ag, Cu, Ni, and In). Most of the PbO contained in the glass was reduced to Pb. More than 90 % of the Au, Ag, Cu, and Ni were recovered in the metallic phase together with Pb, while only about 60 % of indium was recovered with the metals, and the remaining 40 % was included in the glassy phase as glass component oxide. The overall material flow in the experiments is shown in Fig. 7.

3.5 *Chemical Thermodynamics of Oxidation Reduction*

Chemical thermodynamics calculations were carried to determine if the added metals were present as metals or oxides.

When a metal M forms an oxide M_mO_n, with 1 mol O_2, the equilibrium of the oxidation reduction is commonly expressed as

$$\frac{2m}{n}M + O_2 = \frac{2}{n}M_mO_n \tag{1}$$

where m and n are integers.

The change in the standard Gibbs free energy ΔG^0 is expressed as

$$\Delta G^0 = -RT\ln K \tag{2}$$

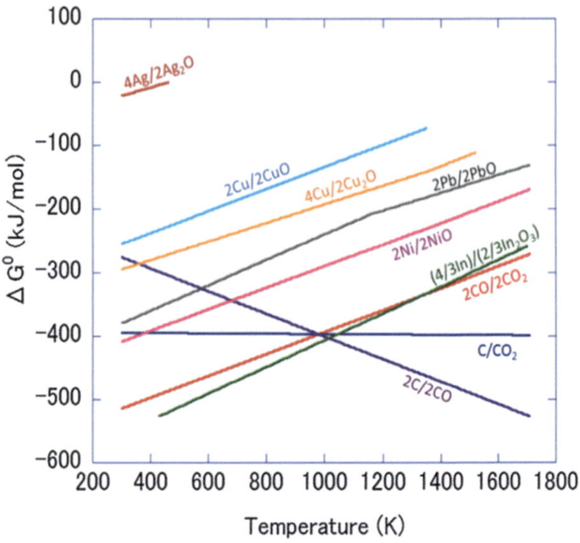

Fig. 8 Ellingham diagram of the added metals, Pb, and C (Except Au)

where R is the gas constant, T is the absolute temperature, and K is an equilibrium constant. The changes in ΔG^0 of an element versus T are shown by an Ellingham diagram. The Ellingham diagrams for Pb, the added metals, and the reducing reagent C are shown in Fig. 8. It has no data for Au, because Au does not form oxides. Ellingham diagram is a good-enough index to discuss the oxidation-reduction reaction.

With increasing negative values of ΔG^0, the equilibrium reaction (1) moves to the right (oxidation). Oxides with higher plots are reduced by metals or C compounds which lie below. As in Fig. 8, the ΔG^0 negative values of Ag and Cu are smaller than those of Pb, and when PbO is reduced to Pb, Ag and Cu are not oxidized. The 2Ni/2NiO line lies below that of 2Pb/2PbO, but above that of $2CO/2CO_2$. The CO in the reactions here is generated by incomplete combustion of C, and Ni is not oxidized in the presence of CO here. This makes it possible to assume that Au, Ag, Cu, and Ni are present as metals when heated with C (carbon) as the reducing reagent. For indium, the $(4/3In)/(2/3\ In_2O_3)$ line is at a level similar to that of $2CO/2CO_2$, and indium may be assumed to oxidize to In_2O_3 when CO oxidizes to CO_2. Further, the $(4/3In)/(2/3\ In_2O_3)$ line lies below that of 2Pb/2PbO, and the PbO contained in the glass is reduced to Pb by the indium.

$$3PbO + 2In \rightarrow 3Pb + In_2O_3 \qquad (3)$$

The In_2O_3 is dissolved and remains as a component of the glass functioning as a network modifier (NWM) in the silicate glass [16], and In_2O_3 is ionized in glass as below.

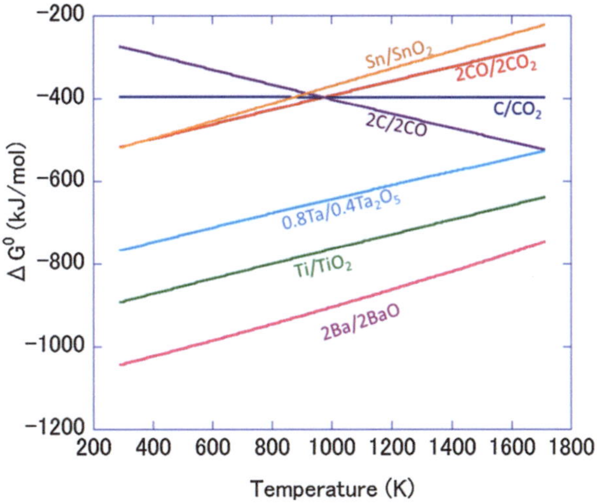

Fig. 9 Ellingham diagram of oxidation of metals not investigated here

$$In_2O_3 \rightarrow 2In^{3+} + 3O^{2-} \quad (4)$$

The net result is that when lead glass is melted with C, the Au, Ag, Cu, and Ni are recoverable in the Pb alloy, while indium will tend to oxidize and dissolved into the glassy phase. This makes it difficult to recover indium by the reduction melting method employed here. Actually, the indium in LCD panels is present as In_2O_3, adding to the difficulty when attempting to reduce and recover indium.

The ΔG^0 was also calculated for metals that were not tested here. Tantalum (Ta), barium (Ba), titanium (Ti), and tin (Sn) were selected from among the metals contained in PCB (Table 1). The Ellingham diagram of oxidation for these metals with 1 mol O_2 is shown in Fig. 9, which shows that the Sn/SnO_2 line lies above that of $2CO/2CO_2$.

It may be expected that Sn can be recovered by the reduction melting method, the plots for Ta, Ti, and Ba lie below that of $2CO/2CO_2$ and they will not be reduced by C and cannot be recovered by this method.

4 Summary

Metal recovery of metals added to a model lead glass by reduction melting was investigated. The results showed that Au, Ag, Cu, and Ni could be recovered with Pb, but that indium remained in the glass as an oxide. These results agree with chemical thermodynamics calculations, which showed that the recoverable metals are more difficult to oxidize than CO. The tendency toward oxidation of indium is similar to that of CO, and the recovery rate of indium was lower. The possibility of recovering elements that were not tested was also examined by chemical thermodynamics calculations. From the calculated values of ΔG^0, it has been suggested that Sn may be recovered, but that Ta, Ti, and Ba would be difficult to recover by this process. The process can be applied to recover several metals from e-waste together in the reduction melting of CRT funnel glass. If several kinds of waste can be treated at smaller plants in widely distributed locations, it would decrease the CO_2 generated by long-distance transport.

Acknowledgments This research was supported by a grant of The Environmental Research and Technology Development Fund (ERTDF, No. 3K-123019) from the Ministry of the Environment, Japan. The authors would also like to thank Ryo Awaji and Kozo Sugiyama for assistance of experimental work.

References

1. Inano H (2009) Pb recovery from the waste CRT glass by reduction melting method. In: Proc. of eco design 2009, 6th Int. Symp. on Environmentally Conscious Design and Inverse Manufacturing (CD-ROM). Sapporo, Japan. 7–9 Dec 2009
2. Association for Home Electric Appliance (2014) Kaden risaikuru nenji houkokusyo (Annual reports of electric home appliances recycling). pp 22–24. [In Japanese]
3. Inano H, Tada T, Okada T, Hiroyoshi N (2013) Recycling technology for waste CRT glass. In: Proc. of eco design 2013, 7th Int. Symp. on Environmentally Conscious Design and Inverse Manufacturing, EcoDesign 2013 (USB memory). Jeju, Korea. 4–6 Dec 2013
4. International Lead and Zinc study Group. http://www.ilzsg.org/static/statistics.aspx?from=1
5. Industry Council for Electronic Equipment Recycling (ICER) (2003) New approach to Cathode Ray Tube (CRT) recycling
6. Inano H, Hashimoto Y, Kudo K (2005) lead separation from used CRT glass by reduce melting. Rep Hokkaido Ind Res Inst 304:71–77 [in Japanese]
7. Inano H (2012) Effect of alkali metal oxide on Pb recovery from the waste CRT glass by reduction melting method. In: Matsumoto M, Umeda Y, Masui K, Fukushige S (eds) Design for innovative value towards a sustainable society, Proc. of Eco Design 2011, 7th int. symp. on environmentally conscious design and inverse manufacturing. Kyoto, Japan. Springer, Heidelberg, pp 896–900
8. Okada T, Inano H, Hiroyoshi N (2012) Recovery and immobilization of lead in cathode ray tube funnel glass by a combination of reductive and oxidative melting processes. J Soc Inf Disp 20:508–516
9. Kaji H, Abe K (2004) Toku Kai (Japanese unexamined patent application publication) 2004–162141 [in Japanese]
10. Mongi H, Naruoka T, Igura T, Yoshioka T, Fujimori T (2014) Toku Kai 2014–94366 [in Japanese]
11. Inano H, Tada T, Okada T, Hiroyoshi N (2012) Lead recovery from waste CRT glass by reduction melting/chloride volatilization hybrid method. Ceram Jpn 47:96–100 [in Japanese]
12. Chen M, Zhang FS, Zhu J (2009) Lead recovery and the feasibility of foam glass production from funnel glass of dismantled cathode ray tube through pyrovacuum process. J Hazard Mater 161:1109–1113
13. Yoda S, Ohtake K, Takebayashi Y, Kanda T, Toku Kai 2002–346500 [in Japanese]
14. Sasai R, Kubo H, Kamiya M, Ito H (2008) Development of an eco-friendly material recycling process for spent lead glass using a mechanochemical process and Na2EDTA reagent. Environ Sci Technol 42:4159–4164
15. Inano H, Tomita K, Tada T, Hiroyoshi N (2014) Metal recovery mediated by molten Pb glass. In: Sigen-Sozai kouensyu (Proc. of fall conference of The Mining and Materials Processing Institute, Japan) [in Japanese]
16. Smets BMJ, Krol DM (1984) Group III ions in sodium silicate glass. Part 1. X-ray photoelectron spectroscopy study. Phys Chem Glasses 25:113–118

Part V
Strategy for Sustainable Society

Rethinking the Ecodesign Policy Mix in Europe

Carl Dalhammar

Abstract In the European Union (EU), there are a large number of mandatory and voluntary policies applied to improve the environmental life cycle impacts of products. Policies are enacted at both the EU level and among the EU member states, and there is a need to develop a better coordination between the various policies. This contribution outlines the various policies and examines various conceptual approaches to coordination. It further outlines some future policy challenges.

Keywords Product policy • Life cycle • Life cycle thinking • Policy mix • European Union

1 Introduction

The number of policy instruments for inducing ecodesign and product life cycle improvements has increased in the last decades [1, 2], and the European Union has become a leader in standard setting [3]. Early policies included extended producer responsibility rules, chemical content obligations, and eco-labeling schemes. Over time these rules have developed and been complemented with rules related to ecodesign and energy efficiency, public procurement, and energy labeling. Also the focus of these rules has changed over time. For instance, the Ecodesign Directive [4] has mainly been applied to set minimum requirements for product energy efficiency – like the Top Runner program in Japan – but has lately been used to set criteria to increase product longevity and induce resource efficiency [5]. While there is a need for many different types of policies, the large number of policies can be confusing. The fact that some policies and laws are enacted at the EU level whereas some are applied at the member state (national level) can add to this confusion. An important policy issue is therefore how the various policies should be coordinated and for what purposes. In this contribution, we will outline the various types of policies. We will then discuss various approaches toward

C. Dalhammar (✉)
IIIEE, Lund University, Lund, Sweden
e-mail: carl.dalhammar@iiiee.lu.se

M. Matsumoto et al. (eds.), *Sustainability Through Innovation in Product Life Cycle Design*, EcoProduction, DOI 10.1007/978-981-10-0471-1_37

coordination at the conceptual level and provide examples of how this is done in practice. Finally, we will provide some general recommendations on how the coordination could be improved.

2 The Policy Mix in the European Union

2.1 *Types of Policies*

When it comes to binding regulations, these are set at the EU level and can be categorized into three main categories. One category concerns chemicals in products. The EU regulates this for specific types of products, with regulations for several product groups including electrical and electronic equipment (EEE), toys, packaging, and building products. There is also horizontal legislation in place that covers a wide number of product groups, most notable through standard set in the main EU chemical regulation, the REACH [6]. REACH will have an indirect impact on chemicals in products as it sets standards for chemicals in several ways. If a chemical is phased out entirely or its use is highly restricted, this will influence also if and how the chemical is present in products. Exposure scenarios can also be important as they can account for and address exposure of chemicals from products. Further, REACH has some specific rules relating to chemicals in products. A second category of rules concerns recycling of waste products.

The EU has rules that set targets for recovery and recycling for several product groups (cf. Table 1). Further, there are rules that divert certain types of waste from landfills. Together, the rules provide a policy framework for increased recycling and energy recovery.

The rules related to chemicals in products and recycling complement each other. For instance, by removing hazardous substances from products, the recycling operations will be safer and more cost effective. However, the rules can sometimes also lead to conflicts. One example is when manufacturers prefer virgin materials to recycled ones, as recycled materials can contain some toxic substances which mean their products will not comply with chemical rules [7]. Thirdly, there are rules for the energy efficiency of products, so-called minimum energy performance standards (MEPS), which are set for specific product groups under the Ecodesign Directive [4]. In addition to these three main types of regulations, rules related to product durability and resource savings have started to emerge under the Ecodesign Directive [2, 5]. Examples include durability standards for vacuum cleaners and lighting products and the use of duplex printing as default for printers. More standards to promote durability and recycling are under discussion [8].

The mandatory regulations on product performance and characteristics are complemented by an increasing number of other instruments. They include mandatory energy labeling for some product groups, as well as voluntary energy labels

Table 1 EU rules for collection, recycling, landfilling

Waste stream	Target years	Minimum recovery	Minimum recycling	Collection rate
Packaging	2008	60 %	55 %	
Cars	2015	95 %	85 %	100 %
Electronics	2006	75 %	50 %	4 kg/person
Batteries	2011	50–75 %		
	2012			25 %
	2016			45 %
Tires	2006	Zero tires going to landfill		
Biowaste diverted from landfills	2006	Reduction to 75 % of the 1995 level		
	2009	Reduction to 50 % of the 1995 level		
	2016	Reduction to 35 % of the 1995 level		
New targets in waste framework directive	2015	Separate collection paper/metal/plastic/glass		
	2020	50 % municipal waste		
	2020	70 % construction and demolition waste		

(e.g., the Energy Star). When it comes to eco-labels, there is one EU-run eco-labeling scheme (the Flower), as well as several successful national (e.g., the German Blue Angel) and regional (e.g., the Nordic Swan) schemes. There are also other types of widespread labeling – e.g., TCO labeling [9] – that include environmental criteria.

The main rules for how to conduct public procurement are set at the EU level, but green public procurement (GPP) is executed at the national level. Most EU member states have their own GPP criteria and often establish their own guidelines for procurement. The EU procurement rules have widened the scope for environmental requirements allowed in GPP in the last 10 years [10]. Some EU member states have also worked with technology procurement and public procurement for innovation (PPI), which governments can make use of to trigger the development of new, more environment-friendly products on the market [11]. Typically, manufacturers can hesitate to develop better products if they are uncertain about the demand, and governments can make use of various tools to encourage the development of new products. For instance, Sweden has successfully used procurement to induce the design of more energy-efficient appliances such as fridges and freezers.

Governments also fund R&D and demonstration projects to trigger fundamental research into new product solutions, but generally it takes several years until the solutions reach markets.

Subsidies for consumers and industries can be used to increase the market uptake of energy-efficient products, which lead to larger market shares and – over time – lower prices for energy-efficient products, due to economies of scale and learning. Such schemes are set up at the national level.

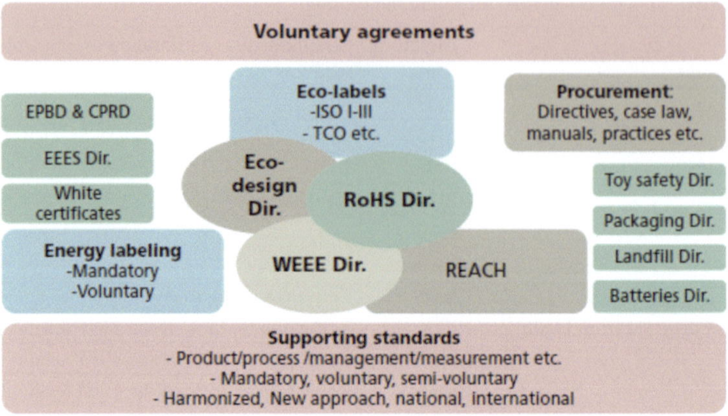

Fig. 1 Scope of ecodesign

2.2 Example: Electrical and Electronic Products

Figure 1 outlines the policy mix for electrical and electronic equipment (EEE): the Ecodesign Directive [4] sets rules for energy efficiency, whereas the REACH [6] and the RoHS Directive [12] regulates hazardous content. For EEE that are classified as toys or contain batteries, additional rules apply. Waste EEE is regulated through the WEEE Directive [13]. Many EEE, such as white goods and vacuum cleaners, are also subject to mandatory energy labeling rules. Manufacturers choose whether to apply voluntary tools, such as voluntary eco-labels and energy labels, on their products. Various standards are important for the setting of requirements and showing compliance. Some rules regulate the way voluntary instruments are applied. For example, EU procurement rules influence what environmental requirements governments can apply in their procurement programs [10]. Other policies (not further discussed here) can also impact EEE.

3 Approaches for Coordination

3.1 Introduction

Obviously, the various policies address different life cycle aspects. This is however not the only important dimension when it comes to a successful policy mix. Below, we discuss some important conceptual approaches to policy coordination and provide examples of their application.

3.2 *Policy Mix for Innovation Among All Manufacturers on the Market*

One way of looking at the interaction of standard set in various policies is to view them as having slightly different functions in relation to more and less progressive manufacturers and more and less environment-friendly products on the market (cf. Fig. 2).

The figure depicts the "ideal" situation for a given product group, with a focus on product energy efficiency. Mandatory standards (MEPS) are used to cut off the worst performers from the market, whereas eco-labeling, procurement, and other policies are used to stimulate the design, marketing, and diffusion of products with a better than average environmental performance. Over time, the interactions of the different instruments will improve the environmental performance of the whole product group and allow for the setting of stricter mandatory standards on product performance (represented by the figure at the bottom); also eco-label and GPP criteria will be strengthened over time. The different instruments tend to interact over time, in different ways, for instance [14]:

- Voluntary initiatives and economic instruments will often lead to legislation at a later stage, as progressive companies may lobby for this in order to avoid other enterprises being free riders.
- Non-regulatory initiatives might also cause or have the potential to cause trade distortions within the EU, thereby triggering a need for EU-wide regulations to promote a more level playing field.
- Voluntary initiatives will prove that certain measures are possible at reasonable costs and thereby combating industry arguments that regulations will be too burdensome. Thus, they can pave the way for legal standards at a later stage.
- Voluntary systems may provide benchmarks for legislation. For instance, criteria in eco-labeling schemes may over time be incorporated in laws and become the legal standard.

The main role of the mandatory standards is to eliminate the least environmental-friendly products and to push eco-innovation among the laggards,

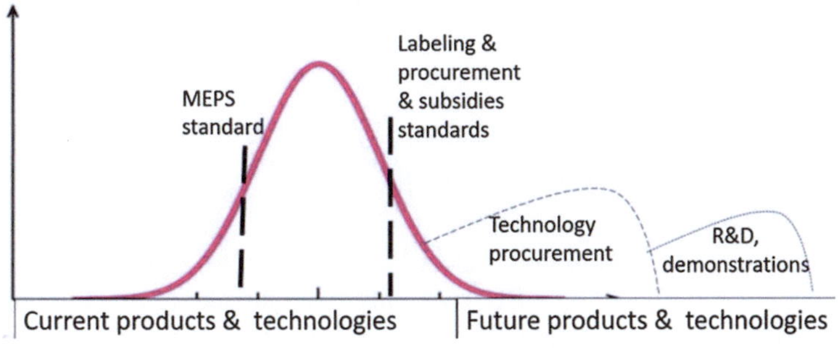

Fig. 2 Interactions of policies for product energy efficiency

whereas other instruments may stimulate eco-innovation among front-runners. An "optimal" way to strengthen policy is to coordinate the stringency of requirements among policy instruments and to regularly update criteria in both ends in order to push developments and provide continuous incentives for eco-innovation and ecodesign.

Looking at existing EU standards for energy efficiency, it is important that mandatory requirements (MEPS) are coordinated with standards applied those in labeling and procurement. Otherwise, we may run into problems. For instance, if requirements in energy labeling are not updated often enough, a product may get a high ranking in the energy label scheme though it does not comply with mandatory standards (if the MEPS are recently adopted). This could lead to confusion and undermine consumer confidence in the policies. For this reason, the European Commission has started to coordinate the process of setting requirements in the EU Ecodesign and EU Energy labeling directives. But there is less coordination of MEPS and eco-labeling and procurement criteria. A study on dishwashers found that both the EU Ecolabel and the Nordic Swan Ecolabel (run by governments in the Nordic countries) had failed to update their standards to keep up with mandatory energy standards [15]. This implies that eco-labeling schemes must be better at updating their standards to keep up with mandatory standard setting.

When it comes to chemicals in products, there are also possible ways to combine mandatory requirements on product content and mandatory information requirements. The RoHS Directive regulates the chemical content of EEE. The Ecodesign Directive has complemented these rules by regulating – for specific product groups – that when products contain certain chemicals, this must be communicated on the product packaging. This also means that consumers and professional purchasers can get and act upon this information.

Regarding recycling policies, a proposal for combining mandatory and voluntary standards for new types of resource- and recycling-related requirements has been suggested by Fabrice and Mathieux [16]: Their proposal is that there can be legal requirements on "maximum disassembly time" for some product groups in order to make the product faster and thus more cost effective to take apart and recycle. If eco-labels require even shorter "maximum disassembly time" than the legal standard in order to obtain the label, a differentiation between mandatory and voluntary instruments is provided.

3.3 Coordination for Addressing Life Cycle Aspects in Various Life Cycle Phases

Some environmental and social aspects are difficult to address through legal requirements set in the EU, as EU laws only have jurisdiction in Europe. These may typically include environmental issues related to the extraction and production phases, such as release of chemicals, cutting of rainforest, and an unsafe working environment. If such issues are difficult to address by mandatory laws, GPP and labeling schemes and other voluntary approaches can be used to improve the situation [14].

3.4 Coordination for Radical Innovation

Another type of interaction concerns radical innovation. Most traditional policy instruments cannot stimulate radical innovation under normal conditions. For instance:

- Legal standards can only address specified, identifiable entities. This means that they can address established product groups but are often ill suited to address product groups where the speed of innovation is quick. It is also difficult to set rules for environmental impacts of services. Further, it is difficult – for legal and political reasons – to mandate the use of functional solutions, such as the use of "thin clients" instead of PCs, even if such solutions exist and are preferable from the environmental viewpoint.
- Eco-labels are also bound by the need to establish certain criteria for defining the product group, which means that more radical solutions may not be eligible for labeling even if they are better from a sustainability viewpoint. Another problem may be that the criteria set (e.g., on biodegradability, recycling etc.) may exclude new, better solutions.

This means that procurement policies, R&D support, and competitions and awards may be crucial complements to the traditional policies, to support radical solutions.

3.5 Horizontal and Vertical Legislation

Another way to look at the interaction is to look at policies that are directed toward single product groups (vertical laws) as a complement to horizontal rules (which cover several product groups). REACH [6] is the main EU instrument to address chemicals. It will influence chemicals in products as it regulates chemicals throughout supply chains and has specific information requirements related to products (see Articles 7 and 33 in REACH). However, REACH's rules on chemicals in products are deficient, and the rules on provision of information on request (Art. 33) do not properly include some actors such as recyclers. Rules for specific product groups could be used to address (to some extent at least) such shortcomings. As changes to REACH are very cumbersome process, product-specific rules set through specific directives or the Ecodesign Directive can also be more flexible to implement.

3.6 EU Policies and National Policies

An important issue concerns the relation between member state policies and EU policies. While the EU can control the coordination of standards under its control, such as the mandatory standards, energy labeling, and the EU eco-label, it has less influence over member state standards such as national eco-labeling schemes and GPP practices and standards. This means that member states must adjust their policies to make sure the interactions with EU policies work well. The example of labeling was discussed above.

However, it should be noted that there should be some scope for member state action. While industry often argues that it's better to have EU-wide policies in order to have a level playing field, there is a need for dynamics in order to set new policies and provide continuous incentives for ecodesign to manufacturers [17]. If the EU does not adopt progressive standards in certain areas of ecodesign, the member states should be granted some scope to do so. Further, member state actions often force the European Commission to act and consider the implementation of new EU standards [17].

4 Regulating Products in Systems

A problematic issue concerns how to deal with products that are part of larger systems. In the Ecodesign Directive, some standards make use of an "extended product" approach, e.g., by regulating an electric motor and the variable speed drive connected to the motor. This allows for the setting of stricter standards than if laws only cover a "narrow" product definition.

In the future, we can expect more focus on the "system dimension." The expected regulation of products and components in buildings – such as windows, insulation material, and various equipment – makes it important to deal with such issues. Some of the main concerns are:

- How do we define the regulated entity in terms of components, product, function, and system? A system approach may be necessary, since the performance of a given product depends on how it functions together with the system, and maintenance and user behavior influence the performance. But a system approach requires different scoping methods and methodologies and may encounter barriers related to legal design (it is easier to regulate a well-defined unit, such as a single product).
- How do we deal with a situation where two very different types of products affect each other's performance (e.g., washing machines and detergents) but where legal standards are set at different forums without much coordination? For some product groups, the capacity to improve the technology is limited, unless changes take part in other product groups.

- The use of some products will mean that other products are required, which will lead to more environmental impacts. For instance, the use of gas ovens often require more ventilation and thus more energy in the system; this should be considered when requirements are set and imply that some products should perhaps be "penalized" in some way. This has been suggested for gas ovens. But some stakeholders argue that such regulations are problematic as not every home with gas hobs also has ventilation, and nor do only homes with gas hobs make use of ventilation [18].

Regarding appliances in buildings, there are many different issues that need to be dealt with. For instance, "built-in" intelligence in appliances and how these function together with the building are of great concern. The need to have appliances designed to suit the needs of passive houses and other types of energy-efficient buildings (current appliances may not fit well) is also important. The fact that different EU Directives use different metrics may be problematic. For instance, the Energy Performance of Buildings Directive (EPBD) has energy requirements related to square meters, which is different from the metrics applied in the Ecodesign Directive. This may cause problems when choosing the right appliance for the building. Further, both these Directives should function together with the Energy Efficiency Directive.

Interviews with people in industry [14, 19] have shown that there are differing views regarding the benefits and drawbacks of a "system orientation," and therefore many industry associations abstain from making concrete recommendations on whether and how to implement a systems approach. Further, producers of "plug-in appliances" seem to have fewer concerns with legal requirements as these products are often well defined. It further seems that the "system approach" is controversial for some product groups, but less so for others [20].

System approaches may however provide certain benefits to producers [14]. More and more money tend to be made on the services related to products (maintenance, repair, new business models, etc.), and most manufacturing firms have increased the business offerings related to services in the last decades. In fact, the transition to a mindset of "servicization" may constitute a huge challenge for many traditional manufacturing firms. A system approach can help to bring about new business models and new revenue streams. However, it can also constitute a challenge for some firms. Industry seems somewhat split on the issue of "regulating systems"/"adopting a systems perspective in regulation." Whereas some industries prefer a more system-oriented approach, others see this as risky and prefer the "single product" approach [19].

The example of windows is very interesting from a systems perspective. Recent research shows how problematic it is to set energy labeling requirements for windows, in order to provide a differentiation between various products for consumers and building companies [21]. The energy efficiency of the window does not only depend upon the design and the climate where it is used, but also very much on how the building where the window is inserted is designed (with shading, solar inflow, shutters etc.). In the EU, which has several climate zones and very varying building traditions, the issue of how to regulate windows and make an appropriate labeling scheme for the consumer is very complex. In labeling, it is also important to strike a compromise between accuracy and simplicity of the information provided: simplicity – e.g., by providing easy categories and a user-friendly design – enables the use of the label by more actors, but the information may become less nuanced. Finally, in the case of windows, a selection is often made by a contractor, and not consumers, which needs to be taken into account: there may be reason to consider which type of information that is useful for different actors.

5 Concluding Remarks

The increasing number of product policy instruments is necessary: without policy there is inadequate incentive for ecodesign among manufacturers [17]. But the ever-increasing number of policies makes it difficult for market actors to get an overview of all relevant requirements. It is therefore crucial that various policies are coordinated. In this paper we have discussed some conceptual approaches to coordination and provided examples of concrete situation where coordination is crucial. In order to provide continuous incentives for innovation and ecodesign among all market actors, the coordination of mandatory and voluntary policies must be well thought through.

The national schemes complement EU policies and can provide dynamics on the market, which is positive. But it is crucial that national policymakers consider EU policies when designing national policy schemes and adjust national policies to make sure the coordination works between the two policy levels.

In practice, the coordination of various mandatory and voluntary standards tends to be suboptimal at both the EU and domestic levels. To some extent, this reflects the nature of policymaking: actors have to fight hard and long to implement policies at all, and the coordination and interaction between various public bodies are often poor. This applies both within the European Commission when implementing EU policies and among (and within) authorities at the national level. But the coordination has improved over time, as policymakers learn to deal with the complexities involved.

The planned regulation of more and more products that are part of larger systems, such as buildings, increases the complexity. The fact that the larger entity (e.g., the building) is also regulated through other policies increases the complexity and the risk for double regulation. We therefore need to undertake more research to find out how we can improve the methods for regulating the "system dimension."

Acknowledgments This research was funded by the Swedish Energy Agency (Energimyndigheten) through the project "Promoting ecodesign for the advancement of innovation, competitiveness and environmental improvements," and by the Swedish Foundation for Strategic Environmental Research (MISTRA) through the research program "Resource-Efficient and Effective Solutions based on circular economy thinking" (MISTRA-REES).

References

1. Van Rossem, C, Dalhammar C, Toulouse F (2009) Designing greener electronic products: building synergies between EU product policy instruments or simply passing the buck? Report: European Environmental Bureau (EEB), Brussels
2. Dalhammar C (2014) Promoting energy and resource efficiency though the ecodesign directive. Scand Stud Law 59:147–179
3. Selin H, VandeVeer S (2006) Raising global standards: hazardous substances and e-waste management in the European Union. Environment 48(10):6–17
4. Directive 2009/125/EC of the European Parliament and of the Council 21 October 2009 establishing a framework for the setting of ecodesign requirements for energy-related products, OJ 2009 L 298/10
5. Bundgaard AM, Remmen A, Overgaard Zacho K (2015) Ecodesign directive version 2.0: from energy efficiency to resource efficiency. Report, Miljostyrelsen [Danish Environmental Protection Agency], 2015
6. Regulation (EC) No 1907/2006 of the European Parliament and of the Council of 18 December 2006 concerning the Registration, Evaluation, Authorisation and Restriction of Chemicals (REACH), establishing a European Chemicals Agency, OJ 2006 L 396/1
7. Tojo N, Thidell Å (2012) Material recycling without hazardous substances – experiences and future outlook of ten manufacturers of consumer products. Report, Swedish Chemicals Agency
8. Jepsen D, Spengler L et al (2015) Delivering resource-efficient products. How Ecodesign can drive a circular economy in Europe. Report, European Environmental Bureau
9. tcodevelopment.com
10. Dalhammar C (2015) The application of 'life cycle thinking' in European environmental law: theory and practice. J Eur Environ Plan Law 12:97–127
11. COWI (2009) Bridging the valley of death: public support for commercialisation of eco-innovation. Report to the European Commission
12. Directive 2011/65/EU of the European Parliament and of the Council of 8 June 2011 on the restriction of the use of certain hazardous substances in electrical and electronic equipment, OJ 2011 L 174/88
13. Directive 2012/19/EU of the European Parliament and of the Council of 4 July 2012 on waste electrical and electronic equipment (WEEE), OJ 2012, L 197/38
14. Dalhammar C, Machacek E, Bundgaard A, Overgaard Zacho K, Remmen A (2014) Addressing resource efficiency through the ecodesign directive: a review of opportunities and barriers. TemaNord 2014:511. Nordic Council of Ministers, Copenhagen
15. Bundgaard AM, Overgaard Zacho K, Remmen A (2013) Product policies on the environmental performance of washing machines Investigating the synergies and coherence between policy instruments. Report: Nordic Council of Ministers
16. Ardente F, Mathieux F (2014) Identification and assessment of product's measures to improve resource efficiency: the case-study of an energy using product. J Clean Prod 83:126–141
17. Dalhammar C (2007) 'Product and life cycle issues in European environmental law: a review of recent developments'. Yearbook of European Environmental Law, vol 7. Oxford Univ. Press, Oxford
18. DEFRA (2012) Re: ecodesign and energy labelling proposals for: ovens, hobs, grills, range hoods, coffee machines and horizontal labeling. Position statement 29th May 2012
19. Teknikföretagen [Association of Swedish Engineering Industries] (2011) Ecodesign directive. How to create a more efficient legislation. Report
20. CSES/Oxford Research (2012) Evaluation of the ecodesign directive (2009/125/EC). Final Report to the European Commission, 2012
21. Janssens C (2015) Towards an EU energy labelling scheme for windows: insights and learning of the development phase of the first label for an energy-related building product. Proceedings form ECEEE2015

Global Initiative on UPCYCLE Carbon Footprint Certification and Label Systems for Creative Waste Management and Greenhouse Gas Reduction

Rattanawan Mungkung, Singh Intrachooto, Tananon Nudchanate, and Kannika Sorakon

Abstract Upcycling aims at turning scraps and wastes into new materials or products with equivalent quality or better than original through creative design. The upcycling processes are based on the direct use or the technology and production processing with consideration of potential environmental impacts and greenhouse gas. This is in line with the national policy on waste management by applying the principle of 3Rs (Reduce, Reuse, and Recycle), as well as the reduction of greenhouse gas emissions associated with waste management. In addition, upcycling is seen as one way to add values on wastes and promote the development of creative economy. Under this context, the UPCYCLE Carbon Footprint certification and verification system was initiated and developed in June 2015 under the leadership of the Department of Environmental Quality Promotion, Ministry of Natural Resources and Environment. There are five criteria in the UPCYCLE Carbon Footprint certification scheme: (1) scraps and wastes, (2) upcycling process, (3) product quality, (4) creative design, and (5) carbon footprint. For the requirement of carbon footprint, the avoided GHG emissions of upcycle materials or products shall be higher than their life cycle GHG emissions. A case study of upcycled glass tiles was used to demonstrate how to calculate the associated carbon

R. Mungkung (✉)
Center of Excellence on enVironmental strategy for GREEN business (VGREEN), Kasetsart University, Bangkok, Thailand

Department of Environmental Technology and Management, Kasetsart University, Bangkok, Thailand
e-mail: fscirwm@ku.ac.th

S. Intrachooto
Creative Center for Eco-Design, Department of Building Innovation, Kasetsart University, Bangkok, Thailand

Department of Building Innovation, Kasetsart University, Bangkok, Thailand

T. Nudchanate • K. Sorakon
Center of Excellence on enVironmental strategy for GREEN business (VGREEN), Kasetsart University, Bangkok, Thailand

M. Matsumoto et al. (eds.), *Sustainability Through Innovation in Product Life Cycle Design*, EcoProduction, DOI 10.1007/978-981-10-0471-1_38

footprint. It aims to be used as a communication and marketing tool to ensure that the certified upcycled materials or products are made from wastes/scraps, fit for use, have a good quality, and friendly to the environment.

Keywords Carbon footprint • Creative design • Greenhouse gas reduction • Upcycle

1 Introduction

Continuous increase in waste volumes within the manufacturing sector has led to the exploration of turning waste into products. Upcycling activities in Thailand emerged as a business enterprise in 2006 with the success of Osisu Company. The upcycling has then been expanding into industry especially small enterprises. In 2008, the concepts of 3Rs (Reduce, Reuse, and Recycle) were implemented with 37 SMEs for creative economy under the "Waste-to-Wealth" project initiated by the Industrial Technology Assistance Program (iTAP), National Science and Technology Development Agency, Thailand's Ministry of Science and Technology. Several kinds of wastes (such as wood offcuts, leather remnants, wasted papers, packaging wastes, and dead-stock fabrics) were used to make more than 500 furniture and home décor products. In 2014, the concepts of upcycling (i.e., creative transformation of wastes for a new purpose with better-than-original environmental and economic values through design and doesn't require extensive processing) were introduced, and the "Upcycling" project was launched as a continuation of Waste-to-Wealth to stimulate creative waste management solutions and innovative design among SMEs [1]. Despite being creative with resource efficiency, these manufacturers were questioned about their upcycled products regarding their environmentally friendliness and qualities.

In parallel, the Department of Environmental Quality Promotion (DEQP), Ministry of Natural Resources and Environment has developed a project titled the "Promotion of environmentally-friendly products" by developing its own "G (reen)-production" certification with the "G-" label to encourage the application of cleaner production, resource efficiency, and pollution prevention among small enterprises and to promote the G-label as a marketing tool. Hundreds of upcycled products were produced from these small enterprises in various parts of Thailand.

With the growth and popularity of upcycled products and materials in Thailand, in 2014, the "G-Upcycle" certification scheme was initiated by DEQP and used to certify upcycled products in the country. The five criteria of G-Upcycle are (1) source and amount of scrap, (2) product quality according to relevant industrial standards, (3) production processing using cleaner technology, (4) creative design, and (5) innovative solutions for waste management.

2 Development of Upcycle Carbon Footprint Certification and Labeling Systems

Having recognized the national policies on waste management as well as climate change mitigation policy, the "UPCYCLE Carbon Footprint" certification and labeling system has been initiated in 2015 as a national certification scheme under the cooperation of DEQP and Thailand Greenhouse Gas Management Organization (TGO). The draft standard has been prepared by a national technical committee, using the G-Upcycle manual [2] and the Carbon Footprint national guideline [3] as the key references as well as related international standards, related mainly to reuse and recycling but none on upcycling. The draft document has been reviewed by experts in the field of creative design, certification and labeling, and environmental technology. The standard document was presented to the stakeholders for consultation and finalized by the appointed national steering committee.

3 Requirements of Upcycle Carbon Footprint Certification

Products shall be described that they are one of this following product categories: (i) Materials (e.g., artificial stone, blocked cement, ceiling, wall, yarns, paper, tile, etc.); (ii) Furniture (e.g., desk, chair, shelve, etc.); (iii) Home decorative (e.g., lamp, carpet, pot, etc.); (iv) Fashion (e.g., bag, belt, necklace, shoes, hat, etc.); and (v) Others.

There are five criteria for the UPCYCLE Carbon Footprint certification [4], which are:

3.1 Scraps and Wastes

Scraps and wastes are referred to as wasted materials, reclaimed materials, or dead-stock materials or products that are environmental burdens for disposal by land-filling or incinerating.

Use of wastes and scraps shall be based on these following requirements:

3.1.1 Ratios of Scraps and Wastes

At least 20 % of reclaimed material contents (by weight) shall be used for upcycled materials or products

3.1.2 Sources of Scraps and Wastes

Sources of scraps and wastes shall be identified (e.g., postconsumer wastes, postindustrial wastes, or wastes from temporary uses) and transported with consideration of potential environmental impacts and greenhouse gas emissions. Post-consumer wastes are referred to as postconsumer wastes or community wastes. They are dispersed and not easy to collect for manufacturing such as snack packages, plastic bags, milk pouches, coffee bags, straws, styrofoam containers, fruit juice packages, or construction debris. Post-industrial wastes and scraps are referred to as manufacturing offcuts, leftover materials or products generated by manufacturers or service providers, material dead stocks, as well as agricultural wastes. They are abundant and highly redundant and easy to collect for manufacturing, such as leather and fabric remnants, glass offcuts, steel scraps, or wood offcuts.

Wastes from temporary use are referred to as wastes that remain after short-term uses from exhibitions and fairs and public-relation activities, such as carpets, papers, or advertising boards.

3.1.3 Preparation of Scraps and Wastes

Level of effort and resources used to prepare scraps and wastes for remanufacturing shall be minimized to encourage the direct use or the processes that minimize amounts of water, energy, and new resources by applying these following methods:

3.1.3.1 Preparation of scraps and wastes by using renewable energy, alternative sources of energy, or human labor

3.1.3.2 Preparation of scraps and wastes that avoids toxic chemicals such as chlorine, solvents, etc.

3.1.3.3 Use of production process with energy-saving policies

3.2 Upcycling

Upcycling processes shall be based on the direct use or technology and production process with consideration of potential environmental impacts and greenhouse gas, as follows:

3.2.1 Upcycling processes with environmental management of chemicals, energy, pollutions, and wastes

3.2.2 Upcycling processes with safety management for healthy workplaces

3.2.3 Upcycling processes with labor management for social responsibility by hiring elderly people, imprisoned persons, disables, or the locals

3.2.4 Upcycling processes with skill development management by supporting personnel in various capacity building such as attending seminars, training, or taking field trips on product design, waste management, or related know-how

3.3 Quality

Upcycled materials or products shall have equivalent or better quality than their original and fit for use, through quality control systems to ensure that their quality meets specifications or complies with the required industrial standard (if any).

3.4 Creative Design

Upcycled materials or products shall be developed with creative design to distinguish itself from others and enhance the reclaimed materials.

3.5 Carbon Footprint

The avoided GHG emissions of upcycled materials or products shall be higher than their life cycle GHG emissions.

The calculation of life cycle GHG emissions of upcycled materials or products must be based on the requirements as specified in the national guideline for product carbon footprinting as well as the method specified in the Product Category Rules (PCRs) of a specific product or group products.

The calculation method for the avoided GHG emissions associated with the disposal of wastes is based on the assumption that if the scraps and wastes are not used for upcycled products, then they will be disposed by landfilling and emit GHGs. The calculation method of the avoided GHG emissions associated with the disposal of wastes shall be based on the IPCC Guideline for National Greenhouse Gas Inventories – Volume 5: Waste, on waste disposal in a shallow landfill (Table 1).

The calculation of the avoided GHG emissions associated with the production of virgin materials is based on the assumption that if the scraps and wastes are used for upcycled products, then the GHG emissions associated with the production of virgin materials will be avoided. The GHG emissions of production of virgin materials shall be based on these following sources, in decreasing order of preference:

- Source 1: National Life Cycle Inventory (LCI) databases on the TGO's website (the latest version) and list of emission factors (EFs)
- Source 2: LCI data from theses and research projects conducted in Thailand
- Source 3: Peer-reviewed journals, technical reports, or theses in the context of Thailand
- Source 4: Databases available in LCA software
- Source 5: Industrial databases and databases specific to individual country

Table 1 GHG emissions from a shallow landfill

Waste type	GHG emissions (tCO_2e per ton of waste (dry weight))
Paper	2.93
Fabric	2.00
Food	2.53
Wood	3.33
Garden wastes (i.e., leaves, grass)	3.27
Paper diaper	4.00
Rubber and leather	3.13

- Source 6: Publications from international organizations (e.g., The Intergovernmental Panel on Climate Change (IPCC), The United Nations (UN), The Food and Agriculture Organization of the United Nations (FAO), etc.)

In case of upcycling more than one time, the summation of avoided GHG emissions from each time of upcycling must be added up.

For the calculation of life cycle GHG emissions of upcycled materials or products, seven Kyoto gases shall be included in the carbon footprint calculation. The gases are carbon dioxide (CO_2), methane (CH_4), nitrous oxide (N_2O), hydrofluorocarbons (HFCs), perfluorocarbons (PFCs), sulfur hexafluoride (SF_6), and nitrogen trifluoride (NF_3). The unit of analysis shall be defined as the mass of upcycled material or product per piece.

All sources shall be included, except for the acquisition of wastes/scraps. If the GHG emissions from research and development are significant, it shall be included. Capital goods are referred to as the buildings, machines, and equipment that have the life span of more than 1 year and shall also be excluded.

The unit of analysis shall be defined as the mass of upcycled material or product per piece. The calculation method of the life cycle GHG emissions of upcycled materials or products shall be based on the prototype production (one piece, at least) or three batches of production. Once there is a real massive production by industry, there shall be the data verification at the production site again to ensure the representativeness of data.

All data used for the calculation of life cycle GHG emissions of upcycled materials or products shall be properly recorded for verifying within 2 years (at least) or over the life span of upcycled materials or products with the labels which are available in the markets.

When comparing between two upcycled materials or products, it is necessary to take into account the quality, performance, and life span of such upcycled materials or products. Related technical aspects shall be defined, such as upcycling technology or processes to be taken into consideration when comparing between two upcycled materials or products that perform the same functions.

The required data for carbon footprint assessment are:

(i) Activity data

Input data are the inputs required for upcycling processes, which are raw materials, energy, chemicals, water, and others. Output data are the outputs from upcycling processes, which are main product, coproduct, wastes, and emission to environment.

Activity data shall be collected from the primary data if they are direct activities and accessible (such as suppliers to supply wastes/scraps to upcyclers or designers). In case there is no data access, then the secondary data can be used based on these following sources, in decreasing order of preferences (similar as above).

For activity data, the required inventory data are:

1. Amount of scraps and wastes including other raw materials.
2. Amount of inputs and outputs associated with packaging production. The primary data for packaging of the product will be required. In the case where this data is not available, secondary data can be used. In the case where the carbon footprint of the packaging of a product is less than 5 % of the total carbon footprint of the product, it can be neglected and excluded.
3. Amount of energy used for preparation of wastes/scraps. The GHG emissions from fuels shall include:

    3.1. Emissions from the production of fuel
    3.2. Emissions from fuel use (e.g., $kgCO_2e/kg$ fuel, $kgCO_2e/MJ$ electricity or heat) based on the source of energy used

        3.2.1. Emissions at site of energy use (e.g., coal and gas combustion)
        3.2.2. Emissions from energy carrier acquisition
        3.2.3. Emissions from fuel use in transport
        3.2.4. Upstream GHG emissions (mining, transport of fuel to electricity production site) including biomass acquisition for using as fuel
        3.2.5. Downstream emissions from waste treatment and management from energy conversion

In terms of distribution or transport, the required data for road transport are the type of vehicle, type of fuel, distance, loading rate, and return trip. The distance shall be based on the Department of Highway's website (http://map-server.doh.go.th). For the sea transport, the required data are the type of vessel, name of port in Thailand, and port in overseas. The distance between two ports shall be based on the international container shipping's website (http://searates.com).

For activity data associated with the use of upcycled materials or products (including installment, use, and maintenance during use), these are the following sources:

- Source 1: The product's application shall be based on published technical papers, such as its ISO 14025 or PCRs.
- Source 2: The product's application shall be based on the instruction recommended by manufacturers (including installment, use, and maintenance during use).

- Source 3: The product's application shall be based on the typically use method by general consumers, which is a set of rules that describe a product's operations and certain assumptions about the product.

For the final disposal stage, the required inventory data are the amount of wastes after reaching its life span. Transport of wastes to dispose by landfilling shall be based on the transportation by 10-wheeler, 16-ton truck over a distance of 40 km (normal roads) with an empty return trip.

(ii) Emission factors (EFs)

EFs are referred to as the amount of GHG emissions associated with the production of main materials, minor materials, energy, chemicals, transport by various types of vehicles, etc.

In the case where EFs of some inputs are not available, the EFs of substances having similar physical and chemical properties shall be used as the substitutes. If no such substitute data is available, the highest EF of the inputs in the particular life cycle stage shall be used as the substitute.

The calculation of GHG emissions, i.e., carbon footprint, of upcycled materials or products must be based on these following steps:

(i) Converting the primary and secondary data of inputs/outputs to EFs by multiplying their loadings with respective emission factors
(ii) Converting GHG emissions into CO_2e by multiplying the individual EF by the relevant GWP
(iii) Finalizing the carbon footprint result, expressed in terms of CO_2e per product unit

Items that contribute lesser than 1 % of the total carbon footprint can be cut off. However, the total cutoff cannot be more than 5 % of the anticipated GHG emissions. In case of a cutoff, the assessed emissions shall be scaled up to represent 100 % of the EFs associated with the product unit defined.

It should be noted that the UPCYCLE Carbon Footprint toolkit, based on the national guideline of carbon footprint of products, shall be applied to calculate the carbon footprint to ensure that the avoided GHG emissions of upcycled materials or products are higher than their life cycle GHG emissions.

The symbol of UPCYCLE Carbon Footprint (Fig. 1) consists of:

(i) UPCYCLE in the symbol indicates that these materials or products are upcycled by using scraps and wastes for designing.
(ii) CO_2 in the symbol indicates that these upcycled materials or products have the avoided GHG emissions higher than their life cycle GHG emissions.

Using the symbol of UPCYCLE Carbon Footprint shall be based on these following rules [5]:

1. The symbol of UPCYCLE Carbon Footprint shall not be used in the way that will lead to misunderstanding or cause any damage.

Fig. 1 The symbol of UPCYCLE Carbon Footprint

2. The symbol of UPCYCLE Carbon Footprint shall not be used as a part of trademark, tradename, co-marking, or others that is not related to the verification scope of UPCYCLE Carbon Footprint.
3. The symbol of UPCYCLE Carbon Footprint, as shown in the figure, shall be used in any color that is appropriate and suits the packaging material. The label can be put on packaging (or website, product catalogue, display shelve, etc.) at a readable size to ensure communication with consumers.
4. The symbol of UPCYCLE Carbon Footprint can be used for 3 years [4]. There will be a warning to renew the license in advance, at least 5 months before the license expiry date. The symbol of UPCYCLE Carbon Footprint can still be used after the expiry date for another 3 months only.
5. The license of UPCYCLE Carbon Footprint shall be renewed by applying for verification and certification again.

It is worth mentioning here that the revision of requirements will take place once every 2 years to ensure that the requirements are up to date with the real situation.

4 Case Study

To demonstrate the implementation of UPCYCLE Carbon Footprint certification scheme, an example of a case study of glass tiles (with the brand "Granite Glass®") was performed.

Fig. 2 Granite Glass® made from wasted glasses

4.1 Product Description

The studied product was Granite Glass®, which was made from discarded glass bottles (Fig. 2).

4.2 UPCYCLE Carbon Footprint Compliance

4.2.1 Scraps and Wastes

The proportion of discarded glass bottles in Granite Glass® was 98 %. The discarded glass bottles were collected from local sources, within the radius of 40 km from the factory, and transported to the factory by 10-wheeler trucks. The preparation of discarded glass bottles was kept minimum by only sorting and washing them with water with no use of chemical.

4.2.2 Upcycling

The upcycling processes started with grinding in two steps: rough and fine grinding, by using grinding machines (Fig. 3). The grinded pellets were passed through sieves to ensure homogeneous sizes. Pigments and glue were added with grinded pellets and then formed into a shape (98 × 98 × 9 mm). Firing at the temperature 900 °C, 4 h, was the last stage. They are packed in corrugated boxes for sales.

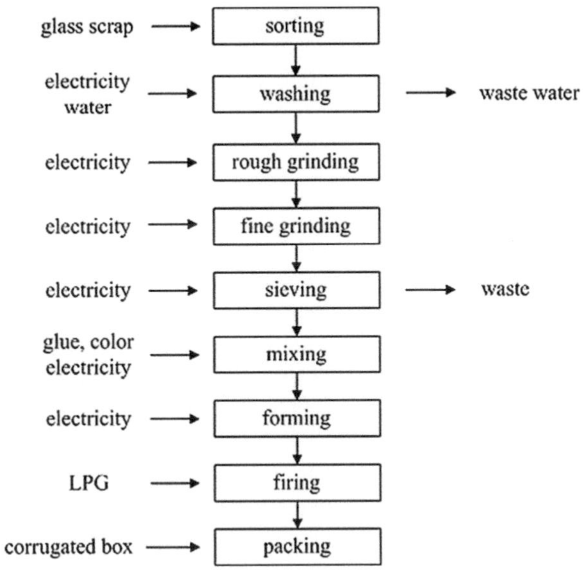

Fig. 3 Upcycling processes of granite glass tiles

4.2.3 Quality

To ensure the quality of Granite Glass®, they were tested in terms of water absorption, breaking strength, and Young's modulus according to the related industry standard of Thailand Industry Standard Institute (TISI). The results confirmed the quality was in compliance with the standard values.

4.2.4 Creative Design

Using discarded glass bottles to make Granite Glass® offered a creative way for waste management. Their initiative on the upcycling of discarded glass bottles was captured and disseminated by the media.

4.2.5 Carbon Footprint (CF)

GHG Emissions of Virgin Material Production

If the discarded glass bottles were not used for making Granite Glass®, then there will be a need to produce Granite Glass® from virgin material. This, thus, helps avoid 26.0 $kgCO_2e$ per m^2 of Granite Glass® [6].

GHG Emissions of Waste Disposal

If the discarded glass bottles were not used for making Granite Glass®, then there will be a need to dispose emptied glass bottles (after use) by landfilling. This, thus, helps avoid 0.0778 kgCO_2e per m^2 of Granite Glass® [3].

GHG Emissions of Granite Glass® from Discarded Glass Bottles

The calculation of life cycle GHG emissions, i.e., carbon footprint, was referred to the national guideline of carbon footprint of products [3]. The required inventory data associated with each subprocess were identified and collected from the company's recording systems. The emission factors of electricity and water, including transport, were sourced from the national databases. Other inputs (e.g., pigment and lubricant) were gathered from international databases.

It was found that 25 kg of discarded glass bottles, 1.59 kWh of electricity, 5.2 kg of LPG, and 0.5 kg of pigment were required to produce 1 m^2 of Granite Glass®. The carbon footprint value of upcycled Granite Glass® was 22.9 kgCO_2e per m^2 of Granite Glass®. The key hot spot was linked to the use of LPG in the firing process.

5 Results and Discussions

Discarded glass bottles from post-consumption were used to upcycle to make the Granite Glass®, which has offered a new way of waste management through creative design. At least 98 % of discarded glass bottles were used, and the preparation of pellets did not require intensive processes. The results of quality test have proved that Granite Glass® is fit for use.

The avoided GHG emission as the consequences of using discarded glass bottles for upcycled Granite Glass®, by avoiding the GHG emissions from the production of virgin materials and the disposal of wastes, was 26.1 kgCO_2e per m^2 of Granite Glass®, which was higher than the CF value of Granite Glass® (Fig. 4). As a result,

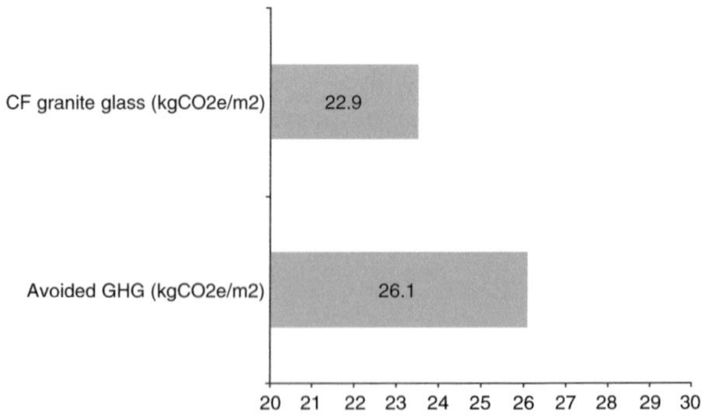

Fig. 4 Carbon footprint of upcycled granite glass compared to the avoided GHG emissions

the upcycled Granite Glass® is in compliance with the carbon footprint criterion of UPCYCLE Carbon Footprint certification scheme.

6 Outlook

At present, there are various upcycled products available in Thailand. It is expected that the UPCYCLE Carbon Footprint certification scheme will be an effective green marketing tool to promote more upcycling activities and upcycled products into mainstream markets. Simultaneously, it will stimulate innovative ideas for waste management and GHG reduction. Through the market demand and supply mechanism, it supports the move toward sustainable consumption and production and will contribute to the low-carbon society.

Acknowledgments This research is supported by the National Science and Technology Development Agency (NSTDA).

References

1. Intrachooto S (2013) Upcycling: creative scrap utilization. National Science and Technology Development Agency, Ministry of Science and Technology, Pathum Thani, Thailand, 200 pp. http://www.nstda.or.th/eng/#
2. Department of Environmental Quality Promotion (DEQP), Ministry of Natural Resources and Environment (2014) Manual of G-upcycle certification and verification, 1st edn. DEQP Publisher, Bangkok, 27 pp
3. Thailand Greenhouse Gas Management Organisation (TGO) (2015a) National guideline of product carbon footprint, 5th edn. Bangkok, 35 pp
4. Department of Environmental Quality Promotion (DEQP), Ministry of Natural Resource and Environment (2015) UPCYCLE carbon footprint: certification and verification, 1st edn. DEQP Publisher, Bangkok, 27 pp
5. Thailand Greenhouse Gas Management Organisation (TGO) (2015b) Principles and requirements of using the carbon footprint symbol, 2nd edn. Amarin Printing and Publishing Publisher, Bangkok, 25 pp
6. Kellenberger D, Althaus HJ, Jungbluth N, Künniger T (2007) Life cycle inventories of building products. Final report ecoinvent data v2.0 No. 7. Swiss Centre for Life Cycle Inventories, Dübendorf

Sustainable Energy Strategy Primarily Involving Renewable Resources in Japan

Haruki Tsuchiya

Abstract This paper focuses on the future energy strategy that primarily involves the use of renewable resources, such as hydropower, geothermal, biomass, photovoltaic and wind power, in Japan to decrease CO_2 emission. We estimated the national energy demand in 2050 by considering population decrease, macroeconomic prospects and energy efficiency improvement. The results of a survey on the supply potentials of domestic renewable energy are used to allocate renewable energy supply to end-use energy demand. We analysed three scenarios of energy supply systems. Scenario 'A' involves providing supply for pure electricity demand only, and scenario 'B' involves providing supply for pure electricity demand, electricity to EVs (electric vehicles) and hydrogen to FCVs (fuel cell vehicles). Scenario 'C' involves providing supply for pure electricity demand and hydrogen and thermal energy for all types of energy demand. We performed dynamic simulation of the annual electricity supply with hourly weather data for three scenarios and discussed the scenario structures.

Keywords End-use energy • Renewable energy • Solar energy • Wind energy

1 Introduction

Renewable energy is essential for building safe, self-sufficient and sustainable energy systems without carbon dioxide emission. As a result, much attention is focused on a renewable-rich energy future. There have been studies on providing global energy in 2050 with 100 % renewable energy, largely wind, hydropower and solar power, by Stanford University [1] and WWF International [2]. The National Renewable Energy Laboratory in the USA published a report in 2012 that US electricity could be supplied by 90 % renewable energy [3].

The Japanese government declared the official target of CO_2 emission is '80 % reduction' in 2050 from the 1990 level [4]. However, the details on how to achieve the target are not clear. There was a renewable-rich scenario study for 2030

H. Tsuchiya (✉)
Research Institute for Systems Technology, Tokyo, Japan
e-mail: tsuchiya@systemken.com

M. Matsumoto et al. (eds.), *Sustainability Through Innovation in Product Life Cycle Design*, EcoProduction, DOI 10.1007/978-981-10-0471-1_39

produced by Japanese government in 2013, which showed that renewable energy, including solar photovoltaic technology and wind power, will supply 26–37 % of electricity in 2030 [5]. However, the government revised the long-term energy prospects in 2015, with renewable energy supplying 22–24 % in 2030 [6]. WWF Japan published a 100 % renewable energy scenario in 2011 [7]. The study showed that Japanese energy demand will decrease by efficiency improvement and will be provided with 100 % renewable energy in 2050.

In this paper, we studied the actual process of introducing a large share of renewable energy to the national energy supply in 2050. In the study of renewable energy expansions, we recognised that there are three scenarios for introducing a large share of renewable energy according to the types of energy end use and developing stages. This paper considers these three scenarios with dynamic simulation of the electricity supply including large-scale solar and wind power. In addition, we included discussions on the three scenarios.

2 Future Energy Demand

First, we estimated the future energy demand. Analysis revealed that future energy demand will decrease mainly because of population decrease and energy efficiency improvement in 2050. The official reports of the Japanese population forecast for 2050 show that it will decrease to 76 % of the year 2013 level [8]. The population decrease will inevitably cause the decline of energy demand. The GDP growth rate is forecasted officially as 1 % for 2020–2030. The society will shift to a more service-oriented and reduced material production.

We examined the energy end use and the future activity level to estimate the energy demand in 2020, 2030, 2040 and 2050, considering population, number of households, GDP and industrial production of steel, paper, cement, chemical materials and material resource index, as shown in Table 1.

We estimated the future energy demand in each end use using the following equation:

$$\begin{aligned}\text{Future Energy Demand} = \text{Energy}\ &\text{Demand in 2013} \times \text{Future Activity level} \\ &\times \text{Energy Efficiency Improvement}\end{aligned} \quad (1)$$

An appropriate macroeconomic index is used to represent the future activity level in each end use. The efficiency improvement will be achieved by a twice efficient light-emitting diode for lighting, heat pump technology with high COP (coefficient of performance, present 3.0–6.0 in 2050), insulation for houses and buildings that will decrease heating demand to one third and inverter control motors in factories with efficiency improvement of 30 % or more.

The material use in the industrial sector will decrease due to the use of recycling systems. In particular, the recycling rate of steel will be almost 75 % in 2050, which is much higher than 30 % today. Energy demand for steel production will decrease

Table 1 Macroeconomic index ($1 = 120 yen)

Macroeconomic index	2013	2020	2030	2040	2050
Population (million)	127.29	124.10	116.19	107.17	97.07
Population ratio to 2013	1.00	0.975	0.913	0.842	0.763
Real GDP (year 2000 trillion dollar)	4.42	5.47	6.09	6.91	7.09
Real GDP ratio to 2013	1.00	1.238	1.377	1.562	1.603
Number of households (million)	55.95	54.46	52.69	47.40	45.19
Number of households (ratio)	1.00	0.973	0.942	0.847	0.808
Industrial production index (year 2010 = 100)	98.9	114.4	125.8	139.8	144.0
Crude steel production (million ton)	111.52	114.58	105.95	88.70	88.70
Ethylene production (million ton)	6.76	7.05	6.87	5.71	5.71
Cement production (million ton)	58.82	55.64	53.15	45.09	41.69
Paper and paperboard production (million ton)	26.66	30.85	30.58	29.00	28.27
Material resource index (ratio)	1.000	1.021	0.965	0.827	0.807

because energy demand for an electric furnace for recycled steel production is less than that for a blast furnace, which is used in the iron ore reduction process. In the transportation sector, EVs (electric vehicles) and FCVs (fuel cell vehicles) will replace conventional vehicles driven by internal combustion engines, thereby drastically reducing energy demand to almost one third because of the use of high-efficiency electric motor drives.

The energy demand in 2050 is estimated to be nearly 55 % of that in 2013. The energy demand structure of each sector is shown in Fig. 1. We refereed the BAU scenario that is published as the Asia/World Energy Outlook by the Institute of Energy Economics in Japan [8].

Thus, we assume that the annual electricity demand in 2050 will be 63.7 % of the year 2013 level. The end-use demand is divided into pure electricity, electricity for fuels (mainly for EV and FCV) and fuel for heat and transport. End-use pure electricity demand is 984 TWh in 2013. Thermal end-use demand is 246 MTOE (million ton oil equivalent) in 2013 (1 TOE = Electricity 11,630 kWh).

3 Renewable Energy Resources

We performed a survey of the renewable energy potentials. Although Japan has already developed hydropower and geothermal power, these sources are expected to be developed further. Solar power has expanded recently and is expected to further expand up to 2050. Wind power is currently not heavily developed but has large potential.

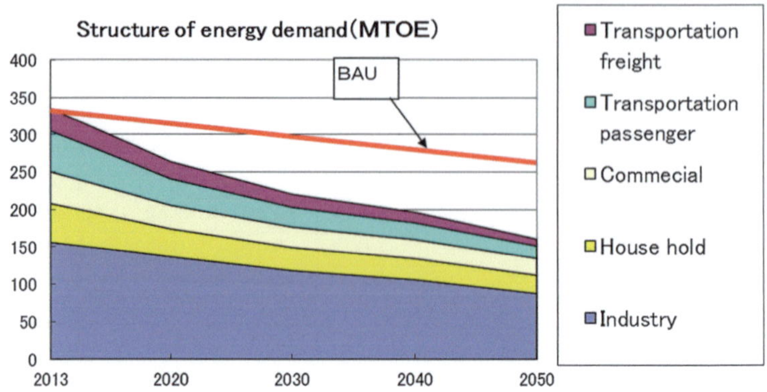

Fig. 1 Future energy demand structure

3.1 *Renewable Energy Potentials*

Several studies on renewable energy resources in Japan exist in the literature. Table 2 shows the renewable energy capacities of the present situation, maximum potential and this study.

The capacity of hydropower is 20.7 GW today. We assumed the capacity will be 33 GW in 2050. The geothermal plant is 540 MW today, and we assumed it will be 14.2 GW in 2050; the maximum is 33 GW according to the energy vision of the Japan Geothermal Association.

The waste biomass potential is estimated to be 20 MTOE. As 68 % of Japan's national land is forest, the stored wood biomass in the forest is estimated to be 5 Gm ^{3}. This wood biomass potential is as large as 30 MTOE annually if it is harnessed every 40 years. However, it is unknown if this potential is practically achievable or not, as the cost of collecting biomass is high and cannot compete with present oil prices.

3.2 *Solar Photovoltaic Power*

Solar photovoltaic power systems were introduced with the aid of subsidies in the 1990s and are rapidly increasing because of the government mandated feed-in-tariff system in 2011; nearly 23 GW was introduced cumulatively until 2014. The price of a photovoltaic system is analysed using a learning curve, and the past price records show that the price decreased to 82 % every time the cumulative production doubled in Japan [9].

NEDO reported that solar PV usage will be as follows: 200 GW in the household sector, 150–200 GW in the commercial sector, 150 GW in industries and 150–200 GW for electric vehicles in the transportation sector [10]. We estimated

Table 2 Renewable energy resources (GW)

Sources	2014	Max potential	Scenario 'C'
Hydropower	20.7	37	33
Geothermal	0.54	33	14
Photovoltaics	23	700	477
Wind power	2.7	1856	111
Biomass	4	12	8

the maximum photovoltaic capacity will be 700 GW when the conversion efficiency will reach 25 % in 2050.

There is the Extended AMeDAS weather data, which has hourly weather data of 842 sites nationwide in Japan [11]. We used it to calculate solar photovoltaic power and wind power.

The calculation method for solar power is to split the hourly solar radiation into the direct beam and the scattered irradiance and to combine both sources of sunlight on the panel oriented to the south with a tilt angle [12]. The optimum tilt angle to achieve the maximum annual generation is set at 'latitude −5' degrees at each construction site, as is known experimentally. The inverter of 94 % efficiency is included, with a partial load factor curve. The solar capacity factor is approximately 12 %.

3.3 *Wind Power*

There are 2790-MW wind turbines already introduced as of 2014. A study of the energy potential performed by the Ministry of Environment showed very large wind potentials, 283 GW onshore and 1572 GW offshore [13]. The Japan wind power association reported their vision to 2050; the goal is a total of 75 GW in 2050, which is composed of 38 GW onshore, 19 GW fixed offshore and 18 GW floating offshore. Japan is surrounded by sea and has long coastlines. Offshore wind power will be a large source of energy in the future. However, in this study, AMeDAS wind data sites do not include offshore sites. As a 70-MW wind turbine is being built offshore in Fukushima, many of them will be built offshore in the future. As offshore wind has higher wind speed and will have larger capacity factor and our estimates used onshore wind data only, the results will be very conservative.

In this study, a 2-MW wind turbine with horizontal axis and three blades is used to calculate the wind power in the respective sites. The blades start to rotate at a cut-in wind speed of 3 m/s and stop at a cut-out wind speed of 25 m/s. The conversion efficiency is assumed to be 40 %. The diameter of rotor is 80 m, and the hub height is 56 m. The wind speed of meteorological data is extended by a power factor law to calculate the wind speed at the hub height [14].

As we excluded the wind power sites having a capacity factor less than 18 %, 99 sites were selected for wind sites among the 842 sites. The site selection of wind power resulted in the average capacity factor of 27 %.

The hourly generation amount by single unit of a 100-kW solar system for the respective site is calculated. The necessary unit numbers of 100-kW solar PV systems are allocated by considering the annual electricity demand of the respective areas. These unit numbers of solar and wind turbines are multiplied by the hourly electricity generation to calculate the electricity supply throughout the dynamic simulation.

4 Three Scenarios

We propose that three scenarios are suitable to explain the future situation according to the stages when we expand the share of renewable energy resources.

We will call the case that the renewable sources supply energy for only pure electricity end use as scenario 'A'. The promising renewable energy technologies today are electricity supply technologies, such as solar photovoltaic technology and wind technology. In the initial stage, only electricity supply technologies among the renewable energy technologies will be strongly promoted. Scenario 'A' involves providing supply based only on the electricity specific demand, such as lighting, motor, air conditioning and communication.

When we increase the share of renewables, it is anticipated that the intermittent nature of renewable electricity generation will cause periods of lack and excess of energy compared to the energy demand. The lack of electricity is compensated by backup power and storage from pumped hydro and battery storage systems. Excess electricity will be supplied to other energy end-use applications, such as vehicles in the transportation sector. We refer to this case as scenario 'B'. In this case, excess electricity will be used for EVs and for water electrolysis to produce hydrogen for FCVs.

Scenario 'C' involves providing supply to satisfy the total energy demand, such as electricity, vehicles and thermal fuel demand. In this case, excess electricity will be supplied to EVs, FCVs, hydrogen cogeneration and heat pump. Other thermal demand will be supplied by solar heat and biomass. The features of three scenarios are summarised in Table 3.

Pure electricity demand is the same for three scenarios. However, the amounts of total electricity generation are different, depending on the allocations of excess electricity to the energy demand. Solar PV and wind power have larger deployments for scenarios 'B' and 'C'. The ratio of the total electricity generation to electricity demand is 126 % for scenario 'B' and 168 % for scenario 'C'.

Table 3 Features of the three scenarios of energy demand and supply

Scenario	'A'	'B'	'C'
Energy demand/supply	Electricity	Electricity + EV + FCV	All: electricity + EV + FCV + heat
PV (GW)	147	306	477
Wind (GW)	37	79	111
Electricity demand (TWh)	627	627	627
Total electricity generation (TWh)	634	791	1050
Total electricity/demand (%)	101	126	168

5 Dynamic Simulation of Three Scenarios

Scenario 'A' is a basic supply scenario. Scenarios 'B' and 'C' are extensions of scenario 'A'. Here, we will explain three scenarios by introducing a dynamic simulator of electricity supply [15]. The simulator is a computer programme that selects combinations of energy supply options suitable to electricity demand every hour throughout a year.

5.1 Characteristics of Solar Power and Wind Power

The dynamic simulator provides the hourly electricity generation of solar PV (photovoltaic) systems at 824 sites and of wind power at 90 sites.

Figure 2 shows the monthly solar PV and wind generations throughout a year. Wind is low in summer and high in winter, whereas solar has the opposite characteristics. As a result, these energy sources are complementary to each other. We find that the optimal ratio of electricity generation by solar PV systems and wind power systems is approximately 2:1, as electricity demand in summer is larger than in winter [16], and this ratio is used in three scenarios.

Figure 3 shows the hourly generation of power for solar PV systems and wind systems. The hourly wind power is relatively flat over 24 h, whereas the hourly solar has a bell-shaped pattern that peaks during the daytime.

5.2 Dynamic Simulation

We used the standard 2000 weather data of Expanded AMeDAS published by the Architecture Association of Japan, which includes the hourly data of solar radiation and wind speed throughout a year in 842 sites nationwide. The allocation of solar

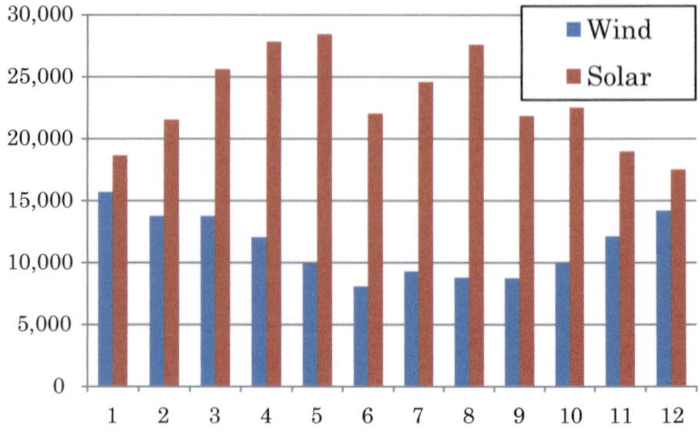

Fig. 2 Monthly solar and wind power (GWh)

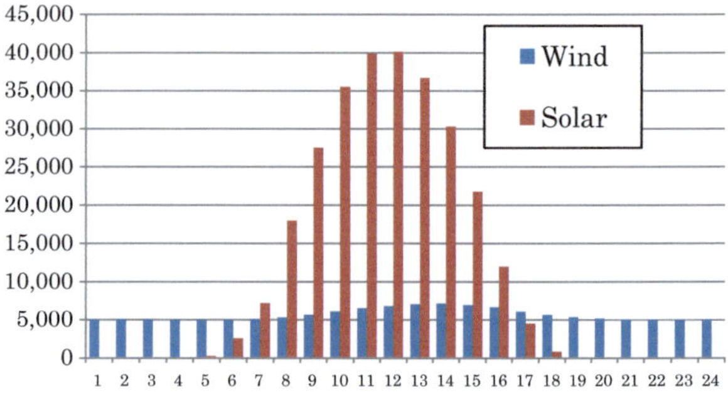

Fig. 3 Hourly solar and wind power throughout a year (GWh)

PV and wind power in scenario 'B' is shown in Table 4, which is based on ten areas of the present electric power companies.

The supply areas of the ten electric companies are Hokkaido, Tohoku, Kanto, Chubu, Hokuriku, Kansai, Chugoku, Shikoku, Kyushu and Okinawa from north to south. However, we treated the transmission line as being integrated in a single grid in the dynamic simulation.

The monthly electricity consumption data of ten electric power companies and the hourly pattern for a day are referenced for the respective monthly simulations. Figure 4 shows the dynamic simulation of the hourly electricity supply for 3 days. The electricity demand grows in the morning and peaks at approximately 14:00. Geothermal and biomass systems are constant supplies of power. Wind power systems exhibit supply fluctuations. PV systems exhibit bell-shape supply patterns with the peak occurring during the daytime. Hydropower supply is under the schedule of peaking at approximately 18:00.

After settling the load and scale of renewable resources, computer calculations were performed in the following order. The hourly electricity load (L) was subtracted by the renewable supply as follows:

1. Electricity load (L) is supplied by geothermal power (G), which is constant throughout the year.

Table 4 Allocation of solar PV and wind power in ten areas (scenario 'B')

Area	PV (MW)	PV (GWh)	PV CF (%)	Wind (MW)	Wind (GWh)	Wind CF (%)
Hokkaido	12,936	12,466	11.00	10,400	27,648	30.35
Tohoku	32,140	31,045	11.03	10,767	23,699	25.13
Kanto	95,097	109,423	13.14	11,499	29,529	29.31
Chubu	42,367	49,188	13.25	7380	17,680	27.35
Hokuriku	11,195	10,884	11.10	2394	4514	21.52
Kansai	51,095	55,411	12.38	4181	8840	24.14
Chugoku	21,401	23,272	12.41	13,924	29,717	24.36
Shikoku	9400	11,184	13.58	3236	8840	31.18
Kyushu	28,396	32,712	13.15	13,533	33,103	27.92
Okinawa	2603	2966	13.01	1651	4514	31.21
Total	306,630	338,550	12.60	78,966	188,084	27.19

MW capacity, *GWh* annual generation, *CF* capacity factor

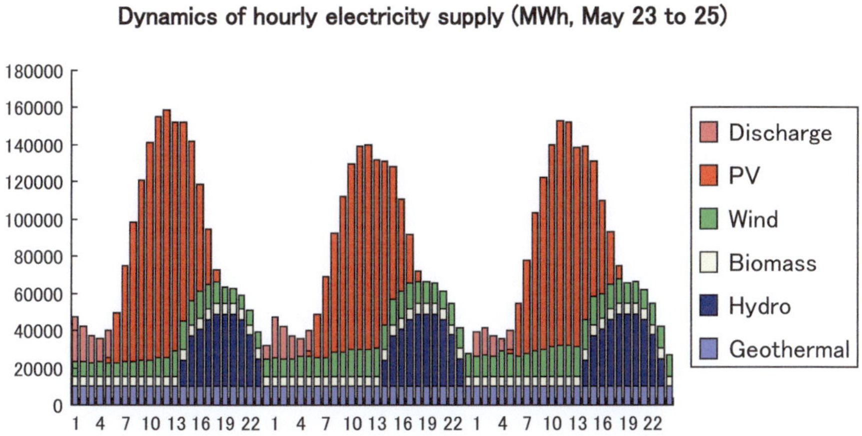

Fig. 4 Supply of PV, wind, discharge, biomass, hydro and geothermal powers for 3 days

$$L = L - G \tag{2}$$

The updated L is the load for the next step.

2. Hydropower (Hy) is supplied according to the time-dependent supply pattern.

$$L = L - Hy \tag{3}$$

The hourly operation of hydropower is designed to function effectively during the difficult time of the peak load and at the end of the daily solar radiation period. The hydropower system is programmed such that the maximum hydropower supply is achieved at approximately 18:00, starting at 15:00 and ending at 22:00 every day.

3. The next step depends on the updated load.

   3.1. If the amount of solar (S) and wind power (W) are less than the load, i.e., L > S + W, then

$$L = L - (S + W) \tag{4}$$

   The updated load is supplied from biomass power (Bi).

$$L = L - Bi \tag{5}$$

   If biomass power is not sufficient for the load, then discharge from battery or pumped hydro (Bd) can supply power.

$$L = L - Bd \tag{6}$$

   If the load remains positive, then the load is supplied by backup energy (Ba).

$$Ba = L \tag{7}$$

   3.2. If the solar and wind supply are larger than the load, i.e., S + W > L, then the charge to battery or pumped hydro (Bc) is supplied until fully charged.

$$Bc = (S + W) - L \tag{8}$$

   If the battery or pumped hydro are fully charged, then the rest of the electricity is accounted as excess electricity (Ex).

$$Ex = \text{rest of}(S + W) - L \tag{9}$$

Throughout a dynamic simulation, the energy balance is confirmed at every hour using the following equation:

$$Ex = S + W + Hy + G + Bi + Ba - L + Bd - Bc - Bs \tag{10}$$

where Bs is the loss of battery or pumped hydro.

The simulation starts on April 1 with batteries half charged and ends on March 31 of the next year, for which the meteorological data of the same year are used.

The necessary capacities of electricity generation in 2050 are PV 147 GW, wind 37 GW, hydro 28 GW, geothermal 14 GW and biomass 8 GW in scenario 'A'.

The electricity storage system is also analysed. Japan has the pumped hydropower supply of 26 GW over 10 h. The storage capacity is 260 GWh.

A large battery capacity is required to reduce the backup power. When increasing battery storage incrementally, we find 300 GWh is sufficient to allow for zero backup power. Table 5 shows the scale of electricity generation for three scenarios.

Table 5 Summary of the simulations

Simulation items	Unit	Scenario 'A'	Scenario 'B'	Scenario 'C'
Solar photovoltaics capacity	MW	147,637	306,630	476,980
Wind power capacity	MW	36,851	78,966	110,552
Pumped hydro/battery capacity	GWh	260/300	260/300	260/300
Electricity demand	GWh/year	626,945	626,945	626,945
Average electricity demand	MW	71,569	71,569	71,569
Peak electricity demand	MW	109,965	109,965	109,965
Total electricity generation	GWh/year	633,841	790,623	1,050,310
Total share of generation	%	101.1	126.11	167.53
Solar photovoltaics	%	25.77	53.53	83.26
Wind	%	13.66	29.28	40.99
Hydro	%	27.76	21.58	21.57
Geothermal	%	17.08	13.88	13.88
Biomass	%	9.83	7.83	7.82
Backup power	%	6.99	0	0
Peak back-p power	MW	44,525	9047	0
Battery charge	GWh/year	3616	18,080	5717
Battery discharge	GWh/year	3586	17,076	5302
Battery loss	GWh/year	370	1805	566
Battery loss/demand	%	0.06	0.29	0.09
Max charge level	%	51.38	100	100
Average charge level	%	1.4	88.78	97.33
Pumped hydro charge	GWh/year	29,337	73,740	55,507
Pumped hydro discharge	GWh/year	22,667	56,705	42,612
Pumped hydro loss	GWh/year	6800	17,012	12,784
Pumped hydro loss/demand	%	1.08	2.71	2.04
Pumped hydro max charge level	%	89.65	100	100
Pumped hydro average charge	%	7.66	50.91	73
Excess electricity	GWh/year	190	145,637	410,052
Excess electricity/demand	%	0.03	23.23	65.4
Peak excess electricity	MW	1792	184,569	312,912

Electricity is supplied by a combination of renewable energy sources. Battery and pumped hydro are used to store excess electricity and discharge when the supply is not adequate.

6 Scenario 'B'

Scenario 'B' involves the renewable energy supply to address the pure electricity demand, the electricity for EVs and hydrogen for FCVs. Renewable electricity generation is assumed 126 % of the pure electricity demand. Here, we first discuss the energy demand for EVs and FCVs.

6.1 Energy Demand for Vehicles

The number of passenger vehicles was 55.5 million in 2010 and will be decreasing to 49 million in 2050. Table 6 shows the energy demand for passenger vehicles for approximately 2050. We assumed half of the vehicles are FCVs, and the other half are EVs. The annual driving distance is 10,000 km, and the fuel economy is 195 Wh/km.

We introduced an on-board PV panel to add some energy to improve vehicle's fuel economy. Figure 5 shows the concept of a future vehicle with 600 W of photovoltaic supply at vehicle's rooftop area of 3 m^2. The vehicle has regenerative braking and recovers electricity from the rotation energy of wheel with 50 % efficiency. In the case of a FCV, there is a hydrogen sub-tank, which is used for storage of the hydrogen produced on board.

The photovoltaic panel on the vehicle's rooftop is assumed to supply 20 % of the annual driving demand, 39 Wh/km for each vehicle [17]. If the PV cost decreases according to the learning curve, then the use of the PV panel will be a realistic option in the near future. As the PV panel for the vehicle's rooftop does not require a support frame structure, as in mega solar plants on the ground and the automobile production process is highly automated, the cost per kW is expected to be relatively lower than that of a PV plant. The use of a PV panel for the vehicle's rooftop will be particularly effective in areas with abundant solar radiation.

Table 6 Electricity demand of future vehicles

Electricity demand	Unit	FCV	EV
Driving load/km (A)	Wh	195	195
Rooftop PV/km (B)	Wh	39	39
Net demand/km (A–B)	Wh	156	156
Charge efficiency	%		90
H2 production efficiency	%	85	
FC generation efficiency	%	60	
Annual driving distance	km	10,000	10,000
Annual electric demand/vehicle	kWh	3059	1733
Number of vehicles	Million	24.5	24.5
Annual electric demand	TWh	74.9	42.5

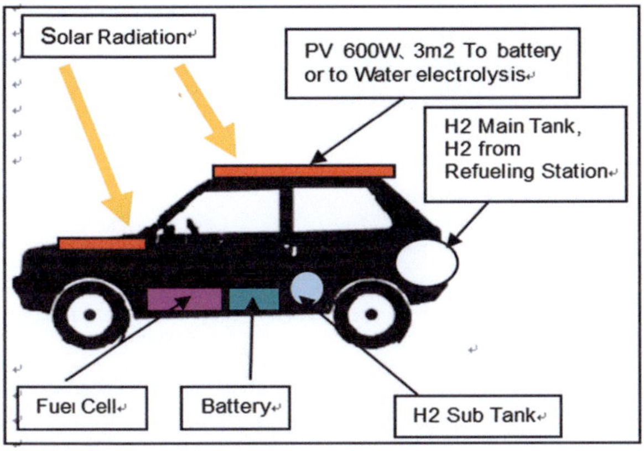

Fig. 5 EV and FCV with a rooftop PV panel

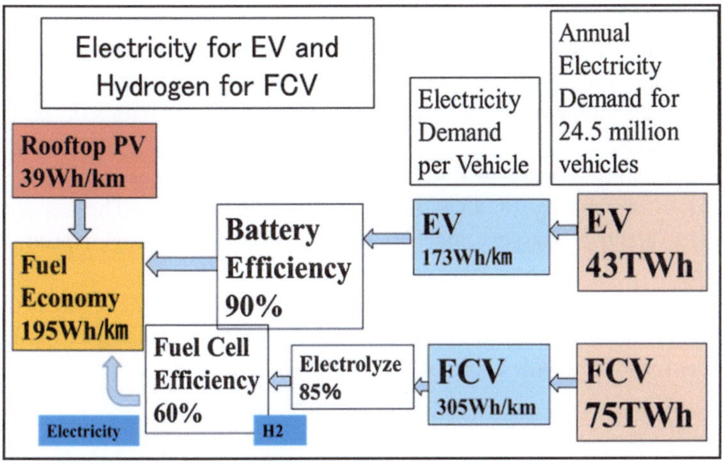

Fig. 6 Energy flow for an EV and a FCV

Figure 6 shows the energy flow for an EV and a FCV. Electricity for the EV is supplied with 90 % battery efficiency. Electricity for the FCV is supplied with 85 % electrolyse efficiency and with 60 % generation efficiency of the fuel cell on board. The hydrogen demand for the FCV and the electricity demand for the EV are estimated using the conversion efficiency shown in Table 6. The annual electricity demand for an EV is 1733 kWh and that for a FCV is 3059 kWh.

6.2 Excess Electricity and Hydrogen

The excess electricity is the difference between generation and demand. The results of electricity show fluctuations, especially during the summer season. There are periods of very small excess electricity in June. Solar PV is not large in this rainy

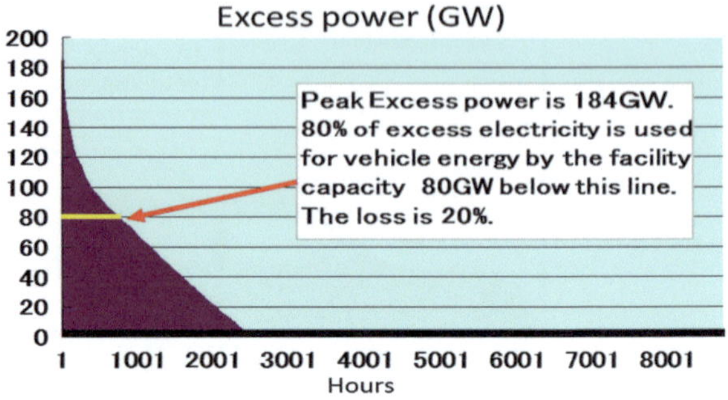

Fig. 7 Load duration curve of excess electricity (GW)

month. The amount of wind power is not as high in the summer, and the peak of solar generation does not occur in summer, but in April and May. It may be not possible to adequately supply energy to EVs and FCVs in June. It is necessary to solve this problem via a variety of design methods, including energy storage systems and demand response.

If we attempt to use all of the excess electricity, then high capacities of electric chargers and hydrogen production plants are required, which will not be used frequently. This situation is not economical because the capacities of the facilities become too large.

The load duration curve in Fig. 7 shows that 80 % of excess electricity can be recovered by a capacity of 43 % of peak excess electricity. The capacities are 29 GW for electric chargers for EVs and 51 GW for hydrogen producers for FCVs [18].

The total electricity generation is 782 TWh, of which 627 TWh is for pure electricity demand, 147 TWh is for passenger vehicles, 43 TWh is for EVs and 75 TWh is for FCVs, with a loss of 29 TWh, as shown in Fig. 8.

7 Scenario 'C'

Scenario 'C' aims to supply energy to all end-users in 2050. This scenario requires the addition of thermal energy supply to the 'B' scenario. By increasing the solar and wind powers, the electricity generation is assumed 168 % of the electricity demand. Here, we discuss the fuel demand for heat in the industry, commercial and household sector and fuel for transport.

We assumed that biomass will be used for trucks, cargo ships and air flights in the transportation sector. In addition, biomass will be used for thermal demand for low-to-medium temperature heat in the industrial sector. Some part of the thermal demands in the industrial sector and the building sector will be supplied by solar heat, with the assistance of heat pumps. If we have enough biomass to supply energy to all of the thermal end-use applications, then it is simple to achieve the goals of the 'C' scenario. However, the available scale of biomass supply is not yet

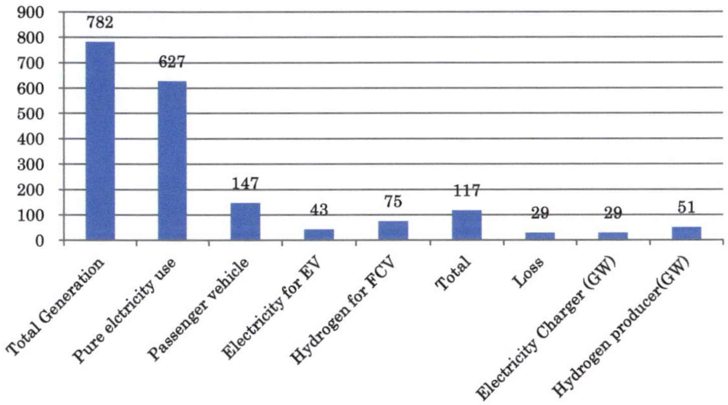

Fig. 8 Effective use of excess electricity (TWh)

precisely known. As a result, we considered that some part of the thermal demand will be supplied from excess electricity and solar heat.

Scenario 'C' involves not only supplying electricity and energy for vehicles but also converting excess electricity to hydrogen for cogeneration and thermal energy demand. Other than EVs and FCVs, fuel cell drives are used for passenger ships because of its silent power characteristic. Figure 9 shows the end-use energy demand for the following categories: pure electricity, electricity for fuels (EV, FCV, hydrogen cogeneration and heat pump) and fuels (for industrial heat, hot water, heating demand and transportation).

Cogeneration in this scenario is used to supply electricity and heat for the hot water demand of residential and commercial buildings using the hydrogen produced from excess electricity at a conversion efficiency of 85 %. Hydrogen will be converted to electricity using a cogeneration fuel cell with efficiency of 40 % into exhaust heat 40 % and into loss 20 %. Solar heat will be harnessed by domestic solar hot water system and will be used for hot water demand in households and industry. Industrial solar hot water will be used for the heat input of an efficient heat pump and supply energy for drying, washing and cleaning in industrial processes.

Figure 10 shows the total energy supply structure. We assumed the amounts of biomass energy of 20 MTOE from the wastes of agriculture and the construction industry and 36 MTOE from energy crops and forests in 2050. Solar heat will be developed to supply 18 MTOE in 2050. PV on vehicle's rooftop will supply 4 MTOE.

There are some difficult points in supplying energy from renewable sources to address all of the energy demand. We could not confirm the technology to produce steel using hydrogen by 2050. If such technology is not available until 2050, then some amount of gas or coal will be used in the iron and steel industry. However, the demand will be not so large, as the recycling rate of steel will be as high as 75 % in 2050.

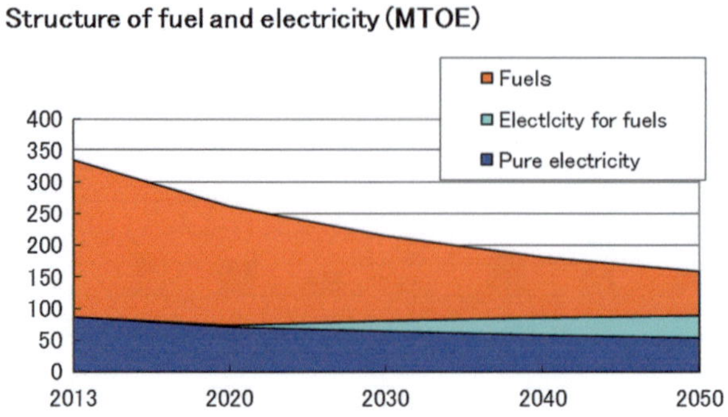

Fig. 9 Types of end-use energy demand

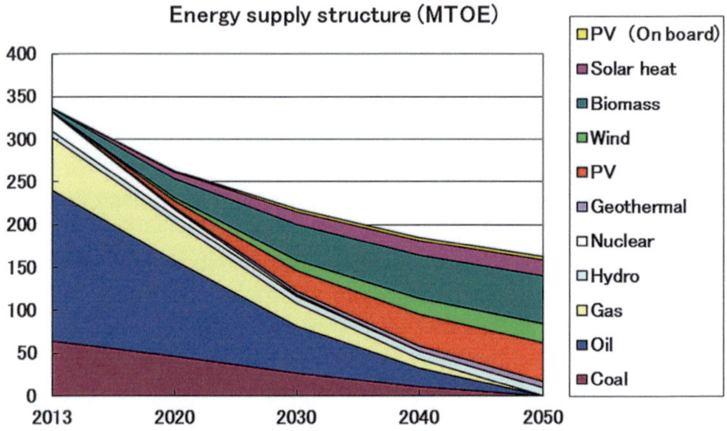

Fig. 10 Total energy supply structure in scenario 'C'

8 Discussions

We found there are three renewable-rich future scenarios in our study of renewable energy introductions, which are categorised according to introduction stages. When the share of renewable sources increases, the intermittent nature of renewable electricity generation will cause periods of lack and excess of energy compared to the energy demand. The lack of electricity is supported by backup power and storage of pumped hydro and batteries. Scenario 'A' involves the supply of electricity only. This scenario is simple to implement, but the share of renewable sources is larger, resulting in periods of lack and excess of electricity. Scenario 'B' is developed to address such situations. In this scenario, energy is supplied to meet the demands of both pure electricity and transportation. The excess electricity is used for electricity to power EVs and for hydrogen to power FCVs.

Excess electricity will be used to address energy demand that is not restricted by time. In scenario 'B', as shown in Table 5, the lack of electricity decreases due to the larger scale of solar PV power and wind power, and the amount of excess electricity increases. Scenario 'B' is effective for ensuring a stable electricity supply and for addressing other end uses of transportation demand at the same time. In scenario 'C' shown in Table 5, the lack of electricity disappears, but a large amount of excess electricity appears. The excess electricity will supply all of energy end-use needs together with biomass and solar heat with heat pump,

9 Conclusion

We studied three types of renewable-rich future energy scenarios. The promising renewable energy technologies today are electricity supply technologies, such as solar photovoltaic technology and wind technology. In the initial stage, the electricity supply among renewable energy technologies will be strongly promoted. However, when the share of renewable electricity increases, the intermittent nature of renewable energy will cause periods of lack and excess of energy compared to the energy demand. The lack of electricity is addressed by backup power and storage. Excess electricity can be used to address energy demands that are not restricted by time.

We analysed three scenarios. Scenario 'A' involves supplying pure electricity demand only. Scenario 'B' involves supplying both pure electricity demand and transportation energy demand for EVs and FCVs. Scenario 'C' is an enhancement of scenario 'B'. In scenario 'C', excess electricity is used to address some part of the thermal demand. Additionally solar heat and biomass are used to meet the entire thermal energy demand. The idea behind these three scenarios is to provide a response to the developing stages of renewable energy expansion. We hope that the approach is helpful to understand the developing steps of renewable-rich future energy systems.

References

1. Jacobson MZ et al (2011) Providing global energy with wind, water, solar power, part 1 & 2. Energy Policy 39:1154–1190
2. The energy report, 100% renewable energy by 2050, WWF international. Feb 2011
3. Renewable electricity futures study, National Renewable Energy Laboratory, United States (2012)
4. The 4th basic environmental plan 4th Apr 2012, Japanese Government
5. Revolutionary energy scenarios for 2030, Energy strategy committee, Japanese Government, 2013
6. Japanese long term energy prospects for 2030, METI, Japanese Government, Jul 2015

7. Energy scenario proposal toward post carbon society, Part 1–4, World wide fund for nature Japan, 2011–2014
8. (2010) Handbook of energy and economic statistics in Japan, The energy data and modelling centre, Institute for Energy Economics, Japan
9. H Tsuchiya (1999) Learning curve cost analysis for model building of renewable energy in Japan, Proceedings IEA workshop on experience curve for policy making – the case of energy technologies, Stuttgart, pp –76
10. Photovoltaic Roadmap: PV2030+, NEDO (New Energy and Industrial Technology Development Organization), (2010)
11. Extended AMEDAS (2008) Hourly and daily weather data, Japan architecture association
12. Solar energy utilization handbook, Japan solar energy society, (1985)
13. Study of potential of renewable energy (FY2010), May 2011, Ministry of environment, Japan
14. Handbook for building wind energy systems, NEDO (New Energy and Industrial Technology Development Organization), Japan, (2005)
15. Tsuchiya H (2012) Electricity supply largely from solar and wind resources in Japan. Renew Energy 48:318–325
16. H Tsuchiya (2014) Optimal mix of solar and wind power for electricity supply matching with demand, grand renewable energy conference, Tokyo, July 2014
17. Tsuchiya H (2008) Innovative renewable energy solutions for hydrogen vehicles. Int J Energy Res 32:427–435, John Wiley
18. H Tsuchiya (2014) Hydrogen and electricity supply for future vehicles by excess electricity from solar and wind power, grand renewable energy conference, Tokyo, July 2014

Participatory Design as a Tool for Effective Sustainable Energy Transitions

Benjamin C. McLellan, Yusuke Kishita, and Kazumasu Aoki

Abstract Sustainable energy systems and transitions towards such systems are often discussed among experts or high-level stakeholders, but rarely involve individuals from the general community. This paper argues that it is important for such stakeholders to be engaged in the visioning or formation of plans for future energy systems in order for effective transitions to take place. The results of a number of related studies and techniques for achieving such visions are presented. Furthermore, research has indicated that there is a need to effectively nurture niche actors in order to provide the seeds of future sustainable regimes with the possibility to emerge. A discussion of how policy could promote this will also be described.

Keywords Participatory design • Energy scenarios • Transition theory • Policy

1 Introduction

The challenge of achieving sustainable energy systems is one of the most critical endeavours of society, without which it will not be possible to mitigate potential drastic climate change. The transition to sustainable energy systems is an important consideration, trying to identify the most appropriate and effective pathways and policies to enable such major restructuring to occur. Various technologies, policies and scenarios are discussed – in academia [1, 2], government [3, 4] and non-governmental organisations of various types [5–7]. However, it is apparent that most such exercises depend on technical assessment by experts and rarely incorporate the opinions of individual consumers – let alone enable participation in the development of the scenarios themselves [8]. This approach is feasible under business-as-usual conditions, with a highly centralised system of energy

B.C. McLellan (✉)
Kyoto University, Graduate School of Energy Science, Kyoto, Japan
e-mail: b-mclellan@energy.kyoto-u.ac.jp

Y. Kishita
National Institute of Advanced Industrial Science and Technology, Tsukuba, Ibaraki, Japan

K. Aoki
University of Toyama, Toyama, Japan

M. Matsumoto et al. (eds.), *Sustainability Through Innovation in Product Life Cycle Design*, EcoProduction, DOI 10.1007/978-981-10-0471-1_40

infrastructure, low demand-response flexibility and relatively low number of controllable loads in the system. However, the advent of the prosumer, liberalisation of markets, increases in distributed generation and renewable energy and increasing "smartness" of the grid make this pattern of development appear to be theoretically flawed, if not obsolete. Moreover, with the improvement in knowledge about alternatives and rapid dissemination of information with new communication technologies, and likewise with the ability to gain almost instantaneous feedback (automated or manually) through the same technologies, as highlighted by the societal use of social networking services (SNS), there is an ever-increasing ability to engage with stakeholders and obtain their input.

Participation in the design phase of sustainable energy scenarios or plans is not merely a concept that can lend legitimacy to the outcome. The process of participatory design can enable the ongoing education of stakeholders and the expansion of understanding – which is an important element in take-up of new technologies or engagement with new operating regimes. Moreover, particularly in the case of energy – a ubiquitous utility verging on being a definable human need – the consumer base that makes up the general stakeholder body has a very large potential to influence the operation and effectiveness of the system (e.g., in Japan, the residential sector accounts for around 30 % of electricity consumption alone [9]). This demand, however, is diffused across many point loads, making effective collective action difficult. Without the engagement of such diffuse actors, it is virtually impossible to encourage support for effective collective decisions, except under drastic circumstances.

This paper discusses the theoretical background and practical experience of utilising a variety of processes to engage stakeholders in the participatory design of energy system scenarios.

2 Methodology

2.1 Participatory Process Precedents

Participatory processes have been used in a variety of decision-making and design contexts. Among the most prominent techniques, participatory technology assessment (PTA) has a long history of utilisation, particularly in Europe; design "charrettes" have been applied to post-disaster reconstruction [10], low-energy building design [11] and other contexts. Participation has been applied to the selection of policy – for example, in the Japanese [4] and French contexts [12]. Considering sustainable design, participation and consideration of various stakeholders is considered an important aspect in industrial operations [13] – and processes that integrate such stakeholders as participants in a "codesign" process are vital. While these processes are gradually becoming more common in the design of single industrial operations, where the problems being addressed are typically more technically and spatially bounded, the use in broader, less-bounded situations is more unusual. In both academic and political arenas, this is the general rule, where even the most inclusive scenario development often uses technical experts or multi-criteria assessment as a proxy for real participation. There is thus a challenge for the mindset of those responsible with developing energy scenarios.

In current industrial design practice, generally the most participatory component of the design process (with regard to non-technical stakeholders) is the period of consultation or comment allowed on an Environmental Impact Statement (EIS), at which time general stakeholders can provide feedback on a design [14]. This is generally too late in the design cycle to create effective and cost-effective changes in the final project [15] – but may be sufficient to stop a project at the final regulatory hurdle. This is often a unidirectional approach where feedback is taken and responses not given directly. Such a process can be useful, but often not as potentially useful as a process containing appropriate feedback loops and iterations [15]. In the case of larger policy initiatives or energy plans, it is common that the only feedback enabled is (in democracies) through elections – which may come too late and conflagrate too many issues to have a clear decision on one strategy.

2.2 *Criteria for Appropriate Methods*

In order to effectively integrate participatory processes in the planning of sustainable energy futures, a number of key elements should be considered. Firstly, sustainability being multifaceted and stakeholders being diverse, there needs to be some well-defined and structured process of obtaining the information required by the technical team. In the case of industrial processing operations, this has been developed and tested elsewhere [13, 16] and has shown to promote effective improvement of the sustainability contribution of these organisations. Secondly, in the case of future-oriented processes where it is aimed to determine preferences on yet-to-be-promoted technology, or in the case where "opportunities" are being sought (not just risks), it is important that these structured processes can prompt innovative thinking to get beyond the easily identifiable standard solutions or barriers.

2.3 *Methodology Used Here*

Based on an understanding of the energy system – technically, institutionally and with regard to both the supply and demand-side requirements of current and emerging technologies – a methodology was developed to enable the project team to obtain sufficient data to undertake energy scenario modelling.

The methodology developed over a series of four case studies in 3 years, involving one group of general public participants and three groups of interdisciplinary students in the first semester of a graduate degree in energy science. All the case studies were undertaken in Japan – in the first case this was undertaken in Japanese, while the remaining three studies were in English. The first case involved mostly Japanese participants, while in the other three studies, the participants were a mixture of Japanese and international students. In all cases, the focus was on building a sustainable future energy system plan for Kyoto.

The process utilised a workshop format, with a combination of structured and semi-structured tasks led by a facilitator. In the first case, only one workshop was undertaken, due to time constraints for the attendees. In the remaining three cases, two or three workshops were undertaken, enabling an iterative improvement of the plans and increasing understanding of the desired outcomes.

Three main aspects were sought from these workshops:

1. Understanding the context of a sustainable energy system within a sustainable, preferable and resilient societal system – particularly identifying the improvements needing to be made for the future. Defining what a sustainable city or state would look like or require.
2. Understanding the preferences for supply-side technologies – particularly in the controversial situation for nuclear power in Japan at present.
3. Understanding energy service requirements and the ability to shift these.

The main steps in each of the workshops (on the basis of three sessions here, although in some cases this was squeezed into two) are shown in Fig. 1.

Workshop 1 was used to introduce the scope of the study – in this case producing a vision of a sustainable Kyoto city, with a focus on energy. Various information about the city was introduced, and the scope also enabled consideration of more than just the city when energy generation was concerned. The first major activity of the

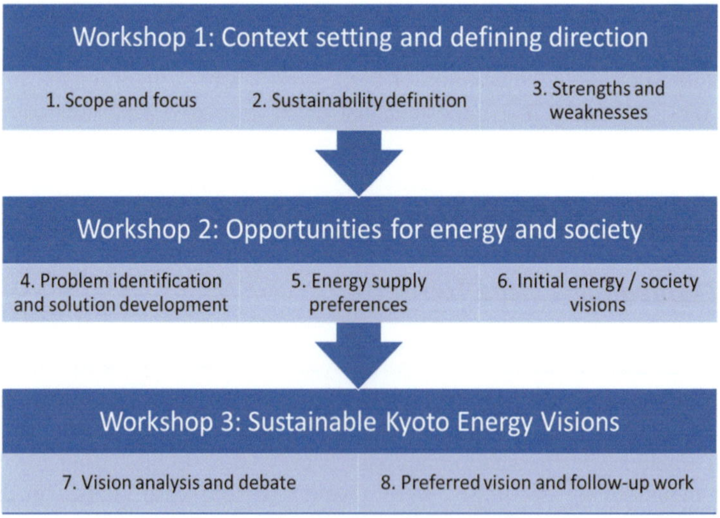

Fig. 1 Outline of workshop process and content

workshop was to derive a definition of sustainability within the context of Kyoto – a brainstorming session started with identifying different elements of sustainability, which were pulled together as a guiding statement for the remainder of the activities (this approach has been found useful elsewhere [17]). As an example, the definition derived in one of the studies was:

> Maintaining or improving the quality of life within environmental limitations, while maintaining the cultural / historical heritage but enhancing a vibrant economy and society.

This definition reflects the specific context of Kyoto, which is highly influenced by its cultural heritage, while focusing on key areas perceived to be important current vulnerabilities – demographic shifts, economic stagnation and self-sufficiency within environmental limitations. For each workshop, the derived definition was different but reflecting much the same type of concerns – quality of life, environmental preservation and a vibrant economy – while retaining cultural heritage value.

The next major stage of the process was to consider the strengths and weaknesses of Kyoto – loosely framed as "what Kyoto should keep" and "what Kyoto should change" and in an additional activity "what is better or worse about Kyoto than my home town", with most of the student group being in Kyoto for only a few years. This activity, involving small groups discussing and producing suggestions, was useful in identifying areas that were immediately obvious and highly relevant to the stakeholders in their everyday activities as well as at the level of the community.

At the end of the first workshop, the participants were provided with a primer on energy in Kyoto – from supply to demand – which they were able to read more about between workshops. They were also provided with a summary of the workshop outcomes in the interim.

In the second session, the participants undertook an exercise to identify specific problems and potential solutions for a sustainable Kyoto, based on a "Five Capitals" framework [18]. They worked on a single capital in small groups and then rotated between the stations to augment or respond to the previous groups' suggestions, then finally discussing with all participants. Also in the second workshop was a facilitated discussion of alternative energy supply technologies – using a visual technique for displaying the technologies in comparison to each other and a baseline, as regards to their preferability. Some typical results are shown in Fig. 2. These discussions required all participants to come to a consensus regarding the relative preference of each technology. Interestingly, the energy science students gave a tighter grouping of technologies, whereas the general stakeholder group were more willing to place selected technologies in extreme positions – particularly a negative response to nuclear power. Such information was able to be combined later with data from a large-scale survey of Japanese consumers regarding this and an expanded list of issues associated with energy and policy.

At the end of the second workshop, the participants were presented with some typological scenarios for sustainable energy futures for consideration. These scenarios or, rather, end-point characterisations of the potential energy-based lifestyles – e.g. a

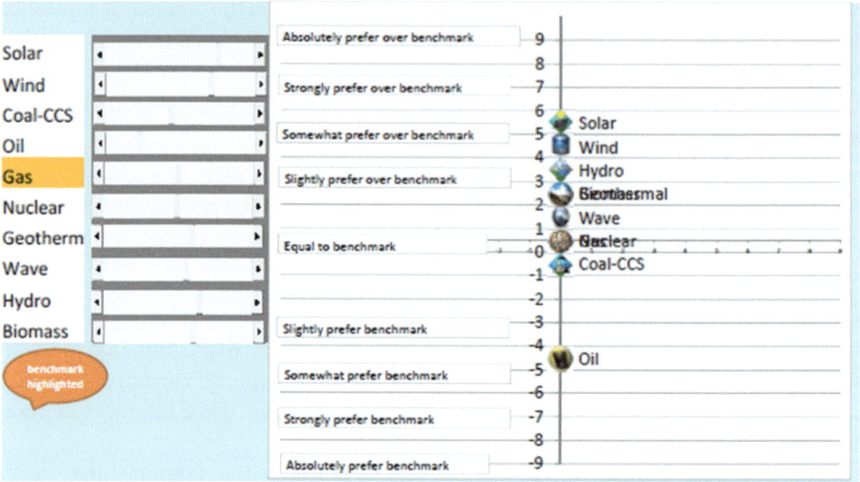

Fig. 2 Relative ranking of preference for alternative electricity supply technologies

"smart and sophisticated" future in which high levels of electrification are matched with renewable energy or a "simple" future in which a reversion to more agricultural and mechanical end-use technologies – are utilised. In the final session, stakeholders help to shape and select elements of their preferred scenarios (and in the case of the students, they were required to develop their own scenarios and put together a full report on it).

A variety of tools were developed for use in the workshop process and the decision-making process. On the key agreed criteria of:

- Self-sufficiency
- Vibrant economy
- Job creation
- Local environmental quality
- Global environmental impact
- Diversity and youth

The results in Fig. 3 show the relative performance of each of the energy technologies with regard to each element. This data, from the first student workshop, was then weighted using their own preferences to identify a single-indicator "Sustainable Kyoto Index" (weightings shown in Table 1). Scores (from −5 to 5) were given by each individual participant for each technology against the six criteria listed above. These scores were averaged to give a group score for each criteria and technology. The criteria were then combined using an another scoring survey, in which a score of 1–9 was given for each headline criteria, which was then normalised before multiplying the individual criteria and summing for the "Sustainable Kyoto Index". It is apparent that based on the preferences of the workshop, solar power is most preferable for Kyoto's future energy system – interesting

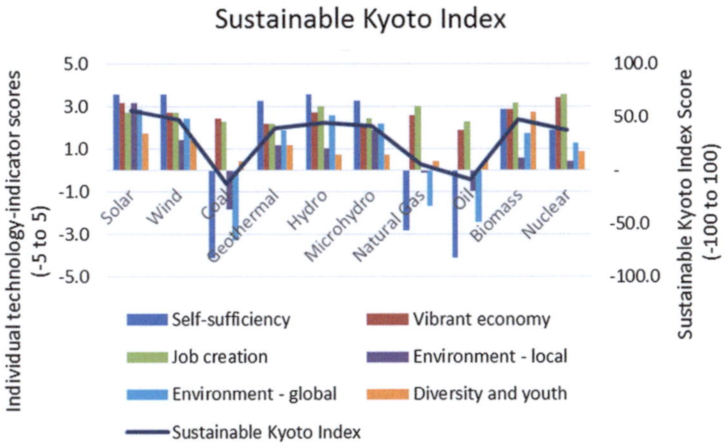

Fig. 3 Comparative weighted performance of supply-side energy technologies

Table 1 Criteria weighting towards "Sustainable Kyoto Index"

Headline criteria	Weighting
Self-sufficiency	0.18
Vibrant economy	0.14
Job creation	0.19
Local environmental quality	0.16
Global environmental impact	0.14
Diversity and youth	0.20

moreover because there are specific restrictions on residential solar systems in Kyoto to avoid impacts on cultural heritage.

In addition to understanding the relative supply-side preferences for energy, the process resulted in a large number of potential problem-solution parings for problems identified on the Five Capitals framework. These could be used directly to contribute elements of a future sustainable energy vision.

2.4 *National Consumer Survey*

To supplement such small-group workshops, data was also obtained from a large-scale national survey undertaken in various regions of Japan in March 2014 (2581 respondents). This survey was focused in part on parallel questions (e.g. supply-side energy preference and technologies or actions appropriate for improving sustainability of the energy system); it was also used as a tool to try and understand the

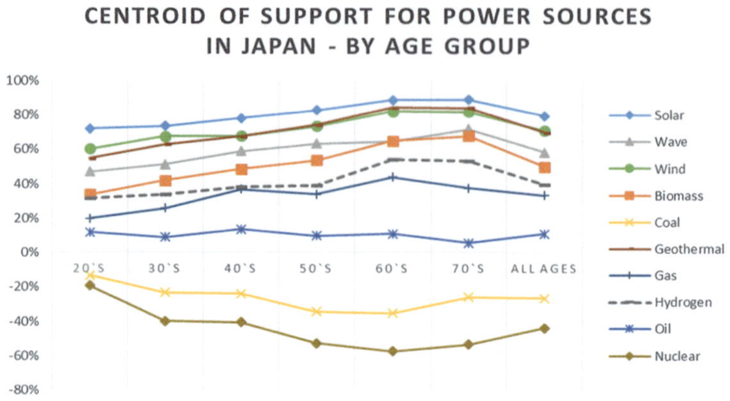

Fig. 4 Support for power sources in consumer survey by age group

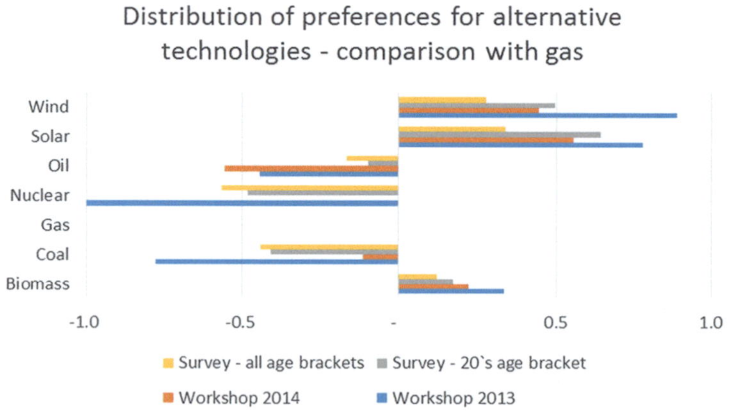

Fig. 5 Preferences for selected energy generation technologies – comparing techniques

potential transitions that may be taking place or are identified as favoured by the populace.

Of the various questions asked in the consumer survey, one key question is chosen here for comparison – the preference for alternative energy sources (due to relative ease of comparability). In Fig. 4, the support for different alternative power sources was converted from a five-band scale plus uncertainty (strongly support, support, neutral, oppose, strongly oppose, uncertain) using weightings of 2, 1, 0, −1, −2, 0, respectively, to obtain a "centroid of support" for each power source. The difference between each of these sources and the comparator (natural gas) were then normalised using the largest possible value and compared with similarly normalised workshop values for selected technologies (Fig. 5). It is apparent that the survey provided a relatively clear preference for technologies, which varies by age group but mostly with regard to the magnitude rather than direction or order of preference.

Figure 5 shows similarly that the two techniques are relatively consistent with direction of support or opposition, but not necessarily to preference order. In particular, in the 2014 workshop nuclear was equally preferred to gas.

2.5 Acknowledging Fundamental Limitations

An important aspect of building feasible or achievable scenarios is to understand the limiting physical, economic, social and technical constraints. In the case of the survey and workshop for visioning, described above, these were not directly incorporated (apart from a willingness to pay or willingness to undertake behavioural changes – described below). For example, it is easy for consumers to state a preference for solar energy, but if the cost, the available space for installation, the ability to import or manufacture PV cells and the available sunlight are insufficient, then the scenario may be unfeasible. It is important that any model utilising such preferences be able to take these constraints into account when the progressive and final energy mixes are developed.

In the current case study, no model was directly utilised; however, in future studies – and in the general case – it would be most useful to have a "live" process, whereby the participants can see the influence of their preferences within the fundamental limitations of the system. This should not just enable consideration of supply-side preferences, but also of the demand-side choices.

3 Results

3.1 Scenario Development

By way of example here, we discuss the ways in which a scenario for supply-side energy generation (particularly the electricity mix) could be developed using the data presented above.

In the first instance, aiming to model supply of electricity requires an understanding of demand. In this case, we assume that demand is stable over the upcoming period to 2050. This will inevitably change if we incorporate the preferences and opportunities developed in the workshops, but for the sake of this paper, this simplification is considered useful. Japan's demand for electricity has been largely stable over the last decade or so [9] and, despite some anticipated efficiency and demographic-led reductions, is expected to be at similar levels (around 1000 TWh) out to 2030 [4].

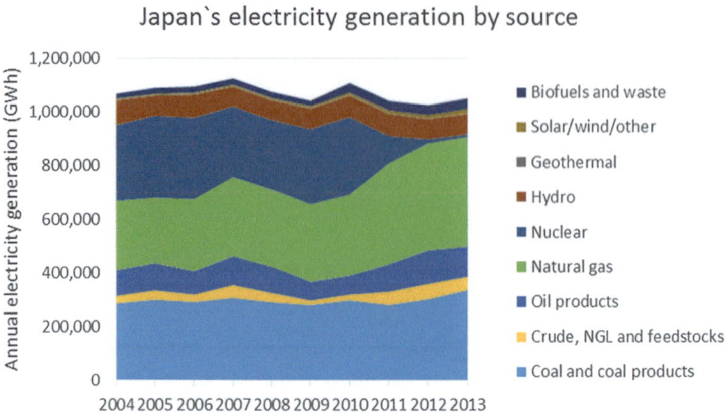

Fig. 6 Japan's electricity generation by source [9]

More importantly to our understanding of potential scenarios and future developments is the mix of electricity generation by source (Fig. 6). Since the closure of nuclear power plants in the wake of the Fukushima disaster of 2011, the electricity mix has included a larger share of gas, oil and coal making up for idle nuclear capacity. In developing a future scenario, the current situation can be considered either as the existing system boundary or as an aberrance from the normal system state (which normally includes more nuclear power). In this study, we applied the electricity mix of 2010 and 2013 as alternative starting points. Regarding the demand for electricity in the future, for the sake of demonstration, we have used two alternative methods of estimating – one assuming a constant 1000 TWh of demand out till 2050 and one assuming that demand remains constant as a per capita rate of usage. We also estimate population changes by extrapolating recent patterns of demographic shift estimated at 10-year intervals [19]. The trend in population is shown in Fig. 7.

With regard to supply, the preference of consumers for alternative technologies was used to set the final state of the electricity supply.

With regard to understanding the priorities of consumers for the characteristics of the energy system, from the Kyoto respondents to national survey (111 respondents), it was identified that the priorities were (1) no blackouts, (2) safety, (3) maintaining or lowering costs, (4) reducing the environmental impact (particularly global warming) and (5) promoting the use of local resources. The top five items were the main priorities for each of the age groups; however, the order or preference changed somewhat according to the age group, as shown in Fig. 8. It was interesting to note that the requirement for no blackouts was more important for the youngest age groups, while the eldest age group was most concerned about safety.

Another important question regarding the potential to develop future scenarios was the consideration of behavioural change and investment decision. In the consumer survey, applicants were asked to choose as many options as they wanted out of a variety of behavioural change and investment-requiring changes for

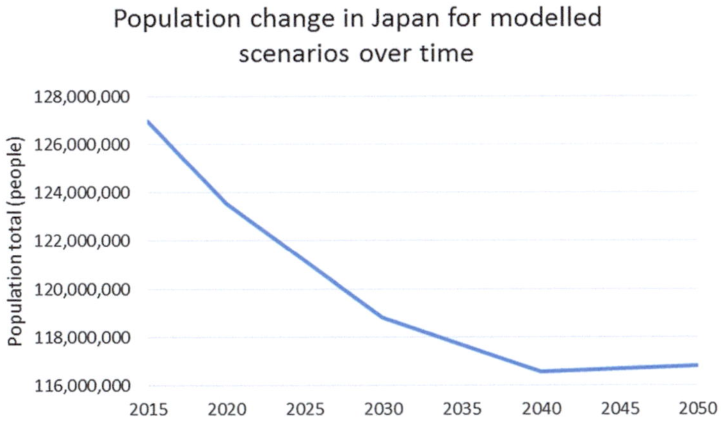

Fig. 7 Population trend used for the scenario

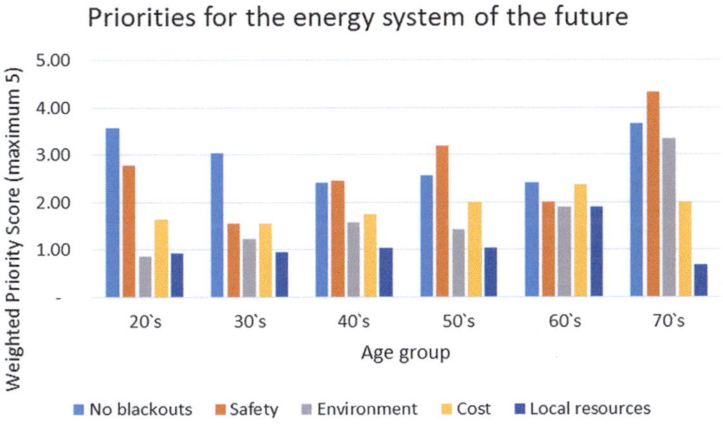

Fig. 8 Kyoto consumer priorities by age group

reducing energy use or environmental performance of energy (responses shown in Figs. 9 and 10, respectively). It is immediately apparent that there are a greater number of respondents willing to undertake behavioural change rather than invest money in significant infrastructural, refurbishment or relocation changes. There was more interest in installing solar energy systems than in using fuel cells or other generators, and in other questions we noted that the main barrier to actualisation of this was cost [20].

Regarding behavioural change, although participants were willing to walk for short trips and reduce the use of air conditioning, it appears that some activities

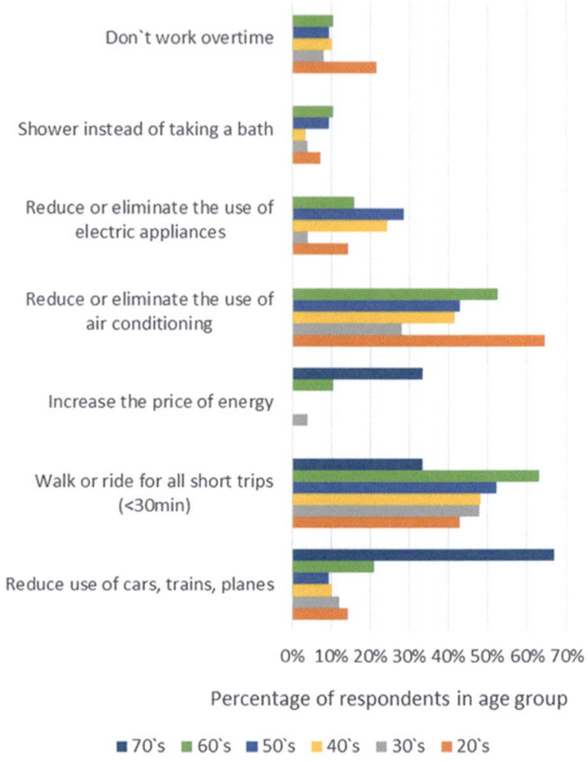

Fig. 9 Acceptability of behavioural-only changes for Kyoto consumers

they were either unable or unwilling to adjust – such as bathing or reducing overtime at work – and accepting an increase in the price of energy was very unpopular. Understanding such preferences can assist in setting boundaries and conditions for scenarios. For the sake of this paper, these behavioural changes are not considered, but will be incorporated in future work as discussed below.

As part of the outcomes of a consensus-based visioning process, it would be ideal for the participants to agree on not only the ideal vision, but on their own role and commitment within it. For example, in order to use the most renewable energy at a reasonable cost, demand-side usage reduction may be necessary. If the participants are willing to commit to such activities and have a clear awareness of the practical limitations of the scenarios, then it may be argued that the decision is more legitimate.

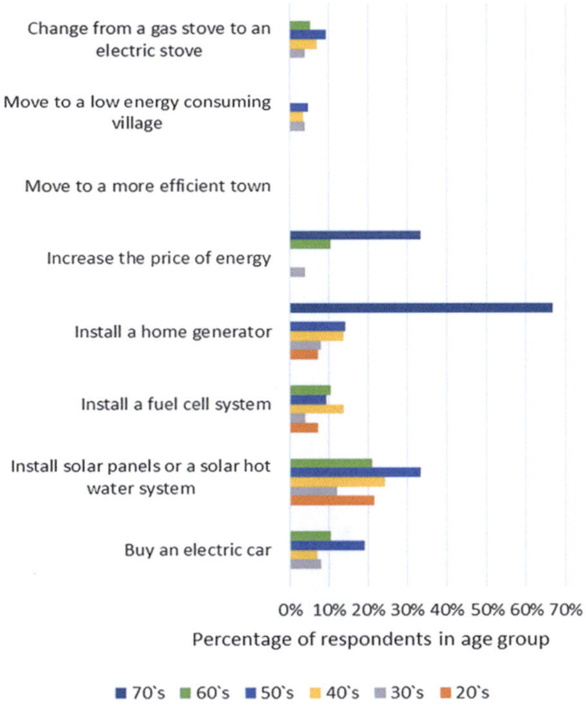

Fig. 10 Acceptability of investment-requiring lifestyle changes for Kyoto consumers

4 Discussion

4.1 Simple Scenarios

In this case, two simple scenarios for the future of electricity generation in Japan are shown in Fig. 11 – (a) with steady total demand and the change in generation starting with a 2010 mix of electricity generation and (b) with steady per capita demand and starting with a 2013 electricity generation mix. The preference of consumers in this case was used to develop the final energy mix, by assuming that the ratio of the level of preference should equal the final ratios of electricity generation in the mix. The two technologies that had an overall negative preference (nuclear and coal) were given a final mix percentage of 0 %. Although an over-simplification, the scenario here has used exponential growth or reduction to model the mix change for individual technologies based on the overall required growth out to 2050. To balance the mix with demand, in (a) nuclear and coal are used equally up to 2040 to fill any lack in supply from other technologies, while in (b) gas and coal are utilised. Hydropower is assumed to be stable in both scenarios, as the current capacity is close to the practical maximum [21].

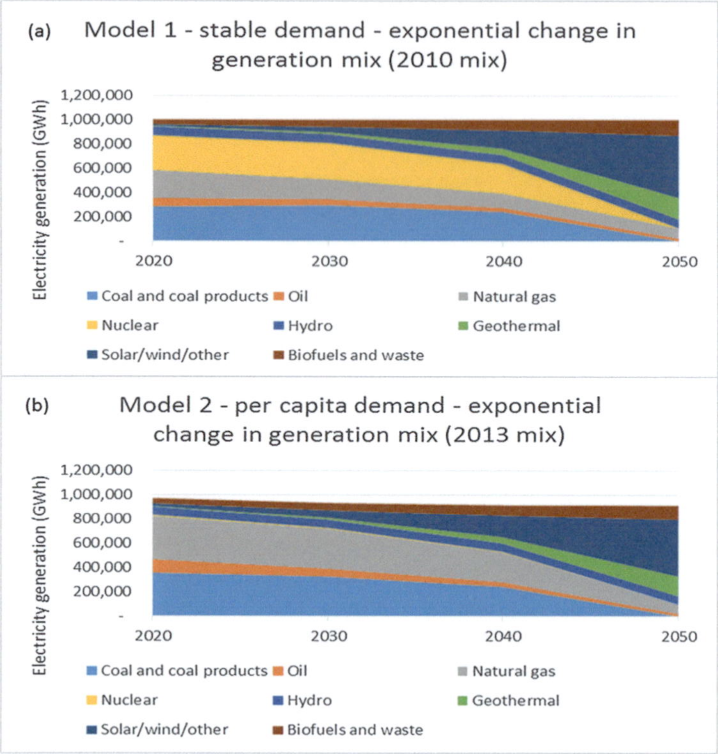

Fig. 11 Two alternative scenarios of electricity generation

4.2 *Feasibility and Indicators of Performance*

In the case study here, the feasibility of the preferred mixes is not explicitly tested. However, scenarios performed by others elsewhere indicate that these scenarios are within the bounds of conceivable possibility [22–24]. Regarding decision-making on the ultimate desirability of a scenario, it would be useful to have performance indicators – such as cost, CO_2 emissions – on which to compare alternatives. Alternatively, indicators such as the “Sustainable Kyoto Index” highlighted above could be employed. This work is ongoing for the current scenarios, but is an important and crucial consideration for processes such as this to be undertaken in the future. In the workshop process, the discussion of such indicators and a comparison of alternatives and the sensitivity of parameters would be undertaken as a final session in which the consensus decision would be made.

4.3 Improvements

These scenarios are drastically oversimplified due to time constraints; however, advances are being undertaken to improve the models being applied. The key novelty here is the application of consumer preference – which is unusual if not unique in existing models. The work being undertaken will further include the demand-side behavioural changes as described in the previous section, as well as clarifying more sophisticated methods for incorporating the supply-side preferences.

4.4 Educational Outcomes

One of the direct outcomes of the visioning process employed in this study was the ability to strengthen educational outcomes for students trying to understand sustainable energy systems. In the case of general public participants, there is also benefit in the ability to prompt questions and enable stakeholders to better understand the limitations and potentials of certain technologies or approaches.

4.5 Evidence for Energy Transitions

Part of the survey described here was undertaken to identify whether or not a transition is likely to be occurring in the Japanese energy system and what type of future energy system may actually be emerging – or at least representative of consumer preferences at present. This study has been more thoroughly discussed elsewhere, but the key conclusion was that though there may be a transition underway, it is unlikely that this transition will lead to a more distributed generation system. The current system transition appears to be more likely to emerge as a greater mix of renewable energy technologies, but remaining largely centralised (e.g. mega-solar plants, wind farms).

This type of transition is not reflected in the scenarios shown here – as they do not identify whether the electricity is generated from decentralised or centralised power plant. This is an area requiring additional model development for future assessments.

5 Summary

This paper has covered a number of areas in the discussion of energy scenario development through participatory processes. In current scenario-design methods, there is rarely feedback or input from general stakeholders or consumer preference surveys. However, the changing nature of energy production – from centralised to distributed – and the increase in information and interactive capability make stakeholder interaction both more easily achieved and more essential.

This paper has demonstrated that such an approach can be possible. However, this work is in its infancy, and further work is being undertaken to use this available data and information in a more sophisticated model.

The study has shown a number of useful and important outputs which can feed in to the scenario development process and which are currently not typically used. These include:

1. Identifying stakeholder preference for alternative technologies, which may indicate potential for uptake in future
2. Identifying preferable and unacceptable behavioural changes and investments that could be utilised to shift energy usage
3. Identifying energy system characteristics most desirable and least acceptable – to match proposed energy systems to these

It is apparent that some useful approaches can be developed to enable the input of participant stakeholders in developing future sustainable energy scenarios.

References

1. Zhang Q et al (2012) Scenario analysis on future electricity supply and demand in Japan. Energy 38(1):376–385
2. Elliston B, MacGill I, Diesendorf M (2013) Least cost 100% renewable electricity scenarios in the Australian National Electricity Market. Energy Policy 59:270–282
3. MEJ (2011) Potential for the introduction of renewable energy (FY 2010) [in Japanese]. Ministry of Environment of Japan, Tokyo
4. National Policy Unit (2012) Options for energy and environment – the energy and environment council decision on 29 Jun 2012. (cited 22nd Aug 2012). Available from: http://www.sentakushi.go.jp/english/
5. WEC (2007) Transport technologies and policy scenarios to 2050. World Energy Council, London
6. Beyond Zero Emissions (2010) Zero carbon Australia: stationary energy plan. Melbourne Energy Institute, University of Melbourne, Melbourne
7. IEA (2013) World energy outlook. (cited 24 May 2013). Available from: http://www.worldenergyoutlook.org/
8. McLellan BC et al. (2013) Assessing sustainable regional energy systems: a case study of Kansai. In 4th international conference on sustainable future for human security. Kyoto
9. IEA (2013) Energy balances of OECD countries. International Energy Agency, Paris

10. Zhang H, Mao Z, Zhang W (2015) Design charrette as methodology for post-disaster participatory reconstruction: observations from a case study in Fukushima, Japan. Sustainability 7(6): 6593
11. Parrish K, Singh R, Chien S-C (2015) The role of international institutional partnerships in delivering low-energy building design: a case study. Sustain Cities Soc 14:383–389
12. Mathy S, Fink M, Bibas R (2015) Rethinking the role of scenarios: participatory scripting of low-carbon scenarios for France. Energy Policy 77:176–190
13. Corder GD et al (2012) Engineering-in sustainability through the application of SUSOP®. Chem Eng Res Des 90(1):98–109
14. McLellan BC et al (2009) Incorporating sustainable development in the design of mineral processing operations – review and analysis of current approaches. J Clean Prod 17(16):1414–1425
15. Corder GD, McLellan BC, Green SR (2012) Delivering solutions for resource conservation and recycling into project management systems through SUSOP®. Miner Eng 29:47–57
16. Tuazon D et al (2012) A practical and rigorous approach for the integration of sustainability principles into the decision-making processes at minerals processing operations. Miner Eng 29:65–71
17. Corder GD, McLellan BC, Green S (2010) Incorporating sustainable development principles into minerals processing design and operation: SUSOP®. Miner Eng 23(3):175–181
18. Forum for the Future (2011) The five capitals. 2011. (cited 12 Sep 2011). Available from: http://www.forumforthefuture.org/project/five-capitals/overview
19. MIAC (2015) Japanese government statistics. Ministry of Internal Affairs and Communication, 2015. (cited 23rd Jul 2015). Available from: http://www.stat.go.jp/english/data/index.htm
20. McLellan BC, Aoki K, Kishita Y (2014) Transitioning to sustainable future energy systems: surveying lock-in. In EcoDesign products and services symposium, Ecodesign Promotion Network: Tokyo
21. WEC (2010) Survey of world energy resources. World Energy Council, London, p 608
22. Zhang Q et al (2012) Integration of PV power into future low-carbon smart electricity systems with EV and HP in Kansai Area, Japan. Renew Energy 44:99–108
23. Esteban M, Portugal-Pereira J (2014) Post-disaster resilience of a 100% renewable energy system in Japan. Energy 68:756–764
24. Esteban M, Zhang Q, Utama A (2012) Estimation of the energy storage requirement of a future 100% renewable energy system in Japan. Energy Policy 47:22–31

A Fuzzy Monte Carlo Simulation Technique for Sustainable Society Scenario (3S) Simulator

A.M.M. Sharif Ullah

Abstract Sustainability means meeting the current needs without jeopardizing the potentials of fulfilling the future needs. Sustainable Society Scenario (3S) Simulator is one of the effective computing frameworks for assessing the sustainability of an issue from a greater context. The 3S Simulator uses the sampling, forecasting, and/or backcasting related information while performing the simulation for some given scenarios. In this study, a fuzzy Monte Carlo simulation technique is introduced for its use within the context of 3S Simulator. The technique uses the numerical data related to sampling, forecasting, and/or backcasting and induces the underlying possibility distributions (fuzzy numbers). The simulated states are then stochastically generated from the induced fuzzy numbers. The effectiveness of the proposed simulation technique is demonstrated by providing numerical results that deal with the forecasting of CO_2 emissions from a growing population of passenger vehicles having variable mileage and fuel efficiency.

Keywords Sustainable society • Scenario simulator • Fuzzy logic • Monte Carlo simulation • Uncertainty

1 Introduction

According to the United Nations, sustainability means meeting the current needs without jeopardizing the potentials of fulfilling the future needs [1]. To assess the impact of some regulations, plans, systems, and/or products on the sustainability, one can use the following indices: emissions of greenhouse gases, depletion of resources/materials, health hazard potentials, retainment of ecological integrity, stability of economic growth, and fulfillment of human/social needs. The scale of assessment starts from the individual level and goes up to the global level. Such entities as data, information, knowledge, logical structure among various regulations, plans, activities, solutions, assumptions, and expectations are needed to assess the sustainability. In addition, a set of scenarios are needed to facilitate the

A.M.M.S. Ullah (✉)
Department of Mechanical Engineering, Kitami Institute of Technology, Kitami, Japan
e-mail: ullah@mail.kitami-it.ac.jp

M. Matsumoto et al. (eds.), *Sustainability Through Innovation in Product Life Cycle Design*, EcoProduction, DOI 10.1007/978-981-10-0471-1_41

assessment [2]. A computing framework called *Sustainable Society Scenario* (3S) *Simulator* [3] has been proposed to systematize the sustainability assessment process. See the works [2, 3] for a comprehensive description on the 3S Simulator. The 3S Simulator simulates the future states of some sustainability indicators taking the input data from the *data level*, which is linked to the *structured scenario level* [3, 4]. The structured scenario level evolves from the *problem definition level* (storylines) [3, 4].

Numerous authors have been using the 3S Simulator. For the sake of better understanding, some of the works are briefly described below. Kishita et al. [5] have developed a 3S Simulator for assessing the gain in the sustainability by gradually increasing the number of users of fuel cells in a location in Japan. One of the simulation challenges in [5] deals with the uncertainty in predicting the number of fuel cells used at a given time in the future. Mizuno et al. [6] have studied the sustainability of Japanese manufacturing sector taking 2010 as the base year and 2050 as the target year. They have customized the 3S Simulator and introduced a framework called Scenario-based Industrial Structure Analysis (SISA). Using SISA [6], some of the indices associated with the stability of economic growth and fulfillment of the human/social needs have been assessed, such as the employment status in various manufacturing sectors, the contribution of manufacturing in the GDP of Japan, and the CO_2 emissions from the manufacturing sector. The simulation is performed based on the hypothetical values of the input parameters without considering the energy and material depletion. The inclusion of the real-life field data might pose some challenges for the simulator as pointed out in [6]. Iwai et al. [7] have customized the 3S Simulator for the six scenarios of renewable energy-driven products and studied the diffusion potentials of the products by the year 2030 in a location in Japan due to some unforeseeable factors such as a change in the energy policy. They have also considered the decrease in the population [7].

Nevertheless, from the viewpoint of simulation, the block diagram shown in Fig. 1 can schematically illustrate the structural modules of the 3S Simulator. As

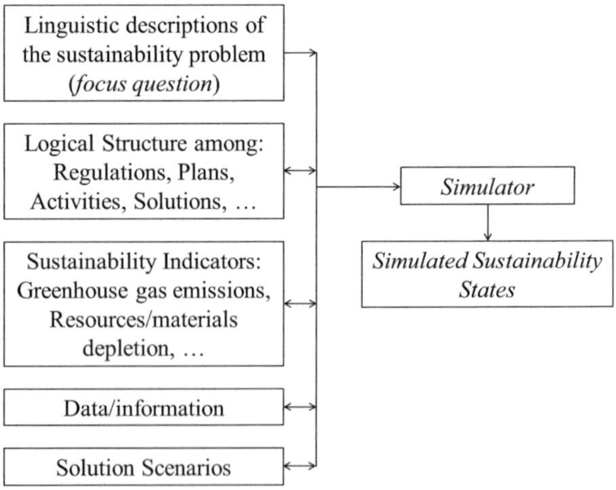

Fig. 1 3S Simulator from the context of simulation

seen from Fig. 1, the 3S Simulator consists of six modules, namely, focus question, logical structure, sustainability indicators, data/information, solution scenarios, and simulator. Focus question means a linguistic description of the sustainability problem. The simulator simulates the sustainability states for making a decision. Except the first and last modules, the other modules exhibit a network structure and do not follow a hierarchy per se. This also means that these modules (except the first and last modules) are defined interactively for a given focus question. The elements within a module are organized on a hierarchical basis, however. It is worth mentioning that the other sustainability assessment approaches use the same strategy (simulate the sustainability states), but do not explicitly show the presence of the modules underlying the 3S Simulator.

However, to perform a simulation, simulation-relevant data/information is needed. Since the data/information has two main categories, namely, crisp and granular information, the simulator of the 3S Simulator may encounter these two categories of data/information. Note that the crisp information means the sharp numerical values and the granular information means the values with sharp or imprecise boundaries. To understand the differences between the crisp and granular information, consider the examples listed in Table 1. See [8, 9] for more details on the categories of information.

To model the granular information for the sake of computation, both probability and possibility distributions can be used [10]. See [11] for the definition of probability and possibility distributions. The possibility distributions are popularly known as fuzzy numbers [11]. Appendix A also provides the definition of a fuzzy number. Fuzzy numbers have extensively been used in sustainability assessment [12]. Some authors have integrated both fuzzy numbers and probability distributions in simulating the sustainability states [13]. However, fuzzy numbers used in a simulation are user-defined and, thereby, bring a great deal of subjectivity in the assessment process. The subjectivity can be avoided if a set of numerical data is used to induce the fuzzy numbers, as shown in [10]. Since the sustainability assessment process from the context of 3S Simulator entails forecasting, backcasting, and sampling [2–7, 14], it is important to see how the fuzzy numbers can be constructed using the numerical data underlying the forecasting, backcasting, and sampling. From this contemplation, this article is written.

The remainder of this article is organized as follows. Section 2 describes the proposed framework. Section 3 describes the data modules. Section 4 describes the fuzzy number induction process. Section 5 describes the proposed simulation process. Section 6 describes how to simulate the CO_2 emission states using the proposed simulation technique. Section 7 provides the concluding remarks of this study.

Table 1 Categories of information

The number of users of the eco-product is	Categories of information
100	Crisp
Between 100 and 200	Crisp granular
About 100	Fuzzy granular
Very low	
Normally distributed with mean 100 and standard deviation 5	Probability granular
Most likely about 100	Fuzzy-probability granular

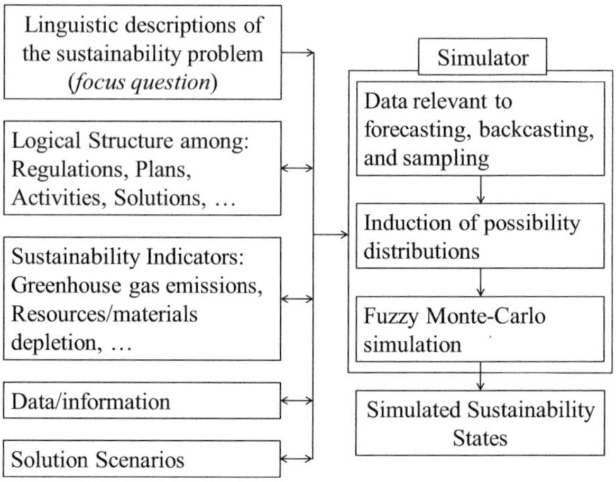

Fig. 2 Proposed simulation framework

2 Proposed Simulation Framework

This section describes the proposed simulation framework, as schematically illustrated in Fig. 2.

The proposed simulation framework consists of a simulator having three modules. The first module is the *data module* that acknowledges the data relevant to forecasting, backcasting, and sampling. The second module is the *inductor module* that induces the possibility distributions (fuzzy numbers) using the data stored in the first module. The other module is the *simulator* itself that simulates sustainability states using a fuzzy Monte Carlo simulation process. Needless to say, the induced fuzzy numbers are used in the simulation process. The following three sections describe these three modules.

3 Data Module

This section describes the data module. In particular, three patterns are considered, namely, forecasting, backcasting, and sampling. The structure of the data depends on forecasting, backcasting, and sampling. Therefore, this section first describes these concepts and then introduces the data structure.

3.1 *Forecasting*

Figure 3 schematically illustrates the concept called forecasting. As seen in Fig. 3, forecasting means determining the future states from some of the given states in the past.

Let $x_p(t)$, $t = n1, n1 + 1, \ldots, n2$, be the states of a variable in the past for the time interval $[n1, n2]$, $n1 < n2$. Let $x_f(t)$, $t = n2 + 1$, $n2 + 2, \ldots, n3$, be the forecasted states of the variable in the future for the time interval $(n2, n3]$, $n2 < n3$. In forecasting, $x_p(t)$ are given and $x_f(t)$ are simulated states. Thus, the forecasting is given by Eq. 1:

$$x_p(t)t \in [n1, n2] \overset{\text{Forecasting}}{\longrightarrow} x_f(t)t \in (n2, n3] \tag{1}$$

3.2 *Backcasting*

Figure 4 schematically illustrates the concept called backcasting. As seen in Fig. 4, the backcasting starts from some expected or desirable states denoted as $x(T)$ at a point of time T in the future. The expectation is defined by a possibility distribution (fuzzy number) denoted as F in the universe of discourse of $x(T)$ denoted as U_T, i.e.,

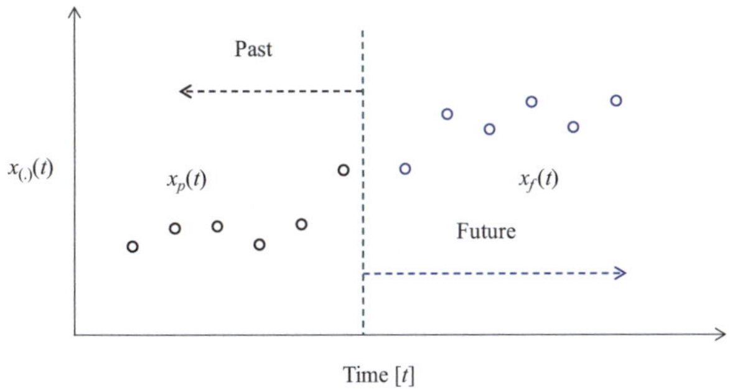

Fig. 3 Forecasting

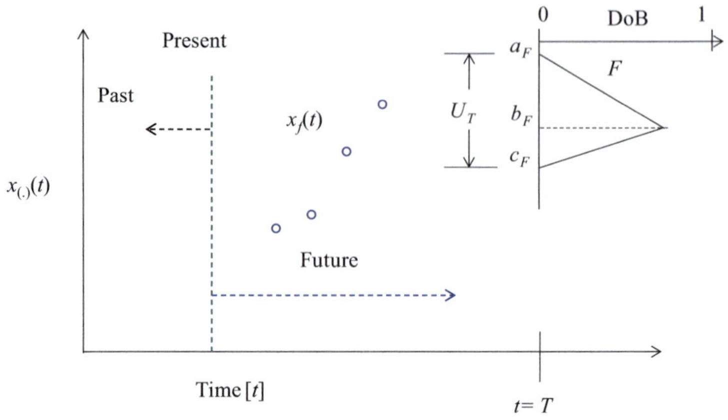

Fig. 4 Backcasting

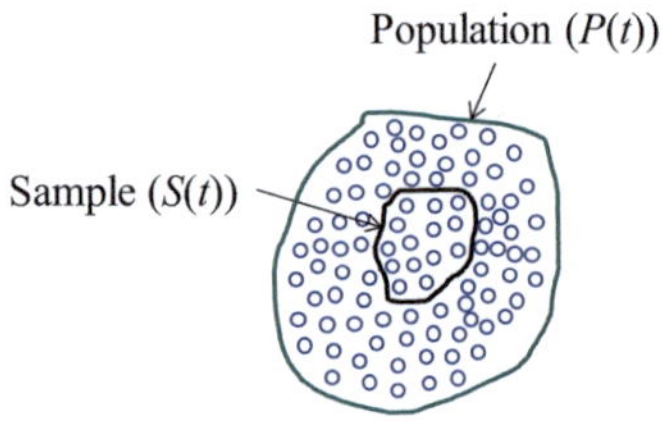

Fig. 5 Sampling

$x(T) \in U_T$. The backcasting simulates the future states denoted as $x_f(t)$, $t = n1$, $n1-1,\ldots,n2$, $n1 > n2$, in the backward direction of time.

Thus, backcasting is given by Eq. 2:

$$x(T) \in U_T, F \overset{\text{Backcasting}}{\rightarrow} x_f(t) \quad t \leq T \tag{2}$$

3.3 Sampling

Figure 5 schematically illustrates the concept of sampling. Sometimes it is not possible to observe (or keep a record of) all pieces of information about an issue. In this case, a finite set of observations is recorded known as sample $S(t)$ from the whole, i.e., the population $P(t)$, at a point of time t. Here, a sample denoted as $S(t)$ contains a set of observations $x(i,t)$, $i = 1,\ldots,n$, at a point of time t, and, thereby, given by Eq. 3:

Table 2 Input data structure

Item	Data structure
Forecasting	$(t,x_p(t))$, $t = n1,n1+1,\ldots,n2$
Backcasting	$x(T) \in U_T$, F, T is a point of time in the future
Sampling	$(i,x(i,t))$, $i = 1,\ldots,n$

$$P(t) \overset{\text{Sampling}}{\rightarrow} S(t) = \{x(i,t)\ |i = 1, \ldots, n\} \tag{3}$$

3.4 *Data Structure*

The above three concepts called forecasting, backcasting, and sampling produce three types of data structures, as listed in Table 2. The data underlying these three data structures serve as the inputs for executing the next phase (Fig. 2).

As listed in Table 2, forecasting underlies a time series, i.e., an ordered pair $(t, x_p(t))$ $t = n1,n1+1,\ldots,n2$. The backcasting underlies a universe of discourse ($U_T \in \mathfrak{R}$) and a fuzzy set F showing the expected/desired state of a variable at a future time point T. Sampling also underlies an ordered pair $(i,x(i,t))$, $i = 1,\ldots,n$, showing the observed states of a variable at a time t.

4 Induction of Possibility Distribution

This section describes the induction of possibility distributions underlying the forecasting and sampling. The backcasting does not require an induced possibility distribution (Fig. 4).

4.1 *The Case of Forecasting*

In the case of forecasting, a fuzzy number is induced using the ordered pair $(t,x_p(t))$, $t = n1,n1+1,\ldots,n2$. The induction process is schematically illustrated in Fig. 6. As seen in Fig. 6, the induction process needs the growth (or decay) rate $d(t) = dx_p(t)/x_p(t)$, $dx_p(t) = x_p(t+1) - x_p(t)$, $t = n1,\ldots,n2-1$. From the growth rate, a point cloud is created. The point cloud means here a plot of ordered pairs $(d(t),d(t+1))$, $t = n1,\ldots,n2-2$. The probability-possibility transformation described in [10] is used here to induce a possibility distribution (in this case a triangular fuzzy number denoted as G). The support and core of G are denoted as $[a_G,c_G]$ and b_G, respectively.

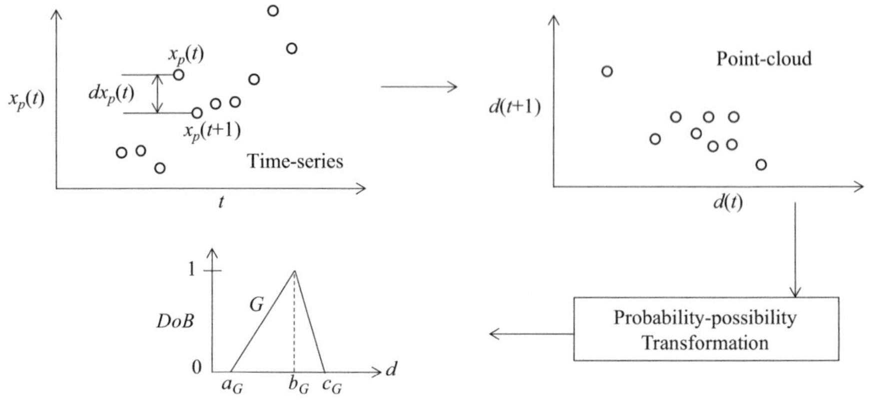

Fig. 6 Inducing possibility distribution for forecasting

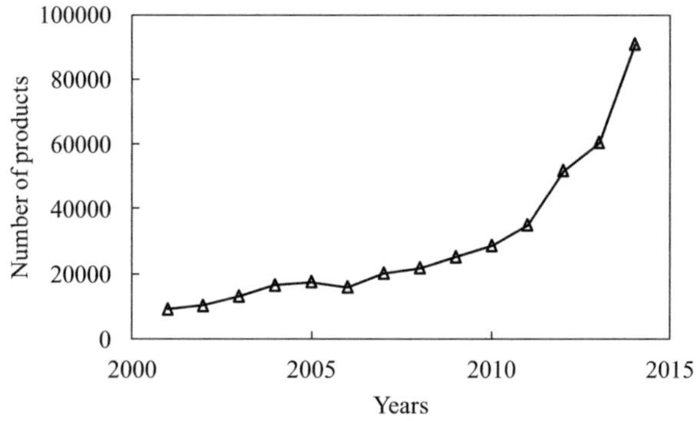

Fig. 7 A trend of a product population

To get more insight into the possibility distribution induction process for the case of forecasting, let us consider the following numerical example. Figures 7, 8, and 9 show the time-series plot, growth point cloud, and induced fuzzy number, respectively.

Figure 7 shows the population of an eco-product over a period of 14 years, from 2001 to 2014. The time-series plot in Fig. 7 can be converted into a triangular fuzzy number following the steps shown in Fig. 6. Accordingly, the point cloud of the product growth (*d*(*t*)) is created, as shown in Fig. 8. The probability-possibility transformation described in [10] has been applied to the point cloud (Fig. 8) to induce the triangular fuzzy number *G* shown in Fig. 9. The induced triangular fuzzy number has the following support and core: $[a_G, c_G]$ = [0.056,0.256] and $b_G = 0.154$.

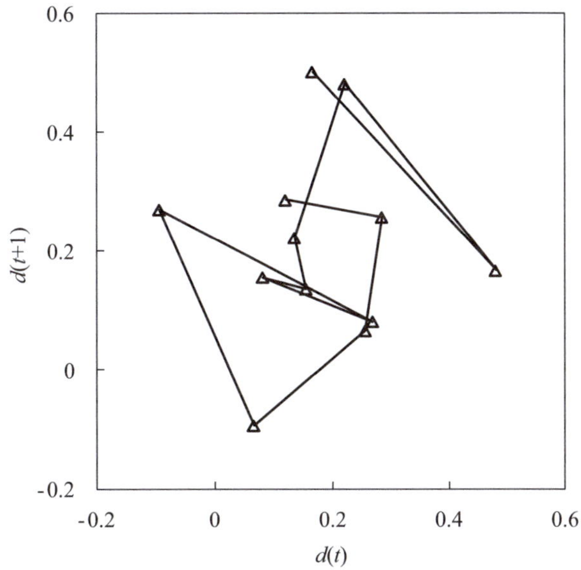

Fig. 8 Point could of product growth

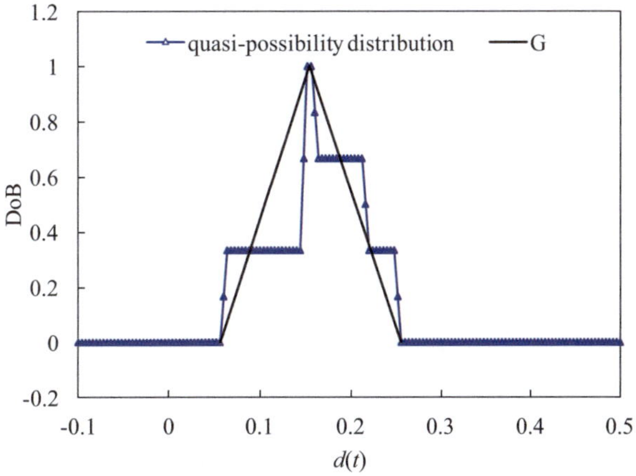

Fig. 9 Induced possibility distribution for forecasting

4.2 *The Case of Sampling*

In the case of sampling, the sequence of the data points is not important, unlike the previous case. Therefore, here a fuzzy number is induced after sorting the data points in the ascending order. The induction process is schematically illustrated in Fig. 10. As seen from Fig. 10, the sorted data points are denoted as $y(j,t) \in \{x(i,t) \mid i = 1,\ldots,n\}$ where $y(j,t) \leq y(j+1,t)$, $\forall j = \{1,\ldots,n-1\}$. The point cloud consisting of the points $(y(j,t),y(j+1,t))$, $j = 1,\ldots,n-1$, are used to induce a possibility distribution (in this case also a triangular fuzzy number denoted as H). The support

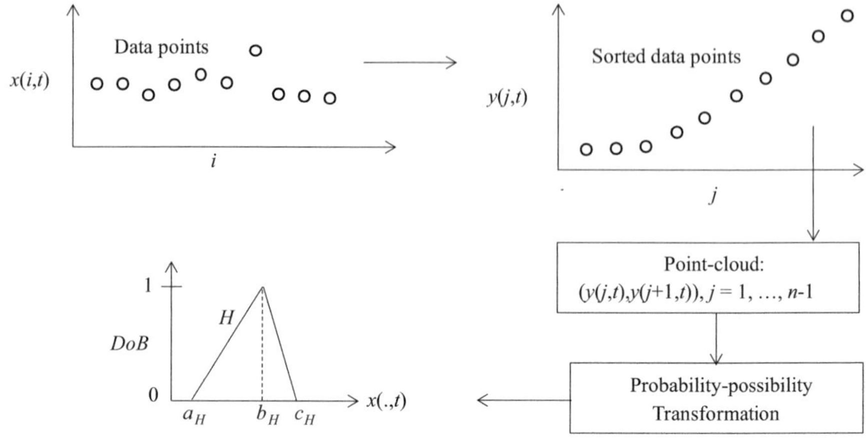

Fig. 10 Inducing possibility distribution for sampling

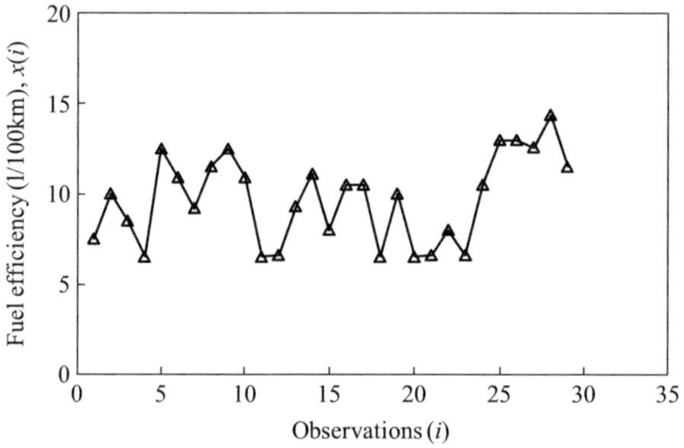

Fig. 11 Sampled data points as observed

and core of H are $[a_H,c_H]$ and b_H, respectively. To get more insight into the possibility distribution induction process for the case of sampling, the fuel efficiency data of 30 randomly selected passenger cars is considered [9]. Figures 11, 12, 13, and 14 show the data points in the sample as observed, the data points after sorting in the ascending order, the point cloud of the sorted data points, and the induced fuzzy number, respectively. The observed data points shown in Fig. 11 can be converted into a triangular fuzzy number following the steps shown in Fig. 10. Accordingly, the observed data points are sorted in the ascending order (Fig. 12) and then represented by a point cloud (Fig. 13). The probability-possibility transformation described in [10] has been applied to the point cloud (Fig. 13) to induce the triangular fuzzy number H shown in Fig. 14. The induced triangular fuzzy number has the following support and core: $[a_H,c_H] = [6.4,13]$ and $b_H = 10.2$.

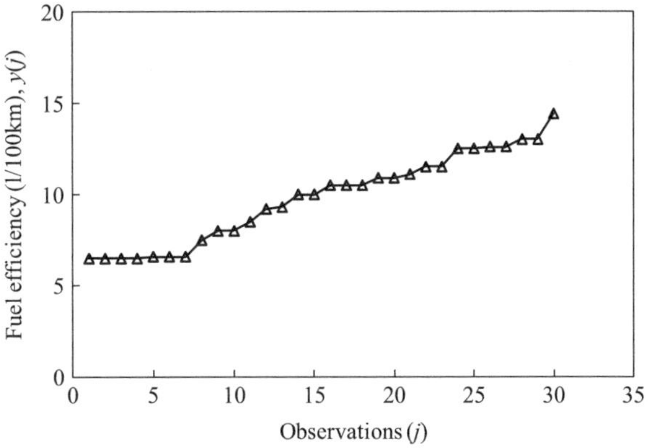

Fig. 12 Sampled points in the ascending order

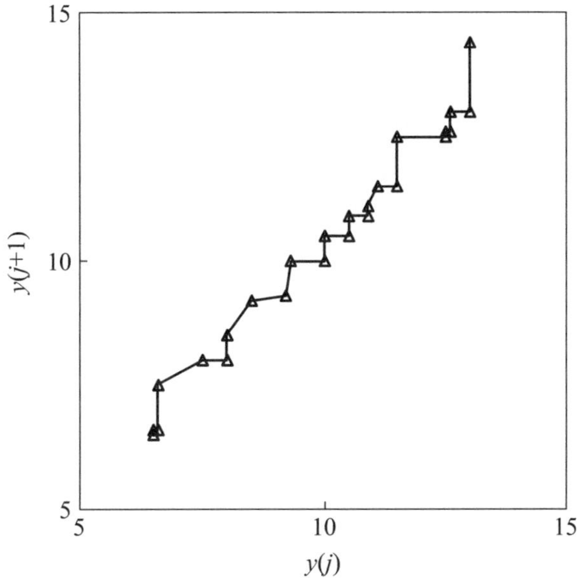

Fig. 13 The point cloud of sorted data points

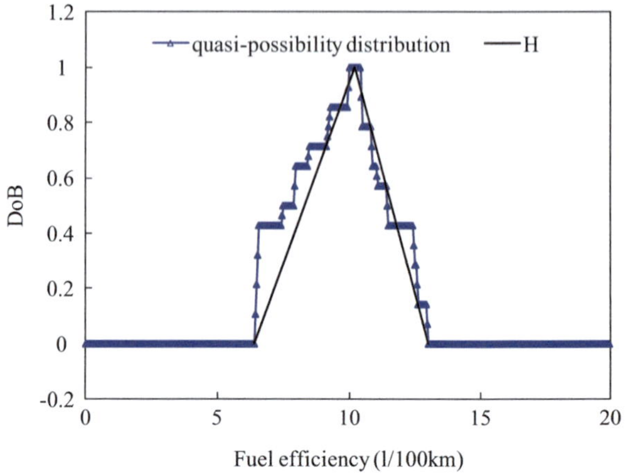

Fig. 14 Induced possibility distribution

5 Simulation Process

This section describes the simulation process. The main point is to get an alpha-cut of a triangular fuzzy number and use the boundary of the alpha-cut in the simulation. Figure 15 schematically illustrates the process of getting the boundaries of an alpha-cut from a triangular fuzzy number. As seen from Fig. 15, let A be a triangular fuzzy number. The support and core are denoted as $[a,c]$ and b, respectively. Let r be a random number in the interval [0,1], i.e., $r \in [0,1]$. This random number is the alpha of the alpha-cut of a fuzzy number [11]. The lower and upper limits, denoted as x_L, x_H, respectively, are calculated by using Eq. 4. A simulation process can be defined in such a way so that the alpha-cuts are repeatedly generated and the upper/lower limit of the alpha-cuts are recursively used in generating the next state, $x(t+1)$, from the present one, $x(t)$:

$$\begin{aligned} x_L &= a + r(b-a) \\ x_H &= c - r(c-b) \end{aligned} \tag{4}$$

5.1 Simulation Process for Forecasting

For the case of forecasting, the fuzzy number G is induced in the universe of discourse of the growth rate $d(t)$. Therefore, the relationship between the present and next states is as follows: $x(t+1) = x(t)(1+d(t))$. This yields a simulation process as defined in Eq. 5:

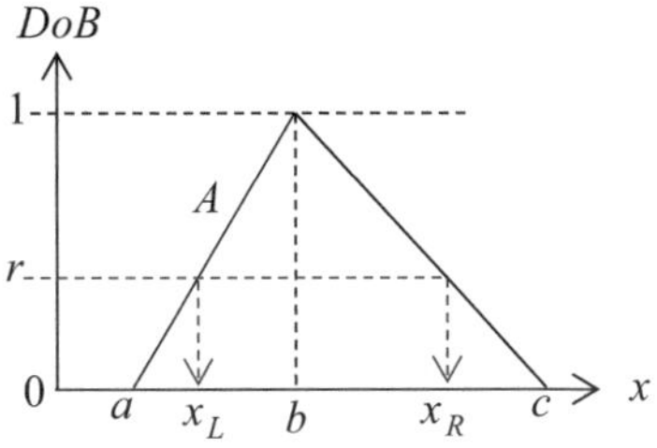

Fig. 15 A random alpha-cut at alpha $= r$

$$
\begin{array}{llll}
\text{Algorithm : Forecasting} & & & \\
\text{Define} & G(a_G, b_G, c_G), x(n2), n3 & & \\
 & \text{For} \quad t = n2, \ldots, n3 - 1 & & \\
 & & r_t \leftarrow [0, 1] & \\
 & & d_L(t) = a_G + r_t(b_G - a_G) & \\
 & & d_H(t) = c_G - r_t(c_G - b_G) & \\
 & & d(t) \in \{d_L(t), d_H(t)\} & \\
 & & x(t+1) = x(t)(1 + d(t)) & \\
 & & x_f(t) = x(t+1) & \\
 & \text{End For} & & \\
\text{Output} & X_f = \left\{x_f(t) | t = n2 + 1, \ldots, n3\right\} & &
\end{array}
\tag{5}
$$

The output of the simulation process denoted as $X_f = \{x_f(t) \mid t = n2 + 1, \ldots, n3\}$ is the set of all forecasted states. It is worth mentioning that while performing the simulation, the growth rate $d(t)$ can be chosen to be either $d_L(t)$ or $d_H(t)$. If one sets $d(t) = d_L(t)$, a set of states called *lower bound* denoted as X_{fL} is created. On the other hand, if one sets $d(t) = d_H(t)$, a set of states called *upper bound* denoted as X_{fH} is created.

5.2 *Simulation Process for Sampling*

The simulation process for sampling is very similar to that of the forecasting except the fact that in sampling the simulated states are meant for a point of time t. The simulation process for sampling is expressed by Eq. 6. According to Eq. 6, the output of the simulation process denoted as $S(t) = \{x(i,t) \mid t = 1, \ldots, N\}$ is the set of all simulated sample states. In addition, if one sets $x(i,t) = x_L(i,t)$, a set of states called lower bound denoted as $S_L(t)$ is created. On the other hand, if one sets $x(i,t) = x_H(i,t)$, a set of states called upper bound denoted as $S_H(t)$ is created:

$$\begin{array}{lll} \text{Algorithm : Sampling} & & \\ \text{Define} & H(a_H, b_H, c_H), N, t & \\ & \text{For} \quad i = 1, \ldots, N & \\ & \quad r_i \leftarrow [0, 1] & \\ & \quad x_L(i,t) = a_H + r_i(b_H - a_H) & \\ & \quad x_H(i,t) = c_H - r_i(c_H - b_H) & \\ & \quad x(i,t) \in \{x_L(i,t), x_H(i,t)\} & \\ & \text{End For} & \\ \text{Output} & S(t) = \{x(i,t) | i = 1, \ldots, N\} & \end{array} \tag{6}$$

6 Case Study

This case study deals with the problem of CO_2 emission from the usages of passenger vehicles (cars) in a given geographical location. The CO_2 emission at a point of time t in the future denoted as $C(t)$ is given as

$$C(t) = K \sum_{i=1}^{P(t)} M(i,t) \cdot E(i,t) \tag{7}$$

In Eq. 7, $P(t)$ is the car population at a point in time t (year); $M(i,t)$ and $E(i,t)$ are the i-th (simulated) mileage and fuel efficiency of a car in km/year and liter/100 km, respectively; and K is the CO_2 emission factor in kg-CO_2/liter (usually 2.3 kg-CO_2/l of fossil fuel [9]). To simulate $P(t)$, the simulation process (Eq. 5) for forecasting is needed. In this case, $P(t)$ underlies a growth rate given by a fuzzy number of type G. The fuzzy number is already shown in Fig. 9 and repeated in Table 3. To simulate $M(i,t)$, the simulation process (Eq. 6) for sampling is needed, i.e., $M(i,t)$ underlies mileage given by a fuzzy number of type H. The fuzzy number is shown in Table 3. The process defined in Fig. 10 induces the fuzzy number. To simulate $E(i,t)$, the simulation process (Eq. 6) for sampling is needed, i.e., $E(i,t)$ underlies a growth rate given by a fuzzy number of type H. The fuzzy number is already shown in Fig. 14 and repeated in Table 3. Therefore, the simulated results of $P(t)$, $M(i,t)$, and $E(i,t)$ can be integrated as defined in Eq. 7 to forecast the lower and upper limits of $C(t)$. (Note that the sets of data for mileage and fuel efficiency are taken from [9].) The simulated lower and upper limits of CO_2 emission ($C(t)$) in Mt are shown in Fig. 16. As seen from Fig. 16, the forecasted CO_2 emission lies in the range of 0.9–2.5 Mt after 10 years (2025) from now (2015). This is for a simulation instance. If one repeats the simulation process for quite a large number of times, the lower and upper limits might change. Therefore, the forecast at given point of time can be given by another possibility distribution that induces from all a large number of simulated instances. In this case, the possibility distribution takes the form of a

Table 3 Possibility distributions for simulation

Car population growth rate (type G fuzzy number) ($a_G = 0.056$, $b_G = 0.154$, $c_G = 0.256$) 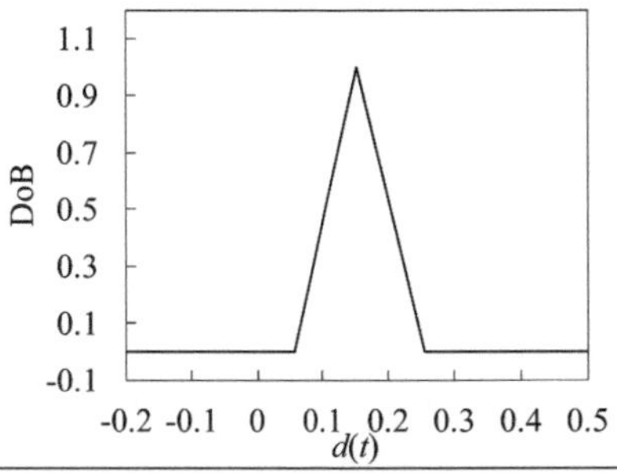	Mileage (km) (type H fuzzy number) ($a_H = 9732$, $b_H = 25898$, $c_H = 50088$) 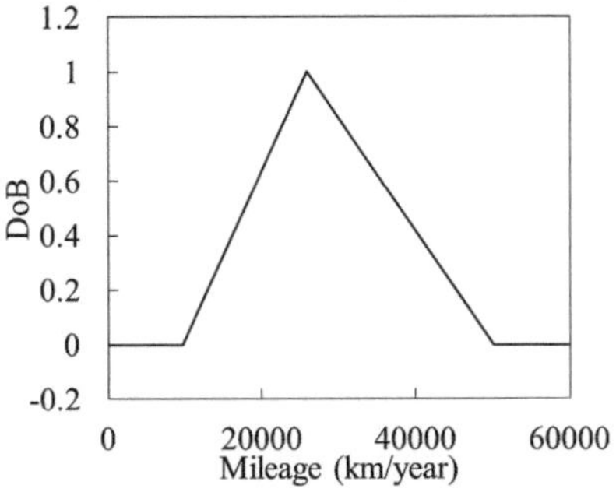
Fuel efficiency (liters/100km) (type H fuzzy number) ($a_H = 6.4$, $b_H = 10.2$, $c_H = 13$) 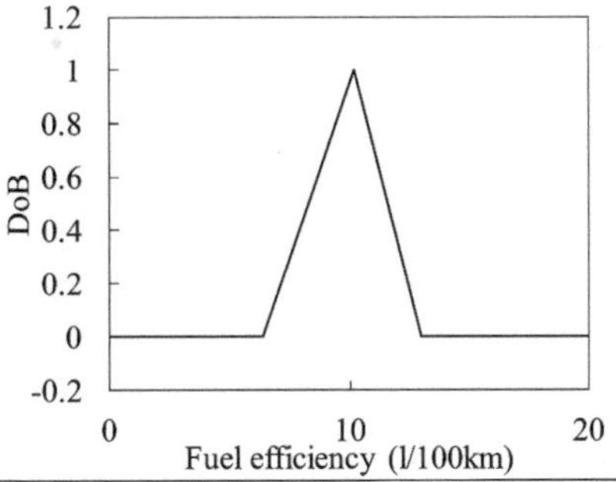	
[l = liters]	

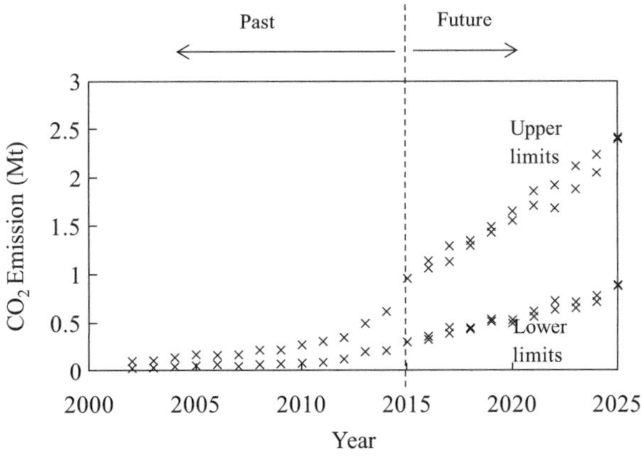

Fig. 16 Forecasted CO_2 emission limits

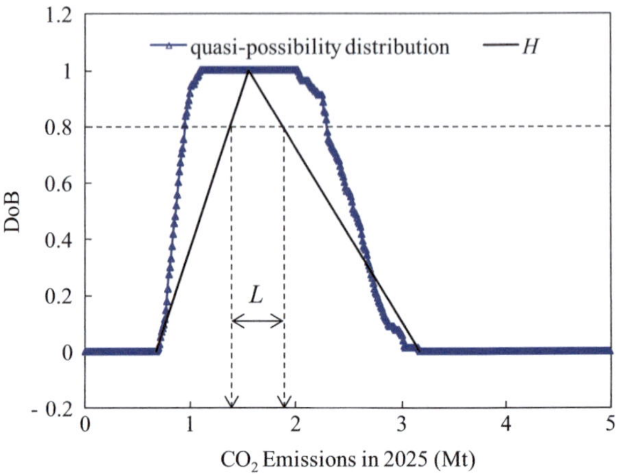

Fig. 17 Possibility of CO_2 emission in 2025

fuzzy number of type *H* (i.e., the fuzzy number meant for sampling data). Figure 17 shows such a fuzzy number.

In particular, the fuzzy number shown in Fig. 17 corresponds to 100 simulation instances of the lower and upper limits of CO_2 emission in 2025. Instead of saying that the CO_2 emission in 2025 is in the range of 0.9–2.5 Mt, the fuzzy number in Fig. 17 can be used to find either a range or a crisp value of CO_2 emission possibility in 2025. In particular, the fuzzy number *H* shown in Fig. 17 corresponds to a support [0.68, 3.17] Mt and core 1.56 Mt. If an alpha-cut is taken at alpha = 0.8, denoted as *L* in Fig. 17, then the alpha-cut = *L* becomes a range [1.384,1.882] Mt. Therefore, instead of mentioning a rather long range, one can say that the most possible value of CO_2 emission in 2025 is about 1.56 Mt. Alternatively, there is at least 80 % possibility that the CO_2 emission in 2025 is in the range of [1.384, 1.882] Mt.

It is worth mentioning that the alpha-cut where the alpha is less than 0.5 may not be considered for forecasting because the statements are false more than they are true in the case of alpha = 0.5.

7 Concluding Remarks

In most cases, both the determination of probability distribution from the sampled data and the regression analysis have been the main approach for simulating the future states within the context of 3S Simulator. Since what will happen in the future is just a matter of possibility, possibility distribution must play an explicit role in simulating the future states. This is what exactly done in this paper. Although the described case study deals with the forecasting using the possibility distributions underlying both forecasting and sampling, one can use the same approach for the backcasting, too. This issue is open for further research. Nevertheless, fuzzy

Monte Carlo simulation uses the equivalent probability distributions of a given possibility distribution. This work uses a different approach where the possibility distributions directly participate in the simulation process, making it (the approach) an original one for performing the fuzzy Monte Carlo simulation.

However, some research works lie ahead. For example, in this work the triangular fuzzy number replaces the quasi-fuzzy numbers in the simulation process. Is it the right thing to do? This question remains open for further research. Nevertheless, the triangular fuzzy number remains almost the same even though the set of data changes to an extent. In other words, employing the triangular fuzzy number instead of the (steplike) quasi-fuzzy number creates a somewhat robust simulation result.

Appendix A: Fuzzy Number

For a fuzzy number A, there exist four numbers, $a_1,\ldots,a_4 \in \mathfrak{R}$, and two functions $L_A, R_A : [0,1] \rightarrow \mathfrak{R}$, where L_A is an (left-side) increasing function and R_A is a (right-side) decreasing function. The membership function (or the degree of belief function) denoted as $DoB_A(x)$ ($x \in \mathfrak{R}$) is given as

$$DoB_A(x) = \begin{cases} 0 & x < a_1 \\ L_A & a_1 \leq x < a_2 \\ 1 & a_2 \leq x \leq a_3 \\ R_A & a_3 < x \leq a_4 \\ 0 & x > a_4 \end{cases} \tag{A.1}$$

The boundary $[a_1,a_4]$ and $[a_2,a_3]$ are called the support and core of the fuzzy number, respectively. The fuzzy number exhibits the properties of normality and convexity [11]. It is also semicontinuous having a bounded support. See [11] for the explanation. For a triangular fuzzy number, $a_2 = a_3$ is true.

References

1. Report of the World Commission on Environment and Development (1987) Annex to general assembly document A/42/427. Nairobi, Kenya. United Nations, New York
2. Fukushige S, Yamamoto K, Umeda Y (2012) Lifecycle scenario design for product end-of-life strategy. J Remanufacturing 2(1):1–15. doi:10.1186/2210-4690-2-1
3. Umeda Y, Nishiyama T, Yamasaki Y, Kishita Y, Fukushige S (2009) Proposal of sustainable society scenario simulator. CIRP J Manuf Sci Technol 1(4):272–278. doi:10.1016/j.cirpj.2009.05.005
4. Mizuno Y, Kishita Y, Fukushige S, Umeda Y (2015) Proposal of a computational design support method for sustainability scenario design. Trans JSME 81(822):14-00269–00214-00269. doi:10.1299/transjsme.14-00269

5. Kishita Y, Nakamura Y, Kegasa A, Hisazumi Y, Hori T, Fukushige S, Umeda Y (2014) Scenario analysis of the diffusion of fuel cells in the residential sector. Procedia CIRP:294–299. doi:10.1016/j.procir.2014.06.007
6. Mizuno Y, Kishita Y, Fukushige S, Umeda Y (2014) Envisioning sustainable manufacturing industries of Japan. Int J Autom Technol 8(5):634–643
7. Iwai N, Kurahashi N, Kishita Y, Yamaguchi Y, Shimoda Y, Fukushige S, Umeda Y (2014) Scenario analysis of regional electricity demand in the residential and commercial sectors – influence of diffusion of photovoltaic systems and electric vehicles into power grids. Procedia CIRP:319–324. doi:10.1016/j.procir.2014.06.076
8. Zadeh LA (2005) Toward a generalized theory of uncertainty (GTU) – an outline. Info Sci 172 (1–2):1–40. doi:http://dx.doi.org/10.1016/j.ins.2005.01.017
9. Khozaimy O, Al-Dhaheri A, Ullah AMMS (2011) A decision-making approach using point-cloud-based granular information. Appl Soft Comput 11(2):2576–2586. doi:10.1016/j.asoc.2010.10.007
10. Ullah AMMS, Shamsuzzaman M (2013) Fuzzy Monte Carlo simulation using point-cloud-based probability–possibility transformation. Simulation 89(7):860–875. doi:10.1177/0037549713482174
11. Ullah AMMS, Harib KH (2006) Zadehian paradigms for knowledge extraction in intelligent manufacturing. In: Kordic V, Lazinica A, Merdan M (eds) Manufacturing the future. Pro Literatur Verlag Robert Mayer-Scholz, Mammendorf, pp 291–308. doi:10.5772/5052
12. Suganthi L, Iniyan S, Samuel AA (2015) Applications of fuzzy logic in renewable energy systems – a review. Renew Sustain Energy Rev 48:585–607. doi:10.1016/j.rser.2015.04.037
13. Wieland R, Gutzler C (2014) Environmental impact assessment based on dynamic fuzzy simulation. Environ Model Software 55:235–241. doi:10.1016/j.envsoft.2014.02.001
14. Wada H, Kishita Y, Mizuno Y, Fukushige S, Umeda Y (2013) Proposal of a design support method of backcasting scenarios for sustainable society. Trans JSME Ser C 79(799):845–857. doi:10.1299/kikaic.79.845

The Minerals-Energy Nexus: Past, Present and Future

Benjamin C. McLellan

Abstract Minerals and energy have been highly intertwined throughout history, but are becoming even more so with two key emerging trends: the gradual declining grade and increasing depth and complexity of ores and the increasing demand for high-performance functional materials in energy technologies. This paper examines these trends from the perspective of eco-design, focusing on the competing requirements of energy for producing and processing minerals and the demand for minerals in energy technologies. One key element examined is recycling – its impact on supply and the opposing implications of miniaturisation for recyclability of energy technologies. Peak minerals are addressed in conjunction with the implications of nonconventional resources such as deep ocean deposits.

Keywords Minerals • Criticality • Energy • Sustainability • Resources

1 Introduction

Minerals and energy are highly intertwined – not just through fuel minerals (coal, uranium) but also through the use of metals as functional materials in clean energy technologies and other energy infrastructure. Moreover, the requirement of energy for extracting and processing minerals is a significant proportion of global energy consumption. This nexus of minerals and energy is thus a highly important area of concern for sustainable energy. This paper aims to present a brief review of the important interconnections of minerals in energy and vice versa, examining a number of scenarios for critical minerals used in clean energy technologies as examples of the importance of considering these two fields together.

While important to consider the current status of the nexus, it is also critical to consider the trends that currently shape the future and the implications for eco-design. In particular, in this paper the considerations of recovery of critical or

B.C. McLellan (✉)
Graduate School of Energy Science, Kyoto University, Yoshida honmachi, Sakyo-ku, Kyoto 606-8501, Japan

Sustainable Minerals Institute, The University of Queensland, St Lucia, QLD 4072, Australia
e-mail: b-mclellan@energy.kyoto-u.ac.jp

M. Matsumoto et al. (eds.), *Sustainability Through Innovation in Product Life Cycle Design*, EcoProduction, DOI 10.1007/978-981-10-0471-1_42

rare metal components from clean energy technology waste are considered. These critical metals have been of increasing interest due to political and economic shifts and are often the most important functional materials in clean energy technologies. Moreover, their use is of particular relevance to, and has the potential to be strongly influenced by, eco-design.

The approach taken here has been to examine the statistics of minerals [4] and energy usage [1] in isolation and in conjunction. This provides background on the past and current status of the minerals-energy nexus across a variety of its manifestations and is described in Sects. 2 and 3. A survey of the use of critical minerals in energy technologies was also carried out [5] and is briefly presented to show the importance of these minerals and to highlight some challenges for energy and for eco-design of energy technologies. Based on this data, the development of scenarios for alternative supply and demand patterns of selected minerals was undertaken. Finally, the comparison of some emerging routes for sourcing key materials is applied – including recycling and deep ocean mineral deposits. Such alternative deposits may be required, and the implications of their exploitation should be examined.

2 Energy Usage for Minerals

2.1 Energy Use in the Minerals Industry

The first consideration in the nexus is the usage of energy for the production of minerals that are vital to society. Figure 1 shows the total global energy usage in the

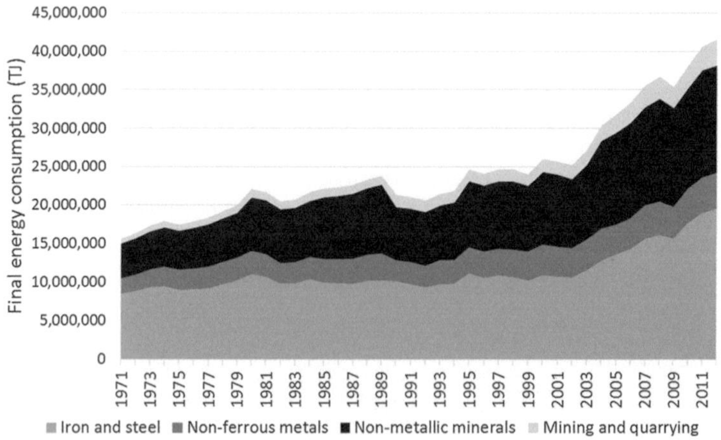

Fig. 1 Energy usage in the minerals industry by sector [1]

minerals industry – separated by sector – over a period of 42 years to 2012 [1]. It is apparent that the energy usage is dominated by iron and steel and non-metallic minerals sectors – these two sectors being essential for bulk infrastructure associated with industrialisation and economic development. Over the period, the total energy use in the sector has rapidly increased; however, when considered on the basis of the produced mineral product (Fig. 3), the specific energy usage has declined. This decline has not been consistent – partly this is attributable to the availability and reliability of data from key sources [1, 6], but the major variability in non-ferrous minerals may also be attributable to the changes in the mix of produced mineral products. Figure 2 shows further the amount of energy estimated to be used per tonne of product in the mining and quarrying sector – these figures are likely to have a higher level of error associated with them, as the mined and quarried material are not consistently reported on a national or international scale. However, if the data is considered to be reasonably accurate – or at least the methodology is internally consistent – then, contrary to the other sectors, which

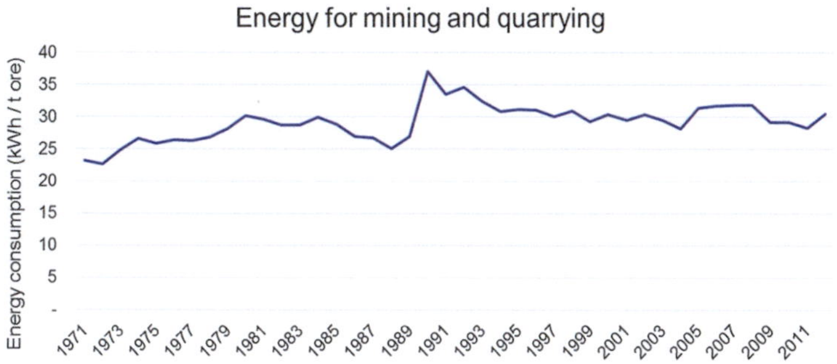

Fig. 2 Energy per tonne of mined material

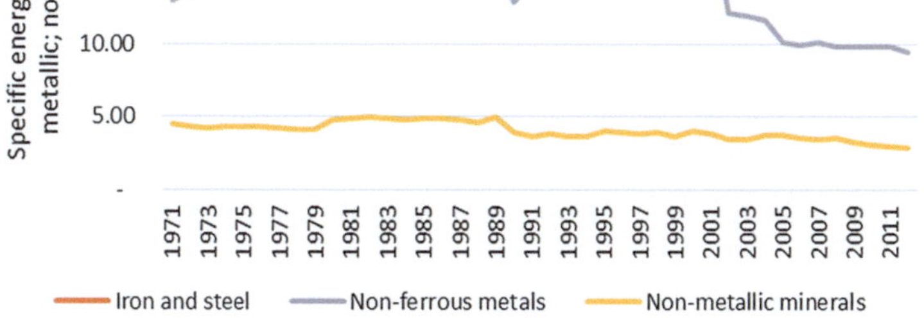

Fig. 3 Energy per tonne of product mineral

show some level of efficiency increase, the mining sector is showing an efficiency decrease per tonne of product. One reason that this could logically be occurring is the fact that mining is now being pursued at greater depths and lower ore grades than previously, both of which are necessitating greater energy consumption.

With regard to the overall global environmental impact of the minerals industry, one key indicator associated with energy is the amount of renewable energy usage. Figures 3 and 4 show the calculated utilisation of direct (fuel, heat, reagent) and indirect (electricity consumption) energy from renewable sources in the global industry [1] compared with the same figure for total final consumption of energy across all sectors of the global economy. This shows that energy used in the minerals industry is significantly lower in renewable energy percentage than the total global average. As has been shown elsewhere [7], this may be largely due to the high requirements for high-temperature thermal energy in the industry, making substitution of this energy relatively difficult. The most significant improvement in the renewable energy contribution to a subsector is in the non-ferrous metals sector, which has a large proportion of electricity usage [7] – and with many operations in countries with large-scale hydrothermal (e.g. Canada) or geothermal (Iceland) and low emission intensity [8].

The growth in renewables within the sector should be compared to the total growth in the sector itself (see Table 1, Fig. 5). It is apparent that the renewable energy increase has been much higher than the overall energy use in the downstream minerals sectors and that, combined with energy efficiency gains, this is a likely contributor to lower overall emissions for these sectors.

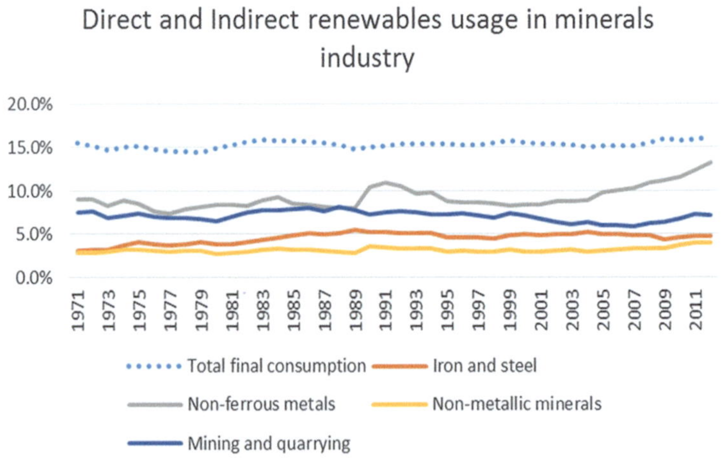

Fig. 4 Direct and indirect usage of renewable energy in the minerals sector

Table 1 Growth by subsector in production and energy consumption

Category	Growth in subsector			
	Iron and steel	Non-ferrous metals	Non-metallic minerals	Mining and quarrying
Production (tpa)	167 %	228 %	380 %	322 %
Energy (TJ)	130 %	139 %	207 %	455 %
Renewable energy (TJ)	244 %	249 %	334 %	428 %

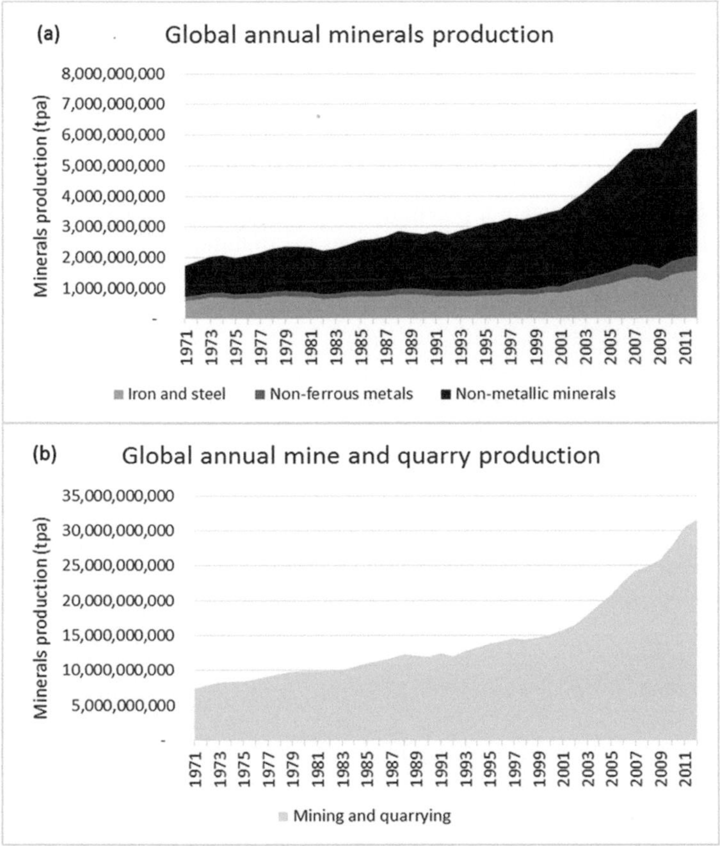

Fig. 5 Global minerals production: (**a**) ferrous, non-ferrous and non-metallic subsectors; (**b**) mining and quarrying

3 Minerals Usage for Energy

3.1 *Fuel Minerals*

The most straightforward usage of minerals in the energy industry is the use of fuel minerals – particularly coal and uranium. As global demand for energy – especially electricity – has grown over the past 40 years or so, the production of coal and uranium has likewise grown (Fig. 6). Coal production is also closely related to the need for metallurgical coal used in the production of steel. On the other hand, uranium is almost exclusively utilised in energy generation (some usage for producing weapons or medicinal isotopes aside). The expanded production of coal experienced two particularly rapid growth periods – in the 1970s with a likely influence from the oil crises of that decade and in the 2000s with the rapid development of countries such as China. Important in the considerations of the future potential of these fuels are the available reserves and resources. The World Nuclear Association [9] reports resources (less geological and economic confidence than reserves) that indicate a potential for 100 years of uranium consumption at current levels, while BP's total coal reserves [3] (bituminous and non-bituminous) give a potential global life of reserves of 113 years.

3.2 *Nonfuel Functional Minerals*

Apart from mineral fuels, there are many minerals that perform important functions in the energy system; many of these are listed as critical minerals for major consuming nations of the world [10]. Minerals are generally listed as critical or

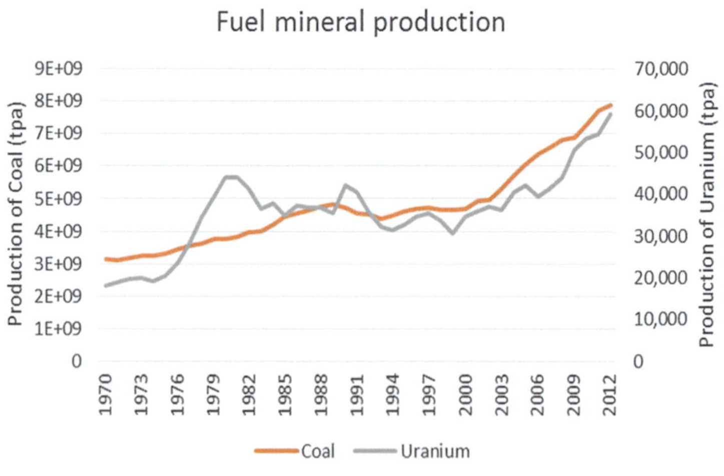

Fig. 6 Fuel mineral production (various sources [2, 3])

Table 2 Some critical minerals in energy technologies

Technology	Critical minerals	Function
Wind turbines	Dysprosium, neodymium	Permanent magnets
	Copper	Generator windings and wiring
Photovoltaics	Indium, gallium, selenium, tellurium	Photoactive materials (total PVA reported)
	Copper	Electrical connections and power electronics
Fuel cells	Platinum	Electrodes/catalysts (PEM FC)
	Yttrium, lanthanum	Electrolyte and electrode materials (SOFC)

After [10]

strategically important if there is a threat of supply shortages (via economic, physical or trade-imposed scarcity) and if such shortages would lead to significant adverse economic or social outcomes. In the case of energy, some of these critical minerals are relatively mundane (copper for electrical wiring and windings in generators), while others are more exotic (rare earths, platinum group metals and by-product metals for fuel cells and advanced energy-efficient technologies).

A summary of some important critical minerals in selected clean energy technologies is shown in Table 2. As mentioned above, metals such as copper are ubiquitous in electronics, electrical generators and motors and electrical transmission networks.

The potential for future supply restrictions is important with regard to these minerals, and understanding the rapid expansion of demand and the rate at which supply can be increased is highly important. This equation can become complicated in the case of minor coproduct metals such as selenium, tellurium, gallium and indium – as the estimates of available resources are largely tied to the estimates of the parent metal groups (typically copper, lead and zinc deposits).

4 Minerals-Energy Nexus and the Future

Following on from the previous sections which give an overview of the unidirectional flows of minerals and energy into the other sector, this section expands on the important considerations of the minerals-energy nexus.

4.1 Critical Minerals

Critical minerals are one of the foremost considerations of the minerals-energy nexus – specifically in relation to those technologies and minerals that are considered vital for the low-carbon future considered to be essential for avoiding catastrophic climate change. On the demand side, understanding how such minerals are utilised, what their current functions are, and how they might be substituted with alternatives are all vital factors to estimate future growth. In some situations, it is recognised that there are few adequate alternatives or that the alternatives offer worse performance than the current mineral component. In the case of rare earths, the recent supply restrictions due to China's export quotas prompted significant rapid expansion of global mine production as well as technological advances to reduce the use of rare earths significantly [11].

4.2 Peak Minerals

The concept of peak minerals has been an area of increasing research interest in recent years [12]. Following on from the concept of peak oil developed by Hubbert in the 1950s, the aim to identify potential peaks for minerals production – which may lead to significant supply-side constraints in an otherwise exponentially increasing demand scenario – is important to consider. Although there are various challenges to the approach – particularly the availability of data about past production and ultimately recoverable resources. Figure 7 shows one such peak curve, for tellurium in this case, whereas Fig. 8 shows a variety of supply scenarios and a demand scenario for tellurium (demand is for thin-film solar cells). In Fig. 8, scenario D uses the peak curve model, while DE and DF also use a peak model but include recycling at different rates. In this particular example, it is apparent that tellurium demand rapidly outstrips supply – making a large influx of thin-film CdTe solar cells (even at a low rate of 1–5 % of the total global photovoltaic market) likely to be unfeasible.

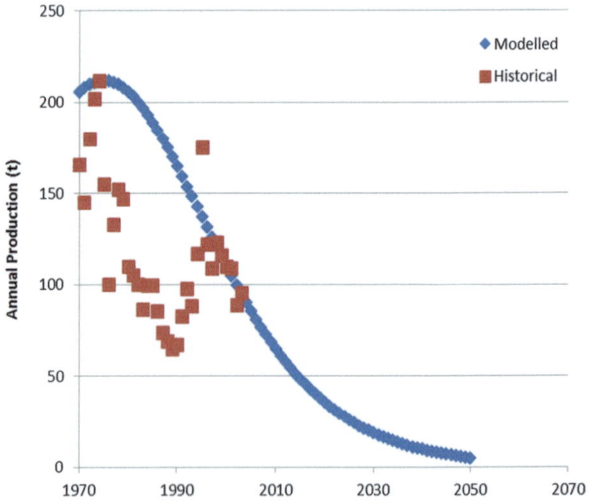

Fig. 7 Example modelled peak curve for tellurium

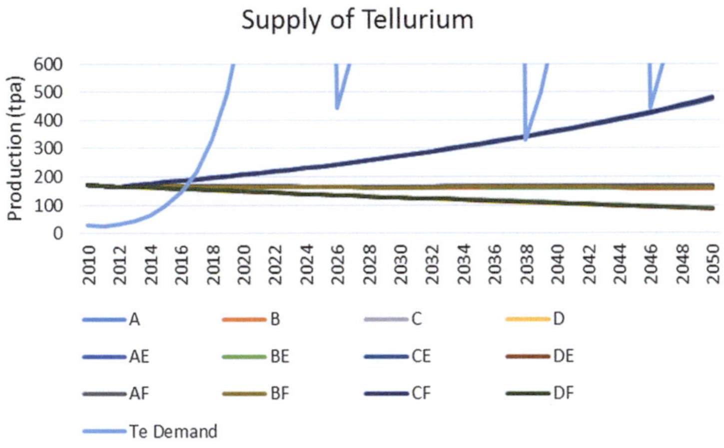

Fig. 8 Supply and demand of tellurium (example scenarios)

Not only does the peak curve consideration have implications for the potential supply of minerals, but it can also indicate the proximity of increasing energy and environmental costs of extraction. In considering the life cycle impacts, it could be useful to allow for such considerations – as the peak is passed, it is likely that a declining grade and increasing financial and environmental cost will be observed.

The scenarios shown in Fig. 8 were calculated based on the use of CdTe thin-film PV cells to provide for the increase in PV estimated following existing trends under the following constraints for the supply side:

- A Constant 2012 production
- B Linear trend growth extrapolated from 2002 to 2012 onwards
- C Year-on-year trend growth extrapolated from 2002 to 2012 onwards
- D Hubbert peak curve estimate

Two recycling scenarios were then applied to scenarios A–D with the following recycle rates:

- E Current recycling rate (20-year lifetime/lag for materials to be recycled)
- F Improvement in recycling rate growing to 90 % by 2050 (20-year lifetime/lag).

The demand for tellurium in solar cells was modelled according to recent rates of PV expansion with a limit of 30 % of the total energy production, in the scenario shown in Fig. 9.

4.3 Unconventional Resource Impacts

As terrestrial deposits of minerals decline in profitability and in overall resource availability, mining companies have moved to consider alternative resources. One

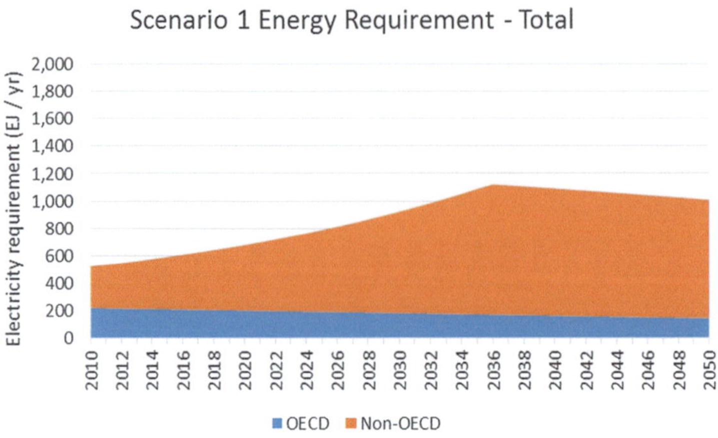

Fig. 9 Total energy demand in scenario

such resource is the ore found at the bottom of the ocean. Deep sea mining is potentially going to start for the first time in Papua New Guinea in the next few years, with the company Nautilus Minerals currently building its equipment [13]. The economic feasibility and environmental impacts of this mining are yet to be fully demonstrated through a full-scale operation, but the grade of minerals indicates that it could be beneficial on both categories on a life cycle/supply chain basis [14]. The mining phase, however, is comparable to the deepest onshore mines, making the impact of the mining phase itself likely to be high. Sufficient consideration of the entire life cycle is therefore an important comparative factor to use in deciding on production from such deposits and the incorporation of these into products.

4.4 Considerations for Eco-Design

Eco-design, aiming to reduce the environmental impact of products and services, is a vital tool for improving the outcomes of future societal advances. With regard to the minerals-energy nexus, it is essential to consider a variety of impacts across the full life cycle. In design, a number of frameworks have been developed to enable the integration of sustainability principles in industrial processing operations [15], in chemical processes (green chemistry) [16] and generally in engineering (green engineering) [17, 18] (a more thorough review of techniques is available elsewhere [19]). It is vitally important to consider a number of key principles with regard to functional materials and their overall life cycle impact or benefit. While some of these principles have been discussed for some time, it is worthwhile considering them specifically from the perspective of the nexus.

Firstly, recyclability, resource recovery and design prerogatives are considered. Particularly in the case of photovoltaics and fuel cell technologies, the use of nanothin films of functional minerals in order to enhance certain properties as well as (often most importantly) to reduce the cost is often in direct opposition to the recyclability and recoverability of the metal components. With the energy requirement and other environmental impacts typically increasing exponentially with respect to a decrease in grade of mineral, it is apparent that recycling of devices with such minor components is likely to lead to poor recovery and/or high environmental load. With regard to the energy-minerals nexus, the demand for some metals may outstrip conventional supply in the near term even if only energy technologies are considered as demand (e.g. Fig. 8). In order to supplement some of that gap in demand, it will be important to recycle – but if the energy of recycling is higher than the primary metal, there is a feedback with regard to the required global energy need – thus requirements for energy could increase dramatically (above the already expanding unit energy) in the case of certain critical-metal-using technologies.

As shown in Table 3, the comparative grades of some energy system urban ores are below those of terrestrial deposits that are currently considered economic to mine. However, the table also shows that enabling disassembly – at least the removal of certain major components – can significantly enhances the grade of the material. In some cases, such design for disassembly may enable reuse or minimum refurbishment before incorporation in another product, rather than expending the additional energy to recycle. Understanding the end-of-life requirements and balancing the product cost with the broader societal and environmental consideration of resource recovery are therefore important.

Secondly, it is important to consider the viability of using alternative materials – or materials requiring lower levels of purity – particularly in competitive markets. In the case of the coproduct metals used particularly in photovoltaics, the availability of mineral is uncertain and the quantity produced each year is small. This can affect the market dynamics, making the prices fluctuate, although likely to rise in future as demand increases faster than supply. It is also important to consider from a societal perspective that the priority of end utilisation is not necessarily translated

Table 3 Comparative grades of energy system urban ores with terrestrial deposits

	Grade (wt%)		
	Energy system urban ore		
Critical minerals	Component	System	Terrestrial deposits
Dy, Nd	30 % (magnet) (28 % Nd/2 % Dy)	0.03 %	0.05 % Nd/0.02 % Dy ~0.8 % Nd/0.5 % Dy
Cu	~100 % (wire/windings)	0.8 % (wind)	>0.5 % (typical cut-off grade)
		0.1 % (PV)	
In, Ga, Se, Te (as total)	0.5 % (panel)	0.3 %	In (<0.01 %) in Zn ores
	14.5 % (panel without glass)		Se (8 %) in Cu slimes
			Se (<0.00001 %) in Cu ores
			Te (1 %) in Cu slimes

After [10]

effectively into market power or market share – the use of neodymium and dysprosium in magnets for coal-fired power plant may be considered equal to wind power or electric vehicles in the market, although the end impact is very different and the ability to pay for the material is also variable.

Thirdly, regarding the systems of recovery – although these fall more broadly under the "industrial ecology" label than eco-design perhaps – the recovery rates of end-of-life products are highly affected by the available collection and recovery infrastructure. It is likewise important to consider that the recyclability of a product may be only a theoretical concept in places that do not have the infrastructure or economic incentives to recycle [20]. Affecting the ability to recycle – even from large-scale clean energy technologies – is the quantity of waste and the timescale at which that becomes available [5]. In the case of large-scale "megasolar" or wind farms, the scale of deposit is typically still quite low compared to existing terrestrial deposits. Although the avoidance of overburden and the requirement to dig out the deposit may be advantageous, the small deposit size (typically more than one order of magnitude smaller than standard mines) makes the attraction of the deposit specifically for recovery operations lower. This is exacerbated in the case of distributed generation – for example, household solar systems [5].

Finally, with the advent of new mines at the bottom of the ocean or exploiting recycling, it may be an opportunity to further expand the environmental and country-of-origin labelling of minerals in order to enable consumers to make better-informed, socially responsible purchasing decisions.

5 Summary

This article has described some of the key energy-minerals relationships that are an important consideration for the energy-minerals nexus. The study has presented data indicating that the growth of demand for minerals is continuing, but the use of energy in extraction of these minerals is growing at an expanding rate, despite improvements in efficiency for the processing and smelting to metals. This is an important potential impact for the production of primary ores and subsequent metals.

Vitally, with regard to sustainable energy futures, there is a need to consider the future availability and recyclability of minerals when products or services are being designed. The study showed that in example technologies used for clean energy, there is a potential for recycling to present worse "ore grades" than primary mines, but that design for disassembly can drastically enhance the potential benefits. Moreover, considering the nature and size of future natural and urban ore deposits is essential to understand the transitions of energy-minerals nexus impacts in future.

Acknowledgments This research was undertaken using a Grant-in-Aid from the Japan Society for the Promotion of Science, Research Project Number: 26701014.

References

1. IEA (2013) Energy balances of non-OECD countries. International Energy Agency, Paris
2. BGS (2010) World mineral production 2004–2008. British Geological Survey, Keyworth
3. BP (2013) BP statistical review of world energy 2013. Cited 24 Sept 2013. Available from: http://www.bp.com/statisticalreview
4. USGS (2013) Mineral commodity summaries 2013. United States Geological Survey, Washington, DC
5. Giurco DP et al (2015) Mineral-water-energy nexus: implications of localized production and consumption for industrial ecology. In: Brau YNAL (ed) The International Sustainable Development Research Society (ISDRS) 21st annual conference – the tipping point: vulnerability and adaptive capacity. ISDRS, Geelong
6. USGS (2015) Mineral commodity summaries 2015. United States Geological Survey, Washington, DC
7. McLellan BC et al (2012) Renewable energy in the minerals industry: a review of global potential. J Clean Prod 32:32–44
8. McLellan BC (2011) Optimizing location of bulk metallic minerals processing based on greenhouse gas avoidance. Minerals 1(1):144–156
9. WNA (2013) Uranium production figures, 2005–2012. 2013 May. Cited 21 June 2013. Available from http://www.world-nuclear.org/info/Nuclear-Fuel-Cycle/Mining-of-Uranium/World-Uranium-Mining-Production/#.UcOe5y9-OP8
10. McLellan BC (2015) Sustainability and design considerations for critical minerals in clean energy technologies. In Shibata J, Ahn JW (eds) The 13th Japan/Korea international symposium on resources recycling and materials science. The Resource Processing Society of Japan and Korea Institute of Geoscience and Mineral Resources, Kyoto
11. McLellan B, Corder G, Ali S (2013) Sustainability of rare earths—an overview of the state of knowledge. Minerals 3(3):304–317
12. May D et al (2012) Peak minerals: theoretical foundations and practical application. Nat Resour Res 21(1):43–60
13. Nautilus Minerals (2013) Nautilus Minerals Inc. – Solwara I Project. Cited 8 Apr 2013. Available from http://www.nautilusminerals.com/s/Projects-Solwara.asp
14. McLellan BC, Motoori R, Tezuka T (2015) Comparison of lifecycle impacts of seafloor mining with onshore equivalents. In: Chung JS (ed) ISOPE 2015. Kona, Hawaii
15. Corder GD, McLellan BC, Green SR (2012) Delivering solutions for resource conservation and recycling into project management systems through SUSOP®. Miner Eng 29:47–57
16. Anastas PT, Breen JJ (1997) Design for the environment and green chemistry: the heart and soul of industrial ecology. J Clean Prod 5(1–2):97–102
17. Allen D, Shonnard D (2002) Green engineering: environmentally conscious design of chemical processes. Wiley, New York
18. Anastas PT, Zimmerman JB (2003) Design through the 12 principles of green engineering. Environ Sci Technol 37(5):94A–101A
19. McLellan BC et al (2009) Incorporating sustainable development in the design of mineral processing operations – review and analysis of current approaches. J Clean Prod 17 (16):1414–1425
20. Corder GD, Golev A, Giurco D (2014) "Wealth from metal waste": translating global knowledge on industrial ecology to metals recycling in Australia. Miner Eng 76:2–9

Estimation of Reduction in CO_2 Emissions by Using ICT Throughout Japan

Tomomi Nagao, Minako Hara, Shinsuke Hannoe, and Jiro Nakamura

Abstract Many countries, including Japan, have taken measures to reduce the load placed on the environment, thus, helping reduce the effects of global warming. One measure, the spread of information and communications technology (ICT) has increased the effects of global warming. However, ICT is expected to be of benefit regarding environmental load reduction, specifically by making it possible to reduce CO_2 emissions. Such technology can reduce the need for human physical effort and enhance the efficiency of physical distribution and other industrial activities. The use of ICT to estimate the reduction in CO_2 emissions is important if the government is to adopt a policy of promoting ICT and addressing environmental issues. Since 2003, we have continuously estimated the reduction in CO_2 emissions throughout Japan while updating the estimation factors. In this paper, we report on the results of an estimation of the 2013 data. The reduction in CO_2 emissions in 2013 was less than that for 2012, which was predicted in 2008. By analyzing the estimation factors, we reveal that these results are due to the ICT penetration rate. We also discuss a future prediction based on the penetration rate obtained for 2013.

Keywords ICT • CO_2 emission reduction • ICT penetration rate

1 Introduction

Nations throughout the world have focused on efforts to meet CO_2 emission reduction targets with the goal of reducing the effects of global warming [1]. The Japanese government is currently discussing new CO_2 emission reduction targets for Japan for COP21 while considering energy problems. Japan has been working to reduce CO_2 emissions since the adoption of the Kyoto Protocol in 1997 [2]. At that time, Japan set a CO_2 emission reduction target of 25 % by 2020 (compared with the

T. Nagao (✉) • M. Hara • S. Hannoe
Network Technology Laboratories, NTT, Tokyo, Japan
e-mail: tomomi.nagao@lab.ntt.co.jp

J. Nakamura
Device Technology Laboratories, NTT, Kanagawa, Japan

M. Matsumoto et al. (eds.), *Sustainability Through Innovation in Product Life Cycle Design*, EcoProduction, DOI 10.1007/978-981-10-0471-1_43

1990 figure). The Japanese government declared that it is necessary to take full advantage of information and communications technology (ICT) in order to achieve this target [3]. Such technology has increased the power consumed by PCs, terminals, and network facilities, which is a negative impact on the environment. However, ICT also has a positive impact on the environment. For example, ICT-based services can reduce the need for human physical effort and enhance the efficiency of physical distribution and other industrial activities. The widespread use of ICT throughout society may lead to reduced CO_2 emissions. Using ICT to reduce the environmental burden is called "Green by ICT" [4]. By estimating the reduction in CO_2 emissions achieved using ICT throughout Japan, we can estimate the contribution of ICT to meet the government targets for the reduction of CO_2 emissions.

In this paper, we discuss the estimation results related to the reduction in CO_2 emissions achieved using ICT throughout Japan based on 2013 data and review the ICT usage scenarios of this past estimation. We also add future ICT usage scenarios. We used statistical data to develop estimation models in relation to these scenarios. We also updated the estimation factors in the conventional estimation models. We discuss ICT usage scenarios that are expected to make a large contribution to CO_2 emission reduction in the future by comparing the future prediction made in 2008 with the 2013 estimation results.

2 Background

2.1 Previous "Green by ICT" Work

The Japan Forum on Eco-Efficiency released the guideline for information and communication technology (ICT) eco-efficiency estimation for estimating the reduction in the CO_2 emissions of individual ICT services in 2006 [5]. Subsequently, a method for estimating the amount of CO_2 emitted by ICT services internationally was standardized in 2012 in ITU-T Recommendation L.1410: "Methodology for the assessment of the environmental impact of information and communication technology goods, networks and services" [6]. This recommendation includes a method for estimating the amount of CO_2 emitted by ICT services that Japan proposed and established in 2006.

A method for application to company-wide ICT services has also been developed based on the estimation method for individual ICT services [7, 8]. Information and communication technology companies want to inform society of their environmental efforts by announcing the reduction in CO_2 emissions that they have achieved by using ICT as part of their corporate social responsibility program. Some ICT companies, including NTT, have set quantitative targets for reducing CO_2 emissions [9, 10]. In 2014, the "Life Cycle Assessment (LCA) for the Organization of Offering ICT Services" study group of the LCA Society of Japan released beta version guidelines for a method designed for estimating CO_2 emission reduction at the enterprise level [11].

2.2 *Previous "Green by ICT" Work in Japan*

A method for estimating the reduction in energy consumption by using ICT throughout Japan has been studied since 2002. The reduction in Japan's total energy consumption was forecast for 2010 on the basis of statistics [12]. This estimation method was subsequently improved upon in three ways, namely, by classifying the ICT usage scenarios of the estimation targets, using continuously published statistical data, and using the ICT penetration rate [13]. With reference to these estimation methods, the Study Group on ICT Policy for Addressing Global Warming, a subsidiary group of the Ministry of Internal Affairs and Communications in Japan, discussed the reduction in CO_2 emissions since 2007 due to ICT. They estimated the amount of reduction in fiscal year (FY) 2006 and also forecast the reductions for FYs 2010 and 2012 by forecasting the penetration rates until 2012 using statistical data [14]. In 2009, the ICT Policy Task Force for a Global Era, a subsidiary group of the Ministry of Internal Affairs and Communications in Japan, forecast the amount of CO_2 emissions reduction until 2020. They considered adding new estimation objects and the impact of the presence and policy introduction of regulation.

As 5 years has passed since the last estimation, some changes can be inferred in the use of ICT. We must update the ICT contribution by using current statistical data. We must also add appropriate scenarios for current usage. The updated estimation results can be compared with the predictions for 2012. The ICT promotion strategies to date can be reviewed by referring to this comparison. Furthermore, this new estimation can help us develop strategies for the future.

3 Method

The concept of estimating CO_2 reduction by using ICT throughout Japan is based on the following assumption. If ICT had not been introduced, it is assumed that there would be activities having the same functionalities as ICT services. The introduction of ICT is assumed to reduce CO_2 emissions as the result of reducing such activities. CO_2 reduction is estimated by identifying such activities and making the estimation models for those of activities. This concept is also adopted regarding estimation methods for individual and company-wide ICT services. The basic estimation model is shown in Fig. 1. We define "penetration rate" as the number, percentage, or market share with respect to the introduction or utilization of each service based on ICT throughout the country. "Scale of activity" is defined as the frequency or quantity of activities considered to have the same functions as ICT services had ICT not been introduced. "Reduction rate" is defined as the

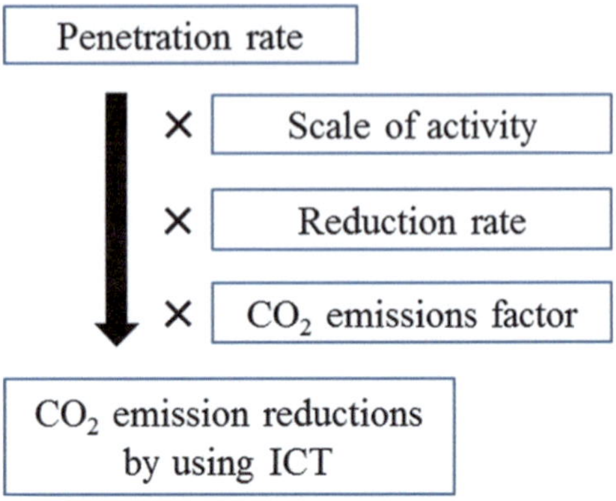

Fig. 1 Basic estimation model

percentage by which the scale of an activity is reduced by introducing ICT. The "CO_2 emission factor" is defined as CO_2 emissions per activity. We now describe the basic model for estimating the reduction in commuting through the use of "telework" by using usage scenarios as an example (Fig. 2). The "penetration rate" is the number of teleworkers, and the "scale of activity" is the average distance each means of transportation travels for the purpose of commuting. With telework, the "reduction rate" is 20 % because only 52 days/year are assumed to be telework days [15]. The "CO_2 emissions factor" is the CO_2 emission per unit moving distance of each means of transportation. By multiplying these factors, we can estimate the CO_2 emitted from activities that are assumed to have the same function as ICT had not been introduced. In this paper, we define this estimation result as the reduction in CO_2 emissions derived from ICT use.

The estimation target is seven assessment fields and 23 ICT usage scenarios. We added four new ICT usage scenarios, "electronic medical record services," "telemedicine services," "online banking services," and "electronically recorded monetary claims services," to the 2013 data estimation. The targets of the estimation in 2013 are shown in Table 1. The four new scenarios were selected considering their possible future spread. The possibility of these services spreading is referred to as the "smart Japan ICT growth strategy" and includes other policies for promoting ICT usage established by the Japanese government [16]. "Electronic medical record services" and "telemedicine services" can be expected to account for a considerable amount of CO_2 emission reduction among the promoted ICT services. As part of the development of an estimation method for company-wide ICT services, we also constructed estimation models related to "online banking services" and "electronically recorded monetary claims services."

For the four added scenarios, we investigated actual cases in which each service is used based on previous studies. We identified the main activities that we can expect to be reduced by using ICT. To estimate the contribution of each service throughout Japan, we obtained the data for each service, namely, penetration rates, scales of activity, reduction rates, and CO_2 emission factors, from published statistics. We constructed estimation models of the four scenarios based on the

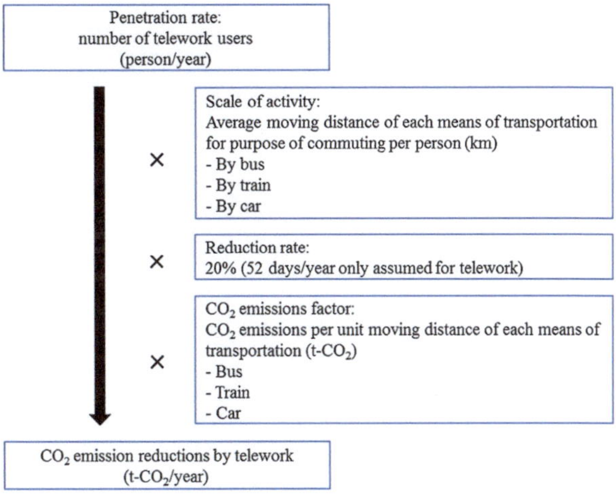

Fig. 2 Estimation model of telework as example

Table 1 2013 estimation targets

Assessment field	Usage scenarios
Electronic commerce for consumers	Online shopping
	Online issuing of air tickets
	Purchase of tickets at convenience stores
	Putting in place of ATMs
	Online banking
Electronic commerce for corporations	Online trading
	Electronically recorded monetary claims
	Supply-chain management
	Reuse market
	Electronic medical record
	Telemedicine
Shift to information of materials	Music-related contents
	Image-related contents
	PC software
	Newspapers and books
Movement of people	Telework
	TV conferencing
	Remote management
Intelligent transport systems	ITS
Electronic government and municipalities	Electronic bidding
	Electronic applications (tax return filling)
	Electronic applications (online reception)
Energy control	BEMS/HEMS

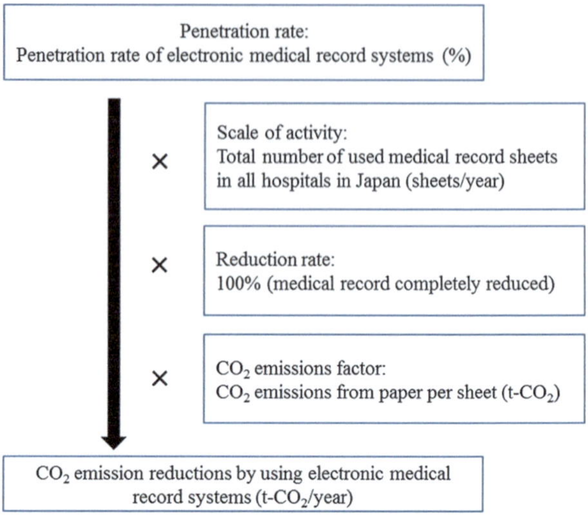

Fig. 3 Estimation model of electronic medical record services

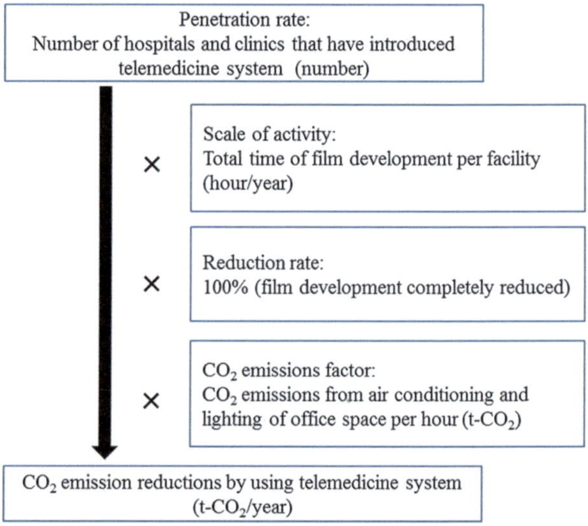

Fig. 4 Estimation model of telemedicine services (time of film development reduced)

basic estimation model. The estimation model of "electronic medical record services" is shown in Fig. 3, and those of "telemedicine services" are shown in Figs. 4, 5, and 6. Figure 4 is a model for the reduction of time of film development. Figure 5 is a model for the reduction of time from diagnostics to response. Figure 6 is a model for the reduction of film transport. The estimation model of "online banking services" is shown in Fig. 7, and that of "electronically recorded monetary claims services" is shown in Figs. 8 and 9. Figure 8 is a model for the reduction of transport. Figure 9 is a model for the reduction of time of creation.

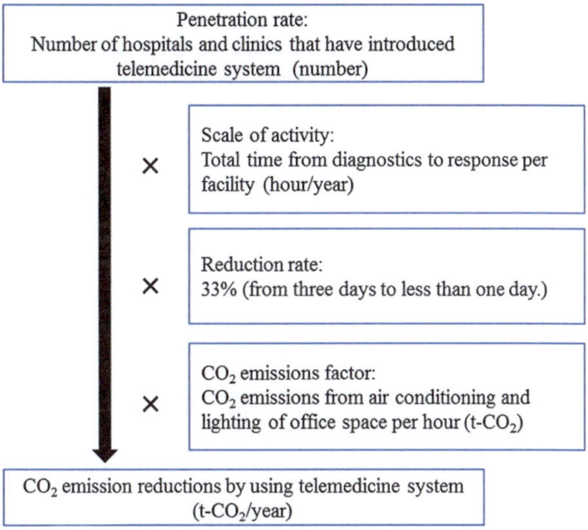

Fig. 5 Estimation model of telemedicine services (time from diagnostics to response reduced)

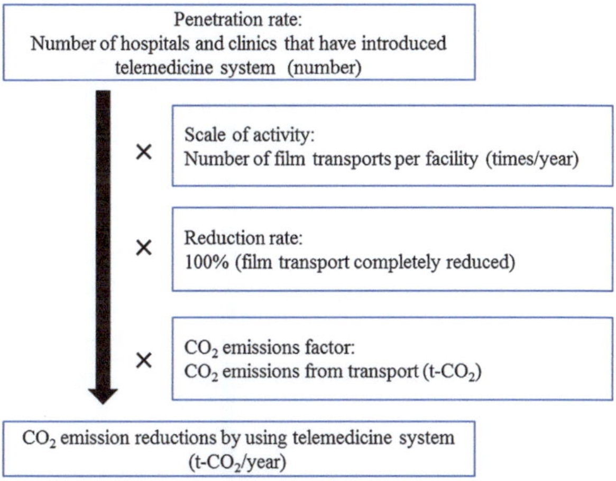

Fig. 6 Estimation model of telemedicine services (film transport reduced)

Nineteen conventional scenarios are described in a report on the results on eco-efficiency case collection and estimation standard of ICT services [17]. Certain factors have been updated in these scenarios. Some factors were replaced when more appropriate statistical data became available. In addition, we modified the estimation models according to the appropriate data where necessary.

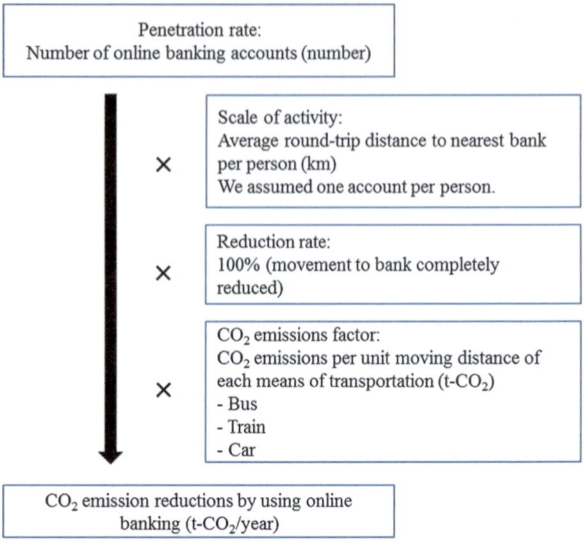

Fig. 7 Estimation model of online banking services

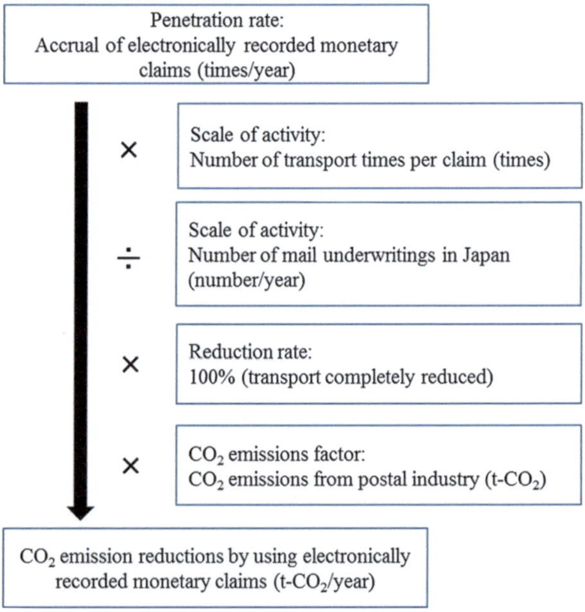

Fig. 8 Estimation model of electronically recorded monetary claims services (transport reduced)

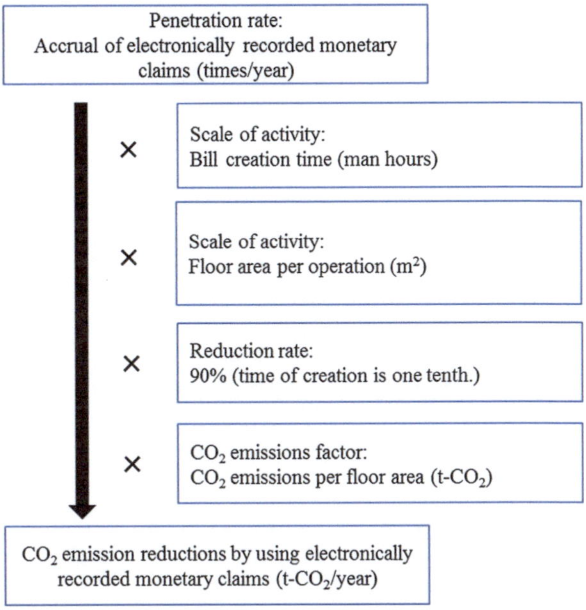

Fig. 9 Estimation model of electronically recorded monetary claims services (time of creation reduced)

4 Results and Discussion

4.1 Results

Each factor of the estimation models for the 4 added scenarios and the 19 conventional scenarios was updated with the latest available statistical data. We estimated the reduction in CO_2 emissions of each scenario according to each estimation model and totaled the results for all the scenarios. The total CO_2 emission reductions for 2013 of all the scenarios amounted to 40.86 million ton-CO_2. The estimation results for each scenario are listed in Table 2. The revised factors are shown as black dots in the table. In this estimation, we constructed estimation models with CO_2 emissions because the purpose was to estimate reductions in CO_2 emissions. Until recently, estimation models for estimating energy consumption have been used. These models were restructured to estimate CO_2 emissions. Therefore, all the updated estimation models use a different CO_2 emission factor from past estimation data. Also, when more appropriate statistical data become available during the course of an investigation, the factors and estimation models will be improved using the appropriate data. There are usage scenarios that have not been updated in this estimation. "Intelligent transport systems (ITS)" has conventionally been estimated mainly using the penetration rate of "vehicle information and communication system (VICS)," which is part of ITS. We did not renew the penetration rate of VICS because it has been published that VICS migrates to new services. "Automated teller machines (ATMs)," "electronic applications," and "building

Table 2 Results of 2013 estimation

Assessment field	Usage scenarios	Reduction in CO_2 emission (million t-CO_2)	Penetration rate	Scale of activity	Reduction rate	CO_2 emissions factor
Electronic commerce for consumers	Online shopping	6.08	●	●		●
	Online issuing of air tickets	0.07	●	●		●
	Purchase of tickets at convenience stores	1.73	●	●		●
	Putting in place of ATMs	2.61				
	Online banking	1.27	–	–	–	–
Electronic commerce for corporations	Online trading	1.26	●	●		●
	Electronically recorded monetary claims	0.15	–	–	–	–
	Supply-chain management	9.11		●		●
	Reuse market	2.36	●	●		●
	Electronic medical record	0.01	–	–	–	–
	Telemedicine	0.93	–	–	–	–
Shift to information of materials	Music-related contents	0.12	●	●		●
	Image-related contents	0.08	●	●		●
	PC software	0.61	●	●		●
	Newspapers and books	0.30	●	●		●
Movement of people	Telework	4.32	●	●	●	●
	TV conferencing	2.02	●	●		●
	Remote management	0.07	●	●		●
Intelligent transport systems	ITS	3.08				

Electronic government and municipalities	Electronic bidding	0.00	●			●
	Electronic applications (tax return filling)	0.00				
	Electronic applications (online reception)	0.00				
Energy control	BEMS/HEMS	4.68				
Total	–	40.86	–	–	–	–

energy management system (BEMS)/home energy management system (HEMS)" were not estimated, and the 2006 results were used.

4.2 *Discussion*

According to the estimation results, the largest reduction in CO_2 emissions was achieved by using "supply chain management (SCM)." Estimation scenarios of SCM use suppress unnecessary production and reduce factory and warehouse size. The efficiency of wholesale and retail distribution through logistics management is also an estimation scenario. Expanding ICT in business activities can greatly contribute to reducing CO_2 emissions. However, in this estimation, it should be noted that the 2006 penetration rate was used. Since the statistical document used in conventional estimation has changed, we could not obtain the current penetration rate of SCM. In recent years, SCM has often been used in conjunction with other systems as part of the core system of a company. In the future, we should estimate core systems including SCM.

Among the added scenarios, "online banking" accounts for a large CO_2 emission reduction. Since "online banking" is widely used by general users, the reductions are large. The other three scenarios are used between companies or between medical institutions, so reductions are relatively small.

4.3 *Future Prediction*

The reductions in CO_2 emissions in the 2013 data were compared with those predicted from the 2008 data. The reductions in CO_2 emissions for 2012, which were predicted in 2008, were 68.04 million ton-CO_2 [18]. The 2012 penetration rates were forecast by subtracting the regression line for each past penetration rate. The 2012 penetration rates were replaced with the penetration rates for 2006, and CO_2 emission reductions were predicted. We focused on the penetration rate used for prediction in 2008. The reductions in CO_2 emissions and the penetration rates used for the 2013 estimation and those used for the estimation predicted in 2008 are listed in Table 3.

According to Table 3, in all the scenarios except for "purchase of tickets at convenience stores" and "newspapers and books," the 2013 penetration rates are smaller than those of 2008 predictions. However, ITS was not estimated. In the 2008 prediction, scenarios that were expected to have a relatively large impact (more than 10 %) on the total reductions in CO_2 emissions of 2012 were "online shopping," "online trading," "SCM," "reuse market," and "BEMS/HEMS." A comparative analysis of the penetration rates revealed that the 2013 penetration rates achieved only about half the values expected in 2008. Also, scenarios that were expected to have a relatively large impact on the total CO_2 reduction amount

Table 3 Comparison of estimation results in fiscal 2013 and forecasts of 2012

Assessment field	Usage scenarios	Penetration rate	Penetration rate in 2013 (actual)	Reduction in CO_2 emission in 2013 (actual) (million t-CO_2)	Penetration rate in 2012 (forecasted in 2008)	Reduction in CO_2 emission in 2012 (forecasted in 2008) (million t-CO_2)
Electronic commerce for consumers	Online shopping	Penetration rate of e-commerce for B2C	3 %	6.08	7 %	7.12
	Online issuing of air tickets	Utilization rate of ticket reservation systems	70 %	0.07	83 %	0.06
	Purchase of tickets at convenience stores	Utilization rate of ticket reservation systems	16 %	1.73	15 %	0.64
	Putting in place of ATMs	Number of ATMs, number of reduced stores	73,380	2.61	81,100	3.19
	Online banking	Number of online banking accounts	60,258,506	1.27	–	–
Electronic commerce for corporations	Online trading	Penetration rate of e-commerce for B2B	26 %	1.26	40 %	8.36
	Electronically recorded monetary claims	Number of accrual of electronically recorded monetary claims	1,156,224	0.15	–	–
	Supply-chain management	Penetration rate of SCM	50 %	9.11	80 %	18.39
	Reuse market	Reuse rate	1 %	2.36	2 %	11.97
	Electronic medical record	Penetration rate of electronic medical record systems	20 %	0.01	–	–

(continued)

Table 3 (continued)

Assessment field	Usage scenarios	Penetration rate	Penetration rate in 2013 (actual)	Reduction in CO_2 emission in 2013 (actual) (million t-CO_2)	Penetration rate in 2012 (forecasted in 2008)	Reduction in CO_2 emission in 2012 (forecasted in 2008) (million t-CO_2)
	Telemedicine	Number of hospitals and clinics that have introduced the telemedicine system	2,453	0.93	–	–
Shift to information of materials	Music-related contents	Electronic distribution rate in the record market	17 %	0.12	27 %	1.33
	Image-related contents	Market share of electronic distribution in the video market	11 %	0.08	26 %	0.25
	PC software	Penetration rate of SaaS and ASP service	22 %	0.61	30 %	0.61
	Newspapers and books	Market share of electronic distribution in the publishing market	4 %	0.30	2 %	0.95
Movement of people	Telework	Number of teleworkers	9,300,000	4.32	16,300,000	0.63
	TV conferencing	Market scale of TV conferencing	42,000,000,000 yen	2.02	66,000,000,000 yen	3.07
		Substitution rate to the TV conferencing	37 %		–	
	Remote management	Number of vending machines	2,560,000	0.07	2,670,000	0.05
Intelligent transport systems	ITS	Penetration rate of VICS	18 %	3.08	21 %	4.01

Electronic government and municipalities	Electronic bidding	Implementation number of electronic bidding	53,460	0.00	480,000	0.02
	Electronic applications (tax return filling)	Utilization rate of tax return filing	49 %	0.00	50 %	0.08
	Electronic applications (online reception)	Utilization rate of online reception	97 %	0.00	100 %	0.01
Energy control	BEMS/HEMS	Payment amounts of BEMS	47,000,000,000 yen	4.68	104,700,000,000 yen	7.30

for 2013 were "online shopping," "SCM," "telework," and "BEMS/HEMS." Although "BEMS/HEMS" was not estimated, it may help in reducing CO_2 emissions by promoting the spread of "online shopping," "telework," and "SCM."

5 Summary

We added four ICT usage scenarios to take into account recent ICT spread and constructed estimation models for the scenario based on the basic model. We estimated the reductions in CO_2 emissions by using statistical data for 2013. We also estimated 19 conventional scenarios. We totaled the results for 23 scenarios and estimated the 2013 CO_2 emission reductions as 40.86 million ton-CO_2.

We reported this result to the Japanese government as information regarding the contribution of ICT. The penetration rates have hardly increased compared with the 2008 predictions. We propose reviewing efforts to promote ICT utilization. The government can position ICT as an important factor for meeting new CO_2 emission reduction targets.

We should review the 19 conventional ICT usage scenarios. These usage scenarios may have changed as SCM and ITS changed. We need to modify the estimation models and factors in response to these changes. In addition, we should review the reduction rates. According to Table 2, the reduction rate data have hardly changed. An assumed reduction rate is used in conventional estimations because it is difficult to obtain actual data. To obtain data, it is necessary to demonstrate the degree to which activity is reduced by using ICT. Since the reduction rate is a requirement when measuring the contribution of ICT, we must consider ways of obtaining such data.

References

1. Summary and Estimation of COP 19/CMP 9 (2015) The 19th Conference of Parties to the UNFCCC and the 9th session of the conference of the parties serving as the meeting of the parties to the Kyoto Protocol) (2015) Ministry of foreign affairs of Japan
2. Kyoto Protocol to the United Nations Framework Convention on Climate Change (1998) United Nations
3. ICT Policy Task Force for a Global Era (2009) T Ministry of Internal Affairs and Communications in Japan
4. 2010 WHITE PAPER Information and Communications in Japan (2010) Ministry of Internal Affairs and Communications in Japan
5. The Guideline for Information and Communication Technology (ICT) Eco-Efficiency Estimation (2006) The Japan forum on eco-efficiency
6. L.1410: Methodology for environmental life cycle assessment (LCA) of information and communication (ICT) goods, networks and services (2012) International Telecommunication Union Telecom

7. Nagao T, Hannoe S, Takei Y, Saito K (2014) Green ICT – method for estimating CO_2 emission reduction of ICT services. In: Proceedings of going green CARE INNOVATION 2014, Vienna, 1.2.3
8. Hannoe S, Nakamura J, Nagao T, Takei Y, Sakurai A, Saito K (2014) Analysis of influence of ICT services including application-service and network-service on CO_2 emission. In: Abstract of eco balance 2014, Tsukuba, 28D4-4
9. The Green Vision 2020 (2012) NTT CSR report, NTT Corporation
10. Green Policy 2020 (2008) Fujitsu Corporation
11. The Guideline for Business of Information and Communication Technology (ICT) CO_2 emission reduction Estimation (β version) (2014) The LCA Society of Japan "Life Cycle Assessment (LCA) for the Organization of Offering ICT Services" Study Group, in Japanese
12. Nakamura K, Nishi S, Aoki T, Yano Y, Setoguchi Y, Yoshida M, Kii T (2002) Analysis of ICT development and energy consumption, In: Proceedings of 18th conference on energy, economy, and environment, vol 16, no 1, p 391
13. Nakamura J, Honjo K, Tatemichi H, Tanaka T, Ibata Y, Nishi S (2003) Estimation of environmental impact of the spread of the information communications services in Japan, In: Proceedings of ISIE meeting and LCA forum extended, joint SETAC Europe, p 73
14. MIC Communications news (2008) Ministry of Internal Affairs and Communications in Tokyo, vol 19, no 8
15. Energy Saving in Society by IT Solutions – how to quantify the contribution of 'Green by IT' Practical Guide (2013) Survey and Estimation Committee Green IT Promotion Council, p 29
16. Smart Japan ICT Strategy (2014) Ministry of Internal Affairs and Communications in Japan
17. Report of study result on eco-efficiency case collection and estimation standard of information communications technology (ICT) services (2004) Japan Environmental Management Association for Industry, pp 74–153, in Japanese
18. Report of Study Group on ICT Policy for Addressing Global Warming (2008) Ministry of Internal Affairs and Communications in Japan

The Role of Industrial Design in Effective Post-disaster Management

Areli Avendano Franco, Liam Fennessy, and Judith Glover

Abstract In the aftermath of a natural disaster, post-disaster management (PDM) stakeholders tackle both the planning for and delivery of the subsequent recovery phase to reinstate permanent, safe and sustainable livelihoods to affected communities. Such complex scenarios demand particular capabilities from a range of knowledge fields. Access to all types of required expertise can be difficult to arrange, and as such PDM practitioners face many 'gaps' that limit their efficacy in practices. This is often the case where specialised design expertise is required. Two aspects that arise when PDM skills are examined from an industrial design (ID) perspective are (a) the need to take a more holistic and human-centred design approach to augment the focus on civil infrastructure or urban scale and (b) the role of designedly systems and participatory approaches to assist in leveraging affected communities in the re-establishment of materially and technologically mediated daily activities. By framing various ID disciplinary methods inside the PDM practice, this paper explores the opportunities for a greater inclusion such methods in the development of a novel master's course.

Keywords Sustainability • User centred • Product-service system • Education • Design thinking

1 Introduction

This paper outlines how industrial design (ID) might play a significant role in post-disaster management (PDM) – alongside other professions – in actualising sustainable recovery and resilience of communities impacted by a natural hazard. By describing the general context and principles that are used in PDM, we highlight some of the gaps that have been identified as challenges within the current practices. Then, we suggest that the inclusion of ID methodologies in PDM education can contribute to the effective and sustainable management of material and immaterial resources. A key element of taking an ID approach is the capability to work in

A. Avendano Franco (✉) • L. Fennessy • J. Glover
Department of Industrial Design, RMIT University, Melbourne, Australia
e-mail: areli.avendano@rmit.edu.au

M. Matsumoto et al. (eds.), *Sustainability Through Innovation in Product Life Cycle Design*, EcoProduction, DOI 10.1007/978-981-10-0471-1_44

cooperation with communities and relevant stakeholders by deploying a systems approach through participatory methods. Finally, we describe the development of a new master's course based on specialised knowledge and practical skills drawn from a range of practitioners, academics and industry leaders.

2 Post-Disaster Management (PDM) Context

2.1 Response and Recovery Phases

In the literature, disaster management (DM) refers to the management of resources and lines of responsibility for dealing with the humanitarian aspects of emergencies to lessen their impacts [1]. DM can be divided into four phases: mitigation, preparedness, response and recovery. The first two phases take place before a disaster occurs; the latter two phases come after it has occurred [2]. Here, we use the term post-disaster management (PDM) to refer to the response and recovery phases.

2.2 Complexities in a PDM Situation

After a country has been stricken by a natural hazard, emergency services carry out short-term relief activities such as the provision of temporary shelter and basic needs relief, as well as medical aid. When there is a serious disruption of the functioning of a community or a society due to human, material, economic or environmental losses and recovering from it exceeds the ability of the affected community or society to cope using its own resources, it is said that a disaster has occurred [3]. A greater challenge in PDM is the subsequent recovery phase in which permanent, safe and sustainable restoration of the livelihood of the communities affected is essential [4].

PDM is an intricate process initiated by governments in cooperation with different stakeholders involved in the various activities that need to take place. Stakeholders include all levels of government, political decision makers, humanitarian agencies, activists and emergency or disaster recovery managers as well as engineers, architects and urban designers. For aid agencies, the management of disaster recovery actions is highly complex and contingent on the scale and category of disaster; the capabilities of other involved stakeholders; and the geographic, cultural, political and economic conditions of those stricken. Management of disaster recovery actions and efforts of the many people engaged in PDM is complex, since the particular context and impacts of a disaster are required to be addressed on a case-by-case basis [5].

2.3 *Skills Required in PDM Situations*

For those involved in making key decisions in a PDM situation, identifying the various 'gaps' in coordinating long-term recovery actions that effectively respond to the needs of communities in a sustainable manner is critical [6]. The latter generally involves an efficient coordination and communication among the hundreds of stakeholders, particularly when they are often competing for scarce resources [2]. Experts in the PDM area point to the lack of practical skills of those out in the field to effectively address the most relevant issues according to the communities themselves [7]. Furthermore, it is highlighted that those involved in such situations require a greater level of understanding of the appropriate set of skills needed in specific situations. That is, a willingness to take on assigned roles and interchange them as needed as opposed to taking on a profession-specific role that might not be relevant to specific situations [8]. Research in the area highlights four key capabilities for effective PDM [2, 4–7]: (**1**) a clear understanding of the roles and capabilities of the diverse stakeholders involved, (**2**) the adequate understanding of the context and impact of the disaster, (**3**) the ability to successfully create permanent and sustainable recovery strategies and (**4**) the effective response to long-term needs of communities through effective collaboration with members of affected regions. The above underlines that PDM requires processes to define and judge problems as well as needs on the macro- and micro-scales in interconnected sociocultural, technical, environmental, political and economic systems.

3 Looking at PDM Through an Industrial Design Lens

3.1 *ID Perspectives and Methods*

It is uncommon to find examples of the ID profession explicitly engaging in the area of PDM, even when its methods and processes clearly present a useful addition to the other design- and engineering-based fields attempting to work across PDM. Two main aspects that arise when education towards PDM is examined from an ID perspective are (a) the need to take a more holistic and human-centred design approach to augment the focus on civil infrastructure or urban scale and (b) the role of designedly participatory approaches to assist in leveraging affected communities in the re-establishment of materially and technologically mediated daily activities. ID perspectives offer a way to assess and react to complex situations that involve both technical and human-centred approaches. Design, situated as an advocate for improved human experiences, mediates the often competing needs of individuals and the complexities of their environments. As a practice, ID draws together technical and creative problem-solving with mixed methods from ethnography and systems thinking to deliver new products, services and systems. Then, ID thinking offers a quite specific set of skills for PDM including the ability to quickly assess, visualise and prototype solutions, to identify and solve ill-defined problems, to create innovative approaches to socio-technical concerns, and to design permanent and sustainable product and service solutions [8].

ID as a discipline has moved in the past three decades from a previous role strongly characterised by the aesthetic and technical resolution of client-based commercial problems to a practice in which the designer understands their primary responsibility is to the end users and environments in which their 'design' is to operate. This transformation is largely due to the development of discourse in critical subfields such as design for sustainability (DfS), product-service systems (PSS) and user-centred design (UCD). DfS allows for long-term social and economic development alongside environmental considerations [9]. PSS perspectives introduced methods to map complex systems of users and artefacts [8]. A UCD perspective privileges the empathy and understanding of users and puts the role of the designer as the mediator of technology and people to ensure human-centred outcomes [8].

3.2 Design for Sustainability (DfS)

Over the period of 2000–2015, thousands of disasters associated with natural hazards have occurred around the world. Earthquakes, floods, tsunamis and tropical cyclone were among the most frequent hazards [10]. It has been suggested that the increasing changes in climatic conditions heighten the likelihood of disasters taking place [11]. Population growth is another factor that increases the vulnerability of communities. Currently, more than one billion people in developing countries are concentrated in high-risk areas living in often deplorable conditions [11]. It is mostly this population – whose basic needs are not always provided for – who is at most risk from being affected by cyclones, floods, earthquakes and so on. The difference in level of impacts in these areas compared to more urbanised areas is in part explained by the level of disaster preparedness and the quality of housing, infrastructure and services [5]. Depending on the type of disaster, the impacts might include physical harm, human casualties, partial or total loss of livelihoods, environmental degradation, infrastructure damages and/or the destruction of entire communities [6]. That these factors are unique to every disaster scenario underlines the need for adaptive methods to rethink how the international community undertakes PDM activities.

Approaches commonly taken by ID, namely, socially responsible design, design for affordability, design for change and design for social innovation, are helpful in addressing problems with complex and intertwining systems of sociocultural, technical, environmental and economic issues. As a discipline that has had such a bearing on the specification of goods for manufacture and consumption, sustainability discourses and the implications of such a practice are now well entrenched within disciplinary values and methods [12]. ID has a long history of contribution through NGOs via the development of often low-cost mass-manufactured goods that provide poor communities with improved amenities and pathways towards income generation and options to avoid needless environmental impacts in the transition from agrarian ways of living and working [13]. The contribution of design in the sustainability field takes four main forms:

(a) industrial design deployed as a set of processes by which local craft and production practices might be modernised or made less socially and environmentally problematic;
(b) industrial design and design research used to define new products and services that transition communities to new socio-material practices;
(c) new products designed to bring critical technical infrastructure into communities to increase opportunities for social and economic development;
(d) service and systems design used to tackle complex sociotechnical issues including education, sustainability and social justice.

3.3 User-Centred Design (UCD)

In the PDM context, recovery is defined as 'the coordinated process of supporting disaster affected communities in the reconstruction of the physical infrastructure and restoration of emotional, social, economic and physical well-being' [1]. It requires the adequate coordination and the creation of comprehensive strategies based on the long-term needs and interests of the affected communities [5]. However, recovery is often merely understood as the provision of permanent housing and other infrastructure without consulting communities [6]. This understanding is due to the fact that buildings are one of the most evident ways for agencies, organisations and governments to show that resources are being spent and aid being delivered [5]. The recovery phase does however require far more than civil infrastructure, shelter, governance and health and social services.

A key part of the PDM process is not so much rebuilding infrastructure but keeping further disruption for those affected to a minimum [14]. For example, housing reconstruction is likely to take place before key public services are; therefore, recovery might take several years after the disaster has occurred. As a result, short-term interventions are frequently required to assure the availability of basic services and safety of households [6], that is, helping the most vulnerable and affected communities to return to normal (or improved) ways of living. People live their lives and do their work through interactions with material and technological artefacts – products and objects that provide a service – be it utilitarian, generative or social. Understanding these interactions and designing for them is the domain of industrial design and requires a UCD approach [15]. Such an approach privileges stakeholder and participatory design processes useful in PDM where there is often a lack of community participation in the process of determining what 'recovery' will result in. Communities hold the information about their own needs, expectations and desires; however, in PDM situations, they are often removed from the design, financing and planning of their own recovery process [14].

3.4 Product-Service System Design (PSSD)

The response phase in PDM takes place in the early hours after the disaster has occurred and depending on the magnitude and type of disaster they will vary [2]. Emergency activities include rescue, relief and rehabilitation [3, 5]. Rescue is the first action taken after disaster, initiated by local residents and then supported by trained individual from governmental departments. Rescuing survivors can take up to 7 days and demands a complex mix of human labour, expertise technologies and products. Preventing further infrastructure damage and immediate relief of the needs of those affected can take up to 3 months [1, 5]. The rehabilitation part can take up to 5 years. One of the most evident activities of the response phase relates to the provision of temporary shelter, basic needs relief, provision of medical aid as well as the planning of and rebuilding homes and infrastructure [11]. Since all of these activities require the provision of material goods, they could also be supported by service systems solutions in which the role of contextually fit-for-purpose design sits as a fundamental inclusion. PSSD promotes shifting from creating products to fulfilling needs by delivering functions through a mix of products and systems while fulfilling the needs of stakeholders with less environmental impact.

PDM guidelines and/or programmes often assume a centralised project planning, management and decision-making processes in order to speed up procedures and deliver solutions [5, 11]. By the very nature of work that PDM is tasked to do, and the political nature of decisions made, a centralised structure approach provides a pragmatic robustness. However, it risks aggravating issues when action is taken with limited information available. PDM research shows that the complexity of a post-disaster situation requires a systems approach and a move away from focusing decisions only on the technical [7] or macro view aspects. Drawing on mixed methods from anthropology, ethnography and sustainability, ID can provide an applied 'ecosystems' approach to solutions in which the interactions between people and things (material artefacts, technologies and service schemes) are considered both for their short- and long-term humanitarian value and for their implications. The PSS model enables new types of stakeholder relationships and/or partnerships, in which systemic resources optimisation is possible while benefiting all involved [16]. This may provide PDM with alternative scenarios by which recovery actions can be directed at the human and service scale in ways that sit outside the realm of expertise of other design disciplines.

4 Towards an Inclusion of ID Methodologies in PDM Education

4.1 Moving Away from Preconceptions About Design

The contribution of design in PDM is often understood as the specification of an environment to be rebuilt [2, 5, 6]. However, design can offer much more. It provides a means by which people can significantly improve their lives and satisfy their needs through the creation of material solutions, services and systems that make environments amenable and sustainable [8]. Applying ID methodologies such as DfS, UCD and PSSD is potentially new to training for PDM. Such methodologies go beyond the technical and logistics aspects of the immediate response phase and seem a critical inclusion within complex PDM projects if the creation of sustainable solutions and the recovery of livelihoods are to be adaptive and generative. PDM professional training and university degrees are becoming more commonly available. Where there is the integration of design content and method in these programmes, it is typically located in the recovery phase.

Discourses from urban design, as a form of master planning, and architectural design, as a means approaching the need for an improved built environment, are common and present design in PDM at the urban scale. ID offers PDM alternative ways for thinking through defining and acting on problems at the human-artefact-interaction scale. Inextricably tied to design for manufacture as a means of economic development, industrial design perspectives are attuned to the notion of enterprise and livelihood in ways that are quite different to design disciplines that centre on built environment concerns.

4.2 Newer Set of Skills for PDM Practitioners

Over the past year, the authors have embarked on the development of a new course that contributes to a novel new Master of Disaster, Design and Development at RMIT University. The full master's program is a combination of humanitarian architecture, urban design, urban planning, international and community development and policy courses. Many of these courses are designed to be undertaken by students while engaged in aid and community development work. The course – 'Industrial Design in Humanitarian Scenarios' – attempts to translate methods of thinking, researching and operating through design processes common in industrial design to students from a wide array of disciplinary backgrounds that are concurrently engaged in PDM work. It uses a mix of field appropriate technological platforms for students to gather, reflect upon and to convey to peers their learning through design investigations in real-world PDM scenarios. The course asks students to identify a problem through designedly activities on the ground with affected communities. Problems could be the provision of post-disaster or development resources or an operational problem that limits the efficacy of response, recovery or development aims of the agencies that they work for.

As discussed previously, the use of design in PDM has traditionally focussed on the provision of technical solutions at the urban infrastructure scale. While the projection of a prototypical solution is important for design thinking (and for design education) inside industrial design, it is merely a means by which the nub of a problem can be identified through participative research and creative thinking. The course, which commences by the end of 2016, will provide learners with new tools to do this, and frameworks for possible solutions will be drawn from DfS, UCD and PSSD as well as other ID approaches including design thinking, community-situated design and service design as explained in the next sections.

4.3 Design Thinking

In recent years, the once rather opaque areas of design thinking have developed as ways of describing the problem-defining and problem-solving techniques and methods of designers [17]. Framed for an entrepreneurially oriented reading of design methods – that have sat inside industrial design discourse since the mid-twentieth century – design thinking is inherent in the methods of designers no matter what theoretical or practical field they are exploring in context and application [17].

Design thinking is based on knowledge gained from many fields and disciplines; yet, the essential principle is that 'design' has the ability to conceive, plan and present ideas or create scenarios in the form of products or systems that respond to people's real needs. It has been defined as an 'essentially human-centred innovation process that emphasizes observation, collaboration, fast learning, visualization of ideas, rapid concept prototyping, and concurrent business analysis, which ultimately influences innovation and business strategy' [17]. This emerging concept of design thinking emphasises the integrative character of the design disciplines that intersect with other disciplines either directly or indirectly. In the context of PDM, design thinking can be used to understand concerns from all stakeholders considering their point of view and the issues that concern them and determine the intervention that best serves them all. It can also bring the necessary understanding of the unique characteristics of the context in which a disaster has occurred and envision scenarios for sustainable recovery [18].

Design decision-making processes and strategic planning can contribute to address complex systems or environments. Here, there are clear overlaps with engineering, architecture and urban planning, which are concerned with the functional analysis of buildings or civil infrastructure. However, design is more concerned with sustaining and integrating human needs into the broader 'ecological and cultural environments, shaping these environments when desirable and possible or adapting to them when necessary' [19]. This situates design in a position in which to quickly assess the 'wicked' nature of PDM situations – that is – ill-defined situations that lack clear definitions and solutions towards the visualisation and creation of innovative, permanent and sustainable solutions. For design to actualise its potential, it is imperative that it plays a role in defining the 'design problem' [20].

Problem definition refers to the analytic phase in which the 'design problem' or list of requirements is identified in order that elements that could solve such a problem are documented, and then a series of design concepts that fit those requirements are proposed. 'Problem solution' relates to the synthesis phase in which the design concepts proposed are assessed; one is selected, which subsequently becomes the 'solution' in the form of objects, products or systems [20]. Design thinking brings creativity and design methods to address problems based on users and other stakeholders' insights to then articulate their main concerns through design innovation. Innovations do not necessarily involve the creation of a product but rather provide a strategic approach to create systems or services [21]. Attention should be put to the designers' capacity to contribute to the solution of local problems by using and adapting models, criteria and methods borrowed (at least in concept) from industrial contexts. Designers seek to understand different points of view from stakeholders rather than imposing 'proven solutions' to problems. Design methods also crossover with those used in sociology and anthropology including observations, ethnography, co-creation and discussions. While elevated in fields of new product and service development, government services and policy design; the designedly ways of thinking inside DfS, UCD and PSSD have so far remained largely unexplored as part of the PDM education and practice.

4.4 Community-Situated Practices

Community-situated industrial design (CSID) activity constitutes a particular form of practice [22] that is well suited to PDM education. If the intentionality to solve problems through the design of new product or services is put to one side (even momentarily), a CSID engagement with stakeholders resembles a rather loose form of participatory rural appraisal (PRA) [23, 24]. Here, design thinking, visualisation and prototyping are used as a means by which communities' needs (problems) and desires (preferences) can be brought into the open and then leveraged with the designer towards real outcomes that matter on the ground.

Design and designing thus becomes a transactional mechanism – a language – that provides a bridge between current realities and preferred futures. In other contexts, this approach is variously termed 'participatory design', 'co-design' and 'co-creation' and is deployed frequently in the inclusive design, social design and social innovation domains [19]. While ostensibly a process of listening and drawing out threads of collective and individual need, this form of design-based PRA comes as quite a natural practice for industrial designers. Indeed in many ways, it mirrors approaches used by industrial designers inside mass-manufacturing contexts to leverage the capabilities of a production workforce in order to resolve problems and specify high-quality production outcomes. While not universal in its application inside industrial design curriculums, it is a common curricula approach in developing world contexts and progressive industrial design undergraduate and post-graduate programs. When combined with methods and knowledge from DfS and UCD such a grounded – or non-bounded – framework for designedly thinking presents very real benefits for PDM education.

4.5 *Service Design*

Service design is also relatively young stand-alone design discipline that has been split out from industrial design and combined with visualisation methods from communications design. It tends to manifest in two main ways: as a means by which commercial operations can induce more effective consumer experiences and in the design of social enterprises or social services [25]. Service design adopts a constructivist approach to service systems innovations that draw heavily from approaches in the area of PSS – which has sustainability at its core – which service design may or may not. Service design has the major concern of understanding, mapping and the communication of user's experiences. It has been suggested that there are four main dimensions in which service design operates: (1) product-service system (PSS) innovation, (2) process innovation, (3) organisational innovation and (4) external relational innovation [25]. In the context of PDM situations, service design can play a role of organising resources with different stakeholders to contribute to the co-creation of final solutions [25]. In this view, service is regarded as 'value creation' in which the focus is put on the intangible nature of such value defined by users and less focus is put on the physical goods that provide such value. The major focus here is on a human-centred approach to solutions.

Service design methodologies, while dealing with immaterial notions of design, could assist in the development of people-based recovery scenarios and are structured so as to achieve effective stakeholder engagement and consultation while identifying key elements to deliver sustainable and enduring solutions [15]. A co-creation approach could give deep understanding of the context in order to generate socially innovative service and policy solutions that meet the expectations of communities affected. Service design, when done well, demands holistic approach from designers: by looking at the bigger picture of a system or service, considering the points of view of involved communities and projecting future and ideal scenarios, a path towards appropriate solutions can be defined.

5 Summary

The most evident opportunities for design intervention in PDM are characterised by a sense of urgency and complexity due to critical conditions. They often emerge in situations that are not covered by policy or governmental programmes. Even when abundant, international aid and financial support are provided, this as such is unable to provide valid and long-lasting solutions to such problems in PDM situations. As a result PDM practitioners and educational programmes require a combination of managerial robustness, creativity and thresholds for forms of contextual complexity that are unlike those required in any other fields.

This paper attempts to describe how the inclusion of industrial design methodologies into PDM education might provide new strategies to develop holistic, human-centred and flexible methods of service delivery in PDM scenarios. The development of a novel master's course that uses digital technologies as a platform to combine problem-based learning, design thinking and post-disaster and development fieldwork as particular form of situated learning offers a mechanism by which the actual fit and value of ID inside PDM education can be further examined. How effective such an inclusion might be both to meanings of practice and to pedagogy in PDM remains to be seen, and, as the new course unfolds and is taken by PDM practitioners, the authors will evaluate its efficacy.

References

1. Haddow GD, Bullock JA, Coppola DP (2014) Introduction to emergency management, 5th edn. Butterworth-Heinemann, Waltham
2. Hidayat B, Egbu CA (2010) A literature review of the role of project management in post-disaster reconstruction. In: Egbu C (ed) Proceedings 26th annual ARCOM conference, Leeds, 6–8 Sept 2010, pp 1269–1278
3. Moe TL, Pathranarakul P (2006) An integrated approach to natural disaster management: public project management and its critical success factors. Disaster Prev Manag Int J 15(3): 96–413
4. Davis I (2007) Learning from disaster recovery-guidance to decision makers. UNISDR, Geneva
5. Johnson C, Lizarralde G (2012) Post-disaster housing and reconstruction. In: Susan JS (ed) International encyclopaedia of housing and home. Elsevier, San Diego
6. Jha AK, Barenstein JD, Phelps PM, Pittet D, Sena S (2010) Safer homes, stronger communities: a handbook for reconstructing after natural disasters. World Bank
7. Pholeros P (2015) Role of design in post-disaster shelter, disaster, design and development symposium presentation, RMIT Melbourne
8. Morelli N (2006) Developing new product service systems (PSS): methodologies and operational tools. J Clean Prod 14(17):1495–1501
9. UNEP (2009) Design for sustainability: a step-by-step approach. The United Nations Environment Program, sustainable consumption and production, Delft University of Technology, Faculty of Industrial Design Engineering, Design for Sustainability Program, Delft

10. Provost C (2011) A decade of disasters-get the key data. The Guardian-UK edition, 18 March. [online]http://www.theguardian.com/globaldevelopment/datablog/2011/mar/18/world-disastersearthquake-data
11. McClean (2010) World disasters report: focus on urban risk. International Federation of Red Cross and Red Crescent Societies, Geneva
12. Manzini E (2003) Scenarios of sustainable wellbeing. Des Philos Pap 1:5–21
13. Ramirez M (2012) Ethics and social responsibility: integration within industrial design education in Oceania. In: Proceedings DRS2012 international conference of the design research society, Bangkok
14. Donovan J (2013) Designing to heal: planning and urban design response to disaster and conflict. CSIRO, Australia
15. Vasantha GVA, Roy R, Lelah A, Brissaud D (2012) A review of product–service systems design methodologies. J Eng Des 23(9):635–659
16. Manzini E, Vezzoli C (2002) Product-service systems and sustainability: opportunities for sustainable solutions. United Nations Environment Program, pp 1–31
17. Lockwood T (2009) Design thinking: integrating innovation, customer experience, and brand value. Allworth Press, New York
18. Crouch C, Pearce J (2012) Doing research in design. Berg, Bloomsbury, London
19. Frascara J (2002) People centred-design: complexities and uncertainties. In: Frascara J (ed) Design and the social sciences: making connections. Taylor & Francis, London, pp 33–39
20. Buchanan R (1995) Wicked problems in design thinking. In: Margolin V, Buchanan (eds) The idea of design. MIT Press, Cambridge, MA, pp 3–20
21. Cooper R, Junginger S, Lockwood T (2009) Design thinking and design management: a research and practice perspective. In: Lockwood T (ed) Design thinking: integrating innovation, customer experience, and brand value. Allworth Press, New York, pp 57–63
22. Varadarajan S, Fennessy L (2012) Educating for design activism. In: Live projects: designing with people. RMIT University Press, Melbourne, pp 200–213
23. Chambers R (1997) Whose reality counts? Putting the first last. Intermediate Technology Publications, London, p 1
24. Singh K (2001) Handing over the stick: the global spread of participatory approaches to development. In: Edwards M, Gaventa J (eds) Global citizen action. Lynne Rienner Publishers, Boulder, pp 163–175
25. Sangiorgi D, Prendiville A (2014) A theoretical framework for studying service design practices: first steps to a mature field. DMI J 1(9):61–73

Undergraduate Students Designing Environmental Concern Products: A Case Study in Design Education

Edilson S. Ueda and Fumio Terauchi

Abstract This paper analyzes experimental learning based on pedagogical case studies that propose practical designs with a sustainable approach. The research has limited resources, time, and scale of context. According to the activities and attitudes of the student team, they expressed strong interest in exploring the possibilities of sustainable design rather than in eco-design principles. They showed interest in focusing on sustainable consumption and consequently leaned toward a sociocultural rather than a technological eco-design approach in their works. Students faced real-world sustainability issues, providing a great opportunity for them to discuss between themselves as well as with the public.

Keywords Sustainable design • Design education • Eco-design • Proposal design

1 Introduction

The activities and attitudes of the student team toward sustainability as well as their inclination while designing sustainable products are analyzed in four phases: understanding, analyzing, proposing, and evolving. This case study stimulated discussion and cooperation within the student team on the common theme of eco-design. In addition, it provided an opportunity for students to assess real-world sustainability issues and confront their ideas publicly.

The pedagogical experience and activities of students contribute to an initial discussion of the principles of the design education for sustainability (DEfS) context. This research is justified by the lack of reference literature that pertains to DEfS. This fact is confirmed in the papers published between 2003 and 2013 by the International Symposium on Eco-Design [1] and between 1997 and 2011 by the Special Issue of the Japanese Society for the Science of Design [2, 3]. Few papers have discussed this issue from the pedagogical viewpoint of design education, especially in industrial DEfS. One example that focused on design education is the research paper entitled "Sustainability Curricula in Design Education," which

E.S. Ueda (✉) • F. Terauchi
Department of Design, Faculty of Engineering, Chiba University, Chiba, Japan
e-mail: shindi2010@hotmail.com

M. Matsumoto et al. (eds.), *Sustainability Through Innovation in Product Life Cycle Design*, EcoProduction, DOI 10.1007/978-981-10-0471-1_45

underlined the current DEfS learning and practice. This paper introduced the development of a sustainability curriculum for design courses for both bachelor's and master's students [4].

2 Method

This pedagogical research is an experimental design that explores the activities, attitudes, and inclination of the student team when designing sustainable products. It is characterized by practice-based research of new knowledge that is generated by a combination of practical design activities and the reflections of the student team.

It is based on experimental learning theory, which describes learning as a process whereby knowledge is created through the transformation of experience [5–7].

According to experts, this method is useful for providing a structure and better understanding about the project-based approach regularly used in design education [8]. Final-year undergraduate students of an industrial design course were invited to participate in the design project. The design project was supervised by a faculty who has the equivalent of graduate research requirements. The pedagogical design project focused on a common theme: eco-design. This was chosen because the term was acknowledged by the public and students. It had not been proposed in previous design projects, meaning that students had not yet had the opportunity to research it. The project lasted for a period of 8 months with a total of 150 h.

3 Design Education for Sustainability (DEfS)

DEfS is based on design for sustainability (DfS) principles, which aim to take the environmental, social, and economic aspects (triple bottom line) of the total life cycle of a product or service into account during the development stage [9].

According to experts [10, 11], DfS is now well established, although it can still be considered to be an emerging field. However, the term eco-design is recognized by academic fields and the public. The terms eco-design and DfS both take into account environmental issues; however, eco-design principles consider environmental factors in each stage of the product life cycle [9]. DfS principles consider the environmental (e.g., resource use, end of life impact) and social impacts of a product (e.g., usability, socially responsible use, sourcing and designing to address human needs) [9].

The term DEfS in its widest sense could also be defined as a practice of education and research design that contributes to sustainability [12]. The newest research [13] on DEfS is related to existing design courses (both bachelor's and master's degrees) that integrate sustainability and eco-design concepts into curricula in the UK, Finland, Italy, the US, South Korea, Japan, India, the Netherlands, and Australia.

4 Design Education for Sustainability Approaches

This section summarizes four examples of approaches that can influence the direction and principles of DEfS.

(a) Design and sustainability context:
According to Fletcher and Dewberry [13], two extremes exist for implementing sustainability into design education programs [4]. On one side is a design context where sustainability is understood within the frame of reference of current design activities and priorities, i.e., conceiving of and producing economic goods for competitive markets. On the other side of the spectrum is a sustainability context. Here, design is viewed as a dimension of sustainability rather than sustainability being viewed as a dimension of design as at the other extreme. This draws on sets of backgrounds, expectations, priorities, and outputs in accordance with the goals of sustainability [14].

(b) Production and consumption
According to experts [5, 15, 16], DfS is a balance between production (quantitative production issues) and consumption (qualitative consumption sphere) approaches, in which both are considered to be notions of eco-efficiency and sufficiency, respectively. These two notions suggest a proposal that can move from product-focused to practice- or use-focused (practices, habits, and lifestyles) solutions.

(c) Redesigning and rethinking
From the perspective of the development of products, industrial designers can redesign and rethink traditional products by reconsidering how objects are conceived, developed, produced, distributed, used, reused, recycled, and disposed [4, 11]. From an educational perspective, experts express that DEfS should support "rethinking" approaches that go beyond "redesigning" [5, 15, 16].

(d) Technological and sociocultural
A technological approach refers to innovative products and processes that reduce environmental impacts. This is often used in conjunction with eco-efficiency and eco-design. Leaders in many industries have been developing innovative technologies in order to work toward sustainability. However, these are not always practical or enforced by policy and legislation. The sociocultural approach considers eco-innovation in terms of usage rather than merely in terms of product [17]. It suggests new environmental scenarios corresponding to new lifestyles, which require the highest level of social acceptance [18].

5 Students' Profile

This research has limited resources, time, and the scale of context. The case study was composed by a student team with four students on a final-year design program run by the Faculty of Engineering, Design Department. All participants were female and their average age was 21. Students did not have experience of eco-design and sustainable design activities and had not received lectures on eco-design or sustainable design over the past 3 years.

The main motivation of students to participate in this design project was the challenge to understand the complexity of sustainability issues and to put into practice a solution through design possibilities. In particular, they expected to break down the barriers of their own fields of study research.

6 Learning Sustainable Design

Those sections describe the results of the activities, attitudes, and inclination of the student team when designing sustainable products as well as the important considerations of the experts.

6.1 DfS Sustainable Design Principles of the Student Team

The sustainability issues and DfS principle described in Sect. 5 was introduced to the study team. During teaching those principles, the student team learned the different key characteristics of sustainable design principles.

The student team expressed strong interest in exploring the possibilities of sustainable design rather than in eco-design principles. They had the possibility to intervene in and offer a new solution and a broader context of current product design. This intervention field referred to the goals around needs, involving different elements to create system change, reevaluating design conventions, multifunctional outcomes, and ecological understanding in appropriate ways.

On the complexity of the DfS principles and different characteristics of product categories in marketing, the student team decided to define products that were everyday products as an important field. According to experts, the consumption of those everyday products has most caused environmental damage [19–21].

6.2 Everyday Eco-products in Marketing

The student team reflected on the DfS principle and decided to intervene in everyday products called "eco-products" in marketing. The sample was found in relevant literatures: *Eco-Design Handbook* [22], *The Total Beautiful Sustainable Products* [23], and *Yearbook Good Design Award* [24].

A total of 256 eco-products were selected and reviewed based on their description in brochures, catalogs, and books. Each eco-product was evaluated based on the basic principle of the life cycle of the product with implementation of sociocultural factors.

By analyzing the characteristics of those eco-products, the student team expressed that for most it was difficult to understand the real significance of the new feature improvement of environmental aspects. Those eco-products with new explanations of features were focused on technical aspects. In other words, the main strategy adopted by companies focused on the production and technological approach.

The reduction of environmental impact during the use phase of product and optimization of the end-of-life system were the main focuses of designers in their proposal designs.

Students felt that consumers went far beyond their satisfaction level, such as enlightenment and awareness toward the environment.

Other aspects noted by them were also the price of eco-products, which was higher than conventional products. They did not appreciate the double price compared with conventional products.

According to experts [25], many so-called eco-products have failed in the market, while market promotion has also concentrated primarily on fuel saving, energy saving, safety, and quality.

6.3 Defining Objectives for Design for Sustainable DfS Products

In this phase, the student team expressed a strong need to define a common goal, exploring a common interest to translate the theoretical study into the practical design. In order to clear the area of intervention of everyday product categories, the students, based on the previous section, chose keywords that were not explored in depth or were not described elaborately in these eco-products.

The student team noted little consideration in relation to user focus, such as aspects on enlightenment and awareness of the environment.

The student team noted that there was also little evidence of eco-products described from environmental consciousness and awareness in order to change users' behavior toward a sustainability approach.

Given those facts, they decided to intervene in the product categories that could be raised and center on users' environmental enlightenment and awareness aspects.

According to experts [26–29], from a pedagogical viewpoint, the activities, attitudes, and inclination of the student team in this stage showed a proactivity toward sustainability. They were interested in solutions to the sociocultural stimuli of products than the physical appeal of the product.

6.4 Experimental Product Design Based on the Eco-feedback Approach

In this stage, the student team put into practice their product development knowledge learned in the previous years of industrial design, such as the design process, material planning, visual communication, and prototyping. Each student proposed a different product category, but each one had a common point between them: user enlightenment and an awareness of the environmental approach.

The student team defined their proposed design based on the following approaches: longevity and attachment of product through use by users, sustainable consumption of products and services through the accessibility of environmental information by users, energy concern through users touching and feeling the product, and ecology concern through users enjoying and using the media.

Figure 1 shows the possible user segments and the relationship with each proposed design.

(A) Nature
Enjoying and learning knowledge of elements of nature. Through the picture book and calendar, by using colorful images with fewer words, the student team highlighted the importance of water to get users interested in nature (both adults and children). The proposed design was aimed at possible user groups.

(B) Light
Intensifying users' experience of energy conservation. Through a small and portable desk light, made with nature and soft materials, users can touch and feel the warmth of light. Through the possibility of touching the light, users can enjoy turning on or off the switches.

(C) Table
Memories and attachment of users toward the long life of the product. This is a proposal that through its everyday use connects family members or old friends and stimulates the memory of users while using it. Through its modular design, it can be increased according to the number of users added over time.

(D) Community eco-service
Promotion of eco-action in the local community. This is a terminal with which users can get easy and fast information about public eco-services (eco-conscious events and guidelines for eco-activities, transportation,

Fig. 1 Proposal design of student team [30]

workshops, garbage collection, reuse, rental, sharing of products between users). It provides motivation to participate in eco-activities easier by facilitating proper and smooth communication with users, enterprises, and administrators regarding environmental problems.

6.5 Effectiveness of the Design Proposal

The objectives of this phase were to allow the student team to present their proposals to the public and to stimulate the student team to review their proposals. The four design proposals were exhibited at an event: undergraduate study of design exhibition. During the 3 days of exhibition, the proposals were displayed to the public through prototypes, panels, and video demonstrations and orally.

Altogether, 254 people visited the proposed design, of which 203 of them expressed their opinions. The similar opinions were grouped in the following viewpoints: (a) Each proposal was educational to focus on enlightenment and awareness (57 %); (b) I do not feel a strong ecological image of products. It could give more emphasis to ecological product message (32 %); (c) it seems that the combination of approaches of each proposal design with a specific segment is irrelevant. All user segments can make use of all proposals (11 %). In general, positive comments were received. However, the student team felt that their proposed designs were lacking in detail, such as the technical aspects of production, material, and cost.

7 Discussion

This experimental research focused on undergraduate students designing environmental concern products were an introduction of DEfS. It was an initial discussion of the principles and complexity of the DfSB context.

After the exhibition of proposal design, the student team noted that more studies on sustainable behavior were necessary, in order to explore the possibilities of use phase of eco-products. In the other words, strategies could influence and encourage user's attitude during the use phase of products in order to decrease the environmental impact.

Facing this fact, the student team identified a new research area that has emerged in the design field: design for sustainable behavior that is described as an intersection of sustainable design and interaction design, applying insights from multiple disciplines to the problems of influencing more environmentally friendly use of products, services, and environments [31].

8 Conclusion

In this research, the student team expressed strong interest in exploring the possibilities of sustainable design rather than in eco-design principles. They showed interest in focusing on sustainable consumption and consequently leaned toward a sociocultural rather than a technological eco-design approach in their works. Students faced real-world sustainability issues, providing a great opportunity for them to discuss between themselves as well as with the public.

9 Future Research

This research was an introduction of DEfS through experimental learning based on a pedagogical case study. Based on the discussion and conclusions above, we will conduct research focusing on exploring a practical way with appropriate content and language of DEfS for students of industrial designers and professors. A possible practical way is to adapt eco-design strategy tools for design education.

References

1. Ecodesign symposium third~eighth international symposium on environmentally conscious design and inverse manufacturing, Ecodesign. (2003, 2005,2007,2009,2011, 2013) IEEE Computer Society 2003;1-838.2005;25-973.2007;29-978.2009;43-1246.2011;42-1272.2013; CD-ROM
2. Japanese Society for the Science of Design (JSSD) (1997) The consideration of the relationship between design and ecology, recycle and PL-law. Special Issue. 4(3):1–51
3. JSSD (2011) Ecodesign. Special Issue. 18–3(71):1–74
4. Casais M, Christiaans HHCM, Almendra R (2012) Sustainability curricula in design education. In: Proceedings of the 14th international conference on engineering and product design education (E&PDE 2012) Belgium. http://repository.tudelft.nl/view/ir/uuid:b2c0c3ae-1a8c-4b4e-9114-1b6cfc9a4645/. Accessed Feb 2015
5. Cooper T (2003) Product development implications of sustainable consumption. Des J 2003 (3):46–57
6. Adams AB, Kayes DC, Kolb DA (2004) Experiential learning in teams. (Working paper, George Washington University, Case Western Reserve University, USA). Simul Gaming 2005. 36:330–345
7. Galea C (2007) Teaching business sustainability. Vol 2: cases, simulations and experiential approaches. Greenleaf Publishing, Saltaire, UK, pp 9–10
8. Green CR (2011) Sustainability in the product design curriculum; evaluating the effectiveness of integration in project based learning. International conference of engineering and product design education. The Design Society, London, pp 343–348
9. Bhamra T, Lofthouse V (2007) Design for sustainability, a practical approach. Gower Publ Limited, Hampshire, UK, 4(39):65–99
10. United Nations General Assembly (1987) Report of the World Commission on Environment and Development: our common future, pp 16–17
11. The United Nations (UN) (2014) Decade of education for sustainable development, UNESCO. http://www.unescobkk.org/index.php?id=961. Accessed Feb 2015
12. Costa F, Prendeville S (2014) Sustainability and ecodesign in Welsh higher education-recommendation based on international best practice. Ecodesign Centre (EDC). Waste Strategy Branch, Welsh Government. http://www.ecodesigncentre.org/sites/default/files/resources/Final_Sust%26Ecod%20in%20Welsh%20HE.pdf. Accessed Dec 2014
13. Fletcher K, Dewberry E, Demi E (2002) A case study in design for sustainability. Int J Sustain High Educ 3(1):38–47
14. Boks C, Diehl JC, Wever R (2006) Sustainable product design, engineering, and management education for industrial design engineering. In: 13th CIRP international conference on life cycle engineering, pp 161–165
15. Marchand A (2009) Sustainability and design education; from products to practices. International conference on engineering and product design education, UK, pp 436–440
16. Reisch LA, Scherhorn G (1999) Sustainable consumption. In: Dahiya SB (ed) The current state of economic science, vol 3. Spellbound Publications, Knutsford, pp 657–690
17. Fussler C, James P (1996) Driving eco-innovation: a breakthrough discipline for innovation and sustainability. Pitman Publishing, London, p 364
18. Manzini, E (1992) Limits and possibilities of Ecodesign; from the ecological re-design of products to the suggestion of new environmental scenarios. Change Design (formerly EcoDesign Foundation), a project of the Society for Responsible Design, Sydney, Australia | www.edf.edu.au, http://www.changedesign.org/Resources/Manzini/Manuscripts.htm. Accessed Oct 2014
19. Aoyagi M (2014) Changing consumption patterns in Japan. National Institute for Environmental Studies. www.eolss.net/sample-chapters/c16/e1-57-03-00.pdf. Accessed Oct 2014
20. MOEJ (2012) Basic policy on promoting green purchasing. www.env.go.jp/en/laws/policy/green/2.pdf. Accessed Dec 2014
21. Asian Productivity Organization (APO) (2014) Eco-products. http://apo-ecoproducts.com. Access Dec 2014

22. Fuad-Luke A (2009) Eco-design handbook – a complete sourcebook for the home and office. Thames & Hudson, Paris, pp 10–352
23. Datschefski E (2001) The total beautiful sustainable products. Rotovision, Brighton, UK, pp 10–176
24. Japan Institute of Design Promotion. Yearbook Good Design Award (2008, 2009, 2010, 2011, 2012) p 10-304(2008), 10-304(2009), 10-712(2010), 10-728(2011), 10- 784(2012)
25. Yim HJ (2007) Consumer oriented development of ecodesign products, Vulkan –Verlag, Germany, p 7
26. Bakker C (1995) Environmental information for industrial designers. Doctoral Thesis, Delft University of Technology, pp 35–56
27. Ueda ES (2013) Aptitudes of industrial designers towards eco design – a survey of Japanese industrial designers during the early activities of ecodesign. Bulletin of JSSD (Accepted in January 2013) pp 1–10
28. Ueda ES (2013) Industrial designers working towards and eco-innovation approach. Bulletin of JSSD. (Accepted in January 2013) pp 1–10
29. Brezet HC (1997) Ecodesign: a promising approach to sustainable production and consumption. UNEP, Arendal, Norway, pp 140–142, 346
30. Saito M, Mizoguchi A, Tanaka Yu, Watanabe M, Ueda E (2011) Study for eco-design method based on users volition. Proceeding of the annual conference of JSSD 58(0):324–325
31. Lockton D, Harrison D, Stanton N (2012) Models of the users: designers' perspectives on influencing sustainable behaviour. J Des Res 10(1/2):8–11

Part VI
Eco-innovation Strategy

The Future of Design for Sustainable Behaviour, Revisited

Casper Boks, Debra Lilley, and Ida Nilstad Pettersen

Abstract At the 2009 Ecodesign conference, the results of a survey on the future of design for sustainable behaviour (DfSB) were presented. In this paper, the survey is revisited, and responses from both surveys are compared and discussed. The contribution of theoretical fields, research priorities, integration in business, and the location and position of DfSB are discussed. The current discourse on behaviour- versus practice-oriented research is addressed, and the paper concludes with thoughts on how DfSB may further mature as a research area.

Keywords Design research • Sustainable behaviour • Sustainable practices

1 Introduction

In recent years, increasing attention has been directed towards the environmental impacts caused during the use phase of products, and the significance of the way people behave and interact with products has been acknowledged. This interest has resulted in a research field investigating how it may be possible to influence people's everyday activities through design, to reduce their environmental burden. This field is often referred to as design for sustainable behaviour (DfSB), although the correctness of this term is debated by, for example, practice theorists. A growing range of case studies have been published (an overview of 28 case studies is presented in Daae and Boks [1]), and ethical aspects of designers and industry influencing user behaviour have also been addressed [2, 3]. Within Northern Europe, researchers in this field have found each other in a research network that meets annually, and several dedicated special issues on the topic have been published in academic journals (see Fig. 1).

C. Boks (✉) • I.N. Pettersen
Department of Product Design, NTNU Norwegian University of Science and Technology, Trondheim, Norway
e-mail: casper.boks@ntnu.no

D. Lilley
Loughborough Design School, Loughborough University, Loughborough, UK

M. Matsumoto et al. (eds.), *Sustainability Through Innovation in Product Life Cycle Design*, EcoProduction, DOI 10.1007/978-981-10-0471-1_46

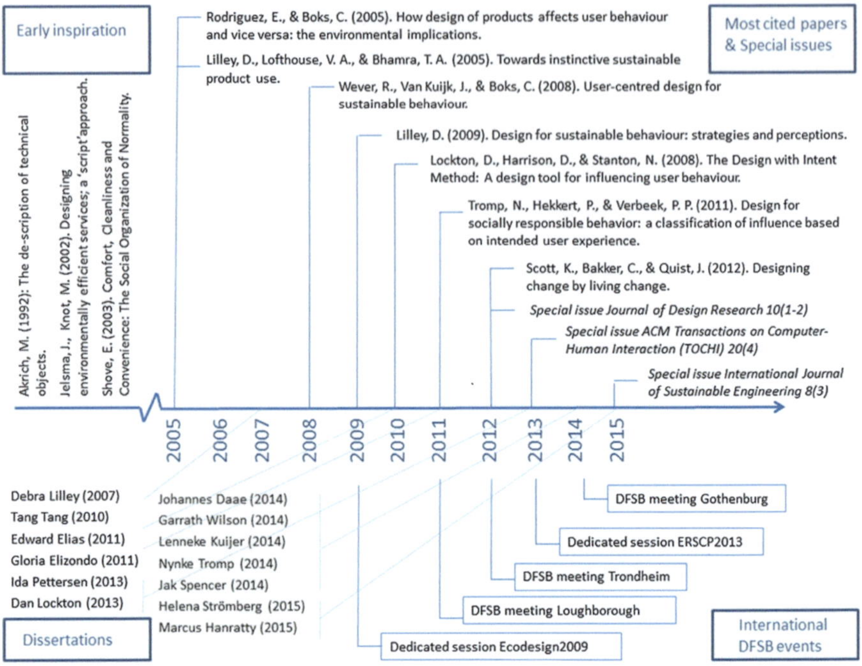

Fig. 1 Visualisation of the important milestones in DfSB research [4–16]

In the proceedings of the 2009 Ecodesign conference, the results of a survey on the future of design for sustainable behaviour were presented [17]. The survey was designed to map the opinions of researchers most active in this emerging field. The study addressed expectations on which academic fields were to contribute significantly to the further development of the field and which topics were most likely to receive external research funding, as well as priorities that should and would be given to theoretical, applied and design-oriented and management- and organisation-oriented research. Another part of the survey addressed questions related to which design strategies would turn out to be most relevant for various (household) practices and behaviours in general, including an assessment of the importance of various factors on the adoption and appropriate use of DfSB strategies. Based on this, a new adapted version of the survey was sent out to scholars within this field, to assess how the field has developed since then, to address future developments and as such document research progress within DfSB research and to potentially fuel discussion on a common future research agenda.

The paper is structured as follows. First, the development of the DfSB field is briefly sketched. Second, the results of the 2015 survey are presented and where appropriate compared to the results of the 2009 survey, leading to a discussion of to what extent expectations of theoretical development and industrial application have come true or not and what expectations exist for the near future.

2 The Past and the Present

In the first decade of ecodesign research, from about 1995 to 2005, there was relatively little focus on the use phase of products and thus on human and social aspects. The vast majority of common ecodesign strategies focused mainly on material aspects, design for disassembly and recycling [18]. Although many ecodesign researchers patted themselves on the back for having a life-cycle perspective, in reality they did nothing more than contribute with end-of-'life-cycle' solutions themselves, focusing on the means to consumption (the product) instead of the practices involved in consumption itself. Research into sustainable consumption has traditionally had relatively little connection to sustainable design and the product level, in terms of research community; these topics have attracted interest from scholars with distinctly different backgrounds and academic perspectives. Strategies related to usage had of course been considered from the early days of ecodesign, such as in tools like the Life Cycle Design Strategy Wheel. But most of these strategies were likewise based on indirect material and end-of-life considerations; lifetime extension for example appealed to postponing the end-of-life stage and avoiding the need for material use. Reduction of energy use focused on using technologies requiring less energy consumption. As such, academic research in the sustainable product design domain has often been done in the 'design engineering' tradition, usually with limited intent to make truly interdisciplinary connections, as in building or extending scientific theory in, from, with or for other scientific domains which may include social sciences, natural sciences or management sciences.

A lot has changed in the past decade however. Nowadays, many scholars see the potential that design research offers to transdisciplinary perspectives into the development of sustainable solutions. The rapidly emerging field of design for sustainable behaviour, which serves as an example of a transdisciplinary enquiry, which investigates, at various levels, how to influence the sustainability impact of consumers' activities, through studying their behaviours and practices, developed over time and in space. As a result, we have seen an academic network develop, which organises international workshops and other ways of scholarly cooperation, several dedicated special issues of acknowledged scientific journals [14–16] as well as doctoral dissertations devoted to this theme. Figure 1 visualises the (arguably) most important academic events related to the development of DfSB as a field of academic interest. It should be noted that the figure represents a very Northern European perspective, which the authors choose to justify by the fact that the bulk of DfSB literature originates from a limited number of Northern European Universities. Adjacent fields, such as sustainable human-computer interaction (HCI), critical design and persuasive technology, focus on similar research questions but do not affiliate themselves with DfSB and are therefore not within the focus of this overview nor this paper.

3 Method and Limitations

Just like in 2009, we received 10 completed surveys, though this time fewer surveys were sent out; surveys were only sent to scholars with a completed or almost completed PhD with a strong relation to design for sustainable behaviour or sustainable practices and to a number of professors that have supervised at least one or more PhD students within DfSB. Only scholars from Northern Europe were addressed, more specifically those that represent or have represented relevant research groups at Loughborough University (UK); the Royal College of Art (UK); Delft University of Technology (NL); University of Twente (NL); Chalmers University of Technology in Gothenburg, Sweden; and the Norwegian University of Science and Technology in Trondheim, Norway. This is obviously a limiting factor, but this was justified by the fact that the majority of scientific publications in the area of DfSB have come from these schools, as mentioned above, and it is this area that we want to investigate in this paper.

The spread among the respondents was 60 % female/40 % male, and the respondents indicated on average 7.3 years of research experience within DfSB.

4 Survey Results by Theme

4.1 Contribution to Theoretical Fields

Respondents were asked their opinion of to what extent 20 different theoretical fields have contributed to the development of DfSB, on a scale from 1 (strongly disagree that it has contributed) to 5 (strongly agree). Figure 2 shows that based on the 2015 survey, three theoretical fields clearly stand out: user-centred design, psychology and persuasive technology. The figure ranks the fields based on their perceived importance in 2015. It can be observed that some clear shifts in opinion have occurred: fields such as sociology of consumption, user experience, human factors, practice-oriented design and co- and participatory design scored considerably lower in 2015 compared to 2009. We should mention however that practice theory was not explicitly mentioned as a separate field and respondents may not have associated sociology of consumption with it. Another observation is that now after 6 years, many more fields received a lower score compared to 2009 (four fields scoring lower than 2).

A second question addressed which fields respondents expect to deliver important future contributions towards the further development of DfSB should receive more attention. The results in Fig. 3 suggest that expectations about user-centred design and PSS/service design have gone up, whereas behavioural economics and industrial ecology are now regarded as less promising. It is also interesting to note that some of the fields mentioned as scoring lower in 2015 than in 2009 for the question on theoretical contributions so far – user experience, co-design and

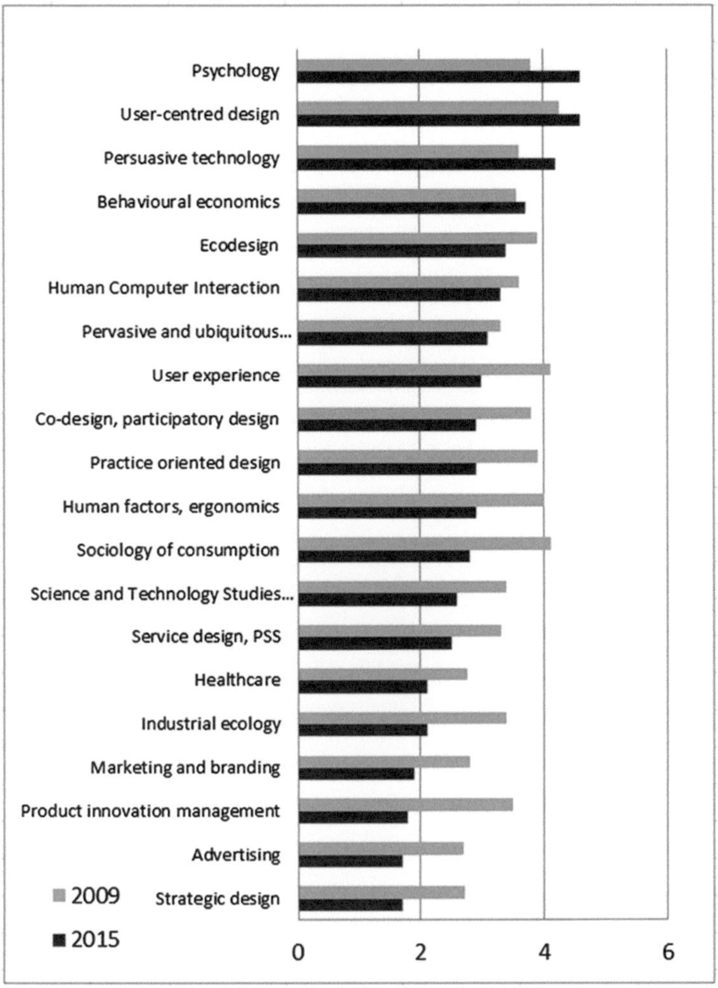

Fig. 2 Importance of past contributions of fields

participatory design and sociology of consumption – are still the ones expected to deliver the most important contributions in the future. Obviously, the respondents expect the most contributions from areas that are traditionally at the core of design research and that already have a strong user focus. The integration of disciplines that are more distant to design research, such as those that are closer to the market (marketing, branding, advertising) and closer to environmental sciences (industrial ecology) or strategy and management (product innovation management, strategic design), is not considered as pressing (yet). Once sufficient evidence will become available that DfSB-inspired products are effective in their goal to contribute to sustainable behaviour, these fields may become more relevant to integrate into DfSB research (see also Sect. 4.3 and Fig. 5).

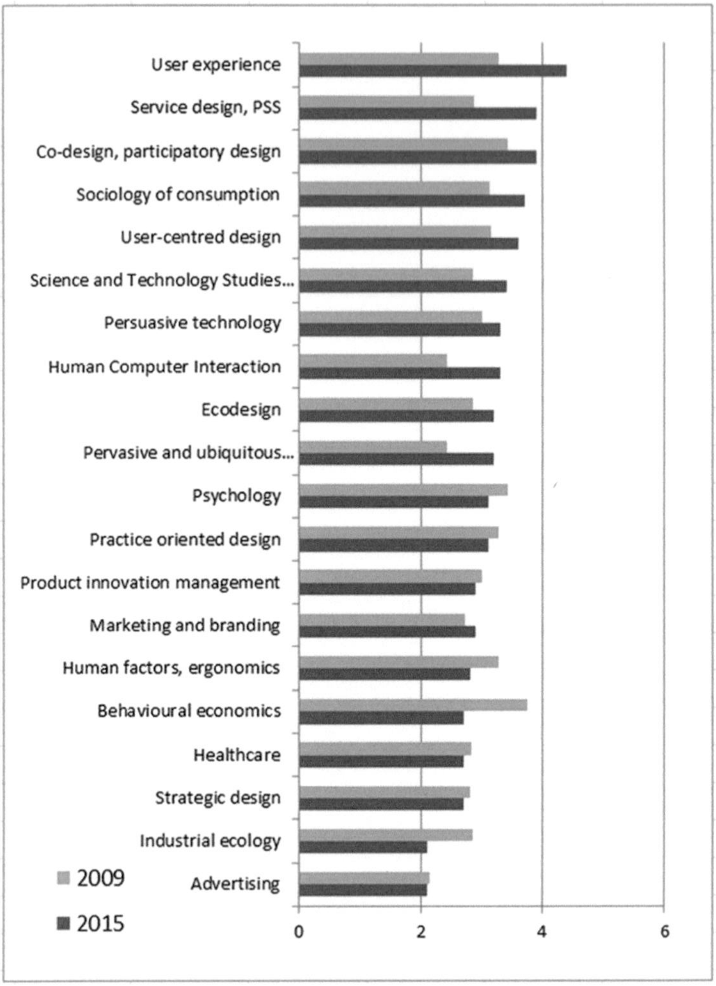

Fig. 3 Importance of expected contributions of fields

4.2 Research Priorities

Another question addressed perceived research priorities; respondents were asked to rank a number of different topics as unimportant or unappealing theoretical topics (score 1) to very appealing and pressing ones (score 5). Also here some interesting developments can be observed (Fig. 4). Evaluation criteria to facilitate the decision making of designers scored highest, with integration into industrial product development practice coming in second place. Perhaps surprisingly, metrics, as in establishing ways of measuring behaviour-related impacts and performance in sustainable product design, scored less compared to 6 years ago, even

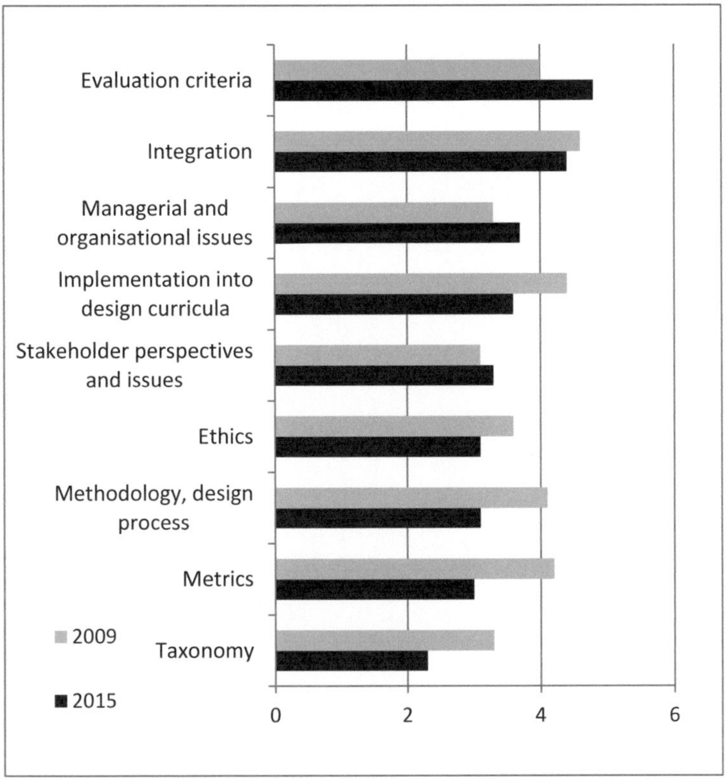

Fig. 4 Perception of importance of topics

though it can be argued that these would support the topic of establishing evaluation criteria, the highest scoring topic. However, it is possible, on reflection, to consider that these two topics were not related to each other by respondents.

Figure 4 also suggests that additional theoretical work on DfSB taxonomies (as in classification and categorisation of design strategies) is clearly something that the respondents see as less pressing than before. This may be a response to the observation that a considerable amount of DfSB research has been devoted to this topic over the past years. A new question in the 2015 survey singled out four of these taxonomies, selected from a previous overview listed in Zachrisson and Boks [19], and asked the respondents whether they would (1) commonly use selected taxonomies as a main reference in/for their research and (2) which one they would prefer to be used as a common reference within DfSB research, both on a scale of 1–5, with 1 = not used/least preferred and 5 = usually used/most preferred. Table 1 illustrates that the taxonomy from Zachrisson and Boks [19] is on average preferred over others; six out of ten researchers (from three different research groups) score this taxonomy with a 4 or 5 on the second question.

Table 1 Preference for various taxonomies

Taxonomy	Descriptions of user-technology interaction								In your research	As common language
Nr. 1 (in [8])	Eco-feedback		Scripts and behavioural steering				Intelligent products and systems		2.4	2.3
Nr 2 (in [20])	Eco-information	Eco-feedback	Eco-spur	Eco-choice	Eco-steer		Eco-technical intervention	Clever design	2.0	2.4
Nr. 3 (in [11])	Thoughtful		Shortcuts			Pinballs			1.7	1.8
Nr. 4 (in [19])	Information	Feedback	Enabling	Encouraging	Guiding	Steering	Forcing	Automatic	3.3	3.8

Respondents replied very similarly compared to 2009 regarding their preferred division of attention for more theoretical research, more applied design research or more management-/organisation-oriented research: preferences for more applied design research has somewhat increased (from 46 to 51 %) at the cost of more theoretical research (from 32 to 27 %) with attention for management-/organisation-oriented research stable at 22 %. This reflects the lack of longitudinal case studies in which strategies have been applied and evaluated versus the aforementioned development of strategies. It also, arguably, points to the need to create an evidence base to prove the success of behaviour-changing interventions in order to make a persuasive argument for their use by industrialists who might require firm confirmation of success in practice or otherwise to be convinced that their use could provide them with competitive advantages, given the absence of related regulations, standards and conventions.

4.3 Integration in Business

Two questions related to the integration of DfSB in industry were included in the survey, just as they were in 2009. Here, respondents were asked about success factors and obstacles for industrial engagement in design for sustainable behaviour, in the context of the 'average consumer electronics multinational'. Though responses were quite similar in both surveys, some interesting shifts could be noted. In 2015, competitive advantage, legislation and innovation were seen as the three main drivers, all scoring higher than in 2009 (Fig. 5). Cost savings was no longer seen as being among the main drivers.

Figure 6 represents how respondents assess a number of potential barriers to integration of DfSB in business. The legend for Fig. 6 is provided in Table 2.

Lack of awareness of 'technology-user interaction'-related sustainability impacts and time and cost pressures in daily product development were seen as the main barriers in both 2009 and 2015 and even more so in 2015. Lack of consensus on acceptable levels of product influence and lack of available tools and methods are seen as less as obstacles now compared to 2009, suggesting that research has made advances in these areas.

4.4 Behaviour Versus Practice

Within the research community there is an interesting ongoing debate about the value of practice theory versus more interaction and behaviour-oriented research approaches. Behaviour-oriented researchers prefer a behavioural psychology perspective to understand (un)sustainable behaviour, placing individual human beings at centre stage. Social practice theorists draw on sociology and advocate a broad, relational perspective, not studying individuals but practices consisting of material

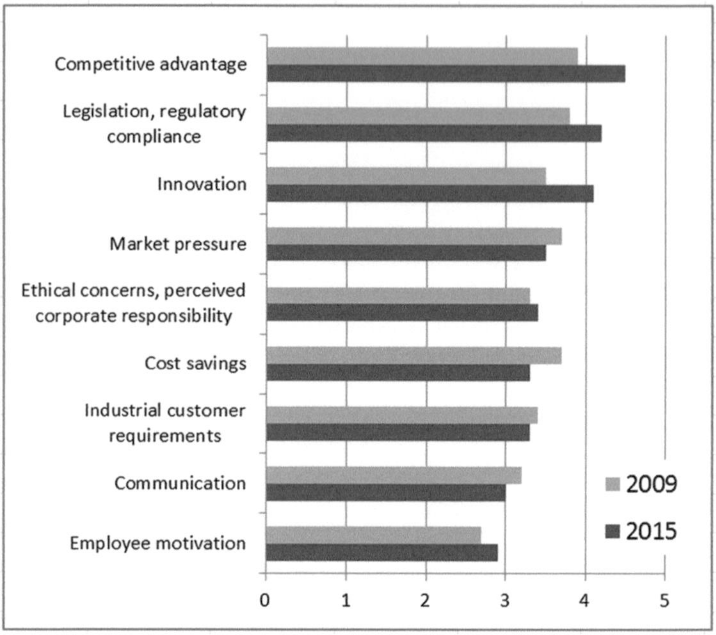

Fig. 5 Main drivers for integration in business

elements, competence and meaning [21] and how they develop in space and time following shared ideas about what is normal. The 2015 survey included a number of statements that respondents were asked to disagree or agree with, using a 5 point Likert scale (5 = strongly agree, 1 = strongly disagree). These statements were loosely formulated and considered to reflect some debates that have been going on, both in literature [13, 20, 22–24] and at conferences. In Table 3 the statements and their replies are presented.

Interestingly, almost all statements were met with both strong agreement and strong disagreement among the respondents. The replies suggest that there is support for the statement that both approaches are complimentary, although they are very different, whereas there are less clear shared opinions about whether research should focus on integrating both. Some respondents added interesting opposing thoughts on this issue. There are some that point out fundamental concerns with behaviour-oriented approaches, such as (1) a focus on incremental savings that tend to disappear in larger trends, (2) a risk of failing to achieve the intended behaviour change, (3) a strong rhetoric of right and wrong behaviours and (4) a risk of missing opportunities on larger scales of change [24]. Other respondents advocate that these include contrived arguments and that positioning practice- and behaviour-oriented approaches against each other is counterproductive; that designers are inherently magpies, picking up whatever little bits of knowledge and theory might help to gain meaning and develop successful solutions; and that both theoretical viewpoints can be useful to achieve this and each can inform the other.

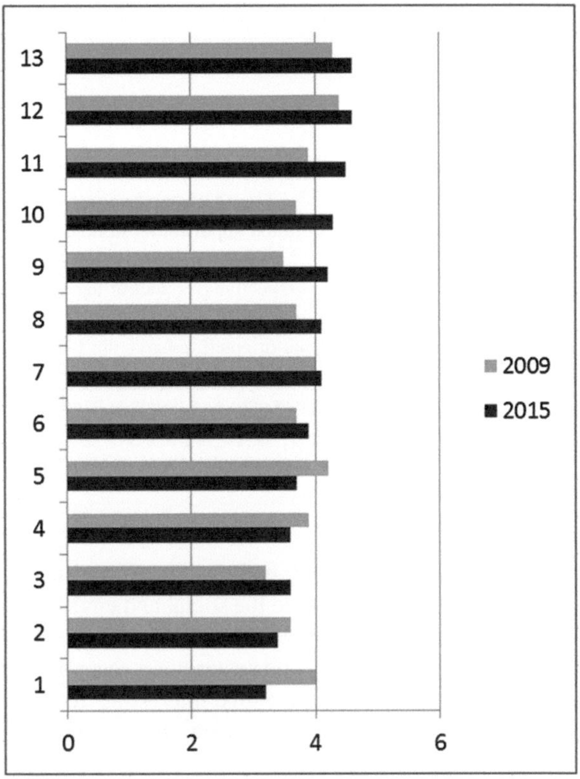

Fig. 6 Main obstacles for integration in business

Table 2 Legend for Fig. 6

1	Lack of awareness of 'technology-user interaction'-related sustainability impacts
2	Time and cost pressures in daily product development
3	Lack of perceived company benefits
4	Limited influence of designers on strategic and managerial issues
5	Lack of perceived company responsibility
6	Lack of know-how
7	Lack of perceived market demand
8	Lack of attention in legislation
9	Lack of appropriate tools and methods
10	Increasing product complexity
11	Lack of proof as to effectiveness of technological influence on behaviour
12	Fear of harming customer relationships
13	Lack of consensus on acceptable levels of product influence

Based on the annotations made by the majority of the respondents, who, granted, mostly affiliate themselves with behaviour rather than practice-oriented research, it is safe to say that most of them interpret and adapt these and other theories in a

Table 3 Scores for statements on behaviour- versus practice-oriented research

Statement	Avg. score	Std dev	Max. score	Min. score
Behaviour-oriented and practice-oriented research are very complimentary	4	1.1	5	2
Behaviour-oriented and practice-oriented research are inherently different	3.6	1.3	5	1
Practice-oriented design has limited applicability in industry contexts	3.5	0.8	5	2
It is beneficial for a broader perspective of issues at hand to integrate both approaches	3.4	1.3	5	1
Behaviour-oriented and practice-oriented research are two sides of the same continuum and meet half way	3.3	1.4	5	1
It is beneficial for a broader perspective of issues at hand to keep exploring both approaches separately	2.8	1.2	5	1
Behaviour-oriented research leads to suboptimal or even counter-sustainable design	2.5	0.8	4	1

broader holistic sense than, perhaps, the disciplines where they originated. If the end goal of DfSB is to reduce the social and environmental impact of activities that individuals engage in through interaction with the world around them, it is implicit that the designed artefacts and systems are acceptable, usable and even desirable to users and therefore commercially interesting for firms to offer. One respondent wrote that a practice level intervention may have a large potential for resource saving, but opportunities for 'radical' change may not be desirable to users for any number of reasons (e.g. comfort, financial, logistical, perceptual, etc.). For that matter, the exact same holds for a behavioural or technological intervention. DfSB, just like any other design discipline, is not beholden to fully adopt any other academic field's knowledge paradigm, particularly in a subject as complex as the fields of human behaviour. These theories are abstractions, simplified ways of viewing an incomplete picture. What is relevant is their usefulness, if any, to the designer and/or design researcher; any theory is only as useful as it proves itself in practice. This debate is likely to be ongoing and indeed should form the basis of further academic discussion.

5 Location and Position of DfSB

Design for sustainable behaviour is obviously an inter- or transdisciplinary research and application field. Although it provides an opportunity for different groups with different backgrounds to work together, the reality is that researchers in this field are often (physically) located in a department in which DfSB is only a research niche. In 2009, the respondents were divided about what the ideal environment would be for a DfSB research group, with 45 % preferring a design for sustainability group and 36 % preferring an interaction design group. Now in 2015, a vast majority of 90 % sees broader design for sustainability research as a preferred research environment. Many respondents were relatively outspoken about the fact that the context in which research is to be performed needs to be design and needs to be sustainability, in order for DfSB to develop in the right direction, provided that the field can interact with a great variety of research environments. It was stated that even though embedding the topic in an interaction design environment could provide a great deal of opportunities, especially when framed as sustainable HCI, its current position which is mostly outside interaction design and HCI has provided a unique position to develop and has led to a distinct canon of knowledge and lens on technologically mediated behaviour. One respondent stated that as a field and compared to interaction design in many aspects, DfSB is more comfortable with the concepts and potential for the Internet of things and physical computing, as it has developed a rich understanding of the agency designed products can exert on our lives. To an extent, being positioned in the area of sustainability also allowed space for the ethical concerns to be explored at a remove from commercial pressures. Similarly, a development outside sociology and psychology has created a very explanatory and descriptive focus which is often inherent to design research.

6 Some Thoughts About the Future

At the end of the survey, some statements were given that probed into the respondents' expectation on how the field will further develop. Interestingly, the analysis of the responses suggests a lack of consensus on a number of issues. There was an equal number of respondents that agreed, disagreed or were unsure about (1) whether the number of PhD students that will work on DfSB themes at Northern European universities will double in the next 5 years, (2) whether DfSB research will turn out to have been a research niche, becoming less visible in 5 years and (3) whether academic DfSB research so far has contributed to commercial DfSB solutions that can be seen on the market today. There were, however, few that expect the launch of a dedicated scientific journal devoted to DfSB.

But there is a strong agreement (save one exception) among the respondents that design for sustainable behaviour and/or practices will be a significant element of their jobs 5 years from now. The senior respondents with tenured positions in academia expect to continue there, whereas most PhD students were unsure whether this would be in academia or elsewhere. This suggests that DfSB is an exciting and inspiring field to work with that is expected to provide challenging research opportunities also within the next years. These could include a broader application of current DfSB approaches. As is reflected by both the 2009 and 2015 survey, so far the (implicit?) background context or goal for much of the DfSB research has been the eventual application of DfSB approaches in industry, resulting in design interventions, on both smaller and bigger scales (but perhaps mostly smaller scale as in products – as illustrated by most case studies available).

In the future, DfSB could focus on the role of other actors relevant to design and use (energy utilities, the public sector, NGOs, etc.) and the development of interventions involving multiple actors. Another approach could be to take DfSB approaches with firms, or other organisations, themselves as actors, as in how these could be informed or forced or most interestingly seduced to integrate sustainability into their activities or study the practices of designers and others, to identify opportunities for influencing the direction in which they develop. But, as the results from the questions about the importance of various theoretical fields discussed in Sect. 4.1 suggest, this is not high on the agenda. Perhaps a re-revisitation of the survey in, say, 2021 will show a different picture.

Acknowledgements We would very much like to thank our colleagues that kindly responded to the survey. In addition we would like to express our appreciation for being part of a very interesting research community.

References

1. Daae J, Boks C (2015) Opportunities and challenges for addressing variations in the use phase with LCA and design for sustainable behaviour. Int J Sustain Eng 8(3):148–162
2. Lilley D, Wilson GT (2013) Integrating ethics into design for sustainable behaviour. J Des Res 11(3):278–299
3. Pettersen IN, Boks C (2008) The ethics in balancing control and freedom when engineering solutions for sustainable behaviour. Int J Sustain Eng 1(4):287–297
4. Akrich M (1992) The de-scription of technical objects. In: Bijker W, Law J (eds) Shaping technology, building society: studies in sociotechnical change. MIT Press, Cambridge, MA, pp 205–224
5. Jelsma J, Knot M (2002) Designing environmentally efficient services; a 'script' approach. J Sustain Prod Des 2:119–130
6. Shove E (2003) Comfort, cleanliness and convenience: the social organization of normality, vol 810. Berg, Oxford
7. Rodriguez E, Boks C (2005) How design of products affects user behaviour and vice versa: the environmental implications. In Environmentally conscious design and inverse manufacturing. Eco Design 2005. Fourth international symposium on, pp 54–61. IEEE
8. Lilley D, Lofthouse VA, Bhamra TA (2005) Towards instinctive sustainable product use 2nd International conference in sustainability, creating the culture, 2–4 November, Aberdeen, Scotland
9. Wever R, Van Kuijk J, Boks C (2008) User-centred design for sustainable behaviour. Int J Sustain Eng 1(1):9–20
10. Lilley D (2009) Design for sustainable behaviour: strategies and perceptions. Des Stud 30 (6):704–720
11. Lockton D, Harrison D, Stanton NA (2010) The design with intent method: a design tool for influencing user behaviour. Appl Ergon 41(3):382–392
12. Tromp N, Hekkert P, Verbeek PP (2011) Design for socially responsible behavior: a classification of influence based on intended user experience. Des Issues 27(3):3–19
13. Scott K, Bakker C, Quist J (2012) Designing change by living change. Des Issues 33 (3):279–297
14. Wever R (2012) Editorial: design research for sustainable behaviour. J Des Res 10(1–2):2012

15. Pierce J, Strengers Y, Sengers P, Bødker S (2013) Introduction to the special issue on practice-oriented approaches to sustainable HCI. ACM Trans Comput-Hum Interact. 20(4) Article 20:1–20
16. Bhamra T, Lilley D (2015) Editorial IJSE special issue: design for sustainable behaviour. Int J Sustain Eng. 8(3):146–147
17. Pettersen IN, Boks C (2009) The future of design for sustainable behaviour. Ecodesign 2009: sixth international symposium on environmentally conscious design and inverse manufacturing, 7–9 Dec, Sapporo
18. Boks C, McAloone TC (2009) Transitions in sustainable product design research. Int J Prod Dev 9(4):429–449
19. Zachrisson J, Boks C (2012) Exploring behavioural psychology to support design for sustainable behaviour research. J Des Res 10(1–2):50–66
20. Bhamra T, Lilley D, Tang T (2008) Sustainable use: changing consumer behaviour through product design, Proceedings of changing the change: design visions, proposals and tools, Turin
21. Shove E, Pantzar M, Watson M (2012) The dynamics of social practice: everyday life and how it changes. SAGE Publications Ltd, London
22. Pettersen IN, Boks C, Tukker A (2013) Framing the role of design in transformation of consumption practices: beyond the designer-product-user triad. Int J Technol Manag 63 (1–2):70–103
23. Kuijer SC (2014) Implications of social practice theory for sustainable design. Doctoral dissertation, TU Delft, Delft University of Technology
24. Kuijer L, Bakker C (2015) Of chalk and cheese: behaviour change and practice theory in sustainable design. Int J Sustain Eng 8(3):219–230

From Eco- to Sustainable Innovation: Approach and Methodology to Guide Design Initiative into the Innovation World

Andrea Gaiardo and Paolo Tamborrini

Abstract The awareness of the fundamental role of innovation in the challenges of global and national society is well known and has been widely discussed. What is less obvious today is which innovation strategy has to be adopted to develop and drive the project onto this path. That is why researchers are looking for, studying and experimenting different types of methodologies. The innovation design lab does the same and believes that the design process initiative is a key to aspirations for a prosperous, innovative future. Innovation and the related design process of creative destruction lead to new ideas, new entrepreneurs and new business models, thus contributing to the establishment of new markets and the creation of new jobs in a sustainable way. Sustainable innovation is, therefore, the key to enable green and growth to go hand in hand, and design has gained the multidisciplinary expertise to drive it in the right way. In this article, we are setting this scenario with the explanation of the state of the art of our methodology, through the presentation of one case study developed inside our innovation design laboratory.

Keywords Sustainable innovation • Environmental education • Sustainable society and scenario

1 Introduction

The awareness of the fundamental role of innovation in global and national environmental, social and economic challenges is well known and has been widely discussed in recent years. The interest in sustainability and its relationship with innovation is not new; however, what is less obvious today is which innovation process has to be adopted to develop and drive the project onto this path. Actors involved in this challenge have understood that innovation comes out from a result of a complex system of relationships among interacting firms, public bodies, institutions, markets and technological opportunities. The outline is an "innovation ecosystem" which is very hard to tackle and handle. Furthermore, the rapid

A. Gaiardo (✉) • P. Tamborrini
Department of Architecture and Design, Polytechnic University of Turin, Torino, Italy
e-mail: andrea.gaiardo@polito.it; paolo.tamborrini@polito.it

M. Matsumoto et al. (eds.), *Sustainability Through Innovation in Product Life Cycle Design*, EcoProduction, DOI 10.1007/978-981-10-0471-1_47

evolution of the worldwide scenario – together with the rise of environmental and social issues, rapid technological development and new business models opportunities – calls for radical answers which have to be more durable, more equitable and more sustainable, with positive impacts on the society. On one hand, this concern has grown and is increasingly acknowledged as essential for all organisations aiming to stay alive [1]. On the contrary, companies increasingly see these factors as opportunities to drive production efficiencies, stimulate growth, reduce cost, improve brand positioning and enhance their business [2]. On this path, many of them have started to use the term "eco-innovation" to describe the contributions of sustainable business development while improving competitiveness. The idea of eco-innovation is fairly recent, but it seems not to be exhaustive because most publications still report modest results in terms of application effectiveness and scope and also in terms of limited effective integration of eco-innovation initiatives compared to the innovation world [3–7]. The main problem which emerges nowadays is how to transfer the theory and eco-design guidelines into the "real world". With regard to eco-design research, whether academic or applied, the key issue continues to be "how to make it happen?" [8]. Although theories and methods are available, in practice, it appears that implementing sustainable design is not an easy task, possibly due to the lack of a holistic approach to the implementation process, from a theoretical and empirical point of view [7]. In general, solutions coming out from R&D laboratories, academic studies and research suffer when the time comes to transform them into tangible and marketable projects. This issue exists for many reasons. The first one is connected to how much the user needs the proposed solution. The second one is related to the context of use, and the third one to the actual capability of the user to use the given solution [9]. Additionally, from the perspective of the enterprises, to modify the traditional productive process and strategy to reach eco-innovative goals seems to be complex and unsustainable [10]. We are witnessing a shift, from the question "how do we do something?" to "why are we doing that?" At present, it seems that users and their context are the starting point and that the path to tread goes backwards, towards technology and business feasibility. All these considerations are increasing attention to the field of design, after not being recognised for years as a relevant factor in the debate on sustainability [11]. Design seems to be the relevant lever to drive, activate and mend the gap with sustainability perspective, technology and business purpose. On this line, our research team founded an innovation design laboratory to research, experiment and develop projects focused on sustainability. The purpose is to find answers to new meanings and emerging needs for those who create innovation in order to progress and prosper (enterprises) and for those who need it to improve their quality of life (user/subject), trying to get closer to the equation Innovation=Sustainability. Accordingly, the article is organised as follows. The first part describes the passage from eco-innovation to sustainable innovation. The second part explains the need and the methodology developed in the laboratory to deal with the case study presented in the last section.

2 From Eco-innovation to Sustainable Innovation

2.1 Definitions

The general definitions of innovation are neutral and open to many directions. The term usually refers to the introduction of a novelty and a change in the way of doing something. Actually, definitions differ when they go more deeply into the description of the characterisation into levels (incremental or sustaining, radical or disruptive innovation) [12], types (product, process, model) and nature (bottom-up, top-down) [13] rather than depending on the content of change. In terms of content of change, it is essential to start and talk about eco-innovation. To achieve it also, several attempts have been made in the literature [14], but setting a univocal definition is not an easy task. Eco-innovation is defined as an innovation that designs new products or processes, which provide value to users and businesses but significantly decrease environmental impacts [3]. Therefore, in a broad sense, the literature on eco-innovation aims to provide a better understanding of the relationship between businesses and environment [15]. However, even these statements do not seem to put an emphasis and give a satisfying innovation content towards a whole sustainable concept. The aim of this concept is to innovate, taking into account the three pillars of sustainability, that is, to say not only environmental issues but also the balance between them and the economic and social aspects of the system into which the innovation takes place. The definition of sustainable innovation seems to be more exhaustive in that case. Sustainable innovation does not only embrace the attributes that characterise a certain type of innovation; it also offers a vision in order to drive progress and give a clear direction to strive for developing sustainable and innovative projects. The term "sustainable innovation" applies to product/service/model and process design able to deliver essential goods and services to pursue the social goals of human health, equity and environmental and economic balance [16]. There has been a long debate on how to transform this definition into an operational way and on how to conceive sustainability more as a heuristic idea. As Carry suggests, "Sustainability is not a fixed ideal, but an evolutionary process of improving the management of systems, through improved understanding and knowledge" [17]. This is the reason why those who are working in this area are striving for finding the right tool or methodology to manage this process within different fields, contexts and times of intervention.

2.2 Sustainability as a Key Driver for a New Innovation Scenario

Nowadays, many firms, companies, institutions or individuals look at sustainability as an inconvenient factor for an unfair global competition, an additional cost for changing established paradigms and as a trouble generator, because of laws and regulations. That is why, from a business perspective, many executives, decision-makers or start-ups consider the possibility to start sustainable initiatives as some sort of divorce [18]. Plus, the system where they act every day is more and more complex and driven by unpredictable aspects. The global growth of markets, new technologies opportunities and emerging cultures are rewriting a system where the old paradigms apparently do no longer work when applied into the contemporary context. The opening up of new possibilities in terms of product and service are breaking the boundaries in every field of human knowledge with astonishing speed, forcing everyone to view itself in a global perspective. The result is a world where local actions have a global influence, and not only vice versa. Whether they like it or not, organisations and societies have to face an apparently distant and elusive international reality, which positive or negative consequences, anyway, are reflected on the local context [19]. Starting from these considerations, it is possible to outline a scenario – built into a sustainability approach – that joins the local with the global (cosmopolitan localism) [20], where sustainability initiatives yield bottom-line and top-line returns in terms of commercial, social and environmental aspects. Moreover, the future scenario will be more and more influenced by bottom-up factors. Nowadays, it has become obvious that implications and changes need to overcome the technical dimension [21]. In fact, in this digital age, we are witnessing a deep-seated need for meaningful solutions, because technological innovations are often complemented and pushed up by non-technological factors. If technology is the "how" and it is, beyond any doubt, a strategic asset, then the "why" is still an open question to be answered. All these aspects increase attention to design as a field of study and as a tool to overcome or mitigate those kinds of issues, giving relevant answers in order to promote sustainable innovation.

3 Design Methodology for Sustainable Innovation

3.1 Design Research and Sustainable Innovation

When it comes to the necessity to innovate "something", theory and practice are different things that have to be blended. If theory sets methodologies and approaches to innovation, practice has to set up tools and processes able to transform intangible purposes into tangible initiatives. The purpose of design research is to support this process. That typically means that research is integral to the design work itself, and inquiries are part of designing, not about the design itself [22]. Design research both inspires imagination and intuition through a variety of methods with related aims: to expose patterns underlying the rich reality of people's behaviours and experiences, to explore reactions to probes and prototypes and to shed light on the unknown through iterative hypothesis and experiments [23]. In order for design to be successful, it must serve the needs and desires of actual humans. Moreover, that is an optimal match point with the concept of sustainability. There is a multifaceted and acknowledged interest in integration, meaning the ability to accommodate a diverse range of activities and individuals with a sustainable view [24]. Furthermore, design research is the strategic tool to offer not only this periscope view but also to make it real. Applying design research thoughtfully, it is possible to understand through which system an innovation will be introduced, its context, its interaction, its relationship with people, territory, product and service and its correspondence with the real needs. Through this process, it is possible to bring together needs, priorities and goals of the actors involved in the system with a real sustainable balance. On account of this, we embrace the definition of innovation as the practice of creating a new viable value/benefit through a sustainable proposal, following the studies of Larry Keely [25] and Roberto Verganti [26]. In this definition, the aim of innovation is to focus on delivering new, viable and concrete sustainable final activities with a tangible value/benefit for all the actors involved. Design with a sustainable perspective opens to new opportunities creating a new economic model, new social engagement and lifestyle improvements with a reduced input of resources into the overall system. Sustainable endeavours are changing, first of all, the vision of how and why we are designed. The context, the focus on reduction and reuse, or the smart allocation of resources, the delivery of better and more suitable product/service to users, the creation of new partnerships with the actors involved in the system, they are all pieces of the innovation system that has to be taken into account to make additional revenues or create new lines of business, solving at the same time social issues such as unemployment, social exclusion, health enhancement, education and so on. Sustainable innovation is, therefore, the key to enable green and growth to go hand in hand, and design [27] has gained the multidisciplinary expertise to drive it in the right way.

3.2 The Need of Methodology

Opportunities coming out from this scenario are encouraging innovators to stand out and to rethink the entire socio-productive system in more sustainable terms before designing a product or a service to be placed in it. But to approach and tackle a sustainable innovation project, it is necessary to count on the right and thorough methodology, in order to prevent and minimise the risk not to find the light at the end of the exhausting innovation tunnel and to create useless and unsuccessful outcomes. Indeed, the design process initiative is a key to aspirations for a prosperous innovative future. Sustainable innovation and the related design process of creative destruction lead to new ideas, new entrepreneurs and new business models, thus contributing to the establishment of new markets and the creation of new jobs. In order to do that, the design methodology of our laboratory implemented the traditional department approach: the systemic design and the experience on eco-design initiatives used and taught in design courses at the Department of Architecture and Design of the Politecnico di Torino applied into the well-known innovation design process to guide the stage of product/service/model design innovation. The core of our approach starts from the analysis of the complex interactions and relationships among different actors (individuals, society, enterprise, culture, territory, etc.) and the related cultural, economic and community area or territory in order to unlock and exploit the innate value of the context as a starting point. The resulting methodology tries to add market value and increase environmental and social acceptance starting from enhancing and mixing the assets of the context where innovation reaches value and great impact results.

3.3 Contexts and Circumstances

The field of possibility within which people establish their life projects is defined by the context in which they find themselves. According to the characteristics of what that defines, we call it a system. This system is made up of users, environments, social, cultural and economic aspects. To understand and summarise all these aspects, we analyse what we define as context and circumstances (or relations). The proposed methodology is therefore based and rooted within the insight of the user context and circumstances. The context embraces all the characteristics and aspects of a particular place where some circumstances are going on. With the term "circumstance", we want to refer to the characteristic actions or relations which take place in a certain context of use, of territory or both. Understanding and knowing these two elements is crucial, strategic and essential in order to obtain successful design outcomes. What context and circumstance do is to determine existing components in term of resources and capabilities which are useful as starting points in order to create or mix new functions and new meanings. These two elements may be different across regions, time-scales, target groups and so on. If we do not consider that there might be some different circumstances across different contexts, we take the risk of thwarting our favourite and successful solutions already used in previous projects and in other contexts. Secondly, knowing these elements in depth, with a contingent statement of what causes what and why, could give the designer the possibility to better formulate the problem or the addressed need, increasing the potentialities of designing the solution in the most appropriate way. Furthermore, this would allow to revalue land resources, to remix or use them in parallel, to enable sustainable innovation initiatives. In practical terms, the first step of the methodology concerns how to make the overall system where the innovation will take place visible and understandable. The interpretation of the state of the art and the recognition of the right lever for innovating are blended with our methodology and the relevant data-driven process.

3.4 Data-Driven Process

Nowadays, more than ever, the new information and communication technologies, and in particular the ability to gather, analyse and disseminate large swaths of data and connect large numbers of people over broad areas, have also enabled greater understanding of complex systems. There is now a range of tools for mapping and better understanding systems that can give a useful insight in order to lead those systems towards innovation [28]. Rearranging numeric data, reinterpreting qualitative information, locating information geographically and building visual taxonomies, we can develop a diagrammatic visualisation – a sort of graphic shortcut – to describe and unveil the hidden connection of complex systems. Visualisation must be open and inclusive, and it must preserve multiple interpretations of complex phenomena [29]. This approach is a strategic way to face the design innovation activity, defined as the process combining all essential steps that lead to innovations, generally involving all the internal source of knowledge generation and learning and the organisational structure and processes of people/organisations committed to the process [30]. The better you know the current state of things and why they are like that, the better you will be positioned to innovate. How is it possible to reach this deep knowledge? We can only understand the complex system through research. Data-driven research is the only tool we can use to build useful information and to interpret the iteration with context, users and circumstances as the main key factors to develop successful designs.

3.5 Design Methodology Process

The design methodology process is described in a general view in the picture (Fig. 1), going through all the stages of the innovation design process. The preliminary step is the definition of the target project, with the description of the topic and context action boundaries. The setup of these boundaries derives from what we would like to achieve starting from an intuition, an idea or an issue to solve. The meta-design phase, or rather the design expectations that will provide indications without the definite solutions to implement the project [31], starts and ends with two iterative processes: the research and the design steps. In this phase, the assignment of the designer is to answer the question about why he is adopting a certain solution, what the solution is and how the adopted solution works. The research step begins with the exploration and collection of broad and tangential information in order to examine the vast array of issues, features and relationships surrounding the system topic. The analysis of this general picture drives into a deep understanding of the topic outlining the real role of all actors involved within their scope, their development and their relations, in their operational context and circumstance. The amounts of data generate the design analysis/data documentation. This output is a fundamental underlayer leading into the second stage of design, where the iterative

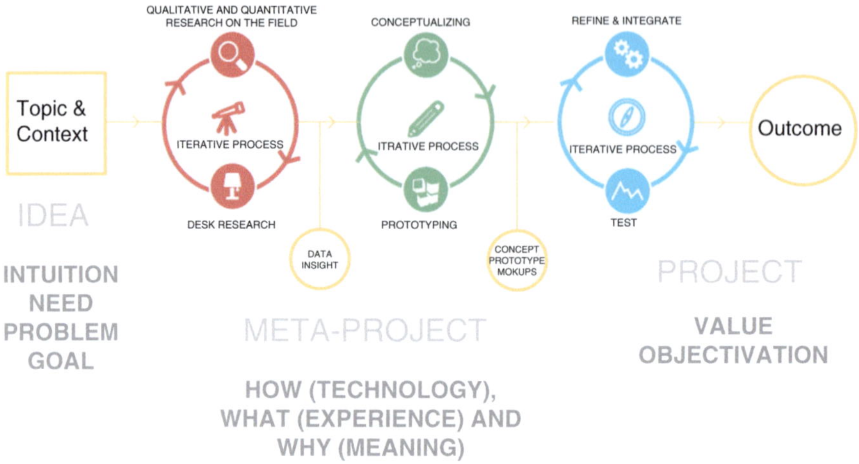

Fig. 1 Design process

process of design starts. At this step, the designer has to develop guidelines and propose solutions according to the previously described exploration research step. All the proposed solutions have to be conceptualised following the sustainable approach and methodology and in accordance with the data insight. The result of this stage has to be a functional prototype or mock-up of the selected concept. The intent of this outcome is to communicate and explain how the concept works and how it answers to the goal and expectation of the project. The last phase is the implementation of the concept in its development and the launch/beta test phase, where the purpose is to produce a valuable objectification, producing and setting useful metrics in order to refine and correct the outcome following the feedback and to measure the impact before delivering the final version of the project outcome.

4 Sustainable Design Innovation: A Case Study

4.1 The Laboratory

Polito Innovation Design Lab [32] is a university laboratory created to research and promote initiatives which help to reach innovative sustainable project and broaden the concept of innovation. The mission of the lab is to manage, coordinate and carry out research projects about products, services or models able to meet real needs of people with regard to territorial potential perspectives of technology, environment, economics, culture and social sciences.

4.2 Case Study: Torino Food Innovation Project

The innovation challenges faced in the laboratory aimed to redefine, reshape and produce impacts and effects with a sustainable perspective. Based on these principles, the Torino Food Innovation (TFI) project was implemented and developed inside the lab. The scope of the project was to train and experiment the ability of the described sustainable design process to enhance a special context or territory, starting from the resources currently present in it. In that particular case study, the TFI scope was focused on studying different districts of Turin and, starting from their problem and resources, introducing a sustainable innovation able to improve the social or the economic or both aspects with the use of food (supply chain and related things) as an enabling factor.

4.3 Torino Food Innovation Development Process

This type of request required a careful contextual analysis to identify the critical situation and the potential for intervention. Indeed, the heart of the project was to transform the existing resources in the food sector, and then it was necessary to know in depth all the infrastructure, the economic relationship, the human value and the cultural assets involved in this system before studying and conceiving any sort of improvement. In order to find out the values of the territories, the different teams analysed different types of qualitative and quantitative information as a quality of natural and built environment, together with identity elements and cultural living conditions, and the economic settlements of the neighbourhoods around the city centre. In particular, this type of research was conducted with a research carried out to investigate the overall context of action in three main fields: environment, socioeconomics and the specific food field. Regarding the first and the second aspects, the project used traditional ethnographic methods to analyse the environment field (natural heritage, infrastructure, public spaces, mobility system, garbage system and quality perceived by residents) and the socioeconomic field (history, economic identity, quality of life, target of residents, culture, events, associations and economic fabric). The food field analysis was conducted instead with the collaboration of Laboratory of Analysis and Territorial and Urban Representations (LARTU) through an experiment testing a digital mapping process (Fig. 2).

The data acquisition was performed in its entirety on the ground, using a laptop and a smartphone app linked to a "collaborative" cloud-based platform allowing users to use, create and share maps, applications and data [33]. The data collected were divided into three main categories:

- Production (urban garden, social garden, etc.)
- Distribution (street food, retail store, lorry owner, etc.)

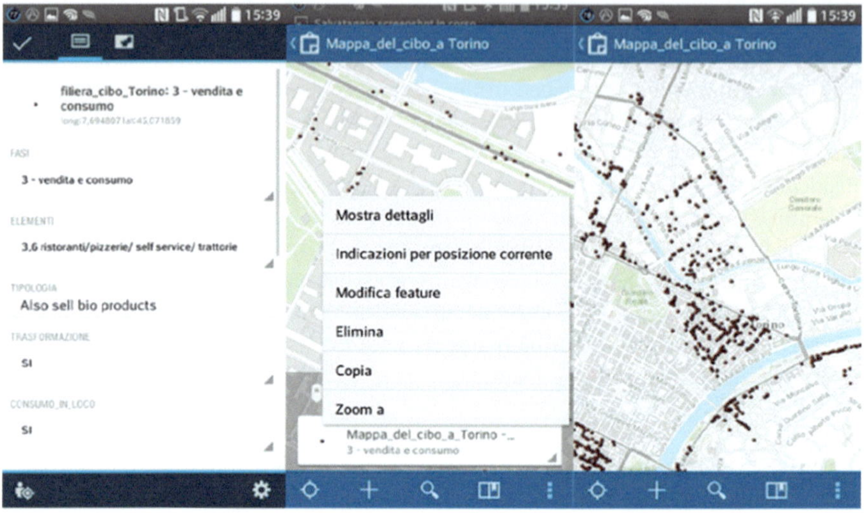

Fig. 2 Screenshot mapping app

- Sale (store, market, hypermarket, etc.)
- Consumption (coffee shop, restaurant, takeaway, etc.)

For each one of them, a distinction was made based on the "nature" of the food in terms of regional, national, ethnic, bio, vegan food, and so on. All the data were put into a map to analyse the distinction and the distribution of the different actors involved inside the food system in the city and in each district (Fig. 3).

The experimentation produced useful data on the chain of production, distribution and consumption of food. It encouraged innovators to reflect on the objectives of the design proposal, with extensive reflections on methods and times of living in the same city. In this way, it was possible to reconnect to large problems, among which:

- Food as a source of well-being and health for different target populations intended both as inhabitants (elderly people, youth, children, foreigners-natives) and as city users (employees, visitors)
- Food as a testimony of culture and traditions rooted in the territory
- Food as a source of innovation in the agroindustrial chain
- Food as a vehicle for interaction and exchange between different ethnic groups and cultures
- Food as an urban polarity and as an "attractor", even against new emerging polarities (universities)
- Food as a symbolic element of sociocultural identity and so on

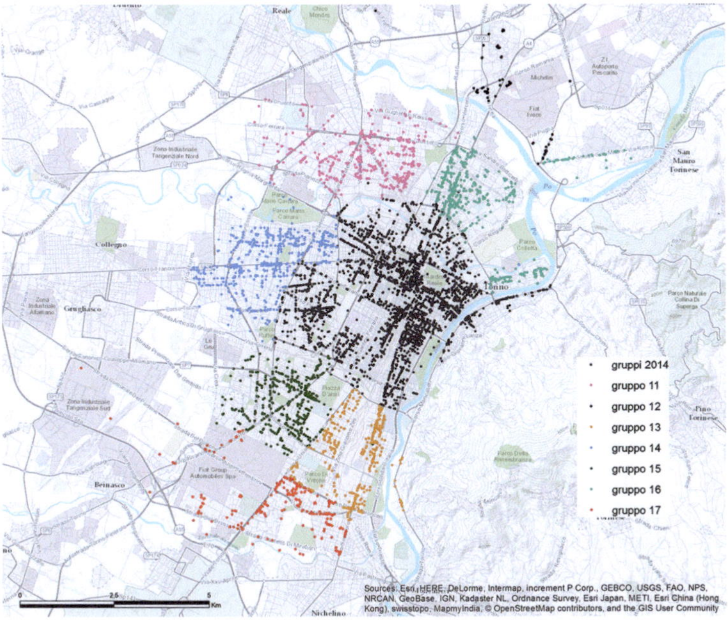

Fig. 3 Overall map results

4.4 Torino Food Innovation Results

Through this full immersion in each district, it was possible to locate the different activities on a geo-database, accompanied and enriched by the quantitative and qualitative information deriving from the ethnographic research. This permitted to obtain an overall picture and knowledge about what, for who and why starting to design solutions with a balance of all the needs inside the reference district context. In other words, acting in a sustainable perspective. This particular project led to the definition and the development of the following specific projects:

- Market Square Borromini: reorganisation and free market with Borgo Po district, city, hill crops, online and off-line for families and the elderly
- TipORTO Park Dora: enhancement of the park through the project of an urban garden, reusing the organic waste of the Valdocco district to produce compost for the garden
- Orabuona, enjoyed the break!: app for service workers and restaurateurs in Cenisia – Cit Turin district for a more organised and varied lunch break, also in relaxed contexts such as parks and gardens
- The Perceived Quality: online tools for the evaluation and promotion of quality perceived by customers in food stores in the Crocetta neighbourhood;

- Bicibo: promotional service in the district of San Salvario, connecting day- and nightlife along the route of a cart with pedals that sells local products and quality aperitifs
- Cibamenti: to integrate and promote healthy and "km 0" food in the district of Vanchiglia and serving students of the Einaudi Campus

To better describe one of the achieved results – the last one – Cibamenti was designed to encourage young people to purchase and have nutritionally correct behaviour experiences thanks to the opportunities offered by the local market. Cibamenti was particularly designed for students as the main target, as one of the biggest campuses of the city present in the district. The aim of the project is double: on one side, to respond to the needs related to student life diet and foster their health, and on the other side, to promote the local economy enhancing the local neighbourhood market (Fig. 4).

The innovation introduced is a service where it is possible to purchase preset shopping bags through an online platform, and they are distributed through a new bank to the local district market. All the preset bags are selected by nutritionists, and they are made with the food sold locally in the market. The intention is to inform and educate users, to simplify the recognition and purchase of products, to provide simple recipes for cooking them. All these elements are marked by a strong social aspect, in order to engage the target audience. The project aims to channel users towards positive eating habits, based on healthy and local food, as selected and quantified according to criteria supported by certifications from experts. In few

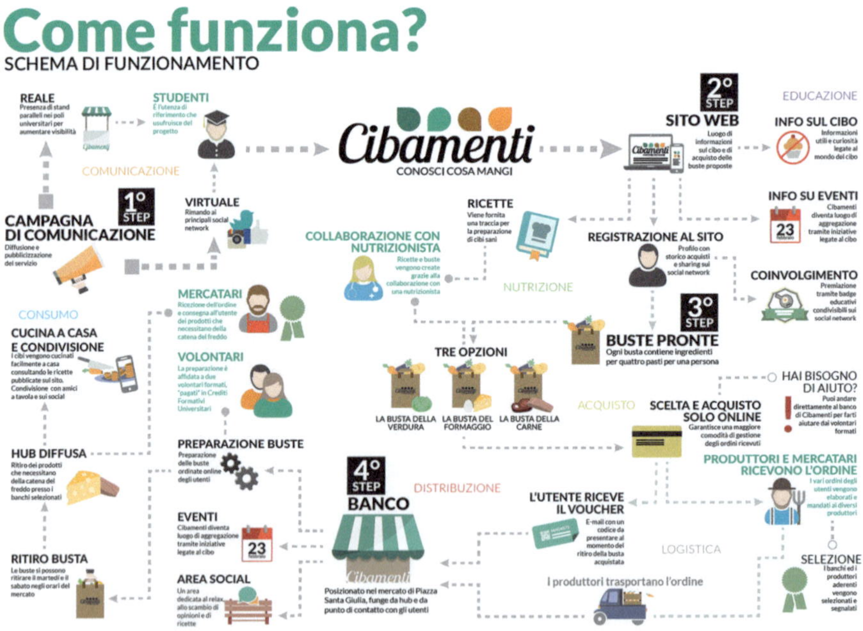

Fig. 4 Cibamenti schema

words, Cibamenti introduces an innovative sustainable service in the area and recombines the possibilities and the resources already present in the context, enhancing the economic aspect (more jobs and more economic activity in the market), the social aspect (food education and social engagement with the local territory) and the environment (less waste of food, regional and local food consumption).

References

1. Nidumolu R, Prahalad CK, Rangaswami MR (2009) Why sustainability is now the key driver of innovation. Harv Bus Rev 87(9):56e64
2. Santolaria M, Oliver J, Gasol C, Morales T, Rieradevall J (2011) Eco-design in innovation driven companies: perception, predictions and the main drivers of integration the Spanish example. J Clean Prod 19:1315–1323
3. James P (1997) The sustainability circle: a new tool for product development and design. J Sustain Prod Des 2:52–57
4. Baumann H, Boons F, Bragd A (2002) Mapping the green product development field: engineering, policy and business perspectives. J Clean Prod 10(5):409–425
5. Deutz P, Mcguire M, Neighbour G (2013) Eco-design practice in the context of a structured design process: an interdisciplinary empirical study of UK manufacturers. J Clean Prod 39 (1):117e128
6. Pigosso DCA, Rozenfeld H, McAloone TC (2013) Ecodesign maturity model: a management framework to support ecodesign implementation into manufacturing companies. J Clean Prod 59:160e173
7. Verhulst E, Boks C, Masson H (2007) The human side of ecodesign from the perspective of change management. In: Proceedings of the 14th CIRP international conference on life cycle engineering, Tokyo, p 107e112
8. Karlsson R, Luttropp C (2006) Ecodesign: what's happening? J Clean Prod 14 (15e16):1291e1298
9. Design-driven innovation – placing user at the centre; design for Europe, European Commission. Mark Nicklas, Deputy Head of Unit, Innovation Policy and Investiment for Growth Intervention. Expo Milan, 24.06.2015
10. Mutlu B, Er A (2003) Design innovation: historical and theoretical perspectives on product innovation by design 5th European Academy of Design Conference, Barcelona
11. Spangenberg JH, Fuad-Luke A, Blincoe K (2010) Design for sustainability (DfS): the interface of sustainable production and consumption. J Clean Prod 18(15):1485e1493
12. Christensen CM (1997) The innovator's dilemma: the revolutionary book that will change the way you do business (Collins Business Essentials). HarperCollins, New York
13. Goffin K (2012) Chapter 6: Sustainability and new product development. In: Cranfield on corporate sustainability. Greenleaf, Sheffield, p 105e118
14. Carillo-Hermosilla J, Del Rio P, Konnola T (2010) Diversity of eco-innovation: reflection from selected case studies. J Clean Prod 18:1073–1083
15. Loorbach D, van Bakel J, Whiteman G, Rotmans J (2009) Business strategies for transitions towards sustainable systems. Business Strategy and The Environment published online
16. Larson A (2012) Entrepreneurship and sustainability v.1.0. http://2012books.lardbucket.org/books/entrepreneurship-and-sustainability/index.html
17. Cary J (1998) Institutional innovation in natural resource management in Australia: the triumph of creativity over adversity. In: Abstracts of the conference 'Knowledge Generation

and transfer: Implications for Agriculture in the 21st Century'. University of California-Berkeley, June 18–19, pp 11–13
18. Nidumolu R, Prahalad CK, Rangaswami MR (2009) Why sustainability is now the key driver of innovation. Harv Bus Rev 87(9):56–64
19. Gaiardo A, Tamborrini P (2015) Systemic Innovation design methodology: the comparison of two cases studies. The Value of Design Research, 11th European Academy of Design Research, Paris
20. Manzini E (2015) Design, when everybody designs, an introduction to design for social innovation. The MIT Press, Cambridge
21. Brones F, de Carvalho MM (2015) From 50 to 1: integrating literature toward a systemic ecodesign model. J Clean Prod 96:44–57
22. Hall E (2013) Just enough research. A Book Apart Publisher, New York. ISBN-13: 978–1937557102
23. Fulton Suri J (2008) Informing our intuition: design research for radical innovation. Rotman Magazine, pp 52–57
24. Tingstrom J, Swanstr€om L, Karlsson R (2006) Sustainability management in product development projects – the ABB experience. J Clean Prod 14(15e16):1377e1385
25. Keeley L, Pikkel R, Quinn B, Walters H (2013) Ten types of innovation. Wiley, ISBN 978-1-118-50424-6
26. Verganti R (2009) Design-driven innovation: changing the rules of competition by radically innovating what things mean. Harvard Business Press, Boston
27. Walker B (2014) A golden age of design. The New York Times. [Online] 22nd September 2014. Available from [Accessed: 15th July 2015]: http://tmagazine.blogs.nytimes.com/2014/09/22/design-golden-age/?_r=0
28. OECD (2014) Data-driven innovation for growth and well-being. Paris
29. Ciuccarelli P (2012) Visual exploration. Online investigation for understanding society, Malofiej, International Infographics Awards, SND-E Society for News Design, Pamplona
30. Filippetti A (2011) Innovation modes and design as a source of innovation: a firm-level analysis. Eur J Innov Manag 14(1):5–26
31. Germak C, Bistagnino G, Celaschi F (2008) Man at the centre of the project, design for a new humanism. Allemandi Press, Turin
32. www.polito.it/innovationdesignlab
33. ArcGIS Online platform (ESRI)

State of the Art of Open Innovation and Design for Sustainability

Ursula Tischner and Lea Beste

Abstract This paper summarizes the state of the art of open innovation and discusses why it is beneficial to connect open innovation activities, methods, and tools with the quest to create radical innovation and design for Sustainability. It is exemplified how this connection can work in practice using the open innovation and design for Sustainability platform www.innonatives.com. The paper concludes with a summary of success and failure factors of open innovation platforms generally and the innonatives platform specifically.

Keywords Open innovation • Radical innovation for Sustainability • Crowdsourcing • Crowdfunding • Success factors

1 Introduction: Crowd-Based Open Innovation and Design for Sustainability

The world is facing a lot of severe Sustainability problems, which seem almost too complex to be solved: from climate change, diminishing biodiversity, via peak oil and resource depletion, to water and food scarcity, poverty, health issues, and social crises. Due to these problems, Sustainability became a well-known notion as being the solution for these issues [1]: it is a topic that is present and discussed around the globe. The willingness to go green is increasing and citizens are aware of problems such as global warming. About 75 % of the German population is aware of climate change, a loss of biodiversity, and the destruction of natural resources [2]. Furthermore, 68 % of Europeans consider climate change to be a serious problem [3]. Citizens want to act in a sustainable manner, but when it comes to picking between certain product attributes versus contributing to a better environment, they almost always decide for the attributes. Therefore innovative ideas have to be generated to make sustainable products as appealing as usual ones [4]. Moreover, there is an

U. Tischner (✉)
Econcept, Agency for Sustainable Design, Cologne, Germany
e-mail: u.tischner@econcept.org

L. Beste
Cologne Business School, Cologne, Germany

M. Matsumoto et al. (eds.), *Sustainability Through Innovation in Product Life Cycle Design*, EcoProduction, DOI 10.1007/978-981-10-0471-1_48

expectation of customers for companies to act in a responsible manner, which includes Sustainability. Out of 100 % participants asked in a survey of the European Commission, 35 % agreed that businesses and the industry are responsible for dealing with the problem of climate change [3]. This clearly shows that not only being innovative but also being sustainable can be used as a competitive advantage to outperform competitors [4].

In the times of a globalized market, it is increasingly difficult for an organization to stand out from competitors. To do so, a competitive advantage is needed, something that contrasts a company favorably from the others. Competitive advantages can be created internally or externally by a firm. Internal changes are often generated by innovation. They not only foster a competitive advantage, but can also be a basis to eliminate competitive advantages of other companies [5]. To answer the need for innovation, the idea was to distribute knowledge from all around the globe via crowdsourcing or so-called open innovation (OI). No one can possess all knowledge alone. Therefore the quest of how to best access and distribute that knowledge led to the implementation of crowdsourcing [6]. "It taps into the knowledge, the creativity, the insight and skill of the world around you. It can help you predict next month's sales, or next season's fashion. It can improve the management of your supply chain. It can enhance your customer's experience of your product and services. It even outperforms the polls in predicting the winners of Presidential elections. And often these improved outcomes can be created with surprisingly modest investment" [6].

The underlying research question of this project was if open innovation and crowdsourcing can be combined to generate radical sustainable solutions and create win-win situations for companies, citizens, and the environment alike.

2 Emerging Tools for Crowd-Based Open Innovation and Design for Sustainability

The project carried out thorough desk, literature, and online research to analyze and evaluate approaches available so far in open innovation literature and online platforms available with the following conclusions.

Various users may have similar problems and demands, as well as very interesting ideas on how to solve them. To avoid the inefficient use of resources and to serve the community, their information and innovations should be broadcasted, freely revealed, and brought together [7]. This is one of the most significant functions of Innovation Communities or so-called open innovation platforms (OIPs). OIPs have been introduced over the last years to facilitate and broaden the innovation process [8]. From the technical view, such online platforms serve as IT-based environments to find new innovators and act as a tool for users to interact [9]. Based on Chesbrough, Komus, and Wauch, and some further sources, Hallerstede [10] created the following definition for OIPs:

> An open innovation platform is defined as a virtual environment that offers digital services, with the aim to allow the creation of innovation by facilitating time- and location independence, voluntary interaction of innovators.

A further reason for the necessity for OIPs is that only through cooperating and multicultural knowledge changes can be made. Especially the needed social innovations toward Sustainability cannot be designed and implemented without the integration of the relevant stakeholders.

Having recognized the trend toward OIPs and the further need for sustainable solutions, the Sustainability Maker project developed and launched the open innovation platform "innonatives.com" (from now on referred to as innonatives), an online platform (OP) promoting the development of open innovation and design projects to solve Sustainability-related problems and implement sustainable solutions as a case study.

Tools that are used in the open innovation context and the innonatives platform are the following:

- Open innovation, a term originally coined by Prof. Henry Chesbrough, meaning "the use of purposive inflows and outflows of knowledge to accelerate internal innovation, and expand the markets for external use of innovation, respectively. [This paradigm] assumes that firms can and should use external ideas as well as internal ideas, and internal and external paths to market, as they look to advance their technology" [11].
- One shortfall of open innovation activities so far were that they are still very much directed toward one company and their benefits but not to benefit society as a whole. Thus, in the open innovation for Sustainability platform innonatives, the term "open innovation" generally stands for innovation activities that involve diverse stakeholders, actors, and experts. For instance, besides of companies, innovation projects can also be initiated by NGOs, municipalities, communities, citizens, networks, etc.
- Crowdsourcing is a term composed by the words "crowd" and "outsourcing" that indicates the act of taking tasks usually performed by contractors (or employees) and outsourcing them to a specific community of people (the Crowd) in systems of mass production [12]. This means in simple words that the general public (the Crowd) is invited and enabled to generate contributions (ideas, innovations, concepts, services, products, etc.) for a company or other organization.
- Crowdvoting, the public or a specific community (the Crowd) votes for challenges, solutions, products, designs to select which of a number of solutions shall be selected. Crowdvoting is also a new form of target group research used by companies to get insight into the public opinion.
- Peer-to-peer (p2p) production is a new form of production (of goods, contents or services) that involves members of communities in an organized way. It is a "coordinated, (chiefly) internet-based effort whereby volunteers contribute project components, and there exists some process to combine them to produce a unified intellectual work" [13].

- Crowdfunding, an approach for generating funds by asking the general public or a specific community (Crowd) to invest in or sponsor activities, such as implementation of new solutions or products, via online platforms, e.g., used by musicians and artists to fund their productions. Different forms of crowdfunding are distinguished such as donation-based or reward-based crowdfunding, crowdlending, and crowdinvestment.
- New business models, emerging via online platforms, e.g., grassroots economies that move the focus from mass production to individualization and production on demand, and on ethical, personal, political, and sustainable values of the goods, mass customization, codesign, and socio-preneurism (social and sustainable entrepreneurism) (see also [14]).

In addition to the new emerging methods and tools for open innovation and crowdsourcing, a growing group of well-educated creative professionals is forming, who are not employed in traditional ways by companies, but are engaged in own creative projects or team up in temporary project groups to carry out divers projects. These professionals use crowdsourcing and crowdfunding sites more and more to realize and implement their ideas outside of the conventional marketplaces, or to be commissioned by traditional companies. More known examples are musicians and artists or designers producing their own pieces of art, music, or designs without involving the mainstream industry and funding production as well as selling them via online platforms. Also the maker and fab lab activities (makezine.com) or craft-oriented do-it-yourself sites like www.craftster.org or www.etsy.com support the very interesting new maker and participatory design movements.

To conclude, most of the individual elements that are used by the new innonatives.com platform are already applied by some actors and in some countries. But none of the previously existing platforms combined all elements and directed all its efforts toward solving Sustainability issues in the holistic triple bottom line approach of people, planet, and profit. The specific innonatives approach developed as a demonstration case study by the project based on the research above is described in the next paragraph.

3 Open Innovation and Design for Sustainability: The Case Study www.innonatives.com

The innonatives platform has been generated as a demonstration case study by the European "Sustainability Maker project" (www.sustainabilitymaker.org), carried out by five core partners and a wider network of universities, companies, and nonprofit organizations. The core partners are "econcept - Agency for Sustainable Design" (manager), "ecosense media & communication," software company "WEBclusive," "Politecnico di Milano," and "Forum soziale Technikgestaltung" (forum for social design of technology). The project is supported by the LIFE financial instrument of the European Union.

The platform "innonatives.com" specializes in innovation and design for Sustainability. From the research carried out in the project it became clear that combining crowdsourcing, crowdvoting, crowdfunding, and an online shop on a single platform to create radical innovation for Sustainability has not been done before. The platform then has been developed by iterative agile software design processes, where parts of the platform have been established, test used, and then improved and then the next parts have been added. Thus the platform is a living model of continuous improvement. The most important functions of innonatives were defined as carrying out innovation Challenges (open innovation, crowdsourcing, and crowdvoting), a crowdfunding module to generate budgets for implementation of sustainable solutions, an online shop to trade sustainable solutions, and an expert and educational system, as well as helpful tools for Sustainability evaluation and development of intellectual property rights concepts. One important aim of innonatives is to enable cross-cultural dialogue of a most diverse group of actors worldwide. The so-called Crowd collaborates for solutions development and exchange of knowledge. Many of the Sustainability solutions needed today exist already somewhere around the world. Often actors in one part of the world simply do not know of the great cases in other parts.

The Challenges (innovation projects) on innonatives had to be designed in a simple and intuitive way (see Fig. 1): To post a challenge, a challenge owner creates and describes his Sustainability-related problem. By proposing this challenge he or she asks the so-called Solvers (community of the platform) to develop solutions. Solvers post ideas and concepts. Crowdvoting and evaluation takes place via the platform. The users who suggested the most promising ideas and concepts will be asked to develop their ideas further and develop a more refined solution. At the end of the Challenge process, all solutions are evaluated by the challenge owner, the experts (internal and/or external experts of the platform), and the Crowd (crowdvoting).

By the use of crowdfunding and the online shop, the winning solutions will then be implemented via the platform, or outside of the platform alternatively, whatever is best and most efficient. All modules can be used independently from each other: the challenge process, the crowdfunding module, and the online shop. However, the platform will always guarantee that all contributions will have a high level of Sustainability.

The platform innonatives defines sustainable innovation in line with the original UN Brundtland definition of Sustainable Development [15] and covers the three dimensions of Sustainability: environment, sociocultural, economic prosperity. Sustainable innovations are defined as new products, services and production-consumption systems that fulfill customer needs and further drastically enhance the social and environmental performance along the whole life cycle and service or production-consumption systems (see also [16]). Sustainable innovation includes not only technological but also social innovation, changing behavior, and how citizens organize daily routines. Social innovation unlike technological innovation does not necessarily need expert knowledge but can be created by nonexpert groups with a good knowledge of the actual situation and the needs and demands of the

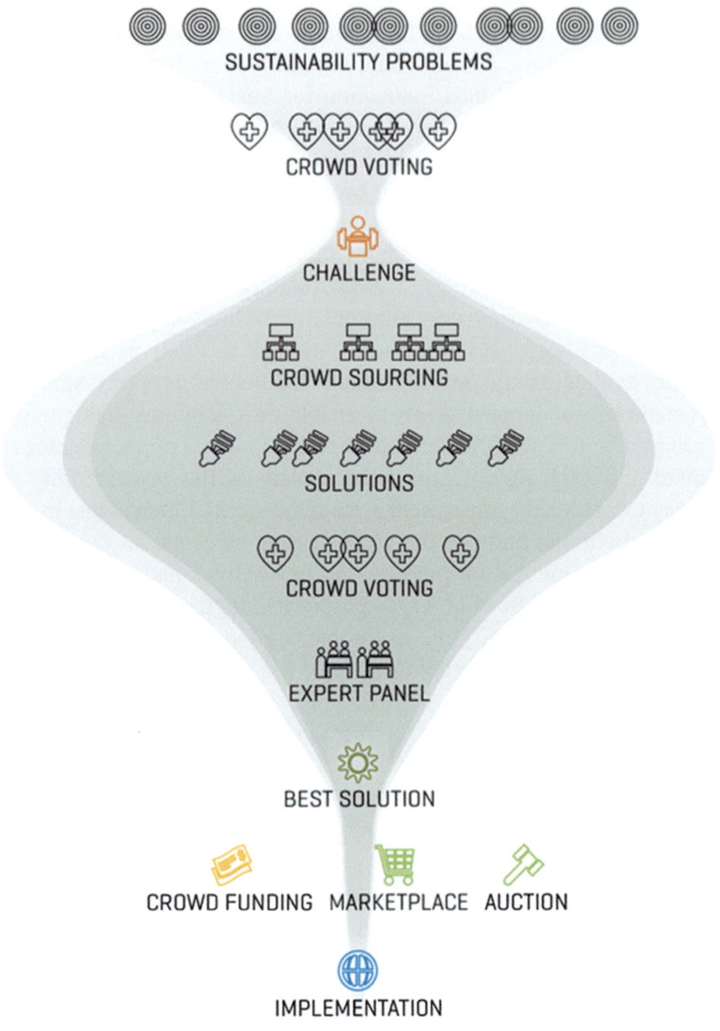

Fig. 1 The standard innonatives process

actors involved. This was one of the assumptions of the project that had to be evaluated by test using it.

A further open question was which business model is "right" for the platform itself. It needed to be sustainable, could be crowd based, and the platform itself could act like a nonprofit organization that invests all surplus gained, beyond the cost of running the platform, in Sustainability solutions generated. It might establish the organizational form of a cooperative also involving the Crowd in decision-making.

4 Why Innonatives.com? About "Wicked Problems" and Possible Solutions

Sustainability-related problems are often referred to as "wicked problems" because they are complex, pose a real challenge to society, and require involvement of many different actors. "Wicked problems" are defined by Rittel and Webber as follows [17]:

- They are difficult or impossible to solve because of incomplete, contradictory, changing requirements difficult to recognize and define.
- They are systemic problems with complex interdependencies and many actors and factors involved.
- It is difficult to analyze wicked problems, and the solution of one aspect may reveal or create other problems.
- There is no ultimate true or false solution; a solution of a wicked problem can only be better or worse.
- Every solution to a wicked problem is a "one-shot operation"; there is no opportunity to learn by trial and error, every attempt counts significantly. Thus, planners are liable for the consequences of the actions they generate.
- Wicked problems cannot be solved by the application of traditional methods but demand creative solutions.

The assumption of the Sustainability Maker project is that by using the open innovation for Sustainability platform innonatives, society might be able to deal with wicked problems because:

- Online open innovation platforms involve a huge group of international diverse and multidisciplinary actors and stakeholders (the Crowd).
- The platform searches for tangible problems within the large complex Sustainability issues that the Crowd can handle, and then another and then another, moving in the right direction.
- Problems to be tackled are proposed, defined, and voted for by the Crowd; thus, there will always be a local and specific problem that a majority of the participants find interesting and well defined.
- Only solutions will be implemented that a majority of participants and an expert panel find worth it and enough actors would like to use. Thus, it is very likely that the state of the art situation is improved.
- Solutions can be tested in niche markets, and then be up-scaled when successful.

The experiences gained so far were positive and lead to the conclusion that indeed some of the effects above can be realized by some of the innovation projects. A more detailed analysis will be available at the end of the EU demonstration project Sustainability Maker in February 2016 (see www.sustainabilitymaker.org).

5 Success and Failure Factors for Crowd-Based Open Innovation and Design for Sustainability

This final paragraph discusses learning outcomes and state of the knowledge about what promotes and might hinder success of open innovation online platforms and more specifically www.innonatives.com.

5.1 Awareness and Accessibility

A successful website has to be known widely and has to be accessible [18]. Anybody having access to the Internet has the opportunity to participate in innonatives any time of the day. The team behind innonatives spreads the word by intensive social media activities and linking to other networks, annual "Sustainability Maker Conferences," newsletters, distribution of flyers, etc. As soon as there is a new challenge planned on innonatives, potential interest groups are contacted via social media platforms. Also the innonatives users are encouraged to create their own communication campaigns. Keeping up with intense media communication constantly is a demanding challenge. It is also necessary to connect local groups with the international community and to create exchange of offline and online communities and allow knowledge transfer. For this, innonatives involves local ambassadors, for instance, schools, where professors and students act as facilitators between local communities and the international online network.

5.2 User Acceptance and Satisfaction

According to Venkatesh et al. [19], three factors are direct determinants of user acceptance and user behavior on the platform. These are performance expectancy, effort expectancy, and social influence. Performance expectancy is outlined to be "the degree to which an individual believes that using the system will help him or her to attain gains in job performance." Effort expectancy is defined as "the degree of ease associated with the use of the system." The social influence is termed as "the degree to which an individual perceives that important others believe he or she should use the new system." Users are not willing to spend more than about one minute of their valuable time to understand how a website works before they leave and will most likely not return. Dickinger and Stangl [20] identified as general success factors for the user acceptance of websites: usefulness, ease of use, enjoyment, website design, trust, content quality, navigational challenges, system availability, satisfaction, perceived value. Thus, design of the website functions and user interface and the establishment of an interesting and positive community of users are essential for innonatives.com. The design team has tried to make the website attractive, e.g., by a color-coding system, as well as intuitively comprehensible, but user testing showed there is still room for improvement.

5.3 Language Issues

In order for participants to be able to communicate with each other, a shared language is critical. At the moment, the majority of communication on the innonatives platform is in English. This has been an obstacle for some users. A temporary solution was the integration of an "online translator" tool. In the long run, the platform shall be available in several languages. Already now, it is possible to open a challenge in another language than English. Concomitant participation is then partly limited to people speaking the language featured in the proposed challenge, excluding all others.

5.4 Transparency and Trust Issues

"Trust is important for information websites, but even more essential for transaction sites" [20]. In an OIP such as innonatives, this refers, for instance, to issues such as payments for crowdfunding and to intellectual property rights (IPR) distribution. IPR management is of outmost importance for the success of an OIP [21]. Therefore innonatives has developed an IPR-contracting module that helps challenge proposers to define fair IPR models and contracts for the challenges. The platform chose for two challenge models: open or closed innovation challenges, with two types of IPR approaches: The open challenges (open to the public) adopt the Creative Commons license "Attribution-NonCommercial-ShareAlike 3.0 Unported (CC BY-NC-SA 3.0)" which users accept upon registration. The principle of this Creative Commons license is that users broadcast their solutions publicly; every innonatives visitor can see them and also use and develop them further, but they have to refer to the originator of the solutions, who owns the IP rights. If an actor wants to use any open solution commercially, a (financial) agreement has to be made with the original inventor.

In closed challenges, specific rules for copyrights and intellectual property rights distribution have to be created, published, and signed before entering into the closed challenge. This process is transparent and accepting it is a precondition for participation. Thus, all users know, what their rights and duties are, when joining any challenge.

5.5 *Motivation of Users*

According to Antikainen [18], several different factors influence the usage of an OP which have been taken into account at innonatives as follows:

- Knowledge exchange and discussion with members to extract knowledge and gain new input or learn something, concerning giving information, advice and experience, professional or nonprofessional: innonatives' challenges ask for exchange of knowledge. Discussions take place and a team can be formed to collaborate.
- Commercial activities: innonatives offers financial rewards in challenges and creates funding opportunities through crowdfunding and the opportunity to trade in the online shop.
- Dating or meeting new people: at innonatives, it is possible to meet new like-minded people from different backgrounds and from all over the world. This enables users to benefit from networking professionally or finding interesting creative people, and creates knowledge transfer as well as encourages cross-cultural understanding.
- Playing and gaming: making it more fun for people to participate: There are some gaming elements integrated in the innonatives platform such as the "reputation management system." Members are ranked according to their contribution level and receive higher levels of recognition the more active they are. This creates also a slight element of competition. Having an interesting avatar (user representation) at innonatives can also add to the fun of the website. Users can choose their display name and picture freely.
- Similarity of users, which "refers to the extent to which potentially relational partners perceive or feel psychologically, morally, or emotionally close or accustomed to each other" [18]. This also counts for similarities in experience and background. Innonatives, being an OIP for Sustainability issues, brings together users that have at least one thing in common: the wish for creating sustainable solutions. They all work toward a similar goal. This is one of the most favorable success factors of the platform. It makes it unique, uniting all users in the same important mission.

5.6 *Picking the Right Crowd and Providing Guidance*

This success factor presented by Hopkins [22] reflects that only approximately 1 % of a crowd will actively participate in crowdsourcing (although the 90-9-1 % Rule is questioned now for social media participation). Considering this, it is crucial to start with a large number of participants, expert and nonexpert audiences. The Crowd should at best have a diversity in thinking which implies specialty knowledge. In innonatives, high value is placed on the diversity of the Crowd and the experts on the platform. The core target group was selected as creative students, young creative professionals, Sustainability experts, and activists, generally actors who care for Sustainability and innovation, SMEs and NGOs. Large companies often carry out open innovation activities already; the smaller companies normally are not able to run own innovation platforms and thus benefit from using innonatives for their innovation projects.

5.7 *Directing, Guiding, and Valuing the Crowd*

Innonatives also takes care of the fact introduced by Hopkins [22] that "crowds require direction and guidance and someone to answer their questions." Users can feel a valued part of the community and are guided by the team and expert groups.

5.8 *How to Formulate Innovation Challenges*

Hopkins [22] also suggests that challenges should be kept simple and be broken down in the smallest manageable components. As many of the users participate in their free time, the tasks should be small, simple enough, and fun enough to be worked on in the spare time available. Thus, challenges on innonatives are selected to have lower complexity and high relevance. Challenge owners are encouraged to define clearly what they want, what they do not want, and what they already know. Meaningfulness is highly important in all innonatives challenges. As Carpenter [23] specified, understanding the case for which a new innovation is needed is crucial to the potential participants. The challenge owner should describe the change that can be made by finding solutions for the given problem. Choosing interesting topics right from the beginning on is essential to encourage active participation. Innonatives challenges are selected to offer highly relevant topics like reducing food waste, banning plastic bags, upcycling waste materials, improving quality of life in poorer communities, etc.

5.9 *Incentives, Rewards, and Achievement*

Having analyzed twelve OIPs, Antikainen and Väätäjä [24] found out that 62.6 % of 100 % respondents state that monetary rewarding encourages their participation on OIPs. 57.2 % state that nonmonetary rewarding would also encourage their contribution. Furthermore, publicizing of rewarded users on the websites is crucial to many respondents, as well as a ranking list showing the quality of ideas. For some innovators, it is more important to have their name and innovation well distributed in a good light. They want to increase their public image [25]. Innonatives acknowledges this by offering financial rewards, royalties, contracts, and other material benefits but also immaterial rewards such as being promoted and receiving reputation points. It seems that Sustainability is a matter of particular interest for the users; they prefer to engage in meaningful projects. Making a change by creating sustainable solutions and doing that together with interesting and creative people offers high satisfaction.

Finally, giving an "opportunity to exercise one's existing skills toward a goal, learn new ones, and receive feedback for improvement" [23] is also an element of innonatives, as the platform offers knowledge exchange with experts through the internal expert system, and educates on using specific tools such as for Sustainability evaluation of challenges and solutions.

6 Summary

This manuscript presented the state of the art and connection of open innovation platforms and radical innovation for Sustainability. It used the platform www.innonatives.com as a case study to discuss success and failure factors for open innovation and design for Sustainability platforms. Preliminary conclusions show that indeed the Crowd can come up with Sustainability ideas and solutions and a platform like innonatives can be an enabler for that. More conclusions and results of the European demonstration project Sustainability Maker will be published on the project website toward the end of the demonstration phase in March 2016 (see www.sustainabilitymaker.org).

References

1. Saarinen J (2006) Traditions of sustainability in tourism studies. Ann Tour Res 33 (4):1121–1140. doi:10.1016/j.annals.2006.06.007
2. Weber, M. (2008). Klimawandel und Klimabewusstsein: zur Schwierigkeit der individuellen Wahrnehmung globaler Umweltprobleme [Climate change and climate awareness: for the difficulty of the individual's awareness of global environment problems]. Social Science Open Access Repository (Ssoar), pp 3602–3615. Retrieved 7 July 2014 from http://nbn-resolving.de/urn:nbn:de:0168-ssoar-155531
3. European Commission (2011). Climate change report. Special Eurobarometer 372. Retrieved 12 May 2015 from http://ec.europa.eu/public_opinion/archives/ebs/ebs_372_en.pdf
4. Ginsberg JM, Bloom PN (2004) Choosing the right green marketing strategy. MIT Sloan Manag Rev 46(1):79–84
5. Grant RM (2008) Contemporary strategy analysis. Blackwell, Malden
6. Chesbrough HW (2011) Foreword. In: Sloane P (ed) A guide to open innovation and crowdsourcing. Advice from leading experts. Kogan Page Limited, London, pp 76–84
7. von Hippel E (2005) Democratizing innovation. MIT Press, Cambridge, MA
8. Battistella C, Nonino F (2012) Open innovation web-based platforms: the impact of different forms of motivation on collaboration. Innov: Manag Policy Pract 14(4):557–575. doi:10.5172/impp.2012.14.4.557

9. Komus A, Wauch F (2008) Wikimanagement: was Unternehmen von Social Software und Web 2.0 lernen können. [Wiki management: what companies can learn from social media software and Web 2.0]. Oldenbourg, München
10. Hallerstede SH (2013) Managing the life cycle of open innovation platforms. Springer Gabler, Wiesbaden
11. Chesbrough HW, West J, Vanhaverbeke W (2006) Open innovation. Researching a new paradigm. Oxford University Press, Oxford
12. Howe J (2006) The rise of crowdsourcing. Wired Mag 14:1–4
13. Benkler Y (2006) The wealth of networks: how social production transforms markets and freedom. Yale University Press, New Haven, p 277
14. Gassmann O (ed) (2010) Crowdsourcing. Hanser Verlag, München
15. Brundtland GH (1987) Report of the World Commission on environment and development: "our common future". United Nations, New York
16. Belz F-M, Peattie K (2012) Sustainability marketing. A global perspective. Wiley, Chichester
17. Rittel HWJ, Webber MM (1973) Dilemmas in a general theory of planning. Policy Sci 4:155–169. Accessed at http://www.uctc.net/mwebber/Rittel+Webber+Dilemmas+General_Theory_of_Planning.pdf
18. Antikainen MJ (2007). The Attraction of Company Online Communities. A multiple case study. Doctoral dissertation. Retrieved 7 May 2015 from https://tampub.uta.fi/bitstream/handle/10024/67697/978-951-44-6850-6.pdf?sequence=1
19. Venkatesh V, Morris MG, Davis GB, Davis FD (2003) User acceptance in information technology. Toward a unified view. MIS Q 27(3):425–478
20. Dickinger A, Stangl B (2013) Website performance and behavioural consequences: a formative measurement approach. J Bus Res 66(6):771–777. doi:10.1016/j.jbusres.2011.09.017
21. Meige A, Golden B (2011) Multistep dynamic expert sourcing: a novel approach for open innovation platforms. In: Sloane P (ed) A guide to open innovation and crowdsourcing. Advice from leading experts. Kogan Page Limited, London, pp 189–202
22. Hopkins R (2011) What is crowdsourcing? In: Sloane P (ed) A guide to open innovation and crowdsourcing. Advice from leading experts. Kogan Page Limited, London, pp 15–21
23. Carpenter H (2011) Motivating the crowd to participate in your innovation initiative. In: Sloane P (ed) A guide to open innovation and crowdsourcing. Advice from leading experts. Kogan Page Limited, London, pp 76–84
24. Antikainen MJ, Väätäjä HK (2010) Rewarding in open innovation communities – how to motivate members. Int J Entrep Innov Manag 11(4):440–456. doi:10.1504/ijeim.2010.032267
25. von Hippel E, von Krogh G (2006) Free revealing and the private-collective model for innovation incentives. R D Manag 36(3):295–306. doi:10.1111/j.1467-9310.2006.00435.x

An Analysis of the Ecodesign Scientific Network 1994–2014

Jacobus Marthinus van der Bank, Casper Boks, and Johan Braet

Abstract Although the field of ecodesign has been described by many different types of scientific review, an extensive network analysis has not been provided in literature. The authors explore collaborative network study techniques that are able, based on an extensive database of scientific articles, to visualize how networks have emerged, highlighting coauthor relations. The paper answers questions related to what appear to be and have been the main collaboration networks in the past 20 years. It is found that a giant main cluster of coauthors exists, surrounded by several smaller clusters. By highlighting key authors at different time periods, it was also found that a core group of authors exists who contributed to the field from the start and are still active in the field.

Keywords Ecodesign • Scientific network • Coauthorship network

1 Introduction

With ecodesign having become a well-established research field studied at higher education and research institutes across the world, many studies have attempted to characterize this field. Such studies typically address emerging trends in research such as cross-disciplinary research, describing how new scientific fields add to increased understanding, how terms and definitions develop over time, and how new industries, firms, and application areas have been or can be successfully targeted. However, a characterization of the field from a collaborative network perspective has not been attempted yet. Such an approach can potentially provide insights into where the main networks exist and identify key authors and author

Jacobus Marthinus van der Bank is also at the School of Business and Finance, University of the Western Cape, South Africa.

J.M. van der Bank (✉) • J. Braet
Department of Engineering Management, University of Antwerpen, Antwerpen, Belgium
e-mail: jaco.vanderbank@uantwerpen.be

C. Boks
Department of Product Design, Norwegian University of Science, Trondheim, Norway

M. Matsumoto et al. (eds.), *Sustainability Through Innovation in Product Life Cycle Design*, EcoProduction, DOI 10.1007/978-981-10-0471-1_49

groups, how global a research community is, and how these developments may have changed over time.

In a scientific network, a research paper forms the basic research unit. Research papers are selected based on a set of criteria and aggregated into various higher research units, i.e., author(s), institution, journal, keywords, etc. These research units are denoted by nodes to form a network. A link normally represents a type of relationship such as citation, coauthorship, co-citation, bibliographic, or co-word [1]. This study draws from previous scientific network studies and implements methods put forward by them to try and map the scientific and social development of ecodesign over a 20-year period.

2 Paper Structure and Flow

The remainder of the paper is structured as follows: In Sect. 3, we provide a brief literature review on ecodesign, to provide the context and definitions surrounding the study. Furthermore, we provide an overview of work done on characterizing the field. This leads us to a number of research questions we seek answered in our study. In Sect. 4, we introduce network studies and provide a brief overview of work done on coauthorship networks. In Sect. 5, we describe our dataset in detail. We explain issues surrounding author ambiguity and how we dealt with it. In Sect. 6, we perform different statistical analyses on the complete network. In this section, the focus is on both the full network and time periods/snapshot or evolution of the network.

In Sect. 7, we revisit our research questions and attempt to answer them by our analysis of the networks. We conclude the paper, highlighting key results and limitations of the study.

3 Different Ways to Review Developments in Ecodesign

Ecodesign is a concept that integrates multifaceted aspects of design and environmental considerations into the product development process [2]. The emphasis is placed on creating solutions with lower environmental impact by integrating environmental aspects within the product development process while simultaneously adapting a life cycle perspective. There seems to be consensus about the fact that Ecodesign, as a research field, emerged in the early 1990s, when it was introduced as a promising approach to sustainable production and competition [3]. One of the first milestones was provided by [4], who published a manual, cowritten with the United Nations Environment Programme (UNEP), with a step-by-step approach to ecodesign for companies. It is however beyond the scope of this paper to provide a history of the field itself; instead, we wish to sketch how the field has been described in literature.

Review papers are probably the main method used to provide a holistic view of ecodesign research. In such papers, scholars have usually provided either an overview of obstacles and success factors for implementation of ecodesign [5–10] or provided an overview of tools and methods for ecodesign [11, 12], analyzing weaker and stronger points depending on the chosen application. Hernandez Pardo et al. [13] present a framework that intends to characterize and classify 40 ecodesign projects extracted from the literature. Boks and McAloone [14] have sketched the field of ecodesign, or sustainable product design, as having gone through four transitions, each broadening the field and making it more complex: from opportunistic to realistic research, from a product to a systems perspective, from an environmental to a sustainability context, and from a concept development to technology transfer and commercialization perspective. In 2001, Boks et al. [15] took a more bibliometric perspective by categorizing close to 3000 references in the ecodesign field in an attempt to analyze where academic debate occurred at that time (conference proceedings versus journals), what platforms were most popular, etc.

In 2009, Boks [16] provocatively provided a rather subjective description of ecodesign research(ers), categorizing them as "blood types": The analytical blood type, describing this group as "engineering and technology focused," "legislation compliance oriented," focusing on topics like toxicity control, advanced life cycle inventory data collection, and aligning end-of-life technology (development) with product design (requirements); the facilitative blood type, representing a more management-oriented approach to sustainable product design, focusing on performance measurement, stakeholder involvement and supply chain management, corporate social responsibility, and issues of implementation and operationalization; and the creative blood type, representing research borrowing from a wide range of disciplines that share perhaps an innovative ambition and focus.

All of these reviews serve their own purpose, but a number of interesting research questions can be formulated that neither the reviews nor characterizations (provided in this section) can be objectively answered, but for which collaborative network analysis is highly suitable. These questions are:

1. What appear to be the main networks for the scientific ecodesign community?
2. Is it possible to identify key authors or author groups that connect networks within this community?
3. Do answers to these questions vary when looking at different decades?

4 Collaborative Network Studies

With the popularity of network studies, accelerated by the discovery of small-world and scale-free networks, a growing trend has emerged in studying the different types of scientific networks. Social networks have been widely studied from different perspectives [1, 17–22, 25] over the past 50 years. Before the turn of the millennium, social network studies were mostly restricted to small systems and were often viewed as static graphs, whose nodes and links represented various quantifiable social interactions [18, 23]. This took a turn when methods derived from statistical physics moved the focus to large networks and allowed for the systematic development of new methodologies. The different methodologies along with the theoretical and empirical results created new and unimagined possibilities for research and a wealth of applications in many fields ranging from computer science to biology [18]. Barabási [18] identified important results stemming from the above. Firstly, most networks inherit the small-world hypothesis [17, 18, 24]. According to the small-world hypothesis, the average separation between nodes is rather small. Therefore, you should be able to find a short path along the links between most pairs of nodes. Secondly, according to [1, 17–19, 23], real networks display a degree of clustering higher than expected for random networks. Lastly, the degree distribution contains important information about the nature of the network [18, 23]. This has led to interest in studying scale-free power-law distribution of networks.

Newman [19, 20] took an important step in bringing modern network ideas to collaboration networks. Furthermore, the study of Barasi [18] built on the work of Newman argued that collaboration networks should be viewed as a prototype of evolving networks, where the accent is on the dynamics and evolution of the network. They proposed a continuum model for the evolution of dynamic complex networks [23]. This paper follows the same approach in which the coauthorship network constantly expands by the addition of new authors to the database, as well as the addition of new internal links representing papers coauthored by authors that were already part of the database [23]. The topological properties of the network are determined by these dynamic growth processes [18].

4.1 Network Approaches and Types

Researchers have approached scientific collaboration networks from the following perspectives: macrolevel statistics, mesolevel clustering techniques, and microlevel indicators [1]. Macrolevel statistics are useful in identifying the global structure and features of the network. Mesolevel approaches focus on the behavior of a group of actors, and various clustering techniques can be classified into this category. Lastly, microlevel indicators are useful in understanding an individual node's power, stratification, ranking, and inequality in social structures [1, 19]. Furthermore, the interaction of research aggregates (author, topic, year, location) can be examined from different types of scientific networks. Scientific networks, the principal focus of this study, are representative of a type of social network. Scientific networks focus on finding patterns of constructs or intersections between social actors [1]. Scientific networks, such as coauthorship networks, can be seen as a complex dynamic network [18, 23]. In this paper, the main question for the network is "How do they evolve?" [18, 23]. It examines the coauthorship network from all three perspectives, thereby providing multiple lenses to individually zoom in on the different perspectives of micro-, meso-, and macrolevel, providing a holistic view when combined.

5 Dataset Description

The network was constructed using the Scopus electronic online database. Different search terms and combinations of search terms were used to create the dataset for the period of 1994–2014. For a list of search terms used, see Table 1. The dataset consists of 4493 journals and conference papers, spanning the previous 20 years, that contained one or a combination of the search terms either in the title, the abstract, or as a keyword. The papers stem from a variety of fields ranging from environmental studies, management, engineering, and product development. Each paper was provided with a unique number to ensure correct cross-referencing. The dataset of papers included the following: (1) author(s), (2) title, (3) publication date, (4) abstract, (5) keywords, (6) source, and (7) journal article/conference paper. The data in the dataset were processed in several steps. First, all the duplicate papers were identified and deleted from the dataset. The second step was to attain uniformity among the authors. We experienced similar difficulties as [1, 17, 18, 20, 23], with identifying unique authors.

First, the same author might be identified in different ways. For example, Jelle Alexander Meysman might use JA Meysman in one paper, J Meysman in a different paper, and Jelle Alexander Meysman in the third paper. If this issue went unchecked, redundant nodes (three different authors) will form in the network creating inaccurate links. Therefore, to reduce the redundancy and eliminate double nodes within the dataset, we have revised author's names and converted them to a uniform format. A manual process was implemented, whereby author's names were double-checked using the Scopus database. In some cases, two distant authors shared the same initial and surname, making it impossible for the programs (Cytoscape, VOSviewer) to identify them as two different authors. In such a scenario, we added the author's affiliations, thereby creating two unique authors. The error is mainly important for two main reasons. Firstly, this error is important for authors of eastern decent. Secondly, it would influence the final network by inflating the coauthorship network of one of the authors. Newman [17] showed that the error introduced by those problems is of a few percent. Our results are also affected by these methodological limitations; however, we do not expect it will have a significant impact on the results due to the manual editing of the authors' names using Scopus.

Table 1 Search terms employed in the creation of the dataset

Ecodesign	Sustainable product development
Environmental design	Sustainable product innovation
Environmentally conscious design	Environmental product innovation
Green design	Sustainable design
Green product design	Design for environment
Green product development	Design for sustainability
Environmentally friendly design	Design for sustainable behavior
Environmentally friendly product design	Environmental product development
Product service systems	Life cycle design
Sustainable product design	

5.1 *The Construction of Networks*

Using the dataset obtained from Scopus, we procured two different interlinked networks: Publication and Coauthorship networks.

For the coauthorship, matrix multiplication was used to generate the network. Similar methods were used by [1, 23] in their analysis of networks. The publication network was used to create the coauthorship network.

5.2 *Publication Network*

"Which author wrote which paper?" is the main question in the publication network. The network is the most basic of the two and consists of the article that forms the basic research unit.

5.3 *Coauthorship*

The publication network was used as the basis to build the coauthorship network. The aim of the network is to determine which author writes papers with whom [23]. To build the coauthorship network, the following formula was used, in which the coauthorship network is denoted by C, where C (i,j) = 1 if author *i* is an author of paper *j* and 0 otherwise, resulting in a sparse matrix. In our coauthorship network, all authors listed on a paper are connected with one another, and all links are reciprocal (e.g., if a is a coauthor of b, then b is a coauthor of a) as was used in [23].

Table 2 Summary of the results of the analysis of ecodesign scientific analysis

Period	1994–1999	2000–2004	2005–2009	2010–2014
Number of papers	380	421	854	2,838
Number of authors	679	883	1,734	6,094
Coauthorships per paper	1,78	2,09	2,03	2,14
Clustering coefficient c	0.667	0.792	0.754	0.747

6 Data Analysis

Two network visualization and analysis software programs were used, VOSviewer which is based on Clauset, Newman and Moore's [26] algorithm for weighted networks, and Cytoscape for network analysis and visualization of the network. This resulted in very attractive visualizations of global cooperation among authors that have published about ecodesign and adjacent fields, allowing for a reflection and discussion on the network collaboration aspect of this field. Table 2 below provides the reader with a summary of our results.

6.1 Number of Authors

The network consists of over 8000 unique authors. Each period of observation experienced expansion, with new authors joining the field, and contraction as other authors left the network. According to the data in Table 2, the network had around 679 authors during the early days of ecodesign. This number grew to over 6000 individual authors by 2014 and continues to grow.

In our coauthorship network, the average number of coauthors per paper has increased steadily over the last 20 years, from 1.78 to its current value of about 2.14. The largest number of authors per paper was 38, followed by 24 and 23 authors per paper.

6.2 Ecodesign Scientific Network Analysis

In Figs. 1 and 2, we present the start of ecodesign as a scientific network.

The network is scattered with a high number of different clusters. To the left, we see the start of the giant cluster starting to take form. The key authors during the period are G. Keoleian contributing ten documents with 11 coauthors, P. Sheng contributing nine documents with 11 coauthors, and A. Stevels contributing six documents with 14 coauthors. Interesting to mention is the document with the most coauthors during this period had 21 authors. The biggest cluster consisted of 22 authors and five documents linking them together.

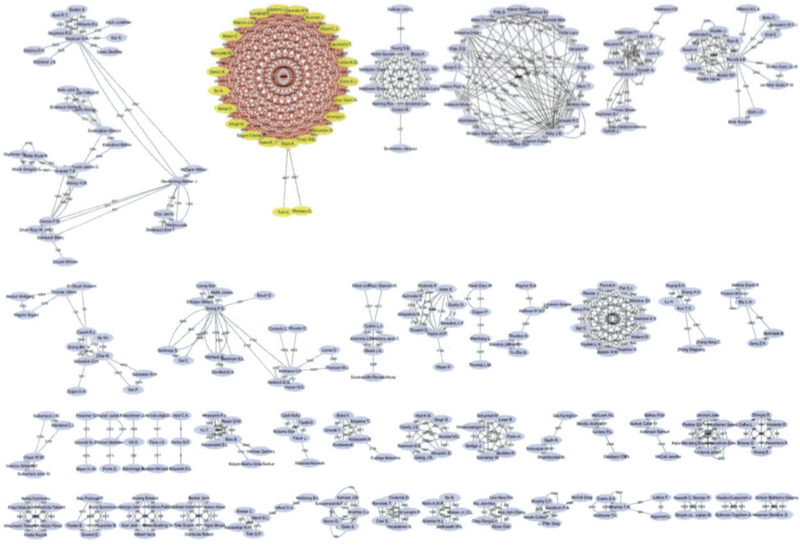

Fig. 1 Ecodesign scientific network 1994–1999

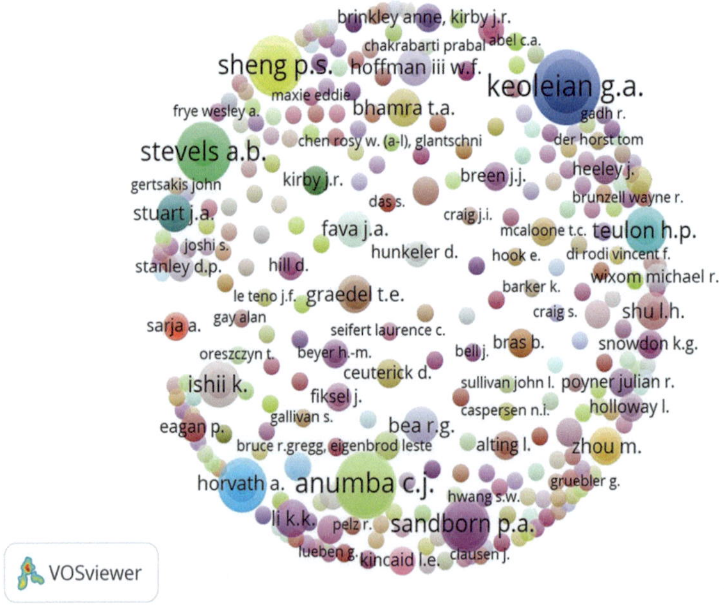

Fig. 2 Ecodesign scientific network 1994–1999 VOSviewer

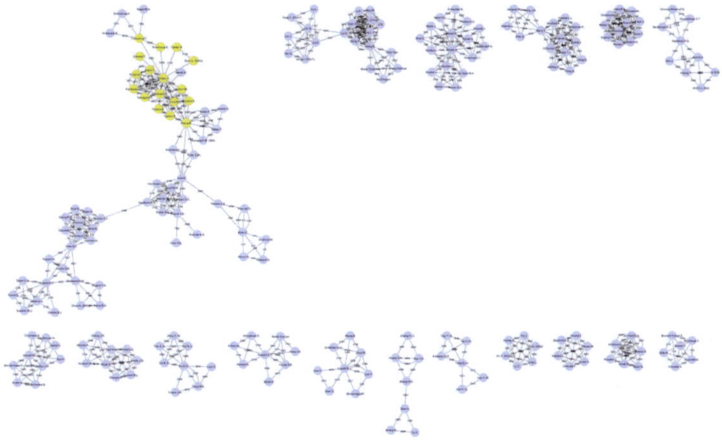

Fig. 3 Ecodesign scientific network 2000–2004 with Y. Umeda coauthorship network highlighted in *yellow*

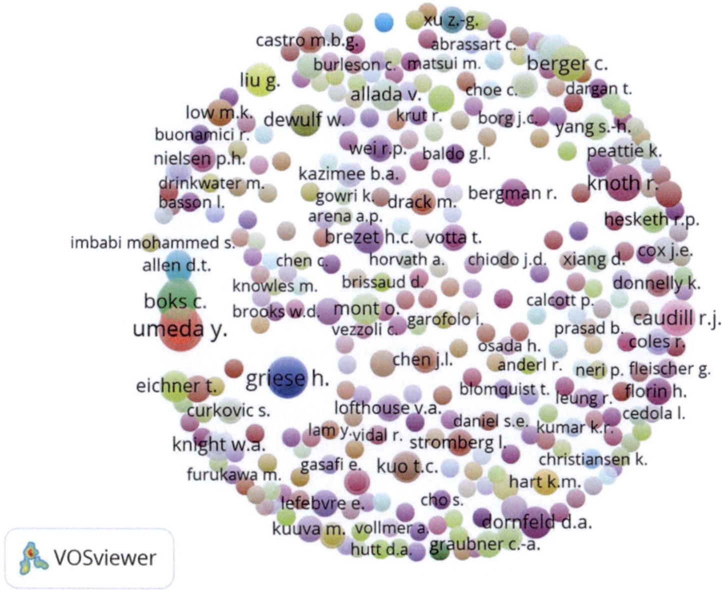

Fig. 4 Ecodesign scientific network 2000–2004 VOSviewer

A number of different clusters form around key authors in the field with little or no connecting link between them.

In Figs. 3 and 4, the scientific network for the period of 2000–2004 is indicated. Two of the key authors for this period are Y. Umeda and H. Griese. Y. Umeda coauthored six and five documents, respectively, with 16 coauthors each. In Fig. 3, the coauthorship network of Y. Umeda is highlighted in yellow. The network

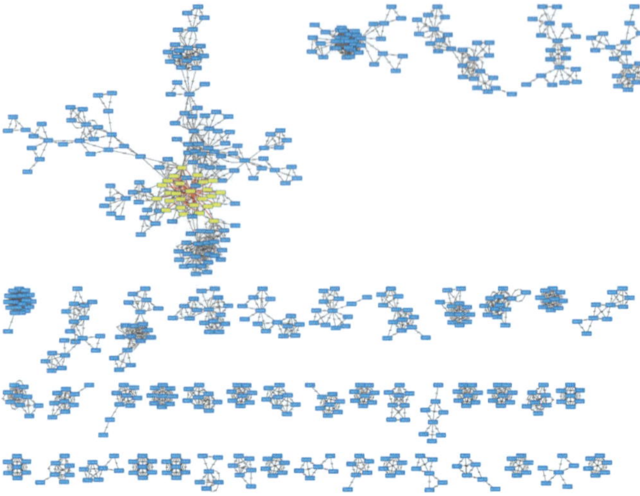

Fig. 5 Ecodesign scientific network 2005–2009 with T. Sakao's coauthorship network highlighted in *yellow*

continues to grow with more clusters forming around the giant cluster. Both key authors are found within the giant cluster.

In Figs. 5, 6, 7, and 8, we show the steady growth of the scientific network form 2005–2009. Ecodesign as an academic research domain grew rapidly as the network doubles in number compared to the previous period of 2000–2004, both in number of new documents and new authors, to the scientific network.

Figure 5 shows the lack of connectedness of the different authors and Fig. 8 is an enlargement of Fig. 7. In the left part of Fig. 5, we see the giant cluster emerging with the key authors for this period in it. The giant cluster consists of 161 authors and 124 documents. The key authors for this cross section are T. Sakao with 19 documents and 26 coauthors, Y. Umeda with 16 documents and 18 coauthors, and C. Boks with 11 documents and 14 coauthors.

The period between 2010 and 2014 saw the field grow at a much higher rate than the previous periods put together. During this period, 6094 authors took part in advancing the field – more than three times the amount of the previous 5 years. The giant cluster in the scientific network (Fig. 9) continues to grow as new authors joined the networks of the more established authors. The giant component consisted of 627 authors and 321 papers. Figure 9 shows how fragmented the field of ecodesign truly is. The majority of authors are unconnected to each other.

The key authors in Figs. 9 and 10 are as follows: D. Millet with a contribution to the network of 23 documents and 27 coauthors, C. Boks contributed 16 documents and coauthored with 14 different authors. In Fig. 9, the coauthor network of C. Boks is highlighted in yellow. During this period, the author moved from the giant cluster (Fig. 6) to the lower left side of the figure. Shimomura Y. is also highlighted in Fig. 10 below as one of the key authors for this time period. It is also interesting to

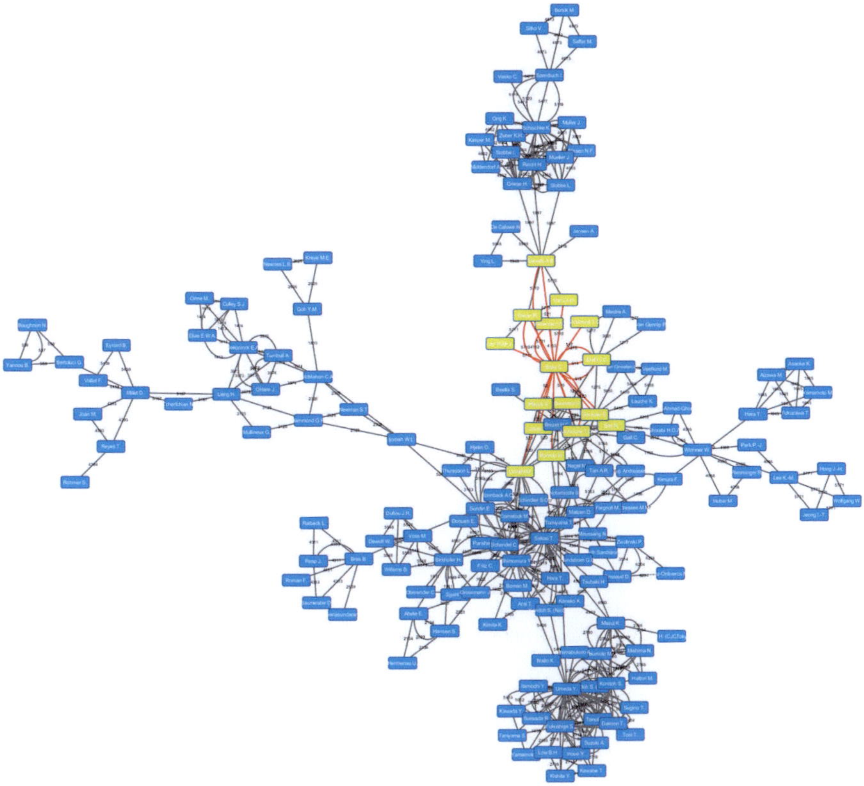

Fig. 6 The coauthorship network of one of the key authors for 2005–2009, C. Boks

see the development and growth of smaller networks to the right of the giant. We can track some of their creation and growth almost from the start of the network in 1994 to where it is now.

The following two figures (Figs. 11 and 12) present the scientific coauthorship network from 1994 to 2014. They show the giant cluster with smaller clusters to the right and below it. The giant cluster consists of 1588 authors with most of the key authors of the field nested within the giant. The second biggest cluster consists of 195 authors.

Figure 12 shows the giant cluster along with the second biggest cluster of the network to its right. It is tempting to say that the giant cluster is the engine room of ecodesign; however, the clusters to the right of it have contributed to the development and evolution of ecodesign and still do. Future research could study the reason why the network is scattered and if it would be interesting to link the two giants with each other. Table 3 provides an overview of the key authors in the field of ecodesign for the time period 1994–2014.

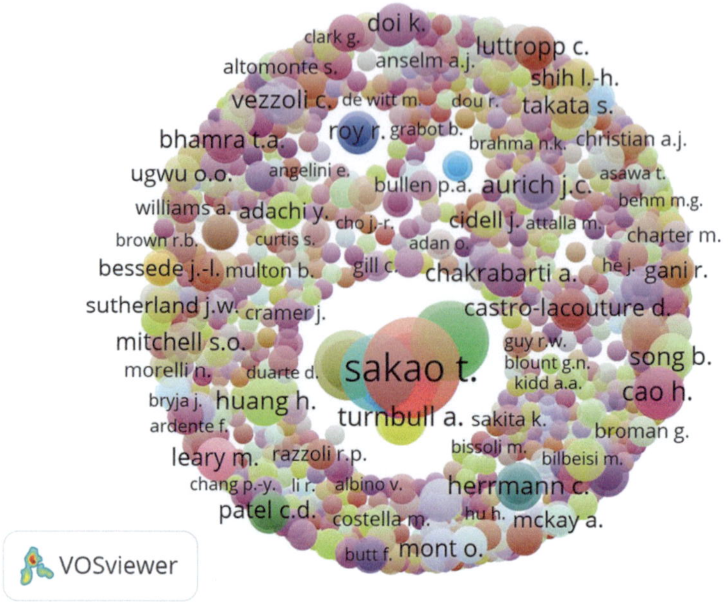

Fig. 7 Ecodesign scientific network 2005–2009 VOSviewer

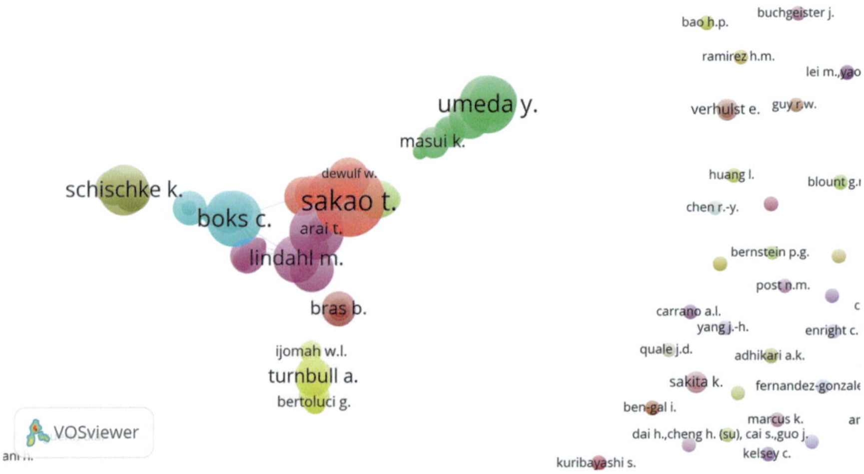

Fig. 8 Ecodesign scientific network 2005–2009 VOSviewer

7 Conclusions and Limitations

In this paper, we studied the scientific network of ecodesign over the past 20 years. In our study, the nodes are authors, and a tie (edges) between two authors represent a coauthorship relationship of one or more papers. To illustrate the growth of the

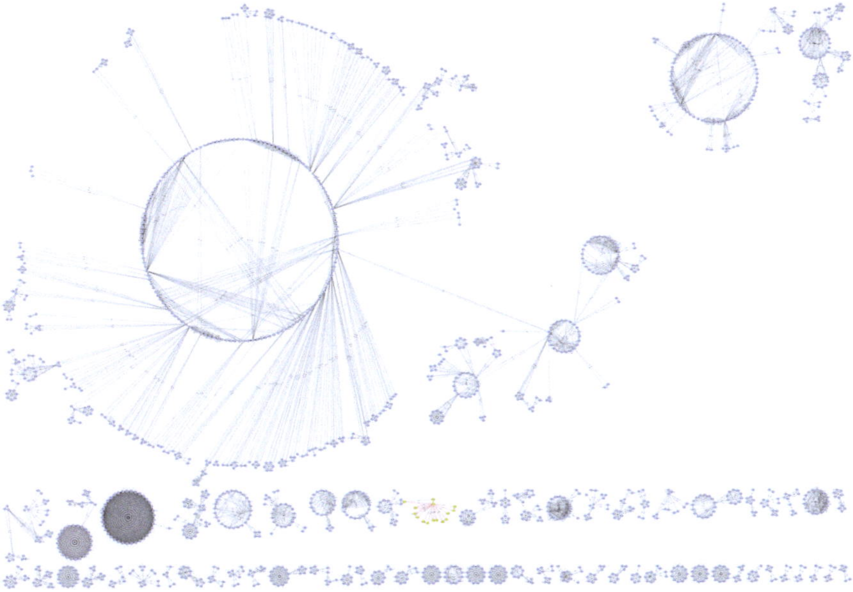

Fig. 9 Ecodesign scientific network 2010–2014

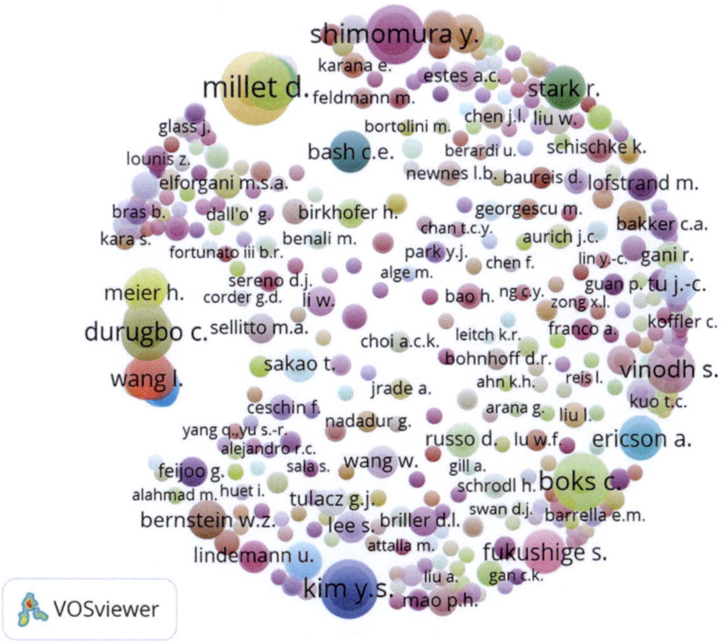

Fig. 10 Ecodesign scientific network 2010–2014 VOSviewer

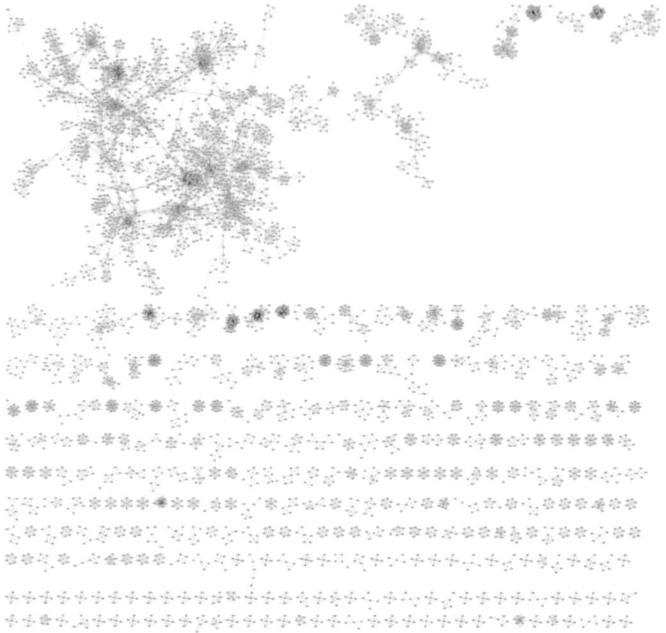

Fig. 11 Ecodesign scientific network 1994–2014

Fig. 12 Ecodesign scientific network 1994–2014 giant cluster

coauthorship network, the network was divided into five time periods (1) 1994–1999, (2) 2000–2004, (3) 2005–2009, (4) 2010–2014, and (5) 1994–2014. For each of these periods, a basic network analysis was done. Furthermore the key authors of the period were identified based on their contribution to the research field by way of published articles or conference papers and the total amount of coauthors they had in their network. We found that the network was scattered with multiple clusters and that most authors are not connected. However,

Table 3 Key authors in the field of ecodesign

Authors	Number of documents	Number of coauthors
Boks C.	34	32
Shimomura Y.	30	28
Y. Umeda	32	41
Sakao T.	29	35
Millet D.	27	30
Kim Y.S.	18	28
Brissaud	17	23
Schischke K.	18	35
Bhamra T.A.	16	27
Stevels A.B.	13	23

the scientific community grew rapidly during the last period, both in amount of documents and in authors. The results are quite interesting for future research and for understanding the field.

In Sect. 3, we asked three important questions. The first question can be answered by viewing Figs. 4, 5, 6, 7, 8, and 9 in which the giant cluster is visible. Also interesting to note is the smaller clusters developing around the giant cluster. Further research could analyze these clusters more in detail to determine why they are not part of the giant cluster. It would also be interesting to see as how the smaller clusters develop as time proceeds.

When analyzing the network, we could identify key authors for each time period. This yielded interesting results, by highlighting the key authors at different time periods. From this, we could determine a core group of authors who contributed to the field from the start and are still active in the field.

Our study is a first step to better understand the field of ecodesign from a network perspective, and as such, there are limitations to the study. For this study, we only used documents contained on the Scopus database, thereby potentially leaving out a number of documents that could have aided us in building a more accurate coauthorship network. We hope, given the popularity of network studies and the need to investigate ecodesign from different perspectives, more studies will follow and provide more detail on other important network characteristics.

Acknowledgments This work was supported by the National Research Foundation of South Africa [grant number 94666].

References

1. Yan E, Ding Y (2012) Scholarly network similarities: how bibliographic coupling networks, citation networks, cocitation networks, topical networks, coauthorship networks, and coword networks relate to each other. J Am Soc Inf Sci Technol 63(7):1313–1326
2. Karlsson R, Luttropp C (2006) EcoDesign: what's happening? An overview of the subject area of EcoDesign and of the papers in this special issue. J Clean Prod 14(15–16):1291–1298

3. Brones F, de Carvalho MM (2015) From 50 to 1: integrating literature toward a systemic ecodesign model. J Clean Prod 96(SI):44–57
4. Brezet H, Hemel VC (1997) Ecodesign: a promising approach to sustainable production and consumption. United Nations Environment Programme, UNEP, The Hague
5. Baumann H, Boons F, Bragd A (2002) Mapping the green product development field: engineering, policy and business perspectives. J Clean Prod 10(5):409–425
6. Johansson G (2002) Success factors for integration of ecodesign in product development – a review of state-of-the-art. Environ Manag Heal 13(1):98–107
7. Tukker A, Jansen B (2006) Environmental impacts of products: a detailed review of studies. J Ind Ecol 10(3):159–182
8. Handfield RB, Melnyk SA, Calantone RJ, Curkovic S (2001) Integrating environmental concerns into the design process: the gap between theory and practice, In: ISEE Transactions on Engineering Management
9. Boks C (2006) The soft side of ecodesign. J Clean Prod 14(15–16):1346–1356
10. de Medeiros MN, Ribeiro JF, Cortimiglia JLD (2014) Success factors for environmentally sustainable product innovation: a systematic literature review. J Clean Prod 65:76–86
11. Byggeth S, Hochschorner E (2006) Handling trade-offs in Ecodesign tools for sustainable product development and procurement. J Clean Prod 14(15–16):1420–1430
12. Unger S, Schneider N, Salhofer F (2008) A review of ecodesign and environmental assessment tools and their appropriateness for electrical and electronic equipment. Prog Ind Ecol Int J 5 (1):13–29
13. Hernandez JGV, Noruzi MR (2011) A Short Note on the Organizational Economics, In: International conference on information and financial engineering, vol. 12, pp 324–326
14. Boks C, McAloone T (2009) Transitions in sustainable product design research. Int J Prod Dev 9(4):429–449
15. Boks C, Stevels A, Koster J (2001) The impact of literature references in the field of applied ecodesign. Ecodesign 2001, second international symposium on environmentally conscious design and inverse manufacturing, 12–15 December, Tokyo, Japan
16. Boks C (2009) "Social perspectives in sustainable product development: The Human Individual as Object of Study". Ecodesign 2009: sixth international symposium on environmentally conscious design and inverse manufacturing, 7–9 December. Sapporo, Japan
17. Newman MEJ (2001) The structure of scientific collaboration networks. Proc Natl Acad Sci 98 (2):404–409
18. Barabási AL (2002) Evolution of the social network of scientific collaborations. Phys A 311 (3):590–614
19. Newman MEJ (2001) Scientific collaboration networks. II. Shortest paths, weighted networks, and centrality. Phys Rev E 64(1):016132, Jun
20. Newman MEJ (2003) The structure and function of complex networks. SIAM Rev 45 (2):167–256
21. Weisstein EW (n.d.) Small world network. Wolfram Research
22. Watts DJ, Strogatz SH (1998) Collective dynamics of 'small-world' networks. Nature 393 (6684):440–442
23. Kas M, Carley KM, Carley LR (2012) Trends in science networks: understanding structures and statistics of scientific networks. Soc Network Anal Min 2(2):169–187
24. Travers J, Milgram S (1969) An experimental study of the small world problem. Sociometry 32(4):425–443
25. Wasserman S, Faust K (1994) Social network analysis: methods and applications
26. Waltman NJ, Van Eck L (2013) A smart local moving algorithm for large-scale modularity-based community detection. Eur Phys J B 86(11):471

Recent Developments in Ocean Energy and Offshore Wind: Financial Challenges and Environmental Misconceptions

Miguel Esteban, Alexandros Gasparatos, and Christopher N.H. Doll

Abstract The first generation of (pre-commercial) ocean energy devices emerged in 2008, but comparatively little practical commercial development has taken place in wave and tidal power since then. Currently offshore wind power is in a sense the only true commercial deployment of an ocean energy technology seeing wide-scale adoption. Recently, several developments related to tidal energy are taking place, though progress in wave power devices remains disappointing, with several companies facing financial difficulties. This paper presents a review of recent developments in ocean energy and discusses the financial challenges and environmental misconceptions regarding such technologies.

Keywords Renewable energy • Ocean energy • Environmental challenges

1 Introduction

Much of the debate on solutions to climate change has centred on a range of technologies such as carbon capture and storage (CCS), geo-sequestration, ocean fertilisation, the so-called next generation of nuclear technologies, solar, wind power and biofuels. However, the majority of these technologies have been associated with some type of environmental problem. For example, it is still not completely clear if CCS is viable, the environmental impacts of large-scale biofuels have been called into question [1], and the Fukushima incident in 2011 in Japan brought such large-scale contamination that makes it unlikely that democratic countries will have an easy time building new installations of this type.

M. Esteban (✉)
GPSS-GLI, The University of Tokyo, Kashiwanoha, Japan
e-mail: esteban.fagan@gmail.com

A. Gasparatos
IR3S, The University of Tokyo, Tokyo, Japan

C.N.H. Doll
United Nations University (UNU-IAS), Tokyo, Japan

M. Matsumoto et al. (eds.), *Sustainability Through Innovation in Product Life Cycle Design*, EcoProduction, DOI 10.1007/978-981-10-0471-1_50

Thus, it is necessary that new ways of producing energy are found, in order to satisfy the world's growing demand for energy. The first generation of commercial ocean energy (OE) devices was introduced in 2008, with the first units being installed in the UK and Portugal (SeaGen and the Pelamis, respectively). These technologies are not only clean (as they do not emit greenhouse gases) but they also have an almost negligible visual impact, especially compared to other renewable sources such as hydropower or onshore wind (provided that offshore turbines are located at a sufficiently long distance from the coastline [2]).

OE has the potential to make an important contribution to the supply of energy to countries and communities located close to the sea, though so far it has not been utilised on a significant scale [3]. The Fifth Assessment Report of the United Nations Intergovernmental Panel on Climate Change (IPCC AR5) highlighted the important role that technology will play in addressing climate change. In fact, large-scale penetration of various combinations of renewable power could theoretically power entire countries, such as Portugal [4] or even Japan [5].

Nevertheless, it is important to remember that there are technical limits to OE penetration, as although it has been reported that its theoretical global potential is between 20,000 and 92,000 TWh/year (compared to the world consumption of electricity of around 16,000 TWh/year), it is unlikely that this technology by itself will be able to solve the energy needs of the planet [6]. Also, there are important feasibility issues. The theoretical potential of tidal energy throughout the world might be 3000 GW, of which less than 3 % is located in areas suitable for power generation [7]. Wave energy has a higher potential, at around 1000–10,000 GW, which is in the same order of magnitude as world electrical energy consumption, though so far it appears to lag in terms of technical development. Another advantage of tidal currents over waves (or wind) is its predictability, as tides can be accurately predicted weeks or even years in advance. Aside from tides and waves, it is worth noting that OE devices can also generate electricity by making use of thermal (ocean thermal energy conversion or OTEC) or osmotic energy [6].

European countries have so far been at the forefront of developments in OE, giving that its potential in the continent is promising [8, 9], with the technically achievable resource estimated to be at least 280 TWh per year [8]. In 2003, the US Electric Power Research Institute (EPRI) estimated the viable resource in the USA at 255 TWh per year (6 % of demand), comparable to the energy currently generated by conventional hydropower [8].

The above suggest that OE technologies have indeed some potential for providing clean and renewable energy in the future. The objective of the current work is to review some recent developments in OE technologies across the world (Sects. 2, 3, and 4). Offshore wind and ocean energy systems are rarely studied together. However, the authors will argue here that as they share the same (marine) environment and have a number of common characteristics, it is worth considering them together. Consequently, the authors conduct a survey of the environmental impact of such technologies, to dispel some misconceptions about the level of expected impacts (Sect. 5).

2 Current Development in Ocean Energy

2.1 Tidal Barrages

Tidal barrages operate in a way similar to dams, by trapping the incoming tide and slowly releasing it to produce electricity. Tidal barrages were built as early as 1966, when the plant at La Rance (France) came into operation. China also started to build a number of barrages around this time, as part of policies from 1958 onwards that emphasised energy independence as a key route to poverty alleviation [10, 11]. Elsewhere, Canada built a barrage at Annapolis, which began service in 1984. In the UK, tidal barrages, such as that proposed for the River Severn, are currently being reappraised. However, opposition from the Environment Agency and other groups appears to make it unlikely that the project will ever reach construction phase [12]. Other smaller schemes such as the Swansea Bay Tidal Lagoon [13] might have more chances of being built.

The environmental impacts of these structures have generally hindered their large-scale application, as they have been known to have some impacts on marine biodiversity [14].

At present, the only country seriously undertaking efforts to construct tidal barrages is South Korea, which recently completed a 254 MW tidal barrage at Sihwa-ho Lake. Another plant almost three times the size is under planning for Ganghwa. The South Korean government claims that the 355bn won (USD 382 million) [11] project will be profitable. In other countries, it is not clear that these plants are economically viable, due to the massive infrastructure investments they require [13] and the fact that often the environmental damage outweighs the benefits of the structure. These traditional or "old" types of ocean energy are thus not without their problems, though modern OE technology is far more environmentally benign [13].

2.2 Modern Ocean Energy Developments

In the first decade of the twenty-first century there was a widespread interest in OE technologies [15, 16]. In fact, 2008 marked the year that the modern OE industry moved from the prototype stage to installation of the first showcase commercial farms. The first operational projects included the Pelamis project (in Portugal) and SeaGen (in Northern Ireland), both having completed installation by the end of the summer of 2008 [17]. However, there have recently been many setbacks to the modern OE industry, with Pelamis Wave Power facing financial difficulties and being put into administration in 2014, Aquamarine Power cutting back to a skeleton staff [18] and a number of other big firms abandoning OE projects or collapsing altogether [19]. However, it is worth noting however that SeaGen continues to operate and has been delivering energy into the national grid since 2008 [18].

Despite these setbacks, many devices that employ hydrokinetic conversion systems that use tidal or river stream flows are currently under development [20], whilst numerous devices using other technologies have already completed prototype testing [18]. Furthermore, several other projects and prototypes are undergoing full-scale testing. For example, the European Marine Energy Centre (EMEC) has four grid-connected berths for wave and five for tidal devices, all of which are either in use, booked or awaiting for support installations to be constructed (such as the WaveHub) [21]. Other promising developments in the UK and France include the installation of the first tidal arrays (4 MW) made up of two OpenHydro turbines in France and Canada [22].

Another type of OE technology that is reaching maturity is OTEC, which has significant potential for countries located in tropical regions. These plants must be located in an environment where the warm surface seawater must differ about 20 °C from the cold deep water that is no more than about 1,000 m below the surface. This is typically in areas between latitudes 20° north and south of the equator [23]. India, for example, has considerable potential for the development of OTEC energy, and some trial plants have already been built there, such as a 1 MW OTEC and a 100 t/day freshwater desalination plant [23]. The potential OTEC resource for Indonesia has been estimated to be able to produce enough electricity for the entire country [19]. However, currently the only fully functioning plant, with a rated capacity of 100 kW, exists in Okinawa, Japan [24].

2.3 *Offshore Wind Energy*

Although this type of energy is not traditionally considered as part of the OE sector, the authors would like to point out that many of the technologies employed (and the constraints faced) by offshore wind and OE industries are similar. Thus, offshore wind energy can be considered to be similar to OE, in that it is located in the same type of environment (marine), and could potentially have many synergies (e.g. joint utilisation of grid connections).

European countries have traditionally dominated offshore wind. Figure 1 shows how the UK currently has more installed capacity than any other country [20], and the rate of installation is expected to expand considerably in the following years, going from an annual installation rate of 194 MW in 2008 to over 400 MW in 2009 and to 800–1,000 MW per annum for the period 2010–2014 [25]. It is believed that the UK's seas could eventually provide enough extra wind energy to power the equivalent of 19 million homes. By having an extra 25 GW of electricity generation capacity (in addition to the 8 GW of wind power already built or planned offshore), offshore wind could be enough to power every household in the UK [26]. According to some estimates, this could provide over a quarter of the UK's electricity needs whilst generating thousands of jobs [26, 27].

However, the development of offshore wind is not limited to Europe. Increasingly Asia-Pacific is emerging as a significant market [28]. China is quickly

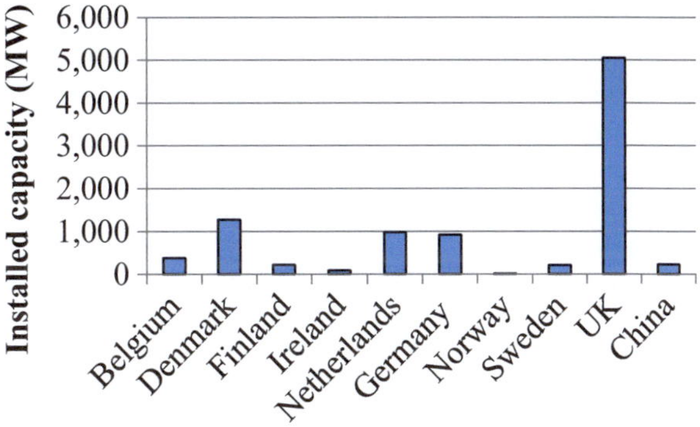

Fig. 1 Installed offshore wind capacity for selected countries in 2015 (in MW)

becoming one of the world leaders in the adoption of offshore wind energy. In China's case, offshore wind energy has the advantage of being located close to the population centres, in contrast to onshore wind. The offshore wind power resource has been estimated at 200 GW, almost ten times the total (onshore) presently installed wind capacity [29].

Particularly interesting recent developments in offshore wind power include research into floating wind turbines, with prototypes currently being tested in Norway [30] and Japan. Such floating turbines represent a promising development for countries where the coastal seas are rather deep and could greatly reduce installation costs.

3 Future Potential of Wind and Offshore Energy

As already discussed, ocean energy is currently at the initial stage of commercialisation of the first-generation technologies (referred to as the pre-commercial stage) (Sect. 2.2). The first commercial devices are now starting to be connected to the grid [31], and, like other power generation technologies, it will eventually be succeeded by a "second-generation technology", which would further improve the current designs using the benefit of experience. There is some evidence that these second-generation technologies are already under development given that devices such as SeaGen will likely soon be placed in arrays.

The cost of energy from initial tidal stream farms has been estimated in the range of 0.15–0.55 USD/kWh for tidal and wave energy farms and 0.11–0.22 USD/kWh for tidal streams [32]. These are expected to decrease to 0.10–0.25 USD/kWh by 2020. Estimated learning factors are believed to be 10–15 % for offshore wave and 5–10 % for tidal stream [32]. A concrete example of the perceived cost of the first generation of modern ocean energy units can be found in Portugal, where the government put in place a feed-in tariff (a legislative price incentive which encourages the adoption of renewable energy through a long-term declining subsidy) equivalent to approximately 0.23 EUR/kWh [33].

To identify long-term cost developments, learning curves have been derived for different technologies, which identity the correlation between cumulative production volumes of a given technology and a reduction in costs. For many technologies, the learning factor falls in the range of between 0.75 for less mature systems to 0.95 for well-established technologies [34]. A learning factor (LF) of 0.9 would signify that costs are expected to fall by 10 % each time that the cumulative output for that technology doubles. Using this theory, OE could be expected to be competitive with oil sometime between 2017 (LF = 0.75) and 2033 (LF = 0.9) [35].

4 Challenges and Large-Scale Implementation of OE

There are a number of challenges that are perceived to hinder the large-scale penetration of OE. Some of these problems are legal or regulatory in nature [36]. Others relate to the financial considerations of the projects [17], as OE is generally more expensive than conventional forms of energy, as explained above.

However, like other intermittent renewable sources (see below), the main criticism voiced about OE is its reliability. It is generally believed that renewable power poses a problem to traditional electricity grid systems due to fluctuations in the electricity produced, which is generally time and season dependent. In particular, the production of wave and offshore wind energy would depend on the wind and waves at a given location. As a result, during times of low wind or waves little electricity would be produced, a problem often referred to as intermittency.

Traditional grids were not originally designed to cope with significant peaks and troughs in electricity supply. Therefore, load levelling and the absorption of fluctuations are believed to represent a problem for the widespread use of renewable power, OE included. However, intermittency can be partly attenuated in large countries such as Japan, as unfavourable weather conditions for one renewable source in one part of the country might be compensated by better conditions elsewhere [37]. It is worth noting here that tidal energy is far more predictable and follows daily fluctuations that can be predicted years in advance and could thus reliably contribute to smoothening the system [38].

The use of electricity storage techniques, together with other sources of renewable energy (such as hydropower and biomass energy), can also be used so that a country such as Portugal could rely 100 % on renewable power [4]. The problem, however, is not only how to effectively store excess energy during periods of high production but the fluctuations in the level of demand. These can broadly be classified into two types: short-term fluctuations between day and night and long-term variations associated with the different seasons of the year [5].

For a 100 % renewable electricity system, short-term fluctuations can be absorbed using batteries in electric vehicles (EV) or specially built storage. Batteries in electric vehicles could hence serve the dual purpose of eliminating CO_2 emissions from transportation and providing storage during fluctuations in renewable power production, particularly if they are produced in great numbers [39]. If such a high penetration of electric vehicles is realised, then their batteries could store the wind and wave power produced during the night. This would be more economically feasible than using batteries dedicated exclusively to electricity storage. The storage potential of electric cars could also be complemented by batteries dedicated exclusively to electricity storage, e.g. similar to a battery unit in the north of Japan that stores part of the electric load of a 50 MW wind farm. A total of 17 batteries (2 MW-NAS) are employed, using a total area of just over 4,000 m^2 of floor space. For the case of long-term storage, hydrogen could be a good storage medium [40, 41], where electrolytically produced hydrogen could be stored in conventional tanks under pressure and then re-electrified during peak demand seasons through various custom-made fuel cells.

5 Environmental Issues

There has been some concern that OE sources could impact coastal ecosystems or habitats at (or near) the seabed. Such ecosystems can harbour important biodiversity or offer breeding areas for fish and birds. However, given the current low level of OE deployment, there is little empirical evidence to quantify the likely environmental effects.

Of the available range of OE technologies, only tidal barrages can be considered relatively "mature" [15]. The environmental impacts of these structures have generally hindered their wide-scale application, as they have been shown to have some impacts on marine biodiversity, e.g. the entrapment of whales for the case of the Annapolis power plant in Canada [15]. Similar to conventional dams, they can also produce changes in sediment loading, salinity and water turbidity [42].

As already discussed, at present there are no large-scale modern OE installations so many of the potential environmental risks are speculative. As the first modern (pre-commercial) ocean energy units are only now becoming operational, there is little track record and indeed understating of the potential environmental problems. In fact, the lack of understanding of the environmental impact of modern OE appears to be a major obstacle to its development [43]. However, it is worth noting that to date Environmental Impact Assessments (EIAs) have shown no evidence that modern OE could have a significant environmental impact [44], though this is obviously context dependent.

A few studies have identified potential effects of OE on marine biodiversity, which can be either positive or negative [45, 46]. Some of these possible effects include the upwelling of water rich in nutrients that could affect marine life, during the operation of OTEC systems that extract cold water from deeper regions of the ocean [42]. Wave energy extraction devices could interfere with sediment transport systems [47], though the potential magnitude of the interference would appear to be rather small [48]. Such effects might affect depositional processes, change current and wave fields and eventfully alter the substrates that form the habitat for benthic organisms. However, it is worth nothing that some benthic organisms living in areas with high potential tidal or wave energy resources are likely to be relatively resistant to the low levels of disturbance caused by modern OE devices [49]. These units could nevertheless affect coastal erosion and result in a loss of habitat area, which could be a significant driver of ecosystem change in regions of high biodiversity.

Furthermore, it is not inconceivable that the noise associated with some types of OE, or the rotary movement of tidal turbines, could have an effect on certain species of marine mammals or fish [42]. Whilst tidal turbines are much slower than wind turbines, they could nevertheless interfere with some species, such as diving birds [49] or fish [50]. Electromagnetic fields [47] could affect sensitive species [42], though these effects are likely to be limited to the vicinity of grid connection cables [49]. Also, there has been some speculation about the local polluting effect and toxicity of lubricants and paints [50]. Furthermore, the installation and decommissioning of OE devices could result in a temporary degradation of habitat and water quality, through increased turbidity in the water column from disturbances to the seabed [51]. Finally, increased vessel movements and noise during these phases could also have an impact on various marine animals, fish stocks and bird populations [51], though these phases are relatively limited in the life cycle of OE units.

The few full-scale (pre-commercial) OE devices installed so far have shown no discernible effects on biodiversity. For instance, the SeaGen device in Strangford Narrows, Northern Ireland, has a nearby seal colony and occasionally basking sharks and seems to have had no influence on them [52]. Studies at this site using acoustic Doppler current profiling have failed to show evidence of any significant changes to the velocity of flow direction due to the installation of the turbine [52]. Finally, there have been no changes in the numbers and distributions of bird species and benthic animals such as lobsters [53].

Studies of offshore wind farms suggest that porpoises will initially avoid newly constructed wind farms, but eventually return to the area the following years [54]. Sakaguchi [55] looked at the environmental impacts of three pre-commercial tidal units that have had 1–3 years of monitoring after installation and concludes that no negative impacts were confirmed for any of the projects.

Conversely, the development of modern OE could indeed have a positive impact on biodiversity and fish stocks, considering that the installation of modern OE units will require these areas to remain out of bounds for fishing and other sea traffic [16]. The delimitation of some marine areas can be beneficial to the preservation of fishing stocks [50] and have other biodiversity benefits, though this is not entirely clear at present, as how changes to water circulation could affect the distribution of species in still not well understood [51].

Sakaguchi [55] highlights the current lack of evidence on whether modern OE farms will have the same outcomes on biodiversity as existing (pre-commercial) isolated units, highlighting the need for further study in this area. However, in the authors' opinion, and given the small size of proposed modern OE developments and the negligible impacts observed up to now, learning about potential environmental problems from the actual installation of (slightly larger) farms appears warranted at present.

6 Conclusions

The first generation of (pre-commercial) ocean energy devices emerged in 2008, with the first units being installed in the UK and Portugal. However, despite intense media attention, comparatively little commercial development has since taken place in recent years in wave and tidal power. In a sense offshore wind power is the only true commercial ocean energy source experiencing wide-scale adoption.

It is interesting to note that despite several tidal devices still operating to date, and the first tidal arrays likely to be built in 2016, many wave energy companies have fared badly. Companies such as Pelamis (considered at one point to be one of the leaders in the sector) have been taken into administration. The reasons for such failures are not clear, though installing devices in areas of high wave energy is likely to pose greater challenges than what was initially thought. Given the uncertainties and harsh environment in which wave devices have to operate, the regular nature of tidal currents appears to make this type of technology more predictable and easier to maintain and operate than wave power.

It is likely that developments in tidal energy will continue in the future, though they will have to compete with other types of renewables, mainly solar, wind and geothermal. The costs of such technologies are rapidly decreasing, and whilst OE is likely to remain expensive for some time, it does have a future if it positions itself as part of a global energy solution that can help to smoothen the intermittency of other renewable sources. It could also prove to be the most appropriate source of energy for small islands, particularly those which have an abundance of this resource, low populations and limited options for other forms of centralised electricity generation.

It is important to note that modern OE (as opposed to tidal barrages) appears to have few environmental problems associated with it. This can substantially help mitigate greenhouse gas emissions whilst also not contributing to damaging local ecosystems.

References

1. Gasparatos A, Stromberg P, Takeuchi K (n.d.) Biofuels, ecosystem services and human wellbeing: putting biofuels in the ecosystem services narrative. Agriculture, Ecosystems and Environment (in press, obtained through private communication)
2. Ladenburg J (2009) Visual impact assessment of offshore wind farms and prior experience. Appl Energy 86:380–387
3. Pelc R, Fujita RM (2002) Renewable energy from the ocean. Mar Policy 26:471
4. Krajacic G, Duic N, Carvalho MG. How to achieve a 100% RES electricity supply for Portugal? Appl Energy 88(2). The 5th Dubrovnik Conference on Sustainable Development of Energy, Water and Environment Systems, held in Dubrovnik September/October 2009. pp 508–517
5. Esteban M, Zhang Q, Utama A, Tezuka T, Ishihara KN (2010) Methodology to estimate the output of a dual solar-wind renewable energy system in Japan. J Energy Policy 38:7793–7802
6. Soerensen HC, Weinstein A (2008) Ocean energy: position paper for IPCC Proc. of IPCC Scoping Meeting on Ren. En. Sources. Lubeck, Germany

7. World offshore renewable energy report 2002–2007, published by Renewables UK, UK Department of Trade & Industry. Authors: Douglas-Westwood Ltd
8. Scruggs J, Jacob P (2009) Harvesting ocean wave energy. Science 323:1176
9. Cornett AM (2008) A global wave energy resource assessment, Proceedings of the Eighteenth International Offshore and Polar Eng. Conference, Vancouver, Canada, 6–11 July
10. Tan M (2009) China's potential on the renewable energy development. Final thesis at the United Nations University Institute of Advanced Studies.
11. Watts J (2009) South Korea lights the way on carbon emissions with its 23bn pounds green deal, The Guardian. Retrieved April 21, 2009, http://www.guardian.co.uk/environment/2009/apr/21/south-korea-enviroment-carbon-emissions
12. Owen E (2008) Environment agency boss opposes Severn Barrage. New Civ Eng J 1:5
13. Gosden E (2015) Cardiff tidal energy lagoon "could power every home in Wales", daily telegraph 2 March 2015. http://www.telegraph.co.uk/news/earth/energy/11445579/Cardiff-tidal-energy-lagoon-could-power-every-home-in-Wales.html
14. Rourke FO, Boyle F, Reynolds A (2010) Tidal energy update 2009. Appl Energy 2010:398–409
15. Leary D, Esteban M (2011) Recent developments in offshore renewable energy in the Asia-Pacific. J Ocean Dev Int Law 42:94–119
16. Leary D, Esteban M (2010) Renewable energy from the ocean and tides: a viable renewable energy resource in search of a suitable regulatory framework. Carbon Clim Law Rev 3:417–425
17. Westwood A (2008) SeaGen installation moves forward. Renew Energy Focus 9:26–27
18. Williams A (2015) Wave and tidal energy in 2015: finally emerging from the labs. Renewable energy world.com. http://www.renewableenergyworld.com/articles/2015/02/wave-and-tidal-energy-in-2015-finally-emerging-from-the-labs.html. Accessed 24 June 2015
19. Deign J (2015) Tidal industry gets boost from MeyGen kick-off. Tidal today. http://social.tidaltoday.com/installation/tidal-industry-gets-boost-meygen-kick?utm_campaign=TID%2004FEB15%20Newsletter&utm_medium=email&utm_source=Eloqua&utm_source=Newsletter&utm_medium=eshot&utm_campaign=Tidal&elq=1b452334c19549f7b51ef47e1921a087&elqCampaignId=852. Accessed 5 Feb 2015
20. Khan MJ, Bhuyan G, Iqbal MT, Quaicoe JE (2009) Hydrokinetic energy conversion systems and assessment of horizontal and vertical axis turbines for river and tidal applications: a technology status review. Appl Energy 86:1823–1835
21. Bryden I (2009) Progress towards a Viable UK marine renewable energy. Coasts, marine structures and breakwaters. Edinburgh, Scotland, 16–18 Sept 2009
22. OpenHydro to install grid-linked tidal arrays in France, Canada in 2015, Tidal Today. http://analysis.tidaltoday.com/openhydro-install-grid-linked-tidal-arrays-france-canada-2015?utm_campaign=TID%2004JUN15%20Newsletter&utm_medium=email&utm_source=Eloqua&elqTrackId=85395252A1D5839328B8D6E5344111AA&elq=cdc66d2b9bfb47e1a69abf998c666144&elqCampaignId=2313&elqaid=5356&elqat=1. Accessed 28 June 2015
23. Ikegami Y, Achiruddin D, Abdullah K (2009) Possibility study of OTEC & DOWA in Indonesia, Proc. of International Symposium on Sustainable Energy and Environmental Protection (ISSEEP) 2009 Yogyakarta, Indonesia, 23–26 Sept 2009
24. http://otecokinawa.com/en/Project/index.html
25. British Wind Energy Agency (BWEA) (2007) UK offshore wind: staying on track. Forecasting offshore wind build for the next five years
26. The Guardian Newspaper. Offshore wind farms could meet a quarter of the UK's electricity needs, published 25th June 2009. http://www.guardian.co.uk/environment/2009/jun/25/offshore-wind-uk-homes
27. Esteban M, Leary D, Zhang Q, Utama A, Tezuka T, Ishihara KN (2010) Job retention in the British offshore sector through greening of the North sea energy industry. J Energy Policy 39:1543–1551
28. Global Wind Energy Council (GWEC) (2009) www.gwec.net. Accessed 22 June 2009

29. Zhau X (2010) The development of modern power system and power system technology. APPEEC 2010 Conference, China, 28 to 31st of May, 2010 (plenary speech)
30. BBC (2009) Floating wind turbine launched. http://news.bbc.co.uk/2/hi/8085551.stm. Accessed 22 June 2009
31. Lemming J, Morthorst PE, Clausen NE, Jensen PH (2008) Contribution to the chapter on wind power in: energy perspectives 2008. Int Energy Agency
32. International Energy Agency (IAE) (2007) World energy outlook 2007
33. Soerensen HC, Naef S (2008) Report on technical specification of reference technologies (wave and tidal power plant), NEEDS project report 2008
34. Teske S (2008) Energy [r]evolution, European Renewable Energy Council (EREC) and Greenpeace International, ISBN 978907336189
35. Esteban M, Leary D (2011) Current developments and future prospects of offshore wind and ocean energy. J Appl Energy 90:128–136. http://www.sciencedirect.com/science/article/pii/S0306261911003813
36. Leary D, Esteban M (2010) Renewable energy from the ocean and tides: a viable renewable energy resource in search of a suitable regulatory framework. Carbon Clim Law Rev 4:417–425
37. Esteban M, Leary D, Zhang Q, Utama A, Tezuka T, Ishihara KN (2010) Job retention in the British offshore sector through greening of the North Sea energy industry. J Energy Policy 2010:1543–1551
38. Carnegie Corporation (2009) Development timeline. www.carnegiecorp.com.au/index.php?url=/ceto/development-timeline. Accessed 8 Apr 2009
39. Guille C, Gross GA (2009) Conceptual framework for the vehicle-to-grid (V2G) implementation. Energy Policy 37:4379–4390
40. Sherif SA, Barbir F, Veziroglu TN (2005) Wind energy and the hydrogen economy—review of the technology. Sol Energy 78:647–660
41. Levene JI, Mann MK, Margolis RM, Milbrandt A (2007) An analysis of hydrogen production from renewable electricity sources. Sol Energy 81:773–780
42. Sathaye J, Lucon O, Rahman A, Christensen J, Denton F, Fujino J, Heath G, Kadner S, Mirza M, Rudnick H, Schlaepfer A, Shmakin A (2011) Renewable energy in the context of sustainable development. In: Edenhofer O, Pichs-Madruga R, Sokona Y, Seyboth K, Matschoss P, Kadner S, Zwickel T, Eickemeier P, Hansen G, Schlömer S, von Stechow C (eds) IPCC special report on renewable energy sources and climate change mitigation. Cambridge University Press, Cambridge
43. Huertas-Olivares C, Soerensen HC, Rousseau N, Villate JL, Lopez JR, Sarmento A, Russell I, Neumann F (2008) First outcome of the WAVEPLAM project: assessment of non-technological barriers and best practices. 2nd International Conference on Ocean Energy (ICOE), 15–17 Oct 2008, Brest, France
44. Soerensen HC, Weinstein A (2008) Ocean energy: position paper for IPCC, Proceeding of IPCC Scoping Meeting on Renewable Energy Sources, Lubeck, Germany. www.needs-project.org. Accessed 29 Apr 2009
45. Broadhurst M, Orme CDL (2014) Spatial and temporal benthic species assemblage responses with a deployed marine tidal energy device: a small scaled study. Mar Environ Res 99:76–84
46. Witt MJ, Sheehan EV, Bearhop S, Broderick AC, Conley DC, Cotterell SP, Crow E, Grecian WJ, Halsband C, Hodgson DJ, Hosegood P, Inger R, Miller PI, Sims DW, Thompson RC, Vanstaen K, Votier SC, Attrill MJ, Godley BJ (2012) Assessing wave energy effects on biodiversity: the wave hub experience. Philos Trans R Soc A Math Phys Eng Sci 370:502–529
47. (2009) US Department of Energy, Report to Congress: potential environmental effects of marine and hydrokinetic energy technologies (Public Review Draft). www.ornl.gov/sci/eere/EISARReport.html. 23 Mar 2009
48. Mendes L et al. (2008) Analysis of the impact of a pilot zone for wave energy conversion offshore Portugal. In: Proceedings of the 18th international offshore and polar engineering conference, Vancouver, BC, Canada, 6–11 July

49. Argyll Tidal (2013) Environmental Appraisal (EA) for the Argyll Tidal Demonstrator Project. Environmental Appraisal
50. Esteban M, Webersik C, Leary D, Thompson-Pomeroy D (2008) Joint United Nations University-Institute of Advanced Studies, University of New South Wales, Yokohama National University Report: "Innovation in responding to climate change: Nanotechnology, Ocean Energy and Forestry". http://www.ias.unu.edu/sub_page.aspx?catID=111&ddlID=738
51. OERA (2012) Marine renewable energy: background report to support a Strategic Environmental Assessment (SEA) for the Cape Breton Coastal Region, inclusive of the Bras D'Or Lakes. www.aecom.com
52. Fraenkel PL (2007) Marine Current Turbines: Moving from Experimental Test Rigs to a Commercial Technology. In: 26th international conference on Offshore Mechanics & Arctic Engineering
53. Fundy Ocean Research Center for Energy (2011) Environmental effects monitoring report (OpenHydro). Announced October 2011. Retrieved November 15, 2014 from http://fundyforce.ca/wp-content/uploads/2012/05/FORCE-Environmental-Effects-Monitoring-Report-Oct-2011.pdf
54. Whitford J (2008) Final report background report for the Fundy Tidal Energy Strategic Environmental Assessment. Project No. 1028476
55. Sakaguchi T (2015) Marine spatial planning for an ocean current power farm: a case study of Tokara, Japan. Master Thesis, University of Tokyo

Renewable Energy Policy Efficacy and Sustainability: The Role of Equity in Improving Energy Policy Outcomes

Andrew John Chapman, Tetsuo Tezuka, and Benjamin McLellan

Abstract Australian renewable energy policy is guided by the renewable energy target, which prescribes audacious targets for both large-scale and overall levels of renewable energy electricity generation. In order to achieve these targets, subsidies have been offered for both large- and small-scale generators, the most pronounced of which are the feed-in tariffs offered in each state for household solar photovoltaics. While high installation rates and positive economic and environmental impacts have been realized, little consideration has been given to the social equity impacts of these policies. This study assesses the current energy policy approach through a novel energy evaluation framework which incorporates an assessment of cost and benefit distribution and societal equity impacts. Through a multi-scenario energy policy analysis, an evidence-based alternative scenario is proposed, which can achieve environmental and economic goals while improving societal equity and sustainability outcomes in accord with observed Australian equity preferences.

Keywords Equity • Policy • Renewable energy • Subsidization • FIT

1 Introduction

This study analyzes recent Australian renewable energy (RE) policy from the viewpoints of efficacy, efficiency, and, most importantly, sustainability. Australia, along with governments around the world, has attempted to stimulate the installation of renewable energy at the community level as part of an overall strategy to achieve energy security and to address climate change by reducing greenhouse gas (GHG) emissions [1]. In order to achieve desired installation targets, the Australian Government has used two key stimulatory policies, namely, feed-in tariffs (FITs) and renewable energy certificates (REC). These policies have been successful in increasing installations, most prominently solar photovoltaic (PV) systems in the residential sector. The outcomes of these policies will be assessed from a unique standpoint, considering not only the efficacy (the environmental achievements and

A.J. Chapman (✉) • T. Tezuka • B. McLellan
Graduate School of Energy Science, Kyoto University, Kyoto, Japan
e-mail: chapman.john.27e@st.kyoto-u.ac.jp

M. Matsumoto et al. (eds.), *Sustainability Through Innovation in Product Life Cycle Design*, EcoProduction, DOI 10.1007/978-981-10-0471-1_51

their costs) and efficiency of Australian RE policy but also the associated equity impacts. Equity is defined in this research as being primarily concerned with the distribution of costs and benefits of RE policy implementation within Australian society. The analysis considers the immediate to short-term (until 2020) equity impacts of policy implementation phases. In order to ensure equal treatment across income levels within society, vertical equity – the sharing of costs according to income level, in order to redistribute wealth in society – is applied. This is intended to enforce a user-pays system, fair value of subsidizations, and payments to participants and to limit the burden on nonparticipants in Australian subsidization schemes. The jurisdictional boundary of this research is the Australian National Electricity Market (NEM) encompassing the states of Queensland, New South Wales (and the ACT), Victoria, South Australia, and Tasmania.

Australian RE policy is guided by the renewable energy target (RET), consisting of a small-scale renewable energy scheme (SRES) and the large-scale renewable energy target (LRET). Installations at both the large and small scale attract RECs (separately administered); however, while the overall target for RE installation in Australia is 20 % of all electricity generated, only the LRET has a hard gigawatt hour (GWh) target of 41,000 GWh by the year 2020. Small-scale RE generators (almost exclusively rooftop PV) are eligible for FITs, administered by the states, whose payments are borne by electricity retailers before being passed on to residential electricity consumers through increased electricity tariffs. FITs have varied since their introduction in 2008, in both the amount per kilowatt hour (kWh) offered and the nature of the FIT, either net or gross.

2 Aim

The aim of this study is to measure the sustainability of energy policy, defined as policy which can meet environmental and economic goals, without impairing societal equity. This definition is synonymous with the ideals of sustainability being a subset of economic, social, and environmental factors [2]. Building on an analysis of identified Australian RE policy impacts [3], three scenarios which include a baseline scenario, the current Australian policy (FIT)-based scenario, and an alternative, evidence-based energy policy scenario are analyzed in order to inform economic and environmental outcomes of differing policy approaches in the Australian NEM in 2020. An original equity evaluation methodology is established in order to objectively evaluate the policy efficacy and sustainability of each scenario while also quantifying equity for the purpose of these evaluations. Finally, a policy-applicability contrast is undertaken with the Lorenz curve equality assessment in order to establish the proposed energy policy evaluation framework and equity evaluation methodology's value as a policy development and objective comparison tool.

3 Methodology

In order to measure efficacy, efficiency, equity, and ultimately the sustainability of energy policy, firstly, a framework is developed which identifies economic, environmental, and social equity factors which are broadly agreed upon internationally as constituent parts of sustainability assessment. Secondly, scenario development is undertaken incorporating a baseline scenario, a business as usual scenario incorporating the FIT, and an evidence-based alternative energy scenario. Each of the three scenarios' efficacy will be measured according to the proposed energy policy sustainability framework and stated Australian government energy and environmental goals under the RET in 2020, with equity measured according to the comparative societal distribution of environmental and economic impacts across income levels. Equity outcomes across the three scenarios are compared, and ultimately measured against an "Australian equity preference," determined via an environmental scan.

3.1 Equitable Policy Sustainability Framework

Four international sustainable development frameworks and reports were selected for incorporation into the equitable energy policy sustainability framework based on the international profile of the publishing authorities and the multinational, cooperative nature of their development. These reports include "Indicators of Sustainable Development" developed by the United Nations Commission on Sustainable Development [4], "Inclusive Green Growth: The Pathway to Sustainable Development" developed by the World Bank [5], "Contribution of Working Group III to the Fourth Assessment Report of the Intergovernmental Panel on Climate Change" (IPCC) [6], and "Energy Indicators for Sustainable Development: Guidelines and Methodologies" provided by the International Atomic Energy Agency [7] (IAEA). There are many linkages between each of these documents, ranging from a common platform for the definition of sustainable development between the Organisation for Economic Co-operation and Development (OECD), the United Nations Environmental Programme (UNEP), and the World Bank to joint reporting over time to the United Nations Department of Economic and Social Affairs by both the UN Commission on Sustainable Development (CSD) and the IAEA. Additionally, IPCC reporting shares intergovernmental linkages with both the UNEP and the World Meteorological Organization. These frameworks and reports are analyzed to identify the common factors of economic and environmental evaluation and, further, social equity factors which represent the distribution of these economic and environmental impacts, in order to thoroughly analyze the sustainability of energy policy. Table 1 summarizes the factors which will make up the energy policy sustainability framework proposed in this study, used to effectively measure the economic, environmental, and equity impacts of the energy policy scenarios.

Table 1 Energy policy sustainability framework factors

Environmental equity	Economic equity	Social equity
GHG emissions (CO2-e)	Levelized cost of electricity (LCOE)	The distribution of economic and environmental costs and benefits across income levels
Resource management	Impact of subsidization on electricity price	Distribution of costs
RE technology deployment efficiency	GDP impacts	Electricity price increases
	Market impacts	Allocation of subsidies
		Distribution of benefits
		Employment
		GHG reduction
		Participation

Social equity defined as the distribution of costs and benefits across societal income levels is synonymous with vertical equity and is useful in establishing the gap between the highest- and lowest-income household relative equity level.

3.2 Energy Policy Scenarios

3.2.1 Baseline Scenario

FITs were first introduced in Australia on 1 July 2008; so in order to negate the effect of the FITs introduction, the baseline scenario (and subsequent scenarios) will begin from January 2008 through to the end of 2020 on a business as usual basis, with no FIT stimuli for the installation of renewable energy. Estimates of PV and wind installations in 2020 are based on pre-FIT installation trends from 2001 to 2008.

3.2.2 FIT Scenario

The FIT scenario will utilize outcomes of the Australian RE policy impact analysis and project changes in electricity supply sources within the NEM in 2020 according to the following assumptions:

1. Solar and wind power deployment increases are calculated based on deployment trends in 2014 [8–11]. Generation is determined based on average NEM solar and wind per annum generation levels.
2. In order to determine residential solar PV net FIT payments, electricity export rates are rationalized between 32 % and 50 % depending on the yearly average size of PV systems installed [12].

3.2.3 Alternative Energy Policy Scenario

The alternative energy policy scenario will utilize key learnings derived from the baseline and FIT scenarios in order to optimize social equity outcomes according to Australian equity preferences while achieving RET goals to the highest degree practicable.

Assumptions common to all scenarios are as follows:

1. Electricity consumption will reduce by 0.5 % per annum from 2015 (based on 5-year average consumption trends [11, 13]) due to energy efficiency improvements and the retirement of energy-intensive industry.
2. Generation from hydroelectric sources within the NEM will remain stable at historical average levels, ignoring impacts such as drought or high rainfall years [13], and no further installation will occur before 2020 [14].
3. Liquid fuels' contribution to NEM generation will be locked at one tenth of 1 % of the total generation in each year.
4. Biofuels' growth is forecasted using 2008–2013 data and gas' using 2008–2014 data, both projected forward in 2020.
5. It is assumed that sufficient generation plant will be in place to meet peak demand. Therefore, each GWh generated from renewable sources will offset fossil fuel generation. This offset will be divided across black and brown coal, dependent on the type and location of the installed renewable energy (e.g., black coal for solar PV and bioenergy installation, predominantly installed in Queensland and NSW, and brown coal for wind, predominantly installed in South Australia and Victoria). Reduction in annual electricity generation and increases in gas generation are reduced across brown and black coal according to their market share.
6. The GHG intensity factors of each generation technology type are constant over time, as shown in Table 2.

3.3 *Australian Equity Preferences*

Although a "fair go" is a part of Australian culture, Australia is at the high end of income inequality, and the gap between the richest (20 %) and poorest (20 %) is

Table 2 Generation GHG intensity factors

Generation source	GHG intensity	Ref
Black coal	0.87 tCO2-e/MWh	[15, 19]
Brown coal	1.25 tCO2-e/MWh	
Gas	0.46 tCO2-e/MWh	
Liquid fuels	0.92 tCO2-e/MWh	
Biofuel (combustion)	0.024 tCO2-e/MWh	
Solar (mono, multi-Si, CdTe)	0.036 tCO2-e/MWh	[16]
Hydropower (run of river)	0.0087 tCO2-e/MWh	[17]
Wind (onshore)	0.0093 tCO2-e/MWh	[18]

similar to that of the UK, the USA, Singapore, and New Zealand [20]. When asked about climate change, survey respondents stated that government policy should create fairness and balance in society, based on their belief that climate change affects low-income groups the most [21]. When asked to describe an ideal future for Australia, workgroup respondents across multiple locations identified common factors including access to good education; participatory democracy; freedom; work-life balance; a healthy, climate change-contained environment; sustainable industries; and equitable access to services and resources. All respondents identified a preference for social equity (specifically full employment and wealth distribution) and preservation of the natural environment over economic growth [22]. When assessing intergenerational distribution preferences, respondents understood that those who benefit from the implementation of a policy are unlikely to be the same people who are paying for them. While a small portion (approx. 12 %) of respondents chose the preservation of the societal status quo, the majority chose to favor young, or future generations, even when "nontrivial" amounts of money were involved. They reasoned that investment would help younger generations and their willingness to invest was based on perceived impacts which would affect future generations negatively [23]. This future-oriented conservation focus was reinforced in a survey of acceptable risk and social values of water allocations which again identified strong support for intergenerational equity and a preference for evidence-based policies and plans managed for the public good [24]. In addition, when health-care decision makers were surveyed on desirable allocation of health gains, a majority favored the young, those of poor health, and where preference was specified, those of a lower socioeconomic status [25]. It should be noted that in some cases these preferences are assessed prior to implementation of policies and may be representative of respondent's desires rather than an approximation of their actual actions. Based on the analysis of recent empirical findings, an approximation of a desirable distribution for Australian economic and environmental costs and benefits is detailed in Fig. 1.

The distribution bias is determined by the gap between richest and poorest households' level of benefit or incurred cost, and, based on this distribution, an "equity snapshot" is prepared which expresses the relative equity afforded to each

Fig. 1 Australian preferred cost and benefit distribution

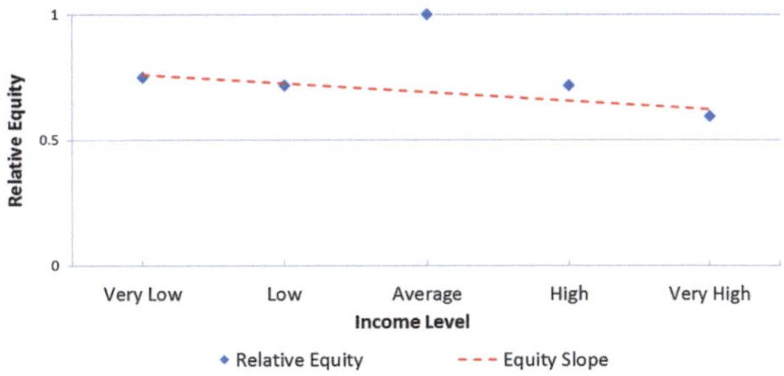

Fig. 2 Australian preference equity snapshot

income level as a result of a given policy. Equity is quantified in this study through the consideration of each income levels' benefit or burden for each of the economic and environmental factors evaluated, and, within the equity snapshot, an equity slope is drawn, demonstrating the balance of equity in society. A positive slope means a reduction in equity (or equity which favors high-income households) and vice versa. Figure 2 clearly demonstrates the reported desire of the Australian public for a more equitable society, where lower-income households benefit from energy policy implementation and the gap between rich and poor is redressed.

4 Results

Each scenario's NEM electricity generation trends in 2020 are summarized along with the scenario-specific economic and environmental outcomes. The distribution of these impacts is then presented concurrently, according to the framework alongside a weighted, comparative equity analysis.

4.1 *Baseline Scenario*

With no exogenous stimulation of RE, generation comes predominantly from fossil fuels, with limited growth experienced in PV and wind power as shown in Fig. 3.

Under the baseline scenario, the major impact on GHG reduction is the annual generation decrease in the NEM, and RE generation goals for both RE generation and large-scale generation are not met. The environmental outcomes of this scenario are outlined in Table 3. The baseline scenario causes minimal negative economic impact, as subsidization is not offered for RE generation meaning that electricity prices are not significantly impacted. However, with limited new RE installation, particularly PV, the learning curves are relatively lethargic. The

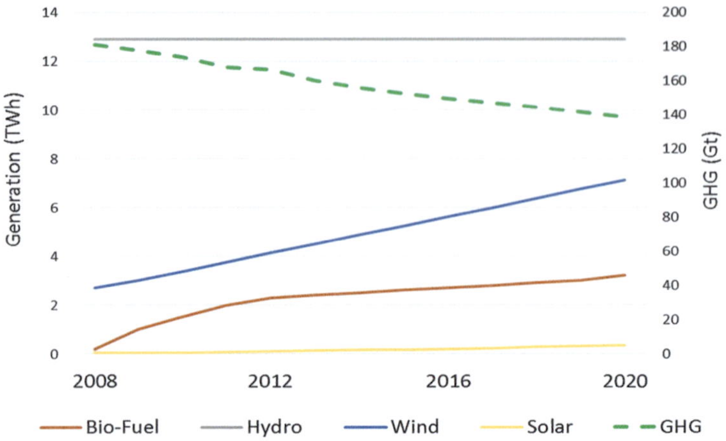

Fig. 3 Baseline RE generation and GHG forecast

Table 3 Baseline environmental outcome summary

Factor	2020 outcomes
GHG emissions	*Gross GHG emissions* **reduced by 23.4 %**
	Generation GHG intensity **reduced by 13.9 %**
Resource management	RE generation increased to **12.8 %** of the NEM total, **63.8 % of the target**
	Large-scale RE installed totals 23,220 GWh, **56.6 % of the target**
RE technology deployment efficiency	*GHG intensity of RE*
	Ideal: 0.009t $CO_{2\text{-}e}$/MWh
	Actual: 0.011t $CO_{2\text{-}e}$/MWh, **126.7 % of ideal**
	Generation efficiency
	Ideal: 2600 MWh/MWp
	Actual: 1791 MWh/MWp, **68.9 % efficient**

levelized cost of electricity (LCOE) is calculated using Australian Treasury modeling studies [26] alongside future Organisation for Economic Co-operation and Development (OECD) generation cost analyses [27]. Direct RE jobs are derived using job multipliers and identified RE job impacts in Australian [3].

The job number represents jobs created in 2020, while hydropower job numbers are sourced from actual Australian 2015 employment numbers, assumed to be constant [28]. Cost reductions for PV and wind in each scenario are based on learning curves in the author's case study and conservative estimates based on additional national and international learning curve data [8, 10, 29, 30], as shown for all scenarios in Figs. 4 and 5.

A summary of environmental impacts of the baseline scenario is provided in Table 4.

Fig. 4 PV market impacts

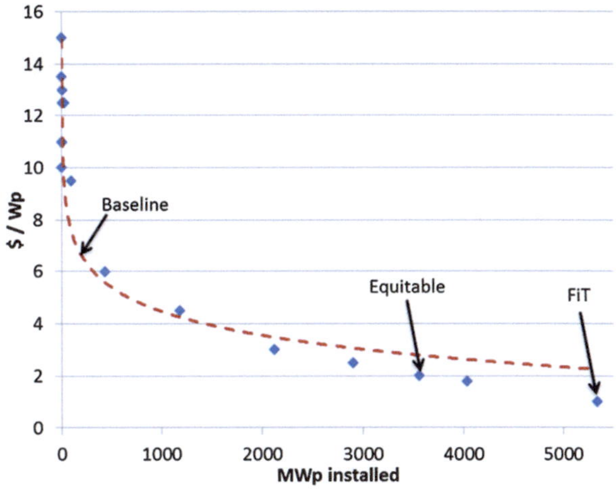

Fig. 5 Wind market impacts

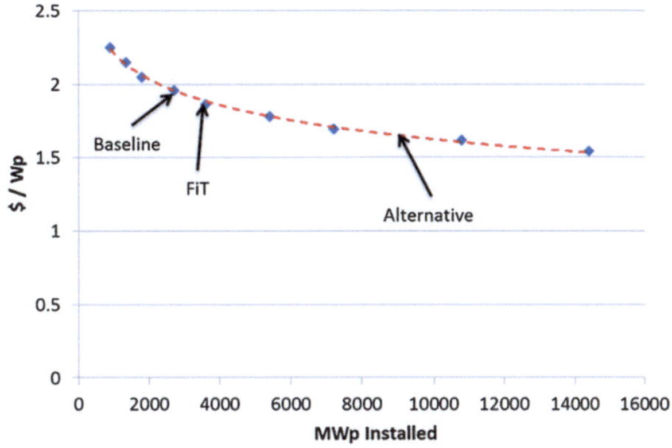

Table 4 Baseline economic outcome summary

Factor	2020 outcomes
Cost of generation	LCOE of *\$92.12/MWh*
Electricity price	*Nonsignificant change*
GDP impact	2573 direct RE jobs in 2020, 28 % growth from 2008
Market impact	Solar PV \$7/Wp, wind \$1.90/Wp (44 % and 17.2 % reductions from 2008)

4.2 *FIT Scenario*

Under the FIT scenario, representative of Australian RE policy from 2008, rapid growth in the PV sector is experienced, increasing overall RE installation compared to the baseline scenario, as shown in Fig. 6.

As with the baseline scenario, FIT scenario environmental and economic outcomes are summarized in Tables 5 and 6.

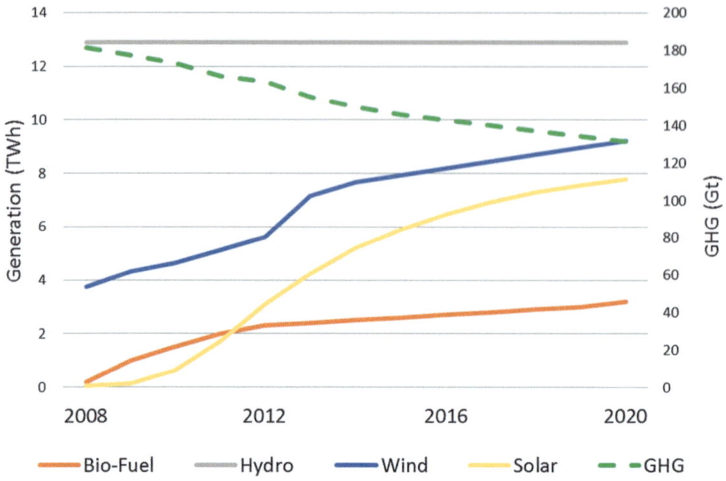

Fig. 6 FIT scenario RE generation and GHG forecast

Table 5 FIT environmental outcome summary

Factor	2020 outcomes
GHG emissions	*Gross GHG emissions* **reduced by 27.6 %**
	Generation GHG intensity **reduced by 19.6 %**
Resource management	RE generation increased to **17.9 %** of the NEM total, **89.6 % of the target**
	Large-scale RE installed totals 25,330 GWh, **61.8 % of the target**
RE technology deployment efficiency	*GHG intensity of RE*
	Ideal: 0.0091t $CO_{2\text{-}e}$/MWh
	Actual: 0.0168t $CO_{2\text{-}e}$/MWh, **185 % of ideal**
	Generation efficiency
	Ideal: 2600 MWh/MWp
	Actual: 2043 MWh/MWp, **78.6 % efficient**

Table 6 FIT economic outcome summary

Factor	2020 outcomes
Cost of generation	LCOE of *$99.45/MWh*
Electricity price	$578.4 million in FIT payments in 2020, $79.01 per non-FIT household
GDP impact	3761 direct RE jobs in 2020, 51 % growth from 2008
Market impact	Solar PV $1/Wp, wind $1.70/Wp (92 % and 24.4 % reductions from 2008)

4.3 Alternative Energy Policy Scenario

Following the vastly different results obtained from the baseline and FIT scenarios, in terms of environmental and economic benefits gained or costs incurred, an appreciation of the impact policy settings have on sustainability outcomes within society is made clear. In order to reduce some inequitable outcomes projected under the current FIT scenario and additionally to fully meet the environmental goals of the RET, learnings from both baseline and FIT scenarios are applied to a new energy policy. Key learnings are summarized in Table 7.

In applying these learnings, the alternative energy policy scenario will, from 2015, prioritize the installation of large-scale wind power as shown in Fig. 7 in order to meet RET goals and end installation of PV in order to reduce the burden of FIT payments on non-FIT households.

Increasing wind power generation has a double benefit, as wind is installed predominantly in the southern, brown coal states. Under this scenario, by 2020, wind generation exceeds brown coal in the NEM. A summary of the environmental and economic impacts of this scenario is shown in Tables 8 and 9.

Table 7 Scenario learnings

Environmental learnings	Economic learnings
Wind power is the best option generating technology, 2nd best GHG/MWh technology (the most efficacious, hydro is maximized)	Solar PV is the cheapest technology per Wp; however, wind power's superior generation per Wp installed makes it a cheaper generating technology
A large (>5000 MWp) installation of residential PV is insufficient to achieve RET environmental goals	Wind power deployment generates the second highest amount of jobs
	Solar PV deployment creates the most RE jobs/MWp but is also the most expensive from an LCOE and price impact standpoint

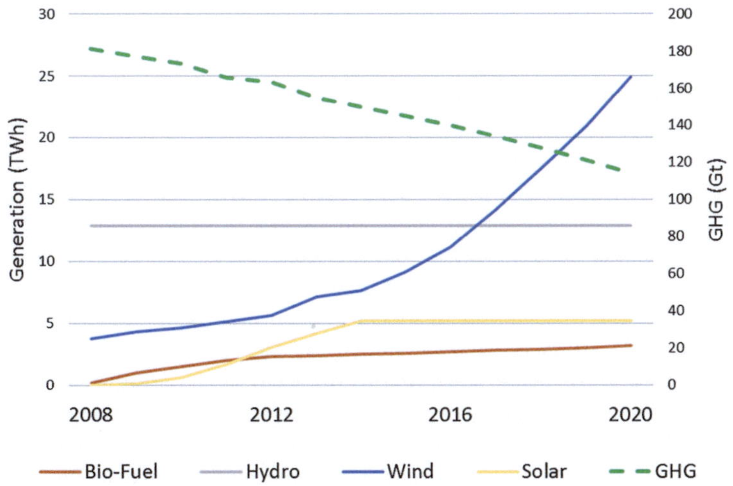

Fig. 7 Alternative scenario generation and GHG forecast

Table 8 Alternative environmental outcome summary

Factor	2020 outcomes
GHG emissions	*Gross GHG emissions* **reduced by 37.1 %**
	Generation GHG intensity **reduced by 29.3 %**
Resource management	RE generation increased to **24.98 %** of the NEM total, **exceeding the target**
	Large-scale RE installed totals 41,000 GWh, **meeting the target**
RE technology deployment efficiency	*GHG intensity of RE*
	Ideal: 0.0091t $CO_{2\text{-}e}$/MWh
	Actual: 0.0132t $CO_{2\text{-}e}$/MWh, **145 % of ideal**
	Generation efficiency
	Ideal: 2600 MWh/MWp
	Actual: 2265 MWh/MWp, **87.1 % efficient**

Table 9 Alternative economic outcome summary

Factor	2020 outcomes
Cost of generation	LCOE of *$96.36/MWh*
Electricity price	$496.4 million in FIT payments in 2020, $64.47 per non-FIT household
GDP impact	5894 direct RE jobs in 2020, 136 % growth from 2008
Market impact	Solar PV $2/Wp, wind $1.65/Wp (84 % and 26.7 % reductions from 2008)

4.4 Comparative Equity Analysis

Each scenario describes a vastly different energy future for the year 2020, achieving environmental and economic goals at differing levels of efficacy and efficiency. The achievement and means of achievement of these goals impact upon societal equity, as each household is impacted differently according to their level of participation and subsequent allocation of subsidies, the amount and distribution of GHG reductions, as well as policy-driven electricity price and employment. In order to derive an equity distribution and slope for each scenario, the distribution of costs and benefits is determined, and their impact is weighted according to the comparative size of each impact, across the three scenarios. Table 10 outlines the precedents and assumptions for these distributions and weighting.

A matrix of the distribution factors, based on precedents and calculations as outlined in Table 10, is initially populated for each scenario. These distribution factors are then rationalized according to the ratio of the absolute values of the weighting factors, simultaneously across all three scenarios. This concurrent comparative analysis identifies the relative cost and benefit distribution bias and equity slope simultaneously for each of the three scenarios as shown in Figs. 8 and 9. Cost and benefit distribution bias is determined for each scenario based on the difference in relative distribution of impacts, between the lowest- and highest-income levels. The equity slope takes an equally weighted, rationalized (with a

Table 10 Equity distribution and moderation factors

Equity factor	Distribution factors	Weighting factors
Participation	Australian participation precedents in [31, 32]	% of nonsubsidized households
GHG reduction	Assumed to be equal	Gt GHG reduced
Employment	Case study job allocation and salaries in [33]	Number of direct RE jobs in 2020
Subsidy allocation	Participation rate multiplied by % of households per income level	Subsidy (FIT) payment amount
Electricity price impact	Electricity price % increase due to subsidization (or LCOE increase) per income level	Actual $ increase per annual average electricity bill

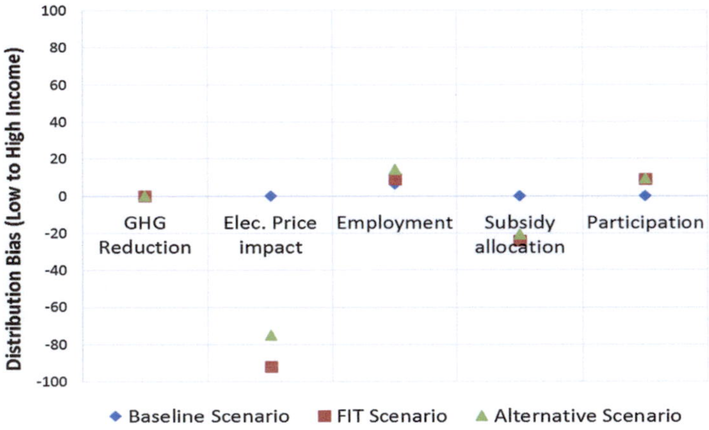

Fig. 8 Per-scenario cost and benefit distribution bias

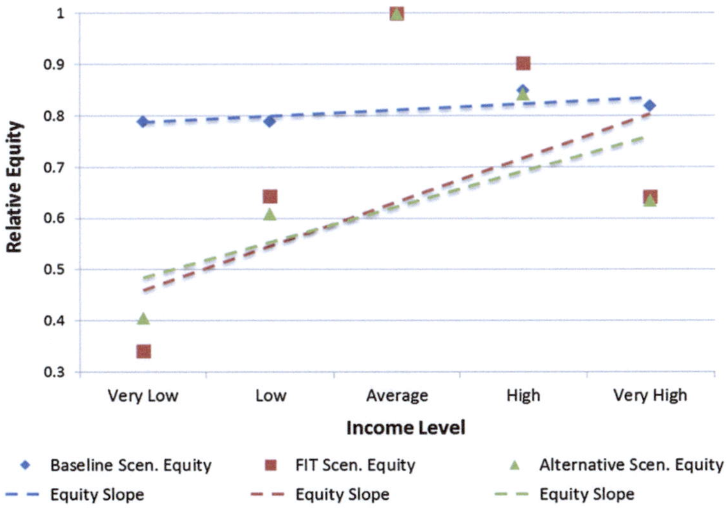

Fig. 9 Per-scenario equity slope

maximum value of 1) assessment of each of the five equity impact factors across the five defined income levels and plots a regression line (least-squares method) between them in order to enable easy comparison of equity outcomes for each of the three scenarios.

The distribution bias and equity slope results demonstrate that the baseline scenario impairs equity the least, while the FIT scenario impairs it the most. The evidence-based alternative scenario is the only scenario under which RET goals are met, and, when compared to the FIT scenario, even with the FIT in place until 2015, equity is improved, evidenced by the milder positive slope and reduced gap between very low- and very high-income household relative equity shown in Fig. 9.

5 Discussion

The method proposed has proven to be useful in assessing existing RE policies and making informed, evidence-based changes in order to derive a more equitable policy based on the best available data and other evidence. This tool may be useful at the evaluation phase for energy policy in order to ensure that negative equity impacts are controlled while meeting policy goals. It is valuable to undertake a comparison with complementary methods which also seek to measure equity. There are some interesting results in the distribution and equity analysis, specifically with regard to subsidy allocation and participation, and a consideration of income and equity factor Lorenz curves may be useful. Figure 10 outlines the Lorenz curve for each of the five social equity factors considered in this study alongside income distribution.

Fig. 10 Equity factor and income Lorenz curves

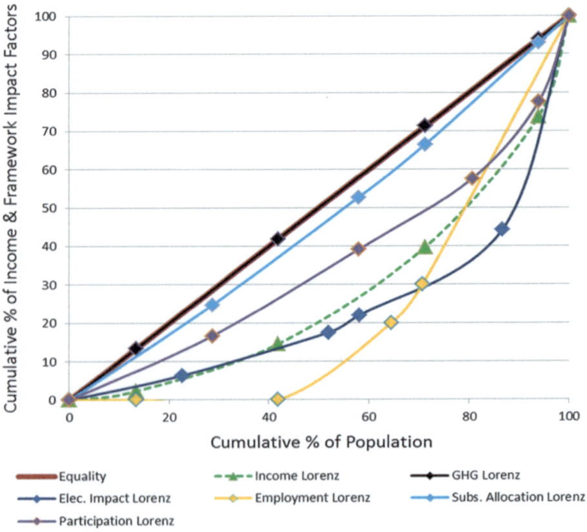

As Fig. 10 demonstrates, GHG reduction is equally shared across society, and the most equal distributions thereafter are participation and income; however, the electricity price and employment distributions are decidedly unequal. While the Lorenz curve is a useful tool for an overall snapshot of the shape of societal equality, it is difficult to pinpoint the level of impact that each factor has on an overall societal equity analysis. The method proposed in this study does provide such an insight through both a distribution and relative equity analysis across income levels.

6 Conclusion

In this study, equity has been quantified, through a consideration of the distribution and level of impact of economic and environmental energy policy factors. This original approach, achieved through policy scenario comparison, evidence gathering, and the proposal of alternative energy policies utilizing the proposed framework and methodology, leads to better equity outcomes – more in accord with Australian equity ideals and an expedient achievement of energy policy goals.

Through a consideration of existing frameworks, a new methodology which is accessible, applicable in multiple jurisdictions, and useful to policy makers has been developed. The combined use of the framework and other equity appraisal tools such as the Lorenz curve may assist the policy maker in not only evidence-based policy development and refinement but also identify equity factor priorities in policy decision-making processes.

By improving economic, environmental, and equity outcomes, it follows that sustainability will also be positively impacted.

References

1. Buckman G, Diesendorf M (2010) Design limitations in Australian renewable electricity policies. Energy Policy 38:3365–3376
2. International Union for Conservation of Nature (2006) The future of sustainability: re-thinking environment and development in the twenty-first century. Prepared by W.M. Adams Professor of Conservation and Development, Department of Geography, University of Cambridge, UK. p 2
3. Chapman A, McLellan B, Tezuka T (2016) Residential solar PV policy: an analysis of impacts, successes and failures in the Australian case. Renewable Energy 86C:1265–1279
4. United Nations Commission on Sustainable Development (2001) Indicators of sustainable development: framework and methodologies. Background Paper No. 3, Department of Economic and Social Affairs
5. World Bank (2012) Inclusive green growth – the pathway to sustainable development. International Bank for Reconstruction and Development/International Development Association, Washington, DC

6. International Panel on Climate Change (2007) Climate change 2007: Mitigation. Contribution of Working Group III to the Fourth Assessment Report of the IPCC. In: Metz B, Davidson OR, Bosch PR, Dave R, Meyer LA (eds). Cambridge University Press, Cambridge/New York
7. International Atomic Energy Agency (2005) Energy indicators for sustainable development: guidelines and methodologies. International Atomic Energy Agency, Vienna
8. Australian PV Institute (2014) Australian PV market since April 2001. Australian PV installations since April 2001: total capacity (kW)
9. Clean Energy Council (2012–2014) Clean energy Australia reports 2011–2013
10. International Energy Agency (2008–2011) IEA wind annual reports 2007–2010
11. Australian Energy Regulator (2009–2014) State of the energy market reports 2009–2014
12. Independent Pricing and Regulatory Tribunal (2012) Solar feed-in tariffs: setting a fair and reasonable value for electricity generated by small-scale solar PV units in NSW. Energy – Final Report March 2012
13. Green Energy Markets (2014) Review of the NEM in 2014. Snowy and Tasmanian Hydro historical generation levels
14. Elliston B, MacGill I, Diesendorf M (2013) Least cost 100% renewable electricity scenarios in the Australian National Electricity Market. Energy Policy 59:270–282
15. Australian Energy Market Operator (2014) Carbon Dioxide Equivalent Intensity Index (CDEII) Available online: <http://www.aemo.com.au/Electricity/Settlements/Carbon-Dioxide-Equivalent-Intensity-Index>
16. Fthenakis V, Kim H (2011) Photovoltaics: life-cycle analyses. Sol Energy 85:1609–1628
17. Varun I, Prakash R (2009) LCA of renewable energy for electricity generation systems: a review. Renew Sustain Energy Rev 13:1067–1073
18. Geuzuraga B, Zauner R, Polz W (2012) Life cycle assessment of two different 2MW class wind turbines. Renew Energy 37:37–44
19. Farine DR, O'Connell DA, Raison J, May R, O'Connor MH, Crawford DF, Herr A, Taylor JA, Jovanovic T, Campbell PK, Dunlop MIA, Rodriguez LC, Poole ML, Braid AL, Kriticos D (2012) An assessment of biomass for bioelectricity and biofuel, and for greenhouse gas emission reduction in Australia. GCB Bioenergy 4(2):148–175
20. Wilkinson R, Pickett K (2010) The spirit level: why equality is better for everyone. Penguin Books, London
21. McManus P, Shrestha K, Yoo D (2014) Equity and climate change: local adaptation issues and responses in the City of Lake Macquarie, Australia. Urban Clim 10(1):1–18
22. Boschetti F, Fulton EA, Grigg NJ (2015) Citizens' views of Australia's future to 2050. Sustainability 7(1):222–247
23. Scarborough H, Bennett J (2008) Estimating intergenerational distribution preferences. Ecol Econ 66(4):575–583
24. Syme GJ (2014) Acceptable risk and social values: struggling with uncertainty in Australian water allocation. Stoch Env Res Risk A 28(1):113–121
25. Mooney G, Jan S (1997) Vertical equity: weighting outcomes? Or establishing procedures? Health Policy 39:79–87
26. Australian Academy of Technological Sciences and Engineering (2011) New power cost comparisons; levelised cost of electricity for a range of new power generating technologies. Report prepared for the Garnaut Climate Change Review, March 2011, Melbourne, Australia
27. West J (2012) A comparative analysis of the future cost of electricity generation in OECD and non-OECD countries. Electr J 25(1):68–83
28. Australian Bureau of Statistics (2015) Statistics: 4631.0 – employment in renewable energy activities. Commonwealth of Australia, Australian Capital Territory
29. Junginger M, Faaij A, Turkenburg WC (2005) Global experience curves for wind farms. Energy Policy 33(2):133–150
30. Melbourne Energy Institute (2011) Renewable energy technology cost review. Technical Paper Series March 2011. The University of Melbourne

31. Higgins A, McNamara C, Foliente G (2014) Modelling future uptake of solar photo-voltaics and water heaters under different government incentives. Technol Forecast Soc Chang 83:142–155
32. Bruce A, Watt ME, Passey R (2009) Who buys PV systems? A survey of NSW residential PV rebate recipients. In: Proceedings from Solar09, the 47th ANZSES annual conference, Townsville, Australia
33. Payscale (2015) Average salary for industry: renewable energy power generation – Australia. Available Online: <http://www.payscale.com/research/AU/Industry=Renewable_Energy_Power_Generation/Salary>

Study on the Diffusion of NGVs in Japan and Other Nations Using the Bass Model

Yue Zhu, Koji Tokimatsu, and Mitsutaka Matsumoto

Abstract In this study, the Bass Model was utilized to estimate the future diffusion of natural gas vehicles (NGVs), in Japan and other countries. The parameters contained in the model, namely, coefficient of innovation (p), coefficient of imitation (r), and the potential market (N), were extracted from the trend as indicated by the actual diffusion curve up to 2014, after which the ultimate diffusion number was calculated based mainly on these parameters. Firstly, three scenarios were established to analyze the future diffusion of NGVs in Japan, with the highest outcome being 8.5 million, accounting for 14 % of the total vehicles. Secondly, as a feasibility study, the Bass Model was utilized to examine twelve countries where NGVs are popularized. The actual diffusion curve pertaining to Iran, Argentina, India, Brazil, Italy, Columbia, Thailand, Ukraine, and Bolivia could ideally fit the Bass Model curve, while the fit was poor for China, Pakistan, and the United States. The reasons for the poor fit were then explained. Finally, the conditions and requirements to facilitate the highest diffusion number in Japan were discussed.

Keywords Natural gas vehicle • Bass model • Diffusion model • Scenario analysis

1 Introduction

1.1 Background

As the realization of a sustainable society progresses, the development of environmentally conscious design products is growing in importance, as is promoting the diffusion of these products. In this study, natural gas vehicle (NGV), as one type of clean-energy vehicle, was emphasized as the focus of our research with the aim to

Y. Zhu (✉) • K. Tokimatsu
Department of Environmental Science and Technology, Interdisciplinary Graduate School of Science and Engineering, Tokyo Institute of Technology, Yokohama, Kanagawa, Japan
e-mail: etsuhector@gmail.com

M. Matsumoto
Human Informatics Research Institute, National Institute of Advanced Industrial Science and Technology (AIST), Tsukuba, Ibaraki, Japan

M. Matsumoto et al. (eds.), *Sustainability Through Innovation in Product Life Cycle Design*, EcoProduction, DOI 10.1007/978-981-10-0471-1_52

justify the creative scenarios and to establish its diffusion number in road transportation in several countries.

Natural gas is the main fuel used for NGVs. Compared with the 1980s, the proportion of natural gas in the current supply of primary energy sources has increased in comparison with that of oil. As is indicated by projections of the New Policies Scenario (World Energy Outlook [WEO] – 2013) [1], the growing demand for natural gas as a primary source of energy will exceed that of any other energy source. The scenario for 2035 indicates that this demand would be approximately 200 % more than it was during the 1980s.

At the same time, in pace with the recent advances in technology, natural gas from shale has become the fastest growing contributor to the total primary energy sources in the United States. Many other countries are following suit and have started to pursue the exploitation of their shale gas deposits. The interest in shale gas has influenced the global market price of natural gas. According to WEO – 2013 [1], the price of natural gas in three representative regions (Japan, Europe, and the United States) is likely to fall somewhat in the near future.

1.2 *Features of NGV*

As a representative product relative to natural gas, a natural gas vehicle (NGV) possibly has distinct economic and environmental advantages compared with the other types of vehicles. Engerer and Horn [2] have pointed out that the performance of the NGV is generally good; while its emission of CO2 is lower compared with that of other vehicles. Moreover, an NGV has the lowest emission of NOx and nearly zero emission of particulate matter [2]. An NGV is regarded as an outstanding, clean-energy vehicle, and the diffusion of these vehicles is taking place worldwide. Currently, an estimated 19.9 million NGVs are in use worldwide [3]. In Japan, the diffusion of NGVs had started in 1990 and 43,600 NGVs are in use at present [3].

As the comparative contents of Table 1 indicate, NGVs have a number of advantages, including a mature production technology and an outstanding performance. NGVs also have low-level emissions of atmospheric pollutants, such as

Table 1 Features of NGVs compared with conventional gasoline vehicles

Life cycle CO_2 emission	Life cycle NOx emission	Atmosphere pollution	Running distance
60–80 %	Very few when compared to GV	Nearly ZERO	50 %
Running cost (fuel price/running distance)	Vehicle price	Infrastructure	Production technology
50 %	140–200 %	Relatively less	Mature

Moreover, future atmosphere pollution from NGV may completely disappear according to improvement of denitrification technology or substitutable fuel as liquefied natural gas nearly zero levels of NOx or SOx. However, the disadvantages of NGVs include a limited number of natural gas stations and a relatively poor running distance.

1.3 Objectives

In this study, scenario analyses were conducted to project the diffusion of NGVs in Japan until 2050. Three research questions were addressed by the study, as follows.

Firstly, what was the possibility of NGVs diffusing in Japan and what would be the extent of diffusion for the period between 2020 and 2050?
Secondly, how should future environmental conditions be changed to help achieve the highest diffusion level for NGVs in Japan?
Thirdly, what would future diffusion number in other countries probably be where NGVs are popularized?

2 Methodology

The Bass Model is widely used for accurately describing the popularizing process of industrial products. The rules include [4, 5]:

1. Customers who initially buy the products can be defined as the sum of "innovators," who have decided independently to purchase, while "imitators" are those who followed the decision of the innovators to make the purchase.
2. The proportion of the number of innovators (during one period), "V_t," in a population who has never made a purchase (during the same period), "Y_t," is fixed and could be shown as "p."
3. The proportion of the number of imitators (during one period), "W_t," in a population who has never made a purchase (during the same period), "Y_t," could be shown as "p_t," where it fulfills the condition "$p_t' = r \cdot n_t$," where "n_t" is the diffusion rate (during the same period) and "r" is constant.
4. The ultimate diffusion number "N" is fixed. This means the population who has never made a purchase "Y_t" and the cumulative population who has made a purchase "$n_t \cdot N$" always fulfills the condition "$Y_t + n_t \cdot N = N$." Thirdly, what would future diffusion number in other countries probably be where NGVs are popularized?

Following the above relationship, new customers during period t, shown as "X_t," and an alternative formula could be indicated as follows:

$$X_t = V_t + W_t \tag{1}$$

$$V_t = p \cdot Y_t = p \cdot (1 - n_t) \cdot N \tag{2}$$

$$W_t = p_t' \cdot Y_t = r \cdot n_t \cdot Y_t = r \cdot n_t \cdot (1 - n_t) \cdot N \tag{3}$$

Hence, in the case of period t, seen as a continuous value, an additional boundary condition settled as "$n_t = 0 = 0$" is given, and the Bass Model with a cumulative diffusion number until period t, shown as "N_t," is formulated as follows [5]:

$$N_t = N \cdot \frac{1 - e^{-(p+t)t}}{1 + r/p \cdot e^{-(p+t)t}} \tag{4}$$

To make the definition of the three parameters meaningful, "N" is the ultimate carrying capacity, "p" is the coefficient of innovation, "r" is the coefficient of imitation, and "N_t" on the left-hand side is the diffusion number at time t.

Generally, we considered two diverse conditions in which the ideal value of the three parameters ("N," "p," and "r") were obtained in different ways when the Bass Model was utilized to project the diffusion estimation. Firstly, if the product has some market share, the values for "p," "r," and "N" are estimated using the actual diffusion trend to date. Secondly, if the product has a limited market share, the parameter values of similar products having some market share are used.

The data we employed were obtained from the Japan Gas Association [3] (originally from The Gas Vehicles Report). These data indicated the rate of diffusion of NGVs in the relevant countries for the period from the 1990s to the end of 2013. The relevant countries chosen for the research are Japan and 12 other countries where the rate of diffusion of NGVs was high. In descending order, these countries are Iran, China, Pakistan, Argentina, India, Brazil, Italy, Columbia, Thailand, Ukraine, Bolivia, and the United States.

3 Results

3.1 Japan

3.1.1 Scenario Description

Scenario 1: The diffusion of NGVs up until March 2014 was assumed to approach market saturation level; therefore, p, r, and N were estimated by using the actual diffusion trend to date.

Scenario 2: From 2020, it was assumed that the NGVs would have a diffusion ratio similar to that of the diesel vehicle (DVs) in Japan in the past. The DV was chosen as a similar product for reasons that will be explained later. Therefore, the diffusion rate was expected to progress rapidly up to 2050.

Scenario 3: For unspecified reasons, the diffusion of NGVs would be one-tenth of the annual increase of Scenario 2.

3.1.2 Result of Scenario 1

Scenario 1 is coherent with the diffusion rate experienced from 1990 to 2013, which was at saturation level at that time.

The diffusive parameters, namely, the coefficient of innovation "*p*," coefficient of imitation "*r*," and the carrying capacity "*N*," were extracted from the diffusion curve trend up until 2013. Moreover, the annual diffusion rate up until 2050 was calculated by using the values obtained for "*p*," "*r*," and "*N*" to extend the diffusion rate curve.

Concerning the parameter settings after fixing the value of "*N*," we minimized the square error between the Bass Model curve and the actual diffusion curve. Once the accumulated least square error was found for the initial value of "*N*," the value of "*N*" was changed as unit of 1000 less or more to confirm the ideal values for "*r*," "*p*," and "*N*." The results were 0.370, 0.00224, and 45,000, respectively. The outcome shows that the diffusion number for 2050 in this scenario was 45,000 (43,600 NGVs were in use in Japan up until March 2014). This amounts to 0.07 % of the total vehicles in 2050 in Japan (the total number of vehicles in Japan in 2050 was expected to be 63,900 according to the Ministry of the Environment, Japan [6]). To make the results be easily recognized, we marked the simulating outcomes as a blue curve in Fig. 1 as diffusion number in Scenario 1. Hence, for comparison to the historical diffusion, the blue curve was plotted together with a red curve which shows the actual diffusion number.

Additionally, the diffusion number estimated in Scenario 1 until 2020 was considered to constitute at an initially slow rate of diffusion, before the conspicuous rapid increase of diffusion after 2020 in the other two scenarios.

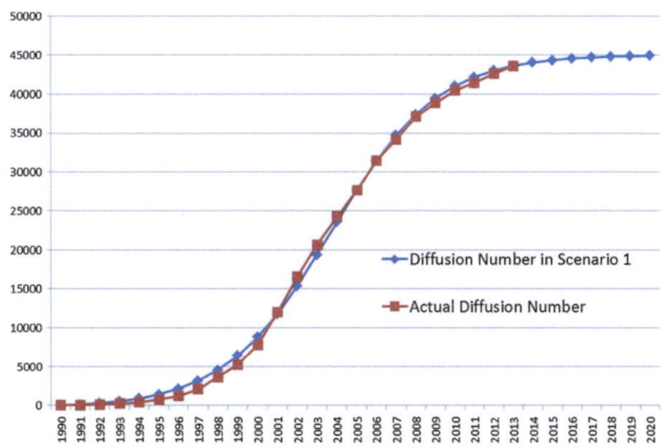

Fig. 1 Results of estimated diffusion number of NGV compared to actual diffusion number in Scenario 1

3.1.3 Result of Scenario 2

In Scenario 2, the conspicuous rapid increase of diffusion after 2020 was assumed comparable with that of DV in Japan between the 1970s and the 1990s. Our reasons for assuming that DV and NGV have significant similar features are as follows:

- NGV and DV both emit less CO2 than does the conventional GV [3].
- NGV and DV both use fuel that is less expensive than gasoline. According to the Tokyo Gas Corporation, the fuel price ratio for one kilometer for three types of vehicles is NGV:DV:GV = 7:10:14 [7].
- Both NGV and DV are popular as heavy-duty vehicles, such as buses or trucks. In contrast, the hybrid vehicle (HV) and the electric vehicle (EV) are popular as passenger vehicles.
- NGV and DV are both equipped with mature technologies, while other green-energy vehicles, such as the EV and the fuel cell vehicle (FCV), still contend with technological obstacles that impede the wide deployment of these vehicles.
- Relevant to the next-generation energy for vehicles (electric power, natural gas, or hydrogen), the EV could be suitable for small, light cars. However, for heavy vehicles such as buses or trucks, NGV would be the sensible choice as a substitute for the DV [3]. The NGV is therefore the obvious choice as replacement for the main role of DVs, which is as heavy vehicles.

The preceding indicates that in Japan the DV would be the optimal similar product for NGV, relevant to the prediction of rapid diffusion.

Data obtained from Isuzu Motors Limited [8] were employed to obtain the diffusion parameters "*p*," "*r*," and "*N*" for Scenario 2. The diffusion rate of the DV was not a good fit for the Bass Model curve, indicating that it was unsuitable for applying directly to predicting the future spread of the NGV. Therefore, we applied the increasing rate of market share of the DV between 1972 and 1995 instead, which we transferred to the Bass Model. As the Bass Model mainly discussed on the correlation between innovating and imitating behavior in one market, it did not make difference when the estimated object was changed from diffusion number to diffusion rate. In this way, the growing rate of the NGVs from 2021 to 2050 could be described.

Therefore, the growing market share of DVs (1972–1995) was considered the extracted diffusion parameters. The total increased market share from 1972 to 1995 was defined as the whole diffusion procedure, indicated by "$N_{occupancy}$," which measured 13.5 %. Hence, the annual increase in market share from 1972 was measured individually, which was then considered as the annual rate of diffusion. Subsequently, methodology similar to that for Scenario 1 was used to find the optimal values for p and r from this diffusion process; the values obtained were 0.0131 and 0.222, respectively.

Once the values of p and r were confirmed, the annual increase in market share of the NGV for 2020–2050 could be calculated by using Eq. 4. Finally, the annual diffusion rate of NGV after 2020 was described as follows:

$$N_t = N_{2020(\text{Scenario 1})} + N_{\text{occupancy}} \cdot \frac{1 - e^{-(p+t)t}}{1 + r/p \cdot e^{-(p+t)t}} \cdot NV_{\text{yearly}} \quad (5)$$

(*: "NV_{yearly}" means annual total number of vehicles.)

The diffusive parameters were set as $N_{2050} = 8{,}446{,}387$, $p = 0.0131$, and $r = 0.222$. The calculations indicated approximately 8446 NGVs for 2050, accounting for 13.22 % of the total number of vehicles [6].

3.1.4 Result of Scenario 3

Scenario 3 represents the diffusion between that of Scenario 2 and that of Scenario 1. The annual diffusion rate beyond 2020 was calculated in a way similar to that of Scenario 2. The adjustment of the annual increased rate of diffusion was one tenth of that of Scenario 2 for the period 2021–2050. The result was $N_{2050} = 885{,}044$.

3.1.5 Summary of Three Scenarios

The overall results from the three scenarios were summarized in the figure below to facilitate comparison. Figure 2 contained three curves and they were all calculated outcomes on the base of the scenarios and the model.

3.2 Nations Other Than Japan

Twelve countries, of which the diffusion rate of the NGVs was already considerable, were chosen for this study. In descending order, these are Iran, China, Pakistan, Argentina, India, Brazil, Italy, Columbia, Thailand, Ukraine, Bolivia, and the United States. Since there was some market penetration in these countries already, the values for p, r, and N were estimated by employing the actual diffusion

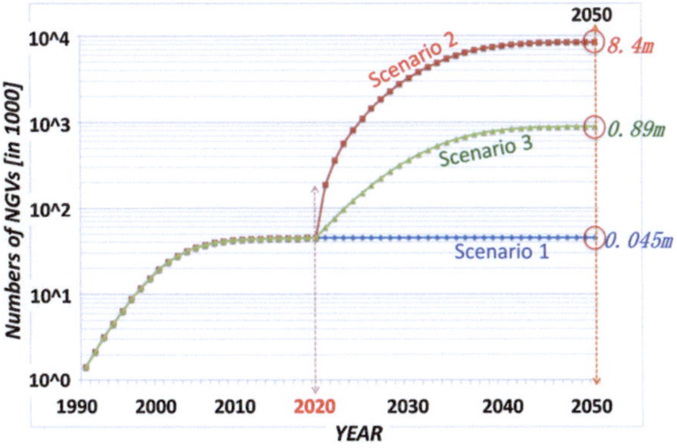

Fig. 2 Results of the diffusion level of NGVs in Japan for Scenarios 1, 2, and 3

trend to date. Furthermore, these countries could be divided into two groups according to whether the Bass Model was a good fit or not.

3.2.1 Nations for Which the Bass Model Was a Good Fit

For nine countries, except China, Pakistan, and the United States, the Bass Model curve ideally fitted the actual diffusion curve. Table 2 summarizes the results (from left to right shows the NGVs' number in 2014, ultimate diffusion number calculated by the Bass Model, value to the coefficient of innovation, value to the coefficient of imitation) clearly.

3.2.2 Nations for Which the Bass Model Was Not a Good Fit

The Bass Model was not a good fit for three countries, namely, China, the United States, and Pakistan. As regards China, the ultimate diffusion number N was calculated as infinite. Figures 3 and 4 indicate that in respect of the United States and Pakistan, the actual diffusion curve was not a good fit for the Bass Model curve.

Table 2 Outcomes of three values for nine nations

Countries	Diffusion status (2014)	*N* (ultimate)	*p*	*r*
Iran	3,500,000	3,900,000	0.0000412	0.668
Argentina	2,331,912	3,000,000	0.00287	0.186
India	1,800,000	2,800,000	0.000495	0.336
Brazil	1,764,137	1,800,000	0.000283	0.539
Italy	823,000	4,300,000	0.0000585	0.0671
Columbia	463,930	600,000	0.0000972	0.350
Thailand	422,812	600,000	$6.08*10^{-6}$	0.511
Ukraine	388,000	700,000	$2.83*10^{-12}$	0.343
Bolivia	273,342	600,000	0.00119	0.295

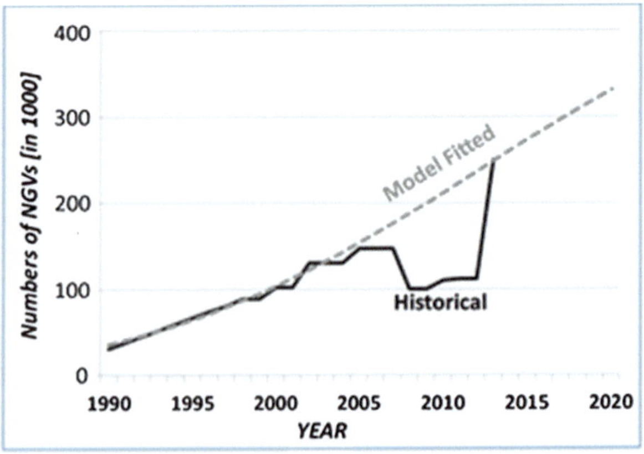

Fig. 3 Comparison of historical data and the model fit results for the United States

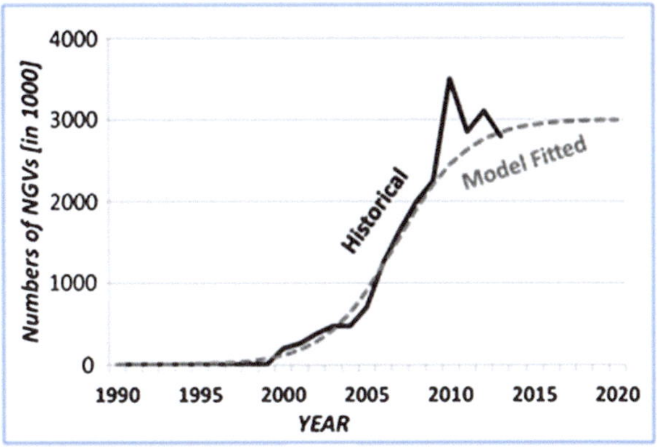

Fig. 4 Comparison of historical data and the model fit results for Pakistan

4 Discussion

4.1 Japan

4.1.1 Comparison with the Report from MOE [6, 9], Japan

In 2009 and 2010, the Ministry of the Environment (MOE) of Japan conducted two diffusion simulations, based mainly on the relationship between the population fluctuation and the preference of various age groups for certain types of vehicle [6, 9]. Table 3 below shows the measured simulation results from the 2010 MOE report for Japan. The NGV diffusion rate by 2050 was calculated as 1289.

Scenario 1 indicates the diffusion of NGVs for 2050 as 45,000, which is markedly lower than is the number calculated by the MOE for Japan in 2010. The scenario would only come true on the condition that there were no change in either the future energy market or the government policy of Japan.

Scenario 3 indicates the NGV diffusion rate for 2050 as 885,000, which is close to the predicted rate of the MOE for Japan of 2009 [9] of 842,000. This can be explained by assuming that the guiding principle of the government of Japan for 2009 was the same as that in our Scenario 3, i.e., a 90 % slower rate than that of Scenario 2.

The diffusion forecast of Scenario 2 was 8446 NGVs in 2050, which is approximately 6.5 times that of the estimate of the MOE of Japan of 2010, which was 1289.

4.1.2 Realizing Method of Scenario 2

Scenario 2 in our research showed an outcome that was approximately 6.5 times that of the MOE [6]. However, the likelihood of Scenario 2 coming true could depend on the number of NGVs substituted for other types of vehicles in future. Among the alternative vehicles other than CDVs, it was most likely that NGVs

Table 3 Predicted rate of diffusion by MOE [6], Japan, 2010

Next-generation vehicle		Diffusion number in 2050 Total as **63,900** (Unit: 1000)	
Light vehicle, track	**EV**	28,441	28,441
Small-sized and normal passenger vehicle	**EV**	5564	31,200
	HV	11,615	
	PHV	11,588	
	CDV*	0	
	FCV	2421	
Normal track and bus	**EV**	289	4259
	HV	1399	
	NGV	**1289**	
	CDV*	936	
	FCV	346	

*CDV** clean diesel vehicle

would replace EVs and PHVs (plug-in hybrid vehicles). The reasons for this statement are explained below.

Firstly, in April 2014, the MOE and the Ministry of Economy, Trade, and Industry (METI) of Japan instituted a policy to promote the establishment of joint fuel stations for both NGVs and FCVs [10]. Therefore, following the introduction of the Toyota FCV Mirai in December 2014, the first joint hydrogen and CNG station successfully activated in the Nerima District of Tokyo. As the predicted number of hydrogen vehicles for 2050 was 1000–5000 (MOE, Japan, 2010 [6]), it could be expected that more such joint fuel stations would be opened. Access to a larger number of fuel stations would obviously be favorable for the diffusion of NGVs.

Secondly, according to the MOE of Japan for 2009 [9], EVs were 200–300 % more expensive than the same class of gasoline vehicle, while NGVs were only 40–100 % more expensive than similar gasoline vehicles.

Thirdly, the information from the MOE of Japan for 2009 [9] indicated that the EVs and PHVs were reliant on the quality and price of their internal batteries. In contrast, the NGVs relied only on the natural gas supply system.

Based on the above reasoning, a hypothesis could be constructed. Future NGV will supposed to exist as a complete substitute for CDV (936,000) (Table 2) firstly, while the rest of 6331 (6331 = 8446 (Scenario 2) – 936 – 1289) could partly replace the market for EV and PHV, where the total number of these vehicles are calculated as 45,882 (45,882 = 28,441 + 5564 + 11,588 + 289) (Table 2). The result shows that in the future NGVs would capture about 1/7 of the EV and PHV market.

Additionally, the number of CNG stations in relation to the substantial diffusion rate of NGVs indicated in Scenario 2 was calculated based on the relation formula from the MOE of Japan for 2009 [9], indicated below:

$$N_{\mathrm{NGS}} = 0.3896 \cdot {N_{\mathrm{NGV}}}^{0.6542} \tag{6}$$

(where N_{NGS} means the number of natural gas stations and N_{NGV} means the number of natural gas vehicles.)

Therefore, for 8446 NGVs, the corresponding number of CNG stations would be 13,244. Elsewhere, the same MOE report for Japan for 2009 [9] indicates that the number of CNG stations in 2050 would be 2934. However, the number of stations obtained in Scenario 2 was approximately 3.5 times larger than was that of the government. The MOE report for Japan for 2009 [9] indicated 13,602 battery charging stations for EV and PHV in 2050. Therefore, to a certain extent, the number of battery charging stations could be replaced by the CNG stations. Additionally, the total number of gasoline stations in 2014 was approximately 34,000 [11], which, to some degree, could also be replaced with CNG stations in view of the expected decline in the number of GVs in future. As a result, the number of CNG stations could be nearly 10,000. Further, as technological development advances, NGVs with a larger natural gas fuel capacity would probably be introduced. This would imply that the projected number of CNG stations required by 2050 would likely be less than the number calculated.

4.2 *Nations, Other Than Japan, for Which the Bass Model Was Not a Good Match*

4.2.1 China

The data relevant to China indicated that NGVs in this vast country appeared to be still at a rapidly diffusing stage. Therefore, when the feasibility study of the Bass Model was utilized, the ultimate carrying capacity N, in this case, was calculated to be infinite, meaning that N was even larger than the population of China (approximately 1368 million). The conclusion was therefore that the Bass Model was not suitable in situations where products were diffusing extremely rapidly.

4.2.2 The United States

In the United States, NGVs showed a slight increase from the 1990s up until 2007; however, the penetration was limited to buses, garbage trucks, and single-unit delivery-truck fleets [12]. In addition, supposedly because of the increase in the number of conventional fuel stations [13], the number of CNG stations has declined between 1996 and 2006. The decline in the number of stations directly affected the rate of diffusion of NGVs for 2008. This is illustrated by the number of NGVs dropping to 100,000, despite being 147,000 in 2007. Following the decline in 2008, the number of NGVs has increased slowly up until 2012, with the diffusion being 112,000 [3].

Fortunately, NGVs have shown a rapid rise in numbers since 2012, with the diffusion reaching 250,000 for that same year [3]. By about 2010, the success of the shale gas revolution had resulted in the increased availability of natural gas; however, the gas market had also become oversupplied [14]. The NGV market was shown to be the best target market for gas suppliers seeking demand, borne out by only 3 % of the primary energy consumption of natural gas in 2011 in the United States being for transportation. Therefore, the NGV market definitely has the most potential to grow [12].

In addition, the US government has published the "Blueprint for a Secure Energy Future" (President Obama, March 2011) and the New Alternative Transportation to Give Americans Solutions Act (Nat Gas Act) of 2011 (US Government, April 2011) to stimulate the diffusion of NGVs [14]. Policies were introduced to, inter alia, substitute the vehicles run in airports with NGVs. As an incentive, a maximum of 4000 dollars was offered as discount for purchasing NGVs instead of conventional medium-sized vehicles [15]. At the same time, vehicle companies such as Honda, Isuzu, and Ford have widened their market share in the United States by promoting NGV sales [14].

4.2.3 Pakistan

The NGV was first introduced in Pakistan during the early 1990s, 70 years after the first commercialization of the technology in Italy. The diffusion rate of NGVs has grown exponentially up until 2010 because of the policy of deregulation of their sales in Pakistan. There were two main motivations for the rapid diffusion of NGVs in Pakistan, namely, (1) a reduction of the dependency on imported oil and (2) the availability of local natural gas and an existing pipeline infrastructure [16].

As a result of the rapid diffusion of NGVs, the proportion of natural gas consumption by NGVs has increased from 0.341 % (2000) to 7 % (2009). Such an obvious dominant market share appears to have displaced directly and indirectly the fertilizer, power, and household sectors that mainly rely on natural gas. The abnormally high consumption of natural gas by transportation has led to severe power disruptions and shortages of fertilizer, a crucial requirement of the agricultural sector. Since 2011, in an effort to remedy the shortage of natural gas, the government of Pakistan has curtailed the diffusion of NGVs in the country. The predicament of Pakistan should serve as a salutary lesson to countries that endeavor to promote the use of NGVs as an alternate to the conventional fuel vehicle option. The judicious allocation of natural gas for transportation has to be considered to avoid the risk of running out of this resource [16].

5 Summary

In conclusion, scenario analyses were carried out to simulate three future diffusion situations for NGVs, based on the introduction of this type of vehicle in Japan. Scenario 2, based on the diffusion of DVs in Japan, was particularly emphasized to discuss the conditions conducive to the diffusion of NGVs. Furthermore, the future diffusion of NGVs in twelve other countries was estimated to show the likelihood of the large-scale diffusion of these vehicles.

However, Bass Model has its limitation that it is hard to insert some parameters like fuel price and number of gas station into the formula. It is because the Bass Model was constructed on the study of behavior on innovating and imitating in one product's market.

Future research should include verifying the influence that social factors (e.g., price fluctuation of the energy market and the implication of government policy) could have on the diffusion of NGVs.

Acknowledgments We wish to express our sincere appreciation to the Japan Gas Association for providing the numbers of NGVs in use in the countries.

References

1. Fatih B, Laura C, Tim G, etc. International Energy Agency. World Energy Outlook 2013. OECD/IEA 2013, pp 45–47, 61–65
2. Engerer H, Horn M (2010) Natural gas vehicles: an option for Europe. Energy Policy 38:1019–1021
3. The Japan Gas Association. About diffusion of natural gas vehicles, 2014 edition, pp 3, 10. http://www.gas.or.jp/ngvj/catalog/book_spread.html. Access date 29 Sept 2015
4. Matsumoto M, Kondoh S, Fujimoto J, Umeda Y, Tsuchiya H, Masui K, Lee H-Y (2008) A diffusion model for clean energy vehicles. J Jpn Soc Energy Res 29(3):49–55
5. FM Bass (1969) A new product growth model for consumer durables. Manag Sci 15–1, p 215–227
6. Ministry of the Environment, Japan (2010) Strategy for diffusion of vehicles friendly to environment. 3, pp 23, 97, 115. http://www.env.go.jp/air/report/h22-02/index.html. Access date 29 Sept 2015
7. Tokyo Gas Corporation. Frequently asked questions, Q9, A9 http://eee.tokyo-gas.co.jp/product/ngv/faq.html. Access date 29 Sept 2015
8. ISUZU MOTOTS LIMITED. Diffusion of diesel vehicles in Japan http://www.isuzu.co.jp/technology/d_databook/status/status_01.html. Access date 29 Sept 2015
9. Ministry of the Environment, Japan (2009) Strategy for diffusion of vehicles on next generation, 2009. 5, pp 76, 97 134, 147–148. http://www.env.go.jp/air/report/h21-01/index.html. Access date 29 Sept 2015
10. Ministry of Economy, Trade and Industry, Japan. New policy promotion about the joint fuel station both for NGV and FCV http://www.meti.go.jp/press/2014/04/20140421002/20140421002.html. Access date 29 Sept 2015
11. Japan Oil Association. Data for movement of the number on gasoline stations in areas, Japan. http://www.sekiyu.or.jp/topics/data_a.html. Access date 29 Sept 2015

12. Sector Effects of the Shale Gas Revolution in the US. Alan Krupnick, Zhongmin Wang, and Yushuang Wang, http://www.rff.org/RFF/Documents/RFF-DP-13-21.pdf. Access date 29 Sept 2015, p 3; p 20
13. CNG vehicle rise spurs filling station projects. http://analysis.fc-gi.com/natural-gas-vehicles/cng-vehicle-rise-spurs-filling-station-projects. Aaccess date 29 Sept 2015
14. The contribution of natural gas vehicles to sustainable transport, OECD/IEA2010, p 57
15. Area reports about diffusion of NGV in the U.S., Kimura Makoto. http://www.jetro.go.jp/ext_images/jfile/report/07001452/us_natural_car.pdf. Access date 29 Sept 2015
16. Running out of Gas? Lessons from the NGV market in Pakistan, Abrar Chaudhury, http://papers.ssrn.com/sol3/papers.cfm?abstract_id=2391524. Access date 29 Sept 2015

Key Success Factors of Green Innovation for Transforming Traditional Industries

Yu-Chen Huang, Jui-Che Tu, and Tsai-Wei Lin

Abstract Through expert interviews this study obtained the influence factors of the transformation of traditional industries and set the framework of the expert questionnaire. Then the study used the Delphi method and the expert questionnaire to uncover the ranking of the influence factors and the key success factors of transforming traditional industries. The purposes of this research are as follows: (1) according to the literature reviews and expert interviews to analyze and evaluate green industry indicators; (2) based on the green industry indicators above to establish the ANP model of green industry; (3) through the ANP model to obtain and assess the weight ranking of green industry; and (4) finally, to uncover the key success factors of green industry. Furthermore, this study provided assessment check items for the transforming traditional industries toward green industries. Based on the research results, it can be expected to give clearer guidelines for the industrial transformation and sustainable development strategy, and to create innovative and sustainable green business strategies.

Keywords Traditional industries • Green innovation • Green industries • Sustainability • Key success factors

1 Introduction

In order to respond to environmental deterioration and rapid scientific and technological development, traditional industries in Taiwan are faced with operating pressure. As a result, some traditional industries have begun to lose their competitive advantages. However, such serious challenges also bring the opportunities of industrial transformation, reorganization, development advantages, and key technologies.

Y.-C. Huang (✉)
Department of Creative Design and Architecture, National University of Kaohsiung, Kaohsiung, Taiwan
e-mail: ych3993@nuk.edu.tw

J.-C. Tu • T.-W. Lin
Graduate School of Design, National Yunlin University of Science and Technology, Yunlin, Taiwan

M. Matsumoto et al. (eds.), *Sustainability Through Innovation in Product Life Cycle Design*, EcoProduction, DOI 10.1007/978-981-10-0471-1_53

Therefore, in addition to facing industrial competition in market and environmental changes, traditional industries must change their internal operational model and the development of their core technology, while considering the direction of their sustainable industrial development strategy, in order to deal with environmental and economic changes [1, 2].

The current global economic and business system is gradually moving toward liberalization and globalization. Internationally, the specific division of industries has been the trend. Faced with this trend, traditional industries must strengthen the integration of resources, exert potential advantages, establish a highly efficient enterprise system, and strive for an advantageous position of sustainable development. Consequently, future traditional industries must enhance their industrial transformation and upgrading through science, technology, and innovation [3].

Daniel et al. [4] indicated that “green innovation can assist in obtaining the leading position, while other enterprises remain in the original place.” Enterprise “green innovation” can not only create value for products and society, it can also consider the sustainable symbiosis between human beings and the environment. Moreover, in the face of economic growth and environmental challenge, green innovation considers strengthening economic growth and protecting natural resources, and encourages making effective use of natural resources and increases the strategic elements of green growth in pollution costs. However, regarding green innovation operation, only some products in certain parts and links of the industry are actual green products or conform to one certain regulation which cannot be called green industrial innovation. In most cases, due to the fuzzy cognition of “green”, or a lack of unified standards for different industrial characteristics, enterprises have no idea about how to carry out green strategies. Such varied reasons cause many enterprises to be fearful of green innovation, leading to a wait-and-see attitude toward green innovation transformation.

Transforming to green innovation by traditional industries is not just labeling brands or products as green or offering green information in marketing. Enterprises must make fundamental changes and developments from the inside to the outside, including industrial culture, internal management, product R&D, and selling. In times of rising environmental protection awareness, the innovation of products, processes, and technology is the main driver of environmental competitive advantages. More and more enterprises begin to transfer from simply obeying governmental strategies of environmental protection regulations to actively promoting innovative strategies, and continuously improving industrial performance in order to improve the industrial system and promote self-competitiveness. Thus, enterprises can maintain competition advantages for sustainable operation [5].

After the financial tsunami of 2006, advanced countries actively promoted the strategies of green growth one after another. The development of cleaning technologies and improved resource producing are applied to respond to the challenges of environment changes and master the national development trend of a green and low carbon economy, as well as the 10 tendencies and drivers of science and technology, products, and service innovations. Twenty years in the future, this trend will have a substantial shock on the operation, commerce, and consumption models of all enterprises [6]. Therefore, green innovation is the key to competition at present. Faced with current environmental challenges, every country and industry identifies with this concept. Moreover, industries in Taiwan must actively make breakthroughs and innovations in the direction of low pollution and high performance and strive toward green innovation in order to avoid being eliminated. This shows that industries need to begin to attach more importance to improving self-competition advantages, and industrial transformation and sustainable operation are gradually being regarded as the main operational strategies. Sustainable industrial operations can facilitate industries to advance toward their goals through innovative thinking and breakthroughs.

In order to respond to the current trend of industrial transformation and strategic development of green innovation, the value and ideas of operational decision makers are able to drive the overall industry; however, there should also be relevant support measures and concrete improvement methods. Therefore, to guide all traditional industries in Taiwan to promote new green innovations with international competitiveness, this study conducted in-depth interviews and investigation using the successful case of the most representative traditional manufacturing industry in Taiwan, which is the benchmark among green innovation transformations, in order to draw conclusions and analyze the direct key success factors of green innovation industrial transformation. It is expected that these research findings can provide a reference for enterprises with the goal of green innovation transformation. This study summarized the relationship figures of industrial transformation factors, as shown in Fig. 1.

In response to the current industry trend of transformation and green innovation strategy development, this study investigated the status of the green industry and future development trends, reflecting on industry green business, trying to upgrade traditional industries, environmental responsibility, and reducing the negative impacts of environmental damage. Thereby, we expect to discover the key success factors and traditional industrial transformation of green industry assessment that have driven green industry. The purposes of this research are as follows: (1) according to the literature reviews and expert interviews to analyze and evaluate the green industry indicators; (2) based on the green industry indicators above, to

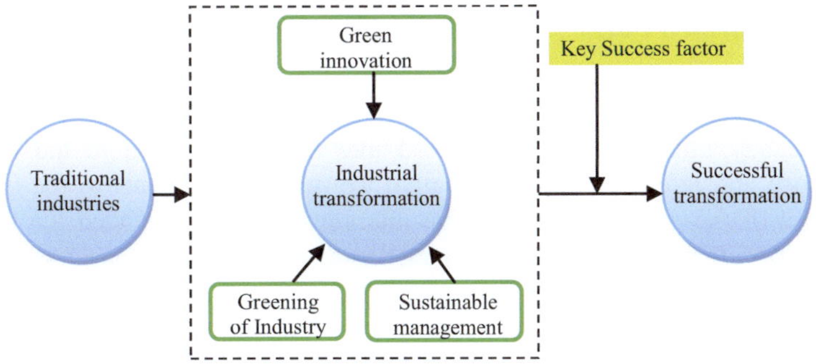

Fig. 1 The relation map of industrial transformation factors

establish the ANP model of green industry; (3) through the ANP model to obtain and assess the weight ranking of green industry; and (4) finally, to uncover the key success factors of green industry.

2 Method

2.1 Research Framework

This study conducted a literature review to explore the related theories of industrial transformation, green innovation, and critical success factors in order to deeply understand the trend and development of traditional industry transformation and green industry. And, through expert interviews, we found the strategy factors of the traditional industry transition to a green industry. In addition, this study acquired the influence factors of traditional industries transforming through expert interviews, set in the framework of an expert questionnaire. Then the Delphi method and analytic network process (ANP) analysis were used to find the influence factor ranking and the key success factors of transforming traditional industries. Furthermore, the study provides assessment check items for traditional industries transforming toward green industries, as shown in Fig. 2.

2.2 Research Subjects

This study achieved a more comprehensive strategy through government, industry, and education. Therefore, this study invited three experts to be interviewed, all of whom are related to a green industrial professional background. This included a manager, R&D manager, and professor (shown in Table 1).

2.3 Research Tools

1. Expert Interview
   This study utilized a semi-structured interview. According to the research purpose and reference-related research, we made adjustments to the question outline of the expert interview. The interview contents included the trend of today's traditional industry, and the factors of green industry business strategy.
2. Questionnaire Survey
   According to the literature review and the experts' advice, the study developed the questionnaire and separated parts 1 and 2 to conduct the questionnaire survey. The survey objects mainly obtained the traditional industry of

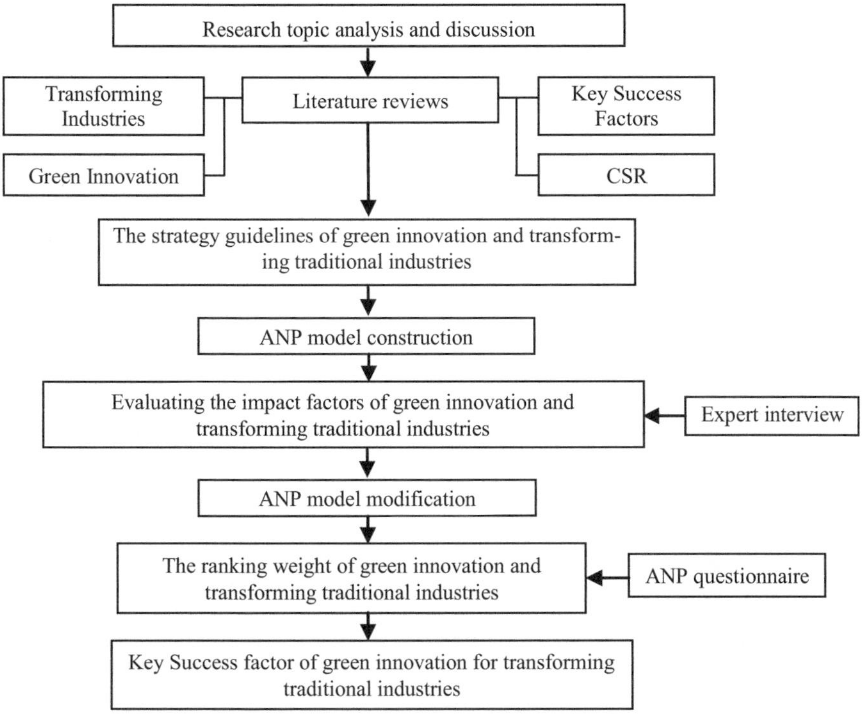

Fig. 2 Research framework

"green paragon" award for 3 years in the ministry of economic affairs of international trade bureau of industry operators. Finally, the results of questionnaire survey were followed through the ANP analysis and the key success factors were obtained.

2.4 Analysis Method

Data analysis was conducted to analyze the correlation among transforming industries, green innovation, and key success factors, and to find the advantage ranking of the important elements and weights of the criteria for establishing a transforming industry procedure in the key success factors. As stated above, ANP was used to analyze issues related to dependence and feedback. It is a mathematical theory that systematically deals with problems and has dependence and feedback [7]. The advantage of ANP is that it is a complete framework that allows clusters in the elements of the network to connect with each other by all means [8]. Therefore, using the basic theory and decision-making procedure of ANP, this study established the ANP model analysis method and the links between the important

Table 1 Research subjects' backgrounds

No.	Subjects	Affiliation	Specialty
1	Manager	©EPEA Intl.UmweltforschungGmbh Taiwan Branch	Green education popularization
			Green industry certification
			Industry management strategy
2	R&D Manager	DA.AI Technology Co.,Ltd	Environmental education popularization
			Laboratory accreditation
			Pharmaceutical analysis chemistry
3	Professor	Industrial Engineering & Business Management, National Taipei University of Technology	Life-cycle assessment
			Role of information and environment report
			Product environment design
			Environmental management strategies
			Cleaning products and processes

elements and decision-making criteria. The ANP decision-making procedure and steps were as follows.

1. It was assumed that the criteria were independent, and the criterion weights were obtained through a pair comparison.
2. The weights of the projects were compared using the criteria. It was assumed that the projects were independent.
3. Mutual dependence among the criteria was investigated.
4. The dependence relationships among the projects were investigated based on the criteria.
5. Based on Steps 1 to 3, the weights of the dependence of the criteria were calculated.
6. Based on Steps 2 and 4, the weights of the projects of the criteria were calculated.
7. Based on Steps 5 and 6, the weights of the projects were calculated to establish the super matrix and limit matrix and analyze the ranking [7].

This study used the Super Decisions software, which was developed by Saaty and his research team at the Creative Decisions Foundation, to calculate decision-making issues related to dependency and feedback. The research and development of Super Decisions software enables decision makers to analyze and make decisions on issues easily without spending time on calculation. Decision makers simply have to construct a relationship network of decision-making issues, and there is no need to clarify what type of matrix is formed in the diagram or to calculate the complex limit super matrix.

After interviews with experts, the geometric mean among all criteria was derived by applying the geometric mean calculation method on the result of the experts' choices. The correlation between all criteria and their weight ranking could be clearly understood by looking at their geometric mean, which is also an important basis for inputting data into the Super Decisions software for analysis in the future.

According to the advice from Saay (1980), the evaluation scale was divided into nine symmetrical scales, and the paired comparison was conducted. If the paired comparison matrix A is an $n * n$ matrix, only simply calculated, the n (n – 1)/2 comparison values of evaluation factors, respectively, conduct A_i and A_j (i,j = 1, 2, 3, ... , $n;i \neq j$) to comparisons and we can obtain the paired comparison A:

$$A = [\alpha_{ij}] = \begin{cases} 1,2,3,4,5,6,7,8,9,\frac{1}{2},\frac{1}{3},\frac{1}{4},\frac{1}{5},\frac{1}{6},\frac{1}{7},\frac{1}{8},\frac{1}{9} & ,i < j \\ 1 \quad ,i = j \\ 1/a_{ji} \quad ,i < j \end{cases} \tag{1}$$

$$A = [a_{ij}]_{n\times n} = \begin{bmatrix} 1 & a_{12} & \cdots & \cdots & a_{1n} \\ a_{21} & 1 & \cdots & \cdots & a_{21} \\ \cdots & \cdots & 1 & \cdots & \cdots \\ \cdots & \cdots & \cdots & 1 & \cdots \\ a_{n1} & \cdots & \cdots & \cdots & 1 \end{bmatrix}_{n\times n} \tag{2}$$

$$a_{ij} = \frac{1}{a_{ji}} \; i,j = 1,2,3, \ldots, n; i \neq j \tag{3}$$

$$a_{ij} = 1, \qquad i = 1,2,3, \ldots, n \tag{4}$$

Finally, the study selected the most competitive advantage of the transforming strategy indicators. Through those steps above, we can obtain the weight ranking of the relationship between each of the selection principles and program. Based on the experts' confirmation and evaluation analysis results we will have the Key Success factor of green innovation for transforming traditional industries.

3 Results and Discussions

3.1 *Establishing the Evaluation Indicators of ANP Network Hierarchy Questionnaire*

According to the literature review for this study [9–15], we extracted the related target indicators and strategy evaluation indicators for the transition of traditional industry. Then through screening and removing the repetition, we further modified

Table 2 ANP model strategy evaluation indicators

Indicator dimension	Strategy evaluation indicators
Industrial transformation	Product innovation
	Diverse development
	Establishing innovation industry image
	Capitalizing professional ability
Green product	Using environmental protection material
	Reduce package or nonpackage
	Recycle
	Life-cycle assessment
	Create green product
Green manufacture procedure	Using reusable material
	Reducing exhaust emission
	Waste disposal process
	Building a green factory
Green marketing	Establishing new image for innovation of industry
	External communication regarding industrial environmental management concepts
	Intensifying the importance of green consumption
	Using electronic systems
Green innovation	Intensifying creative design ability
	Green development innovation
	Product service system
Green management	Establishing green enterprise culture
	Green human resources management
	Executing green audit
	Striving for benchmarked certificates of domestic and international environmental protection
	Implement green purchasing

the vocabulary to construct the key success factors of ANP to a traditional industry transformation questionnaire with the strategy evaluation indicator content, as shown in Table 2.

3.2 ANP Model Modification

This study's framework, according to the advice of the expert interviews, was revised as shown in Table 3. The left side is the ANP original model; the right side contains the expert amendment opinions. The underscored indicators were those that were modified and the bold indicators were the new added evaluation.

The related ANP model framework of analytic hierarchy and network processes is shown in Fig. 3. The expert questionnaire was constructed based on this framework.

Table 3 Comparisons of ANP model

ANP original model		Expert amendment opinions	
Dimension	Strategy evaluation indicators	Dimension	Strategy evaluation indicators
Target dimension	*Industrial transformation*	Target dimension	*Transformation strategy*
	Green product		Green product
	Green manufacture procedure		Green manufacture procedure
	Green management		Green marketing
	Green marketing		Green innovation
	Green innovation		Green management
Industrial transformation	*Product innovation*	Industrial transformation	*Implement product innovation*
	Diverse development		*Diversified development ability*
	Establishing innovation industry image		*Utilizing industrial and professional abilities*
	Capitalizing professional ability		*Creating new brand image*
Green product	Using environment protection material	Green product	Using environment protection material
	Reduce package or nonpackage		Reduce package or nonpackage
	Recycle		Recycle
	Life-cycle assessment		*Developing green product design*
	Develop product		*Using the disassembled and decomposition materials*
			Strengthen the product service extension
Green manufacture procedure	Using reusable material	Green manufacture procedure	Using reusable material
	Reducing exhaust emission		Reducing exhaust emission
	Waste disposal process		Waste disposal process
	Enhance resource-using efficiency		
	Building a green factory		*Using nontoxic materials*
			Curb greenhouse gas emissions
Green marketing	Establishing innovation industry new image	Green marketing	Establishing green industry new image
	External communications on industrial environmental management concepts		External communications on industrial environmental management concepts
	Intensifying the importance of green consumption		Intensifying the importance of green consumption
	Using electronic systems		*Reducing consumables product*

(continued)

Table 3 (continued)

ANP original model		Expert amendment opinions	
Dimension	Strategy evaluation indicators	Dimension	Strategy evaluation indicators
Green innovation	Intensifying creative design ability	Green innovation	Intensifying creative design ability
	Green R&D and innovation		Green R&D and innovation
	Product service system		*Creating green resource management*
			Establishing green supply chain
Green management	*Establishing green enterprise culture*	Green management	*Developing green organizations*
	Green human resources management		Striving for benchmarked certificates of domestic and international environmental protection
	Executing green audit		*Establishing green management strategy*
	Striving for benchmarked certificates of domestic and international environmental protection		*Implement green purchasing*
	Implement green purchasing		

3.3 The Weight Ranking of Evaluation Indicators of Green Industry Transformation

Table 4 shows that companies place great emphasis on the dimensions of "green innovation" and "green management" during implementation. Through preliminary estimation traditional enterprises aim to make innovative breakthroughs for existing industrial problems, including internal to external management and operational ideas. Therefore, they especially place emphasis on green innovation and green management. Relatively, the dimensions of "green products" and "green processes" are obviously ignored.

3.4 Key Success Factors of Transforming Traditional Industries

Daniel [16] mentioned that "most industries have three to six key factors that determine whether or not they can succeed"; however, more recent studies have suggested five to ten items. Hence, based on the overall analysis and results, as investigated by experts, 10 evaluation indicators were selected as the key success

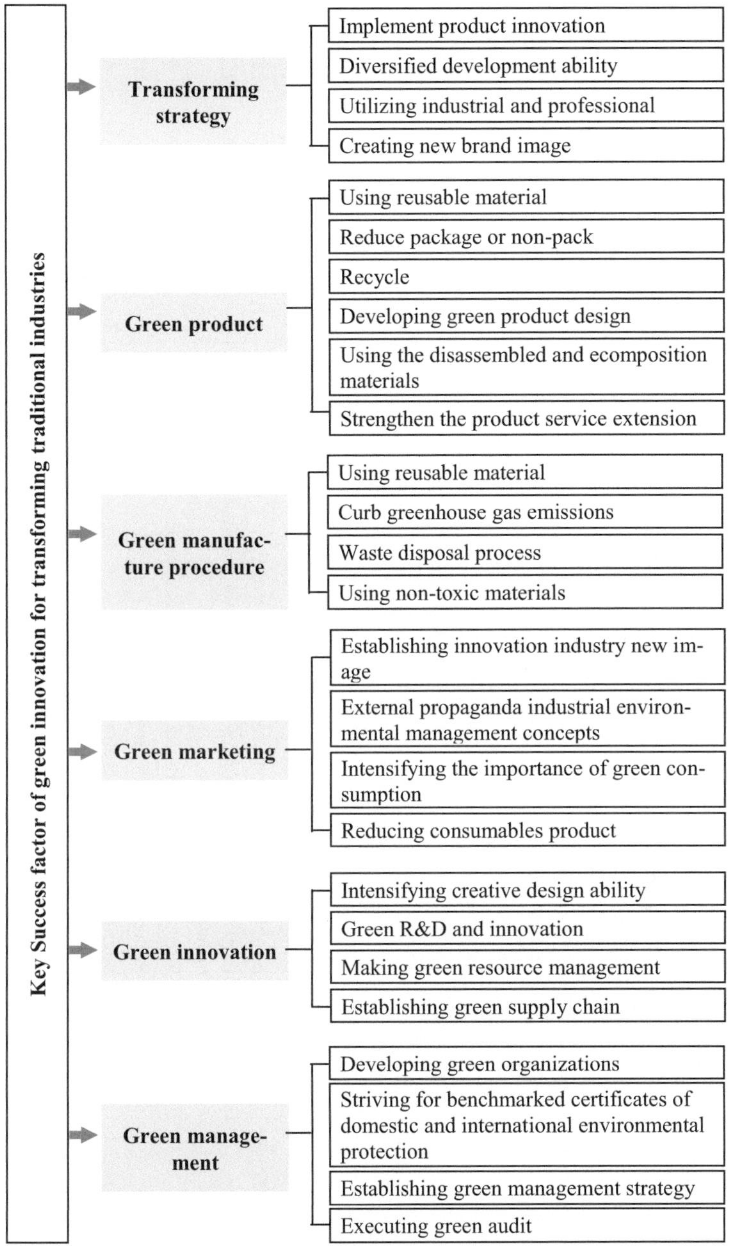

Fig. 3 The ANP model of green industry transformation

Table 4 Industry operators' analysis result of whole evaluation

Target dimension	Dimension weight	Ranking	Evaluation indicators	Hierarchy weight	Hierarchy ranking	Weighting	Final ranking
Transforming strategy	0.09938	4	Implement product innovation	0.2367	2	0.02391	11
			Diversified development ability	0.1932	3	0.01888	16
			Utilizing industrial and professional ability	0.4734	1	0.04699	7
			Creating new brand image	0.0966	4	0.00959	20
Green product	0.08492	5	Using reusable material	0.1718	3	0.01459	17
			Reduce package or nonpackage	0.0866	6	0.00735	21
			Recycle	0.2696	1	0.02290	12
			Developing green product design	0.2431	2	0.02065	15
			Using disassembled and e-composition materials	0.1249	4	0.01060	19
			Strengthen the product service extension	0.1037	5	0.00881	23
Green manufacture procedure	0.03719	6	Using reusable material	0.0882	4	0.00328	25
			Curb greenhouse gas emissions	0.1646	3	0.00612	24
			Waste disposal process	0.1765	2	0.00656	23
			Using nontoxic materials	0.5705	1	0.02122	14
Green marketing	0.09979	3	Establishing innovation industry's new image	0.1263	4	0.01261	18
			External communication of industrial environmental management concepts	0.2260	3	0.02256	13
			Intensifying the importance of green consumption	0.3907	1	0.03899	9
			Reducing consumables product	0.2567	2	0.02562	10
Green innovation	0.37719	1	Strengthening the innovation design	0.1408	3	0.05313	6
			Green R&D and innovation	0.1408	3	0.05313	6
			Creating green resource management	0.2628	2	0.09913	3
			Establishing green supply chain	0.4554	1	0.17177	1

Green management	0.30151	2	Developing green organizations	0.1787	3	0.05390	5
			Striving for benchmarked certificates of domestic and international environmental protection	0.3198	2	0.09642	4
			Establishing green management strategy	0.3632	1	0.10951	2
			Executing green audit	0.1382	4	0.04166	8

Table 5 Weight ranking of key success factors

Ranking	Evaluation indicators	Weighting
1	Establishing green supply chain	0.17177
2	Creating green management strategy	0.10951
3	Establishing green resources management	0.09913
4	Striving for benchmarked certificates of domestic and international environmental protection	0.09642
5	Developing green organizations	0.05390
6	Intensifying creative design ability	0.05313
7	Green R&D and innovation	0.05313
8	Utilizing industrial and professional abilities	0.04699
9	Executing green audit	0.04166
10	Intensifying the importance of green consumption	0.03899

factors out of 26 strategy evaluation indicators, as shown in Table 5 and illustrated as follows.

1. In terms of the experts' degree of attention, "establishing green supply chain" ranks at the top. By means of improvement and development of green technology, industrial technology breakthroughs, and application of high technology, enterprises pursue innovative design and initiate industrial operation methods [17]. By reducing the upstream and downstream costs, the management system of a green supply chain is constructed from green purchasing and manufacturing to logistics and transportation. Enterprises make the commitment together with suppliers to reach the goal of sustainable operation.
2. In terms of the experts' degree of attention, "making green management strategy," "establishing green resource management," and "striving for benchmarked certificate of domestic and international environmental protection," respectively, rank as second, fourth, fifth, and eighth. These four criteria are all in the dimension of green management. Contemporary industries have integrated the concepts of environmental protection into industrial operational management, involving every level, field, aspect, and process of industrial management. It is required that sustainable development should be considered at all times during the process of implementation [4]. In addition, an overall environmental examination should be conducted within industrial operations by establishing a set of perfect environmental management systems.
3. In terms of the experts' degree of attention, "establishing green resources management," "intensifying creative design ability," and "green R&D and innovation," respectively, rank third, sixth, and seventh. These three criteria are all in the dimension of green innovation. Through improvements and developments of green technology, industrial technology breakthroughs, and applications of high technology, enterprises pursue innovative design, initiate industrial operation methods, and intensify industry efforts to develop sustainable operation [17]. Knowledge and ability, including design, innovation, service, and so

on, are applied to satisfy or create market demand and enhance industrial competitive power.

4. In terms of the experts' degree of attention, "making good use of industrial and professional ability" ranks seventh. In order to change or upgrade industrial technology and promote industrial competitiveness, enterprises must make good use of their professional abilities, and fully exert industrial expertise and employ various industrial resources in order to increase operational benefits and ensure the long-term survival and development of industries.
5. In terms of the experts' degree of attention, "intensifying the importance of green consumption" ranks ninth. With service orientation geared toward the appeal of environmental protection, enterprises should narrow the focus of the information provided to consumers, strengthen communications with the public, and assist consumers in buying and using products and services in a manner that values environmental protection, and advocate the concept of environmental protection in order to strengthen consumers' concept of green consumption.

4 Conclusion

This study discussed the evaluation indicators and key success factors for green industries. Using the analytic network process (ANP), this study effectively determined the weight rankings of evaluation indicators and key success factors. The findings can serve as reference for traditional industries to transform into green industries, and develop innovative operational ideas and concepts. The results can be beneficial for the development of sustainable operations; reduced time, cost, and efforts; and enhancing industries' emphasis on environmental protection. The evaluation criteria can be effectively integrated into various aspects of products, processes, marketing, and service, to render industries more suitable to green operational concepts. In addition, it is expected to provide reference for future researchers aiming to conduct in-depth discussions of other industrial aspects related to green industries.

1. According to the literature review and the contents of expert interviews, this study concluded and analyzed the evaluation indicators of green industries, and conducted an expert investigation method. Through in-depth discussions of the three aspects of official, industrial, and academic fields, the six target dimensions as evaluation indicators of green industries were: transformation strategies, green products, green process, green marketing, green innovation, and green management. Then, through the suggestions and amendments of expert interviews, 26 strategic evaluation indicators under each dimension were listed, as follows.
2. In order to determine the key success factors, this study selected the top 10 items from all expert evaluation results, and the overall weight ranking is: (1) establishing a green supply chain; (2) establishing a green management

strategy; (3) creating green resource management; (4) striving for benchmarked certificates of domestic and international environmental protection; (5) developing green organizations; (6) intensifying creative design ability; (7) promoting green R&D and innovation; (8) utilizing industrial and professional abilities; (9) executing a green audit; and (10) intensifying the importance of green consumption.

Although this study has established the basic research of the key success factors of green industry, there are still some questions that need to be investigated by further in-depth studies in the future. Following are some recommendations.

1. The researchers can explore the differences between various industries and can investigate further or make a comparative study of the key success factors of other industries in the future.
2. The researchers can survey the different consumers' points of view, such as corporate transformation after research of purchase behavior and perception.
3. The results of the study presenting green industry and sustainable development are correlated. Using the results of the framework and indicators of the key success factors for practical applications of research, the benefits of a green environment will follow.

References

1. Drucker PF (1993) The five most important questions you will ever ask about your nonprofit organization; participant's workbook [Drucker Foundation]. Jossey-Bass Publishers, San Francisco
2. Shao J, Zhou J (2011) Study on the influences of industry transformation on the sustainable development of resource-exhausted city space. Procedia Eng 21:421–427
3. Bräutigam D, Tang X (2014) Going global in groups: structural transformation and china's special economic zones overseas. World Dev 63:78–91
4. Daniel C, Winston AS, Stella Fred Esty (NRT) (2009) Green to gold: how smart companies use environmental strategy to innovate, create value, and build a competitive advantage. Brilliance Audio, Grand Haven
5. Hashim H, Bakar SMA, Lim JS (2014) Green industry for low carbon economy: palm oil green assessment tool. Energy Procedia 61:2759–2762
6. Lee KH, Min B (2015) Green R&D for eco-innovation and its impact on carbon emissions and firm performance. J Clean Prod 108:534–542
7. Chiu YH, Jheng YM, Hsieh TY (2012) Using fuzzy delphi method and analytic network process in the development assessment of the cultural and creative industries park. J Archit 80:85–109
8. Hsieh KT (2012) Establishing an integrated TRIZ and ANP methods for innovative product development concepts selection-using rear view camera as an example, MA thesis, Graduate Institute of Management Sciences, Tam-Kang University, New Taipei City, Taiwan
9. Lee VH, Ooi KB, Chong AYL, Seow C (2014) Creating technological innovation via green supply chain management: an empirical analysis. Expert Syst Appl 41(16):6983–6994
10. Amores-Salvadó J, Martín-de CG, Navas-López EJ (2014) Green corporate image: moderating the connection between environmental product innovation and firm performance. J Clean Prod 83:356–365

11. Nonaka I, Yamanouchi T (1989) Managing Innovation as a self renewing process. J Bus Ventur 4:299–315
12. Jänicke M (2012) Green growth: from a growing eco-industry to economic sustainability. Energy Policy 48:13–21
13. Gabler CB, Richey RG Jr, Rap A (2015) Developing an eco-capability through environmental orientation and organizational innovativeness. Ind Mark Manag 45:151–161
14. Chun SH, Hwang HJ, Byun YH (2015) Green supply chain management in the construction industry: case of korean construction companies. Procedia Social Behav Sci 186:507–512
15. Lam HL, How BS, Hong BH (2015) Chapter 12 – green supply chain toward sustainable industry development. In: Assessing and measuring environmental impact and sustainability, pp 409–449. doi:10.1016/B978-0-12-799968-5.00012-9
16. Daniel (1961) Conceptualization and measurement of consumer satisfaction and dissatisfaction. Market Science Institute, Cambridge, MA
17. Shih LH (2010) The green innovation and product development. Tsang Hai Publishing, Taichung City

Postmodern Dynamics of Innovation and Knowledge in the Context of Sustainable Energy Development

Harald E. Otto

Abstract The rise of new knowledge economies in developed nations has changed not only economic dynamics and values, but also the nature of society. Knowledge has become the main resource in respect to products, services, processes, and business models. Within sustainable energy development, most programs and instruments were designed and implemented under neoclassical views and paradigms. However, since we are in a period of transition, the additional elements of economic dynamics and social structures and behavior related to rising knowledge societies and the rapidly evolving economics of knowledge need to be taken into account for the entire life cycle of policies and programs. In this study, an analysis is provided of conceptual and structural shortcomings in current sustainable energy development focusing on energy efficiency and conservation. This is approached from a socio-technological and economic viewpoint by taking into account the impact of postmodern science and engineering and the dynamics of rising knowledge societies.

Keywords Energy efficiency • Energy conservation • Sustainable energy consumption • Low-carbon society

1 Introduction

Intensifying concerns about global climate change, energy security, and depletion of fossil fuel resources, amid an almost unprecedented surge in energy demand worldwide, have reinforced the roles of energy efficiency and conservation, which represent long-standing critical elements within sustainable energy development. Empirical evidence from various studies on energy trends since the energy crisis in industrialized nations induced by the OPEC oil export embargo, shortly before the mid-1970s, suggests that there are many opportunities to reduce the demand for energy through energy efficiency and conservation and that these can be tapped into

H.E. Otto (✉)
Department of Industrial Engineering and Mathematical Sciences, Polytechnic University of Marche, Ancona, Italy
e-mail: h.e.otto@univpm.it

M. Matsumoto et al. (eds.), *Sustainability Through Innovation in Product Life Cycle Design*, EcoProduction, DOI 10.1007/978-981-10-0471-1_54

by highly cost-effective investments. However, as socio-technological and economic reality is far from being either rational or perfect, social and economic actors in the industrial, commercial, and private sectors alike seem not to be making much use of this potential. This is despite the fact that, in addition to environmental and social net benefits that might be difficult to discern at first glance, there are also private gains to be obtained through cost-effectiveness, which is quite an obvious economic incentive, and these too remain mostly ignored.

Reasons for this unfortunate situation are presumably manifold. Economic efficiency, operationalized as maximizing net benefits to society, is different from and does not automatically translate into the physical concept of energy efficiency that is related to technology and energy services. The latter is not free of costs. Therefore, the efficiency of economic decisions in respect to energy efficiency is interrelated with the economic behavior of the actors and the economic efficiency of related markets such as the energy market and the capital market. Here, departures from efficient dynamics can be regarded as barriers and failures in both markets and the behavior of actors. Policy-based intervention aimed at correcting those barriers and failures is still mostly designed, implemented, and evaluated with models and parameters used for their predecessors. In most cases, models, as a part of the larger structure of a paradigm, tend to reflect the preconceptions of their builders. Thus, the newly developing structures and dynamics of knowledge generation and diffusion and their related economics, recent unprecedented innovations in information technology, and fast-paced information-based progress in all scientific fields are not fully accounted for. This situation is mirrored not only by the failures of instruments and programs to deliver results as planned but also by the accrual of policy-induced costs that far exceed their estimated net benefits.

In this study, an analysis is provided of a selection of conceptual and structural shortcomings in current sustainable energy development with a focus on energy efficiency and conservation. This is approached from a socio-technological and economic viewpoint by taking into account the impact of postmodern science and engineering and the dynamics of rising knowledge societies. It has been undertaken with the aim of facilitating and supporting concerted efforts by researchers and policymakers to translate insights gained into contributions to the gradual improvement of both design and implementation of programs and instruments.

2 The Economics of Knowledge and Energy Program Needs

From the viewpoint of technological change and economics, energy efficiency relates to investment decisions, which trade off the higher initial costs of energy-efficient technology against uncertain lower future energy operation costs. Within such an approach, less energy is required as input for an energy service to achieve the same results. However, energy demand can also be reduced by considering changes in the processes and behavior of social and economic actors, which are related to energy use. In the case of the former, innovation and technology are directly used to reduce energy consumption, while, in case of the latter, information and knowledge, mostly in concert with information technology, are employed as a means to induce reductions in energy consumption.

A considerable body of scientific work and numerous studies from within energy engineering and economics, providing theories supported by empirical evidence, suggest that potential investments in energy efficiency remain to a large extent unexploited, even though some appear obviously cost-effective. On the other hand, extensive theoretical and empirical work from behavioral economics and environmental psychology suggests that there is considerable potential for reducing energy consumption by achieving behavioral change in energy use regarding both commercial enterprises and private households and that this remains either overlooked or ignored.

This unfortunate situation is referred to somewhat esemplastically in the literature as the energy efficiency gap caused by market and behavioral failures and market barriers. Seen from the neoclassical economic viewpoint, markets and their imperfections represent a violation of the principles of rational behavior, costless transactions, and perfect information. This can be attributed to unpriced costs such as energy security and environmental and knowledge externalities, unpriced goods and services such as technological innovation and education, misplaced incentives such as the principal-agent problem and the involvement of intermediaries, the bounded rationality and heuristic decision-making of economic actors, and a lack of information or imperfect information. Disincentives to the adoption or use of energy-efficient technology and obstacles such as the incompleteness of the energy-efficient market and the low priority of energy issues among consumers are usually not considered as actual market failures, but instead are regarded as market barriers [1–3]. However, these too are contributing to the protraction of the innovation process and are slowing down the diffusion of energy-efficient technology.

The rise of new knowledge economies in developed nations has changed not only economic dynamics and values but also the nature of society. Knowledge has become the main resource in respect to products, services, processes, and business models. This is in contrast to the situation in traditional industries and the concept of neoclassical economics, where land, capital, and labor were the three factors of production. Knowledge has different properties from traditional commodities due to its public goods nature. Knowledge does not diminish by use (it is non-rivalrous), and it is difficult to prevent others from using it (it has only partial excludability). It is a type of both input and output, with current innovations being the input for the next set of innovations. Knowledge also has positive externalities such as information and knowledge spillovers. For example, investment in research and education benefits many. There is also positive feedback reflected in increased returns. For example, in relation to utilization and supply, the more you know and the more you use that knowledge, the easier it is to acquire and use further knowledge.

With these new economic values and new sets of dynamics, social relations, culture, and institutions are undergoing change, resulting in what is now commonly termed the (networked) knowledge society. There are now several different views regarding the nature of currently developing knowledge societies, and these relate to the primacy of scientific knowledge, the rise of knowledge work, and the increasing interconnection of work and social relations enabled by information and communication technology. Within such a scenario, obviously an open culture is not only encouraged, but also technologically supported. In this culture, information is made publicly available. Information technology mediated communication and collective thinking are developed. Social mobilization occurs, and there is decentralized production, distribution, and consumption of knowledge (cf. [4–6]). Many programs were designed and implemented under neoclassical views and paradigms. Of course, this is not wrong in itself, as the capital market, the energy market, and the innovation market still have these dynamics. However, since we are in a period of transition, the additional elements of economic dynamics and social structures and behavior related to rising knowledge societies and the rapidly evolving economics of knowledge need to be taken into account for the entire life cycle of policies and programs from their conception and design to their implementation, and through evaluation and correction, right up to their termination.

3 Dynamics of Knowledge and Sustainable Energy Development

3.1 Knowledge Access and Knowledge Generation

Energy policy mechanisms currently applied in most countries can be subdivided into two basic types of instruments, namely, quantity-based and price-based. The former are instruments such as renewable portfolio standards strongly supported by industry and utilities. The latter are auctions and feed-in tariffs, also referred to as premium payments or minimum price standards, which support a more diverse participation of social actors, thus being also more attractive for decentralized small-scale and community-based projects. Feed-in tariffs are so-called *demand-pull* instruments [7, 8], due to government efforts to create markets for renewable energy technologies. They are currently applied in at least 61 countries [9], though over half of these measures have been enacted only since 2005. Feed-in tariffs are often seen as effective instruments, due to their capacity to remove market barriers, thus enabling a rapid deployment of significant capacity (for a good overview, see [10]). This is also considered an important factor for innovation, as new and novel technologies usually remain too costly compared to their conventional counterparts. Feed-in tariffs help the new technologies to make a swift and timely transition from the niche markets they reside in into the general market [11]. Now, if innovation and development accelerate and a new technology becomes less expensive in terms of production and deployment costs, support from feed-in tariffs should recede and make way for a more natural market competition. However, due to the structure of feed-in tariffs in the first place, which is aimed at providing stability and reduced uncertainty and risk for investors, they paradoxically counteract their own purpose over time, and, if they are successful, they eventually result in cost escalation and thus also undermine public support. An example is the recent revision of feed-in tariffs for solar photovoltaic installations around the world, resulting in tariff cuts of up to 45 % in many developed European countries. This can be seen as a countermeasure to dampen the booming rate of installations [9] and was due to an unprecedented steep decline in price of solar photovoltaic technology that took place mostly in 2009–2010.

There are current developments to secure copyright protection for the digital computerized reproduction and distribution of information, and there are growing efforts to assert and enforce intellectual property rights over technological knowledge, mostly through the use of copyrights, patents, and other novel instruments of legal protection. Because of these moves, the availability of and access to knowledge are about to be severely restricted at the expense of public domain knowledge. This is a typical characteristic of socio-technological systems, where economic factors and commercial interests weigh in during rapid deployment of a new technology, which creates scarcity where there was none before. In this case, it creates an artificial scarcity of technological knowledge and know-how. This effect combines unfortunately with the translation of knowledge from a public good into a commodity, though the characteristics and mechanisms of the production, improvement, and distribution/diffusion of knowledge and those of actual tangible commodities are in most parts incommensurable (see discussions in [12, 13]). This situation may lead to underinvestment in the innovation of efficient energy technology, because commercial and industrial actors fear that they will be unable to fully capture and exploit the benefits of their innovative efforts. Some portion of this benefit will be shifted to other social and market actors, due to the natural characteristics of knowledge as a public good. This tendency has received considerable attention in postmodern economics literature and is often considered to be an innovation market failure. Another aspect associated with innovation market failure is the spillover effects of knowledge and technological know-how. This is especially the case with the latter, which significantly improves through what is called *learning-by-doing*, referring to empirical observations that the costs of production tend to decline as the cumulative production of new technologies increases. In a similar fashion, collective knowledge (as a public good, not a commodity) quickly improves and develops, thus furthering innovation and leading to new discoveries, if it is used, exchanged, and commented upon by many social actors simultaneously. This is made possible due to its characteristics of being non-rivalrous in use and having low excludability combined with multiplicative effects referred to as the 'standing on the shoulders of giants' effect (from Latin *nanos gigantum humeris insidentes*), expressing the meaning of building on knowledge stock and previous discoveries.

The case of feed-in tariffs for solar photovoltaic technology in Europe, as one type of a demand-pull instrument within sustainable energy development, represents a particularly pertinent example that features many deficiencies related to not taking into account the dynamics of postmodern society and the economics of knowledge as discussed earlier. At this point, it should be mentioned that the potential of demand-pull instruments within current energy policies and programs to contribute to technological innovation is in general overlooked as their main goals are different from technology-push-oriented instruments, which address issues of technological innovation and deployment explicitly. The design and implementation of such feed-in tariffs are mostly based on neoclassical economics, ignoring the potential for various opportunities of knowledge generation. For example, most of the potential for advancing technological innovation through learning-by-doing was actually lost to economic actors outside Europe, who provide not only for their own markets but for the majority of the European market. Although this situation, economically seen, represents a considerable trade deficit in new energy technology, it represents an even greater deficit in knowledge generation potential, which is presumably amplified further by chances lost to benefit from knowledge generation related to firm innovation through learning-by-exporting (cf. [14, 15]).

The alternative option to demand-pull instruments in sustainable energy development is directly furthering research and innovation. Within the given context, this is a position that has been argued for by advocates of the so-called *technology-push* approach [16], which stipulates that governments should allocate funding for the innovation of new energy technology and systems driven by knowledge and know-how generation and research and development. For (post)modern industrialized nations, this approach is indeed a viable option. However, considering the characteristics of rising knowledge societies and concomitant phenomena attributable to the likewise rising knowledge economies, several newly developing issues need to be better understood and taken into account. These manifest themselves both as new barriers apparently inhibiting progress and as new opportunities mostly remaining unappropriated. Due to the public good nature of knowledge, technological innovation is prone to research and development spillovers, representing from an economic point of view unpriced goods, which markets tend to underproduce. This situation is reflected in a considerable gap between the private returns firms can capture from their investments in research, development, and deployment and the social returns available. The social returns in this context include externalities related to energy such as environmental benefits in the form of a reduction in environmental pollution, a reduced depletion of fossil energy resources, and energy security. The rate of private returns compared to the rate of social returns is estimated to be about two to four times lower, with the largest gaps to be found in fragmented industries (cf. [17–19]).

To increase incentives for firms to conduct research and invest in innovation, instruments such as subsidies, in the form of tax credits or direct funding of particular research proposals, are usually applied as countermeasures to the knowledge spillover problem. Another option here is to conduct research in the public sector at national laboratories and research facilities or through direct private research contracts. With rapidly evolving social changes in knowledge societies and the rise of knowledge economies, these options and their instruments need to be related more than ever before to the structures, concepts, and dynamics of the various forms of knowledge itself and knowledge generation as well as knowledge access. An example of this is asymmetric knowledge, in the sense that firms specializing in a particular technology usually have more knowledge than outsiders such as national research facilities or private research contractors. This not only puts them in a better position regarding evaluation of the future potential of certain technologies, including their commercial feasibility under given market constraints, but also allows them to conduct research and development faster, more efficiently, and at lower cost. This can be achieved partly by benefiting from their knowledge stock and that keeps marginal costs and transaction costs related to retrieving and accessing knowledge and know-how required as input for innovations at a very low level.

To achieve an acceptable balance between the rate of private returns and the rate of social returns and net benefits, additional issues need to be considered. In many cases a part of the research funding will be consumed by increased expenditure such as wage increases due to temporary shortage of qualified research personnel. This drawback can be turned into a positive stimulus by complementing research subsidies with subsidies for training and education. Here the improvement and promotion of higher education seems to be one favorable option, which in this case implicitly promotes an increase in an unpriced public good as a service. At the same time, this approach of knowledge access through education would be a countermeasure for increasing expenditure in higher education for science and engineering. Another viable option would be to combine this approach with policy instruments related to information provision (discussed in more detail in the next subsection), such as energy audits. Since energy audits, which are usually implemented as government-funded programs, are conducted in many cases by science and engineering faculties of institutions of higher education and by national research laboratories, which also involve supervised students, explicit use should be made of the potential of the benefits and synergies which exist among education, training, and various forms of knowledge generation, and the acquisition of scientific and engineering skills, perhaps with many deemed to be prerequisites for the next generation of energy experts and professionals.

Another issue that needs to be considered in this scenario is related to the properties of knowledge as a commodity and the consequences for technological innovation within developed industrial nations which are in transition due to evolving dynamics in the economics of knowledge. When research is performed by firms, in most cases the results of knowledge generation and technological innovation will be protected by patents and intellectual property rights. This will increase the stock of knowledge of those firms and thus improve the potential for making efficient use of the 'standing on the shoulders of giants' effect while keeping the costs of knowledge access and use at a minimum for them and as a premium to transaction costs for others. This situation will systematically lead toward an increase in commercialization and a concentration of knowledge production and to more uniformity in composition. In the context of sustainable energy development, this trend might form a negative synergy with demand-pull-oriented instruments such as feed-in tariffs for renewable energy, and it might lead toward an artificially created focus on one particular technology, thus impeding recognition and development of alternative, promising, or even superior technologies.

3.2 Provision and Representation of Information

Promotion of energy efficiency requires a certain amount of context-specific information that is meaningful, expressive, easy to comprehend, and represented in an effective as well as efficient manner. This information needs to be provided for as many economic and social actors as possible, and this applies whether the changes are technological changes involving the adoption of energy-efficient technology where new equipment gradually replaces older or whether they are social changes involving behavioral changes in energy use. However, current information programs and instruments are aimed at correcting market failures related to a lack of information or imperfect information, and these are unable to deliver the results expected due to deficiencies related to underestimating or even overlooking the range of potential benefits and synergies which can be obtained by taking into account on a broader base the dynamics and economics of knowledge as outlined earlier.

Adoption of any new technology incurs various expenses for any economic actor, firm, organization, or household. This may be in the form of purchasing new equipment, costs in searching for and collecting information, or costs for learning about new technology and how to adapt it to a current use environment. Energy-efficient technology is no exception to these requirements. Where there is a lack of information or where there is imperfect information about the availability of energy-efficient products and the potential of future energy savings, suboptimal investment in energy efficiency often occurs as a result. This happens especially in the case of small- and medium-sized enterprises, where transaction costs for searching for and obtaining information can become high due to a shortage of trained technical staff. Government programs to organize and fund industrial energy audits and household energy audits are currently one way in which many industrialized nations remedy this situation. Compared to other energy policy instruments, such as subsidies in the form of feed-in tariffs for renewable energy, the energy audits are a low-cost instrument with promising results, as about half or more of all audit recommendations are accepted and implemented. However, the design and implementation of most, if not all, current energy audits are deficient, because they are focused primarily on information provision to promote energy efficiency regarding technology adoption. Many important aspects, and the dynamics of knowledge-related activities, are ignored and thus neither supported nor explicitly funded.

For example, regarding auditing, the potential to generate knowledge through learning-by-doing is in most cases ignored. Also, if the knowledge and insight gained during an audit were documented and made available, it would contribute to the collective knowledge on energy auditing. Moreover, through evaluation by others, it would increase in value and make energy auditing more effective and efficient as knowledge spillovers and increasing returns of knowledge supply and utilization, typical dynamics of knowledge societies, get to work. Efforts toward collecting data in large databases, as is done, for example, by governmental energy departments of developed nations, are by no means adequate. In most cases, these are only a compilation of data from the energy diagnostics and audit reports. One might even ask whether the cost of updating and maintaining those databases is actually in balance with the net benefit they are supposed to provide. Regarding the representation of information, as studies have revealed (cf. [20–23]), an important aspect in this context is detailed personalized information, as that leads to knowledge that translates into informed decisions on how to achieve goals. The number and complexity of interacting factors is small in household energy audits compared to those in industrial energy audits. Here, therefore, user innovation represents another mostly ignored opportunity for knowledge generation, as discussed elsewhere in this study. Users need to be given an environment that will permit such activities. Even though the information technology exists and the communication infrastructure is already in place, unfortunately concerted efforts, which are indispensable to make the best use of these, are still not in sight.

The situation in the case of information provision as a feedback on energy use in households seems to be even less favorable, as the differences between what we know and what is actually done in practice in public programs and policies seem considerable, and they are widening further. For example, all empirical studies on energy consumption feedback have consistently shown that, to be both effective and efficient, information needs to be presented clearly and regularly, in a tailored manner, giving appliance-specific breakdowns (for an overview of recent work, see [24, 25]). This requires an energy infrastructure with smart meters and interactive in-house displays. Most industrialized nations, however, have smart meter rollout programs featuring only a simplified metering technology without any appropriate feedback. They provide sufficient information for energy utilities to improve their demand-side management and thus advance their own economic interests, but they deprive the household of almost any chance of becoming an active part of the process. They are also in conflict with several criteria related to cultural values, such as freedom of choice and privacy. Perhaps, therefore, it is not so surprising that they are struggling with public acceptance. Not only do they oppose social and cultural norms, but they also act as obstacles to technology deployment. Another aspect of this scenario is that, up until now, opportunities have been ignored to involve users on a larger scale as active sources contributing to user innovation. Users could help not only to customize artifacts such as actual feedback displays (both hardware and software) but also in the provision of the information itself. Thus, they would be able to further identify, shape, and improve parameters of the feedback on energy. In this manner, knowledge and insight could also be gained on how energy services, lifestyles, energy consumption behavior, and their interrelationships are perceived and valued in the eye of the beholder. There are many successful examples of user participation in information technology where customization of functions and adaptation of parameters defining the interface between technology and society on a local and individual level have brought about knowledge and innovation. Some of these would have been impossible to achieve by relying solely on traditional methods in academic settings and commercial laboratory-based research and development.

In the given context, information relating to energy efficiency could be provided to the economic actors subject to behavior change, i.e., to customers seen as prospective buyers of energy-efficient goods and equipment. This would provide several opportunities, still neglected, to overcome underinvestment in energy efficiency. Such underinvestment is considered in the economics literature on energy efficiency, which is obviously influenced partly by insights into behavioral economics, and is regarded as a market failure attributable to asymmetric knowledge and information. In practice, this situation is reflected in the struggles of the markets and in policy as well as program instruments designed to overcome the knowledge imbalance among manufacturers, sellers, and customers. Information provision in respect to energy efficiency can contribute to improving this situation, if designed and implemented according to what is scientifically known (see also discussions in [26]). This can be achieved through programs such as labeling, which are currently either in operation or planned for introduction in over 70 countries worldwide. However, more advanced approaches are required to match increasing requirements. Information about the differences in future operating costs between less-efficient and more-efficient products and appliances needs to be provided to encourage private economic decisions to invest. This type of information can steer customers in a more sustainable direction, but it needs to be designed to answer the questions that are important from the viewpoint of those customers, so the information should not be limited to details of cost-minimization in future technology employment. Novel methods of organizing and implementing the provision of information are required, and these must also take into account the costs of doing so (see also discussions on transaction and search costs as reported in [1, 27]). This could be approached by offering products together with novel information provision, in the form of interactive information systems, which should draw on advanced information technology, its infrastructure and characteristics of the knowledge society, and its economics. Information from different knowledge and expertise areas could be compiled, integrated, and provided in addition to the usual product information. This information could include customer service, marketing, servicing, and maintenance engineering for a range of economic actors such as manufacturers, suppliers, and sellers, all of whom would realize that the economic potential, and therefore the value of knowledge, increases when all can agree to share it. If integrated with information networks and online shopping, information-pull would surely promote its diffusion, while knowledge would be generated through learning-by-using from adopters of energy-efficient technology. Knowledge would be appreciated as a positive externality and used to further extend and improve information transfer and provision in the sense outlined, and the chances are that neither asymmetric knowledge and information nor the actual transaction and search costs which are mentioned by economists as obstacles would remain as market barriers in the sense they are seen now. Of course, to make all this happen, it requires concerted efforts and goal setting with a priority on net benefits.

3.3 Knowledge Transformation and Translation

An improved understanding of the relations between fields of knowledge is not only desirable but necessary. So too is the discovery of a better way to integrate both expert views and knowledge from different disciplines. Perhaps even more important, though, is obtaining a firmer grasp on the means of how knowledge produced in the academy, industrial practice, or daily life moves into society and relates to action within that society. Today, research programs in almost every field within postmodern engineering science disciplines are increasingly coming under scrutiny as to how their goals and results, i.e., knowledge and related artifacts, not only conform to the principles of sustainability and science itself but also benefit society. As already pointed out in [28, 29], the social contract is subject to change. Formerly, the scientific community was granted relative autonomy from political and other societal concerns. Now, however, the demand for more accountable government is overriding the old predominant justification of research and other academic pursuits as a means of producing knowledge for its own sake. The new view is that regarding science as being separated from our real physical world and society at large is rather an artificial and unsustainable approach. Therefore, in many of today's research programs and projects, newly allied scientists, practitioners, and policymakers constitute a kind of transformative alliance that has an impact on both the definition of the problem to be solved and the nature of the discourse on how to approach that problem, generate a solution, and eventually translate the solution into practice. However, the current reality is different due to the gap between today's scientific and technological advances and their application: between what we know and what is actually being done. The problem of what is in the literature [30, 31] termed the *know-do gap* exists in most, if not all, fields of policymaking and program design.

As examples, in the literature (cf. [32–34]) the following known phenomena are discussed: issues related to *free riders*, economic actors who would have invested in energy efficiency or considered a behavioral change in energy consumption had there not been any policy-related influence or benefit; *rebound effects*, the increase in energy demand, causing less-than-proportional energy use reductions due to an decrease in marginal costs of energy services induced by energy efficiency; and *free drivers*, economic actors who, considered as nonparticipants in programs, become induced to invest in energy efficiency or assume a behavioral change in energy consumption. In many cases, within current efforts at measuring and evaluating efficiency, effectiveness, and the costs of instruments used within sustainable energy development, the above phenomena still remain inappropriately accounted for or, in the worst cases, even ignored. The case of free drivers is obviously related to dynamic increasing returns generated by learning-by-using. Adopters of a new technology, either individual consumers or firms, create a positive externality for others in the form of knowledge related to the existence, characteristics, and potential or actual success of that new technology. Not only should this be included in the policy design in the first place, but it should also be taken into account as a policy offsetting effect during evaluation in both ex ante and ex post studies. Taking into account additional aspects of the dynamics of the economics of knowledge and networked societies, as discussed elsewhere in this study, would certainly contribute to gradually improving the design of programs and instruments. We also need to consider the access to and diffusion of such increasing returns regarding, for example, speed and scale, and perhaps we need to examine the relationship with socio-demographic parameters. In this unfortunate situation, knowledge translation as a paradigm emerging from within transdisciplinary research is now being seen as a kind of remedy for closing this know-do gap while aiding a more adequate and meaningful policymaking process. However, according to what can be found so far in the current literature and in government-issued reports across all developed economies, there is still a considerable lack of know-how regarding systematic methods and knowledge-based tool applications for closing, or at least reducing, the know-do gap. Both among academia and in practice, work which considers advances in knowledge visualization and knowledge mapping is surprisingly sparse, though it does exist and is recommended as stimulating study material for sustainable energy development processes. Proposals of frameworks as described in [31], aimed at systematically improving the knowledge translation process by exploiting both the benefits of the process of knowledge mapping itself and actually generated knowledge maps, represent examples pointing in the right direction.

4 Conclusions

As studies suggest and empirical evidence confirms, programs and instruments of sustainable energy development in the form of demand-pull approaches that increase a perceived market payoff, and technology-push approaches that directly foster development and diffusion of energy-efficient technology, are able to intervene in a corrective manner, if designed and implemented according to currently known barriers and failures of markets and economic actor behavior. This situation can be further improved by taking into account deficiencies in policymaking that are related to the developing dynamics and structures of both networked knowledge societies and postmodern science, and the economics of knowledge, as those are rapidly on the rise in all industrial nations. Here, in particular, more support is required to initiate generation of unpriced goods and services in the form of collected energy data, the provision of codified knowledge compiled into information available to the public, and the education of future generations of scientist, engineers, and energy experts, all of which tend to be increasingly underprovided by current markets. It is necessary to consider programs and instruments to address the energy efficiency gap and conservation from two perspectives, namely, technological change and behavioral change. This requires a policy portfolio, the efficient mix of which still remains to be determined. In addition, there is a large untapped potential for improving the individual processes of policymaking itself. As was indicated in this study when addressing the know-do gap, at all levels of policymaking, from the makers and strategists at the very top to the implementers, who are usually government employees and elected officials together with their advocacy groups and policy technocrats, there is a need to benefit from employing new knowledge tools and the dynamics of knowledge creation and organization explicitly for the design and implementation of programs and instruments.

References

1. Howarth R, Andersson B (1993) Market barriers to energy efficiency. Energy Effic 15:262–272
2. Levine MD, Koomey JG, McMahon JE, Sanstad A, Hirst E (1995) Energy efficiency policy and market failures. Annu Rev Energy Environ 20:535–555
3. Jaffe A, Newell R, Stavins R (2004) Economics of energy efficiency. In: Cleveland CJ (ed) Encyclopedia of energy, vol 2. Elsevier, Amsterdam, pp 79–90
4. Gibbons M, Limoges C, Nowotny H, Schwartzman S, Scott P, Trow M (2010) The new production of knowledge: the dynamics of science and research in contemporary societies, 10th edn. Sage, Thousand Oaks
5. Castells M (2010) The rise of the network society, 2nd edn. Wiley-Blackwell, Oxford
6. Foray D (2004) The economics of knowledge. MIT Press, Boston
7. Jaffe AB, Newell RG, Stavins RN (2005) Market barriers to energy efficiency. Ecol Econ 54:164–174
8. Nemet GF (2009) Demand-pull, technology-push, and government-led incentives for non-incremental technical change. Res Policy 38:700–709
9. REN21 (ed) (2015) Renewables 2015: global status report, Renewable Energy Policy Network for the 21st Century, Paris, France
10. Couture T, Gagnon Y (2010) An analysis of feed-in tariff remuneration models: implications for renewable energy investment. Energy Policy 38:955–965
11. Menanteau P, Finon D, Lamy ML (2003) Prices versus quantities: choosing policies for promoting the development of renewable energy. Energy Policy 31:799–812
12. Otto HE (2011) Post-modern engineering science: current trends and shortcomings of IT use in knowledge generation. In: Proceedings of 7th international symposium on environmentally conscious design and inverse manufacturing, 30 Nov–2 Dec. Kyoto. Paper-No: B2–01
13. David PA, Foray D (2003) Economic fundamentals of the knowledge society. Policy Futur Educ 1:20–49
14. Clerides SK, Lach S, Tybout JR (1998) Is learning by exporting important? Micro-dynamic evidence from Colombia, Mexico, and Morocco. Q J Econ 113:903–948
15. Salomon RM, Shaver JM (2005) Learning by exporting: new insights from examining firm innovation. J Econ Manag Strateg 14:431–460
16. Frondel M, Schmidt CM, Vance C (2008) Germany's solar cell promotion: an unfolding disaster. Energy Policy 36:4198–4204
17. Council of Economic Advisers (ed) (1995) Supporting research and development to promote economic growth: the federal government's role. Council of Economic Advisers (CEA), Washington, DC

18. Hall B (1996) The private and social returns to research and development. In: Smith LR, Barfield CE (eds) Technology, R&D, and the economy. Brookings Institution and American Enterprise Institute, Washington, DC, pp 140–183
19. Reddy AKN (1991) Barriers to improvements in energy efficiency. Energy Policy 19:953–961
20. Anderson ST, Newell RG (2004) Information programs for technology adoption: the case of energy-efficiency audits. Resour Energy Econ 26:27–50
21. Desmedt J, Vekemans G, Maes D (2009) Ensuring effectiveness of information to influence household behaviour. J Clean Prod 17:455–462
22. Gruber E, Fleiter T, Mai M, Frahm BJ (2011) Efficiency of an energy audit programme for SMEs in Germany: results of an evaluation study. In: Summer study proceedings of European Council for an energy efficient economy, 6–11 June. Belambra Presqu'île de Giens, France, pp 663–674
23. Abrahamse W, Steg L, Vlek C, Rothengatter JA (2007) The effect of tailored information, goal setting, and tailored feedback on household energy use, energy-related behaviors, and behavioral antecedents. J Environ Psychol 27:265–276
24. Fischer C (2008) Feedback on household electricity consumption: a tool for saving energy? Energy Effic 1:79–104
25. Abrahamse W, Steg L, Vlek C, Rothengatter JA (2005) A review of intervention studies aimed at household energy conservation. J Environ Psychol 25:273–291
26. Egan C, Waide P (2005) A multi-country comparative evaluation of labelling research. In: Summer study proceedings of European Council for an energy efficient economy. 30 May–4 June. Mandelieu La Napoule, France, pp 811–822
27. Wilson CM (2012) Market frictions: a unified model of search and switching costs. Eur Econ Rev 56:1070–1086
28. Pielke RA, Byerly R (1998) Beyond basic and applied. Phys Today 51:42–46
29. Frodeman R (2011) Interdisciplinary research and academic sustainability: managing knowledge in an age of accountability. Environ Conserv 38:105–112
30. WHO/KMS (ed) (2005) Bridging the "Know-Do" gap: meeting on knowledge translation in global health. World Health Organization, Department of Knowledge Management and Sharing, 10–12 Oct, Geneva, Switzerland, WHO Document: WHO/ EIP/KMS/2006.2
31. Ebener S, Khan A, Shademani R, Compernolle L, Beltran M, Lansang MA, Lippman M (2006) Knowledge mapping as a technique to support knowledge translation. Bull World Health Organ 84:636–642
32. Eto J, Vine E, Shown L, Sonnenblick R, Payne C (1996) The total cost and measured performance of utility-sponsored energy efficiency programs. Energy J 17:31–52
33. Geller H, Attali S (2005) The experience with energy efficiency policies and programmes in IEA countries: learning from the critics. International Energy Agency, Paris
34. Davis LW (2008) Durable goods and residential demand for energy and water: evidence from a field trial. RAND J Econ 39:530–546

Part VII
Eco-design of Social Infrastructure

Analysis Modeling for Electricity Consumption in Communication Buildings

Minako Hara, Atsushi Sakurai, Hiroki Oka, Yuriko Tanaka, Yohei Yamaguchi, Yusuke Kishita, Yasushi Umeda, and Yoshiyuki Shimoda

Abstract The telecom services industry is working to prevent increases in CO_2 emissions caused by electricity consumption related to the dynamic growth of telecommunication traffic. Communication buildings, which contain information and communications technology (ICT) equipment, air-conditioning equipment, and other equipment, account for the largest amount of electricity consumed by the telecom services industry. Such companies face difficulties in managing the electricity consumption of communication buildings because they can obtain data on the total amount of electricity consumption in each communication building, but not for individual pieces of equipment. We propose an analysis model that can separate the total electricity consumption of a multipurpose building into several components depending on the consumption purpose by employing an analysis methodology related to the energy demands of typical buildings. We also applied the analysis model to communication buildings as a case study. The results showed that the analysis model was applicable to communication buildings and that the pattern of the electricity consumption differed depending on the communication purpose.

Keywords Green ICT • Communication building • Electricity consumption

M. Hara (✉) • A. Sakurai • H. Oka • Y. Tanaka
Nippon Telegraph and Telephone Corporation, Tokyo, Japan
e-mail: hara.minako@lab.ntt.co.jp

Y. Yamaguchi • Y. Shimoda
Osaka University, Osaka, Japan

Y. Kishita
National Institute of Advanced Industrial Science and Technology, Tsukuba, Japan

Y. Umeda
The University of Tokyo, Tokyo, Japan

M. Matsumoto et al. (eds.), *Sustainability Through Innovation in Product Life Cycle Design*, EcoProduction, DOI 10.1007/978-981-10-0471-1_55

1 Introduction

The telecom services industry is working to prevent increases in CO_2 emissions caused by electricity consumption related to the dynamic growth of telecommunication traffic [1, 2]. Communication buildings, which contain information and communications technology (ICT) equipment, air-conditioning equipment, and other equipment, account for the largest amount of electricity consumed by the telecom services industry. Such companies face difficulties in managing the electricity consumption of communication buildings because they can only obtain data on the total amount of electricity consumption in each communication building, but not for individual pieces of equipment.

For example, as regards research on electricity consumption in ICT sector, Lubritto investigated the relationship between electricity consumption and outdoor temperature of base stations for mobile network [3]. Malmodin analyzed the CO_2 emission amount in production and operation of ICT equipment including user terminals [4]. Imaizumi proposed electricity-saving measures for ICT equipment based on their analysis on detailed electricity consumption data that ICT equipment vendors disclosed [5].

We propose an analysis model that can separate the total amount of electricity consumed by a multipurpose building into several components depending on the consumption purpose by employing an analysis methodology related to the energy demands of typical buildings. We also applied the analysis model to communication buildings as a case study.

2 Method

To realize an effective way of reducing electricity consumption in communication buildings, we must separate the electricity consumption of ICT equipment, air-conditioning equipment, and other equipment. In the following sections, we propose an analysis model for the separation of the electricity consumption in communication buildings based on existing practice and assumptions.

2.1 Approach

This paper defines the functions and usage of communication buildings taking their floor plans into consideration. ICT equipment-related space includes ICT equipment, air-conditioning equipment, and other equipment. A communication building may also include general purpose spaces such as offices, meeting rooms, hallways, and lavatories. Since these spaces have similar functions to those in typical office buildings, they are defined as office spaces, which are assumed to consume similar amounts of electricity per unit area to offices. To estimate the electricity consumption in ICT equipment-related space, the assumed electricity consumption in the office spaces was eliminated from the electricity consumption of the entire communication building. To assess the electricity efficiency of communication buildings, the estimated electricity consumption in ICT equipment-related space was divided by the floor space to provide a benchmark.

2.2 Modeling

Figure 1 shows the scheme for the separation and benchmark calculation of the electricity consumption in a communication building. The input data were the electricity consumption and floor space of the entire communication building and the energy efficiency of the air conditioning. The reference data were from the Data-base for Energy Consumption of Commercial Building (DECC) [6] and the average monthly temperature in each area. In this paper, electricity consumption per space was employed as a benchmark. The steps of the scheme are detailed below:

Step 1

To obtain the daily electricity consumption per space in month n in ICT equipment-related space $E_m(i,n)$ [kWh/m^2/day], the daily electricity consumption in month n in office space $E_o(i,n)$ [kWh/m^2/day] was separated from the

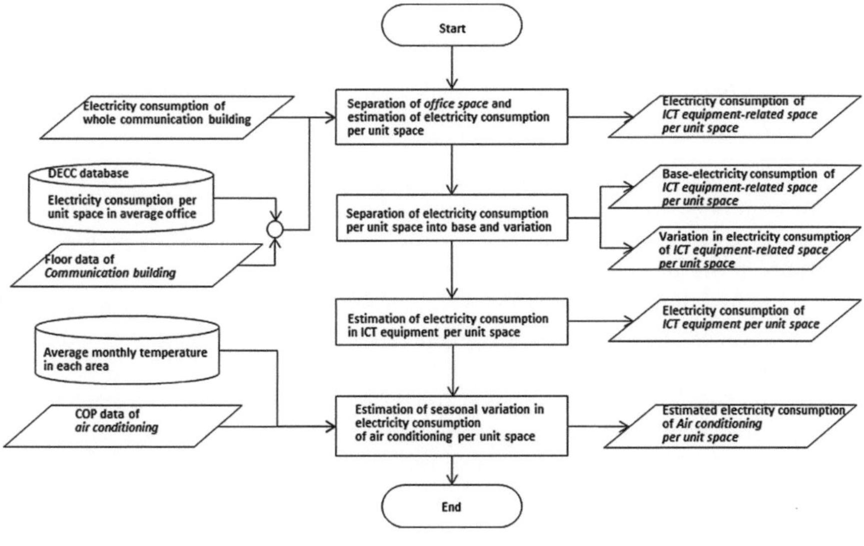

Fig. 1 Scheme of the separation and benchmark calculation of the electricity consumption in a communication building

daily electricity consumption per month of the entire communication building "i" $F(i,n)$[kWh/day]. To estimate $E_o(i,n)$ from $F(i,n)$, we referred to the average monthly electricity consumption of an office building E_o,ave(n) [kWh/m^2/month] derived from DECC. $E_m(i,n)$ was provided from Eq. 1 where $A_m(i)$ [m^2] and $A_o(i)$ [m^2] were the floor of the ICT equipment-related space and office space in the entire communication building "i". E_o, ave(n) was chosen depending on $A_o(i)$:

$$E_m(i,n) = \frac{F(i,n) - E_{o,\text{ave}}(n) \cdot A_o(i)}{A_m(i)} \tag{1}$$

Step 2

To separate $E_m(i,n)$ into base and variation, $E_{m,B}(i)$ [kWh/m^2/day] was defined as the base consumption where $E_m(i,n)$ marked the lowest consumption in month N. Figure 2 shows the concept behind this definition. Then, $E_{m,B}(i)$ and the variation in the daily electricity consumption per space $E_{m,F}(i,n)$ [kWh/m^2/day] in *ICT equipment-related space* are expressed as follows:

$$E_{m,B}(i) = \min\{E_m(i,n)|n\} = E_m(i,N) \tag{2}$$

$$E_{m,\text{F}}(i,n) = E_m(i,n) - E_{m,B}(i) \tag{3}$$

Step 3

$E_m(i,n)$ consists of three components: ICT equipment, air conditioning, and power supply system. These components are assumed to occupy the entire *ICT equipment-related space* in general communication buildings. Thus, $E_m(i,n)$ is expressed as the summation of the daily electricity consumption of ICT equipment per space $E_{\text{mech}}(i,n)$ [kWh/m^2/day], air conditioning per space $E_{\text{air}}(i,n)$ [kWh/m^2/day], and power supply system per space $E_{\text{feed}}(i,n)$ [kWh/m^2/day] in *ICT equipment-related space*:

$$E_m(i,n) = E_{\text{mech}}(i,n) + E_{\text{air}}(i,n) + E_{\text{feed}}(i,n) \tag{4}$$

$E_{\text{air}}(i,n)$ was estimated using the average monthly coefficient of performance (COP) of an air-conditioning system including incidental equipment such as room units and outdoor units and the air-conditioning load which is determined by the heat exhausted from ICT equipment and the heat transferred from outside air. The estimation condition is that the annual electricity consumption of ICT equipment is similar to the heat exhausted from ICT equipment and that the heat transferred between ICT equipment-related space and the outside air can be regarded as heat transmission. Heat transmission is a phenomenon involving heat transfer between spaces divided by a rigid wall. Therefore, the air-conditioning load is larger when the outdoor temperature is higher than the

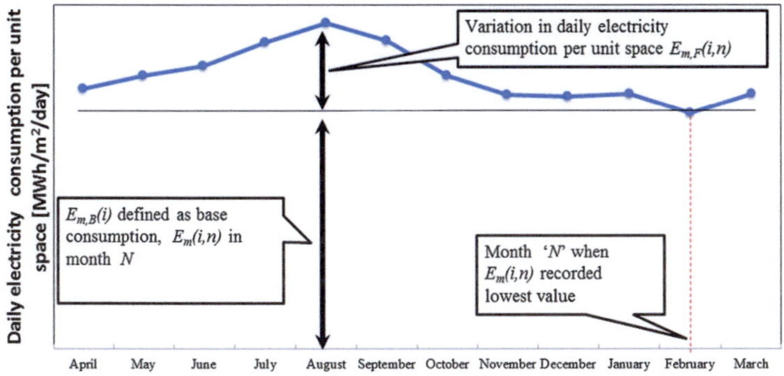

Fig. 2 Explanation of definition of base and varied electricity consumption in ICT equipment-related space

room temperature, and by contrast, the load is smaller when the outdoor temperature is lower than the room temperature because heating is unnecessary in ICT equipment-related space. In addition, $E_m(i,n)$ is expected to mark the lowest consumption in winter as shown in Fig. 2. In this paper, the effect of heat transmission caused by ventilation was neglected to simplify the model.

Considering the daily heat transmission amount per space in month n, $Q(i,n)$ [kWh/m^2/day], the electricity consumption per space of air conditioning $E_{\mathrm{air}}(i,n)$ is expressed as Eq. 5 using the electricity consumption per space of ICT equipment $E_{\mathrm{mech}}(i,n)$ and the COP from the definition in general:

$$E_{\mathrm{air}}(i,n) = \frac{E_{\mathrm{mech}}(i,n) + Q(i,n)}{\mathrm{COP}(i,n)} \tag{5}$$

The electricity consumption per space of a power supply system $E_{\mathrm{feed}}(i,n)$ was calculated by using the definition of the loss coefficient of power supply η as in Eq. 6 in general:

$$E_{\mathrm{feed}}(i,n) = \eta \cdot (E_{\mathrm{mech}}(i,n) + E_{\mathrm{air}}(i,n)) \tag{6}$$

$E_{\mathrm{air}}(i,n)$ and $E_{\mathrm{feed}}(i,n)$ can then be substituted into equation to give the function of $E_{\mathrm{mech}}(i,n)$ alone:

$$\begin{aligned}E_m(i,n) &= E_{\mathrm{mech}}(i,n) + E_{\mathrm{air}}(i,n) + E_{\mathrm{feed}}(i,n)\\&= E_{\mathrm{mech}}(i,n) + E_{\mathrm{air}}(i,n) + \eta\cdot(E_{\mathrm{mech}}(i,n) + E_{\mathrm{air}}(i,n))\\&= (1+\eta)(E_{\mathrm{mech}}(i,n) + E_{\mathrm{air}}(i,n))\\&= (1+\eta)\left(E_{\mathrm{mech}}(i,n) + \frac{E_{\mathrm{mech}}(i,n)+Q(i,n)}{\mathrm{COP}(i,n)}\right)\\&= (1+\eta)\left(\frac{(1+\mathrm{COP}(i,n))E_{\mathrm{mech}}(i,n)+Q(i,\mathrm{n})}{\mathrm{COP}(i,\mathrm{n})}\right)\end{aligned} \tag{7}$$

Then, $E_{\mathrm{mech}}(i,n)$ is expressed as a function of $E_m(i,n)$:

$$E_{\mathrm{mech}}(i,n) = \frac{\mathrm{COP}(i,n)E_m(i,n) - (1+\eta)Q(i,n)}{(1+\eta)(1+\mathrm{COP}(i,n))} \tag{8}$$

Equation applies to month N; namely,

$$E_{\mathrm{mech,B}}(i) = E_{\mathrm{mech}}(i,N) = \frac{\mathrm{COP}(i,N)E_{m,\mathrm{B}}(i) - (1+\eta)Q(i,N)}{(1+\eta)(1+\mathrm{COP}(i,N))} \tag{9}$$

Step 4

Since the components and electricity consumption of ICT equipment exhibit no seasonal change, $E_{\mathrm{mech},B}(i)$ was assumed to be constant. $E_{m,F}(i,n)$ [kWh/m^2/day] is defined as follows:

$$E_{m,\mathrm{F}}(i,n) = E_{\mathrm{air,F}}(i,n) + E_{\mathrm{feed,F}}(i,n) \tag{10}$$

where the variations $E_{\mathrm{air},F}(i,n)$ and $E_{\mathrm{feed},F}(i,n)$ in the daily electricity consumption in month n per floor space of the *ICT equipment-related space* of air conditioning $E_{\mathrm{air}}(i,n)$ and of power supply system $E_{\mathrm{feed}}(i,n)$ were defined by Eqs. 11 and 12, respectively:

$$E_{\mathrm{air,F}}(i,n) = E_{\mathrm{air}}(i,n) - E_{\mathrm{air}}(i,N) \tag{11}$$

$$E_{\mathrm{feed,F}}(i,n) = E_{\mathrm{feed}}(i,n) - E_{\mathrm{feed}}(i,N) \tag{12}$$

Equation 10 was converted as follows:

$$E_{m,F}(i,n) = (1+\eta)\left(\frac{E_{\mathrm{mech,B}}(i)+Q(i,n)}{\mathrm{COP}(i,n)} - \frac{E_{\mathrm{mech,B}}(i)+Q(i,N)}{\mathrm{COP}(i,N)}\right) \tag{13}$$

The system COP introduced in communication building "i" was estimated considering the relationship between outdoor temperature and the system COP based on the air-conditioning system specifications. The average system COP for the air-conditioning system in month n, which is working under similar conditions to the specifications, was defined as $\mathrm{COP_R}$(i,n). In terms of the actual working conditions, system COP is usually lower than the specification due to such factors as deterioration. Therefore, the coefficient $\alpha(i)$ was used in the definition:

$$\mathrm{COP}(i,n) = \alpha(i) \cdot \mathrm{COP}_R(i,n) \tag{14}$$

Substituting the COP of Eq. 13, $E_{m,F}(i,n)$ is given using $\mathrm{COP_R}$ as follows:

$$E_{m,F}(i,n) = (1+\eta)\left(\frac{E_{\mathrm{mech,B}}(i) + Q(i,n)}{\alpha(i) \times \mathrm{COP}_R(i,n)} - \frac{E_{\mathrm{mech,B}}(i) + Q(i,N)}{\alpha(i) \times \mathrm{COP}_R(i,N)}\right) \tag{15}$$

It was possible to estimate the daily heat transmission amount per space in month N, $Q(i,n)$[kWh/m^2/day], by using the heat transmission coefficient of communication building "i", $U(i)$[kW/K·m^2]; the gross area of the building envelope, $S(i)$ [m^2]; *ICT equipment-related space* in a communication building "i", $A_m(i)$ [m^2]; and the temperature difference between the room temperature of *ICT equipment-related space* and the outdoor temperature, $\Delta T(i,n)$ [K], as

$$Q(i,n) = \frac{U(i) \cdot S(i) \cdot \Delta T(i,n) \cdot 24[\mathrm{h/day}]}{A_m(i)} \tag{16}$$

$\Delta T(i,n)$ was estimated by using the temperature recommended in the guideline for ICT equipment as the room temperature of *ICT equipment-related space* and local temperature data as the outside temperature, respectively.

3 Case Analysis

The effectiveness of the analysis model was confirmed using the electricity consumption data of existing communication buildings. We also focused on the characterization of the electricity consumption in communication buildings. DECC has four categories of floor space: space range A, less than 300 m^2; space range B, 300 m^2 or more and less than 2,000 m^2; space range C, 2,000 m^2 or more and less than 10,000 m^2; and space range D, 10,000 m^2 or more. The data we used were the average monthly electricity consumption in communication buildings from April 2012 to March 2013 in the eastern area of Japan. With the focus on regional features rather than construction specifications, the monthly electricity consumption was averaged in each prefecture categorized using the above A, B, and C floor space range. The estimation was implemented under the following assumptions:

- The average monthly outdoor temperature of the prefectural capital was used as the outdoor temperature [7].

- The coefficient of heat transfer was set at 0.39 [kW/K m^2] for walls and 0.42 [kW/K m^2] for roofs assuming that the coefficient of heat transfer of a communication building is similar to that of usual commercial buildings.
- The COP for an air-conditioning system in a communication building was assumed to be 2.57 under a rated condition where the outdoor temperature was 35 °C and the load factor was 100 % [8].
- $\alpha(i)$ was set to minimize the difference between measured $E_{m,F}(i,n)$ and estimated $E_{m,F}(i,n)$ based on Eq. 15.
- The function of the air-conditioning system in ICT equipment-related space was assumed to be limited to cooling. Heating, ventilation, and dehumidification were not included. The conversion coefficient of a power supply system in *ICT equipment-related space* was assumed to be 82 %. This means that η(i) was 18 % [9].

4 Results and Discussion

4.1 Estimation with Analysis Model

The data on electricity consumption in communication buildings were analyzed based on the analysis model. Figures 3, 4, 5, and 6 show the analysis results for sample data in all the buildings and divided by the ICT equipment-related space ranges A, B, and C. The analysis results showed that the proportion of electricity consumption of all the communication buildings was 90 % in ICT equipment-related space and 10 % in office space. In addition, the proportion of electricity consumption as regards all the communication buildings was 50 %, 26 %, and 14 % for ICT equipment, air conditioning, and power supply system in *ICT equipment-related space*, respectively. This electricity use breakdown agreed with data published by NTT West, namely, 55 %, 30 %, and 15 % [9]. This result also showed that ICT equipment accounts for the biggest proportion (as much as 50 %) of the electricity consumed in communication buildings. The relative error for the whole building was from −5.9 to 7.0 % for each month with an annual average of 1.1 %. The relative error for the ICT equipment-related space range A was from −8.8 to 22.9 % for each month with an annual average of 1.1 %.

4.2 Analysis of Model Sensitivity

The sensitivity of the model was analyzed to identify the most effective model parameters. Five parameters were changed for individual estimation as shown in Table 1.

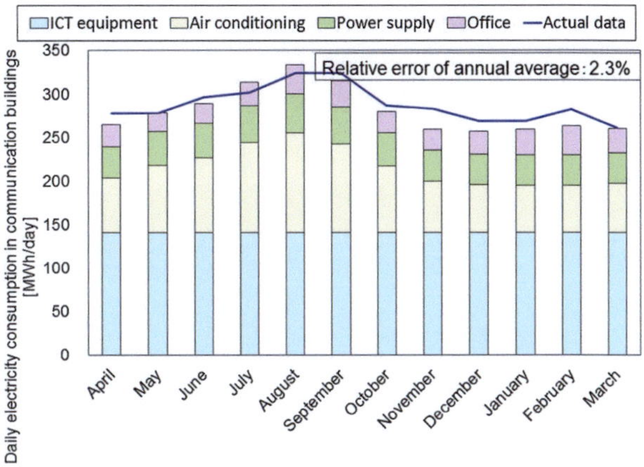

Fig. 3 Daily electricity consumption in all communication buildings

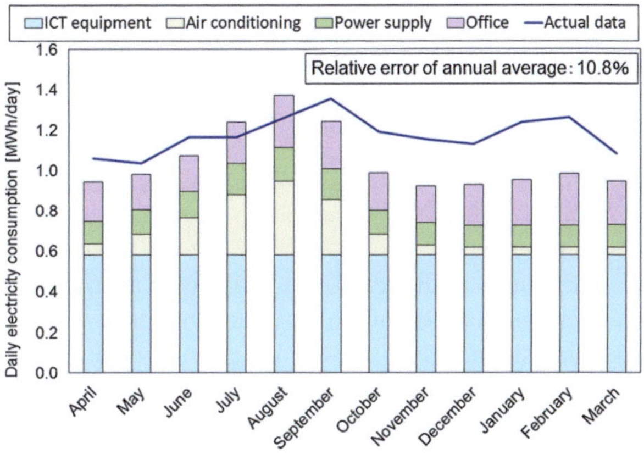

Fig. 4 Daily electricity consumption in communication buildings with the equipment-related space range A

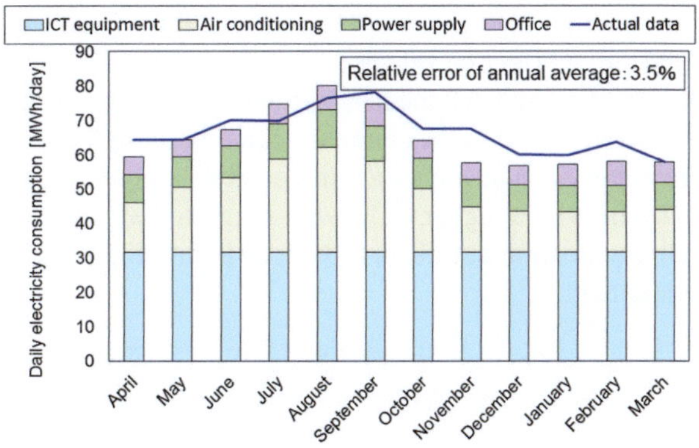

Fig. 5 Daily electricity consumption in communication buildings with the equipment-related space range B

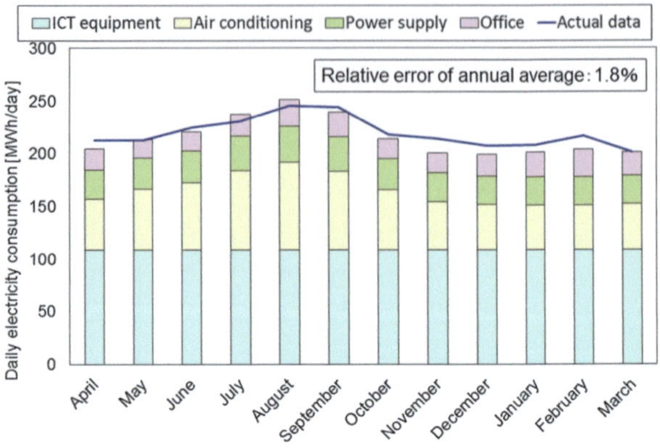

Fig. 6 Daily electricity consumption in communication buildings with the equipment-related space range C

Table 1 Selected parameters for sensitivity analysis

Parameter	Changed amount
Electricity consumption of ICT equipment	E_{mech} decreased by 10 %
Power supply efficiency	η decreased by 10 %
Insulation efficiency	Q decreased by 10 %
Outdoor temperature	T [C°] decrease by 10 %
Air-conditioning efficiency	COP increased by 10 %

The results of the sensitivity analysis of average annual electricity consumption in a communication building revealed that there were no predominant parameters in the analysis model as shown in Fig. 7. A reduction in E_{mech}, η, Q, T [°C], and an increase in COP by 10 % led to reductions of 9.4 %, 0.9 %, −0.7 %, 1.6 %, and 2.5 %, respectively, in the electricity consumption of all the communication buildings. The results showed that improving the electricity consumption of ICT equipment is the most efficient measure for saving energy and improving air-conditioning efficiency. In contrast, improvements in insulation efficiency had a negative effect on power saving in communication buildings in this result. This result suggested that the heat outflow in the summer had a larger effect than the heat inflow in winter in the communication buildings used in this estimation. Therefore, measures such as implementing outdoor-air cooling and replacing air-conditioning systems with a higher COP were suggested as significant ways of saving energy in communication buildings as well as improving the electricity consumption of ICT equipment.

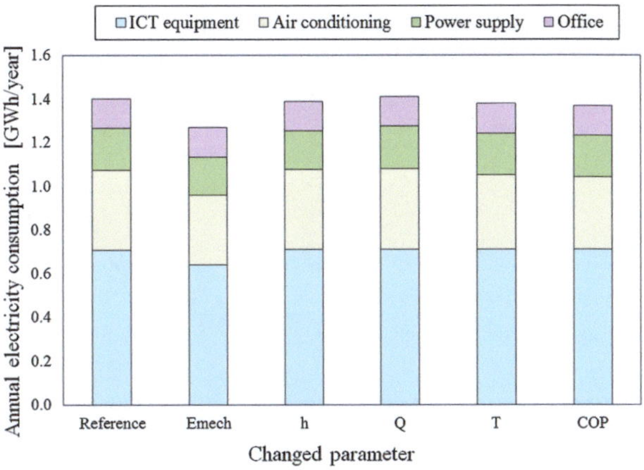

Fig. 7 Annual electricity consumption in communication buildings on sensitivity analysis

4.3 Energy-Efficiency Benchmark in Communication Buildings

With the aim of examining an applicable benchmark in communication buildings, $E_{mech}(i)$ was plotted against ICT equipment-related space as in Fig. 8. A positive correlation was observed between two parameters. This result has considerable importance in the field of energy efficiency in buildings. It indicates that as the scale of a communication building increases, the construction specifications become more energy intensive because the trend is opposite to that of ordinary buildings.

The energy efficiency in communication buildings was evaluated with the deviations from the new regression line as a reference, after eliminating data indicating that there is only a 5 % chance that a deviation this great or greater would occur, as shown in Fig. 9. The results show the highest evaluation in a middle-range group of buildings in a northern-east prefecture and the lowest in a middle-range group of buildings in a northern-central prefecture. Thus, no correlation was observed between the deviations and the spaces or the locations in these evaluation results. The implementation possibility of this benchmark can be examined using a dataset for an individual building in further study.

5 Conclusion

This study improved the analysis model that makes it possible to estimate the electricity consumption of entire communication buildings by separating it into several components including that used for ICT equipment, air conditioning, power supply system, and offices. Estimation with the model rapidly provided the electricity consumption of the individual components.

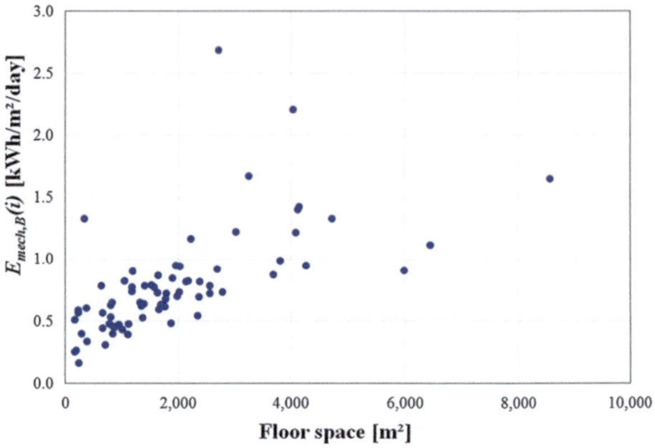

Fig. 8 $E_{\mathrm{mech},B}(i)$ plotted against *ICT equipment-related space*

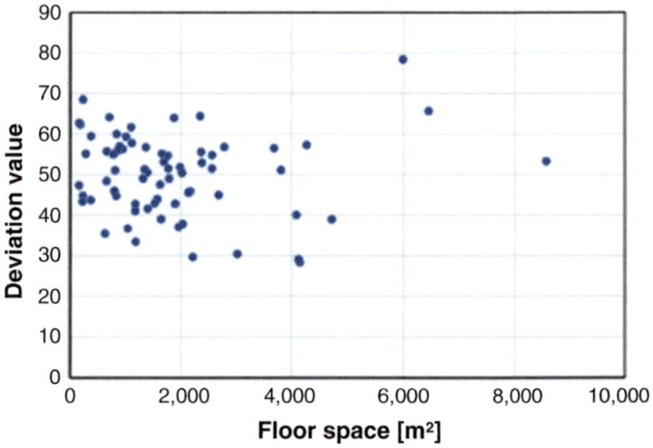

Fig. 9 Deviations from the new regression line of $E_{\mathrm{mech},B,}(i)$ plotted against ICT equipment-related space

The results of the case study using the data on the average electricity consumption of communication buildings in each prefecture showed that the analysis model could provide values comparable to those reported in previous studies.

The results of a sensitivity analysis suggest that a significant way of saving energy in communication buildings is to improve the energy efficiency of ICT equipment and air-conditioning systems.

The case study also suggested that the electricity consumption in ICT equipment correlated most closely to that of the entire communication building. The deviation from the regression line of the daily electricity consumption of ICT equipment per space against floor space of *ICT equipment-related space* is expected to be applicable as a benchmark of energy efficiency for communication buildings.

In further study, the possibility of implementing this benchmark will be examined using datasets obtained in individual buildings.

References

1. Ministry of Internal Affairs and Communications, Japan (2008) Report from Study Group on ICT Policy for Addressing Global Warming. April 2008. http://www.soumu.go.jp/main_sosiki/joho_tsusin/policyreports/chousa/ict_globalwarming/pdf/0804_h1.pdf
2. Kuroda K (2014) Scenario analysis of the electricity consumption in the Japanese Information and Communications Sector toward Green of ICT. In: Proceedings of going green CARE INNOVATION 2014, Schoenbrunn Palace Conference Centre. Vienna, 17–20 Nov 2014
3. Lubritto C et al (2011) Energy and environmental aspects of mobile communication systems. Energy 36:1109–1114
4. Malmodin J et al (2014) Life cycle assessment of ICT carbon footprint and operational electricity use from the operator, national, and subscriber perspective in Sweden. J Ind Ecol 18:829–845
5. Imaizumi H et al (2010) Recent technologies towards green internet. IEICE Tech Rep 110 (49):43–48
6. Japan Sustainable Building Consortium (2013) Data-base for energy consumption of commercial building. Sept 2013. http://www.jsbc.or.jp/decc_english/index.html
7. Japan Meteorological Agency (2015) Climate of Japan. http://www.data.jma.go.jp/obd/stats/data/en/index.html
8. NTT FACILITIES (2010) News release on 15 September 2010. https://www.ntt-f.co.jp/news/heisei22/h22-0915.html
9. NTT WEST (2009) Activities toward green ICT in NTT West. NTT Tech J 21(9):58–61 (in Japanese)

Research on Evaluation Index System and Comprehensive Evaluation of Typical Eco-Industrial Parks

Lei Zhang, Xing Meng, Hongbing Yu, Toru Matsumoto, and Xi Chen

Abstract The eco-industrial park (EIP), as a new industrial park combining the concept of circular economy and industrial ecology principles of construction, has not only become an effective way to achieve sustainable development, improving energy efficiency and the ecological environment quality in many countries, but has also become an important carrier of implementing science-based, adjusting the industrial structure, accelerating the transformation, and upgrading and constructing ecological civilization in China. China began construction of eco-industrial demonstration park pilots in 1999, and subsequently has actively carried out the practice of building eco-industrial parks nationwide. Although the EIPs have made great progress in terms of planning and construction, and achieved some success in improving energy efficiency, reducing environmental pollution, and improving environmental quality, there are many problems in the development process that restrict the stable operation and coordinated development and continuous improvement of the park, such as weak ecological relevance between enterprises, low resources and energy output rate, imperfect organizational structure, unclear construction goals, and insufficient technical innovation. The causes of these problems are mainly due to the government agencies and park management department not paying enough attention to the coordinated development degree of EIPs, lacking a practicable index system and assessment method for assessing the degree of sustainable development and comprehensive development level, which result in their being unable to fully grasp the operational state of the park and also unable to formulate a highly appropriate development strategy. Therefore this

L. Zhang (✉)
The University of Kitakyushu, Kitakyushu, Japan

Sino-Japan Friendship Centre for Environmental Protection, Beijing, China
e-mail: zhanglei0521@126.com

X. Meng • H. Yu
Nankai University, Tianjin, China

T. Matsumoto
The University of Kitakyushu, Kitakyushu, Japan

X. Chen
Sino-Japan Friendship Centre for Environmental Protection, Beijing, China

M. Matsumoto et al. (eds.), *Sustainability Through Innovation in Product Life Cycle Design*, EcoProduction, DOI 10.1007/978-981-10-0471-1_56

chapter, based on the systematic study of the development of EIPs both domestic and abroad, is focused on the theory and methods of constructing the index system of EIPs, establishing the degree of coordination and comprehensive development level index system, and evaluation methods of sector-integrated eco-industrial parks (SIEIPs). Specific research includes the following aspects: (1) reviewing the status of domestic and foreign research and development of EIPs, and at the same time, overviewing several relevant theories and methods of construction of an EIP evaluation index system, including: eco-efficiency, flow analysis, and index systems; (2) drawing on the ideas of the analytic hierarchy process and taking the coordinated development degree and the comprehensive development level as the overall evaluation goal of SIEIPs, from economic development, resource utilization, environmental protection, ecological civilization, park management, and social progress, six subsystems proceeding to consider the current situation, construction effect, stability of development, sustainability and coordination of SIEIP development, and then using frequency statistics and expert consulting methods to select several evaluation indexes initially, further superimposed on the primary indicators analyzed, to screen and ultimately fix the evaluation index system of SIEIPs; (3) applying a multilevel extension comprehensive evaluation method to build a model of coordinated development degree and integrated development level of the SIEIPs, and using this model to evaluate a typical eco-industrial park. The results show that typical EIPs will have to make more effort regarding the stable and sustainable development of circular economy status. In addition, the evaluation results are factor analyzed, proposing strategies and measures for improvement.

Keywords Integrate eco-industrial parks • Index system • Comprehensive evaluation

1 Introduction

Various countries around the world carried out a series of policies and practical activities characterized by resource and energy conservation, green low carbon, and ecological and environmental protection in recent years due to resource shortage, the energy crisis, ecological environment deterioration, frequent natural disasters, and other reasons. A large number of ecological industrial parks were established as one of several productive practical activities, and China also explored and implemented such parks in order to realize sustainable development of resources, energy, economy, society, and ecology. Fifty-nine national ecological industrial demonstration parks have been approved and established in China since the first ecological industrial park in Guigang, Guangxi [1]. Twenty-six parks were qualified in acceptance by the National Ministry of Environmental Protection [2].

However, it is difficult to evaluate accurately the sustainable development degree and comprehensive development level for the process of creating ecological industrial parks and the process of ecologically transforming other parks.

Evaluation concerns the lack of suitable quantitative indicators, therefore it is difficult to form a set of evaluation index systems. Scientific evaluation methods and models cannot be adopted to evaluate the current situation of development, and guidance and constructive suggestions cannot be provided for further developing future parks.

Therefore, it is very important to create a set of index systems capable of accurately evaluating the sustainable development level and comprehensive development level of the park with high practicality and operability.

2 Development Situation of Ecological Industrial Parks at Home and Abroad

2.1 Development Situation of Overseas Ecological Industrial Parks

The ecological industrial park (EIP) was constructed first in Europe; Denmark established the first eco-industrial park in the world in the 1970s, Kalunborg Industrial Symbiont. Denmark's Kalunborg Industrial Park has become a model for ecological industrial parks all over the world after being developed for several decades [3]. Its successful experience is widely focused on by other countries all over the world. Large-scale eco-industrial parks were constructed in Europe in the 1990s. A survey report in Britain shows that there were a total of 60 ecological industrial park projects in Europe by 2005. By-product symbiosis networks have been formed in most ecological industrial parks. Renewable energy recycling technology has been adopted in about 20% of the parks [4].

America is also one of the countries that engages in the practice of national ecological industrial parks at present. Fifteen ecological industrial parks have been planned and constructed within the 3 years since the establishment of four ecological industrial park demonstration pilot communities prior to 1997. The parks are involved in bio-energy development, clean production, waste treatment, recycling, reutilization, and other environmental protection industries [1].

In Asia, Japan has been one of the earliest countries to develop an ecological industry [3]; 10 ecological urban projects and about 30 ecological industrial parks have been built. It ranks at the top among countries in Asia. The more influential parks include Yamanashi-ken Eco-industrial Park and Fujisawa Ecological Industrial Park [4].

In addition, many developing countries are also actively carrying out exploration of ecological industrial parks. For example, a batch of ecological industrial parks with their own characteristics have been planned and constructed in Thailand, Indonesia, Namibia, South Africa, and other countries [3, 4].

2.2 Development Situation and Existing Problems of Domestic Ecological Industrial Parks

Compared with foreign parks, ecological industrial parks started relatively late in China. China launched the pilot work of ecological industrial parks officially in 1999 under the inspiration of successful experience in foreign ecological industrial parks. The first ecological industrial demonstration park was established in China under the support of the State Environmental Protection Administration, Guangxi Guigang Ecological Industrial Park. The park is based on an enterprise for producing sugar with canes, and ecological industry chains of six major systems around the sugarcane field, sugar-making, paper-making, alcohol, cogeneration, and comprehensive environmental treatment were formed. It has become an ecological industrial park with the largest scale and fastest progress in China at present. Ecological industrial parks were planned and constructed in various provinces and cities all over the country in subsequent years. A large number of ecological industrial parks have been constructed in China.

Although ecological industrial parks are constructed and developed faster in China, there is a big gap with developed countries in terms of development quality and comprehensive level of development. Problems are mainly manifested in the following aspects.

1. Management system construction lags. Management personnel in the park have lagged management's thinking mode, management of many parks still remains in the recognition of constructing ecological industrial parks based on the industrial symbiosis system, management of all links in the park development and perfection process is ignored, and most parks do not establish an information-sharing platform matched with the management system in the aspect of park information-sharing construction on the other hand [3].
2. Sustainable development degree of the park is not focused enough. Most parks pay attention to early planning and construction of the ecological industrial park, and do not focus on their own perfection and sustainable development after park construction [3].
3. All subsystems in the park suffer from a low coordinated development degree with poor park stability. Most parks lack considering construction of all subsystems in the construction process as a whole, therefore all subsystems have the disadvantages of low association and poor coordination, thereby affecting the stability of park system development [3].
4. Some parks laterally emphasize construction of the industrial symbiosis system. They have task focus on the exchange of by-products rather than considering fund investment in clean production technology for reducing emissions at the source. Therefore, clean production and industry symbiosis cannot be implemented synchronously in the park, and it is difficult to realize the objective of sustainable development [3].

5. Key technology innovation is not sufficient in the park. Advanced processes and technologies are not sufficiently utilized. For example, energy cascade utilization technology, poisonous and harmful materials alternative technology, green reconstruction, and other technologies are rarely utilized, thereby weakening park competitiveness.

3 Eco-Industrial Park Comprehensive Evaluation Indexes System

3.1 Principles for the Eco-Industrial Park Evaluation Indexes System

Based on the principle of "three priorities and five combinations" put forward by relevant scholars regarding eco-industrial park comprehensive evaluation, the author summarizes the new "three priorities and five combinations" principle through a comparative analysis of related domestic literature.

"Three priorities" means to prioritize the generic statistical indexes, to prioritize the quantitative indexes, and to prioritize the reducing consumption and recycling indexes on the basis of ensuring data availability and evaluation accuracy.

"Five combinations" means to carry out five combinations in selecting and designing evaluation indexes, including the combination of policies and targets, the combination of functions and progressiveness, the combination of systematicness and hierarchy, the combination of dynamics and stability, and the combination of scientificity and operability.

3.2 Establishment of the Eco-Industrial Park Evaluation Indexes

The eco-industrial park is a complicated system composed of several factors, such as economy, resources, environment, ecology, and society, among others. As it is difficult to reflect the development and future trends of the eco-industrial park by one or a few indexes, a comprehensive evaluation is carried out from different aspects and layers according to the type and feature of the eco-industrial park.

With reference to the thought of the analytic hierarchy process (AHP), this thesis divides the eco-industrial park comprehensive evaluation indexes system into four layers: the first layer is the target layer, which makes the coordinated development degree and the comprehensive development level of the park system the general target of the eco-industrial park evaluation; the second layer is the criterion layer: divide the evaluation indexes into six subsystems of economic development, resource utilization, environmental protection, ecological civilization, park administration, and social progress based on the aforesaid principle; the third layer is the state layer: divide the state layer of the eco-industrial park evaluation into two classes by applying the theoretical analysis method and the literature analysis method: Class I refers to the rigid indexes, which includes six aspects of the economic strength, resource consumption, sewage discharge intensity, ecological construction, management level, and employment, and reflects the development status and construction effects of the eco-industrial park [5]; the other class refers to the flexible indexes, which includes six aspects of economic development potentials, recycling degree, pollution control, ecological improvement potentials, infrastructure supporting capacity, and happiness index, and reflects the development stability, sustainability, and coordination of the eco-industrial park; and the fourth layer is the variable layer (the index layer): adopt frequency statistics and expert consultation and select several evaluation indexes [6].

The comprehensive evaluation index system is constructed as shown in Table 1.

4 Eco-Industrial Park Comprehensive Evaluation Indexes System

4.1 Selection of Comprehensive Evaluation Method and Construction of Evaluation Model

The evaluation index system can evaluate the status quo and the development trends of the eco-industrial park well only when it is combined with corresponding evaluation methods. The current comprehensive evaluation methods include the fuzzy comprehensive evaluation method, the AHP, the multilayer extensible comprehensive evaluation method, and the grey correlation cluster evaluation method, among others. This thesis constructs the comprehensive evaluation model based on the AHP and the extenics theory [7].

4.2 Calculation of Weighted Comprehensive Evaluation Indexes

1. Comprehensive efficiency indexes of evaluation objects include economic development index, resource utilization index, environmental protection index, ecological civilization index, park administration index, and the social progress index of the eco-industrial park (i.e., φ_1, φ_2, φ_3, φ_4, φ_5, φ_6). As for the No.i evaluation object (the sample):

Table 1 Comprehensive evaluation index system of integrated ecological industrial park

Target layer O	Criterion layer A	State layer B	Variable layer C
Coordinated development degree and comprehensive development level of the park system	Economic development A1	Economic strength B1, B1	Per-capital industrial added value (10,000 yuan/capital) C1[a]
			GDP average annual growth rate (%) C2[a]
			Percentage of the added value of tertiary industry in GDP (%) C3[a]
			Economic output density (100 million yuan/sq.km.) C4[a]
			Rate of energy output (10,000 yuan/tce) C5[a]
			Rate of water resources output (10,000 yuan/m^3) C6[a]
		Economic development potentials B2	Percentage of scientific research input in GDP (%) C7[a]
			Percentage of high-tech industry output in total industrial output value (%) C8[a]
			Enterprises correlation C9[a]
			Waste chain completeness C10[a]
			Substitutability of raw materials source C11[a]
	Resource utilization A2	Resource consumption B3	Comprehensive energy consumption per unit of industrial added value (tce/10,000 yuan) C12[b]
			Fresh water consumption per unit of industrial added value (m^3/10,000 yuan) C13[b]
			Comprehensive energy consumption elasticity coefficient C14[b]
			Fresh water consumption elasticity coefficient C15[b]
		Recycling degree B4	Repetitiveness of industrial water (%) C16[a]
			Comprehensive utilization ratio of industrial solid wastes (%) C17[a]
	Environmental protection A3	Sewage discharge intensity B5	Wastewater production per unit of industrial added value (t/10,000 yuan) C18[b]
			Solid waste production per unit of industrial added value (kg/10,000 yuan) C19[b]
			COD production per unit of industrial added value (kg/10,000 yuan) C20[b]
			SO_2 emission per unit of industrial added value (kg/10,000 yuan) C21[b]
			COD emission elasticity coefficient C22[b]
			SO_2 emission elasticity coefficient C23[b]
		Pollution control B6	Rate of industrial wastewater discharge compliance (%) C24[a]
			Rate of industrial wastewater discharge reduction (%) C25[a]

(continued)

Table 1 (continued)

Target layer O	Criterion layer A	State layer B	Variable layer C
			Rate of main air pollutants emission compliance (%) C26[a]
			Rate of industrial solid wastes discharge reduction (%) C27[a]
			Rate of household wastes hazard-free treatment (%) C28[a]
			Rate of hazardous wastes treatment and disposal (%) C29[a]
			Rate of sewage treatment plant effluent quality compliance (%) C30[a]
			Average regional environmental noise (db) C31[b]
			Average road traffic noise (db) C32[b]
	Eco-friendly A4	Ecological construction B7	Coverage rate of regulated area of dust and smoke (%) C33[a]
			Rate of secondary air quality standard compliance (%) C34[a]
			Rate of park greenery coverage (%) C35[a]
			Satisfaction of the public toward the environment (%) C36[a]
		Ecological improvement potentials B8	Percentage of environmental protection in GDP (%) C37[a]
			Percentage of clean energy in total energy (%) C38[a]
			Ratio of public transport sharing (%) C39[a]
	Park administration A5	Management level B9	Park environmental report preparation C40[a]
			Completeness of environmental management system C41[a]
			Monitoring on park change C42[a]
			Eco-industrial training C43[a]
		Infrastructure supporting capacity B10	Completeness of supporting infrastructures C44[a]
			Completeness of information system C45[a]
			Percentage of professionals in environmental administration organizations (%) C46[a]
	Social progress A6	Employment B11	Ratio of the number of tertiary industry employees (%) C47[a]
			Ratio of employment added by developing circular economy (%) C48[a]
		Happiness index B12	Engel coefficient (%) C49[b]
			Social security coverage (%) C50[a]

[a]Represents positive index
[b]Represents negative index

$$\phi_{ti} = \sum_{j=1}^{n} \omega_{ij} k(v)_{ij} \tag{1}$$

where φ_{ti} represents the indexes of the No.i evaluation object (the sample); $t = 1$, 2, 3, 4, 5, 6, respectively, indicates the economic development index, resource utilization index, environmental protection index, ecological civilization index, park administration index, and the social progress index of the No.i evaluation object (the sample); $k(v)_{ij}$ represents the correlation function membership of the status data of the No.i evaluation object's (the sample) No.j index; w_{ij} represents the weighted coefficient of the No.i evaluation object's (the sample) No.j index; and n indicates the number of indexes contained in every criterion layer.

2. Coefficient of comprehensive development reflects the general development level and status of each subsystem of the eco-industrial park during the evaluation period:

$$U = \sum_{t=1}^{6} \omega_t \varphi_{ti} \tag{2}$$

3. Coefficient of development coordination is the index to evaluate the intercoordination among subsystems of the eco-industrial park. The following formula can be used to calculate it.

$$H_i = 1 - S_i / \overline{F_i} \tag{3}$$

where S_i represents the efficiency index standard deviation of each subsystem of the No.i evaluation object (the sample); $\overline{F_i}$ represents the average value of the economic development index, resource utilization index, environmental protection index, ecological civilization index, and park administration index of the No.i sample. When φ_1, φ_2, φ_3, φ_4, φ_5, φ_6 are much closer, economic development, resource utilization, environmental protection, ecological civilization, park administration, and social progress of each subsystem of the eco-industrial park are coordinated, and the value is close to 1; otherwise, the indexes are less coordinated, and the value is close to 0.

5 Case Study

5.1 Park Profile

The Tianjin Economic-Technological Development Area (hereafter referred to as the TEDA) was established on December 6th, 1984 with the approval from the State Council. It was one of the first state-level development areas. The TEDA passed the review of relevant NDRC experts on October 29th, 2006 and became one of the first 11 national circular economic pilot parks. Then on May 20th, 2008, the TEDA was formally awarded the title of National Eco-industrial Demonstration Park by the Ministry of Environmental Protection of the People's Republic of China, the Ministry of Commerce of the People's Republic of China, and the Ministry of Science and Technology of the People's Republic of China [8].

The TEDA has gradually transformed into a new industrial district from a single industrial area through its 30-year development. As of the end of 2010, the TEDA has 9546 domestic enterprises with the total registered capital of RMB 177.06853 billion yuan, and more than 3300 foreign-invested enterprises with the total investment of over 15 billion dollars. Presently, the TEDA has nine industrial systems: eight pillar industries of electronic information, biomedicine, food and beverages, automobiles, equipment manufacturing, petroleum and chemicals, aerospace, new energy, and new materials, as well as the modern service industry. Centered on the eight pillar industries, TEDA has formed an eco-industrial prototype characterized by diversified business types, close product chain relations, closed flow of resources, and efficient utilization of resources by continuously promoting industrial structure optimization and upgrading, carrying out industrial ecology, and improving industry aggregation and correlation [8].

5.2 Coordinated Development Degree and Comprehensive Development Level of the Park System

The data of the variable layer quantitative indexes are mainly from the TEDA 2011 Development Report, 2011 Tianjin Statistical Yearbook, industrial park statistical data, relevant regional environmental quality report, governmental research data, relevant literature data and field investigation, and so on. Some data are calculated by basic data.

Please refer to Table 2 for TEDA 2010 status data and evaluation and calculation results.

Please refer to Table 3 for the coordination coefficient and comprehensive cyclic indexes of each subsystem of the eco-industrial park.

6 Summary

6.1 Analysis of Characterization Factor

As seen from the efficiency indexes of the six subsystems of the general eco-industrial park evaluation indexes system, the economic development efficiency index is 0.5536, the resource utilization efficiency index is 0.2601, the

Table 2 Status data and evaluation and calculation results of evaluation indexes

Index layer	Actual value (v_i)	Classical field (a_i, b_i)	Correlation $k(v_i)$	Variable layer weight	Status layer weight	Criterion layer weight
C1	38.41	(15,100)	0.275	0.0641	0.1667	0.3994
C2	25.1	(20.8,30)	0.467	0.1006		
C3	23.1	(43,80)	0	0.1596		
C4	18.36	(2.5,25)	0.705	0.2504		
C5	6.25	(1.25,10.3)	0.552	0.3825		
C6	0.2	(0.034,0.34)	0.542	0.0428		
C7	3.15	(1.8,4.6)	0.482	0.3571	0.8333	
C8	45.3	(20,60)	0.6325	0.3571		
C9	60	(40,90)	0.4	0.1499		
C10	100	(60,100)	1	0.0867		
C11	80	(60,100)	0.5	0.0492		
C12	0.142	(0.05,0.5)	0.204	0.5505	0.75	0.1998
C13	4.05	(0.9,9)	0.389	0.1295		
C14	0.778	(0.06,0.6)	0	0.2601		
C15	0.45	(0.06,0.55)	0.796	0.0599		
C16	87.5	(75,100)	0.5	0.1667	0.25	
C17	90.86	(85,100)	0.391	0.8333		
C18	2.27	(0.8,8)	0.204	0.3813	0.8	0.1998
C19	25.55	(10,100)	0.173	0.1651		
C20	0.24	(0.1,1)	0.156	0.2493		
C21	0.227	(0.1,1)	0.141	0.0996		
C22	0.573	(0.03,0.3)	0	0.0635		
C23	−0.418	(0.02,0.2)	1	0.0412		
C24	100	(95,100)	1	0.1593	0.2	0.1998
C25	5.96	(5,20)	0.064	0.0348		
C26	100	(90,100)	1	0.1593		
C27	9.26	(5,20)	0.284	0.0218		
C28	100	(95,100)	1	0.1593		
C29	100	(100,100)	1	0.1593		
C30	100	(95,100)	1	0.1593		
C31	52.9	(50,60)	0.29	0.0894		
C32	67	(55,70)	0.8	0.0575		
C33	100	(95,100)	1	0.3031	0.25	0.1045
C34	81.37	(75,100)	0.2548	0.4869		
C35	24.7	(30,60)	0	0.1395		
C36	93.3	(90,100)	0.33	0.0705		
C37	2	(1.6,3)	0.286	0.5584	0.75	
C38	48.4	(15,60)	0.742	0.3196		
C39	25.7	(20,40)	0.285	0.1220		

(continued)

Table 2 (continued)

Index layer	Actual value (v_i)	Classical field (a_i, b_i)	Correlation $k(v_i)$	Variable layer weight	Status layer weight	Criterion layer weight
C40	100	(80,100)	1	0.0648	0.2	0.0604
C41	90	(60,100)	0.75	0.5594		
C42	100	(60,100)	1	0.2688		
C43	90	(80,100)	0.5	0.1070		
C44	95	(60,100)	0.875	0.6738	0.8	
C45	90	(40,100)	0.83	0.1007		
C46	43	(75,100)	0	0.2255		
C47	48.4	(40,80)	0.21	0.2	0.8	0.0361
C48	4.79	(3,10)	0.257	0.8		
C49	35.9	(20,40)	0.795	0.75	0.2	
C50	46.9	(30,80)	0.338	0.25		

environmental protection efficiency index is 0.3358, the ecological civilization efficiency index is 0.4363, the park administration efficiency index is 0.6999, and the social progress efficiency index is 0.3342. The park administration efficiency index is the highest, and the resource utilization efficiency index is the lowest. The former is 2.7 times the latter. It is found from Table 2 that the comprehensive development coefficient of the TEDA eco-industrial park is 0.4401, which indicates that the comprehensive development level is not high. It is known from Table 3 that the development of the TEDA eco-industrial park falls into a weak loop, as the upper limit value (positive index) or the lower limit value (negative index) of the classical field in the index system is taken from the advanced level of developed countries or the optimal level of domestic parks of the same kind during the same period. Therefore, the evaluation results conform to the objective facts. It follows that TEDA needs to make efforts to achieve the stable and sustainable circular economic development status and rank among advanced domestic parks or reach world first-class development level.

6.2 *Analysis of Impact Factors*

It's found from formula (1) that the factors impacting the general eco-industrial park coordinated development degree are the subsystem efficiency indexes. When the subsystem efficiency indexes are much closer, the park coordinated development degree is much higher. Then, it is concluded from the above analysis that the resource utilization efficiency index, the environmental protection efficiency index, the ecological civilization efficiency index, and the social progress efficiency index of the TEDA eco-industrial park are relatively low and below 0.5, which is the major factor to lead the weak coordination of the TEDA eco-industrial park system

Table 3 Calculation results of the coordination coefficient of TEDA subsystems in 2010

φ_1	φ_2	φ_3	φ_4	φ_5	φ_6	Standard derivation S_i	Development coordination coefficient H_i
0.5536	0.2601	0.3358	0.4363	0.6999	0.3342	0.1499	0.6567
φ_1	φ_2	φ_3	φ_4	φ_5	φ_6	Average value $\overline{F_i}$	Comprehensive development coefficient U
0.5536	0.2601	0.3358	0.4363	0.6999	0.3342	0.4366	0.4401

development. It is found from formula (2) that the factors impacting the general eco-industrial park comprehensive development level are the weight of each subsystem and its efficiency indexes.

Seen from the weight of each index system layer, the weight of economic development is 0.3994, the weight of resource utilization is 0.1998, the weight of environmental protection is 0.1998, the weight of ecological civilization is 0.1045, the weight of park administration is 0.0604, and the weight of social progress is 0.0361. Economic development has the greatest impact on the eco-industrial park coordinated development degree and comprehensive development level, followed by resource utilization and environmental protection, and then ecological civilization, park administration, and social progress in turn.

6.3 Analysis of Improvement Strategies

It is concluded from the above analysis that the TEDA eco-industrial park should take the following measures to realize coordinated park system development and improve the comprehensive development level: increase scientific research input funds, reinforce scientific innovation and technical R&D, reduce comprehensive energy consumption per unit of industrial added value, improve comprehensive utilization ratio of industrial solid wastes, decrease wastewater production every 10,000 yuan industrial added value and COD production every 10,000 yuan industrial added value, better air quality and increase the ratio of environmental protection in GDP, improve the ratio of professionals in environmental administration organizations, and improve the ratio of employment added by the circular economy.

References

1. Dai Y-w, Liu Y-n, Zheng J, Yu J-h (2006) Comparison of eco-industrial park construction at home and abroad[J]. J Fujian Agric For Univ (Philos Soc Sci) 9(2):48–50
2. http://kjs.mep.gov.cn/stgysfyq/m/201302/t20130222_248379.htm
3. Tian X-g, Xie Q, Fang Z-l, Wang M (2012) Prospect and development of eco-industry park in China [J]. SiChuan Environ 31:128–132
4. Wei W-W (2010) Research on planning and built of eco-industrial park – nanchang high-tech district as an example[D]. Nanchang University, Nanchang
5. Li X-P (2008) Study on the comprehensive assessment system of the eco-industrial park based on circular economy[D]. Tianjin University, Tianjin
6. Lei M (2010) A study on comprehensive evaluation of eco-industrial parks[D]. Huazhong University of Science and Technology, Wuhan
7. Xie Z-M (2012) Research on the resources value flow of circular economy for coal-fired power plant[D]. Central South University, Changsha
8. http://www.teda.gov.cn/html/teda_index2011/TDRY11108/2011-12-30/Detail_577416.htm

The Need to Go Beyond "Green University" Ideas to Involve the Community at Naresuan University, Thailand

Gwyntorn Satean

Abstract University social responsibility (USR), especially in the environmental dimension, is an important engagement as the university is both a polluter and a promoter of environmental quality maintenance. A mechanism which stimulates the university to maintain the environment is the "UI GreenMetric Ranking of World Universities" released by the University of Indonesia (UI) which has embraced the "3E's" (equity, economy, and environment) for sustainable development. Using six criteria, Naresuan University (NU) was ranked 117th in the world out of 360 countries and ranked 10th for universities in Thailand, in 2014. The UI GreenMetric criteria include areas such as infrastructure, water, transportation, waste, and education programs, among others. Green university mobilization is a crucial mission that must be undertaken to achieve the 8th goal under the Eleventh Education Development Plan (2012–2016), so as to become "a university with a friendly environment and appropriate use of resources; a society of peace, generosity, and contribution and efficient operation; a good example for the public and society." However, under the UI GreenMetric criteria, the social dimension can be overlooked by the university, including the involvement of stakeholders both inside and outside the campus. Thus, regarding the leap to become a green university, the following community process should be set out by the university: (1) informing, (2) meeting, (3) decision-making, (4) planning, (5) implementing, (6) monitoring and evaluating, and (7) benefits. These principles embrace the "green university" process which includes G (governance), R (responsibility), E (environment), E (sufficiency economy), and N (network to contribute sustainably to the quality of life of the community surrounding the campus).

Keywords Naresuan University (NU) • UI GreenMetric • Green university

G. Satean (✉)
Department of Sociology and Anthropology, Naresuan University, Phitsanulok, Thailand
e-mail: gwyntorn@gmail.com

M. Matsumoto et al. (eds.), *Sustainability Through Innovation in Product Life Cycle Design*, EcoProduction, DOI 10.1007/978-981-10-0471-1_57

1 Introduction

The UI GreenMetric has become one of the flagship programs of the University of Indonesia (UI). It ranks universities throughout the world according to appointed indicators of campus environmental issues.

The UI GreenMetric model was developed from the Holcim Sustainability Awards; GREENSHIP (the rating system recently developed by the Green Building Council of Indonesia which itself was based on the Leadership in Energy and Environmental Design (LEED) system used in the USA and elsewhere); the Sustainability Tracking, Assessment and Rating System (STARS); and The College Sustainability Report Card. The model mentioned represents the involvement of sustainability assessment systems and academic university rankings.

The implementation of the UI GreenMetric is based on the 3E's (equity, economy, and environment) for sustainable development, covering six main criteria: setting and infrastructure (SI), weighted at 15 %; energy and climate change (EC), weighted at 21 %; waste management (WS), weighted at 18 %; water (WR), weighted at 10 %; transportation (TR), weighted at 18 %; and education (ED), weighted at 18 %. Each area contains 2.7 indicators, for example, area on campus covered in forested vegetation, renewable energy usage policy, sustainable courses/total courses, etc. The participating universities for a green university must prepare the information based on the criteria and submit via the website http://greenmetric.ui.ac.id/. There is no fee for taking part.

The UI GreenMetric World University Ranking was first announced in 2010 with 95 universities worldwide participating. This increased to 360 universities in 2014. On 16 January 2015, UI released the results of its GreenMetric Ranking of World Universities. The University of Nottingham in the UK ranked 1st with a score of 7803, followed by the University College Cork and National University of Ireland (score 7553), with Nottingham Trent University 3rd (score 7530) [1]. Fifteen universities from Thailand participated; Chulalongkorn University was ranked number 41 in the world and ranked number 1 in the country with a score of 6630; Mahidol University was ranked number 72 in the world and ranked number 2 in the country with a score of 6343; Mahasarakham University was ranked number 75 in the world and ranked number 3 in the country with a score of 6326.

Universities participated in the green university program as an effort to promote environmental sustainability on campus and also to involve stakeholders in efforts to create a sustainable environment and university social responsibility (USR). However, the UI GreenMetric-defined criteria are limited only to the campus area and do not cover the social dimension, particularly community participation, which is considered one of the significant social dimensions in the sustainability model (Fig. 1) [2]. Thus, this study focuses on the need to go beyond the green university ideas to involve the community. The case study was conducted in Naresuan University (NU), located in the lower northern region of Thailand and surrounded by both agricultural and commercial communities.

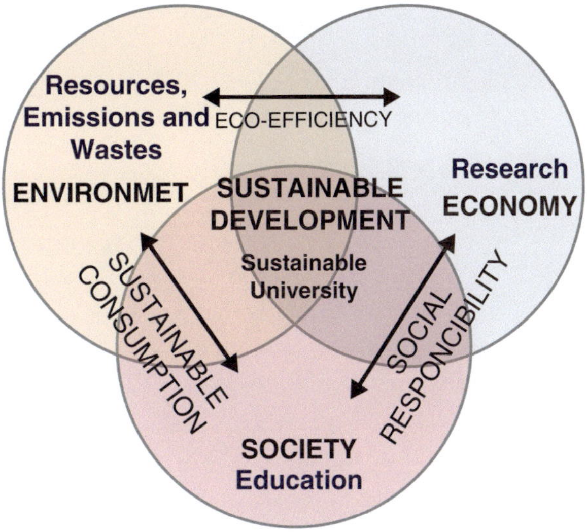

Fig. 1 A sustainability model (Lukman, R., Kranjc, D., Glavic, P. 2010)

NU aspires to become a University of Innovation. Geographically, it is located in Phitsanulok province, a city in the lower northern region. The university's environmental surroundings feature agricultural areas, residential areas, and non-forestry area and non-forested areas with no dominant plant or animal species found because of its wetlands. With irrigation systems, it is suitable as a rice-growing area. However, the area has dramatically changed with the building of student dormitories and commercial buildings. Without regulation, this results in disordered dispersion, especially on the east of Highway No. 117, located on the same side as the university.

Such phenomena cause several problems, including (1) the change of land use from an agricultural to a residential purpose (Fig. 2), (2) the expansion of the community causing waste in infrastructural development, (3) security difficulties, and (4) sewage and wastewater management issues; the most affected areas are Exit Gates 4 (at the rear of NU) and 5 (the side of NU) [3]. The number of students in 2003 was 20,298; this increased to 22,228 in 2014 [4]. Most students reside around the campus in Moo 7 Ban Tha Pho, Moo 8 Ban Kaek, and Moo 9 Ban Nong Lek, Thapoe subdistrict, Muang district, Phitsanulok province. The population of the three villages is around 13,226 people [5].

NU has pledged to fulfill four key missions, including (1) to generate well-qualified students, (2) to produce research, (3) to produce academic service, and (4) to conserve arts and culture [6]. In addition to these four main missions, NU is always aware of its social and environmental responsibility and, therefore, is determined to pursue its ultimate goal for university development, which is to be a university with an appropriate environment, atmosphere, and resource usage, of peace and generosity, and conducive to an efficient operational society, setting a good example and anchoring the people and society (green university) [7]. This reflects the implementation of the 8th goal under the Eleventh Higher Education

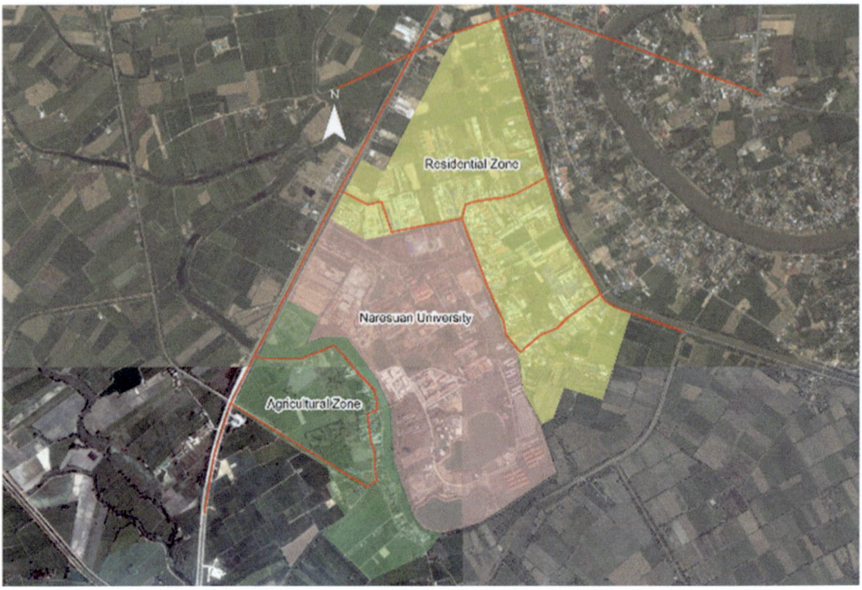

Fig. 2 Land use around Naresuan University was changed from an agricultural to a residential purpose (Environmental Research Centre, NU. 2004) (Source: "Naresuan University." 16°44′46.41″N and 100°11′37.34″E. Google Earth, Esri, DigitalGlobe, GeoEye. 4 April 2015)

Development Plan [8] under Strategy #6 of the Eleventh National Economic and Social Development Plan (2012–2016) [9].

Also, NU is a model of sustainable natural resource and environmental management for society and communities under Strategy #6 of the Eleventh National Economic and Social Development Plan (2012–2016). Thus, NU participated in the UI GreenMetric World University Ranking as a mechanism to enhance the implementation of the green university and develop the reputation of the university to be known internationally and drive toward "Going World Class 2015."

2 Methodology

This is a qualitative research comprising literature review and interviews. The literature review covers: the Guidelines of UI GreenMetric World University Ranking 2014 [10], Seminar Series 2012–2014 Thai Universities on the World Stage published by the Knowledge Network Institute of Thailand [11], reports on data transfer operation, Green University of NU (2013–2014) [12], and research papers related to green universities. The interviews include those with: the chairman of the Green University Committee, enquiring about the ways to operate a green university, and the mayor of Thapoe local administrative organization, Muang district, Phitsanulok province, enquiring about recognition and participation in operating a green university by the local administrative organization. Data collection was conducted from January to April 2015, and the data derived was analyzed using content analysis.

3 Naresuan University Development of Green University

NU participated in the green university ranking in 2013, scored 4567 out of 10,000 points, ranked 198th out of 301 participating universities around the world, and ranked 11th out of 13 participating universities in Thailand. In 2014, to drive the green university more concretely, the Educational Quality Development Division was assigned as a core unit of the operation, appointing advisory committees and the forum of data for green university ranking. As a result, NU was ranked 117th out of 360 participating universities around the world and ranked 10th out of 15 participating universities in Thailand. According to the assessment results of the UI GreenMetric in 2014, NU ranked higher, with a score of 5894 points (Table 1) [12]. Individual evaluation criteria are detailed as follows.

3.1 Setting and Infrastructure

On the setting and infrastructure criteria, NU scored 657 out of 1500 points, representing 43.80 %. The indicators are: (1) the total campus area is 2,054,400 m2, (2) the total ground floor area of buildings is 205,919 m2, (3) the number of students (both part-time and full-time students) in 2013 is 22,792 people, (4) the area on campus covered in vegetation in the form of forest is 218,539 m2 (11 %), (5) the area on campus covered in planted vegetation is 479,636 m2 (24 %), (6) retention—the non-retentive surface on campus as a percentage of the total area for water absorption—is 1,561,655 m2 (76 %), and (7) the percentage of NU budget for sustainability efforts is 7 % which is 104 million baht of the total 1.6 billion baht for 2014.

3.2 Energy and Climate Change

On the energy and climate change criteria, NU scored 1490 out of 2100 points, representing 70.95 %. The indicators are:

1. Energy-efficient appliance usage has partly replaced non-energy-efficient appliances, including replacement of light bulbs and air conditioners with energy efficiency label no. 5 and procurement of energy star computers.

Table 1 UI GreenMetric World University Ranking 2014 (overall) NU (The Green University Planning Committee of NU 2015)

Year	Ranking		Criteria						
	World	Thailand	Total	SI	EC	WS	WR	TR	ED
2014	117	10	5,894	657	1,490	1,200	870	1,100	577
%	58.94	43.80	70.95	66.67	87.0	66.67	87.0	61.11	32.06
2013	198	11	4,567	618	835	1,050	625	1,050	389
%			45.67	41.20	39.76	58.33	62.5	58.33	21.60

2. Renewable energy resources, partial use of clean biomass, and solar and wind power at the School for Renewable Energy.
3. Electricity usage per year is 53,184 kwh (January–December 2013).
4. Energy conservation programs for certain projects, i.e., switching electric bulbs and air conditioners on or off at a set time and adjusting temperatures to 25 °C.
5. Use of green building elements, such as a building with natural ventilation and full-day lighting, i.e., the School for Renewable Energy and the Ekatodsarod Building.
6. Climate change adaptation and mitigation program such as the EV bus system.
7. Greenhouse gas emission reduction policy and 500 bicycles for rent and buses for transporting passengers between the city and campus.

3.3 Waste

On the waste criteria, NU scored 1200 out of 1800 points, representing 66.67 %. The indicators are:

1. Recycling program for university waste: operating in part, for example, double-page paper use, recycling plastic water bottles as small tree pots, reusing PP board, etc.
2. Toxic waste recycling: this is partly treated in some places but mostly taken off campus to a dump, and electronic waste is sold to outside agencies.
3. Organic waste treatment: in a common area of the campus, scraps of woods are gathered to produce compost by the Division of Buildings and Grounds. No report has been made by the faculty.
4. Inorganic waste treatment: this is partially recycled but mostly taken off campus to a dump.
5. Sewage disposal: this is treated before disposal, which is operated by the NU hospital.
6. Policy to reduce the use of paper and plastic in campus: E-document application is adopted to reduce paper use; however, it has not covered all departments in NU.

3.4 Water

On the water criteria, NU scored 870 out of 1000 points, representing 37 %. The indicators are: (1) the water conservation program provides media relations through animation and sticker campaigns for water saving and, (2) as regards piped water, water consumed from piped system is 99 % well water (1,434,240 m^3).

3.5 Transportation

On the transportation criteria, NU scored 1100 out of 1800 points, representing 61.11 %. The indicators are:

1. The number of vehicles owned by the institution is 98 units.
2. The number of cars entering the university daily is 7200 units.
3. The number of bicycles found on campus on average is 1098 units daily.
4. The transportation policy is designed to limit the number of motor vehicles used on campus; students and staff are encouraged to use EV bus.
5. The transportation policy, which is designed to limit or decrease the parking area on campus, is in the preparatory stages.
6. Campus buses are available and service is free.
7. There is no policy for bicycles and pedestrians on campus, but use of these ways is discouraged except for bicycles and pedestrians. There are bicycle and pedestrian ways, and bicycles are provided by NU.

3.6 Education

On the educational criteria, NU scored 577 out of 1800 points, representing 32.06 %. The indicators are:

1. There are 576 courses related to environment and sustainability offered.
2. The total number of courses offered is 4073.
3. Total research funds dedicated to environmental and sustainability research over the last 3 years is 248 million baht.
4. Total research funds over the last 3 years is 555 million baht.
5. The number of scholarly publications on the environment and sustainability published over the last 3 years is 105.
6. The number of scholarly events related to the environment and sustainability is 56.
7. The number of student organizations related to the environment and sustainability is 5.
8. There is existence of a university-run sustainability website (http://green.nu.ac.th/).

All six criteria mentioned above show that NU has been dominant in energy and climate change as displayed by the energy conversation evaluation program as follows:

- NU Public Electric Vehicle – This service has been available for students and staff on campus since 1 June 2012, aiming to reduce the number of accidents, as well as reduce air and noise pollution. In doing so, NU purchased 16 EV bus, 30-seater vehicles, divided into two main lines; service is available from 07:00

to 22:00, every 3–5 min, and it takes an entire 15 min to run an entire trip. There is also a blue line service during the rush hour.

- NU Green Transit – 500 bicycles are available to reduce the number of accidents and air and noise pollution and promote health for drivers. Service is open from 07:00 to 21:00.
- NU Green Building – NU has undertaken structural renovation to reduce energy consumption and improve the flow of air and natural light. The initial implementation of the master plan consists of four stages as follows: Stage 1, the president's office, library office, and multipurpose buildings; Stage 2, humanities building, business building, and dormitories; Stage 3, IT building, international college, and agriculture building; and, lastly, Stage 4, law building, education building, and social sciences building.

Moreover, NU promotes walking and cycling on the main thoroughfare on campus. Initially, NU promoted traffic discipline measures and restricted use of motorcycles (over 9000 units) in the green campus zone during 08:00–17:00. The trial system began on 1 July 2015, before real use began on 1 August 2015 (Fig. 3). As such, the university must provide a motorcycle parking area and four bicycle service points, increase the EV bus with three special routes, prepare bicycle lanes, and plan to grow 25,000 big trees for shady footpaths and bicycle lanes [13]. These measures are part of the transportation criteria regarding transportation policy designed to limit the number of motor vehicles used on campus.

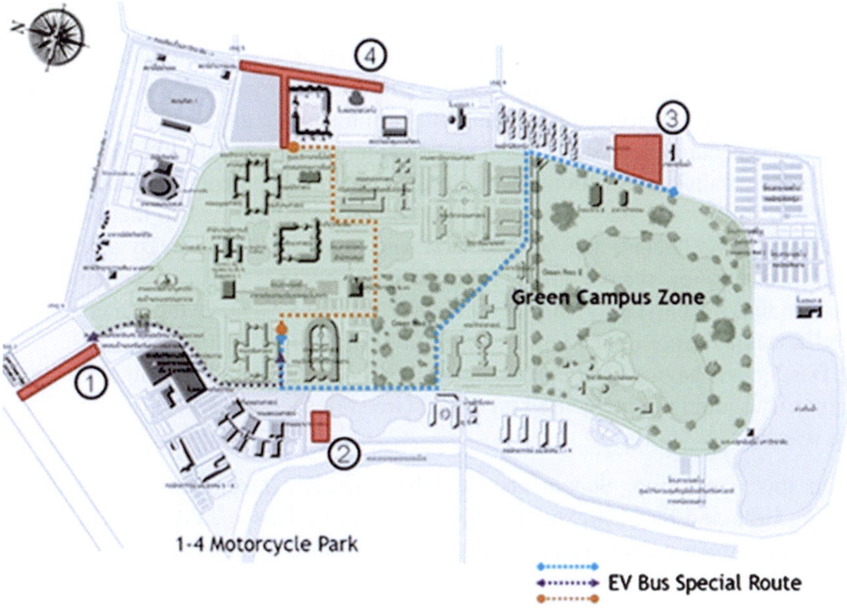

Fig. 3 Green campus zone (Radchapreuk, S. 2015)

However, the restricted use of motorcycles has raised questions as to the availability of internal mass transportation physically, including (1) the adequacy of bike lending, motorcycle parking, CCTV, and security for bicycle riders, (2) the adequacy and wait times for EV buses, and (3) the hot weather and inadequate shade on bicycle lanes. Meanwhile, NU has to answer to (1) operation without considering public opinion or a public hearing of stakeholders including students and personnel on campus and the neighboring public (this includes supporting groups, opposing groups, and neutral groups) and (2) the selected trial used during Phase 1 while vehicles ran normally [14, 15].

Again in 2015, NU is preparing to participate in the UI GreenMetric, with a proactive approach to operations as follows: (1) promote the reuse of food boxes and water glasses to reduce the amount of foam containers, (2) hold the "Green Office Contest" with UI indicators adopted, (3) support student projects on environmental sustainability management, (4) carry out publicity campaigns for energy efficiency, and (5) set up the principal agency for the environmental management of the university.

4 The Need to Go Beyond "The Green University"

NU's effort to mobilize the green university under the UI GreenMetric criteria represents the appreciation and awareness of the university's social responsibility and environmental conservation and promotion. NU consists of more than 27,067 students, faculty, and staff, including the public who use the facilities and services in the campus. Therefore, the university itself is an area where pollution is generated. An obvious example includes solid waste, such as organic waste from campus canteens, municipal solid waste from both administrative units and student dormitories, and hazardous waste from NU hospital and scientific labs. On the contrary, NU is the main organization for environmental management. Consequently, an effective solid waste management system should be established so that this waste can be safely collected, delivered, treated, and finally disposed of. By considering their own situation, the operation of the green university is considered an important mission that the university must continue to carry out to achieve the 8th goal under the Eleventh Higher Education Development Plan (2012–2016).

Despite NU having been ranked higher in 2014, the accomplishment of the green university should not be focused on the UI GreenMetric exclusively for the following reasons:

1. In this evaluation, documents shall be primarily taken into consideration. Suwartha and Sari [16] studied the evaluation by UI GreenMetric as a tool to support green university development. The 2011 assessment ranking reported that the evaluation methods used for the ranking were descriptive and qualitative approaches. A major issue is the reliability of the data given by the participating university. This data was gathered online and inspected by the UI GreenMetric

using a technological approach. Google Earth's facilities were utilized to collect and ensure the validity of the university data, especially about open space areas and campus areas, and field surveys were conducted formally and informally, which is very costly and counterproductive for visiting all the participating universities.

2. The report of the indicative performance measure under the six criteria was descriptive, for example, regarding inorganic waste treatment (trash, discard paper, plastic, metal, etc.), where the options selected by the university that best describe this include (1) burned in the open, (2) taken off campus to a dump site, (3) partially recycled, or (4) fully recycled. The quantitative criteria should be determined as to the percentage reduced each year, so that the participating universities can set target goals and consider the results of operations explicitly. For example, King Mongkut's University of Technology Thonburi has targeted the indicator-based goals clearly in 2017, for instance, reducing waste by 40 %, increasing green space by 20 %, reducing water use by 45 %, and consuming renewable energy by 1 % by the year 2020 [11].
3. The UI GreenMetric has no criteria for the community participation dimension, neither the university community nor surrounding communities. In the case of NU, it was found that traffic discipline measures on restricted use of motorcycles during 08:00–17:00, commenced as a trial system on 1 July 2015, were announced in a message around the campus on 12 May 2015 and publicized on its NU website on 22 May 2015 without public participation, leading to protest among those students who had been affected [17]. The protest asked the executives to suspend the project. Later, the Student Council of NU, the Student Agency, and the Faculty Club organized "Citizens' Dialogue for Green University" to suggest constructive operation and came to the conclusion that "when the university is ready, the student's ready; however, when the university is not ready, how can the students be ready?" [15].

Interviews with the mayor of the local administration indicated that they did not understand the green university and had never been informed of or contributed to the operation. Geng, Liu, Xue, and Fujita's study of creating a green university in China looked at the case of Shenyang University (SU) [18]. Their research suggests that the success of SU in the implementation of a green university included deliberate planning before approval of the university board while operations were supported by broad public participation, including collaboration with university personnel and communication with different stakeholders, e.g., new students, parents, and future employers (the external community). Thus, in the case of the NU's green university operation, it is required that the university create awareness and generate widespread involvement of stakeholders and overlook the barriers of the UI GreenMetric criteria.

5 Ideas to Involve the Community

The concept of green university operation with community participation is due to the incomprehensive UI GreenMetric criteria. The sustainability model (Fig. 1) in the social dimension includes education, based on the indicators related to environment and sustainability, including the number of courses offered, total research funds, number of scholarly publications, events, and student organizations. However, the number of courses, research, and student organizations has not been categorized for activities carried out in the surrounding communities, the secondary stakeholders, while the students, staff, and faculty are primary stakeholders. In addition, satisfaction criteria and the level of participation, such as the opportunity to comment on the green university, are not included. Tsai and Tang's [19] study of the National Taipei University of Technology (NTUT) Development of Ecological Campus demonstrated that NTUT removed walls to create an eco-friendly stream interface and developed urban street space to reach the campus, community, and city links, indicating an interaction between the university and the surrounding community.

The Community Model for Participation in the Green University was developed from the applied concept of public participation, university engagement, good governance, sufficiency economy, and UI GreenMetric with the following details (Fig. 4):

- Community – refers to a group of both primary and secondary stakeholders in the green university operation. Primary stakeholders include executives, faculty, staff, and current students. Secondary stakeholders include student alumni, students' parents, corporate executives, local governments, and citizens around the campus.
- Community Process – refers to the involvement of the community in the green university operation, which was adapted from the public participation model in environmental management proposed by Malikamarn [20] and the participation

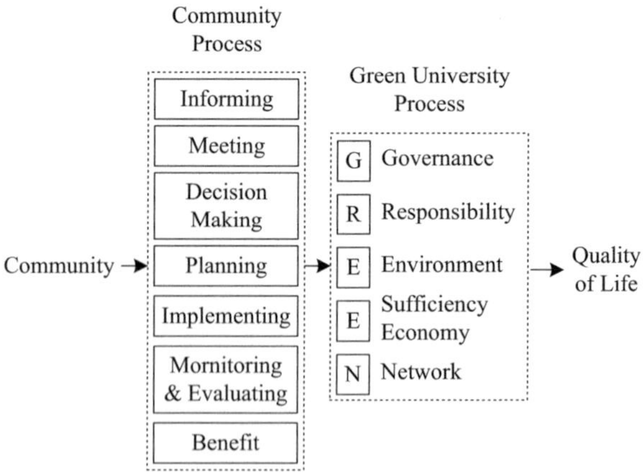

Fig. 4 Community model for participation in the Green University

model in rural development proposed by Cohen and Uphoff [21]. NU encourages the stakeholders to partake in the following activities:

1. Informing is dissemination of information about the project/activity from the beginning of policy development to pursue projects/activities.
2. Meeting is to give an opportunity for stakeholders to make comments so that executives can consider them.
3. Decision-making is to give people the opportunity to voice their opinions if they agree or disagree with the project/activity.
4. Planning is to give the opportunity to plan a work system, analyze priorities, and target goals.
5. Implementing is to give the opportunity to participate in projects/activities or assistance in resources, administration, coordination, etc.
6. Monitoring and evaluating is to have the opportunity to review and monitor the implementation of projects/activities to determine if they meet target goals and objectives.
7. Benefit is acceptance of both positive and negative impacts.

- Green University Process – Governance (G) is an applied principle of management that focuses on:

1. Legitimacy – rules and regulations are modern, fair, and acceptable to the members of the university and community.
2. Fairness – creates values for those working in universities, e.g., discipline.
3. Transparency – disclosure and access to information and data and performance monitoring.
4. Inclusiveness – gives the opportunity to all units of work involved in the implementation of projects/activities voluntarily.
5. Accountability – performs functions and assignments efficiently.
6. Resilience – recognizes the limits of available resources that should be consumed efficiently and cost-effectively.

   All six criteria mentioned contribute to and promote the operation of the green university effectively [22]. The concept of responsibility (R) involves university engagement as transition to working more closely with society. The performance model was previously in the form of academic service and volunteer work, changed into institutional engagement in a long-term partnership seeing the university as stakeholders with communities [23]. Environment (E) refers to six criteria of the UI GreenMetric; they include (1) setting and infrastructure, (2) energy and climate change, (3) waste, (4) water, (5) transportation, and (6) education. Sufficiency economy (E) is a philosophy based on the fundamental principle of Thai culture. It is a method of development based on moderation, prudence, and social immunity, one that uses knowledge and virtue as guidelines in living, including:

1. Moderation: sufficiency at a level of not doing something too little or too much at the expense of oneself or others, for example, producing and consuming at a moderate level to reduce environmental impact.

2. Reasonableness: the decision concerning the level of sufficiency must be made rationally with consideration of the factors involved and careful anticipation of the outcomes that may be expected from such action.
3. Risk management: the preparation to cope with the likely impact and changes in various aspects by considering the probability of future situations [24]. Network (N) refers to building a collaborative network to pursue the green university locally and internationally, e.g., collaboration with provincial administrative organizations, collaboration with networks of lower northern universities, or collaboration with the ASEAN University Network [25].

- Quality of life refers to the well-being of people, both in the university and the surrounding communities both physically and mentally without environmental impact, no exposure to environmental pollution and environmental risks and also benefits from environmental services in the university campus and around university campus. At the same time, people recognize and understand and are of a positive attitude toward the green university [26].

6 Conclusion

In 2015, the UI GreenMetric updated its website with a text invitation for universities worldwide to participate in the green university evaluation; the deadline for submission was 31 October 2015. There is no change in the criteria or indicators [27]. While NU prepares to enter the green university evaluation in 2015, it faces resistance from students and residents against the traffic discipline measures,[1] which have been in force since 7 August 2015, because such measures have affected students and the public both on and off campus and are not in line with the current environment in which most people use motorcycles as the main vehicle for commuting [28, 29]. The protest of students and people who have been affected was successful; NU now allows the use of motorcycles as before 18 August 2015 [30]. The bicycle path was modified only for the united school buildings. However, the implementation of green university continues. In the green university evaluation, transport criteria represent just 18 %, which does not meet the goals of becoming a green university because it affects a lot of stakeholders. However, other criteria of the UI GreenMetric such as energy and climate change (21 %), waste (18 %), water (10 %), and education (18 %) have still been deployed, for example, building management – green office, EV bus transportation system and bicycle, and E-document promotion. However, none of the criteria or indicators of the UI GreenMetric are defined as a policy or action plan concretely by NU to follow university-wide.

[1] Promoting traffic discipline is a measure to support the green university operation based on transportation criteria and indicator #4 and on transportation policy designed to limit the number of motor vehicles used on campus.

With rapid urbanization, NU continues to face various challenges around the university campus such as waste, wastewater, land use change, reduced community green areas, and so on. As a majority of the students and staff reside around the campus, faculty and personnel remain doubtful regarding the urgency of the university in implementing the green university as it relates to the period of data preparation according to criteria in entering the UI GreenMetric World University Ranking 2015. Online questionnaires will be available around July–October of each year.

The results of the UI GreenMetric World University Ranking each year bring forth pride in the university as it has been publicly announced as a green university at the national level. For example, Mahidol University was ranked the 1st green university in 2013 [31], while Chulalongkorn University was ranked the 1st green university in 2014 (Fig. 5) [32].

However, the success of such rankings is merely a mechanism for beginning to be a green university, and it is considered a good start. Identifying as a green university under the UI GreenMetric criteria alone is not sufficient because the indicators only evaluate whether some targeted activities exist or not. The university should set quantifiable performance indicators and short-term (1 year), medium-term (5 years), and long-term (10 years) goals for the implementation of green university.

Furthermore, other factors may affect the success of performance. As Wang et al. [33] proposed, "to achieve the transformation to an ecologically sound society, we need to ask whether we have prepared our students and society-at-large with the necessary knowledge, skills, tools, and experience? Or most importantly, have our mind-changing efforts resulted in helping them and society to make the necessary shifts to the new sustainable development paradigm?" To enable the operations are performed in the right direction, the university should build awareness of comprehensive understanding, seeking participation and acceptance through communication with stakeholders both inside and around

Fig. 5 Chulalongkorn University announced results for the UI GreenMetric World University Ranking 2014 on their website (Chulalongkorn University, 2015)

the campus community, especially the change in the inner environment [34], referring to the attitude and values which affect the behavior and lifestyles of people in a way that supports the environment. Meanwhile, the difference in the organizational culture within the university such as administration policy (vision, mission, and goals), the interdisciplinary fields in the university (Science and technology cluster, Heath Sciences cluster, Humanities and Social Sciences cluster), budget, university and community network, geographical location, community context and environmental issues, means that the university should take these factors into consideration and plan for possible implementation in practice appropriate to current circumstances.

Acknowledgments The author would like to thank Prof. Dr. Sujin Jinahyon (President of Naresuan University) for his brilliant initiative in launching the green university program. My appreciation also goes to Assist. Prof. Dr. Panomkhawn Riyamonkol for her effort during data collection and analysis. I thank Mr. Sakda Homhuan and Mrs. Jiraporn Penglum for their valuable contribution to improvement and discussion of the chapter. I also thank the organizer of EcoDesign 2015 (Association of EcoDesign Societies, Japan, and National Institute of Advanced Industrial Science and Technology, Japan) and the reviewers for their valuable comments to improve the content of this paper.

References

1. UI GreenMetric (2015) Press release, UI GreenMetric World Universities Ranking 2014. Available from: http://goo.gl/uGgZNm
2. Lukman R, Kranjc D, Glavic P (2010) University ranking using research, educational and environmental indicators. J Clean Prod 18(7):619–628
3. Environmental Research Centre, NU (2004) Master plan for landscape architecture and environmental management around NU. Rattanasuwan, Phitsanulok
4. Division of Academic Affairs Naresuan University (2015) Student statistics. Available from: http://reg.nu.ac.th/
5. Taphoe Local Administrative Organization (2014) Local development plan (2015–2017). N. P.: n.p.
6. Naresuan University (2015) Vision & mission. Available from: http://goo.gl/hPLzmB
7. Naresuan University (2015) Green and clean NU history. Available from: http://goo.gl/xj48mJ
8. Division of Planning Naresuan University (2012) The eleventh higher education development plan. Available from: http://www.plan.nu.ac.th/
9. Office of the National Economic and Social Development Board (2015) The eleventh national economic and social development plan (2012–2016). Available from: http://eng.nesdb.go.th/
10. Universitas Indonesia (2014) Guidelines of UI GreenMetric world ranking 2014. Available from: http://greenmetric.ui.ac.id/archieve/
11. Knowledge Network Institute of Thailand (2014) Green campus workshop from Green University to Sustainable Campus. N.P.: n.p.
12. The Green University Planning Committee of NU (2015) NU ranking report (2012–2014). N. P.: n.p.
13. Raksuchon P (2015) Limit the number of motor vehicles Trial Phase 1. Available from: http://www.nu.ac.th/th/green_university.php

14. Radchapreuk S (2015) University without Students: a case study on Green University Policy of Naresuan University. Available from: http://goo.gl/vkk8VQ
15. Achariyach (2015) Review an article of University without Students: a case study on the Green University Policy of Naresuan University. Available from: http://blogazine.pub/blogs/achariyach/post/5405
16. Suwartha N, Sari RF (2013) Evaluating UI GreenMetric as a tool to support green universities development: assessment of the year 2011 ranking. J Clean Prod 61:46–53
17. Dailynews (2015) Students protest against the new policy of NU to prohibit motorcycle in the campus. Available from: http://goo.gl/db1ozc
18. Geng Y, Liu K, Xue B, Fujita T (2013) Creating a "green university" in China: a case of Shenyang University. J Clean Prod 61:13–19
19. Jen-Hui Tsai, Yu-Ting Tang (2012) National Taipei University of Technology Development of Ecological Campus. Design for innovative value towards a sustainable society, pp 1049–1054
20. Malikamarn S (2002) Constitution and public participation in natural resources and environmental protection. Chulalongkorn University, Bangkok
21. Cohen J, Uphoff N (1980) Participation's place in rural development: seeking clarity through specificity. World Dev 8:213–235
22. Graham J, Amos B, Plumptre T (2003) Governance principles for protected areas in the 21st century. Institute On Governance, Ottawa
23. Engagement Thailand (2015) Who are we? Available from: http://www.engagementthailand.org/
24. The Chaipattana Foundation (n.d.) Philosophy of sufficiency economy. Available from: http://www.chaipat.or.th/
25. ASEAN University Network (n.d.) History and background. Available from: http://goo.gl/I30Cw8
26. Organisation for Economic Cooperation and Development (OECD) (2011) Towards green growth: monitoring progress. OECD Publishing, Paris
27. Universitas Indonesia (2015) Criteria & Indicator. Available from: http://goo.gl/cGdiJR
28. JON FERNQUEST, Bangkok Post (2015) Students protest having to walk to class... or bicycle. Available from: http://goo.gl/2HDPmM
29. Bangkok Post (2015) Naresuan backs down on 'green' campus. Available from: http://goo.gl/SyQh1e
30. Phitsanulok Hot News (2015) NU president met student regarding "green university" policy. Available from: http://goo.gl/lVwIyT
31. Mahidol University (2014) Mahidol ranked as Thailand's most eco-friendly university. Available from: http://goo.gl/iNbN1w
32. Chulalongkorn University (2015) Chulalongkorn University Ranked Top Thai University and 40th Global University in UI Green Metric World University Ranking 2014. Available from: http://goo.gl/ceJQeN
33. Wang et al (2013) Moving towards an ecologically sound society? Starting from green universities and environmental higher education. J Clean Prod 61:1–5
34. Dixit, Kunda (1994) The Greenin of the media. Paper presented at AMIC annual conference on "Communication, Convergence and Development". Asian Media Information and Communication Centre, Singapore

Sustainability Assessment of High-Rise and High-Density Urban Structures

Chisato Takahashi, Tomomi Nonaka, and Masaru Nakano

Abstract This study evaluates the environmental and economic effectiveness of high-rises and high-density urban structures in Tokyo. The adopted approach aims to enable production of open spaces and green areas by effective spatial utilization of lands and realize environmental improvement in limited spaces. The proposed evaluation model includes economic and environmental criteria. Economic indicators are composed of the construction costs of high-rise structures and the economic value of the relaxed plot ratio generated by the high-rises. The green spaces created by high-rise structures are evaluated as an environmental indicator. These indicators are adopted and integrated for a sustainability assessment. The proposed model is then tested by applying it to some case studies in the Nihonbashi Tokyo area, and various scenarios are analyzed by changing the parameters used. The results show that the high-rise and high-density urban structures would likely bring about positive economic effects through the relaxation of volume ratio in policy standards.

Keywords Urban planning • Aging society • Sustainability • Building • Social system design • Heat island

1 Introduction

Concentration of population in urban areas is becoming one of the global issues for both developed and developing countries. Excessive concentration of population in urban area causes reduction of green spaces and degradation of dwelling environments. Once the natural greenery is lost by urbanization, it is difficult to reproduce

C. Takahashi
SHIMIZU Corporation, Tokyo, Japan

T. Nonaka (✉)
Department of Industrial and Systems Engineering, Aoyama Gakuin University, Kanagawa, Japan
e-mail: nonaka@ise.aoyama.ac.jp

M. Nakano
Graduate School of System Design and Management, Keio University, Kanagawa, Japan

M. Matsumoto et al. (eds.), *Sustainability Through Innovation in Product Life Cycle Design*, EcoProduction, DOI 10.1007/978-981-10-0471-1_58

it. Hence, urban development should blend parks, walkways, and green roofs in the residential areas.

Sustainability of urban infrastructures has often been evaluated in terms of the economic, environmental, and social parameters [1]. Factors such as energy savings, renewable energy, and efficiency of urban development have also been studied [2–5]. High-rise and high-density urban structures have been studied as influential factors in the formation of urban heat islands [6–8]. However, there have been few studies on the positive effects of creating open and green spaces in high-rise structures.

This study evaluates the effectiveness of high-rise and high-density urban structures in Tokyo based on environment and economic factors. Spatial utilization of land enables production of open spaces and green areas, thereby realizing environmental improvement in limited spaces. The proposed evaluation model includes economic and environmental criteria. Economic indicators include the cost of construction and the economic value of the relaxed plot ratio generated by high-rise structures. The green spaces generated by high-rises are evaluated as an environmental indicator. These indicators are then adapted and integrated for the sustainability assessment. The proposed optimization model minimizes the wall ratio and maximizes the green ratio considering factors such as building height, site area, floor area ratio, construction cost, and the economic effects of capacity relaxation.

The proposed model provides methods to calculate benefits and factors such as (1) the rise in green ratio due to high-rise structures, (2) the economic effect, and (3) the appropriate building height to achieve a balance between the green ratio and the economic effect. The proposed model is applied to some case studies in the Nihonbashi Tokyo area, and different scenarios are analyzed by changing various parameters. This confirms the effectiveness of the proposed method, and the results show that the high-rise and high-density urban structures could bring about positive economic effects by relaxation of volume ratio in policy standards.

2 Evaluation Model of High-Rise and High-Density Urban Structures

In this study, the proposed model assesses sustainability by using the environmental and economic perspectives; these factors are based on the three critical parameters used to assess sustainability: economic efficiency, environmental impacts, and sociality [1]. Earthquake resistance, which is one of the most important social issues in Japan, was excluded from this study, since it is expected to meet national standards to comply with the quake-resistance standards.

2.1 Ratio of Greening and Building Height

If the areas of the site and building are constant, the building needs to be taller in order to maximize the available open area. Green area of the site (the aboveground part only) includes all areas except areas such as pedestrian walkways and car roads.

The building height is defined as H. The proposed model of the structure of the building is shown in Fig. 1. One side of the building is the square root of the area of one floor Ka, and the bottom surface area is Ka. An efficient shape of a building is assumed such that the bottom is a rectangular area. Green space rate Gr, under these conditions, is calculated by Eq. 1 by using the open space ratio Or for the site:

$$Gr = \alpha \cdot Or \tag{1}$$

The a, which represents ratio of green area, is generally set as 0.5 in Japan. The site area Sa, the building area Ka, and open space ratio Or are represented as

$$Or = 1 - \frac{Ka}{Sa} \tag{2}$$

Building area Ka is calculated by Eq. 3 using the number of floors Ts of the building, site area Sa, and the volume ratio Vr:

$$Ka = \frac{Sa \cdot Vr}{Ts} \tag{3}$$

The number of floors Ts of the building is calculated using average floor height Ha and the building height H:

$$Ts = \frac{H}{Ha} \tag{4}$$

Using Eqs. 4 and 1, Gr can be obtained as

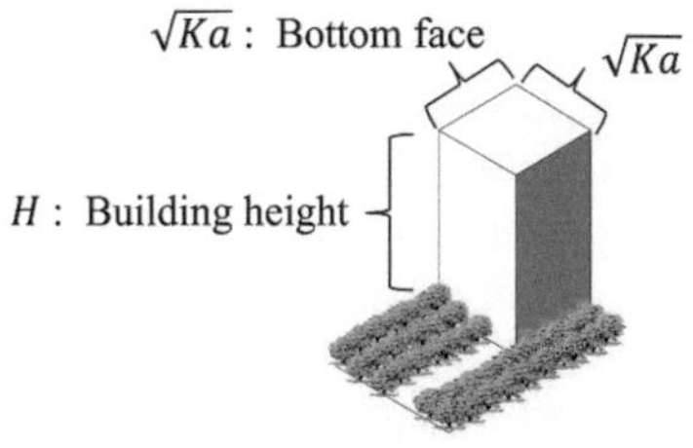

Fig. 1 Model structure of a building

$$Gr = \alpha\left(1 - Vr \cdot \frac{Ha}{H}\right) \tag{5}$$

2.2 Outer Wall Rate and Building Height

Construction costs are highly dependent on the outer wall rate for the total floor area. Outer wall rate Wr is expressed as a ratio between outer wall area Wa and the total floor area $Vr \cdot Sa$:

$$Wr = \frac{Wa}{Vr \cdot Sa} \tag{6}$$

It can also be expressed as Eq. 7 based on the outer wall area Wa, the square root of four outer perimeters, the building height, and the building area:

$$Wr = \frac{4H\sqrt{Ka}}{Vr \cdot Sa} \tag{7}$$

Building area Ka is the quotient of total floor area and the number of floors in the building:

$$Ka = \frac{Vr \cdot Sa \cdot Ha}{H} \tag{8}$$

Outer wall rate Wr is derived from Eqs. 6 and 8 as

$$Wr = 4H\sqrt{\frac{Ha}{H \cdot Vr \cdot Sa}} \tag{9}$$

2.3 Green Ratio and Outer Wall Rate

Based on the building height, we now consider the relationship between the green ratio and outer wall rate. Figure 2 shows the relationship between the building height and the greening rate from Eq. 5 in a commercial area with a volume rate of 600 %. Figure 3 expresses the relationship between the building height and the outer wall rate under the same conditions, from Eq. 9.

Building value Yr is now considered to maximize the greening rate and minimize construction costs. It is modeled as a function of the green space rate Gr and the outer wall rate Wr as follows with weighting coefficients of ω_1 and ω_2:

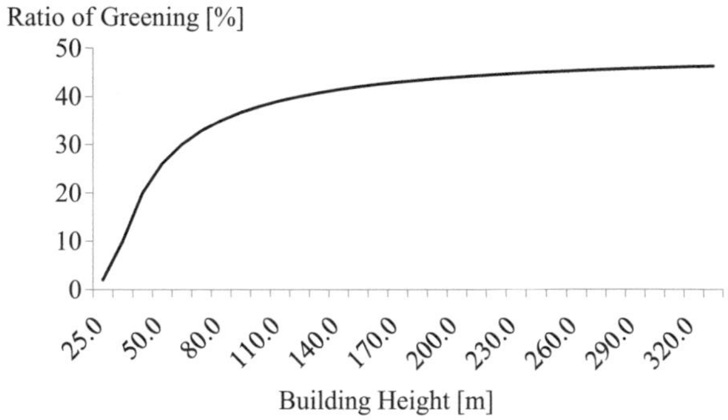

Fig. 2 Building height and ratio of greening

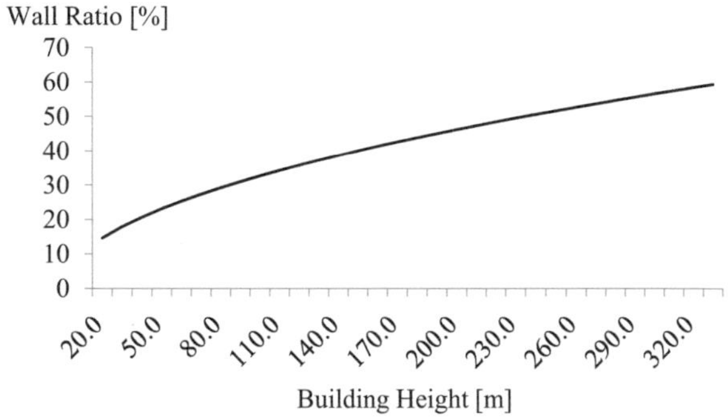

Fig. 3 Building height and wall ratio

$$Yr = \omega_1 \cdot Gr - \omega_2 \cdot Wr \tag{10}$$

In this study, it is assumed that $\omega_1 = \omega_2 = 1$. The building value Yr can be obtained from Eqs. 5 and 9 as shown here in Eq. 11:

$$Yr = \alpha - \alpha \cdot Vr \cdot \frac{Ha}{H} - 4H\sqrt{\frac{Ha}{H \cdot Vr \cdot Sa}} \tag{11}$$

The entire value of Yr is represented by y, and the building height H is represented by x in the following formula:

$$y = f(x) = \alpha - ax^{-1} - bx^{\frac{1}{2}} \tag{12}$$

$$a = \alpha \cdot Vr \cdot Ha \tag{13}$$

$$b = 4\sqrt{\frac{Ha}{Vr \cdot Sa}} \tag{14}$$

The building height x^* that maximizes the overall value y is obtained by $dy/\mathrm{d}x = 0$:

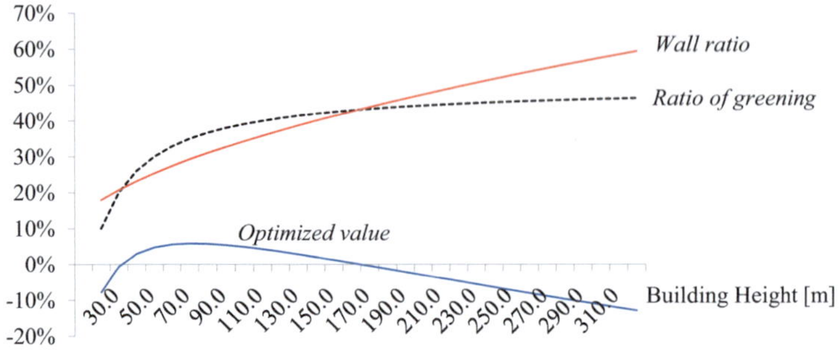

Fig. 4 Overall value of the building height, ratio of greening, and wall ratio (plot ratio: 600 %)

$$x^{*} = \left(\frac{Ha \cdot Sa}{16}\right)^{\frac{1}{3}} \cdot Vr \tag{15}$$

Next, a site area *Sa* of 10,000 [m^2], a volume ratio *Vr* of 600 %, and an average floor height *Ha* of 4[m] are assumed. The building height to maximize the overall value derived by the above equations is 81.4 [m]. The green space rate and the transition of the building height, which maximizes the outer wall rate and the overall value when the volumetric ratio is 600 %, are shown in Fig. 4.

2.4 Economic Value of the Relaxation Volume

A city planning system, which includes special measures for urban renaissance, permits the relaxation of volume rate, depending on the degree of its contribution to urban regeneration. The relaxed volume and revenue amount commensurate with the obtained floor area are considered as a contribution to the public costs. These include installation of public facilities and construction and maintenance of open spaces and sidewalks.

These construction costs are to be absorbed by the expected revenues generated by the building such as rentals. The formula for calculating this is as follows. It is assumed that these construction costs are subtracted from the revenue that is expected for the next 20 years rather than as a new construction provided to the public.

Considering the average floor rental cost per month as *Cct* [yen/month], amortization period as *Yr* [12*Yr* months], the expense rate as *Kr* [yen/month], the rental room occupancy rate as *Rr*, and the construction costs as *C* [yen], then the price *Ct* of relaxation volume can be calculated as in Eq. 16:

$$Ct = Cct \cdot Kr \cdot Rr \cdot 12Yr - C \tag{16}$$

For example, price Ct of relaxation volume per 1 [m^2] will become 726,000 [yen/m^2], based on set parameters if the cost is subject to an amortization period Yr of 20 years, average rental floor rents Cct is 10,000 [yen/month·m^2] [9], the expense ratio Kr *is 70 %*, the rental room occupancy rate Rr *is 70 %*, and the construction cost C is 450,000 [yen/m^2].

Economic value Cc can be calculated from the site area Vt [m^2], the rental floor rents Cct [yen/m^2], and the relaxation volume rate Vk, as shown in Eq. 17:

$$Cc = Ct \cdot Vk \cdot Vt \tag{17}$$

For example, if the site area Vt is set as 10,000 [m^2] and the rental floor rent Cct [yen/m^2] is set as 10,000 [yen/m^2], economic value Cc can be derived as 7.26 billion yen when the relaxation volume rate Vk is 100 %. The economic value Cc will be 21.78 billion yen in case the relaxation volume rate Vk is 300 %. In this case, the price Ct of relaxation volume per 1 [m^2] will be 0 [yen/m^2] if the rental floor rent Cct is 3,827 [yen/m^2]. Thus, economic value is negative in areas where the floor rent Cct is less than 3,827 [yen/m^2]. In addition, when the construction cost is set as 300,000 [yen/m^2], the economic value is negative if the floor rents are less than 2,551 [yen/m^2].

From the above, it can be seen that, while the economic value is higher where floor rent is higher, it is lower in the areas with lower floor rents. Also, it can be said that the economic value becomes positive if it is possible to decrease the construction costs.

3 Case Study

3.1 Targeted Area

As an example, we attempted to calculate the greening rate and economic efficiency of an area in front of the Tokyo Station. The entire area in the Nihonbashi Tokyo Station foreland district is about 102 [ha] (area of residential land only). Nearly 66 % of the buildings are constructed by the old seismic standards (formerly 1981) and 57 % of the buildings are older than 33 years [10]. Thus, it is assumed in this study that nearly 60 % of the buildings require rebuilding.

The building height of the development is assumed to be 120 m. On average, the specified volume rate of the area is 700 %, and it assumes a total volume rate of 900 % of the volume received with 200 % of volume relaxation. During replacement construction, widening of the waste road in front of the building has to be more than 12 m. It is assumed that it is not subject to any volume reduction. Additionally, the site scale is assumed to be about 10,000 m^2. The relaxation volume rate factors in considerations for city planning requirements such as “city block restructuring due to consolidation of land" and "reducing environmental impact and creation of green areas," and it can be assumed to be 200 %.

3.2 *Evaluation*

3.2.1 Appropriate Building Height of the Entire Area

If the site area *Sa* is set as 10,000 m^2, the volume ratio *Vr* as 900 %, and average floor height *Ha* as 4 m, the building height to maximize the overall value is obtained as 122.1 [m] by using Eq. 15.

3.2.2 Greening Rate of the Entire Area

It is assumed that this development can be performed in 60 % of the area of residential land area at Nihonbashi Tokyo Station foreland. With the building height *H* of 122.1 m, a volume ratio *Vr* of 900 %, and an average floor height of the building *Ha* of 4 m, 50 % of the open space of the area can be green. When these are substituted in Eq. 5, a greening rate *Gr* of 35 % of the site and a greening rate *Grn* of the area of 21 % can be obtained. The transition of the building height, which maximizes the overall value, the outer wall rate, and greening rate when the volumetric ratio is set as 900 %, is shown in Fig. 5.

This value is a dramatic increase from the current rate of 3.5 % for the Nihonbashi Tokyo Station foreland. This greening rate is second only to the one in district Tsukiji in Chuo-ku, which has a rate of 22 %. Further, the greening rate will be the same as the current highest rate, which is in Tokyo's Suginami ward and Nerima ward [11].

3.2.3 Economic Value of the Relaxation Volume

It is assumed that this development takes place in 60 % of the residential land area of Nihonbashi Tokyo Station foreland ward. Further it is assumed that the average

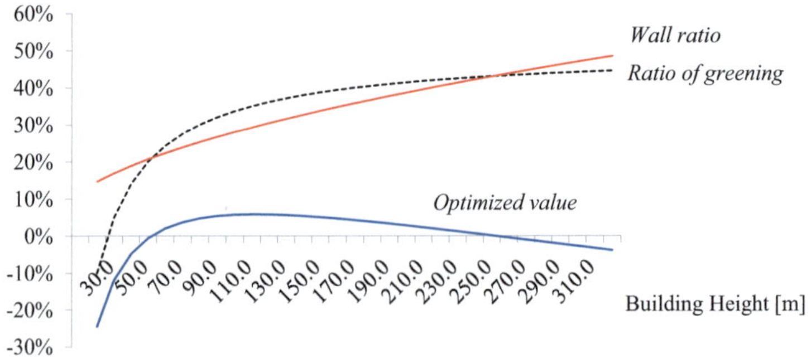

Fig. 5 Overall value of the building height, ratio of greening, and wall ratio (plot ratio: 900 %)

floor rent Cct of 20 years is 15,000 [yen/month·m^2], the expense ratio Kr is 70 %, the room occupancy rate Rr is 70 %, the depreciation period Yr is 20 years, and the construction cost C is 450,000 [yen/m^2]. The price Ct of the relaxation volume per one [m^2] can be obtained as 969,000 [yen/m^2] from Eq. 16. If the target area is set at 102 [ha] and the relaxation volume rate Vk is set at 200 %, an economic value Cc of 1 trillion and 976.7 billion yen is obtained from Eq. 17.

To summarize the results in this section, the key benefits are an improvement of the greening rate from the current 3.5–21 % and that the relaxation volume can produce an economic value as large as 1 trillion and 976.7 billion yen.

3.3 Effect Analysis by Volume Rate Relief for Residential Promotion

In Sect. 3.2, it was considered that the basis of the development is within the scope of the current regulations. It is possible to obtain a large economic value in cities with a high density of high-rises, and this is true especially in inner city areas where high-priced land is used for public facilities such as childcare support and cultural and economic activity-based facilities such as a public contribution expense. Currently few residential structures exist in such areas, and it is difficult to plan further increase in housing. In this section, it is considered that based on a volume rate of 900 %, it should promote housing development by allowing residential and nonresidential development at the same rate. Further, the impact is calculated and analyzed in this case by performing volume relaxation under these conditions.

It is assumed that a nonresidential relaxation volume rate of 400 % and a residential relaxation volume rate of 400 % are added to the volume rate of 900 % (which was verified in the Sect. 3.2), to achieve a total volume rate of 1700 %. The front road width is assumed not subject to road shading limited by widening and waste road. Further, it is assumed that it is not necessary to provide the housing volume relaxation of 400 % to the public.

A transition of the building height which maximizes the greening rate, the outer wall rate, and the overall value when the volume of the rate Vr is 1,700 % is shown in Fig. 6.

3.3.1 Optimization of Greening Rate and Cost

It is assumed that site area Sa is 10,000 [m^2], the volume ratio Vr is 1700 % with the reference volume rate of 700 %, the relaxation volume rate of 1000 %, and the average floor height Ha of the building of 4 m. From Eq. 15, the building height to maximize the overall value is obtained as 230.7 m. Based on this, an outline of the residential promotion plan is shown in Fig. 7.

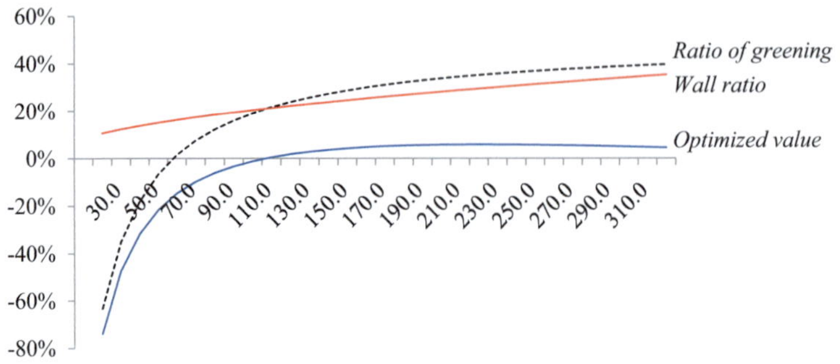

Fig. 6 Overall value of the building height, ratio of greening, and wall ratio (plot ratio: 1700 %)

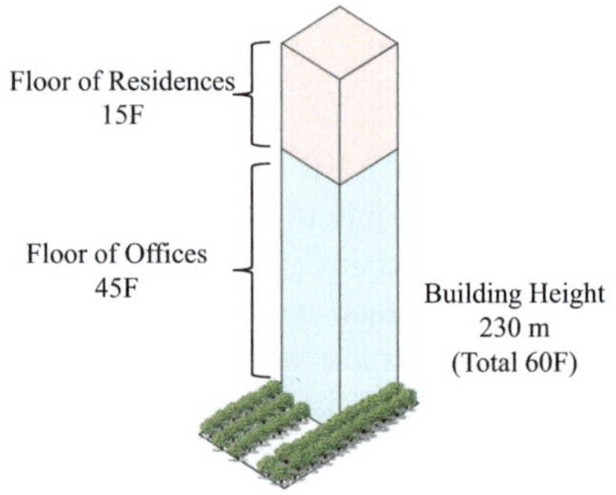

Fig. 7 Outline of the residential promotion plan

3.3.2 Ratio of Greening

In this case, it is considered that the development takes place in 60 % of the residential land area of Nihonbashi Tokyo Station foreland district. Further, it is assumed that the building height H is s 230.7 m, 50 % of the open space the area is to be green, the volume ratio Vr is 1700 % (reference volume rate of 700 %, relaxation volume rate of 1000 %), and the average floor height Ha of building is 4 m. As a result, greening rate Gr of the site can be obtained as 39 % from Eq. 5. Thus, greening rate Gr can be obtained as 39 % in the newly developed site of the above, Nihonbashi Tokyo Station foreland district. Additionally, the overall greening rate of 23 % is a dramatic increase from the current 3.5 % of the Nihonbashi Tokyo Station foreland. Further, it is a higher greening rate than the 22 % achieved by the district Tsukiji in Chuo-ku and the greening rate for Suginami ward and Nerima ward in Tokyo [11].

3.3.3 Economic Value of the Relaxation Volume

The economic value Cc [yen] can be expressed as follows. Housing is assumed to be excluded from the economic value as a provider to the public. In this case, the relaxed volume rate is considered as 600 % by assuming that the housing volume ratio of 400 % is subtracted from the total volume ratio of 1000 %. From Eqs. 16 and 17, economic value *Cc* can be obtained as 5 trillion and 930.3 billion yen, and the price of relaxation volume *Ct* can be obtained as 969,000 [yen/m^2].

The entire economic value of the case, based on a 60 % relaxation volume rate of 600 % of the residential land 102 [ha] of this area (based on 1000 % total relaxation volume and a housing relaxation volume of 400 %), can be obtained as 5 trillion and 930.3 billion yen. The economic value obtained by relaxation volume can be regarded as the value of the "air right"; therefore, it is assumed that it can be used for public contributions.

The key results from this section are the improvement from the current greening rate of 3.5–23 % and that the economic value of the relaxation volume can create an impact as large as 5 trillion and 930.3 billion yen. Since it can lead to increase in housing development, it is also expected that it increases the attractiveness of the cities and improves their energy consumption efficiency.

4 Conclusions

The concentration of population is increasing in the advanced urban cities. Based on the assumption that more high-rise and high-density urban structures are required in the future, this study verified the environment and economic aspects of such urban structures. The findings of our study suggest that the high-rise and high-density building structures can be used to create open spaces and increase the greening rate, thereby contributing beneficially to the environment. The case study finds that there is an optimal height and volume ratio for the high-rise structure based on conditions such as the site area and the floor area ratio. Further, it was found that there is a significant economic effect of increasing the volume relaxation by creating an open space. In addition, it was found that it is particularly effective in large-scale sites with high rental prices.

Acknowledgments This work was supported in part by Grant in Aid for the "Development of simulation models for study of long life housing and sustainable urban infrastructure" from the Association for Sustainable Building Systems in Japan.

References

1. Howes R, Robinson H (2005) Infrastructure for the built environment: global procurement strategies. Elsevier, Oxford
2. Blewitt J (2010) Deschooling society? A lifelong learning network for sustainable communities, urban regeneration and environmental technologies. Sustainability 12(11):3465–3478
3. Ding C (2009) Policy and planning challenges to promote efficient urban spatial development during the emerging rapid transformation in China. Sustainability 11(3):384–408

4. Hal AV, Dulski B, Postel AM (2010) Reduction of CO_2 emissions in houses of historic and visual importance. Sustainability 2(2):443–460
5. Muller MO, Stampfli A, Dold U, Hammer T (2011) Energy Autarky: a conceptual framework for sustainable regional development. Energy Policy 39(10):5800–5810
6. Lo CP, Quattrochi DA, Luvall JC (1997) Application of high-resolution thermal infrared remote sensing and GIS to assess the urban heat island effect. Int J Remote Sens 18 (2):287–304
7. Giridharan R, Ganesan S, Lau SSY (2004) Daytime urban heat island effect in high-rise and high-density residential developments in Hong Kong. Energy Buildings 36:525–534
8. Giridharan R, Lau SSY, Ganesan S, Givoni B (2007) Urban design factors influencing heat island intensity in high-rise high-density environments of Hong Kong. Build Environ 42:3669–3684
9. CBRE, Office Japan (2010), XGATE (in Japanese)
10. Bureau of Urban Development Tokyo Metropolitan Government (2010). Available from http://www.toshiseibi.metro.tokyo.jp/keikaku/shingikai/toshikei181.htm (in Japanese)
11. Tokyo, Chuo City (2005) Dai 4 Kai Chuo-ku Midori no Jittai Chousa, Shi Keikaku Houkoku Shu No.4, (in Japanese)

Analysis of the Energy Consumption of Building Automation Systems

Tamás Iváncsy and Zoltán Ádám Tamus

Abstract Nowadays, the efficient energy usage is more and more important, and reducing the energy usage of buildings lowers the overall environmental effect of energy systems. The building automation systems provide a possible solution to create smart energy distribution in buildings by controlling lighting, heating, cooling and other energy-consuming systems. One possible solution for reducing energy consumption is the use of building automation systems for energy management, and these automation systems can help to optimize the efficiency of heating, air conditioning and the electric energy consumption. In most cases, the overall consumed energy reduction with a building automation system is considered, but without the exact knowledge of the electric energy consumption of the automation system, the overall energy efficiency cannot be evaluated. The aim of our study is to analyze the electric energy consumption of a building management system. The energy consumption of a building automation system is measured in a small office-like environment. The system uses a separate power supply for the sensors and actors on the bus, so the energy consumption of the power supply gives the overall electric consumption of the building management system in different conditions. Based on the result of the analysis, the ratio of the used energy by the system and the energy consumption reduction has been estimated, and an optimal design of these systems is suggested.

Keywords Energy management • KNX system • Building automation • Energy consumption

1 Introduction

Nowadays, from the point of view of sustainability, the energy consumption of office buildings is an important factor. Relating to the buildings' energy management, there are two main points for investigation: The first one is the overall energy

T. Iváncsy (✉) • Z.Á. Tamus
Department of Electric Power Engineering, Budapest University of Technology and Economics, Budapest, Hungary
e-mail: ivancsy.tamas@vet.bme.hu

M. Matsumoto et al. (eds.), *Sustainability Through Innovation in Product Life Cycle Design*, EcoProduction, DOI 10.1007/978-981-10-0471-1_59

consumption and its effect on the environment. The second point is the running cost of the building. The goal is the cutting off the costs by lowering energy consumption, for example. These points are the drivers, which decrease the energy consumption of environmentally friendly buildings as low as possible.

There are several ways for lowering the energy consumption, using different materials for the thermal isolation, using modern highly isolating doors and windows, the use of independent heating for each room or creating rules to switch off the unnecessary lights, to show a few examples.

An advantage of the building automation systems is the independent and automatic control of room heating, cooling and lighting. The temperature can be set according to the needs of the personnel using the building and can follow a program. As an example, temperatures for day, night and weekend could be defined separately.

The illumination level of the different room types can be set according to the needs. The light sources are dimmed and switched according to the information of presence and brightness sensors.

Different sensors are connected to the building automation system, which collects relevant information for the control of the building and all the sensors and actors are using the same bus system. For that reason, the different control systems of the building can work together and so utilize the increased possibilities to reduce the energy consumption further. For example, the lighting system is working together with the shutter system. If the natural light outside is sufficient for reaching the necessary illumination inside, then the shutters are opened, and the lighting is switched off. The intensity of the sunshine can also affect the air conditioning on similar ways, so the shutters and heating or cooling subsystems are working together.

Nevertheless, the use of a building automation system reduces the energy consumption; the system itself also consumes energy for the control of the building. There are many cases wherein the designers and users neglect the electricity consumption of these systems; however, the real energy saving thanks to the installation of building automation system can be estimated only if the energy consumption of the system is also considered.

We were concerned about the electric energy consumption of the building automation and control system, which is necessary to lower the overall energy consumption. In this paper we just study the building automation bus system itself without taking into account the energy consumption of the electric devices switched by the actors.

2 Use of Building Automation Systems

Most of the modern building automation systems are based on bus systems. The sensors, actors and system devices are connected to the bus, which is used for the necessary communication between the devices. The biggest advantage of the bus system is that all of the devices are able to communicate, so the different automated systems can gain information from the other systems and so can work together.

There are different levels of building automation, and with these levels, different energy savings can be realized. These levels differ in the control system functions and in the type of the control. The lowest possible automation level is certainly if there is no automation in the building. In this case the heating, air conditioning, ventilation, lighting and shutter are operated manually. The energy consumption is not limited; the interventions to save energy are delegated to the users of the building. This method, according to the careless behavior of the users, can even increase the energy consumption related to the designed energy performance [1].

A better solution is the automated heating, cooling and ventilation control. In this level the heating and cooling are controlled by electronic thermostats or thermostatic valves in each room separately; the ventilation is timed and also automatic for each room. The lighting and shutter are operated manually, but the lighting has a central, automated switching off option. This solution is capable of increasing the energy performance of the building [2, 3].

The next level is reached; if the control is extended by humidity control in each room, the outer temperature is also measured for the correct setting of the inner temperature. In the control of lighting, the switching on and off is automatic and also the level of the illumination is measured and the light sources are dimmed accordingly. Also the shutters are controlled automatically according to the measured inner and outer illumination values.

To further enhance the automation, the ventilation should be demand and presence driven, and the parts of the automation should work together. In this level the ability for all the components in a bus system to communicate is used as an advantage. That means, for example, the shutter is controlled automatically according to the lighting needs, but if the sun is shining, then the shutter is lowered or raised avoiding or using its heating capability.

For the energy consumption, it would be favorable if the users of the building could not set any parameters; everything would be set automatically. This however would decrease the comfort of the people because the system could not follow the individual needs of users. This affects the energy consumption slightly because the users set the temperature with the thermostat or increase the illumination level infrequently [4]. It is also possible to control the peak electric load of the building with building automation control system. The peak load control results in less comfort in the use of the building. Numerous projects were carried out to collect information about the possibilities of building automation control systems (BACS) in energy consumption control [1].

In the EN 15232 standard [3], four different classes for BACS and technical building management (TBM) are defined:

Class D Nonenergy-efficient building automation and control system
Class C Standard building automation and control system
Class B Advanced building automation and control system and technical building management system

Table 1 BAC efficiency factors for thermal and electric energy for office buildings [3]

Office buildings	A	B	C	D
Thermal energy BAC efficiency factor $f_{BAC,hc}$	0.70	0.80	1.00	1.51
Electric energy BAC efficiency factor $f_{BAC,el}$	0.87	0.93	1.00	1.10

Class A High-energy performance building automation and control system and technical building management system

For each class a minimum requirement of BAC and TBM functions are defined to reach the given level. These levels help calculate the energy performance class of the building. As seen in Table 1, the reference is class C for the factors. It is important to note that these factors can modify the energy performance for a better class, but the determining factor for the energy performance class is not the BACS and TBM. If the energy performance class is better than class D, then the installation of BACS and TBM possibly does not modify the energy performance class [2]. In such a case with just the modification of the building's thermal isolation, changing the air conditioning system for more efficiency can improve the energy performance class.

It is also important to note that, even if a building has a good energy performance class, it does not mean that BACS and TBM are not necessary. With BACS and TBM, remarkable energy savings can be achieved even in a high-energy performance class building, but the energy performance class is not influenced.

3 Energy Consumption of Building Automation Systems

Bus systems of BACS use electric (or electromagnetic) signals for communication and the system components are electric and electronic devices. Three groups of devices are connected to the bus: sensors, actors and system devices. These devices have one thing in common, that is, each one is switched on 24 h a day because the building automation controls the building all day. Each device has its own energy consumption, and besides this, the actors in most cases are switching motors on and off to control the desired parameter of the building (e.g., the motors for raising or lowering the shutters), and these devices also has its own energy consumption.

The energy consumption of the controlled devices cannot be avoided. In an automation class C building, this energy consumption also exists; however, in a class A building, the use of such devices is possibly more frequent, and therefore, the energy consumption could be also higher.

There are different approaches and possibilities to supply the devices connected to the bus system with energy. One possibility is to supply power through the bus cable itself. The advantage of this solution is that the cabling is easier; just installation of the bus cable is necessary for the devices. The disadvantage is that the devices are highly limited in power, because the bus cable is a communication cable and therefore small in diameter, which does not allow high currents. It is easy to measure the used energy of such a system, because the bus system has a small number of power supplies connected to the bus.

If each device has its own power supply, then the communication cable and the low-voltage power cables are also necessary, which means more cabling. The advantage is that the power of the devices is almost not limited. There is also the possibility of using wireless (radio-frequency) communication, which reduces the cabling but increases the energy consumption used for the communication.

In some cases power-line communication is also used. This is advantageous if BACS is installed in the renovation of a building. The measurement of the energy consumption is similar to the previous case; each device should be measured, which means a number of power meters are needed.

Using BACS, the overall energy consumption is reduced to a class A building compared to a class C building, which is the main goal.

4 Odoo House

The Team Odooproject from Budapest University of Technology and Economics, selected from among nearly 50 other applicants, had an opportunity to measure its skills and performance against 20 other teams in the Solar Decathlon Europe 2012 competition in Madrid [5–7]. The goal of the event was to plan and build a house using only solar energy.

To have success in the event, there should be different technologies included in the Odoo house. During the design of the building, it was intended to minimize the equipment, and therefore, passive and active solutions are combined. Air-to-water heat pump supplies the building's thermal energy requirements in heating, cooling and domestic hot-water production. Also artificial ventilation and water re-cooling are built in the house, both custom-made for the Odoo house [6]. To provide electric energy for these systems, photovoltaic modules are mounted on the roof and on the wall of the building. On the roof monocrystalline and on the wall thin-film modules are used.

The temperature, air quality, shutters and the electric energy flow of the PV cells are monitored and controlled by a building automation system. The chosen BACS was the KNX system. KNX is an open standard, which defines the communication protocol between the connected devices on the bus system [8]. The KNX system can use twisted pair cables (TP1), power-line (PL) communication or radio-frequency (RF) as communication channel [9]. The KNX system is a distributed control system, in which each device has a microcontroller running the appropriate software for the device. In other words, the KNX system is a decentralized system; however, it is possible connect computers to the bus system for data acquisition and central status display. For this purpose in the Odoo house, a so-called home server is installed with a central touch screen as user interface. On this screen any information of the control systems in the building can be displayed.

After the contest in Madrid, the Odoo house was assembled again in the garden of the university for further research and educational purposes. In everyday life students are using the house for studying, creating their projects, consultations, etc. This can be treated as a normal office use of the building. This and the installed KNX system made the Odoo house a good choice for our measurements.

5 Energy Consumption of the KNX System in the Odoo House

As mentioned before, we were interested in the electric energy demand of the KNX system. In our case a home server is connected to the KNX system, and this device has a significant energy demand. In the Odoo house the power supply of the KNX system and the home server are connected to the electric power network of the house in different places, so it was possible to measure these two consumptions separately. Although the home server acts as a central managing point in the KNX system, it is not necessary to control any functions of the KNX system. The data acquisition and post processing belong to the management functions of the system and can help to lower the energy consumption of the house by fine-tuning the parameters of the KNX system. The central access to the functions is also convenient, but if the home server is removed from the KNX system, all the control functions remain in properly working state, which comes from the distributed intelligence of the KNX system.

The smallest bus unit of a KNX system is the line segment. On a line segment 64 devices can be connected. On each line segment there is a power supply, which provides electric power for the devices of the bus. Therefore, the bus is used not only for communication but also for supplying power for the connected devices. The voltage of the power supply is 29 V and 640 mA is the maximal output current. That means that each of the 64 devices consumes 10 mA current, which is sufficient for most of the devices. If there are devices with higher-power needs, then the number of devices on the line segment should be reduced or another power supply can be connected, if the length of the bus makes it possible (there is a minimum distance defined between the power supplies [8]).

The KNX system in the Odoo house consists of one line segment with one power supply (with an output rating of 640 mA). On the bus are 48 devices connected. The devices are presence sensors, light sensors, thermostats, air quality sensors, power meters, dimmers, switches and a weather station. All of the devices need a maximum of 10 mA current according the data sheets, so the installed power supply is not used at the maximal possible current rating. Certainly the home server is connected to the KNX bus, but not counted to the devices, because it has its own power source and does not receive power from the KNX bus.

Table 2 The electric energy consumption of the KNX system and the home server during 1 month

	Consumption [kWh]	
Date	KNX	Home server
25/03/2015	0.0	0.0
30/03/2015	1.5	2.2
01/04/2015	2.2	3.2
07/04/2015	4.2	6.1
13/04/2015	6.1	9.0
22/04/2015	9.1	13.3
27/04/2015	10.8	15.8
08/05/2015	14.4	21.1

6 Results

We measured the power consumption separately for the KNX line segment and the home server. A power meter was applied on the 230 V input of the KNX power supply and a separate one on the power supply of the home server. The home server is a small computer running the necessary software tools for the central data acquisition and the user interface for the interaction with the KNX system. If there is no action on the touch screen, the screen is not turned off but a slideshow of photos is running.

The measured overall consumption data is shown in Table 2 and the same is shown in Fig. 1 as a chart. In Fig. 1 a regression line is also shown for the data points, and in both cases 0.999 is the R-square value, which means that the consumption is close to linear. The measured 44 days were weekends and 1 official holiday, and during these days, the house was not used.

The energy consumption of the home server is in average 0.48 kWh daily calculated from the total consumption and the number of days measured. Since the program or the slideshow is running on the home server permanently, it is not supposed to have any significant changes in the electric consumption. The measured results fit these assumptions.

In the measured values of the KNX bus, there are no changes which would show a change in the consumption of the system. According to the data, it can be stated that the power consumption of the system is independent from the usage of the building. The user interaction with the system does not generate more or less energy consumption. The average consumption of the KNX bus system is 0.33 kWh daily.

The average daily consumption values are also calculated for the periods between the readings of the power meters, which show relatively small changes (except the first measured value). For the home server, the mean value is 0.48 kWh as mentioned before, and the average deviation is 0.012 kWh. The mean value for the daily consumption of the KNX bus is 0.327 kWh and the deviation is 0.012 kWh. These results are shown in Fig. 2. In most of the cases, the deviations of the data point are in the same direction and in the same order of magnitude which means that the deviations could result in the same source.

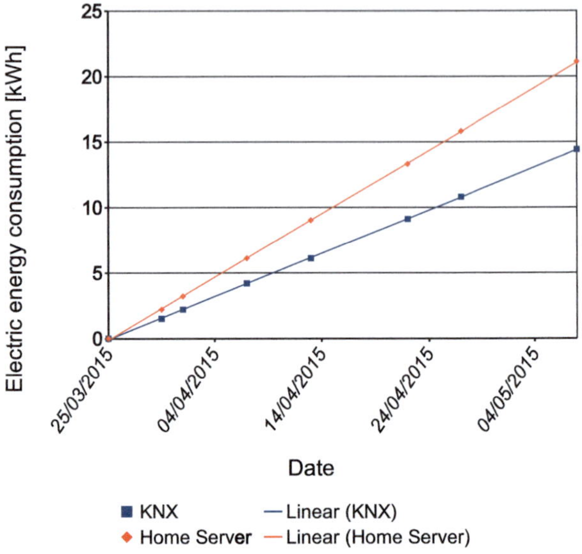

Fig. 1 Electric energy consumption of the KNX system and the home server in about 1 month

The average current ratings are also calculated for the sensors and actors connected to the KNX bus. The average current for the whole measuring period is 9.8 mA for each device. The average currents between the readings are also calculated and the deviation is 0.35 mA. The lowest value is 8.9 mA and the highest is 10.2 mA.

From the results we can conclude that for the calculation of a KNX system's energy demand, it is sufficient to know the number of connected devices. The overall energy demand for a year can be calculated as

$$E_{\text{year}} = 8760[\text{h}] \cdot \sum_{\text{device}} P_{\text{device}} \tag{1}$$

This calculation does not consider the energy needed for the devices switched by the actuators on the bus.

Many examples show that excessive energy savings can be reached with the use of KNX systems [10].

As an example calculation for the energy demand of the KNX system, we choose an office building with 10,300 KNX devices [10]. A saving of 1200 MWh of electric energy was reached with the installation and correct parameter setting of a KNX system. We can calculate the energy consumption of the KNX devices. Each device is consuming 10 mA or even more on 30 V, so each device has at least 0.03 W power. The total system power is calculated as

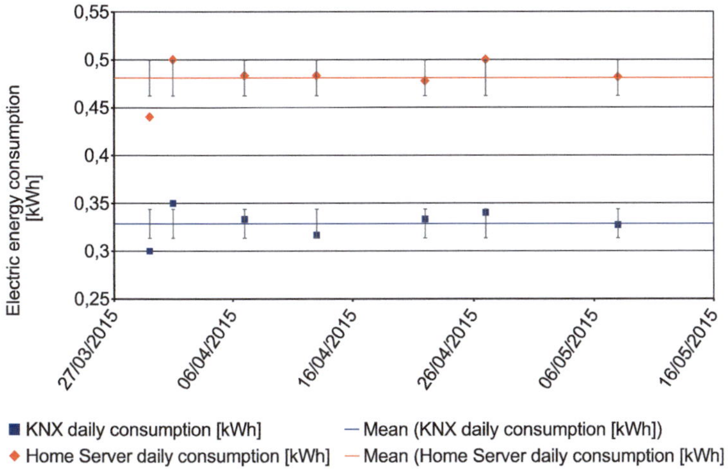

Fig. 2 Daily electric energy consumption of the KNX system and the home server for the periods between the readings of the power meter

$$P_{\text{total}} = 10,300 \cdot 0.03\ [\text{W}] = 3.09[\text{kW}] \tag{2}$$

To calculate the total energy consumption in a year, we can use Eq. 1:

$$E_{\text{KNX}} = 8760[\text{h}] \cdot 3.09\ [\text{kW}] = 27,068\ [\text{kWh}] \tag{3}$$

The result in Eq. 3 is about 2 % of the total reached energy saving in a year (i.e., about 2600 € a year). This seems not too much for an office building, particularly that it is not possible to eliminate this consumption totally, but if in each office building half of this energy is saved, it is a great savings globally.

In smaller installations the KNX system's saving for relative energy consumptions can be about 5 % [11].

One possible solution to achieve lower consumption for the KNX system would be using devices, which would enter standby mode, if they are not sending information on the bus (like sensors in KNX RF systems) [12]. In the KNX RF system, the sensors are using battery as power source, so the device consumption should be as low as possible. In a system about half of the devices are sensors and thus about half of energy consumption.

7 Conclusion

We measured the energy consumption of the BACS in the Odoo house used as an office-like environment. The building automation system is a KNX system extended with a home server for data acquisition and management purposes. We were able to measure the consumption of the KNX bus and the home server separately.

The energy demand of the home server is independent of the usage as it was predictable. We presumed that the energy consumption of the KNX system is strongly dependent on the quantity of communication, in other words on the usage and external condition changes. According to the measurements, the consumed energy does not vary in time significantly. The calculated average current is almost the maximum from the devices and the total power used is the sum of the power need of the devices. To have the energy demand of a BACS more accurate, further investigations are needed. First of all, the previously mentioned switched devices should be taken into account; therefore, it is necessary to calculate and measure the energy consumption. A more detailed measurement of the KNX bus could also make the estimation more accurate. By the detailed measurements, the external factors such as weather and human interventions should also be registered. With an accurate power meter capable of measuring a few watts and recording with time values, daily, weekly and monthly power functions could be recorded and compared to the external factors mentioned before.

The energy consumption of an office building is reduced with the use of building automation system even if the BACS itself is using energy. As the next step, how the energy consumption of the BACS could be lowered should also be considered. In an installation with a few hundred devices connected to the bus, the yearly energy consumption could be lowered even more with an energy-efficient BACS.

Acknowledgments The authors would like to thank Ádám Makk for the valuable help during the measurements.

References

1. Nguyen TA, Aiello M (2013) Energy intelligent buildings based on user activity: a survey. Energy Build 56:244–257
2. Ippolito M, Sanseverino ER, Zizzo G (2014) Impact of building automation control systems and technical building management systems on the energy performance class of residential buildings: an Italian case study. Energy Build 69:33–40
3. E. 15232, Energy performance of buildings – impact of building automation, controls and building management. Comité Européen de Normalisation (CEN), Brussels (2007)
4. Karjalainen S (2013) Should it be automatic or manual – the occupant's perspective on the design of domestic control systems. Energy Build 65:119–126
5. "Odooproject" (2012) [Online]. Available: http://www.odooproject.com. Accessed 28 June 2015
6. Pataky R, Pataky R, Á.ts Á, Á.ts-Leskó Z, Birtalan O (2014) Constructional considerations for the mobile plus-energy house. Energy Build 83(0):195–208
7. Haas-Schnabel G, Szikra C (2014) Dynamic simulation of Odoo. Energy Build 83:117–121
8. I. 14543-3-1:2006, information technology – Home Electronic Systems (HES) architecture – part 3–1: communication layers – application layer for network based control of HES class 1, International Organization for Standardization (2006)

9. Kastner W, Neugschwandtner G, Soucek S, Newmann H (2005) Communication systems for building automation and control. Proc IEEE 93(6):1178–1203
10. Energy Efficiency with KNX. [Online]. Available: http://www.knx.org/media/docs/downloads/Marketing/Flyer/Energy-Efficiency-With-KNX_en.pdf. Accessed 10 Sept 2015
11. Martirano L (2011) Lighting systems to save energy in educational classrooms. In 10th international conference on environment and electrical engineering (EEEIC)
12. KNX Basics. [Online]. Available: http://www.knx.org/media/docs/downloads/Marketing/Flyers/KNX-Basics/KNX-Basics_en.pdf. Accessed 10 Sept 2015

User-Adapting System Design for Improved Energy Efficiency During the Use Phase of Products: Case Study of an Occupancy-Driven, Self-Learning Thermostat

Y. De Bock, A. Auquilla, K. Kellens, D. Vandevenne, A. Nowé, and J.R. Duflou

Abstract Tailoring products to user requirements can improve the energy efficiency without sacrificing user satisfaction. This study considers the case of an occupancy-driven smart thermostat in an office environment. Identifying patterns in past user behaviour enables occupancy prediction to control the heating accordingly. Potential energy savings and related environmental impact reductions, compared to a fixed schedule heating system, are calculated for various heating and building types in three regions (Leuven, Calgary and Tokyo) to account for variations in climate. The obtained energy savings range between 93.2 and 546.5 kWh per year and environmental impact reductions between 2 and 38 EcoPoints.

Keywords Self-learning system • Energy efficiency • LCA • Home automation • User profiling • Usage prediction

Y. De Bock (✉) • K. Kellens • D. Vandevenne • J.R. Duflou
Department of Mechanical Engineering, KU Leuven, Leuven, Belgium
e-mail: yannick.debock@kuleuven.be; karel.kellens@kuleuven.be; dennis.vandevenne@kuleuven.be; joost.duflou@kuleuven.be

A. Auquilla
Department of Mechanical Engineering, KU Leuven, Leuven, Belgium

Department of Computer Science, University of Cuenca, Cuenca, Ecuador
e-mail: andres.auquilla@kuleuven.be

A. Nowé
Computational Modeling Lab, Vrije Universiteit Brussel, Brussels, Belgium
e-mail: ann.nowe@vub.ac.be

M. Matsumoto et al. (eds.), *Sustainability Through Innovation in Product Life Cycle Design*, EcoProduction, DOI 10.1007/978-981-10-0471-1_60

1 Introduction

Energy efficiency of machines, appliances and consumer products receives increasing attention in order to reduce their energy consumption and related carbon emissions and operational costs. This paper discusses the incorporation of automated learning techniques to improve the energy efficiency of heating systems in offices. Typically, space heating is the largest energy consumer in offices in both Europe and the United States. On average space heating accounts for 35 %, or 146 MWh, of the energy consumed by the US office buildings considered in the Commercial Buildings Energy Consumption Survey (CBECS) of 2003 [1]. The incorporation of smart thermostats, which are able to deal with the activity of the inhabitants, can improve energy efficiency in homes by 6–17 % [2]. In an office environment, potential savings could be even higher due to the variety in office hours and habits of different employees, i.e. user profiles, while a fixed heating schedule is typically applied for all offices. Furthermore, as many office buildings are outdated and therefore do not conform to today's insulation standards, these relative savings will have a larger impact in terms of absolute numbers.

Automatically tailoring heating schedules towards the different identified user profiles, i.e. turning the heating off when the office is unoccupied, can substantially reduce energy consumption without compromising user comfort. Aiming at identifying an optimal trade-off between resource utilisation and delivered functionality, the behaviour of users is modelled to allow occupancy forecasting. These predictions facilitate defining a policy oriented towards user comfort and energy efficiency. In terms of satisfying user requirements, this means preheating the house or room in anticipation of the arrival of the user, such that the house or room has reached the desired temperature by the time the user arrives. However, this preheating time must be optimised as heating too long in advance will waste energy. Furthermore, quickly and accurately determining the vacancy of a house or room is important to achieve energy savings.

For this research, a real-life dataset was collected by equipping an office with several sensors to monitor room occupancy. Energy savings, while maintaining user comfort, are identified by comparing the predicted occupancy of the intelligent thermostat to a reference situation of a conventional, schedule-based, heating system. The potential energy and environmental impact savings, for different heating and building types (insulation level and room characteristics) in various regions, are quantified by simulation and a life cycle assessment. The influence of the climate is taken into consideration by computing the environmental impact for three different locations. The reference situation is an office in Leuven, Belgium. Calgary in Canada and Tokyo in Japan were included to account for a colder and warmer climate, respectively.

The research presented in this paper is part of a research project conducted at the Centre for Industrial Management of the KU Leuven, about Personalized Products Emerging from Tailored User Adapting Logic (PERPETUAL).

The remainder of this paper is structured as follows. A discussion of related work is presented in Sect. 2. Section 3 describes the PERPETUAL algorithm for a self-learning thermostat. This is followed by a summary of the different scenarios and their results (Sect. 4). Finally, conclusions are drawn and future work is identified.

2 Background and Related Work

This section first discusses the drawbacks of current thermostats. Then, an overview of occupancy detection and prediction algorithms for heating control is presented, followed by a summary of commercially available thermostats that are labelled smart. Finally, the contributions of our approach are detailed.

Modern thermostats are typically programmable, allowing the user to specify the desired temperature at different times of the day. Consequently, a balance between comfort and energy consumption can be obtained by defining a setback temperature when the house is vacant and preheating the house in anticipation of the arrival of a user. However, these systems are characterised by several complications. Users rely on their personal experience to set the schedule of the thermostat. Therefore, the inferred rules are not optimal; moreover, they are usually inflexible and do not cope with exceptions like coming home early. Furthermore, programmable thermostats are often not properly used due to complex and cumbersome user interfaces [3]. Lengthy manuals, social and practical barriers were also among the user complaints reported in [3]. Peffer et al. prove that only half of the programmable thermostats are used as intended in the United States.

Given these shortcomings and a growing environmental awareness [4], research on energy-efficient heating systems is nowadays mainly oriented towards smart thermostats. Such thermostats are considered intelligent since they are capable of learning the habits of the home occupants and steer the heating accordingly. This intelligent control consists of presence detection and prediction. In the first phase, presence detection enables the system to collect occupancy data. From this information, it is possible to learn common occupancy patterns and construct a predictive model that is capable of forecasting the future occupancy of a room. Then, as new occupancy measurements become available, they are fed into the prediction system which combines these with the knowledge of the common patterns to predict the future occupancy states.

An overview of the most common technologies for occupancy detection and prediction is presented in [2, 5], and a summary of popular approaches is given in the next two paragraphs. A widely used approach for presence detection is employing a wireless sensor network. This is cheap, nonintrusive and easily retrofitted [5]. Camera networks are used to determine occupancy as well; however, such set-ups are expensive and considered intrusive by the user [5]. Also the use of RFID badges, entry-exit logs, a proximity sensor and ultrasonic localisation has been reported in literature [5]. Furthermore, human activity and therefore presence can be derived from the energy consumption of appliances [6]. In [7] a thermal sensor is used in conjunction with a PIR sensor, while GPS data facilitate occupancy detection in [8] and [9].

Since heating systems have a long response time, i.e. it takes time to reach the desired temperature, predicting the future occupancy of a room is essential. In 2014, Kleiminger et al. [2] present a comparative performance analysis of state-of-the-art techniques and classify them into two main categories, namely, schedule-based and context-aware methods. Schedule-based methods rely on historical occupancy schedules, while context-aware strategies try to determine the occupancy state by analysing the context, i.e. location or activity, of the inhabitants. The Neurothermostat [10] uses, as its name implies, a neural network approach for occupancy prediction. To take advantage of the periodic nature of human behaviour, the neural network is combined with a look-up table indexed by the time of day and the current occupancy status. Other well-known techniques for occupancy-driven heating control are the GPS thermostat and time-based presence probabilities [8, 9], the smart thermostat [11], OBSERVE [12] and PreHeat [13]. Still other approaches use logistic regression [14] or fuzzy predictors and time series prediction techniques such as ARMA models [15]. SPOT+ [16] predicts future occupancy using the k-nearest neighbours method from the PreHeat [11] system but also predicts future room temperature and computes the optimal sequence of control inputs.

As human behaviour typically exhibits periodic activities on a daily and weekly basis, the introduction of clustering techniques to extract patterns from a user's outgoing behaviour, i.e. grouping similar days into profiles, could benefit the prediction of future occupancy. This belief has led to several usage profiling techniques that can be used for forecasting the presence of users [17–21].

Finally, methods for forecasting the usage of appliances can be adopted for predicting room occupancy such as [22–24]. This is illustrated by the technique presented in [24] which is an extension of the strategy proposed by Tominaga et al. [21]. A similar relation applies for [20] and [22].

Today several smart thermostats are already commercially available. However, it is noted that the term smart represents variable functionalities. Thermostats that are controlled remotely are already considered smart, while in fact they do not possess any form of autonomous learning and automatic personalisation. Other so-called smart thermostats – in addition to remote control – use geolocation (to facilitate preheating in anticipation of arrival) or presence sensors to switch the heating on and off and/or learn the room characteristics to determine the optimal heating times. Examples of such systems are Momit's smart thermostat, Honeywell's Lyric, Tado,[1] etc. In this paper, a smart thermostat is defined as a thermostat that learns user habits and preferences, i.e. a learning thermostat. By this definition, Google's Nest and Heat Genius are, to the best of our knowledge, the only currently commercially available smart thermostats. The Nest has been the

[1] Momit: www.momit.com/smartthermostat.html. Lyric: lyric.honeywell.com. Tado: www.tado.com

subject of a study conducted by Yang et al. [25]. In this study, an experiment was organised in which 19 households used Nest for at least 1 month and reported their findings. Overall, Nest was well received. However, this was mainly because of its appealing and intuitive graphical user interface and the energy history feature which raises an increased engagement and awareness in usage patterns. The intelligent features however were not that successful as the system could not grasp the intent behind user behaviour and the users did not fully understand how Nest functions.

Compared to the above-presented state of the art, the proposed PERPETUAL method has the following benefits. First, extracting patterns from historical data by clustering enables predictions far in advance and allows for an intuitive graphical user interface showing the expected day pattern. Furthermore, the genericity of the approach allows it to be used for other applications such as dynamical power management, management of electrical car batteries in function of anticipated use and net rates, etc. Finally, to the best of our knowledge, this paper presents the first evaluation of the effects of a smart heating system by quantifying the potential energy savings and related environmental impact.

3 Self-Learning Thermostat

This section describes a set of algorithms, for modelling user behaviour and predicting the future occupancy state of a room, that define the smart nature of a self-learning thermostat. Figure 1 presents an overview of the PERPETUAL approach. This smart system comprises two steps, user profiling and occupancy prediction, respectively, detailed in Sects. 3.1 and 3.2. The former aims to detect occupancy patterns in historical user data, whereas the latter uses those patterns to predict the occupancy state of a room for a certain time frame in the future. Data collection, potential energy savings and environmental impact reductions are discussed in Sect. 4.

3.1 User Profiling

In this first step, historical data regarding room occupancy are analysed in order to identify occupancy patterns.

3.1.1 Data Collection

The historical data were collected by deploying a wireless sensor network (WSN) capable of detecting the presence of a user. Two motion sensors were installed, one near the door and one near the desk in an office environment, to capture movements

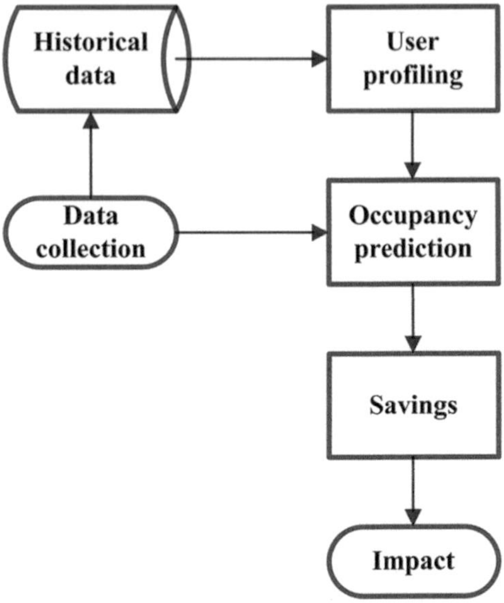

Fig. 1 Overview of the PERPETUAL approach

as accurately as possible. A magnetic reed switch sensor was attached to the door to monitor the position of the door (open or closed). From these sensor measurements, it is possible to derive the presence of an occupant based on a simple decision tree as illustrated in Fig. 2.

Where *motion* and *door* represent the number of movements and times the door was open within a time slot, respectively, and *hour* indicates the time of day. The office is never occupied during the night. This constraint prevents false-positives from occurring during night-time. Occupancy data of the period from the 18th of April to the 10th of July 2014 were used for the reported experiments. Since the heating is not used during the weekends and holidays in the monitored office environment, these days are excluded from the experiment.

3.1.2 Profiling

The historical data are organised as binary vectors – zeros and ones, respectively, indicating absent and present states per time slot of 15 min – representing the occupancy profiles of every day. Patterns are then identified by grouping similar days into clusters. Such a historical pattern, i.e. cluster, is defined as a day template that contains a probability of occupancy for each time slot, as illustrated in Fig. 3.

Human habits and interactions are assumed to occur on a daily basis [26]. Therefore, day patterns are selected. The granularity of a pattern, i.e. the size of the time slots, was set to 15 min. So a pattern consists of 96 time slots.

The Tominaga approach [21] served as a starting point to the developed system. Given the implicit weekly seasonality in terms of occupancy that exists in office

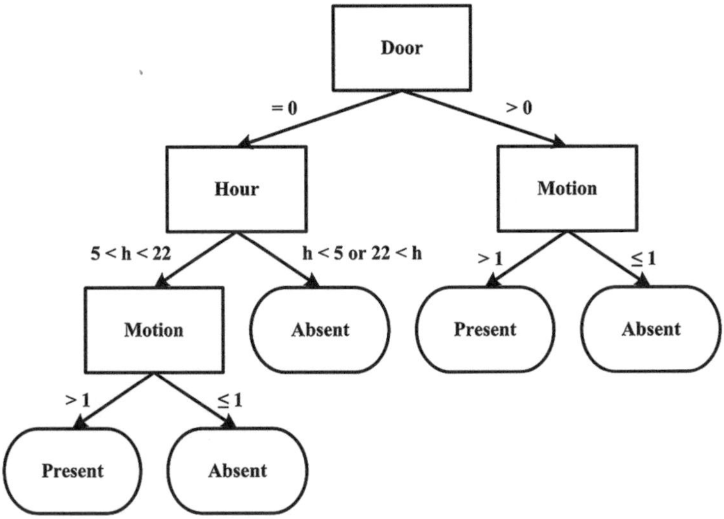

Fig. 2 Decision tree for occupancy detection

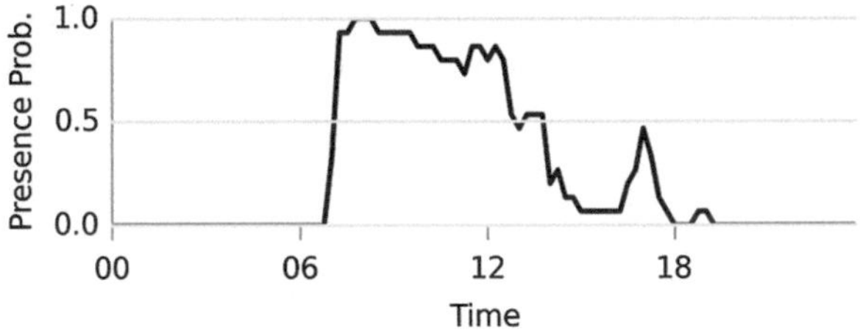

Fig. 3 Example of an occupancy pattern

environments [26], the model is extended with a variable representing the weekday of the observations as introduced by Truong et al. [24]. Although implemented for appliance forecasting, the Truong model can be adapted for occupancy prediction.

Figure 4 depicts the graphical model for user profiling. The directed arrows in the graph indicate the dependencies between the parameters. The observed days, i.e. historical data, and the corresponding weekdays are represented by x_n and q_n, respectively. Both depend on the latent variable z_n, which indicates to which day type the observed day belongs. A possible categorisation of day types could be 8 h workdays, early-off workdays, holidays, etc. This method automatically determines the number of day types, or clusters, K present in the data. The presence probabilities for these day types are estimated for every time slot and represented by $\mu_{k,t}$, where k and t indicate the cluster and the time slot, respectively. The estimation of parameters and number of day types (K) is performed by Dirichlet process mixture (DPM) using a blocked Gibbs sampler. First, its parameters are initialised randomly in such a way that the observed days are allocated uniformly to each day type. Iteratively, this method resamples from its posterior distributions (σ_k,π_k, and μ_k with conjugate priors γ, α, and β, respectively) aiming to maximise the likelihood of the

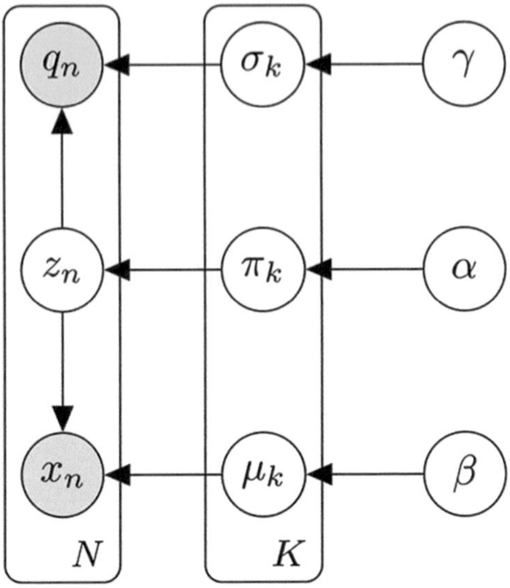

Fig. 4 Graphical model for user profiling [24]

dataset observations. As a result, every observed day has a probability of belonging to a cluster k. The cluster allocation is completed by selecting the cluster with the highest likelihood.

3.2 Occupancy Prediction

The occupancy prediction involves two steps: (1) determine the cluster membership for the next time slots given the current weekday and previous occupancy states and (2) select the occupancy probability associated to the corresponding time slot of the predicted cluster.

As a result of the user profiling step, a vector containing the cluster membership for each day in the dataset was created. With this vector, it is possible to calculate, $\pi_k^* \ (k \in K)$, the probability of each day type k to occur. This value is equal to the number of days belonging to each day type divided by the total number of days. Likewise, the clustering process obtains the day of the week distribution for every day type, $\sigma_{k,w} \ (w \leq 7)$, which allows coping with weekly recurring patterns.

Since the future occupancy status is highly correlated to the previous occupancy states that occurred during the current day, this information is also included in the predictive model. Those previous states are included by using the likelihood of the parameters defined for the graphical model depicted in Fig. 4:

$$p(x_n|\mu_k) = \prod\nolimits_{t=1}^{T} \mu_t^{x_{n,t}} \left(1 - \mu_{k,t}\right)^{1-x_{n,t}} \tag{1}$$

The likelihood function provides an estimate on how well an observed day corresponds to each day type. Thus, the probability that a new observation belongs to a cluster k is computed by combining the probability of each day type (π_k^*), the weekday probability ($\sigma_{k,\,w}$) and the likelihood for the new observations. This value is computed for all clusters. The day type with the highest probability, k^*, is assigned to the new observation.

Finally, the prediction is given by the mean of the occupancy states for the next time slot of the days belonging to k^*. This average is translated to $\{0,1\}$ by defining a threshold τ. If the predicted value is higher than τ, then the new occupancy state is set to present. Without any prior knowledge of the clustering and prediction results, τ can be set to 0.5. This threshold serves as a potentiometer for balancing the performance between savings and comfort.

The presented technique also allows for multi-step-ahead predictions, which is often required for HVAC control as heating a room can take a considerable amount of time depending on its physical characteristics. For N-step-ahead predictions ($N \in \mathbb{Z}$), a prediction has to be made for every $n = [1, 2, \ldots, N]$. Each time a new prediction is made, the vector of previous occupancy values is updated.

4 Impact Quantification

The potential energy savings and the related environmental impact reductions of the discussed approach are illustrated by a real-life use case (reference situation). Moreover, several scenarios were simulated to account for variations in climate, heating type and building characteristics.

The characteristics of the studied room along with the different scenarios are described in Sect. 4.1. This is followed by the computation of the amount of time the heating can be switched off, by using a smart heating system, in Sect. 4.2. Furthermore, the presented strategy is compared to state-of-the-art occupancy prediction methods. Section 4.3 establishes the necessary assumptions enabling the calculation of energy savings. Finally, in Sect. 4.4 the energy and environmental impact reductions are computed for the different scenarios for the three regions and all heating types.

4.1 Room Characteristics

The studied room was a university office occupied by one professor located in Leuven, Belgium. It is a rectangular room with one door and two single-glazed windows and measures 4.5 by 3.3 by 3 m. Adjacent to the office are a hallway and two other offices; the fourth side is an exterior wall mainly consisting of windows (4.5 m^2). The room is on the top floor of a four-storey building. The inner walls are

Table 1 Scenarios

	Volume	Insulation	Floor
Reference	45 m^3	Single-pane windows, R6 walls, R12 roof, average infiltration	Top floor
Scenario V	60 m^3	Single-pane windows, R6 walls, R12 roof, average infiltration	Top floor
Scenario I	45 m^3	Triple-pane windows, R12 walls, R26 roof, good infiltration	Top floor
Scenario F	45 m^3	Single-pane windows, R6 walls, average infiltration	2nd floor

not insulated and the outer wall and roof are estimated to have an R6 and R12 insulation level (US metric), respectively, where R represents thermal resistance. Average infiltration losses, i.e. normal air changes like opening a window, as well as losses through structure construction such as cracks and leaks, are assumed. The heating is scheduled to reach 20.5 °C on weekdays from 7 to 18 h and a setback temperature between 12 and 14 °C is specified. During weekends the heating is turned off.

As physical building characteristics such as envelope materials and insulation levels vary with the climate and age of the building, simulations are performed in different regions for old and modern buildings. Table 1 summarises the different scenarios. In each scenario one parameter is adapted (underlined in the table) with respect to the reference situation. This will allow analysing the influence of the different parameters. In scenario V the length and width, and hence the volume, of the room are increased while keeping the windowed area constant. The insulation level is improved to comply with more modern standards in scenario I. Finally, scenario F assumes the office is on the second floor to parameterise the losses through the roof. For each scenario, savings are calculated for three regions with a moderate (reference: Leuven, Belgium), cold (Calgary, Canada) and warm (Tokyo, Japan) climate.

Furthermore, the type of heating system also influences the energy-saving capacity of an occupancy-driven thermostat. Therefore, simulations are performed with five different heating systems using natural gas, hard coal, fuel oil, wood and electricity.

4.2 Time Savings

The time savings that can be achieved by using a smart heating system are computed for the period defined by the heating schedule, i.e. from 7 to 18 h. These savings represent the amount of time the user is absent during this period, and therefore the amount of time the heating can be turned off.

This is computed by calculating the average number of time slots the model predicted the absence of the user. For this matter, 47 days were used to train the

Fig. 5 Identified day types

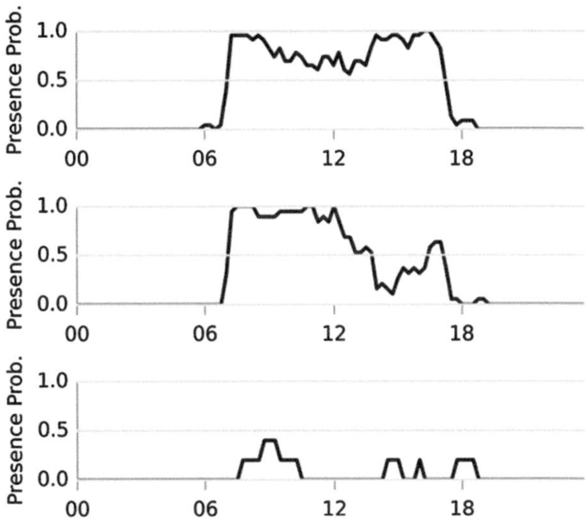

predictive models, and the last 7 days were used for testing. The testing days were not used at any stage in the training phase. The occupancy of the office is predicted half an hour in advance, providing enough time for the heating system to reach the comfort temperature.

For evaluating the system, three metrics are computed: accuracy, miss time and time savings. Accuracy represents the ratio of correct occupancy predictions, i.e. present/absent, to the total number of observations in the test period. The miss time equals the false-negative rate, i.e. the proportion of time slots the model predicts the absence of the user while he is actually present. Finally, the savings measure is the ratio of time slots the model predicted vacancy of the room to the total number of time slots.

The results produced by the profiling algorithm are depicted in Fig. 5. As can be seen, the user's behaviour can be characterised by three kinds of days. These patterns are then used by the predictive model to forecast the occupancy of the room. Figure 6 shows the prediction results for three out of the seven testing days. The results in terms of accuracy, miss time and savings are shown in Table 2.

Our straightforward prediction model achieves a relatively high accuracy of 78 % and an acceptable degree of miss time. The savings that are possible to achieve with our method, i.e. 25 % of the work hours, represent 2.8 h per day on average. These outcomes are similar to the results obtained by state-of-the-art techniques on the same dataset, as shown in Table 2. The BLEMS strategy performs best in terms of accuracy and miss time, while the OBSERVE method achieves the highest savings. However, accuracy can be misleading as it depends on the balance between the classes. In this case the classes are not balanced, as the user is mostly present. Therefore, the miss time metric is more indicative. Improving the accuracy by using more advanced models is part of future research.

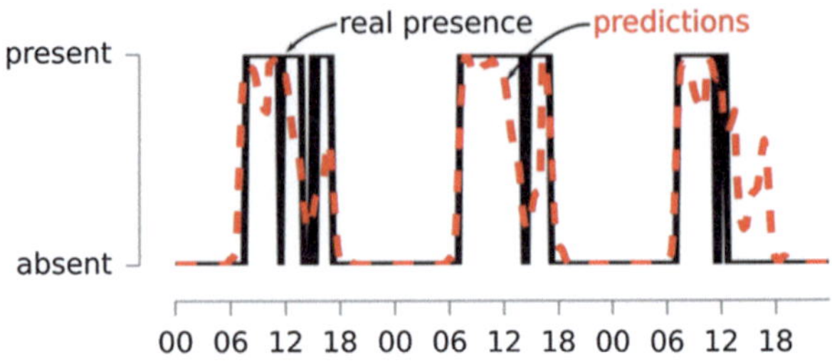

Fig. 6 Predicted vs. actual presence for 3 days

Table 2 Prediction results

Method	Accuracy	Miss time	Savings
PERPETUAL	78 %	9 %	25 %
OBSERVE [12]	80 %	10 %	30 %
PreHeat [13]	79 %	10 %	28 %
BLEMS agent [14]	82 %	8 %	27 %
Fixed schedule	71 %	0 %	0 %

Table 3 Outside temperature and heating days

	Leuven	Tokyo	Calgary
Average temperature	12.0 °C	12.0 °C	4.3 °C
Heating days	243	152	236
Days below 0 °C	2	0	78

4.3 Assumptions

In order to calculate the energy savings from the time savings, the following assumption was made. The average daily savings in time for the studied period are assumed to be the same for the entire year.

To account for the difference in climate, the average outside temperature between 7 and 18 h of the year 2014 was calculated for all regions. Days with an average temperature during office hours above 20.5 °C were excluded. The remaining days are called heating days. Table 3 indicates the average temperature and number of heating days for the different regions.

4.4 Energy and Environmental Impact Savings

Table 4 presents an overview of the heat losses and potential energy savings for the different scenarios and regions. The heat losses per unit of time, calculated using

Table 4 Heat losses [W] and energy savings/year [kWh]

[W/kWh]	Leuven	Tokyo	Calgary
Reference	381/259.2	381/162.2	737/487.0
Scenario V	427/290.5	427/181.7	827/546.5
Scenario I	219/149.0	219/93.2	421/278.2
Scenario F	321/218.4	321/136.6	621/410.4

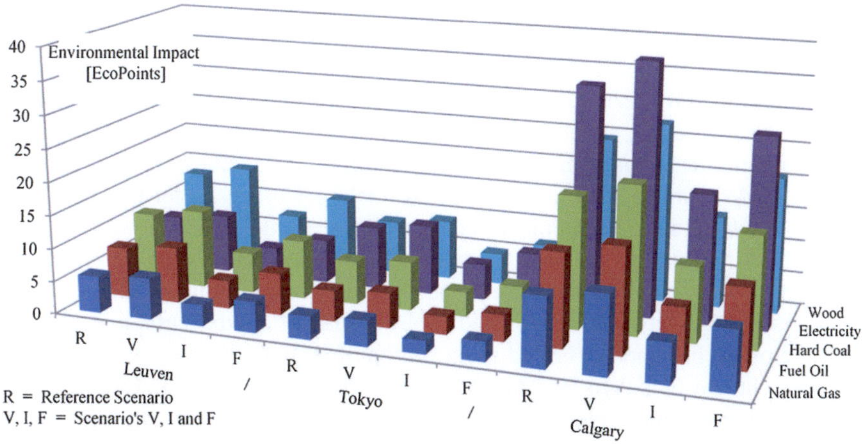

Fig. 7 Potential environmental impact savings [EcoPoints]

the Dimplex heat loss calculator [27], were multiplied with the average time savings of 2.8 h a day and the total number of heating days to obtain the potential yearly energy savings. Finally, as shown in Fig. 7, the environmental impact savings were estimated for different heating types using the Europe ReCiPe H/A LCIA method [28] and ecoinvent3 LCI database[2] [29].

For the reference scenario in Leuven, a potential environmental impact reduction of 5.6–13.0 EcoPoints is estimated depending on the heating type. As expected

[2] The used LCI datasets are:

1. Heat, district or industrial, natural gas {RoW}|heat production, natural gas, at boiler modulating >100 kW
2. Heat, district or industrial, other than natural gas {RoW}|heat production, heavy fuel oil, at industrial furnace 1 MW
3. Heat, district or industrial, other than natural gas {RoW}|heat production, at hard coal industrial furnace 1–10 MW
4. Heat, district or industrial, other than natural gas {RoW}|heat production, softwood chips from forest, at furnace 1000 kW

5a. Electricity, low voltage {BE}| market for
5b. Electricity, low voltage {JP}| market for
5c. Elecbtricity, low voltage {CA-AB}| market for

smaller, better insulated and enclosed rooms experience less heat losses and thus require less heating.

5 Conclusions and Future Work

Smart thermostats that control the heating system according to the predicted occupancy of a room can save a substantial amount of energy. For the reference situation of the professor's office in Leuven, 259.2 kWh could be saved compared to the fixed schedule (7–18 h), resulting in an environmental impact of up to 13 EcoPoints depending on the type of heating system. Evidently, in a region with a colder climate such as Calgary and for larger offices, a lot more savings, up to 546.5 kWh and 38 EcoPoints, can be achieved. As the reference building is an old university building, scenario I, where the insulation level is more in line with modern standards, provides insight in the effect of proper insulation. The effect of heat loss through the roof is accounted for by scenario F. In this case, it is assumed that the office has another heated room on top of it. Both these scenarios and the warmer climate of Tokyo lead to a lower energy-saving potential. The presented occupancy prediction strategy obtains similar results as the current state of the art. It must be noted that the accuracy of 78 % is obtained only during the office hours. Hence, a large and predictable part of the day was not taken into account. The occupancy prediction strategy is in development and part of future research. Furthermore, a more detailed analysis of the energy-saving potential of smart heating systems will be performed in the future. Specifically, the time of day when the heating can be turned off will be taken into account since the amount of energy saved by turning the heating off differs at different moments in time.

Acknowledgments The authors would like to thank IWT Vlaanderen for the financial support and Prof. D. Cattrysse for facilitating a real-life use case.

References

1. EIA (2003) Commercial buildings energy consumption survey. Energy Information Administration, U.S. Department of Energy
2. Kleiminger W, Mattern F, Santini S (2014) Predicting household occupancy for smart heating control: a comparative performance analysis of state-of-the-art approaches. Energy Build 85:493–505
3. Peffer T, Pritoni M, Meier A, Aragon C, Perry D (2011) How people use thermostats in homes: a review. Build Environ 43(12):2529–2541
4. Barnosky AD, Hadly EA, Bascompte J, Berlow EL, Brown JH, Fortelius M, Getz WM, Harte J, Hastings A, Marquet PA, Martinez ND, Mooers A, Roopnarine P, Vermeij G, Williams JW, Gillespie R, Kitzes J, Marshall C, Matzk N, Mindell DP, Revilla E, Smith AB (2012) Approaching a state shift in Earth's biosphere. Nature 486(7401):52–58

5. Nguyen TA, Aiello M (2013) Energy intelligent buildings based on user activity: a survey. Energy Build 56:244–257
6. Kleiminger W, Beckel C, Staake T, Santini S (2013). Occupancy detection from electricity consumption data. In: Proceedings of the 5th ACM workshop on embedded systems for energy-efficient buildings (BuildSys'13), University of Rome, Italy, 13–14 November 2013
7. Beltran A, Erickson VL, Cerpa AE (2013) ThermoSense: occupancy thermal based sensing for HVAC control. In: Proceedings of the 5th ACM workshop on embedded systems For energy-efficient buildings (BuildSys'13). University of Rome, Italy, 13–14 November 2013
8. Gupta M, Intille SS, Larson K (2009) Adding GPS- control to traditional thermostats: an exploration of potential energy savings and design challenges. In: Proceedings of the 7th international conference on pervasive computing (Pervasive 2009), Nara, 11–14 May 2009
9. Krumm J, Brush AJB (2011) Learning time-based presence probabilities. In: Proceedings of the 9th international conference on pervasive computing (Pervasive 2011), San-Francisco, 12–15 June 2011
10. Mozer MC, Vidmar L, Dodier RH (1997) The neurothermostat: predictive optimal control of residential heating systems. Adv Neural Inf Process Syst 9:953–959
11. Lu J, Sookoor T, Srinivasan V, Gao G, Holben B, Stankovic J, Field E, Whitehouse K (2010) The smart thermostat: Using occupancy sensors to save energy in homes. In: Proceedings of the 8th ACM conference on embedded networked sensor systems, ETH Zurich, 3–5 November 2010
12. Erickson VL, Carreira-Perpinan MA, Cerpa AE (2011) OBSERVE: occupancy-based system for efficient reduction of HVAC energy. In: Proceedings of the international conference on information processing in sensor networks 2011 (IPSN 2011), Chicago, 12–14 April 2011
13. Scott J, Bernheim Brush AJ, Krumm J, Meyers B, Hazas M, Hodges S, Villar N (2011) PreHeat: controlling home heating using occupancy prediction. In: Proceedings of the 13th international conference on ubiquitous computing (UbiComp 2011), Beijing, 17–21 Sept 2011
14. Mamidi S, Chang Y-H, Maheswaran R (2012) Improving building energy efficiency with a network of sensing, learning and prediction agents. In: Proceedings of the 11th international conference on autonomous agents and multiagent systems (AAMAS 2012), Polytechnic University of Valencia, Spain, 4–8 June 2012
15. Akhlaghinia MJ, Lofti A, Langensiepen C, Sherkat N (2008) A fuzzy predictor model for the occupancy prediction of an intelligent inhabited environment. In: Proceedings of the 16th international conference on fuzzy systems (FUZZ-IEEE), Hong Kong, 1–6 June 2008
16. Gao PX, Keshav S (2013) Optimal personal comfort management using SPOT+. In: Proceedings of the 5th ACM workshop on embedded systems for energy-efficient buildings (BuildSys'13), University of Rome, Italy, 13–14 November 2013
17. Chang C, Drugan MM, Verhaegen P-A, Nowe A, Duflou JR (2013) Finding days-of-week representation for intelligent machine usage profiling. J Ind Intell Inf 1(3):148–154
18. Vazquez FI, Kastner W (2011) Clustering methods for occupancy prediction in smart home control. In: Proceedings of the 20th IEEE international symposium on industrial electronics (ISIE 2011), Gdansk University of Technology, Poland, 27–30 June 2011
19. Vazquez FI, Kasner W (2010) Usage profiles for sustainable buildings. In: Proceedings of the 15th IEEE international conference on emerging technologies and factory automation (ETFA 2010), Bilbao, 13–16 September 2010
20. Barbato A, Borsan L, Capone A, Melzi S (2009) Home energy saving through a user profiling system based on wireless sensors. In: Proceedings of the 1st ACM workshop on embedded sensing systems for energy-efficiency in buildings (BuildSys'09), San-Francisco, 3 November 2009
21. Tominaga S, Shimosaka M, Fukui R, Tomomasa S (2012) A unified framework for modeling and predicting going-out behavior. In: Proceedings of the 10th international conference on pervasive computing (Pervasive 2012), University of Newcastle, UK, 18–22 June 2012
22. Barbato A, Capone A, Rodolfi M, Tagliaferri D (2011) Forecasting the usage of household appliances through power meter sensors for demand management in the smart grid. In:

Proceedings of the 2nd IEEE international conference on smart grid communications, Brussels, 17–20 October 2011
23. Chang C, Verhaegen P-A, Duflou J-R (2014) A comparison of classifiers for intelligent machine usage prediction. In: Proceedings of the 10th international conference on intelligent environments (IE'14), Shanghai Jio Tong University, China, 30 June–1 July 2014
24. Truong, NC, mcInerney J, Tran-Thanh L, Costanza E, Sarvapali DR (2013) Forecasting multi-appliance usage for smart home energy management. In: Proceedings of the 23rd international joint conference on artificial intelligence (IJCAI'13), Beijing, 3–9 August 2013
25. Yang R, Newman MW (2013) Learning from a learning thermostat: lessons for intelligent systems for the home. In: Proceedings of the 2013 ACM international joint conference on pervasive and ubiquitous computing (UbiComp'13), ETH Zurich, 8–12 September 2013
26. Chang W-K, Hong T (2013) Statistical analysis and modeling of occupancy patterns in open-plan offices using measured lighting-switch data. Build Simul 6(1):23–32
27. Dimplex heat loss calculator. Available online: http://www.dimplex.com/en/customer_support/heat_loss_calculator
28. Goedkoop M, Heijungs R, Huijbregts M, De Schryver A, Struijs J, van Zelm R (Update 2013) ReCiPe 2008, report 1: characterisation. http://www.lcia-recipe.net/
29. Weidema BP, Bauer C, Hischier R, Mutel C, Nemecek T, Reinhard J, Vandenbo C, Wernet G (2013) Overview and methodology. Data quality guideline for the ecoinvent database version 3, Ecoinvent Report 1(v3). The ecoinvent Centre, St. Gallen

A Fully Renewable DC Microgrid with Autonomous Power Distribution Algorithm

Nobuyuki Kitamura, Annette Werth, and Kenji Tanaka

Abstract In this study, we propose an autonomous DC microgrid system for residential community with distributed power exchange control to increase the utilization of renewable natural energy and to ensure minimal energy supply in the event of a large-scale disaster. Assuming small community of 20 houses, each house in the microgrid has its own in-house system that can continue to provide power to appliances from its own batteries and solar panels even when disconnected from the utility grid or power outage. Using solar irradiation time series and individual household power demand records, we examined its performance using real-time simulations. Finally, we propose a sustainable power system based on renewable energy that further minimizes fossil fuel consumption by integrating a renewable auxiliary power supply system such as the hydrogen energy system.

Keywords DC microgrid • Solar photovoltaic • Battery • Renewable energy • Hydrogen • Fuel cell • Power distribution

1 Introduction

The imposed reduction of carbon dioxide emissions in many countries has helped increase the prevalence of renewable energy. However, mega solar and large wind power plants include large equipment and a burden for the power grid to absorb the output fluctuations. In Japan, this is one of the main hurdles that hinder the increased proportion of renewable energy. Indeed, as of 2012, renewable energy except hydropower only represents 1.6 % of the country's electricity generation [1]. For this, self-sufficient power systems that are not dependent on the power grid and that can be deployed similarly to consumer electronics are gaining interest as a

N. Kitamura • K. Tanaka (✉)
Graduate School of Engineering, The University of Tokyo, Tokyo, Japan
e-mail: nobuyuki.kitamura@gmail.com; kenji_tanaka@sys.t.u-tokyo.ac.jp

A. Werth
Graduate School of Engineering, The University of Tokyo, Tokyo, Japan

Sony Computer Science Laboratory Inc., Tokyo, Japan

M. Matsumoto et al. (eds.), *Sustainability Through Innovation in Product Life Cycle Design*, EcoProduction, DOI 10.1007/978-981-10-0471-1_61

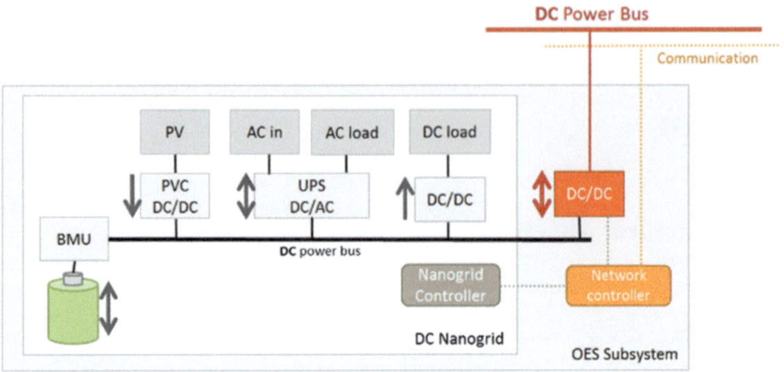

Fig. 1 Elemental subsystem for OES: a DC nanogrid

solution to both increase renewable power sources and to increase resilience in case of disasters [2, 3].

The system we propose uses as building blocks DC nanogrids including PV panels and secondary batteries in each house (Fig. 1). These subsystems are interconnected via a DC power bus, which is used to share energy resources such as PV or also an emergency power supply way even during power outages due to disaster (Fig. 2). An autonomous distributed grid system is thus built for a small community of houses. Furthermore, this system can flexibly expand over several layers by interconnecting not only houses but entire grid systems in a hierarchical way. We named this kind of DC microgrid system open energy systems (OES) [4].

In this paper, we focus on the control strategy of OES and propose an autonomous decentralized power exchange control for OES in order to improve the share of renewable energy within a community.

2 Renewable Energy Use for Homes

In this research we consider solar PV systems as a renewable energy source for the residential community. The utilization efficiency of solar energy can be measured using two indicators: one is the solar replacement ratio (SRR) R_{SRR} which is the percentage of the electricity demand energy that could be replaced with solar energy (Eq. 1). The entire demand energy is given by E_{demand}, the energy from the AC grid is E_{AC}, and the energy corresponding to the battery state of charge (SOC) difference between start and end of simulation is $E_{\Delta SOC}$. Another indicator is the solar operation ratio (SOR) R_{SOR}. This ratio indicates the actual solar generation energy E_{solar} compared to the maximum available energy $E_{unlimited\_solar}$ that could be obtained (Eq. 2). Note that this indicator is inversely proportional to the introduction cost per unit power of solar power generation system:

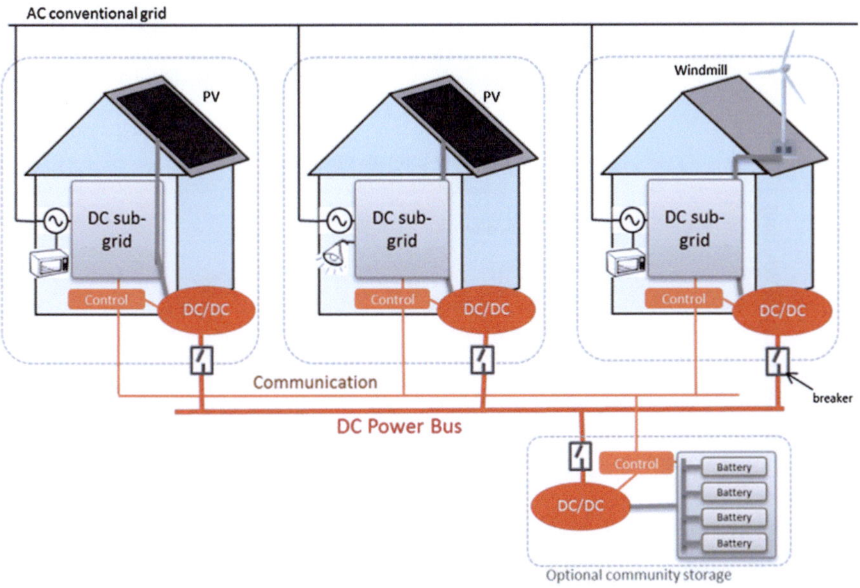

Fig. 2 General system layout for OES

$$R_{\mathrm{SRR}} = \frac{E_{\mathrm{demand}} - E_{\mathrm{AC}} - E_{\Delta\mathrm{SOC}}}{E_{\mathrm{demand}}} \tag{1}$$

$$R_{\mathrm{SOR}} = \frac{E_{\mathrm{solar}}}{E_{\mathrm{unlimited\_solar}}} \tag{2}$$

Solar energy is exposed to significant generation fluctuations that vary greatly throughout the day but also depend on weather and season. On the other hand, the electricity demand also experiences fluctuations according to human activity which usually presents a bottom demand for bedtime hours and peak demands in morning and evening hours. In addition, demand patterns also vary according to the season, depending on the time and day of the week, as well as across households.

For the above reasons, it is necessary to consider the deviation of household demand pattern and solar power generation pattern when examining the energy balance of a residential community.

In a preliminary study, we examined the annual energy balance between 20 houses' demand data (6 kVA contract performance data, Kyushu, Japan) and the solar power generation expectation from NEDO's sunlight database (http://app7.infoc.nedo.go.jp/index.html, 3 kWp, Naha, Okinawa, Japan). As shown in Fig. 3, the demand-response requirements vary not only hourly throughout a day but also monthly throughout 1 year.

In this case, PV system capacity is chosen so that the annual solar generation capacity is approximately 120 % of the amount annual demand with medium sunlight year of the database.

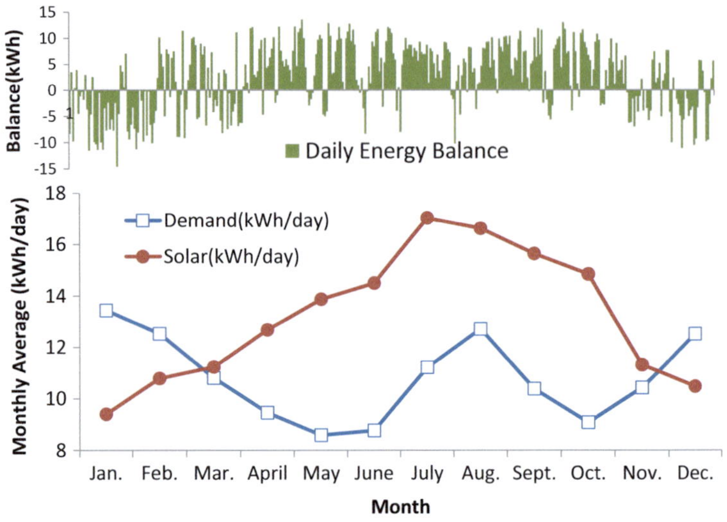

Fig. 3 Demand and solar energy balance for a year

There are two approaches to further improve the utilization of the solar PV: one is to connect each house to the power grid, which is a method for averaging the entire supply-demand balance to avoid full charging and overdischarging of the battery. This approach improves both SRR and SOR. Another approach consists in complete renewable energy system that uses hydrogen and fuel cell. If we assume that it is possible to supply hydrogen to compensate for the gap between the renewable energy and power demand, the micro grid including fuel cell and solar PV can be completely independent from the conventional AC power grid.

3 The Open Energy System (OES) Concept

The OES concept can be seen as an autonomous distributed DC microgrid system. Nanogrids with independent solar panels and secondary batteries in each home are the basic components. Nanogrids are then interconnected through a DCDC converter that serves as an interface to the shared DC bus. OES subsystems are aimed at an easy deployment in homes as well as to be able to continue operation even during utility grid outages or when the subsystem is disconnected from the DC grid. Note that in the future, an auxiliary power supply such as a fuel cell connected to the DC grid could serve as emergency supply during bad weather or disaster. However, in this paper, we only consider the conventional AC grid as power supply when solar energy is not sufficient.

Many current approaches focus on reverse power flow to the AC grid (feeding-in), but this adds a burden to the AC grid resulting into dropping feed-in tariffs or refusal of buying the solar energy by the utilities [5]. Thus, for OES, we do not consider feeding-in to the AC grid and instead provide a distributed control mechanism that allows power exchange within the OES community. We consider only the DC grid in order to achieve the islanding operation as well as full autonomous operation during AC power failure.

4 Feasibility Study on the Real-Time Simulator

We made real-time, multi-domain simulations in MATLAB/Simulink/SimScape. Based on EV technologies, we build a physical model of four subsystems and the interconnections with different kinds of architectures and exchange strategies from conventional AC power grid system presented in JSAE Annual Congress [6]. In this paper, we present a further practical model consisting of 20 subsystems (Fig. 4). Parameters for the model are shown in Table 1.

4.1 Power Exchange Control Method

At first, we examined the impact of the number of houses in an OES community on the solar replacement ratio and consequently the effectiveness of the power exchange between houses. As shown in Fig. 5, the SRR increases as the microgrid size increases. Thanks to the power exchange, demand and supply can be better managed compared to stand-alone houses and thus can help reduce peak demands and enhance solar utilization efficiency. We observe that the SRR stagnates to around 80 % starting from ten houses which represents the minimum number of nanogrids that should be connected for the power exchange to reach its potential to absorb local demand fluctuations.

To obtain these results, we used a model developed in our previous study [8] which is based on a power interchange system for leveling battery levels that uses the DC grid intermittently. This system proved to be an effective strategy for charging the secondary battery [7]. Real demand data and solar generation power estimated from a solar irradiation database are used to verify the power interchange effectiveness throughout 1 year. We compared the three systems. The first system is a conventional PV system without batteries that uses the AC grid for matching the demand, meaning sunlight is only used directly (direct PV). The second system includes batteries in each house, thereby constituting in-house system that could be seen as stand-alone, and one-house DC nanogrid where a PV charger and an internal controller regulate the charging/discharging of the battery (stand-alone nanogrids). The third system adds the power interchange between the nanogrid systems (OES approach). Exchange conditions are shown in Table 1. All three types of system assume an installed solar power system of 3 kWp in each of the 20 households. Each household is subscribed to a 6 kVA contract with an average daily demand of 10 kWh.

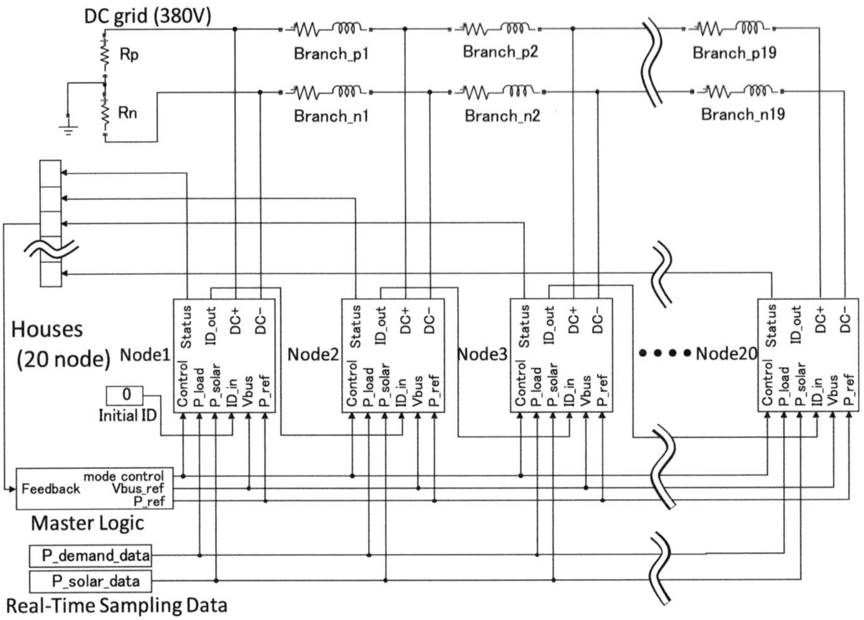

Fig. 4 Physical simulation model sample of the OES

Table 1 Power exchange simulation parameters

Item	Condition	Value
DC power cable (house to house)	8 $mm^2 \times 100$ m	2.3 Ω/km
Power exchange logic (SOC leveling)	ON: $SOC > 90$ % or $SOC < 10$ % or $\lvert SOC\text{-}SOC_{system}\rvert > 25$ %	Discharge 0–1 kW
		($SOC_{system} < SOC$)
		Charge 0–1 kW
	OFF: $\lvert SOC\text{-}SOC_{system}\rvert < 1$ %	($SOC_{system} > SOC$)
DCDC converter loss/efficiency	Min (sleep)	5 W
	Min (operation)	70 W
	Peak efficiency	95 %
Solar PV system	3 kW (peak) (per house)	Average 13.2 kWh/day
Demand electricity	6 kVA (max) (per house)	Average 10.8 kWh/day

For the direct PV system, the SRR for each month is about up to 40 % (Fig. 6) and approximately 35 % of the electricity demand can be replaced by solar energy throughout the year (Table 2). In system 2, a portion of the night power can be provided by the solar energy stored in the battery which increases the SRR to about

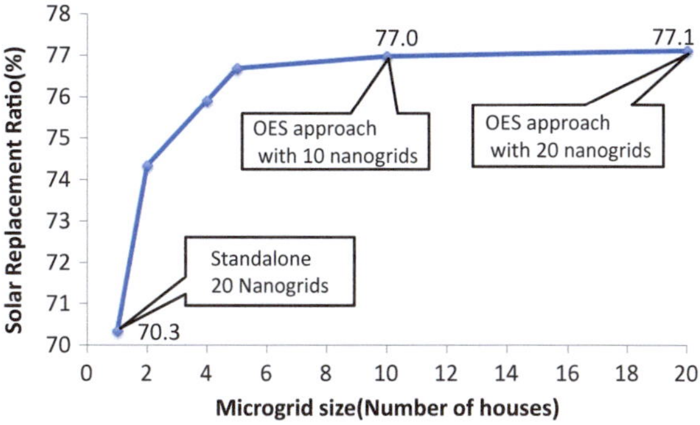

Fig. 5 SRR in function of OES cluster size

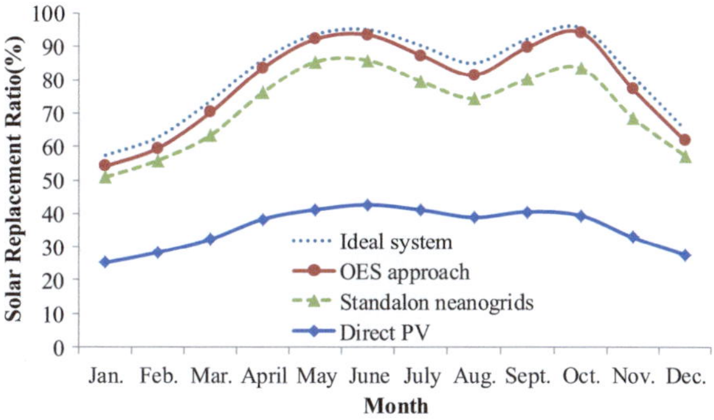

Fig. 6 System-type comparison using monthly SRR

Table 2 Simulation summary (1 year)

System type	Solar PV	Battery	DC power exchange	SOR	SRR
Direct PV	3 kWp	–	–	28.7 %	35.0 %
Stand-alone nanogrids	3 kWp	6 kWh	–	57.9 %	70.3 %
OES approach	3 kWp	6 kWh	Yes	66.7 %	77.1 %

70 % (Table 2). The power interchange used in the third system (OES) could further increase the SRR to 77 %.

Moreover, an increase of the PV system's capacity to 6 kWp results in about 86 % of the demand energy to be replaced by solar energy throughout a year (Fig. 7). But this means more solar panels need to be installed (higher initial costs) and hence the SOR is much lower (Fig. 7).

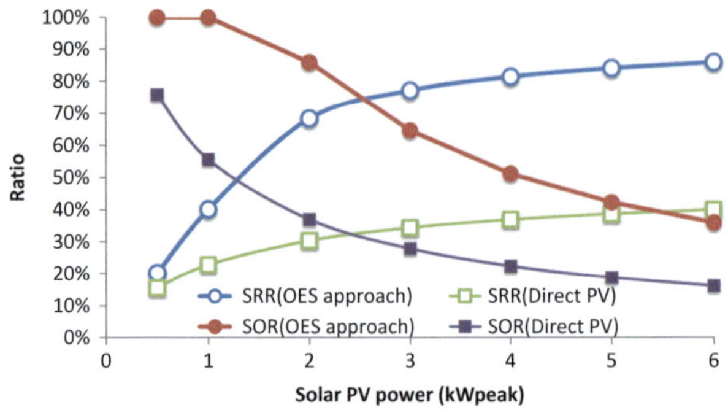

Fig. 7 Solar replacement ratio and solar operation ratio in function of the solar generation capacity

Table 3 Fuel cell system parameters

Item	Condition	Value
Rated power output	Continuous/NET	50 kW
System efficiency (LHV241.8 kJ/mol)	2.5 kW	60 %
	50 kW	50 %
DCDC converter loss/efficiency	2.5–50 kW	95 %
Power distribution and exchange logic	Discharge ON: *SOC* >70 %	Discharge 0–2.5 kW
	OFF: *SOC* <30 %	
	Charge ON: *SOC* <2 %	Charge 0–2.5 kW
	OFF: *SOC* >40 %	

5 Fully Renewable Power System

Maximization of the solar power generation amount by the SOC leveling has been discussed in the previous chapter. In this chapter, we consider a DC power system that uses a fuel cell as auxiliary power supply (APS) in addition to the solar panels and thus provides 100 % renewable energy. In recent years, the fuel cell system could operate even though under −30 °C environment [11], so we consider the DC microgrid system with the fuel cell system. Note that in this paper we do not discuss the sustainability and technical challenges linked to the production and storage of hydrogen.

For our simulations, we assume the fuel cell system parameters and other system performance as shown in Table 3. While the system prerequisites are similar in the previous section, the fuel cell is connected to the DC grid instead of the AC grid, so that it can take the role to generate energy from stored hydrogen to supply homes during power shortages (Fig. 8).

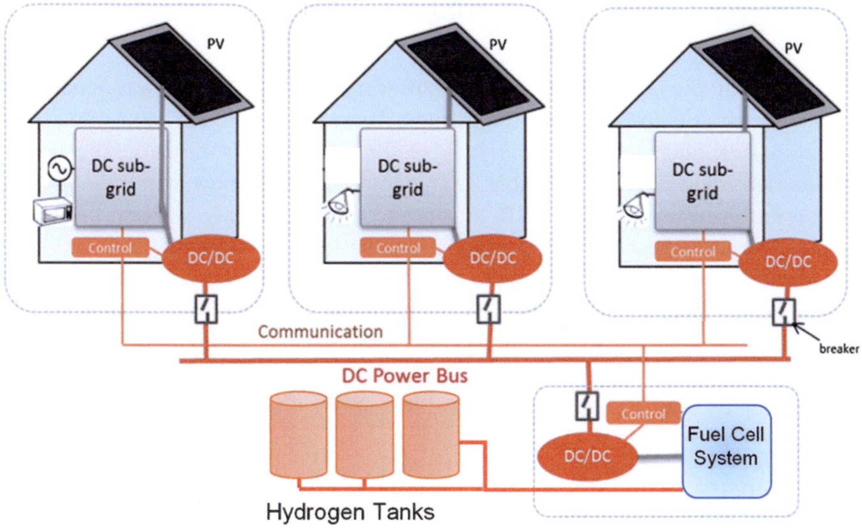

Fig. 8 A fully renewable power system with fuel cell

5.1 FC System Configuration

The considered fuel cell system is a polymer electrolyte fuel cell (PEFC) system similar to the ones mounted in commercially available fuel cell vehicles.

Because of the losses due to air pump load and cross leak hydrogen through the polymer membrane in the idling state of the FC system, the system should be stopped with hydrogen gas purging and system shutdown to avoid such losses. In addition, it is desirable to suppress the counts of start and stop cycle to avoid the degradation of catalyst on the electrode of fuel cells [12].

Therefore, since during evenings and nighttime the SOC decreases because of the absence of solar energy, the fuel cell should be operated continuously to ensure power supply and avoid start and stop operation of the fuel cell system.

Similarly, to above, for the demand data of nodes in the system, we used 20 household data with 6 kVA contracts, and the overall system actual peak demand was about 40 kW or less. Therefore, we considered the fuel cell system of 50 kW shown in Table 3 for this case with 25 % margin. This is approximately half the output of a commercial fuel cell system for vehicle [13].

The fuelcell power generation system's efficiency is 50 % (LHV) at the rated output; the maximum efficiency is 60 % (LHV) at minimum output. The fuelcell system is connected via the DCDC converter with 95 % efficiency to the DC grid, it can provide the power to nodes through the DC grid.

5.2 Power Distribution Algorithm

In this system, the batteries are used for buffering of the fuel cell power generation so that its operation period is not fragmented. Therefore, use of the battery as a solar power buffer, such as in the previous section, is partially limited. The power distribution logic is shown in Table 3. This logic transmits power from nodes that have a recent history of SOC 70 % or more to nodes that have a recent history of empty. The shortage of power is provided by the fuel cell system.

Considering further optimization, the SOC leveling and FC power buffering should be integrated. In addition each node should stop charging from FC system when the SOC reaches to the enough level to keep its power until rising solar energy. The SOC level should be estimated from past demand history data to avoid interruption of solar PV operation caused by battery overflow.

5.3 Simulation

The conditions and the total results for the solar PV – fuel cell hybrid power system simulation – are shown in Table 4. The FC output energy and hydrogen consumption throughout a year are shown in Fig. 9. The FC power peaks and averages are shown in Fig. 10. The number of FC system start/stop cycle and the time-based operation ratio are shown in Fig. 11.

The number of starting and stopping times is on average 2.2 times/day. We observed a power generation peak of up to 37.5 kW, while the average power generation of the fuel cell is about 10 kW. Thus, there is room for system downsizing.

We show that by supplying approximately 580 kg hydrogen per year, we can provide 100 % of the demand energy without using the AC grid. Note that the hydrogen consumption is highly seasonal with peaks of up to 134 kg caused by heating or cooling during winter or summer, respectively. During spring and autumn, the hydrogen consumption can be as low as 10 kg a month.

Thus, we show with this study how to provide 100 % of the power demand using individual solar power generation systems in each home with 3 kWp and 6 kWh of battery. For this we use a DC power bus line that is used both for power exchange between houses and for providing additional power from a fuel cell that needs to be supplied with several hundred kg of hydrogen a year (in the case of 20 houses). If approximately 2.5 (for spring) or ~35 kg (for winter) of hydrogen is always stored, this system can ensure 1 week power supply for 20 houses even in the event of a disaster.

Table 4 Simulation results of OES with fuel cell (1 year)

Number of nodes	Max load	Solar PV	Battery	Fuel cell	SRR	Hydrogen use
20	6 kW	3kWp	6kWh	50kWp	69.3 %	580 kg

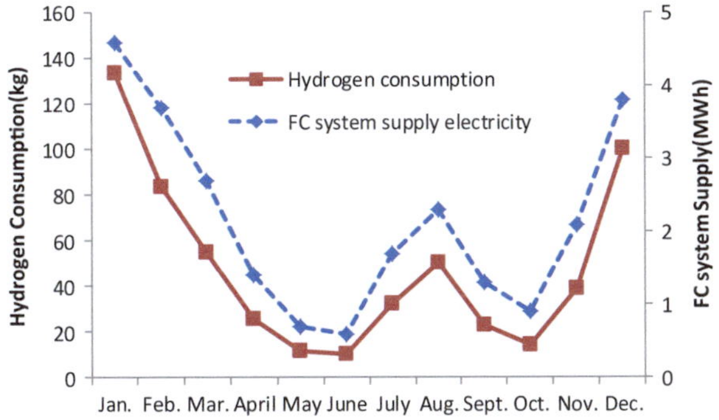

Fig. 9 Hydrogen and electricity consumption

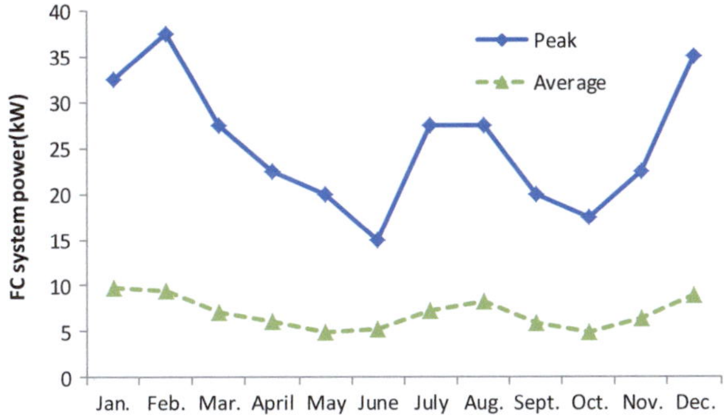

Fig. 10 FC system power output

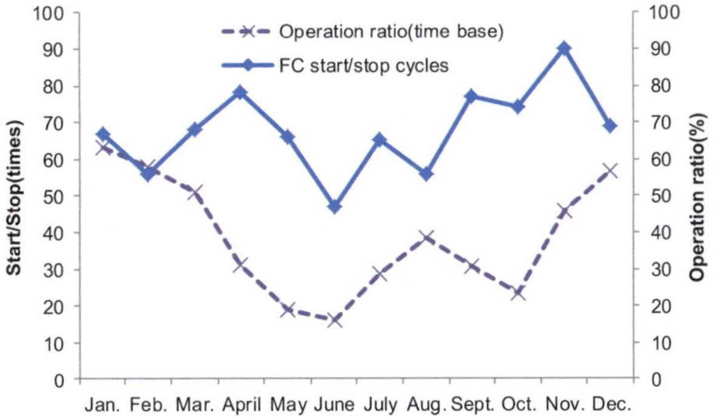

Fig. 11 FC system operation statistics

6 Feasibility Study in the Real World

6.1 Autonomous Distributed Control System

To put the OES approach in practice, we came up with layered software that is able to fully control the power flow between the DC nanogrids and thus allows energy exchange between houses over a DC power bus. The procedure to regulate the bus voltage and current as well as the lower layer software is described in [7]. The highest software layer is responsible for performing the tasks required for the OES to handle power exchanges fully autonomously (without human intervention). The difference with most other systems is that this software is not centralized on one special logical unit, but entirely distributed over the subsystems that communicate with each other over message exchanges. To achieve this, we got inspired by one of the most widely adapted decentralized architectures: pervasive peer-to-peer networks as described in Tannenbaum's book on Distributed Systems [9]. Peer-to-peer systems are horizontally distributed meaning that all processes or systems are equal from a high-level perspective ensuring independence and reducing bottlenecks [9].

We adopted a multi-agent system (MAS) approach which has been shown to work well to control intelligent grid systems made of less-intelligent entities (local controllers) [10]. We split the software in three main parts that are treated by a software agent. A simplified flow chart is shown in Fig. 12.

6.2 Experimental Prototype System

Both hardware and software are being developed and tested on several full-scale prototypes: three-subsystem experimental prototype in the Sony Computer Science Laboratory in Tokyo, another three-subsystem prototype at the Okinawa Institute of Science and Technology (OIST), as well as our full-scale platform at OIST which consists of a community of 19 inhabited houses, all equipped with a DC nanogrid and OES microgrid system (Fig. 13).

We developed an SOC-based autonomous control system as described in the simulation and the previous paragraph. Nanogrids can be added and removed dynamically in a configuration-less plug and play-like way and request or respond to energy exchange request from other houses (third system – OES approach). In practice, the auxiliary DC power supply such as the fuel cell is not implemented. Improvements of the control system, exchange strategy and efficiency are still in progress.

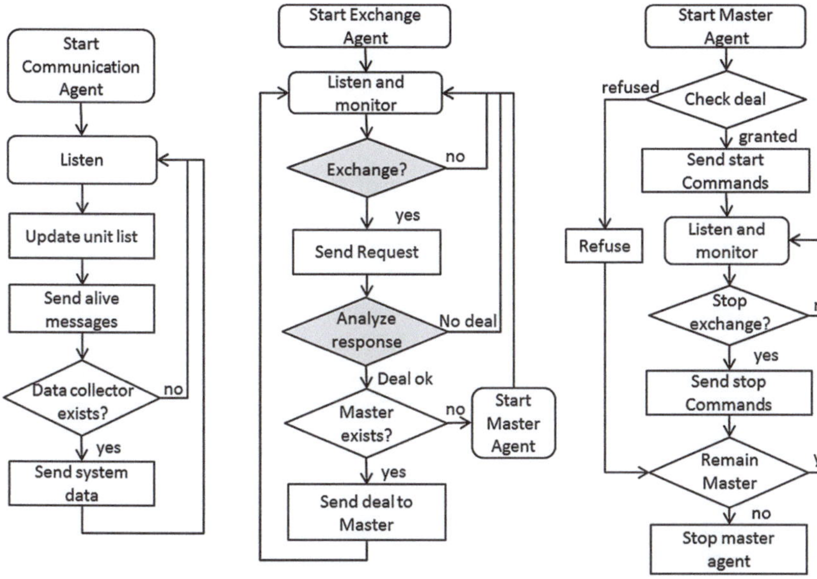

Fig. 12 Simplified flowchart of three main agents for autonomous power transfer

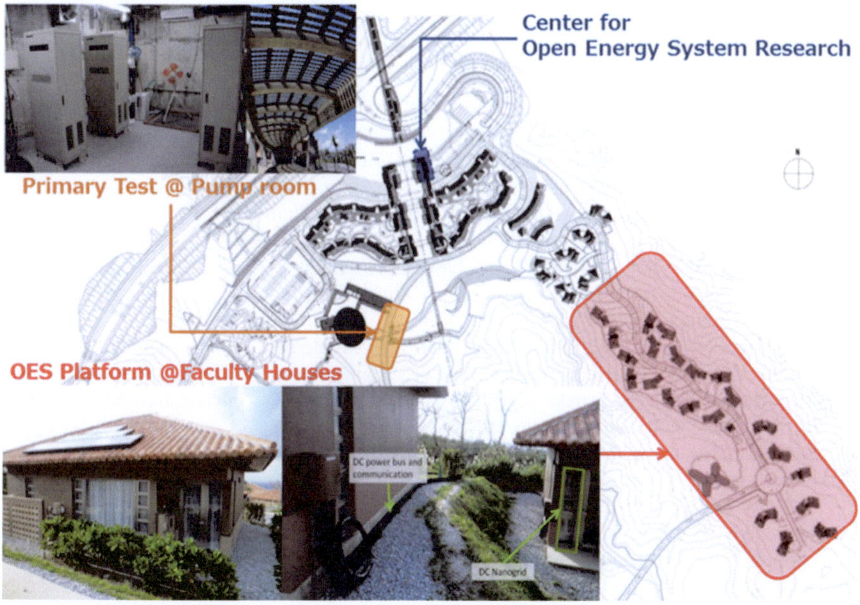

Fig. 13 OES platform at OIST: primary test site and full-scale platform at faculty houses

7 Summary

This ongoing research suggests a new type of DC-based distributed interconnection of DC nanogrids. In this paper, we propose the OES concept that applies to both in terms of hardware and software architecture and show the benefits on 20-node simulations using a physical model. We analyzed this kind of interconnected microgrid system using SRR and SOR and compared it to direct PV systems as well as stand-alone nanogrid systems. We analyzed the impact of the cluster size and the PV capacity throughout a year using real consumer demand data and irradiation data. Finally, we studied the complete renewable power system with fuel cell power supply system based on the OES concept. We further demonstrated the practical feasibility on a full-scale platform. Because of its modular open architecture, it can develop gradually, one subsystem at the time, thus reducing infrastructural investment.

Acknowledgments This research is supported by the "Subtropical and Island Energy Infrastructure Technology Research Subsidy Program" of the Okinawa Prefectural Government and carried out by the research consortium of Sony Computer Science Laboratories, Inc., Okisoukou Co. Ltd., Sony Business Operations Inc. and OIST.

References

1. Agency for Natural Resources and Energy (2014) Strategic energy plan. http://www.enecho.meti.go.jp/en/category/others/basic_plan/pdf/4th_strategic_energy_plan.pdf
2. Abe R, Taoka H, McQuilkin D (2011) Digital grid: communicative electrical grids of the future. IEEE Trans Smart Grid 2:399–410
3. Barnes M, Kondoh J, Asano H et al (2007) Real-world MicroGrids- an overview. In: IEEE international conference on system of systems engineering, pp 1–8
4. Tokoro M (2014) Sony CSL-OIST DC-based open energy system (DCOES). First international symposium on open energy systems, Okinawa Institute of Science and Technology Graduate University (OIST), 14–15 Feb 2014
5. Justo JJ, Mwasilu F, Lee J, Jung J-W (2013) AC-microgrids versus DC-microgrids with distributed energy resources: a review. Renew Sustain Energy Rev 24:387–405
6. Kitamura N, Werth A, Tanaka K (2014) The autonomous DC microgrid system based on electric vehicle technologies. Trans JSAE 45(6):1007–1012
7. Werth A, Kitamura N, Tanaka K (2014) Control scheme for distributed n-to-n power exchange between solar systems including batteries. In: Proceedings going green – CARE INNOVATION 2014, Vienna
8. Kitamura N, Werth A, Tanaka K (2014) The autonomous DC microgrid system for residential community, In: Proceedings going green – CARE INNOVATION 2014, Vienna
9. Tannenbaum AS (2007) Distributed systems, vol. 17. 2nd edn. Pearson Prentice Hall, Pearson Education Inc. Upper Saddle River, NJ 07458 U.S.
10. Dimeas A, Hatziargyriou N (2005) Operation of a multiagent system for microgrid control. IEEE Trans Power Syst 20:1447–1455
11. Kitamura N, Manabe K, Nonobe Y, Kizaki M (2010) Development of water content control system for fuel cell hybrid vehicles based on AC impedance, SAE technical paper, No.2010-01-1088

12. Yousfi-Steiner N, Mocotéguy P, Candusso D, Hissel D (2009) A review on polymer electrolyte membrane fuel cell catalyst degradation and starvation issues: causes, consequences and diagnostic for mitigation. J Power Sources 194:130–145
13. TOYOTA MIRAI Website (2015). http://toyota.jp/mirai/

Part VIII
Sustainability Assessment and Indicators

Sustainability Indicators: Overview, Synthesis and Future Research Directions

Christoph Hollauer, Martin Zäpfel, Daniel Kammerl, Mayada Omer, and Udo Lindemann

Abstract Indicators form an important tool to assess product sustainability. However, an overview over this body of literature and its contents, specifically in regard to indicators for measuring product sustainability, does not exist. The approach applied in this paper is exclusively literature based. In a structured and methodological literature survey, the relevant literature containing various methods for sustainability assessment as well as existing frameworks containing sustainability indicators have been collected. The results of this paper are a list of sustainability impact categories and indicators, spanning the three relevant fields of sustainability. As such, this paper represents a review of existing product-related sustainability indicators.

Keywords Sustainability assessment • Product sustainability • Impact categories • Indicators

1 Motivation and Problem Statement

The consideration of product sustainability is of increasing importance during and after the product development process. This is especially true as products grow to be more complex, and technical systems become larger and more invasive in everyday life and existing economic, ecologic and social ecosystems. Companies are more and more integrating service aspects into their products, thus creating product-service systems or function-oriented business models. They are experiencing further external and internal pressure: For example, product sustainability is becoming an important factor for customer acceptance, and laws and regulations, e.g. regarding CO_2 emissions, are becoming increasingly strict. This results in a constantly rising need to describe and quantify the current degree of sustainability of such products and systems of growing complexity, which is necessary in order to allow the derivation of suitable strategies to increase sustainability of a system or certain parts of it.

C. Hollauer (✉) • M. Zäpfel • D. Kammerl • M. Omer • U. Lindemann
Institute of Product Development, Technische Universität München, München, Germany
e-mail: hollauer@pe.mw.tum.de

M. Matsumoto et al. (eds.), *Sustainability Through Innovation in Product Life Cycle Design*, EcoProduction, DOI 10.1007/978-981-10-0471-1_62

Sustainability measurement and assessment requires suitable sustainability indicators (SI) that cover the ecologic, economic and social dimensions of sustainability. However, no generally approved or standardised set of SI does exist, which might provide sufficient information for the assessment of products considering all three aspects of sustainability [1].

Furthermore, global instructions for enterprises in regard to sustainable behaviour and the selection and application of SI for a sustainability assessment on the product level are still missing [2]. This paper seeks to contribute to the development of a comprehensive set of sustainability indicators by reviewing the existing body of literature concerning sustainability indicators and criteria applicable during the design process in order to assess the sustainability performance of technical products. The paper shows the preliminary result of this research effort.

The structure of the paper is as follows: First, a short overview of the used terms is given in Sect. 2. In Sect. 3, an overview over the research approach is presented. Section 4 contains the results obtained as a consequence of the literature review. Based on these results and the conducted literature review, a research agenda is developed in Sect. 5, containing possible areas for future research.

2 Background

2.1 Definition of Sustainability

Sustainability is a process to ensure a future which is composed of society, institutions, resources and therefore the environment and is based upon specific goals and different ideas [3, 4]. More precisely, sustainability aims at balancing the contemporary needs with the available natural resources and social requirements. Further, the ecologic system has to be conserved as it supports the basis for human life [5]. Lastly, this needs to be achieved along with a strong economic output measured in monetary units and inter- as well as intragenerational equity, the latter of which is not easily measured by any unit [5]. Since these three objectives depend on each other, often involving trade-offs, some authors use the picture of a magic triangle [5, 6]. Because of the involved complexity, [7] developed a multidimensional model on which the following figure was set up (see Fig. 1).

Arrows symbolise the exchange of raw materials and human activities through the boundaries of the individual dimensions which are represented by dashed lines. The largest dimension involved is the ecologic dimension, followed by the social dimension, within which economic operations take place. Within the economic dimension, industrial processes lead to products which are assembled, used and at the end disposed and recycled. Measuring and designing sustainability involves, as described before, a holistic approach spanning the three fields of ecology, economic and social aspects.

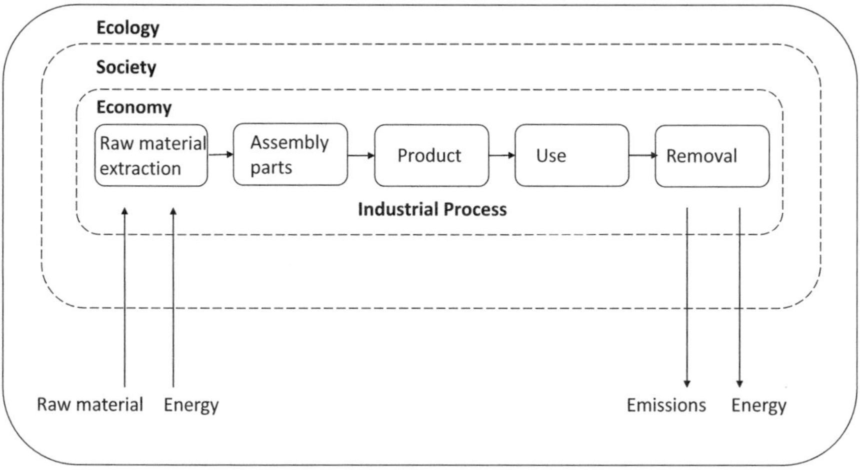

Fig. 1 System of sustainability

2.2 *Definition of Indicators*

An effective sustainability assessment and management requires directly measured and detailed data [8]. Modern information technologies can provide a huge amount of data, but it is time consuming and costly to convert the mass of data into useful information. Therefore, indicators are often used as management tools, which allow practitioners a quick and condensed overview by reducing the complexity of the collected data [9].

Depending on the author and his or her particular intention, indicators are named differently, for example, [10] describes 'performance indicators', [11] use the term 'key performance indicators' and [12] discusses 'core indicators'. To harmonise these concepts, and to be in accordance with [13] as well as [14], this paper concentrates on the general term sustainability indicator.

The information pyramid according to [7] shows the hierarchical structure of primary data, e.g. directly measured gas emissions up to quantifying parameters. Following the direction from raw data to indicators, the amount of information increases, while the amount of specific data decreases. Eventually, a small number of selected SI representing concentrated information in an index allows to decide on sustainability issues (see Fig. 2).

An indicator is always a kind of equivalent variable which cannot be measured directly but strongly depends on other measurements [15]. As an indicator indicates the trend and degree of fulfilment of a specific objective, it needs a reference value, which distinguishes it from a simple variable [16]. Only with such a value can an indicator be used within the scope of sustainable measurements as a SI [14].

In general, different kinds of indicators exist which are described in Table 1 [17–19]. Indicators simplify, measure, analyse and communicate complex information as they are found within sustainability issues. Hereby, trends can be assessed and

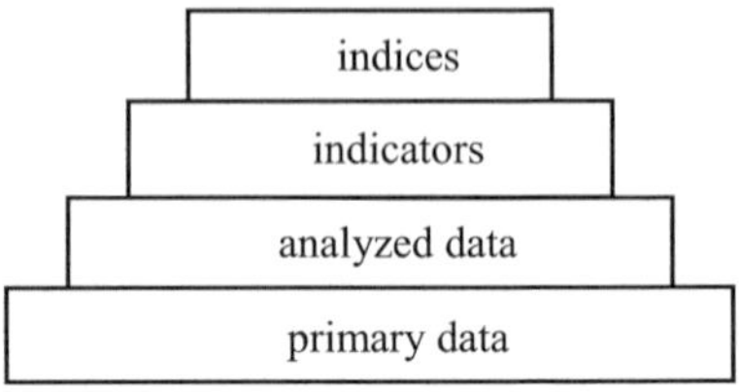

Fig. 2 Information pyramid

Table 1 Indicators

Indicator	Example
Absolute	Raw material in tons
Relative	Raw material related to the produced number of products
Indexed	Data which show the annual percentage change
Aggregated	Summary of data with similar units
Weighted	Rating the data with specific factors
Qualitative	Implemented service systems
Quantitative	kg CO_2 equivalents

areas requiring action can be identified. Because of these benefits, indicators are more and more appreciated and used in sustainable product development [14].

SI can be defined as absolute, relative, indexed, aggregated or weighted indicators in either a qualitative or a quantitative form. For example, an absolute SI could be the mass of raw materials in tons. In relative and therefore often used terms in sustainability measurements, the mass could be related to the number of products which can be created out of it [17].

2.3 *Life Cycle Analysis*

Life cycle analysis takes an entire lifetime into account, from raw material extraction, manufacturing process, use and reparation up to recycling or removal [18]. At best, a life cycle represents a closed loop according to the idea of cradle to cradle, in which recycled products provide the raw material for new products. This closed life cycle is favoured not only by the European Union's *action plan on sustainable production and consumption and sustainable industrial policy* as a benchmark for any life cycle assessment to ensure real sustainability [20]. A complete life cycle analysis is therefore important to avoid transferring problems to different stages of development. Furthermore, it prevents problem displacement to other geographic regions or future generations [18]. Life cycle analysis is often based on the effective use of SI from the three dimensions of sustainability [1, 6].

3 Approach

3.1 Literature Review

In this section, an overview of the research methodology applied in this paper is presented. A graphical illustration of the research methodology is depicted in Fig. 3.

The applied approach is exclusively literature based. In the literature review carried out, to identify SI, Scopus was used in the first step to find relevant papers in German and English. The first results were comprised of papers on SI development methodology, sustainability, sustainable, assessment methods, metrics, measurements, criteria, indicators and data. Similar terms were connected with OR links and the different groups were then connected with AND links (Fig. 4).

After further screening of the abstracts with a focus on a technical and product-related background, the literature review resulted in a total of 212 papers selected for further analysis.

In a next step, the found literature was reviewed and analysed which led to a second iterative literature review based on the first findings. Indicators and impact categories were extracted from the references. The extracted indicators were then collected in a structured list and categorised by economic, ecologic and social dimensions.

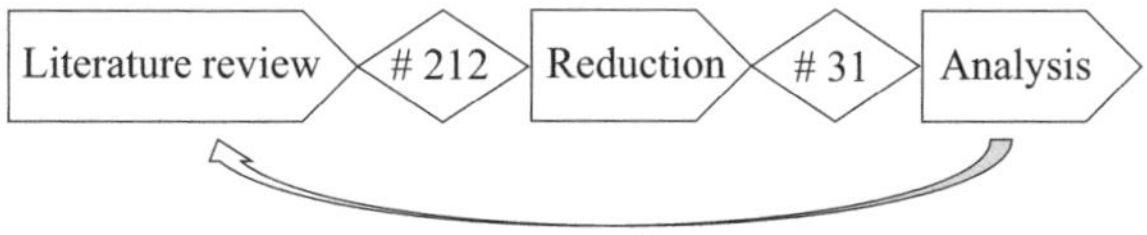

Fig. 3 Research methodology

	aspects			
synonyms (OR)	Bewertung	Nachhaltigkeit	Produktentwicklung	Größe
	Kriterien	Zukunftsfähigkeit	Technik	Anzeige
	Messung	sustainability	technology	Index
	rating	sustainable	„product design“	indi*ator
	criteria		„product engineering“	data
	measur*		produ*t	metri*
	assess*			
	aspect 1	aspect 2	aspect 3	aspect 4
	AND			

Fig. 4 Research strategy

4 Results

4.1 Methods and Frameworks

Based on the literature review, existing methods and frameworks to assess sustainability have been evaluated as well. This is to support the extracted indicators by describing the generation and the appliance of SI.

While frameworks depict a specific structure, methods present a systematic approach. Methods are prescriptive and thereby offer a particular way to meet the objectives [21]. In theory, these terms should be distinguished, but it has been observed that both terms are used inconsistently. For example, [13] as well [22] talk about LCA as a framework, whereas [20] state that LCA is a method defined by ISO EN 14044. In an essay by [23], the terms are used equivalently. Thus, until a general accepted use of both terms is defined, this work concentrates on the term method to assess sustainability.

4.2 Methods to Assess Sustainability

A number of different methods can be used to assess sustainability [14]. Most authors concentrate on one of the three dimensions of sustainability, either on ecologic, economic or social aspects. Further methods, such as the integrating approach of life cycle sustainability assessment (LCSA), exist [24]; however, most authors refer to one of the three methods described below. Hence, this chapter gives a short overview of three methods that are often mentioned in the literature [4, 13, 20], utilise SI and are based on the standard ISO 14040:2006.

4.2.1 Life Cycle Assessment

From a historic point of view, the ecologic dimension and therefore the ecologic life cycle assessment (LCA) are the most important for sustainability assessment [5]. The method shows the impact of a product or service on the environment based on measurable units. This enables the optimisation of a system in regard to ecologic aspects and can lead to decreasing costs, lightweight solutions or extended lifetimes [3]. Furthermore, LCA as it is described by DIN EN ISO 14040:2006-10 forms a basis for the economic and social assessment of sustainability, as it is the only internationally accepted and standardised method for measuring sustainability [20]. The standard for LCA is structured as follows: At first the objective and scope of an assessment are defined, and an inventory analysis is carried out in order to identify all resources used and emissions released. Based on the results of the inventory analysis, the impact on the environment is assessed, the results are categorised and finally the outcome is interpreted and evaluated (ISO 14040,

2006). This procedure is widely accepted and applied. However, no given list of SI exists as the SI need to be composed according to the requirements of the analysis [4].

4.2.2 Life Cycle Costing

Life cycle costing (LCC) is the oldest of the three considered methods. The view on the direct monetary assessment of products during the entire life cycle has changed towards the additional measurement of external and indirect costs such as organisational aspects [13].

The scope of a LCC is to analyse and minimise the costs of products or services. Many methods with specific focuses are used [4], and enterprises even have developed their own guidelines to assess economic sustainability, but no international standardised norm currently exists [18].

Hence, LCC is based to a great extent on the guidelines of ISO 14040 and structured into four stages: Similarly to a LCA, the scope of the assessment is first defined. Similarly to an inventory analysis, all the costs with regard to the measurement are summed up in a next step. Without a specific impact assessment, the costs are categorised and aggregated. This presents the basis for further interpretation of the results [20]. The knowledge of cost drivers and future changes of costs can optimise product development [4].

4.2.3 Social Life Cycle Assessment

Social life cycle assessment (SLCA) considers social aspects during the product life cycle [25]. Compared to LCA and LCC, the four stages of sustainability assessment (definition of scope, inventory analysis, impact categories and interpretation) should be applied similarly [18]. However, some differences exist, e.g. regional relevance of collecting data is more important for SLCA than for LCA [23]. Furthermore, along the entire life cycle, different stakeholders actively influence the products and their social impacts which are difficult to measure. Thus, in contrast to LCA, SLCA uses qualitative indicators in particular, which require more effort and defined guidelines to gain comparable results [23]. On one hand, these guidelines consist of the declaration of human rights and the minimum requirements by the International Labour Organisation (ILO) as a normative basis. On the other hand, they consist of social aspects that can be directly controlled by an enterprise or are affected by the assessed product system [23].

SLCA is still developing and therefore no standards exist and only a limited number of publications are available [4, 23]. Yet, the number of publications has recently been increasing [24] and institutions and enterprises are developing specific approaches [4].

4.3 Synthesised Indicator List

The results of this paper yield a list of sustainability impact categories and indicators, spanning the three dimensions of sustainability. As such, this paper represents a review of existing product-related sustainability indicators, which can be applied during product development for the assessment of product sustainability.

4.3.1 Ecological Indicators

Based on the reviewed literature, ecological SI are mentioned more frequently compared to economic and social SI. Although during the last 5 years, this focus has changed to include social aspects, too. [26] and [4] support the fact that ecologic SI are still discussed more frequently in literature than other aspects of sustainability, as ecologic SI are often measured directly by an inventory of resource use and emissions.

According to [13, 27], Fig. 5 shows an impression of ecologic SI, which are derived from the results of an inventory which lists all resources used and emissions released along an entire life cycle of a product.

The results obtained of a life cycle inventory (LCI) analysis represent factors which directly affect specific criteria, e.g. acidification (midpoint), which itself has a negative impact on the ecosystem. The ecosystem represents one of the so called end points, which provides the required direction of an SI. This direction or trend signals, whether the value of a SI should rise or drop in order to obtain a higher degree of sustainability.

The scope of ecologic SI according to [17] is to compare the ecologic performance of products or enterprises within a certain period of time and to optimise them with regard to ecologic objectives as well as costs. Furthermore, ecologic SI are used in sustainability reports as well as to communicate important facts, e.g. in management systems.

From the literature review, we found 51 unique high-level impact categories for ecologic indicators, such as toxicity, radiation, and raw material, along with a number of indicators and key figures assigned to each of these impact categories.

4.3.2 Economic Indicators

Economic SI represent the monetary flux and the economic impacts of products. Whereas the financial situation of an enterprise is often presented very well in financial reports, the effects of entrepreneurial activities on society are less extensively described [10]. Although these quantitative economic SI are not as meaningful as qualitative SI within sustainability assessment, it is simpler to measure the quantitative SI [11]. In an economic LCI, for example, energy, material or labour costs are also taken into account as shown in Fig. 6 [13]. By relating these aspects,

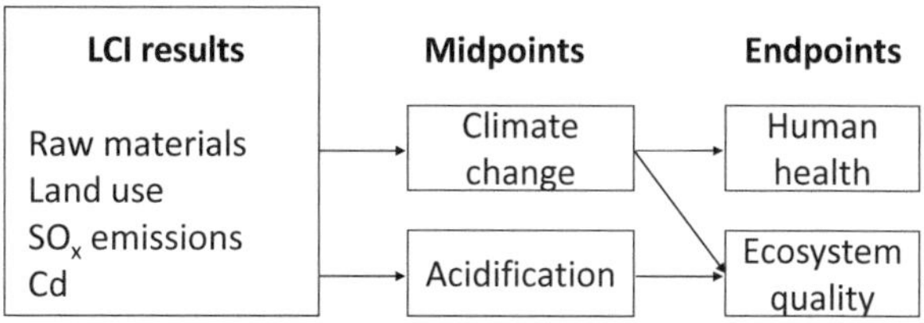

Fig. 5 LCA categories

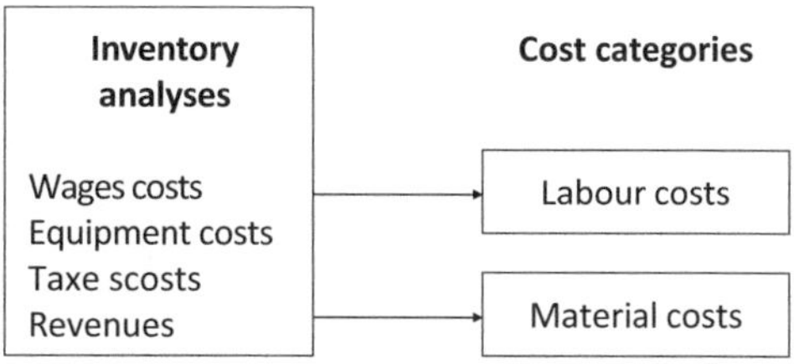

Fig. 6 LCC categories

e.g. to the number of products manufactured, economic SI can be developed (see Chap. 2.2, [17]).

In the economic dimension of sustainability, connections to the environment are often indicated by stakeholder relationships. Stakeholders, for example, are employees, the local community, the global society, costumers and supply chain actors [18]. In addition, [10] names non-governmental organisations, shareholders and investors as additional independent groups of stakeholders, following a multi-stakeholder approach.

Considering the stakeholder relationships, it has long been assumed that resources are always sold to enterprises as long as they could pay a fair price. Nowadays, even CO_2 emissions are monetary appraisable, since CO_2 certificates are traded. However, not every single dependency on nature can be expressed monetarily, and the capability of ecologic systems, e.g. to absorb emissions, is limited. This example shows how complex cause-and-effect relationships could be, making it hardly possible to describe even linear systems in detail. Hence, economic SI can only give an estimation based on experience, since it is impossible within the socio-economic area to predict human behaviour [28].

In our review, we found 110 economic indicators, so far 63 of which have been classified as quantitative and 3 as qualitative. The indicators are structured in 11 main categories and 79 further subcategories.

4.3.3 Social Indicators

The social dimension of sustainability assessment recently has gained an enormous increase of interest [25].

There are complementary kinds of classification methods for social SI. First of all, classifications based on a stakeholder approach, and following the classification established on the basis of impact categories, are widely used [18]. Moreover, the

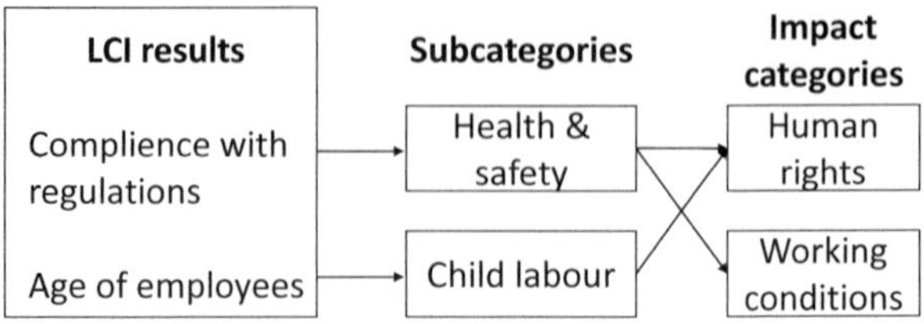

Fig. 7 SLCA categories

well-being approach by Prescott-Allen exists, which additionally takes the ecologic dimension into account [1]. [13] describes impact categories pertaining to the social dimension of sustainability, based on stakeholder categories: human rights, working conditions, health and safety, cultural heritage, governance and socio-economic repercussion (Fig. 7). Results of a social LCI are assigned to further subcategories, for example, child labour which itself is assigned to the impact category of human rights.

The inconsistent use of terms concerning social aspects makes it difficult to classify social SI into categories. For example, [29] clearly describes child labour as an indicator, whereas [30] classify child labour as a key figure. Yet again, [31] refers to the GRI guidelines and classifies child labour as an 'aspect' within the category of human rights. Consequently, there exists no generally applicable definition so that there might be even more options of classification possible.

This may affect the problems enterprises still face when dealing with SI. A huge variety of SI leads to a complex system for sustainability assessment, but enterprises are often interested in a specific selection of SI, as not every single SI is equally applicable throughout the different sectors of industry. Thus, specific guidelines and the consistent definition of terms could increase the application of sustainability assessment [32].

In our review, we found 6 high-level categories, further divided into 46 subcategories.

4.4 Comparison with DTU List

The described list of SI (see Sect. 4.2 and Fig. 8) represents an approach based on the reviewed literature. In this section, the defined system is compared to a reference system that also attempts to collect and structure SI, in this case called environmental performance indicators. As a reference system, the set of SI developed at Denmark Technical University (DTU) was chosen.

Similar to the described table in this work, the table of DTU starts with the name, a symbol and a short characterisation of the SI. The following classification of SI consists of assessment methods, related unities, the desired trend and further information. Moreover the SI are related to the specific phases of the life cycle and to ecologic aspects, e.g. energy, emission or resources. In addition, it is stated whether the considered SI is a qualitative or a quantitative indicator. Finally, the

category	key figure	indicator	Desired Trend	authors	further authors	ecologic	economic	social	quantitative	qualitative
global warming	CO_2, CH_4 emsissions	kg CO_2-equivalents	↓	X		X			X	
product quality	number of complaints	number of complaints per product	↓		X		X			
product service	-	operating service system			X			X		X
...										

Fig. 8 Example of the structure

reviewed literature containing this SI is named as well as the number of authors citing the SI.

In contrast, the work presented in this paper concentrates on the categorisation of SI into key figures and indicators in a holistic way. The focus was to specifically include ecologic, economic and social SI equally, whereas the system by DTU is mainly based on product-related, environmental indicators (e.g. waste water, material or gaseous emissions). Hence, the here presented system of SI contributes to a broad sustainability assessment with regard to all three sustainability dimensions.

Most of the reviewed literature is composed by different authors compared to the findings by DTU. Altogether, around 230 different categories, key figures and indicators are comprised in the here presented list, distributed over the three dimensions of sustainability. The indicators gathered in the DTU list, numbering 234 entries, are more low level and thus more specific, with concrete strategies of gathering the necessary data attached to them. Also, a number of them are represented as absolute as well as relative indicators, e.g. the number of recycled parts and the number of recycled parts in relation to the overall number of parts of a product. The list described in this paper, however, had the intention of structuring the general field of sustainability indicators on a higher level. Thus it would be an important next step to bring both approaches together and generate a database of both high-level categories and low-level indicators, connected by well-defined relationships.

5 Agenda for Further Research

Based on the results presented in this paper, a number of current research gaps have become obvious. In order to present a structured overview about these identified research gaps, they have been grouped into research fields which are shortly described in the following subsections.

5.1 Understanding Sustainable Products and Product Sustainability

Firstly, in order to allow for a more specific discussion about sustainable products, the often unwieldy system property 'sustainability' needs to be broken down into more manageable system properties. For this, the relationships between sustainability (ecologic, economic as well as social) and other system properties, such as robustness, flexibility and safety, need to be further clarified, as already proposed by [33].

5.2 Integrated Product Sustainability Measurement and Design

As stated in section 4, a vast number of sustainability indicators concerning ecologic, economic and social aspects exist in literature. However, we found a lack of support in literature for actually selecting and applying indicators during the product development process in practice. Further, a consistent, expandable system of indicators is needed that supports a targeted top-down approach for the selection of SI as well as an aggregation from low-level data to high-level indicators and indices. This is especially important in the context of the development of complex products and product-service systems, where the choice of optimal sustainability indicators is often not obvious. Choosing the right indicators per product is often based on expert opinions, creating the need to integrate specific experts in the product development process, a fact supported by an interview with a life cycle analysis expert conducted after the literature review has been concluded.

5.3 Data Management and Analytics

Especially during the early phases of product development, such as concept design, developers and engineers are faced with uncertainties in regard to the future properties and the behaviour of the system under development. Such uncertainties are often met by making assumptions or generating necessary data, e.g. through experiments, simulations or the acquisition of commercial databases containing environmental data. However, the life cycle assessment of complex products is dependent on the availability of accurate data, for the collection of which modern products often do contain a considerable number of sensors. To become useful, the raw data then needs to be processed in order to become useful for future development projects to provide a baseline of product sustainability. This allows us to plan for corrective actions to increase future product sustainability. Otherwise, only a limited set of SI can be applied, which is structured as simple as possible and based on easily accessible data, because of the manual effort involved. Hence, we see a need for further sophisticated methods to generate and collect, manage and analyse sustainability-related data, in order to automate portions of the sustainability analysis, e.g. by integrating life cycle assessment and product life cycle management solutions (cp. [34, 35]).

5.4 Model-Based Sustainability Assessment and Design

In order to further the integration of sustainability assessment into design activities, the interface between sustainability and model-based system engineering and its approaches, such as SysML, needs to be further explored. This is derived from the fact that considering the reviewed sustainability indicators, sustainability design is a discipline-spanning effort that needs to be considered on a holistic system level early on in the development process. Since, for example, SysML offers the possibility to integrate parameters and their calculation into models of complex systems, it should be possible to also include the calculation of sustainability indicators in order to enable an efficient, design-accompanying sustainability assessment. Hence, various design alternatives can be assessed during design activities. A further benefit would be the ability to save and reuse the models and their contained information for later development projects as well as for preparing more extensive and accurate life cycle analyses later in the design process.

6 Summary

In this paper, a literature review to capture the currently available sustainability indicators has been presented. The research effort was focussed on generating an overview over existing impact categories and the indicators with which these categories can be further detailed and quantified. For this, existing frameworks and methods – standardised as well as others – have been analysed, and gathered into a single list, which allows a good overview. By conducting a broad review, all three sustainability dimensions have been included. It has been found that a large number of diverse indicators exists for all three dimensions of sustainability. According to our findings, the economic dimension is leading, followed by the ecological and social dimension. However, the indicators from the economic dimension have not been fully condensed yet. Additionally, we found that there are different approaches for structuring indicators and categories, which might make the application far practitioners difficult. Based on the current state of research in sustainability measurement and assessment, a proposal for a research agenda, spanning four fields of research, has been given, which can potentially contribute to the application of sustainability assessment in industry.

However, one of the weaknesses of the SI list presented in this paper is the inherent dependencies between categories and indicators, which cannot be mapped efficiently in a simple list but require a more sophisticated approach such as a database. The literature review itself, although it has been extensive, has not been complete, as the comparison with the DTU list shows. Furthermore, since we focussed on publicly accessible data regarding sustainability analysis and indicators, commercial solutions have not been regarded at this point. The work presented in this paper therefore is of preliminary status and needs to be further refined and extended.

Acknowledgements We thank the German Research Foundation (Deutsche Forschungsgemeinschaft – DFG) for funding this project as part of the collaborative research centre 'Sonderforschungsbereich 768 – Managing cycles in innovation processes – Integrated development of product-service systems based on technical products.'

References

1. Böhringer C, Jochem PE (2007) Measuring the immeasurable—a survey of sustainability indices. Ecol Econ 63(1):1–8
2. Krajnc D, Glavič P (2005) A model for integrated assessment of sustainable development. Resour Conserv Recycl 43(2):189–208
3. Munier N (2005) Introduction to sustainability: road to a better future. Springer, Dordrecht
4. Herrmann C (2010) Ganzheitliches Life Cycle Management: Nachhaltigkeit und Lebenszyklusorientierung in Unternehmen. VDI-Buch/Springer, Berlin
5. Hauff M v, Kleine A (2009) Nachhaltige Entwicklung: Grundlagen und Umsetzung. Oldenbourg, München
6. Grunwald A, Kopfmüller J (2006) Nachhaltigkeit. Campus Einführungen, Frankfurt/Main, Campus
7. Hammond A, Adriaanse A, Rodenburg E, Bryant D, Woodward R (1995) Environmental indicators: a systematic approach to measuring and reporting on environmental policy performance in the context of sustainable development. World Resources Institute, Washington, DC
8. Bennett M, James P (1998) ISO 14031 and the future of environmental performance evaluation. Greener Manag Int 21:71–85
9. Verfaillie H, Bidwell R (2000) Measuring eo-efficiency: a guide to reporting company performance. Retrieved 19 January 2015, from teclim.ufba.br/jsf/ecoeficiencia/wbcsd%20measuring.pdf
10. Global Reporting Initiative (2011) Sustainability reporting guidelines. Retrieved 11 April 2015, from https://www.globalreporting.org/standards/G3andG3-1/Pages/default.aspx
11. Veleva V, Ellenbecker M (2001) Indicators of sustainable production: framework and methodology. J Clean Prod 9(6):519–549
12. NRTEE (2001) Eco-efficiency indicators: calculating eco-efficiency indicators: a workbook for industry. Renouf Publishing, Ottawa. ISBN 1-894737-02-4.
13. UNEP/SETAC Life Cycle Initiative (2011) Towards a life cycle sustainability assessment: making informed choices on products. Retrieved 13 January 2015, from http://www.unep.org/pdf/UNEP_LifecycleInit_Dec_FINAL.pdf
14. Singh RK, Murty HR, Gupta SK, Dikshit AK (2009) An overview of sustainability assessment methodologies. Ecol Indic 9(2):189–212

15. Gladen W (2008) Performance measurement: Controlling mit Kennzahlen, 4th edn. Gabler, Wiesbaden
16. Waas T, Hugé J, Block T, Wright T, Benitez-Capistros F, Verbruggen A (2014) Sustainability assessment and indicators: tools in a decision-making strategy for sustainable development. Sustainability 6(9):5512–5534
17. Jasch C (2000) Environmental performance evaluation and indicators. J Clean Prod 8 (1):79–88
18. UNEP/SETAC Life Cycle Initiative (2009) Guidelines for social life cycle assessment of products. Retrieved 18 February 2015, from http://www.cdo.ugent.be/publicaties/280.guidelines-sLCA.pdf
19. Mitchell G (1996) Problems and fundamentals of sustainable development indicators. Sustain Dev 4:1–11
20. Klöpffer W, Grahl B (2009) Ökobilanz (LCA): Ein Leitfaden für Ausbildung und Beruf. Wiley-VCH, Weinheim
21. Lindemann U (2009) Methodische Entwicklung technischer Produkte: Methoden flexibel und situationsgerecht anwenden, 3rd edn. VDI-Buch/Springer, Berlin
22. Ingwersen W, Cabezas, Weisbrod AV, Eason T, Demeke B, Ma X, Hawkins TR, Lee SJ, Bare JC, Ceja M (2014) Integrated metrics for improving the life cycle approach to assessing product system sustainability. Sustainability 6:1386–1413
23. Dreyer L, Hauschild M, Schierbeck J (2006) A framework for social life cycle impact assessment. Int J Life Cycle Assess 11(2):88–97
24. Klöpffer W (2008) Life cycle sustainability assessment of products. Int J Life Cycle Assess 13 (2):89–95
25. Jørgensen A, Le Bocq A, Nazarkina L, Hauschild M (2008) Methodologies for social life cycle assessment. Int J Life Cycle Assess 13(2):96–103
26. Harmsen J, Powell JB (2010) Sustainable development in the process industries: cases and impact. Wiley, Hoboken
27. Jolliet O, Margni M, Charles R, Humbert S, Payet J, Rebitzer G, Rosenbaum R (2003) IMPACT 2002+: a new life cycle impact assessment methodology. Int J Life Cycle Assess 8 (6):324–330
28. Müller-Christ G (2010) Nachhaltiges Management: Einführung in Ressourcenorientierung und widersprüchliche Managementrationalitäten, vol 1, 1st edn, Nachhaltige Entwicklung. Nomos, Baden-Baden
29. Garbie IH (2013) An analytical technique to model and assess sustainable development index in manufacturing enterprises. Int J Prod Res 52(16):4876–4915
30. Promberger K, Spiess H, Kössler W (2006) Unternehmen und Nachhaltigkeit: Eine managementorientierte Einführung in die Grundlagen nachhaltigen Wirtschaftens, vol 12, Management und Unternehmenskultur. Linde, Wien
31. Daub CH (2007) Assessing the quality of sustainability reporting: an alternative methodological approach. J Clean Prod 15(1):75–85
32. Henle B (2008) Nachhaltigkeit messen: Soziale Indikatoren für Finanzdienstleister, vol 42, 1st edn, Hochschulschriften zur Nachhaltigkeit. Oekom, München
33. Weck OL de, Ross AM, Rhodes DH (2012) Investigating relationships and semantic sets amongst system lifecycle properties (Ilities), third international engineering systems symposium CESUN 2012, Delft University of Technology, 18–20 June 2012, (March 2012):18–20
34. Garetti M, Taisch M (2012) Sustainable manufacturing: trends and research challenges. Prod Plan Control: Manag Oper 23(2–3):83–104. doi:10.1080/09537287.2011.591619
35. Ahmad-Ghorabi O, Rahmani H, Gerhard D (2013) An ontological approach for the integration of life cycle assessment into product data management systems, CIRP Design 2012 – Sustainable Product Development. Springer, London, pp 249–256. ISBN 978-1-4471-4506-6

Strategy Planning Before Urban Mining: Exploring the Targets

Hiroki Hatayama and Kiyotaka Tahara

Abstract This study is a challenge to construct the long-term strategy for recycling. To identify the adequate targets for recycling, we proposed the concept that integrates the criticality assessment and material flow analysis. Prior to the evaluation based on the concept, we developed the framework of the existing criticality assessment for Japan. Furthermore, the information on domestic metal demand and discard were reviewed to evaluate the potential of recycling. By integrating these knowledge, the result of the pilot evaluation was presented with a future perspective of the study.

Keywords Urban mining • Recycling • Material flow analysis • Criticality assessment • Japan • Resource strategy

1 Introduction

Highly developed industrial activities in Japan have been founded on a massive amount of metal consumption. Meanwhile, a large part of the metal supply is depending on the import from foreign countries. Therefore, Japan's governments and industries have been giving much importance on a security of metal resources. As mentioned in the governmental resource securement strategy [1], both common and minor metals are recognized to have a high importance to Japanese industry. As well as a reinforcement of mineral supply from abroad, reducing dependence on imported metals by substitution and recycling is also recognized as the significant measures. All of these measures are more or less effective; however, adequate combination between mitigation measures and target metals should be explored in constructing a long-term resource strategy.

Recycling, also known as "urban mining," has been an important endeavor in Japan's resource management. Developments of technologies and social systems for metal recovery from end-of-life (EoL) products are considered as the effective

H. Hatayama (✉) • K. Tahara
Research Institute of Science for Safety and Sustainability, National Institute of Advanced Industrial Science and Technology, Ibaraki, Japan
e-mail: h-hatayama@aist.go.jp

M. Matsumoto et al. (eds.), *Sustainability Through Innovation in Product Life Cycle Design*, EcoProduction, DOI 10.1007/978-981-10-0471-1_63

measures to secure metal resources. Furthermore, product design and material selection by manufacturing industries would have a great influence on the smooth urban mining. Strategic urban mining must be managed from such a life-cycle perspective. Here, the important questions are as follows: What will be the proper targets for urban mining? What are the metals that should take priority? To answer these questions, it is necessary to assess the supply risk, future demand growth, and domestic recycling potential. Especially, recent studies on the criticality assessment have been seeking to establish the methodology to quantify the relative supply risk and the vulnerability to supply restriction among different metals [2–5]. Those studies pointed out that the criticality should be considered with different components and time scale according to the assessment boundary (i.e., world, nation, or company). However, in Japan, the research and discussion on the criticality have been quite limited. Furthermore, it is not fully examined whether the existing framework for criticality assessment, governmental resource policy, and companies' concern about supply chain risk is corresponding.

The goal of this study is to propose the way to explore how the road map for recycling can be designed in Japan's resource strategy. In the following section, we propose the concept that can identify the preferable target for recycling. In exploring the target based on our concept, the criticality assessment and the material flow analysis (MFA) are the essential analyses. Therefore, the Sects. 3 and 4 explain how these analyses were developed so as to be utilized in our study. Finally, a vision toward a long-term strategy planning is presented in the Sect. 5.

2 Exploring the Targets for Recycling: The Concept

From the aspect of resource security, the targets for domestic recycling from EoL products should meet the following conditions.

First, of course, resource policy and strategy must deal with the metals/materials that are highly important to the domestic industry but exposed to high supply risk. This is what recent works on the criticality assessment try to quantify.

To these critical metals, however, recycling will not always be a silver bullet. Primary metal consumption could be greatly reduced by recycling when a large amount of metal exists in EoL products compared to the amount of consumption. Conversely, even the complete recovery from the EoL products cannot contribute to the significant reduction in primary resource when the metal discard is far smaller than the consumption. Generally, in metal cycles, there is a certain period of time between a growth in metal consumption and the corresponding increase in metal discard that depends on the product lifetimes. Therefore, when demand for a metal grows rapidly and the product lifetime is long, the metal cycle will be less likely to meet this condition. Thus, potential contribution of recycling on criticality mitigation is affected by the balance between domestic metal demand and discard. The balance has been analyzed for a variety of metals using MFA [6–10].

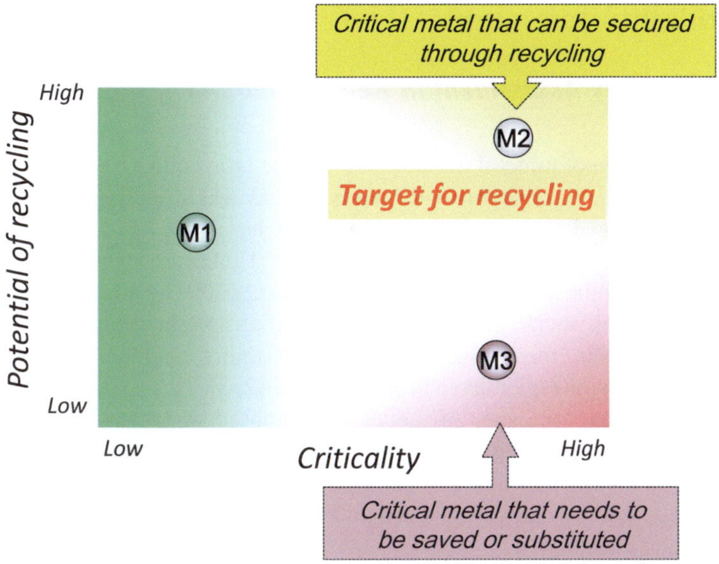

Fig. 1 Concept for exploring targets for recycling based on the criticality assessment and material flow analysis

In exploring recycling targets, our concept can be visualized as a two-dimensional graph that represents the criticality on the horizontal axis and the potential of recycling on the vertical axis (Fig. 1). Metals which criticalities are evaluated as relatively high will be plotted on the right. For metals plotted in the upper right area (yellow area), promoting metal recovery from EoL products would be effective to mitigate the criticality. In contrast, the promotion of metal recovery from EoL products will be less likely to contribute to reinforce the domestic metal supply from EoL products for the metals plotted in the lower right area (red area). For metals in the red area, recycling is not an effective measure to mitigate criticality. Improving material efficiency and developing substitute materials would be more effective than the development of recycling technologies for those metals. Alternatively, it may be required to reinforce the supply of foreign minerals by holding mineral interests.

Thus, Fig. 1 clearly indicates the appropriate measures (not just urban mining) to be taken for each critical metal. Therefore, the concept shown in Fig. 1 would become a powerful tool to explore the target for recycling. However, the concept can be of assistance for Japan's resource management as long as the horizontal and vertical positions of each metal in the figure are evaluated adequately. A lot of information to identify horizontal axis is expected to be provided by the criticality assessment. Meanwhile, the past criticality assessment for Japan does not seem to fully reflect the current resource policy. Therefore, in the next section, the framework of the existing criticality assessment for Japan is developed.

3 Criticality Assessment

3.1 Criticality Assessment in Recent Years

Massive extraction of mineral resources during the last century has brought to light their finite nature. A concern in the early period was a geological availability against increasing demand in the world, which has often been discussed using deletion time or reserves-to-production ratio. However, even the abundant resources in the lithosphere may not easily be accessible for many countries due to the geopolitical, social, economic, and other barriers. In recent studies on the criticality of metals, the pressures on an accessibility to mineral resources for individual country or region have been evaluated [2, 3].

In most studies, criticality of metals is composed of two main constituents: supply risk and vulnerability to supply restriction. Some studies placed the two constituents as horizontal and vertical axes and represent the "criticality matrix" [2, 3]. Other studies calculated a single criticality index by aggregating these different constituents [4, 11].

These constituents are further divided into several components. Components considered in the supply risk are, for example, depletion time, concentration of the production country, and political stability. Components in the vulnerability to supply restriction are GDP of the sectors where the metal is consumed and substitutability. However, it strongly depends on the practitioner of the assessment which components should be included in respective constituents. Furthermore, determination of the adequate weighting factors remains as a critical issue to be discussed even though selection and quantification of the components are fixed.

3.2 Resource Strategy and Criticality Assessment in Japan

Since Japan has been largely relying on the foreign countries for mineral supply, the attempt to identify supply risk has been made for a long time. In the early 1980s, the Ministry of International Trade and Industry in Japan conducted a survey in order to target of stockpiling. They surveyed more than 40 common and minor metals for their amount of reserves, concentration of production country, dependence on imports, etc., which are often employed as a component of supply risk in recent criticality assessments.

The strategy for resource securement [1] announced by the government in 2012 was clearly aware of the criticality assessment. The strategy designated 30 minerals as "strategic minerals" based on each metal's supply risk and economic importance. To reinforce the security for these minerals, the strategy referred to the four countermeasures: acquisition of mineral interests, substitution, stockpiling, and recycling. Especially, the importance of the mineral interests and recycling was mentioned in the Basic Energy Plan of Japan (2010) [12]. The plan defined the self-

Table 1 Criticality components used in the NEDO assessment

Category	Component
Supply risk	Depletion time
	Concentration of reserves
	Concentration of ore production
	Concentration of important trading partners
Price risk	Price change
	Price volatility
Demand risk	Global mine production growth
	Domestic demand growth
	Domestic demand growth for specific uses
Recycling restriction	Stockpiles
	Recyclability
Potential risk	Possibility of usage restrictions

sufficiency ratio of metals based on the amount of metals recycled and mineral resources entitled by the mineral interests and indicated the future goal of the ratio.

Meanwhile several measures are implemented for improving the resource security; the attempt to assess the criticality of metals for Japan in a quantitative manner has been made by the New Energy and Industrial Technology Development Organization (NEDO) [11]. The NEDO assessment figured out the aggregated criticality score for 39 minor metals, where 5 risk categories and 12 components were evaluated (Table 1).

Of course, it is open to debate whether election and quantification of these components are adequate. However, one clear defect of the current framework is that it did not reflect mineral interests. In Japan's resource strategy, holding mineral interests and acquiring entitlements to foreign mineral resources through resource development and investment have been considered an effective measure for stable supply. Actually in 2011, Japan Oil, Gas and Metals National Corporation (JOGMEC) supported the Japanese steel companies to acquire mineral interest of Araxa mine project in order to secure niobium [13]. However, the benefit from this policy implementation will not be appreciated under the framework shown in Table 1.

Not just Japan, but every developed country acknowledges the benefit from the mineral interests and accompanying entitlements to resources. Nevertheless, no criticality assessments have been taking into account the long-term availability of entitled resources. To strengthen the linkage between criticality assessment and policy implementations, "the sufficiency of mineral interests" should be quantitatively evaluated and embedded in the criticality assessment framework.

3.3 Japan's Sufficiency of Mineral Interests

When a company holds mineral interests, the holding company can receive the benefit of entitlement to control mined resources. In general, the percentage of entitled resources to a project's total production is determined in tandem with the investment ratio. To evaluate the degree of the sufficiency of mineral interests, we developed the indicators named "reserve entitlements" and "RE/D (reserve entitlements to demand) ratio."

$$\text{Reserve entitlements of metal } m = \sum_j \left(\sum_i r_{ij} R_{mj} \right) \quad (1)$$

$$\text{RE/D ratio} = (\text{Reserve entitlements})/(\text{Demand for mineral resources}) \quad (2)$$

R_{mj}: Amount of reserves of metal m in mining project j
r_{ij}: Investment ratio in the project j by Japanese company i

The reserve entitlements represent the amount of resources preferentially available to Japanese industry if the companies i maintain the same investment ratio in the future. Furthermore, the RE/D ratio is calculated on the similar concept of geological depletion time. As with the depletion time, RE/D ratio never indicates the duration before the entitlements will be run out. RE/D ratio is rather meaningful in comparing the relative degree of sufficiency among different metals. The RE/D ratio will be increased by both reinforcing mineral interests and by reducing demand for mineral resources through recycling, substitution, etc.

Japanese manufacturing companies and trading companies (called sho-sha) were holding mineral interests in 71 projects in 2012. As a result, Japan's RE/D ratio was calculated as Fig. 2. We found that Japan held relatively high levels of entitlements for niobium, tungsten, and lead. Using this quantified RE/D ratio, it became possible to reflect the mineral interests in the criticality assessment. The details of this section are explained in our paper [14].

3.4 Criticality Assessment for Japan in 2012

By aggregating the RE/D ratio into the existing NEDO framework (Table 1), we evaluated the criticality of 22 metals for Japan in 2012 with 13 components (see our paper [15] for details). The 22 metals were selected from the strategic minerals [1] including both common and minor metals, whereas NEDO assessment [11] focused minor metals only.

The result shown in Fig. 3 clarifies the relatively high criticality for some metals such as neodymium, dysprosium indium, and niobium. Niobium had an advantage for its high RE/D ratio; still the aggregated criticality was high due to the high

Fig. 2 Japan's RE/D ratio for 17 metals, 2012

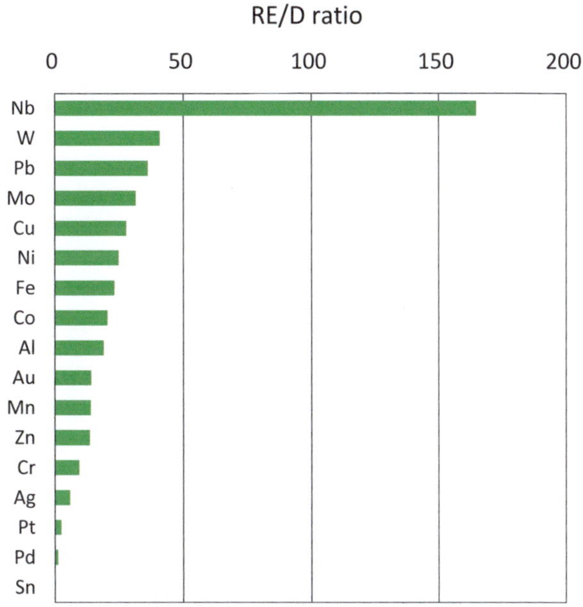

supply risk caused from a concentration of mineral resources in Brazil. It is also indicated that there were few differences in the aggregated criticality scores between the other minor metals and common metals. For common metals such as iron, copper, and zinc, the aggregated criticality was increased by short depletion time and growth in global mine production. The results seemed to be compatible with the Japan's recent resource strategy which designates both common and minor metals as strategic minerals.

4 Material Flow Analysis

The criticality assessment presented in the previous sections would indicate the target that securement should be tackled. However, as mentioned in the Sect. 2, recycling is not always an effective measure to achieve it. Before developing recycling technologies and systems, it should be evaluated how much metals would be recovered from EoL products. Recycling of metals is often hampered by technological and economic constraints. Before these constraints, however, it should be confirmed that there must be maximum amount of metals via recycling.

Table 2 compares the demand and discard of various metals. For the metals with small discard/demand ratio, recycling from EoL products cannot contribute on the reduction of primary metal consumption. For example, most of the indium consumed at the manufacturing processes turns into a process scrap; therefore, the amount of indium contained in the final products becomes quite small. The demand

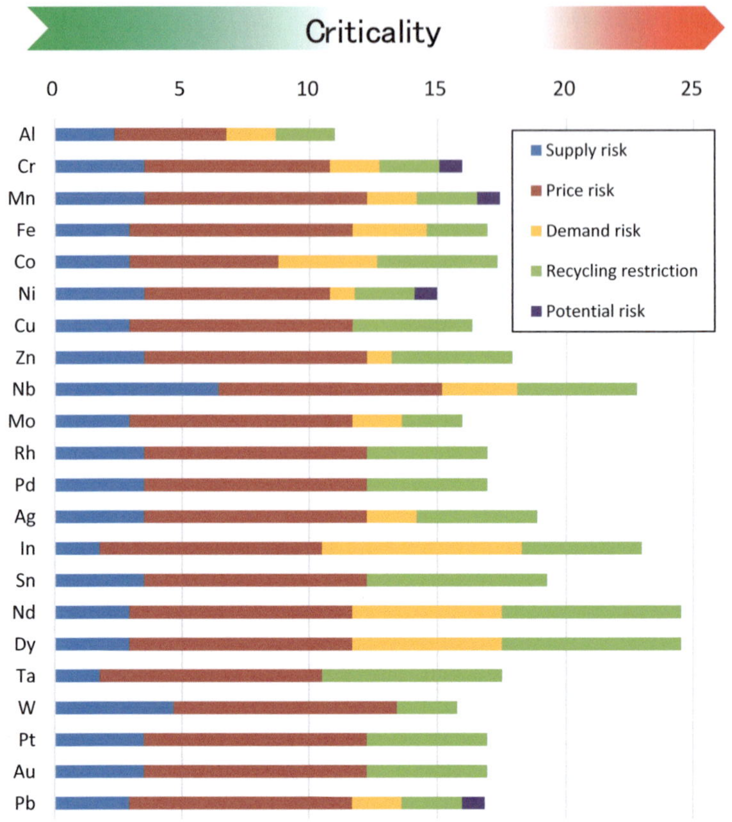

Fig. 3 Criticality of the strategic minerals, 2012

of dysprosium has been increasing in factory automation, home electric appliances, and automobiles, but most of them are still remaining as in-use stock and do not appear in the waste stream.

5 Conclusion: Integrating the Criticality Assessment, Material Flow Analysis, and Dynamic Analysis

On the basis of the criticality assessment and material flow analysis, we allocated some metals on the conceptual chart shown in Fig. 4. By applying the results in Fig. 3 to the horizontal axis and the results in Table 2 (here we regard the discard/demand ratio as "potential of recycling" in Fig. 1), the current situation in Japan was visualized as Fig. 4. Due to the limited availability of criticality assessment and MFA results, only 12 metals are located. Unfortunately among these 12 metals, no metals can be found in the yellow area. However, this snapshot changes toward the

Table 2 Metal demand and discard presented in the past MFA studies

	Demand[a] (kt)	Discard (kt)	Discard/demand	Ref.
Al	3731	1803	0.48	[8]
Cr	552	214	0.39	[7]
Fe	53,077	32,512	0.61	[16]
Co	3138	2581	0.82	[b]
Ni	818	424	0.52	[7]
Cu	860	852	0.99	[17]
Zn	247	168	0.68	[18]
In	861	2.4	0.00	[19]
Dy	275	25.5	0.09	[20]
Ta	319	120	0.38	[b]
W	2800	2160	0.77	[b]
Pb	345	366	1.06	[21]

[a]Domestic final demand
[b]Estimated by the authors

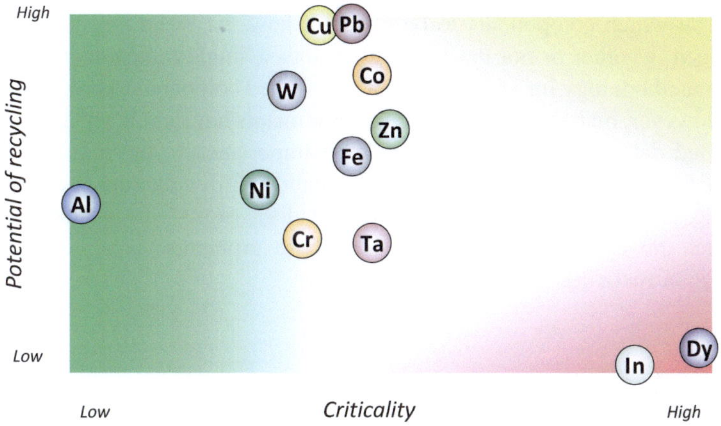

Fig. 4 Snapshot chart representing the criticality and the recycling potential for Japan in 2012

future. The potential of recycling in the future will be affected by the change in metal demand and discard. Especially, the amount of discard of minor metals will be expected to increase. As well, criticality of metals would change in the future in tandem with increase/decrease in geological reserves, metal demands, prices, and so forth. The transition can be illustrated as 3D chart shown in Fig. 5. By introducing dynamic perspective to criticality assessment and MFA, it becomes possible to show adequate timing to enhance urban mining for each metal. The concept developed in this study is a basic framework to move toward the sustainable metal use by employing the right technologies and policies in the right place at the right time.

We are aware that there is much room for the argument on how to define the horizontal and vertical axes. For criticality assessment, the consensus has not been

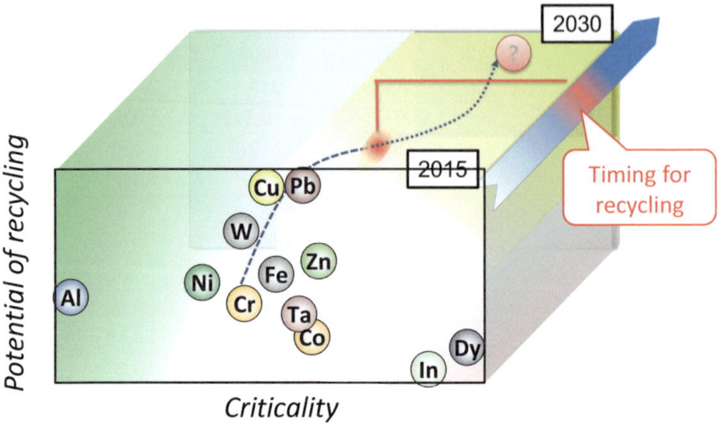

Fig. 5 Long-term strategy planning for recycling

formed on which components to be selected, how the weighting factors will be determined, whether or not the aggregation into a single indicator is appropriate, etc. The methodology for MFA has been established compared to criticality assessment. However, material flows of many minor metals have not been clarified due to the limited data availability despite of their importance to the industry. Furthermore, it may be better to consider the profitability in evaluating the potential of recycling. Huge data, knowledge, and the deep discussions between stakeholders of the metal life cycle are needed to solve these problems. This is a significant challenge of realizing "strategic urban mining."

References

1. Ministry of Economy, Trade and Industry (2010). Strategic energy plan of Japan. http://www.meti.go.jp/committee/summary/0004657/energy.pdf. Accessed 30 June 2015
2. National Research Council (2008) Minerals, critical minerals, and the U.S. economy. The National Academies Press, Washington, DC
3. European Commission (2010) Critical raw materials for the EU. Report of the Ad-hoc Working Group on defining critical raw materials. EC, Brussels
4. Graedel TE, Barr R, Chandler C, Chase T, Choi J, Christoffersen L, Friedlander E, Henly C, Jun C, Nassar NT, Schechner D, Warren S, Yang MY, Zhu C (2012) Methodology of metal criticality determination. Environ Sci Technol 46:1063–1070
5. United Nations Environmental Programme (2009) Sustainable innovation and technology transfer industrial sector studies. Critical metals for future sustainable technologies and their recycling potential
6. Brunner PH, Rechberger H (2004) Practical handbook of material flow analysis. CRC Press, Boca Raton
7. Daigo I, Matsuno Y, Adachi Y (2010) Substance flow analysis of chromium and nickel in the material flow of stainless steel in Japan. Resour Conserv Recycl 54:851–863

8. Hatayama H, Daigo I, Matsuno Y, Adachi Y (2012) Evolution of aluminum recycling initiated by the introduction of next-generation vehicles and scrap sorting technology. Resour Conserv Recycl 66:8–14
9. Kuriki S, Daigo I, Matsuno Y, Adachi Y (2010) Recycling potential of platinum group metals in Japan. J Jpn Inst Metals 74(12):801–805
10. Seo Y, Morimoto S (2014) Comparison of dysprosium security strategies in Japan for 2010–2030. Resour Policy 39:15–20
11. New Energy and Industrial Technology Development Organization (NEDO) (2009) Trend report of development in materials for substitution of scarce metals, Kanagawa
12. Ministry of Economy, Trade and Industry (2010) Basic energy plan of Japan. http://www.enecho.meti.go.jp/category/others/basic_plan/pdf/100618honbun.pdf. Accessed 30 Jun 2015
13. Japan Oil, Gas and Metals National Corporation (JOGMEC) (2011) JFE steel, Nippon Steel, Sojitz, JOGMEC, POSCO and NPS form Japan-Korea partnership group to invest in Brazilian producer of niobium, critical alloying element for high-grade steel products. http://www.jogmec.go.jp/english/news/release/release0067.html. Accessed 30 June 2015
14. Hatayama H, Tahara K (2015) Evaluating the sufficiency of Japan's mineral resource entitlements for supply risk mitigation. Resour Policy 44:72–80
15. Hatayama H, Tahara K (2015) Criticality assessment of metals for Japan's resource strategy. Mater Trans 56(2):229–235
16. Hatayama H, Daigo I, Matsuno Y, Adachi Y (2010) Outlook of the world steel cycle based on the stock and flow dynamics. Environ Sci Technol 44(16):6457–6463
17. Daigo I, Hashimoto S, Matsuno Y, Adachi Y (2009) Material stocks and flows accounting for copper and copper-based alloys in Japan. Resour Conserv Recycl 53(4):208–217
18. Tabayashi H, Daigo I, Matsuno Y, Adachi Y (2009) Development of a dynamic substance flow model of zinc in Japan. ISIJ Int 49(8):1265–1271
19. Yoshimura A, Daigo I, Matsuno Y (2013) Global substance flow analysis of indium. Mater Trans 54(1):102–109
20. Shi W, Daigo I, Matsuno Y, Adachi Y (2010) Substance flow analysis of dysprosium in Japan. J Jpn Inst Metals 74(11):758–765
21. Ando S, Murakami S, Yamatomi J (2010) Material flow/stock analysis of lead in Japan. J MMIJ 126:482–489

Evaluation of Resource Efficiency of Electrical and Electronic Equipment

Tomoaki Kitajima, Yuji Sasaki, and Nozomu Mishima

Abstract Recently, resource efficiency is one of the high concerns in industrialized and developing countries. High resource efficiency is considered to be key of sustainable production. This paper proposes an index to evaluate resource efficiency of a small-sized electrical and electronic equipment (EEE). The index has value of the product based on the functionality, product reusability, and component reusability, as the numerator. And it has the total amount of material consumption measured by TMR (total material requirement) minus recoverable environmental impact as the denominator. The paper carries out practical measurement of the material compositions of different models of smartphones. Then, it investigates the functional data in order to calculate actual resource efficiency value. Through this effort, the paper indicates the effectiveness of the proposed index and clarifies criteria of resource-efficient products.

Keywords EEE • Resource efficiency • Product functionality • Reusability • TMR

1 Introduction

Although modern life heavily depends on natural resources, the supplies of such resources are limited. In addition, demands for natural resources are rapidly increasing because of the growing economies in developing countries. For example, reserves-to-consumption ratios [1] of some critical metals are very small [2]. If we cannot develop technologies to reduce the consumptions of such critical metals or to replace by common metals, modern lives based on such high-tech products will be endangered.

Basically, if the society is more resource efficient, many problems regarding limited natural resources will be eased. High resource efficiency will be effective in reducing cost and enhancing production efficiency at the same time. In considering the resource efficiency, the first step is to evaluate the resource efficiency quantitatively. Some public organizations or private enterprises have proposed resource

T. Kitajima • Y. Sasaki • N. Mishima (✉)
Graduate School of Engineering and Resource Science, Akita University, Akita, Japan
e-mail: nmishima@gipc.akita-u.ac.jp

M. Matsumoto et al. (eds.), *Sustainability Through Innovation in Product Life Cycle Design*, EcoProduction, DOI 10.1007/978-981-10-0471-1_64

efficiency indexes [3, 4]. However, one of the problems is that there is no common definition of resource efficiency.

Even in the G7 summit declaration [5], resource efficiency is an important issue to be studied and enhanced for the future sustainable society. Although it would be difficult to agree on practical index of resource efficiency, at least, a common concept will be necessary. Thus, regarding this situation, the purpose of this study is to propose a concrete index to evaluate resource efficiency of products. Then, the paper also focuses on evaluating actual products based on the index to know the effectiveness of the index in evaluating efficiencies of small-sized electronic products as an example.

2 Proposal on the Resource Efficiency Index

2.1 Basic Concept of the Index

2.1.1 Eco-efficiency

Resource efficiency is based on an eco-efficiency [6] idea which is commonly defined by Eq. 1:

$$\text{Eco-efficiency} = \frac{\text{Value of the product or serive}}{\text{Environmental burden of the product or service}} \quad (1)$$

2.1.2 Resource Efficiency Index

Frequently, LCA result is used for the denominator of the abovementioned equation of eco-efficiency, to calculate the index concretely. However, sometimes carrying out LCA is not so easy. And we feel that the resource efficiency index should focus on criticality of resources than the total environmental burden. Thus, we modified Eqs. 1 and 2 which we think is more suitable in evaluating resource efficiency:

$$\text{Resource-efficiency} = \frac{\text{Value of the product or serive}}{\text{Resource usage index of the product or service}} \quad (2)$$

2.1.3 How to Evaluate the Value

As shown in the former section, the numerator of the resource efficiency index is "*value*" of the product or service. Since there are many functions in a product or many business values in a service, the total value will be the sum of individual functions based on the importance of the functions on the product. In other words, the numerator of the index will be weighed sum of functions. Equation 3 is the numerical expression:

$$Value = \sum_i w_i \cdot \frac{\text{specific value of function } i}{\text{reference value of function } i} \tag{3}$$

w_i: relative weight of the function i

However, as it was mentioned in the introduction, efficient reuse of resources should be evaluated. Thus, in this paper, the value of the products is evaluated by values of the original product calculated by Eq. 3 plus product reuse value and component reuse value.

2.2 How to Evaluate the Resource Usage

In focusing on resource consumption, one good index is total material requirement (TMR [7]). TMR is defined as the total amount of crude metals, ores, soils, removed surface soils, etc. to obtain a unit amount of refined metals. A large TMR value means that a huge amount of ore has to be extracted from earth environment to get the material. It is equivalent that the material has a large environmental burden. Table 1 shows values of TMR of some well-known elements. Using TMR, the denominator of the index can be calculated. Reuse of materials in the aspect of efficient use of resources should be also evaluated. So, in this paper, we evaluate the resource consumption by Eq. 4:

$$\text{Resource usage} = \sum_j TMR_j - \sum_k TMR_k \tag{4}$$

TMR_j: TMR of the material j
TMR_k: TMR of the recyclable material k

2.3 Proposing the Index for Resource Efficiency

Considering all the aforementioned aspects, the paper proposes the resource efficiency index indicated by Eq. 5. Two items to evaluate product reuse value and component reuse value in the equation are expressed by Eqs. 6, 7, and 8:

Table 1 TMR of major elements

Element	TMR
Cu	360
Fe	8
Mg	70
Ni	260
Cr	26
W	190
Al	48
Zn	36
Pb	28
Ag	4800
Au	1,100,000
Pd	810,000
Pt	520,000

$$\begin{aligned}&\text{Resource efficiency}\\&=(\text{Original product value}+\text{Product reuse value}+\text{component reuse value})\ /\\&\left(\sum_j TMR_j - C_3\sum TMR_k\right)\end{aligned} \tag{5}$$

C_3: recycling rate

$$\text{Original product value} = \sum_i w_i \cdot \frac{\text{Specific value of function } i}{\text{Reference value of function } i} \tag{6}$$

$$\text{Product reuse value} = C_0 \cdot C_1 \cdot \frac{V_2}{V_1} \tag{7}$$

C_0: *collection rate*
C_1: *reusable rate*
V_2: value (price) of the second-hand product
V_1: value (price) of the original products

$$\text{Component reuse value} = \frac{C_0}{V_1} \cdot \sum_n C_{2n}(V_{3n} - Lc_n) \tag{8}$$

C_{2n}: reusable rate of component n
V_{3n}: value of the reused component n
Lc_n: labor cost to detach component n

3 Quantification of the Product Value

3.1 Original Product Value

Since basic resource efficiency index has been proposed until now, the next step is to input concrete values in the equations to calculate the index value and examine the effectiveness of the index. Firstly, the paper tried to evaluate the value of the products by a practical example. Taking a mobile phone as an example of the small-sized electronic products, since we are strongly focusing on it, a questionnaire to 83 students to know the relative importance of the functions of a mobile phone was carried out. In the questionnaire, five functions which were listed up through the brainstorming among the authors were asked. The five functions were "continuous life of the battery," "size of the LCD," "pixels of the camera," "capacity of the storage memory," and "the speed of data transfer." The questionnaire is based on conjoint analysis [8]. Combinations of the functions of a mobile phone are combined based on the L8 orthogonal array [9] as conjoint cards. Each correspondent looks at the conjoint cards shown in Fig. 1 and answers which card is most

Fig. 1 The set of conjoint cards

Type A
Battery life: 1 Day Storage memory:16GB Data trans. Speed:30Mbps LCDsize: 2.6 inches Camera res.: 400 mill.

Type B
Battery life: 1 Day Storage memory:16GB Data trans. Speed:30Mbps LCD size: 3.7 inches Camera res.: 800 mill.

Type C
Battery life: 1 Day Storage memory:32GB Data trans. Speed:60Mbps LCD size: 2.6inches Camera res.: 400 mill.

Type D
Battery life: 1 Day Storage memory:32GB Data trans. Speed:60Mbps LCD size: 3.7 inches Camera res.: 800 mill.

Type E
Battery life: 2 Day Storage memory:16GB Data trans. Speed:60Mbps LCD size: 2.6 inches Camera res.: 800 mill.

Type F
Battery life: 2 Day Storage memory:16GB Data trans. Speed:60Mbps LCDsize: 3.7 inches Camera res.: 400 mill.

Type G
Battery life: 2 Day Storage memory:32GB Data trans. Speed:30Mbps LCD size: 2.6 inches Camera res.: 800 mill.

Type H
Battery life: 2 Day Storage memory:32GB Data trans. Speed:30Mbps LCD size: 3.7inches Camera res.: 400 mill.

preferable. Then, the correspondent chooses the second one, and then. Finally, preferable orders from one to eight are given to all the conjoint cards.

By averaging the ranking of card A, B, C, and D, the average of the specification "Battery life: 1 day" can be calculated. Then, by comparing it to those of card E, F, G, and H, the difference in "Battery life" that is important for the respondent can be calculated. The importance of "Storage capacity" can be clarified by comparing A, B, E, and F versus C, D, G, and H. The other functions are the same. As the result of the analysis, relative importance of each function can be calculated by Eq. 9 shown below:

$$\text{Relative importance of function } i = \frac{D_i}{\sum_{i=1}^{5} D_i} \quad (9)$$

D_i: difference of the ranking of low level and high level

Then the relative importance of five functions were calculated as Table 2. However, since the last function is not the function that depends on the design of the mobile phone itself, but depends on the infrastructure, this item was eliminated from the functions. Figure 2 shows the relative importance of other four functions. Table 3 is the reference specifications of the four functions. Finally, original product values of mobile phones can be calculated by Eq. 10:

$$\begin{aligned}\text{Original product value} = {} & 0.4031 \times \frac{\text{Battery life}}{488.7} \\ & +0.2179 \times \frac{\text{LCD size}}{3.15} - 0.2022 \times \frac{\text{Camera resolution}}{717.9} \\ & +0.1768 \times \frac{\text{Storage capacity}}{2020}\end{aligned} \quad (10)$$

3.2 Product Reuse Value

The next step is to quantify the product reuse value. An existing survey [10] identifies how the consumers handle their used mobile phones. Figure 3 shows

Table 2 Relative importance of the five functions of a mobile phone

	Data trans. speed	Battery life	LCD size	Camera res.	Storage capacity
Relative importance	0.3818	0.2492	0.1347	0.1250	0.1093

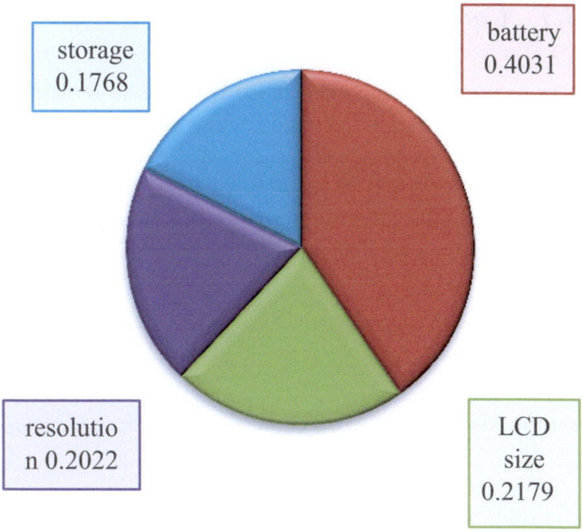

Fig. 2 Relative importance of the functions of a mobile phone

Table 3 Reference specifications of functions of a mobile phone

LCD size [inch]	Resolution [million dot]	Storage capacity [number of photos]	Battery life [h]
3.15	717.9	2020	488.7

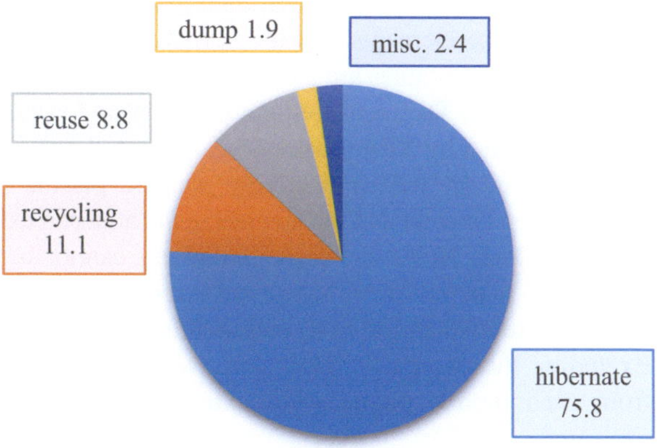

Fig. 3 End-of-life treatment of used mobile phone in percentage [9]

the result. So, 0.088 is used for the reuse rate. Based on Eq. 7, product reuse value can be expressed in Eq. 11:

$$\text{Product reuse value} = 0.088 \cdot \frac{V_2}{V_1} \tag{11}$$

3.3 *Component Reuse Value*

The next item, component reuse value, is also estimated from Fig. 2. Among the used products sent to recyclers, some products are manually disassembled to reuse components. Because only successful example of component reuse is LCD, it is said [11] that from 30 to 40 % of used mobile phones, a reusable LCD can be extracted. Labor cost is estimated to be about 1007 JPY per hour in average. So, finally, component reuse value is calculated by Eq. 12:

$$\text{Component reuse value} = \frac{1}{V_1} 0.039\left(V_{3n} - 1007 \cdot \frac{t}{3600}\right) \tag{12}$$

t: time to disassemble LCD (second)

4 Resource Efficiency of Mobile Phones

4.1 *TMR of Mobile Phones*

Although there are many elements included in the used mobile phones, not every element can be recycled under current system and technologies. Furthermore, different types of smelters have different list of recoverable elements. Tables 4 and 5 show the elements and the corresponding TMR contained in two types of smartphones. The tables are also showing recoverable elements among the long list of the elements. The measurement of material compositions and included amount were carried out by XRF (X-ray florescence [12]) analysis. Only Au (gold) amount was measured by ICP (inductively coupled plasma [13]). Based on the survey, recoverable elements that the smelter in the area can extract are copper, silver, gold, and palladium. However, since palladium cannot be detected in these samples, other three materials were taken into account as recyclable materials. The mobile phones were dismantled first, and then a battery and an LCD unit were detached. Excluding the two parts, other parts were pulverized and mixed. Three samples from each mobile phone were extracted and the material compositions were measured.

4.2 *Resource Efficiency of the Mobile Phones*

Based on the discussion in the former sections and Eqs. 10, 11, and 12, the final equation to express resource efficiency is shown as Eq. 13. As the results of the calculation, resource efficiencies of the two models of mobile phone can be

Table 4 Material composition of a mobile phone

Smartphone type A				
Composing materials	Weight [g]	TMR	TMR of materials used in a mobile phone	TMR of recyclable materials
Na	1.55	50	7.75.E-05	
Mg	0.528	70	3.70.E-05	
Al	4.40	48	2.11.E-04	
Si	7.18	34	2.44.E-04	
Ca	1.27	90	1.14.E-04	
Ti	0.12	36	4.3.E-06	
Cr	0.105	1500	1.58.E-04	
Mn	0.0142	14	1.99.E-07	
Fe	0.475	8	3.80.E-06	
Co	0.0020	600	1.2.E-06	
Ni	0.18	260	4.68.E-05	
Cu	2.57	360	9.25.E-04	1.02.E-04
Zn	0.106	36	3.82.E-06	
Ga	0.0011	14,000	1.5.E-05	
Br	0.0107	1500	1.61.E-05	
Sr	0.010	500	5.0.E-06	
Y	0.0002	2700	5.00.E-07	
Zr	0.005	550	2.90.E-06	
Nb	0.008	640	5.00.E-06	
Ag	0.021	4800	1.00.E-04	1.1.E-05
Sn	0.267	2500	6.68.E-04	
Nd	0.010	3000	3.0.E-05	
Au	0.0184	1,100,000	2.02.E-02	2.22.E-03
Pb	0.0046	28	1.3.E-07	
Misc.	70.2			
Total	89.090		2.29.E-02	2.33.E-03

calculated as Table 6. Specifications of each mobile phone were found in the carrier's web page [14] (Table 7):

$$\text{Resource efficiency} = \frac{0.2179 \cdot \frac{S_1}{3.15} + 0.2022 \cdot \frac{S_2}{717.9} + 0.4031 \cdot \frac{S_3}{488.7} + 0.1768 \cdot \frac{S_4}{2020}}{\left(\sum_j TMR_j - 0.11 \cdot \sum TMR_k\right)} + \frac{0.088 \cdot \frac{V_2}{V_1} + \left[\frac{1}{V_1}\left\{0.039\left(V_3 - 1007 \cdot \frac{t}{3600}\right)\right\}\right]}{\left(\sum_j TMR_j - 0.11 \cdot \sum TMR_k\right)} \tag{13}$$

S_1: size of the LCD of evaluating model [inch]

Table 5 Material compositions of a smartphone (contd.)

Smartphone type B				
Composing materials	Weight [g]	TMR	TMR of materials used in a mobile phone	TMR of recyclable materials
Na	0.0971	50	4.86.E-06	
Mg	0.152	70	1.06.E-05	
Al	3.48	48	1.67.E-04	
Si	3.56	34	1.21.E-04	
Ca	1.73	90	1.56.E-04	
Ti	0.0585	36	2.11.E-06	
Cr	0.75	1500	1.1.E-03	
Mn	0.0319	14	4.47.E-07	
Fe	2.10	8	1.68.E-05	
Co	0.0067	600	4.02.E-06	
Ni	0.316	260	8.22.E-05	
Cu	5.05	360	1.82.E-03	2.00.E-04
Zn	0.176	36	6.34.E-06	
Ga	0.0035	14,000	4.9.E-05	
Ge	0.0005	1500	6.E-05	
Br	0.0149	500	2.24.E-05	
Sr	0.0224	2700	1.12.E-05	
Y	0.0004	550	1.08.E-06	
Zr	0.0278	640	1.53.E-05	
Ag	0.018	4800	8.64.E-05	9.46.E-06
Sn	0.25	2500	6.2.E-04	
Nd	0.015	3000	4.5.E-05	
Au	0.0124	1,100,000	2.36.E-02	2.60.E-03
Pb	0.0113	28	3.16.E-07	
Misc.	71.0			
Total	89.843		2.80.E-02	2.81.E-03

Table 6 Values of two examples

Model	Original product value	Product reuse value	Component reuse
A	1.047	0.039	0.00272
B	2.604	0.027	0.00158

Table 7 Result of resource efficiencies

Model	Total TMR	Resource efficiency
A	0.0207	52.6
B	0.0277	95.0

S_2: resolution of the camera of the evaluating model [mill. pixels]
S_3: battery life of the evaluating model [h]
S_4: storage capacity [photos]

V_1: value of the original product [Yen]
V_2: value of the second-hand product [Yen]
V_3: value of the reused components [Yen]
t: time to detach LCD [s]
TMR_j: TMR of the material j
TMR_k: TMR of the recyclable material k

4.3 Interpretation of the Result

From the abovementioned results, it is clearly indicated that type B has higher resource efficiency than type A. The factor which mostly affected this result most was the original product value. Since smartphone B was launched 1 year after smartphone A to the market, B has higher values regarding the four selected specifications. Totally, it was calculated that the original product value of type B is 2.5 times of type A.

However, still there is a problem in resource efficiency. Since the purpose of resource efficiency evaluation is to enhance sustainability of the product, increasing resource efficiency only by enhancing performances cannot be a perfect answer. Actually, total TMR of type B is 30 % higher than that of type A. This is mainly due to densities of gold observed in Tables 4 and 5, since most of the environmental impact due to element depends on the amount of gold used. It is well known that gold is often used in ICs to achieve high performances. However, in the aspect of sustainability, a new design which uses less amount of gold but is having equivalent or higher performances is desired. Furthermore, there are some problems in this resource efficiency evaluation because of some reasons shown below.

- To calculate original product value, reference data is necessary. The paper used specifications of the newest product as the reference. This means that when a newer product appears, the value of the index changes.
- Conjoint analysis to university engineering students was carried out in order to calculate relative importance of each specification. However, there might be some tendencies due to respondents.
- In this analysis, the price of a second-hand product was used for product reuse value, and the price of used LCD unit was used for component reuse value. However, these prices would not be known in advance. These values are difficult to use in considering resource-efficient designs.
- It was difficult to know the practical flow of used mobile phones. Result of the questionnaire used in the analysis might have some difference with the actual situation.
- Measurement of the material compositions contains some errors. The resource efficiency values are affected by accuracy of the measurement greatly.

To propose the resource efficiency index, finally, abovementioned points should be well examined and solved.

5 Conclusions

In order to provide a design guideline of resource-efficient product, an index to evaluate resource efficiency was proposed. As a case study, the index was applied to evaluate resource efficiencies of smartphones.

The numerator of the proposed index consists of original product value, product reuse value and component reuse value. The denominator of the index consists of TMR amount of the product reduced by TMR amount of recyclable materials. Four functional specifications were chosen to measure original product values. To know the relative importance of the four specifications, conjoint analysis was carried out and it was found that the life of the battery is the most important feature of the smartphones.

Material compositions of the two smartphones were measured by XRF and ICP. Through the measurement, TMR of the smartphones and TMR of recyclable materials were calculated. According to the result, actual resource efficiency values were calculated and compared. Since the index highlighted the enhanced function of the products and amount of critical elements used in the product, the paper concluded that the proposed index can be a useful index, if the authors don't spare efforts to increase the preciseness if the index.

As for future work, much more case studies should be carried out to evaluate resource efficiencies by using the proposed index. And of course, concept proposals and implementations of resource-efficient designs are necessary to establish sustainable production.

References

1. Nishiyama T, Adachi T (1995) Resource depletion calculated by the ratio of the reserve plus cumulative consumption to the crustal abundance for gold. J Nonrenew Resour 4(3):253–261
2. Halada K, Shimada M, Iijima K (2008) Decoupling status of metal consumption from economic growth. Mater Trans 49(3):411–418
3. http://www.toshiba.co.jp/env/jp/products/ecp/factor_j.htm. Accessed 31 Jan 14, in Japanese
4. http://jp.fujitsu.com/about/csr/eco/products/gproducts. Accessed 31 Jan 14, in Japanese
5. https://www.whitehouse.gov/the-press-office/2015/06/08/annex-g-7-leaders-declaration. Accessed 29 June 15
6. DeSimone L, Popoff L (1997) Eco-efficiency – the business link to sustainable development. MIT Press, Cambridge, MA
7. Nakajima K, Yamamoto K, Nakano K, Kuroda K, Halada K, Nagsaka T (2006) Recycle-flow Analysis on Used Cellular Phone Based on Total Materials Requirement. J Life Cycle Assess Jpn 2(4):341–346
8. Green PE, Srinivasan V (1978) Conjoint analysis in consumer research. Stanford University Research Paper 435
9. Fowlkes WY, Creveling CM (1995) Engineering methods for robust product design: using Taguchi methods in technology and product development. Addison-Wesley Publishing Company, Reading
10. http://www.env.go.jp/recycle/index.html. Accessed 31 Jan 14, in Japanese

11. Murakami S, Osugi H, Murakami R, Mukouda A, Tsujimura H (2009) Average lifespan of mobile phones in in-use and hibernating stocks in Japan. J Life Cycle Assess Jpn 5(1):139–145
12. Beckhopff B, Kanngießer B, Lnghoff N, Weddell R, Wolff H (2007) Handbook of practical x-ray fluorescence analysis. Springer Science & Business Media, Berlin, Germany
13. Hou X, Jones BT (2000) Encyclopedia of analytical chemistry – inductively coupled plasma – optical emission spectrometry. Wiley, Chichester, pp 9468–9485
14. Home page of NTT Docomo Co. Ltd., https://www.nttdocomo.co.jp/. Accessed 23 June 15

Regionalized Input-Output Life Cycle Sustainability Assessment: Food Production Case Study

Sergiy Smetana, Christine Tamasy, Alexander Mathys, and Volker Heinz

Abstract Sustainability assessment requires the use of evaluation of regional processes as a basis. Regional statistical data on economic, social, and environmental aspects is used for the construction of regional sustainability assessment system as a basis for regionalized input-output life cycle sustainability assessment performed for a few regions in Germany. The results indicated the capability of a system to estimate the relative efficiency of resource use for diverse regions and separate pillars of sustainability. It was identified that sustainability of food production rates is comparable with rates of regional development, which indicates the promoting or regressing importance of separate production fields.

Keywords Regional sustainability • Input-output • Life cycle sustainability assessment

1 Introduction

Sustainability indicators are referred as measures of system behavior [1–5], which aim to assess important values for human development [4, 6]. They should provide meaningful and simplified information about the complex dynamic systems and assist in decision making in the modern world of regional dependency on global supply chains. Sustainability of regional development in its complex nature is hardly assessable and comparable owing to the lack of universal methodology.

S. Smetana (✉)
Institute for Spatial Analysis and Planning in Areas of Intensive Agriculture, University of Vechta, Vechta, Germany

German Institute of Food Technologies (DIL e.V.), Quakenbrück, Germany
e-mail: s.smetana@dil-ev.de

C. Tamasy
Institute for Spatial Analysis and Planning in Areas of Intensive Agriculture, University of Vechta, Vechta, Germany

A. Mathys • V. Heinz
German Institute of Food Technologies (DIL e.V.), Quakenbrück, Germany

M. Matsumoto et al. (eds.), *Sustainability Through Innovation in Product Life Cycle Design*, EcoProduction, DOI 10.1007/978-981-10-0471-1_65

Input-output (IO) table analysis methods [7–9] are well known for the identification of links for national and even global economies. It allows the transparency of data and results and, furthermore, multiple combination and recombination procedures, which assure flexibility of regional sustainability assessment methodology (RSAM) and its independence from specific data availability. IO tables' combination for complex environmental and economic assessments indicated the capabilities for complex systems analysis through correlation with life cycle assessment (LCA) methods [10].

We argue that IO analysis methods are suitable for assessment of complex regional systems and specific cases' production. Proposed regional sustainability assessment methodology (RSAM) is based on input-output analysis of natural, social, and economic capital transfers in the regions.

The methodological apparatus of the RSAM concept is presented with a comparison of selected regions on the basis of resource cycling and system element interactions. Regional scale of such assessment could provide sufficient and detailed data for sustainability analyses, yet preserving the holism of the system.

RSAM is related to the other LCA-based models and approaches with a regional specification and resource cycling estimation. The most developed regionalized LCA is based on a specific geographic information system (GIS) approach developed by Mutel (2012). However, GIS-LCA regionalized specific methodology does not account for social and economic factors. Regionalization of life cycle sustainability assessment methodology is not yet established, as the methodology itself is under development [11–13].

Other environmental and social approaches have similar limitations, i.e., assessing only separate issues of sustainability. Numerous indicator assessment methodologies consider multiple issues of various sectors, but they are lacking the connectivity options of a holistic system. In case of change of a single element, other indicators should be reassessed with a separate system. On the contrary, RSAM includes multiple indicators with interconnections and indicators of environmental, social, and economic nature with assessment at regional scale.

2 Conceptual Considerations

Regional development is dependent on natural (environmental), social, and economic capital [14–20]. That is why RSAM includes the assessment of regional resources in the context of their interdependency and interconnection. It indicates a region as a complex system with multiple connections of different levels and, therefore, addresses the analysis of regional development as a typical problem of complex system analysis, when the change of one element results in disproportional changes in others [21–23]. System approach used for the RSAM indicates the utmost importance of data quality and availability for the analysis.

As complex social and environmental systems have nonlinear development characteristics, various approaches of complex fluctuation reviews were published.

Most authors agree that the development goes through the stages of initiation, development, stability, and regression or reorganization [24–26]. This paper follows the aggregated overall tendency representation, which consists of multiple components interacting within the system.

RSAM includes assessment of two main sectors, social and environmental, as most resources can be grouped according to the social interactions versus environmental conditions. Social assessment is carried out via separate input-output resource tables, which characterize consumers, service suppliers, regulators, economic activities of producers, and other aggregated types of social and economic resources. The environmental sector includes input-output resource matrices (land, water, mineral, and biotic resources) and associated categories (energy, waste).

3 Methodology

RSAM follows a simple algorithm for the assessment of resource use in the regions. It required data collection for the compared regions, input-output tables' construction, estimation of resource cycling in regions, and comparison and analysis of results. We collected data for the assessment from official statistical sources (Regionaldatenbank Deutschland). Methodologically, RSAM is based on extended input-output tables and flow cycling analyses [7, 9, 27] within an outlined regional area. The IO matrixes reflect the distribution of resources between the analyzed categories. The counting units are diverse in initial matrices and, therefore, cannot be compared and combined in initial "raw" view. However, the assessment of resource efficiency use in separate sectors was performed.

RSAM exploits a number of methods and approaches:

1. Previous data compilation into adapted and extended input-output (IO) tables [7–9]
2. Element data analysis and calculations via transformation of initial resource data with a price index (real regional current prices for the resources) into monetary values for further aggregated system analysis.
3. Connectivity data analysis, which estimates the rate of resource cycling in the system. It is based on a few approaches, which aimed to calculate the length of flows, the magnitude, and the return abilities of the flows to the initial element – Han, Finn, and comprehensive cycling indices [28–33]. Further analysis is performed with evaluation of trend analysis for the identification of system dynamics.

4 The Comparison of Regional Sustainability

We identify regional sustainability as the effectiveness of resource use (social, economic, and environmental) in identified regional borders. Social, economic, and environmental capital-based IO tables, used for the construction of RSAM, are applicable for the diverse regions. The diversity of regions could be reflected in size, border shape, natural resource condition, economic and social variations, etc. The amount of accounting factors is limited only with data availability and accuracy requirements. That is why we selected very economically similar "rural districts" (Landkreis) of level NUTS 3 (nomenclature of territorial units for statistics) for comparison and higher-level NUTS 2 regions, which are more economically and physically diverse. Selected regions are related and situated in the similar climate conditions. Economic development of higher-level regions (NUTS2) includes a higher diversity of production fields. Additionally, a region which has different regional structures and resources was added for comparison (Region H). In total five regions were compared in this case study (Fig. 1, Table 1).

The results indicate that naturally and economically similar regions showed similar results of resource cycling and effectiveness of their use (Figs. 2 and 3). Adapted HCI (Han cycling index – a ratio of resources involved in the cycles to the total available resource stock [34]) demonstrated that regions of Vechta and Cloppenburg (V and C) have a slightly higher rate of resources cycled in the system than the other regions.

FCI (Finn cycling index – a ratio of average path cycling inflows to straight inflows in the system) [29] demonstrated similar trends, but indicated that a region (LS) with a higher complexity of a system is characterized with higher resource cycling (Fig. 3). This index is foreseen as more efficient to indicate the cycles of the resources [28]. Moreover, FCI showed to be more indicative and sensitive for the estimation of sustainability levels of complex systems in our case.

It is evident that the region (H) we used for comparison as a control region (differences in types of industries and natural conditions) clearly demonstrated higher levels of resource use efficiency (Figs. 2 and 3).

Additional assessment technique of RSAM is the relation of external flows to total flows of the system (Fig. 4). This measure indicated the high dependency of regional development on external flows for four regions (LS, WE, V, C), while the control region (H) indicated lower dependency on the external flows. It also demonstrated the dynamic changes for regions (V and C), indicating the abilities of the RSAM for dynamic change assessment.

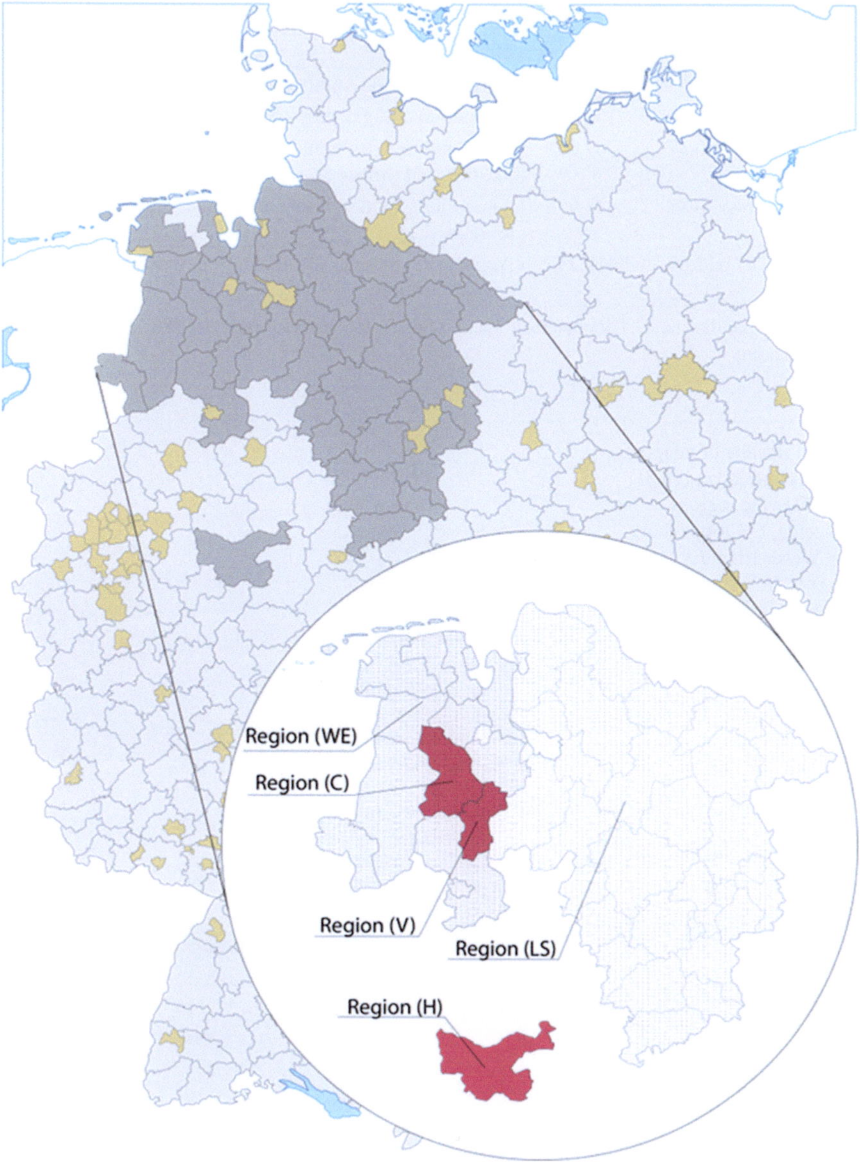

Fig. 1 Study regions of Germany

5 Food Industry

The indication of regional development sets the basis for analysis of separate industries (e.g., pork production, food industry) in relation to the regional development estimates. We indicated the performance of biotic resource flows in regions

Table 1 Main characteristics of the studied regions (Regionaldatenbank Deutschland)

Regions	Area, km^2	Population	Main industries
(V)	812,55	135,374	Agriculture, food, mining, plastics
(C)	1,418,34	158,194	Agriculture, biogas production
(H)	1,960,17	267,601	Metal and foundry industry, machinery, electricity, plastics, and timber
(WE)	14,970,85	2,477,975	Agriculture, food industry, mining, car manufacture, tourism
(LS)	47,612,88	7,918,293	Agriculture, mining, manufacturing, tourism, trade, communication

(*V* Vechta Landkreis, *C* Cloppenburg Landkreis, *H* Hochsauerlandkreis, *WE* Weser-Ems, *LS* Lower Saxony

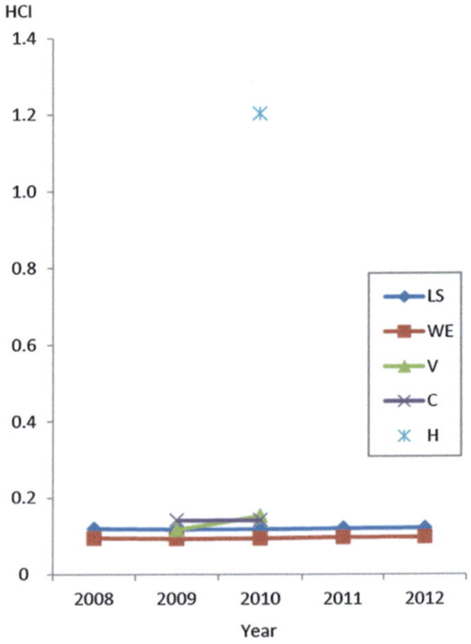

Fig. 2 Results of adapted Han cycling index (HCI) applied to the regions. Note: *V* Vechta, *C* Cloppenburg, *WE* Weser-Ems, *LS* Lower Saxony, *HCI* Han cycling index

for 2010. Biotic resources are characterized with a higher rate of flows cycling in regions with higher organization (WE and LS). At the same time, data weighting with monetized values leveled the difference. All the regions have higher exporting rates than importing for biotic resources.

In the next step pork production in relation to the capacity of region (WE) was identified. Hertwich et al.'s approach was used to connect production data with the components of RSAM [10]. Food (pork) production is evaluated as a part of biotic resources (Table 2), with a comparatively high efficiency of resource use for the region (WE). It reflects the important value of pork production for region (WE),

Fig. 3 Results of adapted Finn cycling index (FCI) applied to the regions. Note: *V* Vechta, *C* Cloppenburg, *WE* Weser-Ems, *LS* Lower Saxony, *HCI* Han cycling index

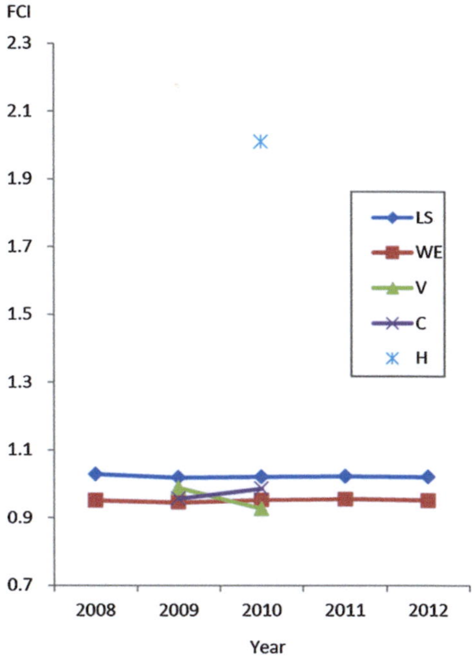

Fig. 4 Results of external flows in relation to the total amount of flows in the region. Note: *V* Vechta, *C* Cloppenburg, *WE* Weser-Ems, *LS* Lower Saxony

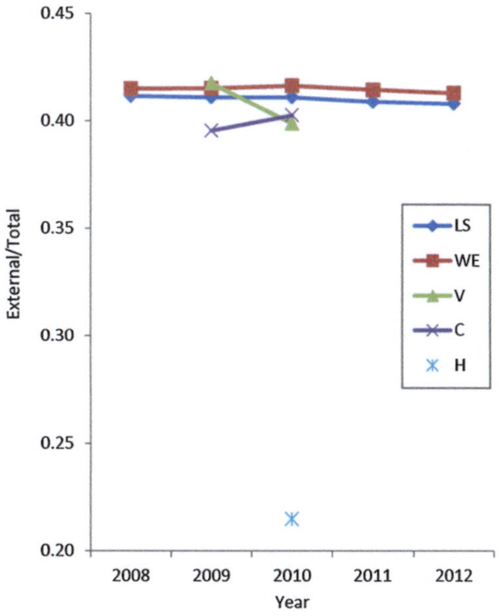

Table 2 Biotic resources comparison for regions in 2010

Regions	HCI	FCI	CCI	I/O	I/total	O/total
Absolute values comparison						
(V)	4.47	3.52	4.02	0.37	0.17	0.46
(C)	5.85	5.56	6.35	0.25	0.12	0.48
(WE)	6.05	7.55	8.63	0.21	0.12	0.57
(LS)	4.11	11.12	12.70	0.12	0.07	0.60
Monetized values comparison						
(V)	5.86	1.30	1.48	0.60	0.30	0.51
(C)	6.48	2.02	2.31	0.42	0.24	0.57
(WE)	7.05	1.86	2.13	0.50	0.28	0.55
(LS)	6.03	2.67	3.05	0.40	0.23	0.57

Note *V* Vechta, *C* Cloppenburg, *WE* Weser-Ems, *LS* Lower Saxony, *HCI* Han cycling index, *FCI* Finn cycling index, *CCI* comprehensive cycling index, *I/O* relation of inputs to outputs, *I/Total* relation of inputs to total flows, *O/Total* relation of outputs to total flows

Table 3 Biotic resource comparison for regions in 2009

Categories	HCI	FCI	E/total
Monetized values comparison			
Region (WE)	0.093	0.945	0.42
Pork production	0.148	1.279	0.44

Note: *WE* Weser-Ems, *HCI* Han cycling index, *FCI* Finn cycling index, *CCI* comprehensive cycling index, *E/Total* relation of external flows to total flows

which triggers the resources to cycle in the region while keeping comparatively low rates of external exchange (Table 3).

6 Summary

RSAM comparison of the regions allowed indication of regional resource holding capacity in the activities, dependency on connections with other regions ("weak sustainability"), and own resources. RSAM is a system-oriented, holistic assessment of regional development trends and separates regional issues in terms of social, economic, and environmental resource application. It has a great potential for the development and application provided that further comparison apparatus is developed. Data availability is one of the most important limitations of RSAM (as well as for other systems based on LCA and IO analyses). The existence of detailed data could expand RSAM, which would then become more holistic and more precise. The indicated limitations of RSAM are preliminary, as they were identified at the stage of theoretical model construction and could be resolved with practical case study application, which is planned as the next step for the research.

The possibilities of RSAM extension should be further tested with dynamic time series application for separate issues of production in multiple production and social spheres. Dynamic stage of RSAM application is tested for food production industry. Case study of regional pork production is analyzed with RSAM as its regional sustainability score. The extension of RSAM should improve the specific fields of environmental and social impact assessment. It could be previously concluded that extension of RSAM would have a probabilistic application unless the complete spectrum of social, environmental, and economic factors is included. The complete coverage of regional interaction elements is limited due to the extended data requirements. That is why RSAM is foreseen to apply a risk assessment approach for the evaluation of environmental and social impacts with regard to probability and magnitude.

References

1. Jordan SJ, Hayes SE, Yoskowitz D, Smith LM, Summers JK, Russell M et al (2010) Accounting for natural resources and environmental sustainability: linking ecosystem services to human well-being. Environ Sci Technol 44:1530–1536. http://dx.doi.org/10.1021/es902597u
2. Ness B, Urbel-Piirsalu E, Anderberg S, Olsson L (2007) Categorising tools for sustainability assessment. Ecol Econ 60:498–508. http://dx.doi.org/10.1016/j.ecolecon.2006.07.023
3. Ramos TB, Caeiro S (2010) Meta-performance evaluation of sustainability indicators. Ecol Indic 10:157–166. http://dx.doi.org/10.1016/j.ecolind.2009.04.008
4. Singh RK, Murty HR, Gupta SK, Dikshit AK (2012) An overview of sustainability assessment methodologies. Ecol Indic 15:281–299. http://dx.doi.org/10.1016/j.ecolind.2011.01.007
5. van Zeijl-Rozema A, Martens P (2010) An adaptive indicator framework for monitoring regional sustainable development: a case study of the INSURE project in Limburg, The Netherlands. Sustain Sci Pract Policy 6:6–17
6. Meadows D (1998) Indicators and information systems for sustainable development. A report to the Balaton Group
7. Leontief W (1951) Input-output economics. Sci Am 185:15–21
8. Olsen JA (1992) Input-output models, directed graphs and flows in networks. Econ Model 9:365–384. http://dx.doi.org/10.1016/0264-9993(92)90019-X
9. Yi I, Norihiro I, Inaba A, Matsumoto K (2007) Development of the interregional I / O based LCA method considering region-specifics of indirect effects in regional evaluation. Int J Life Cycle Assess 12:353–364
10. Hertwich EG (2005) Life cycle approaches to sustainable consumption: a critical review. Environ Sci Technol 36:4673–84
11. Mutel CL (2012) Framework and tools for regionalization in life cycle assessment. DISS. ETH NO. 20604. ETH. http://dx.doi.org/10.3929/ethz-a-007586392
12. Zamagni A (2012) Life cycle sustainability assessment. Int J Life Cycle Assess 17:373–376. http://dx.doi.org/10.1007/s11367-012-0389-8
13. Sala S, Farioli F, Zamagni A (2013) Life cycle sustainability assessment in the context of sustainability science progress (part 2). Int J Life Cycle Assess 18:1686–1697. Springer, Berlin/Heidelberg. http://dx.doi.org/10.1007/s11367-012-0509-5
14. Cooke P, Clifton N, Oleaga M (2005) Social capital, firm embeddedness and regional development. Reg Stud 39:1065–1077. http://dx.doi.org/10.1080/00343400500328065

15. Blažek J, Hampl M (2009) Types and systems of actors in regional development: their function and regulatory potential. Eur Spat Res Policy 16:75–92. http://dx.doi.org/10.2478/v10105-009-0005-5
16. Burkhard B, de Groot R, Costanza R, Seppelt R, Jørgensen SE, Potschin M (2012) Solutions for sustaining natural capital and ecosystem services. Ecol Indic 21:1–6. http://dx.doi.org/10.1016/j.ecolind.2012.03.008
17. Dicks L, Bardgett R, Bell J, Benton T, Booth A, Bouwman J et al (2013) What do we need to know to enhance the environmental sustainability of agricultural production? A prioritisation of knowledge needs for the UK food system. Sustainability 5:3095–3115. http://dx.doi.org/10.3390/su5073095
18. Florida R, Mellander C, Stolarick K (2008) Inside the black box of regional development – human capital, the creative class and tolerance. J Econ Geogr 8:615–649. http://dx.doi.org/10.1093/jeg/lbn023
19. Birch K, Levidow L, Papaioannou T (2010) Sustainable capital? The neoliberalization of nature and knowledge in the European "Knowledge-based Bio-economy". Sustainability 2: 2898–2918. http://dx.doi.org/10.3390/su2092898
20. Devine-Wright P, Fleming PD, Chadwick H (2001) Role of social capital in advancing regional sustainable development. Impact Assess Proj Apprais Taylor & Francis 19: 161–167. http://dx.doi.org/10.3152/147154601781767096
21. Allen CR, Fontaine JJ, Pope KL, Garmestani AS (2011) Adaptive management for a turbulent future. J Environ Manag 92:1339–1345. http://dx.doi.org/10.1016/j.jenvman.2010.11.019
22. Martin R, Sunley P (2007) Complexity thinking and evolutionary economic geography. J Econ Geogr 7:573–601. http://dx.doi.org/10.1093/jeg/lbm019
23. Voinov A. (2008) Understanding and communicating sustainability: global versus regional perspectives. Environ, Dev Sustain. http://dx.doi.org/10.1007/s10668-006-9076-x
24. Simmie J, Martin R (2010) The economic resilience of regions: towards an evolutionary approach. Camb J Reg Econ Soc 3:27–43. http://dx.doi.org/10.1093/cjres/rsp029
25. Garmestani AS, Allen CR, Gunderson L (2009) Panarchy: discontinuities reveal similarities in the dynamic system structure of ecological and social systems. Ecol Soc 14(1):15. [online] URL: http://www.ecologyandsociety.org/vol14/iss1/art15/
26. Allen CR, Cumming GS, Garmestani AS, Taylor PD, Walker BH (2011) Managing for resilience. Wildl Biol 17:337–349. http://dx.doi.org/10.2981/10-084
27. Hendrickson C, Horvath A, Joshi S, Lave L (1998) Economic input-output models for environmental life-cycle assessment. Environ Sci Technol 32:184A–191A. http://dx.doi.org/10.1021/es983471i
28. Allesina S, Ulanowicz RE (2004) Cycling in ecological networks: Finn's index revisited. Comput Biol Chem 28:227–233. http://dx.doi.org/10.1016/j.compbiolchem.2004.04.002
29. Finn J (1976) Measures of ecosystem structure and function derived from analysis of flows. J Theor Biol 56:363–80
30. Bellamy MA, Basole RC (2013) Network analysis of supply chain systems: a systematic review and future research. Syst Eng 16:235–246. http://dx.doi.org/10.1002/sys
31. Liu GY, Yang ZF, Chen B, Zhang Y (2011) Ecological network determination of sectoral linkages, utility relations and structural characteristics on urban ecological economic system. Ecol Model, Elsevier BV 222:2825–2834. http://dx.doi.org/10.1016/j.ecolmodel.2011.04.034
32. Fath BD, Halnes G (2007) Cyclic energy pathways in ecological food webs. Ecol Model 208: 17–24. http://dx.doi.org/10.1016/j.ecolmodel.2007.04.020
33. Ulanowicz RE (2004) Quantitative methods for ecological network analysis. Comput Biol Chem 28:321–339. http://dx.doi.org/10.1016/j.compbiolchem.2004.09.001
34. Han B-P (1997) On several measures concerning flow variables in ecosystems. Ecol Model 104:289–302. http://dx.doi.org/10.1016/S0304-3800(97)00137-3

Spatiotemporal Tools for Regional Low-Carbon Development: Linking LCA and GIS to Assess Clusters of GHG Emissions from Cocoa Farming in Peru

Giancarlo Raschio, Sergiy Smetana, Christian Contreras, Alexander Mathys, and Volker Heinz

Abstract Understanding spatial and temporal variations of greenhouse gas (GHG) emissions at local level is important for both companies in the agri-food sector working to improve sustainability in their supply chains and local governments following a low-carbon development pathway. For this reason, it is necessary to count with methodologies that do not only estimate GHG emissions, but provide analytics about their location and temporal variations. Factors such as spatial distribution of farms, variations on the requirement of inputs depending on the age of the farms, as well as transportation distances for the main and complementary products, should be assessed at farm level. To accomplish this, spatial and temporal analytics can be incorporated into life cycle assessment (LCA) by joining it with geographic information systems (GIS). This paper explores how spatial statistics tools can be applied to identify spatial and temporal trends in GHG emission results obtained from an LCA conducted on cocoa farming in the region of San Martin in Peru. Results indicate that it is possible to identify temporal and spatial trends of statistically significant clusters of farms with high GHG emissions (hot spot analysis).

Keywords LCA • GIS • GHG emissions

G. Raschio (✉)
German Institute of Food Technologies (DIL e.V.), Prof.-von-Klitzing-Str. 7, 49610 Quakenbrück, Germany

Ecosystem Services LLC, 1250 24th Street, NWSuite 300, Washington, DC 20037, USA
e-mail: g.raschio@dil-ev.de

S. Smetana • A. Mathys • V. Heinz
German Institute of Food Technologies (DIL e.V.), Prof.-von-Klitzing-Str. 7, 49610 Quakenbrück, Germany

C. Contreras
Ecosystem Services LLC, 1250 24th Street, NWSuite 300, Washington, DC 20037, USA

M. Matsumoto et al. (eds.), *Sustainability Through Innovation in Product Life Cycle Design*, EcoProduction, DOI 10.1007/978-981-10-0471-1_66

1 Introduction

Suppliers provide businesses with agri-food products from specific areas and not from all producing areas in a country at once. Also, impacts generated from agri-food production activities such as farming and transportation of products (main and complementary crops) vary based on factors such as the spatial distribution of farms and the length and condition of road infrastructure network [1]. This is also true for cocoa production in Peru and its supply to international and domestic chocolate producers.

Cocoa is the perennial crop with the second largest cultivated area in the Peru's Amazon forest, and its production is gaining momentum: the cultivated area increased in 76,300 ha in the past 10 years [2], and it is projected to increase in another 28,000 ha by 2016 [3]. Peru ranks 8th among cocoa producers and 12th among cocoa exporters; however, it is expected that the country will rank among top ten exporters in the upcoming years [4].

The positioning of Peru's cocoa in the international markets, together with a growing global demand and an expected shortage by the year 2020 [5], set the conditions for Peru to keep expanding its cocoa production. This will trigger not only socioeconomic development for low-income farmers but also higher greenhouse gas (GHG) emissions from agricultural and land-use change activities under current agricultural practices. In Peru, GHG emissions from land-use change (35 %) and from agriculture (21 %) account for 56 % of the total GHG emissions in the country; GHG emissions from agriculture have increased in 10 % in the last 10 years [6]. Cocoa in Peru is produced mainly by smallholders whose farms on average are smaller than 3 ha [4]. Smallholders are identified as the main drivers of deforestation in the country given they continuously expand the agricultural frontier at the expense of highly biodiverse forest ecosystems [7].

Environmental impacts, such as global warming as a result of increasing GHG emissions, can be estimated using life cycle assessment (LCA). LCA is a tool that identifies environmental impacts throughout the life cycle of goods and services by using country-level or global average information and taking a snapshot in time of a process. By itself, LCA doesn't account for the diversity of farming practices for a crop in a region and for nonproductive and/or low or high productive years [8].

Some studies have used LCA to assess the environmental impacts from farming practices of perennial crops such as cocoa; however, these studies do not account for the spatial variability of farms, the different requirements of inputs of cocoa at each growing stage, nor GHG emissions from the production of complementary shade crops [9], such as banana for the case of cocoa farmers in Peru [10]. Other studies on perennial crops recognized the need to include spatial and temporal analytics into LCA to account for variations and effects at a local scale [11, 12].

It is then necessary to come up with a method to incorporate spatial and temporal analytics into LCA studies to understand how local dynamics affect the trends of environmental impacts. This can be achieved by coupling geographic information system (GIS) tools with LCA.

Besides identifying where and when impacts occur, it is necessary to identify the variables behind their location and temporal trends. The spatial statistics tools found in GIS can provide useful data for decision-makers in the development of effective low-carbon interventions at local or regional level.

This study explores how LCA and GIS can complement each other by applying spatial statistics to identify and assess spatial trends of high GHG hot spots over time. Results are presented with application to a case study of cocoa farming in the Tocache Province, region of San Martin, which is one of the highest cocoa-producing regions in Peru.

2 Methodology

We used data from a census conducted in 2011 by DEVIDA (National Commission for Development and Life Without Drugs) [13] in the Tocache Province, region of San Martin in Peru (Fig. 1).

The census consisted of a pluri-annual dataset with information on 1187 farms of different implementation date (from 2006 to 2010), thus at different production stages at the moment of the census. For example, at the time of the census, farms implemented in 2006 were at their fifth year (third year of cocoa production), whereas farms implemented in 2010 were at their first year (implementation year), thus not producing cocoa yet, only banana (complementary shade crop). For this study, we only considered farms under cocoa production in 2008, 2009, and 2010 (Fig. 2).

This study is based on attributional LCA approach, as the aim of LCA is to estimate the differentiation of cocoa-farming impacts of, based on agricultural techniques, temporal and spatial conditions. Therefore, product allocation was conducted to account for the production of banana (complementary shade crop) only in the first year of cocoa production (third year of a farm lifetime). We based the allocation on a mass (weight) basis of products. GHG emissions from banana production were allocated to the second and third years of cocoa production by using a factor of 0.1 of the total GHG emissions generated from cocoa farming in that year; this factor was calculated based on secondary information for the study area [14]; GHG emissions from transportation of banana were calculated separately using a GIS database.

We used LCA to calculate GHG emissions via the environmental impact category global warming potential (GWP100) (tCO2eq). The GWP100 was calculated for each of the first 5 years of cocoa farming using as a functional unit 1 hectare (ha)/year. This reference unit is recommended when the focus of a study is on land use [15] and its associated social perspective of preserving landscape [16]. The LCA was conducted using OpenLCA, NREL US inventory data, and IPCC 2004 impact assessment method. GWP100 results were then exported as a shapefile into a GIS system.

Fig. 1 Location of the study area in Peru

2.1 GHG from Transportation

To calculate the distances from each farm to the marketplace, we used publicly available official information on the road network in the study area, generated by the regional government of San Martin in a shapefile format.

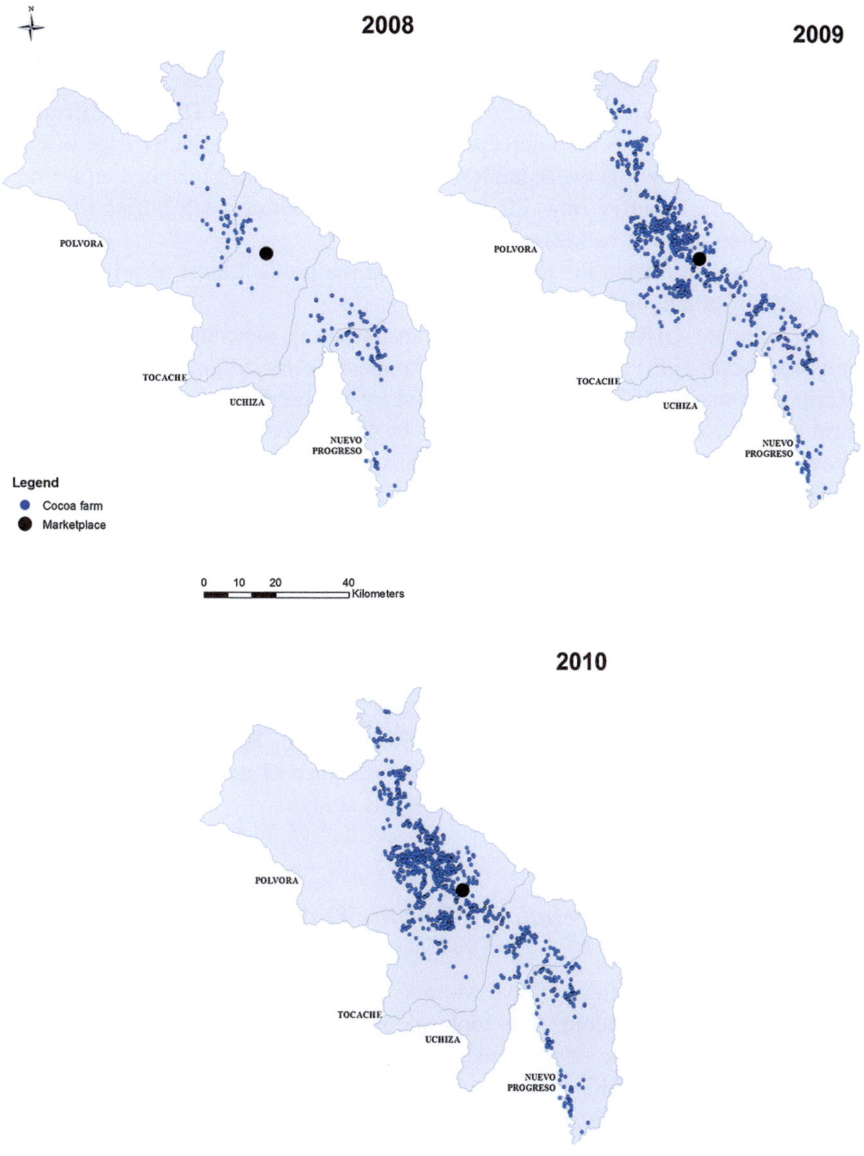

Fig. 2 Farms under cocoa production in 2008, 2009, and 2010

Distances were estimated for each one of the 1187 farms in a three-step process: (1) distance from farms to the road network, (2) distance from the road network to the marketplace, and (3) total travel distance from each farm.

First, we estimated the distance from each farm to the road network via multiple buffer rings with 0.5 km separation distance starting from the road network. Then, we identified the farms within each buffer ring and assigned such distance as (d1) to each farm.

Second, we estimated the distance from the road network to the marketplace via multiple distance buffer rings of 0.5 km from the marketplace. Then, we calculated the travel distance from the interception of the roads with the outer edge of each buffer ring by using a network analyst tool. In the end, we calculated an average distance for each buffer ring, identified the farms within each buffer ring, and assigned such distance as (d2) to each farm.

Finally, we estimated the total distance that the product must travel from each farm (d3) by adding (d1) and (d2).

To calculate GHG emissions from transportation, we multiplied (d3) by an emission factor (EFt) of 0.72061 kgCO_2e/tonnes*km for a light commercial truck obtained from the NREL database [18] and by the production of each individual farm in each year. The EFt was assumed to be similar for all the farms based on our knowledge of the study area. GHG from transportation for the first year of production of each farm was allocated between the transported weight of cocoa and that of bananas, by using the same allocation GWP100 factors.

2.2 GHG from Cocoa Production

Results from GWP100 (tCO_2e/ha) were used to calculate the GHG emissions of each farm producing cocoa in 2008, 2009, and 2010. In cocoa production, we considered only farming and transportation activities (Table 1); GHG emissions from deforestation are not accounted for in this study.

2.3 Spatial Autocorrelation (Moran's I) Analysis

The spatial autocorrelation (Global Moran's I) tool evaluates if a dataset is clustered, dispersed, or random. The tool calculates both a *z*-score and *p*-value to evaluate the significance of that index. We applied this analysis to understand if there is clustering of GHG emissions in each of the assessment years.

The *z*-score refers to the standard deviation of the variable under consideration, and it can be used as an indicator of the degree of clustering: the higher the *Z*-value in a given year, the higher the clustering. The *p*-value refers to the probability that the

Table 1 Total GHG emissions by production year

Production year	Number of farms producing cocoa	Total GHG emissions from cocoa production (tCO_2e)
2008	155	56.12
2009	909	420.67
2010	1187	495.44

Table 2 Total GHG emissions by production year

Significance level (*p*-value)	Critical value (*z*-score)	Confidence level
0.10	<1.65–1.96>	90 %
0.05	<1.96–2.58>	95 %
0.01	>2.58	99 %

observed spatial pattern is the result of a random event. Threshold values for both the *z*-score and the *p*-score depend on the confidence level we aim for [17] (Table 2).

The results from this analysis indicate if there data points are clustered or dispersed, and it is the previous step before deciding to undertake a hot spot analysis.

2.4 Incremental Spatial Autocorrelation

Incremental spatial autocorrelation measures spatial autocorrelation for a series of distances and optionally creates a line graph of those distances and their corresponding *z*-scores. *Z*-scores reflect the intensity of spatial clustering, and statistically significant peak *z*-scores indicate distances where spatial processes promoting clustering are most pronounced. These peak distances are often appropriate values to use for tools such as hot spot analysis [18].

The peak distance shown by this tool represents the distance at which our date showed the highest clustering for a given year. We used this distance to decide the neighborhood size that will define the conceptualization of the spatial relations of the farms, which is needed to perform a hot spot analysis.

2.5 Hot Spot Analysis

Hot spots represent low- or high-value clusters of impact sources with significant statistical evidence to prove, at a certain confidence level, that their aggregation is not the result of a random event [18]. In this study, we focus only on high-value clusters based on the GHG emission values of each farm producing cocoa in 2008, 2009, and 2010. Only high-value hot spots with a confidence level of 95 % or higher were selected for the analysis.

2.6 Directional Distribution (Standard Deviation Ellipse)

By using directional distribution, it is possible to summarize the spatial characteristics of geographic features: central tendency, dispersion, and directional trends. The tool allows elongated distribution identification to estimate an orientation [18]. For this study, the analysis was conducted at a district level.

3 Results

3.1 Spatial Autocorrelation (Moran's I)

Results from the spatial autocorrelation analysis were evaluated based on the parameters indicated in Table 2. Results indicated the presence of strong clustering of high GHG emission values in each of the 3 years of analysis (2008–2010) as shown in Table 3.

3.2 Incremental Spatial Autocorrelation

Results from the incremental spatial autocorrelation analysis indicate that for each of the years of analysis, there is a distance of maximum clustering (Table 4).

The peak distance in 2008 is significantly higher than that of the following years because in that year there were less farms producing cocoa and these were farther apart than in 2009 and 2010.

3.3 High GHG Emission Clusters: Hot Spot Analysis

Results from the hot spot analysis indicated that in 2008 and 2009, high GHG emission clusters existed only in the districts of Polvora and Tocache, whereas in 2010, hot spots were identified also in the district of Uchiza. Hot spots were not identified in the district of Nuevo Progreso in any of the years of analysis (Fig. 3).

Table 3 Spatial autocorrelation results for GHG emissions from cocoa production from 2008 to 2010

Year of GHG emissions from production	Moran's index	*z*-score	*p*-value
2008	0.478653	3.823666	0.000131
2009	0.216964	9.224875	0.000000
2010	0.127221	6.879963	0.000000

Table 4 Incremental spatial autocorrelation for GHG emissions from cocoa production from 2008 to 2010

Year of GHG emissions from production	Peak distance (m)
2008	14661.68
2009	8272.52
2010	5127.71

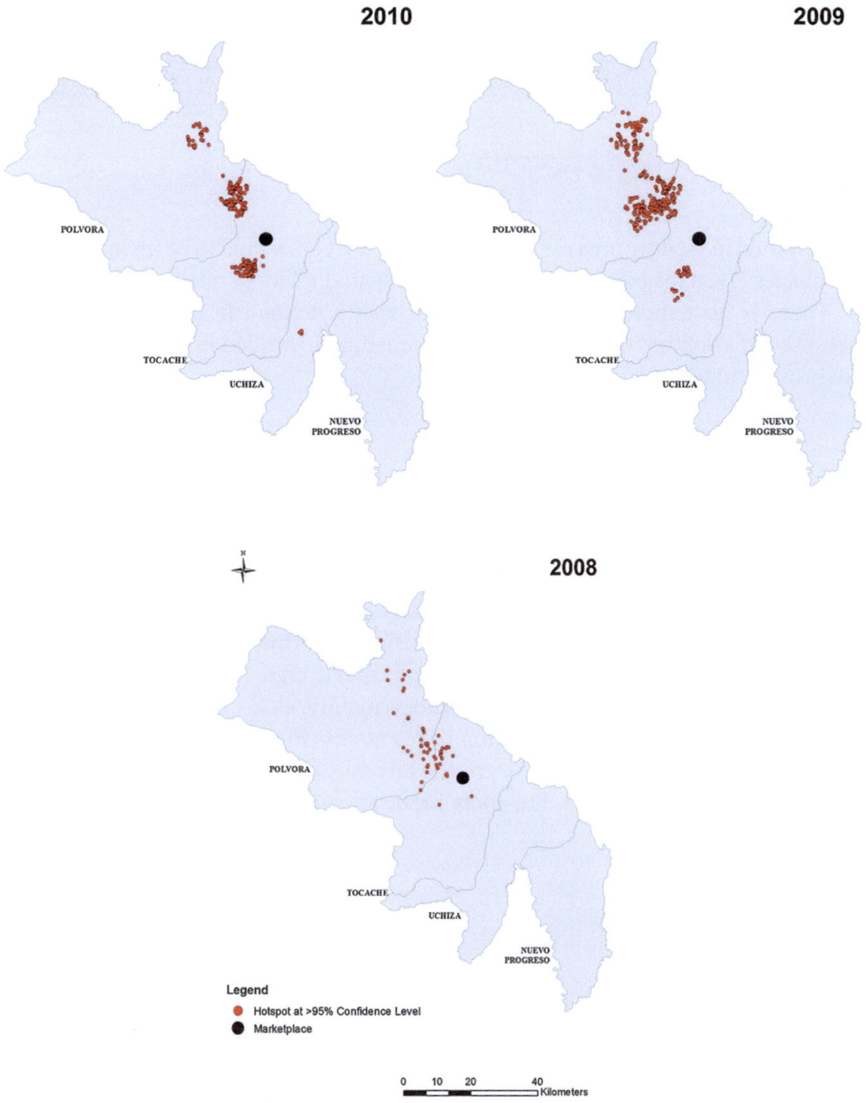

Fig. 3 High GHG emission hot spots from cocoa production (farming plus transportation) in 2008, 2009, and 2010

It can be observed that the share of GHG emissions from hot spots regarding the total GHG emissions from cocoa production decreases over time in the assessment period (Table 5).

The decrease in the number of farms within clusters does not mean that farms have lower GHG emissions over time or that their clustering is a random event. Fewer farms within clusters might respond to the fact that some farms have significantly high GHG emissions (farms in their third year of cocoa production); thus,

Table 5 Temporal variation of the number of farms belonging to high GHG emission hot spots and their relative share of GHG emissions to the total between 2008 and 2010

Production year	% from total number of farms	% of total GHG emissions
2008	35.5 %	48.5 %
2009	37 %	42.8 %
2010	20.7 %	25.9 %

fewer farms in vicinity area or "neighborhood" are considered to be included within the cluster given the conceptualization of the spatial relation, selected for the farms.

There is no temporal trend for the hot spot identified in the Uchiza District because emission hot spots in the district were only identified in the last year of the analysis period.

3.4 *Directional Distribution and Temporal Trends in Hot Spot Location*

Results from the directional distribution analysis indicate that high GHG emission clusters maintain a relatively similar direction over the years. Overall standard deviation appears to be lower along the x-axis but higher along the y-axis. This trend reflects an occupation of the territory mainly along primary roads, which facilitate the access and transportation of resources (Fig. 4).

In Polvora District, hot spots seem to shift toward the northeast of the district, whereas in Tocache District, hot spots seem to shift toward the southwest of the district.

4 Conclusions

- Although coupling LCA and GIS tools is not a novel approach, ongoing LCA studies can provide better understanding on the dynamics of environmental impacts at local level by incorporating spatial and temporal analytics.
- GIS facilitates the generation of data to estimate GHG emissions from transportation at individual farm level, which otherwise would not be possible for LCA alone.
- GIS facilitates the incorporation of the temporal dimension into LCA, thus facilitating accounting for GHG emissions at the farm level and at different production stages. This could allow for environmental impacts not to be estimated as averages or snapshots in time, but providing more accurate results in the estimation of GHG emissions from agricultural supply chains.
- Spatial statistics tools provide statistically robust evidence to identify areas with high levels of GHG emissions from farming and transportation.

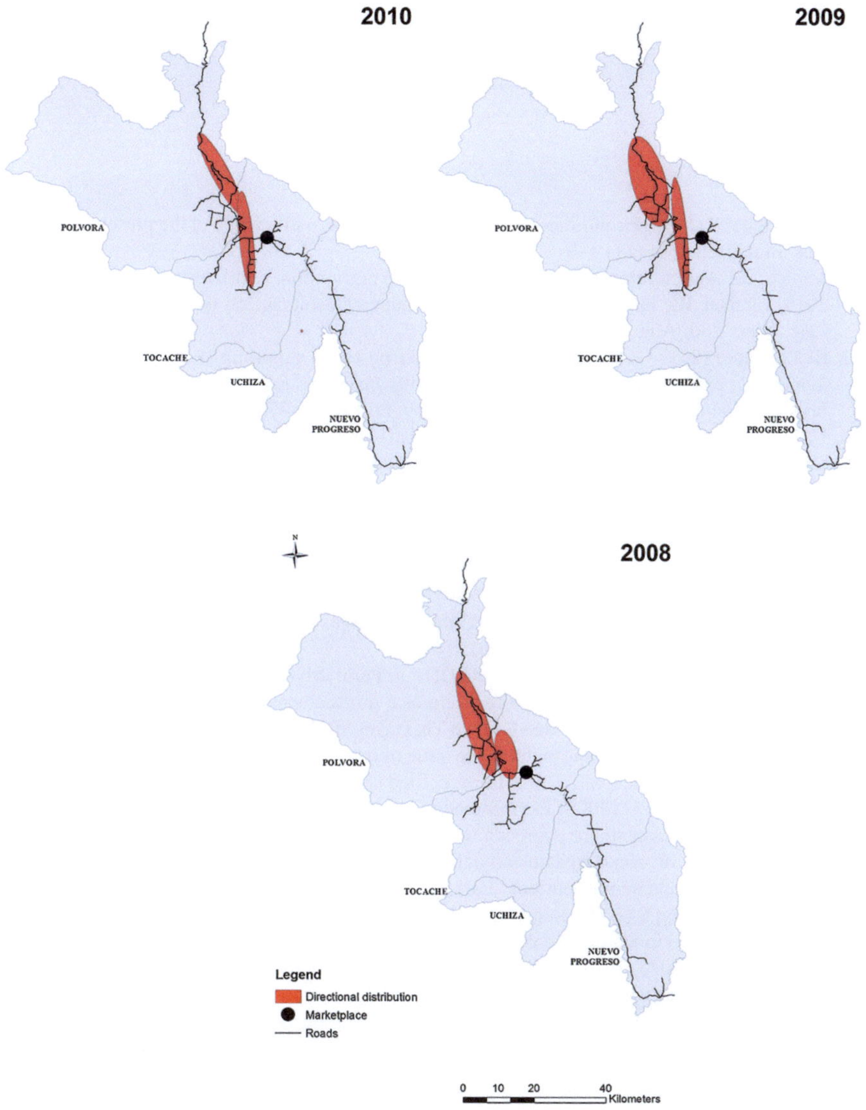

Fig. 4 Directional distribution for high GHG emission hot spots for cocoa production in 2008, 2009, and 2010

– Hot spot analysis can provide insights about the underlying process of land occupation to understand if such process follows a random pattern or if it is associated with spatial variables.
– Hot spots in the study area tend to locate over time toward the forest edge. This should be further analyzed to determine if there is a correlation between high GHG emission clusters and deforestation from implementing cocoa farms.

Acknowledgment This research is possible thanks to the generous support of the Alexander von Humboldt Foundation.

References

1. Schaltegger S (1997) Economics of life cycle assessment: inefficiency of the present approach. Bus Strateg Environ 6(1):1–8
2. Ministry of Agriculture of Peru (MINAG) (2015) Agriculture and livestock dynamics 2004-2013, Chapter III. http://sinia.minam.gob.pe/documentos/dinamica-agropecuaria-2004-2013. Last access 28th Aug 2015
3. El Comercio (2014) Peru could be world leader on cacao exporting [Perú podría ser el líder mundial en exportación de cacao]. Comercio. http://elcomercio.pe/economia/peru/peru-podria-lider-mundial-exportacioncacao-noticia-1704459. Last access 27th June 2015
4. USDA Foreign agricultural sector (2014) Global Agricultural Information Network (GAIN) report. http://gain.fas.usda.gov/Recent%20GAIN%20Publications/Cocoa%20Update%20and%20Outlook_Lima_Peru_10-28-2014.pdf. Accessed on 27th June 2015
5. The Guardian (2014) The cocoa crisis: why the world's stash of chocolate is melting away. http://www.theguardian.com/lifeandstyle/2014/nov/21/cocoa-crisis-world-chocolate-stash-melting-away. Last access 27th June 2015
6. Ministry of the Environment of Peru (MINAM) (2014) First biannual report on Peru's update to the United Nations' Convention on Climate Change. http://unfccc.int/resource/docs/natc/perbur1.pdf. Last accessed 27th June 2015
7. Piu HC, Menton M (2013) Context of REDD+ in Peru: drivers, stakeholders, and institutions [Contexto de REDD+ en Perú: motores, actores e instituciones] (vol 90). CIFOR
8. Basset-Mens C, Benoist A, Bessou C, Tran Th, Perret SR, Vayssières J, Wassenaar T (2010) Is LCA-based eco-labelling reasonable? The issue of tropical food products. LCAFood2010. VII International conference on LCA in the agri-food sector
9. Ntiamoah A, Afrane G (2008) Environmental impacts of cocoa production and processing in Ghana: life cycle assessment approach. J Clean Prod 16(16):1735–1740
10. USDA (2014) Peru: cocoa update and outlook. Global Agricultural. http://www.fas.usda.gov/data/peru-cocoa-update-and-outlook. Last accessed 27th June 2015
11. Mutel CL, Hellweg S (2009) Regionalized life cycle assessment: computational methodology and application to inventory databases. Environ Sci Technol 43(15):5797–5803
12. Bessou C et al (2013) LCA applied to perennial cropping systems: a review focused on the farm stage. Int J Life Cycle Assess 18(2):340–361
13. DEVIDA (2011) Alternative crops monitoring 2011–2012. DEVIDA's Alternative Development Project
14. Silva PA, Castillo LE, Toledo JR (2012) Exportación de nutrientes (N, P, K) en el cultivo de banano orgánico de la Universidad EARTH. Tierra Tropical
15. Fornasiero P, Graziani M (eds) (2011) Renewable resources and renewable energy: a global challenge. CRC press, Boca Raton, p 200
16. Bellon S (Ed.) (2014) Organic farming, prototype for sustainable agricultures. Springer, p 224
17. U.S. Life Cycle Inventory Database (2012) National renewable energy laboratory, 2012. https://www.lcacommons.gov/nrel/search. Last accessed 28th Aug 2015
18. ESRI ArcGIS Help 10.1 – spatial statistics toolbox. http://resources.arcgis.com/en/help/main/10.1/index.html#//005p0000000n000000. Last accessed 27th May 2015

Potential for Greenhouse Gas Mitigation at a Typical Roughage Production System in the Japanese Dairy System

Tatsuo Hishinuma, Kazuyoshi Suzuki, and Yutaka Genchi

Abstract Increasing domestic roughage production is an important measure in designing a sustainable dairy production system in Japan. However, roughage production is one of the agricultural activities that have a significant environmental impact. The objectives of this study were to reveal potentials of greenhouse gas (GHG) emissions in typical roughage production systems using the life cycle assessment method (LCA). Comparative LCA was conducted to evaluate GHG emissions between the domestic roughage production system and imported hay utilization system. The functional unit of LCA was defined as roughage production at 10a, and GHG/TDN ratio was applied comparing roughage production and GHG emissions of domestic and imported roughages. The results showed that the domestic roughage production system produced 20 % less GHG emissions than the imported hay utilization system. GHG/TDN ratios were associated with the LCA results to compare GHG emissions with production of different roughage production systems.

Keywords Roughage • Greenhouse gas emissions • Life cycle assessment

1 Introduction

Dairy production is an important sector for the food supply system. In Japan, dairy production contributed 26 % of the agricultural production value in the livestock production sector (8 % of total agricultural production value) over the last decade. As of 2014, there are 18,600 dairy farms and 1.35 million head of dairy cattle in Japan [1]. The type of dairy farming varies, because the dairy production system

T. Hishinuma (✉)
Department of Agricultural Environmental Engineering, Utsunomiya University, Tochigi, Japan
e-mail: thishinuma@cc.utsunomiya-u.ac.jp

K. Suzuki
Chiba Prefectural Livestock Research Center, Chiba, Japan

Y. Genchi
National Institute of Advanced Industrial Science and Technology, Ibaraki, Japan

M. Matsumoto et al. (eds.), *Sustainability Through Innovation in Product Life Cycle Design*, EcoProduction, DOI 10.1007/978-981-10-0471-1_67

involves several subsystems such as roughage production, livestock management, and manure treatment and utilization.

Mostly, dairy farmers who utilize roughage feed both domestic roughage and imported hay. Imported hay is widely used because of its improved handling ability and labor savings in the roughage production and livestock management systems. On the other hand, the agricultural area devoted to roughage production in Japan has increased slightly. From the standpoint of material circulation and environmental impacts, increasing domestic roughage production is an important measure in designing a sustainable dairy production system [2].

However, roughage production accounts for some of the dairy farming system's most significant environmental impacts. Roughage production systems consume lots of input fuels and materials, such as diesel fuel, petrol, composted manure, chemical fertilizers, herbicides, and other input materials. It has been reported that total energy consumption differs between the grassland dairy production system and the dairy production system with dent corn silage production at mixed upland crop farms [3]. Moreover, it has been reported that direct CH_4 and N_2O emissions from paddy fields and agricultural soils are associated with agricultural activities [4, 5]. Therefore, it is important to evaluate the roughage production system, which includes the potential for greenhouse gas (GHG) emissions, to design a dairy production system with low environmental impact.

In this study, a life cycle assessment (LCA) method was applied to assess a typical roughage production system (hay, maize silage, whole crop rice silage, and pasture) in Japan. We discussed environmental aspects of the domestic roughage production system compared with a utilization system of imported hay from the USA.

2 Materials and Methods

The LCA is a method to analyze environmental aspects and potential impacts of a production system (considered as a process chain) throughout a product's life cycle [6]. The method used for comparing GHG emissions from several roughage production systems was the LCA, which describes GHG emissions from resource material consumption and utilization of agricultural land distinctly and quantitatively.

The LCA consists of four phases: goal and scope definition, life cycle inventory analysis, life cycle impact assessment, and interpretation. The first phase of the LCA is to define the goal and scope of the analysis. The goal definition describes objectives and motives of the analysis and is summarized in Sect. 1 of this paper. The scope definition mainly describes the product system flowchart, the system boundary, and functional unit. The life cycle inventory analysis phase collects input and output data associated with the product system and compiles the data on the basis of the functional unit. The life cycle impact assessment phase evaluates the environmental impact quantitatively. Basically, the data from the life cycle inventory analysis are interpreted in terms of environmental impact, using the indicator of impact category (environmental problem field). The interpretation phase identifies hot spots and potentials to mitigate environmental impacts of the product system, mainly from the results of the life cycle inventory analysis and impact assessment.

2.1 System Description

This study compared the LCA of domestic roughage and imported hay to reveal their potentials to mitigate GHG emissions (Fig. 1).

The domestic roughage production system uses dent corn wrapped silage (DCS), hay, pasture management, and whole crop rice silage (WCRS), which are typical roughage production systems in Japan. The imported hay utilization system

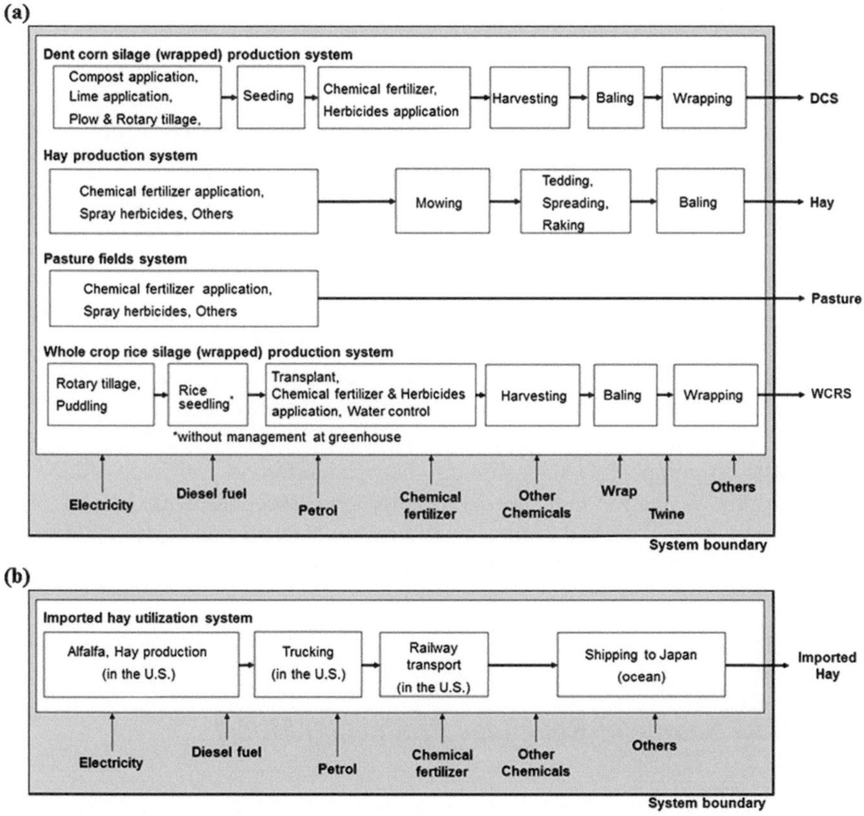

Fig. 1 Description of the domestic roughage production system and the imported hay utilization system. (**a**) Process flow and system boundary of the roughage production system. (**b**) Process flow and system boundary of utilizing imported hay

Table 1 GHG emissions associated with roughage production systems

Greenhouse gases	Emission source
CO_2	Fuels, materials
CH_4	Fuels, materials, paddy fields
N_2O	Fuels, materials, nitrogen fertilizer input, arable land

consumes hay harvested in the USA and transported to Japan. The life cycle of the evaluation system model of roughage production ran from plow or rotary tillage (for preparing cultivation) to harvesting. In particular, the evaluation system model of the imported hay utilization includes the transportation process (i.e., trucking and railway for internal transport and ocean shipping for international transport). Every evaluation system includes production of chemicals and energy materials.

The analysis is based on the functional unit defined as roughage production at the 10a agricultural field. The analysis included estimation of CO_2, CH_4, and N_2O emissions at the phase of life cycle inventory analysis (Table 1). The GHG emissions were evaluated using GWP_{100} (CO_2:1, CH_4:34, N_2O: 298) [7].

2.2 Evaluation Indicator for Comparing Greenhouse Gas Emissions from Roughage Production Systems

To compare GHG emissions between the domestic roughage production system and the imported hay utilization system, we focused attention on a total digestible nutrient (TDN). TDN is a value for the percentage of energy content of feedstuff based on raw materials or dry matter. TDN was calculated from the amount of crude protein (CP), crude fat (C.fat), nitrogen-free extract (NFE), and crude fiber (C.fib), as well as the digestibility of feed nutrition. TDN is used widely in livestock feed and nutrition management in Japan.

In order to compare GHG emissions between different roughage stuffs, we defined as the evaluation indicator GHG emissions associated with 1 kg of TDN production as the GHG/TDN ratio [kg CO_2eq/kg TDN]. Lower GHG/TDN ratio indicates positive trends for GHG emission mitigation in the roughage production system.

2.3 Data Source of Roughage Production Model

The data for inventory analysis (such as fuel and material consumption and chemical fertilizer) of the roughage production system were obtained from the interviews and site investigation. We conducted a survey on material consumption in the roughage production systems at two farms, which were the Utsunomiya

University (UU) farm and roughage production contractor (RC) in Chiba Prefecture. The UU produces DCS, hay, and pasture. The RC harvests DCS and WCRS.

2.3.1 Dent Corn Silage (Bale Wrapped Silage)

Table 2 shows yields, fuels, and material consumption of the two DCS production systems. The UU DCS production system was harvested at three dent corn fields of different hectarage (UU-1, UU-2, and UU-3). Several tractors were utilized at the UU agricultural activities. In particular, the 100PS tractor was mainly used for field work. The manure compost was applied at the dent corn fields with the 100PS tractor and tractive manure spreader, during the cultivation process. The machinery used in the harvesting process were the corn harvester (two rows), roll baler, and bale wrapper (to make bale wrapped silage). TDN content of produced DCS at the UU was estimated using data from standard tables of feed composition in Japan (2001) [8].

The data on fuel and material consumption in the cultivation process of DCS at the RC was obtained from interviews. We analyzed GHG emissions from the cultivation process of DCS at two types of fields: upland field and converted paddy field. We investigated the harvesting activities at the 51.3a of dent corn field and collected the data of fuel and material consumption in the DCS harvesting process. Capacities of machinery for the DCS production system at the RC were about the same as machinery utilized at the UU. The machinery for wrapping dent corn roll bales was crawler type and the fuel was petrol at the RC. We obtained the data of TDN content of DCS produced by RC from site investigation.

2.3.2 Hay

Table 3 shows yields as well as fuel and material consumption of hay production activities in the UU hay production system, harvested at three fields. The grass

Table 2 Roughage yield, fuel, and material consumption (main item) in two DCS production systems

	Unit	UU*	RC
Harvested area	[ha]	5.1	0.5
Seeds	[kg]	85.0	20.0
Diesel fuel	[L]	738.5	192.4
Petrol	[L]	–	7.0
Chem. fertilizers N	[kg]	568.4	35.0
Herbicide	[ml]	29,250	3850
Wrapping materials	[kg]	322.7	86.1
Yields of dent corn	[t]	100.8	26.1

* The values are total amount of each input materials, and total yields, of the three DCS production fields (UU-1, UU-2 and UU-3).

Table 3 Roughage yield, fuel, and material consumption (main item) in hay production system at the UU

	Unit	UU-4	UU-5	UU-6
Harvested area	[ha]	2.8	1.2	1.6
Seeds	[kg]	–	–	–
Diesel fuel	[L]	373.0	236.0	228.9
Petrol	[L]	–	–	–
Chem. fertilizers N	[kg]	274.4	67.2	163.2
Herbicide	[ml]	1120	160	280
Twine materials	[roll]	12.3	4.6	5.9
Yields of hay	[t]	22.2	8.7	10.6

Table 4 Roughage yield, fuel, and material consumption (main item) in pasture production system at the UU

	Unit	UU-7	UU-8	UU-9
Harvested area	[ha]	1.5	1.3	1.3
Seeds	[kg]	–	–	–
Diesel fuel	[L]	25.0	14.2	16.4
Petrol	[L]	–	–	–
Chem. fertilizers N	[kg]	84.0	111.2	114.4
Herbicide	[ml]	40.0	120.0	40.0
Pasture intake (estimation)	[t]	24.1	36.4	33.3

variety of the hay fields was mixed: orchard grass (*Dactylis glomerata* L.), tall fescue (*Festuca arundinacea* Schreb), perennial ryegrass (*Lolium perenne* L.), and white clover (*Trifolium repens* L.). The hay was harvested three times at each hay field and mowed in May, July, and October. The drying process (includes tedding, spreading and raking) before the baling process was important for the quality of storage and nutrients of the hay. In May, the hay production system at the UU had a tedding process of 11–16 sets. The 50PS tractor was mainly used for tedding and several other processes. TDN content of hay was estimated using data from the literature [8].

2.3.3 Pasture

The input and output data of pasture field management were collected from site investigation at the UU (Table 4). The UU had three pasture fields for dairy cattle; the variety of grasses was the same as the hay fields. The length of grazing period was half a year, from May to October. Pasture field management included chemical fertilizer application, routine mowing, and herbicide application. The 50PS tractor was utilized for pasture management. The TDN content of the pasture was estimated using data from the literature [8], including difference of nutrients in the grasses.

2.3.4 Whole Crop Rice Silage (Bale Wrapped Silage)

The data for analyzing the WCRS production system were obtained from the literature [9, 10] and site investigation (Table 5). WCRS was the roughage production system based on the rice production with paddy fields. The WCRS production system, in many cases, has separate work processes between rice farmers and dairy farmers. Rice farmers cultivate whole rice crops at the paddy fields using their own machinery for rice production. Dairy farmers or roughage production contractors also harvest entire rice crops at the paddy fields, but they apply the machinery of hay production, mower balers, or other specific machines for WCRS harvesting.

The data of the WCRS cultivation process (from preparing paddy fields to planting and water control) were based on the case of a typical paddy farmer with a side job. In this farm, the rice-planted area was 1.4 ha, of which 0.7 ha was rice for WCRS. Major machinery used included tractor, rice transplanter (walking type), and rotary harrow.

The data of the WCRS harvesting process (harvesting, baling, and wrapping) were analyzed with regard to the utilization of two types of machinery. Type 1 was a specific harvester for WCRS at paddy fields. Type 2 was hay production machinery (tractor, mower, baler, bale wrapper, and others) to harvest WCRS (Table 6). Type 2 was medium-sized machinery for hay production.

TDN content of WCRS was estimated from the data, which was used to analyze feed nutrients by the Chiba Prefectural Livestock Research Center in 2005 and 2007. The data had variety of rice crops for roughage feedstuff.

Table 5 Roughage yield, fuels, and material consumption (main item) in WCRS production system

	Unit	Case 1	Case 2
Harvested area	[ha]	1.3	1.2
Seeds	[kg]	40.9	37.7
Diesel fuel	[L]	260.4	255.3
Petrol	[L]	26.9	24.9
Chem. fertilizers N	[kg]	58.5	54.0
Chemicals	[kg]	32.0	29.6

Case 1 used specific machinery for harvesting (type 1); case 2 used middle size of hay harvesting machinery (type2) for harvesting

Table 6 Machinery use in WCRS harvesting (two types)

	Type 1	Type 2
Harvested area (WCRS)	1.3 ha	1.2 ha
Harvested area (other)	–	1.0 ha (straw)
Machinery	WCRS harvester (combine type)	Tractor (3)
	Bale wrapper (crawler)	Rotary disk mower
	Wheel loader	Rotary rake
		Trailed roll baler
		Bale wrapper (crawler)
		Wheel loader

2.4 Data Source of Imported Hay Utilization Model

In this study, the imported hay utilization system model is composed of the production process and the transportation process.

2.4.1 Production Process

The process inventory of hay production was analyzed from the literature, which shows fuel and material consumption of alfalfa hay and other hay production systems in the USA [11, 12]. The data included the following hay production areas: western area (Colorado), midwestern area (South Dakota, Missouri, and Wisconsin), south area (Georgia), and east area (New York). In the USA, almost all states produce hay, but material consumption, harvest periods, and yields differ by production area. We applied average data (519.3 g CO_2eq/kg) at seven areas in the USA for comparative analysis (Fig. 2).

The TDN content of imported hay was calculated with digestibility ratio of dairy cattle [3] and nutrient content data [13]. The nutrient data of imported hay was obtained from a report that analyzed actual nutrient conditions of imported hay at dairy farms in Chiba (Table 7). Grass nutrient data of four varieties (alfalfa, timothy, Sudan grass, and oat hay) utilized widely in Chiba were applied for TDN calculation.

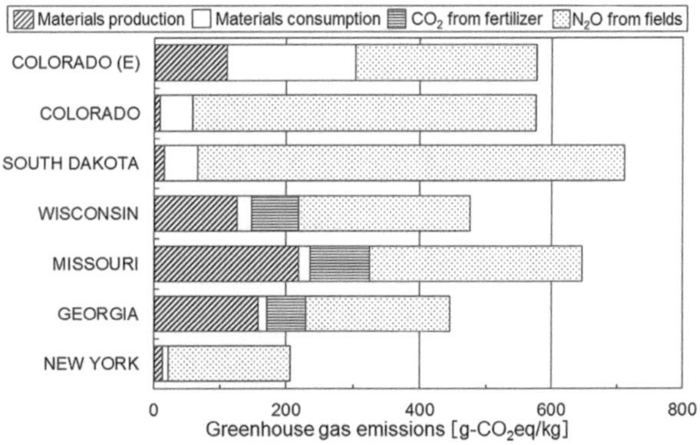

Fig. 2 Greenhouse gas emissions associated with hay production process in the USA N_2O emission from hay fields was defined as 0.78 kg/ha/year as a lower order

Table 7 Nutrient content of the imported hay [8]

[DM-%]	CP	C.fat	NEF	C.fib
Alfalfa	57.4	2.1	43.4	27.0
Timothy	51.8	1.8	46.4	33.3
Sudan grass	62.8	2.4	47.7	35.4
Oat hay	56.6	2.2	53.5	32.5

Fig. 3 Greenhouse gas emissions associated with transportation process of imported hay from the USA

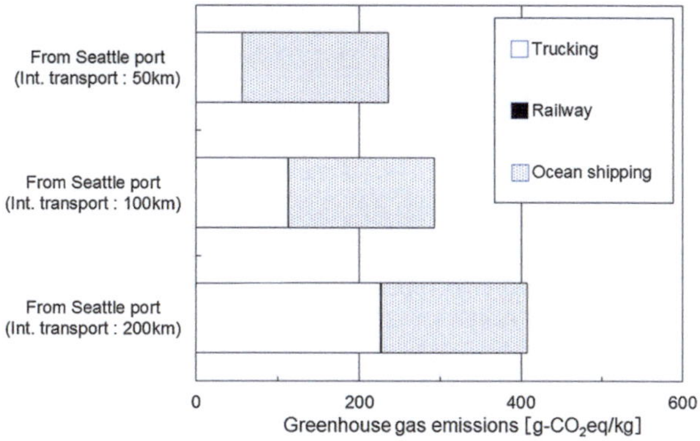

2.4.2 Transport Process

GHG emissions from the transportation process were estimated from hay weight and energy consumption of the transportation means (trucking, railway, and shipping) (Fig. 3). Japan imports a large amount of hay from the USA and Canada. Imported hay from the USA accounted for 70 % of the total weight (2006kt) in 2008, based on Japanese trade statistics. The departure port of the imported hay was assumed to be a major seaport on the west coast of the USA, such as Seattle. The ocean transport distance was defined as 9000 km, which was from the west coast of the USA to Kashima in Japan.

Trucking and railway were used for internal transportation in the USA. The distance of internal transportation was assumed to be 50 km, 100 km, and 200 km, and the contribution of trucking and railway for internal transport was 56.3 % and 43.7 %, respectively, based on a report by Yamaka et al. [14].

2.5 Data Source of Background Data

Most of the underlying data for inventory analysis (which calculates GHG emissions from input materials) are from the database of the LCA software JEMAI-LCA [15], which mostly represents Japanese production. Some data that were missing from the LCA software were obtained from CO_2 emission factor data based on input-output analysis, which was the data of the producer's price basis [16]. The emission factors of CH_4 or N_2O associated with field work, such as irrigation of paddy fields, cultivation, and fertilizer application in the Japanese roughage production, came from the National Greenhouse Gas Inventory Report in Japan [17]. N_2O emissions from arable land and paddy field were calculated using N_2O emissions reported by IPCC (2006) and Akiyama et al. (2005) [18, 19].

To estimate GHG emissions from fuel and material consumption with hay production in the USA, we utilized the emission factor at GREET 1.7 [20]. N_2O

emissions from the USA hay fields were calculated from the N_2O emissions reported by Del Grosso et al. (2006) [21], which gave county-level mean data per hectare for each state.

The GHG emissions of transportation means were estimated utilizing the database of the JEMAI-LCA. Trucking (10 t) was assumed to be the transportation means of internal transportation. Ocean transport was the containership; capacity was of two types: over 4000 TEU (20-ft equivalent unit) and less than 4000 TEU. The ratio of the two types of containership was 65 % for over 4000 TEU and 35 % for less than 4000 TEU [22].

3 Results and Discussion

3.1 Greenhouse Gas Emissions in the Domestic Roughage Production System

3.1.1 Dent Corn Silage (Bale Wrapped Silage)

The amount of GHG emissions from DCS production systems at the UU were 756, 803, and 746 kg CO_2eq/10a (Fig. 4a). The N_2O from managed land in the compost and tillage process was the largest impact, approximately 50 % of total GHG emission. The impact of N_2O emissions associated with compost application was large, contributing 20–30 % of total GHG emissions. The contribution of the chemical fertilizer production was 10–20 % of the total. The TDN yields of DCS at the UU were 325, 194, and 503 kg/10a, respectively.

GHG emissions were 876 kg CO_2eq/10a for upland field and 918 kg CO_2eq/10a for upland field converted from paddy field, in the DCS production system of RC (Fig. 4a). The N_2O from managed land contributed approximately 40 % of total GHG emission at each system. The GHG emissions were large in the DCS production system of upland field converted to paddy field because the amount of required nitrogen fertilizer was high. The TDN yield of DCS produced by the RC was 931 kg/10a.

3.1.2 Hay

The GHG emissions from hay production systems were similar (520, 489, and 522 kg CO_2eq/10a, respectively) between the three fields at the UU (Fig. 4b). Contribution of N_2O from managed land was calculated as 70 % of the total GHG emission. The second largest impact on GHG emissions was chemical fertilizer application, contributed 12–20 % of the total. Without N_2O emission from managed land, GHG emissions were 114–147 kg CO_2eq/10a.

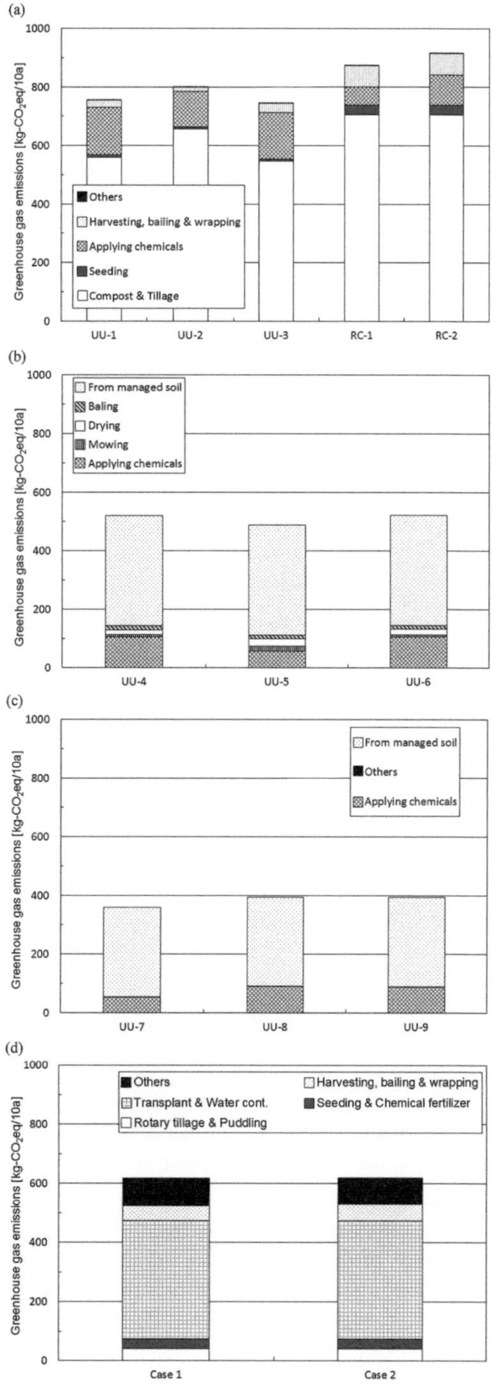

Fig. 4 Greenhouse gas emissions associated with domestic roughage production system. (**a**) Greenhouse gas emissions of dent corn wrapped silage production system at the UU and RC. RC-1 and RC-2 are a case of DCS production at upland field and upland fields converted from paddy fields. (**b**) Greenhouse gas emissions of hay production system. (**c**) Greenhouse gas emissions of pasture production system. (**d**) Greenhouse gas emissions of whole crop rice silage (wrapped) production system

The TDN yields were calculated using a mixed ratio of grass varieties: orchard grass 30 %, tall fescue 30 %, perennial ryegrass 30 %, and white clover 10 %. The TDN yields of hay at the three fields were 493, 468, and 412 kg.

3.1.3 Pasture

The GHG emissions from the pasture production system were 361, 396, and 395 kg CO_2eq/10a (Fig. 4c). The N_2O from managed land contributed 77–85 % of total GHG emission. Agricultural processes associated with the pasture production system at the UU were limited to the application process of chemical fertilizers and herbicides.

The TDN yields from pasture fields were 213, 373, and 328 kg, which were calculated by the amount of feed intake by grazing dairy cattle.

3.1.4 Whole Crop Rice Silage (Bale Wrapped Silage)

The amount of GHG emissions from WCRS production systems at the RC were 617 and 620 kg CO_2eq/10a (Fig. 4d). The impact of N_2O and CH_4 caused agricultural field was 74 % of the total. CH_4 from water control of the paddy field, which included emissions from the "transplant and water control process," had significant impact (60 % of the total GHG emissions). The direct N_2O emission from paddy field, which was calculated as "other process," contributed 14 % of the total emissions in all cases. The difference of fuel consumption in the two cases utilizing two types of machinery for harvesting was ignorable impact in the total GHG emissions of WCRS production system.

The TDN yields of the WCRS production system at the RC were 583 and 582 kg in the two cases.

3.2 GHG/TDN Ratio of Roughage Production System

The GHG/TDN ratio for each roughage production system was analyzed from the results of LCA and calculation of TDN production, which used the average data of each roughage production system (Fig. 5).

The GHG/TDN ratio of the production system of hay and pasture was low (hay as 1.12 and pasture as 1.26), because these systems used low material input in their roughage production. On the other hand, the GHG/TDN ratio in the DCS production system at the RC was lower than the system of the UU, in spite of large GHG emissions, because the TDN yield of the DCS production system at the RC was higher than the system of the UU. Therefore, GHG/TDN ratio results indicated that low yield was an issue in the DCS production system at the UU. The GHG/TDN ratio in the WCRS production system was 1.06. The GHG emissions in the WCRS

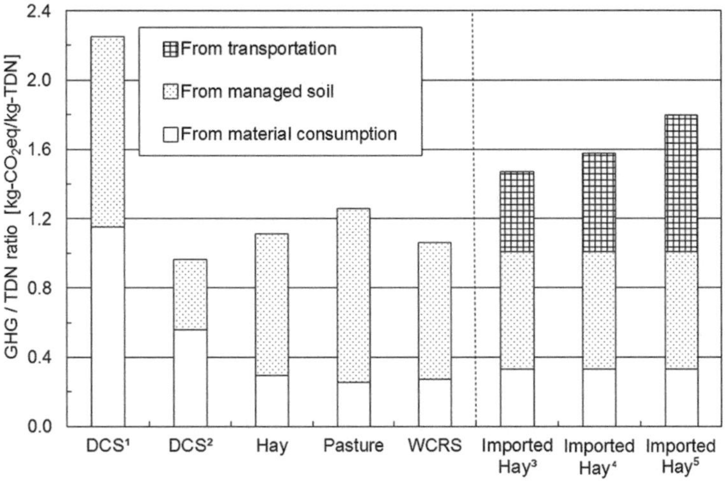

Fig. 5 Comparison of greenhouse gas emissions between domestic roughage production system and imported hay utilization system. *1* The DCS production system at the UU. *2* The DCS production system at the RC. The bar of imported hay represents average value of greenhouse gas emissions associated with 1 kg TDN production of hay at eight areas in the USA. *3* The internal transport distance was 50 km. *4* The internal transportation distance was 100 km. *5* The internal transportation distance was 200 km

production system were larger than the emissions of other roughage production systems (hay and pasture). TDN yield in the WCRS production system was 583 kg, higher than the yield in the hay and pasture production system.

Within the GHG/TDN ratio in the WCRS production system, an impact of material consumption was similar to the hay and pasture production system, which were comparatively low. And, that impact to the GHG/TDN ratio in the DCS production system at the RC was two times higher than the others. Besides, within the GHG/TDN ratio in the DCS production system at the RC, an impact of managed soil was lower than that of the other roughage production systems. These results indicated that the DCS production system at the RC was an intensive roughage production system based on material consumption.

3.3 Comparison of Greenhouse Gas Emissions Between Domestic Roughage Production System and Imported Hay Utilization System

From the results of LCA in the imported hay production system and transportation system, the GHG/TDN ratio of the imported hay utilization system was calculated as 1.47–1.79 (Fig. 5), due to GHG emissions from the internal transportation range (50, 100, and 200 km). Within the GHG/TDN ratio in the imported hay utilization system, the material consumption was 0.33 and the managed soil was 0.68. The impact of the transport process was 0.46–0.79 in the GHG/TDN ratio in the imported hay utilization system; it depended on the difference of internal transport distance. The impact of the managed soil was contributed 38–46 % of the GHG/TDN ratio in the imported hay utilization system.

The GHG/TDN ratio in the domestic roughage production system except for the DCS production system at the UU was 14–46 % lower than the ratio in the production process of imported hay. However, the impact of the material consumption within GHG/TDN ratios was comparable between the imported hay (0.33) and the hay, pasture, and WCRS production systems (0.26–0.30).

Comparing the GHG/TDN ratios of the domestic roughage production system with those of the imported hay utilization system shows that the domestic roughage production system offers potential mitigation of GHG emissions. However, low yield or poor management in the domestic system led to high GHG emissions (such as direct CH_4 emissions from paddy fields), which causes a larger impact than utilizing imported hay in the dairy production system in Japan.

These results suggest that if the GHG/TDN ratio—which was the evaluation indicator of this research—includes a factor of nutrient components of feedstuff and results of LCA, then it has the potential for integrated assessment of environmental impacts and feed quality in the livestock management system. Our study includes uncertainty of N_2O emission from managed soil at roughage production process. While more research is needed to uncertainty analysis from land management, our study showed tendency of GHG emissions in the domestic roughage production system in Japan.

4 Summary

To assess the GHG emissions associated with the domestic roughage production system, the LCA method was applied to typical roughage production systems (dent corn silage, hay, pasture field, and whole crop rice silage) in Japan. A comparative LCA was conducted to compare the domestic roughage production system and the imported hay utilization system, in order to analyze the potential of each to mitigate GHG emissions. To enable evaluation of GHG emissions and roughage quality, the GHG/TDN ratio was suggested as the environmental indicator. The results showed that the domestic roughage production system produces 10 % fewer GHG emissions than the imported hay utilization system, at least in Japan.

References

1. MAFF (2015) Monthly statistics of agriculture. Forestry and Fisheries, Ministry of Agriculture, Forestry and Fisheries, Tokyo
2. Nekomoto K, Hoshiba S, Tamura Y, Kawakami H, Matsumoto K, Morita S (2003) Reduction of nitrogen load by material circulation between dairy farms and dry field crop farms in the mixed production area. J Soc Agric Struct Jpn 34(3):193–198 (in Japanese)

3. Kawakami H, Hoshiba S, Morita S, Noda N, Ikeguchi A (2004) Comparison of two dairy production districts having different feed production systems by multiple evaluation indexes. J Soc Agric Struct Jpn 35(2):103–111 (in Japanese)
4. Yagi K, Minami K (1990) Effect of organic matter application on methane emission from some Japanese paddy fields. Soil Sci Plant Nutr 36(4):599–610
5. Akiyama H, Yan X, Yagi K (2006) Estimations of emission factors for fertilizer-induced direct N2O emissions from agricultural soils in Japan: summary of available data. Soil Sci Plant Nutr 52:774–787
6. ISO (2006) ISO 14040 environmental management – life cycle assessment – principles and framework. International Organization for Standardization, Geneva
7. IPCC (2013) Climate change 2013, the physical science basis. International Panel on Climate Change. Cambridge University Press, New York, pp 712–714
8. NARO (2002) Standard tables of feed composition in Japan (2001). National Agricultural Research Organization edited. Japan Livestock Industry Association, Tokyo
9. Suzuki K, Iguchi M, Uchida K (2005) Cost performance of ensiling systems on whole crop rice silage in Chiba prefecture. Bull Chiba Prefectural Livest Res Cent 5:23–27 (in Japanese)
10. Suzuki K, Somei H (2006) Economical analysis on the custom cultivation model of whole crop rice in Hikata area (Note). Bull Chiba Prefectural Livestock Res Cent 6:59–60 (in Japanese)
11. Heichel GH, Martin NP (1980) Alfalfa. In: Pimental D (ed) Handbook of energy utilization in agriculture. CRC Press, Florida, pp 155–161
12. Chick S, Krummel J (1980) Energy inputs in hay production. In: Pimental D (ed) Handbook of energy utilization in agriculture. CRC Press, Boca Raton, pp 163–168
13. Aoki D, Yonemoto S (2005) Composition of nutrients in commercial hay in Chiba area (the second report). Bull Chiba Prefectural Livestock Res Cent 5:65–71 (in Japanese)
14. Yamaka T, Shibasaki R, Amma K (2004) An analysis on transportation hinterland of international freight in North America continent in which centered on the transportation between North America East Asia. Technical Note of National Institute for Land and Infrastructure Management No. 191 (in Japanese)
15. JEMAI (2005) JEMAI-LCApro. Japan Environmental Management Association for Industry, Tokyo
16. NIES (2002) Embodies energy and emission intensity data for Japan using input-output tables (3EID). National Institute for Environmental Studies, Tsukuba
17. GIO (2006) National greenhouse gas inventory report of Japan 2006 ver. 2. Greenhouse Gas Inventory Office of Japan, National Institute for Environmental Studies, Tsukuba
18. IPCC (2006) N_2O emissions from managed soils, and CO_2 emissions from lime and urea application. Guideline for national greenhouse gas inventories. International Panel on Climate Change. Institute for Global Environmental Strategies, Japan, pp 5–18
19. Akiyama H, Yagi K, Yan X (2005) Direct N_2O emissions from rice paddy fields: summary of available data. Glob Biogeochem Cycles 19, GB1005
20. Argonne (2007) The greenhouse gases, regulated emissions, and energy use in transportation (GREET) model (GREET 2.7). Center for Transportation Research Argonne National Laboratory, Chicago
21. Del Grosso SJ, Parton WJ, Mosier AR, Walsh MK, Ojima DS, Thornton PE (2006) DAYCENT national-scale simulations of nitrous oxide emissions from cropped soils in the United States. J Environ Qual 35:1451–1460
22. Funabashi K, Takahashi H (2004) Port calling League of World Container Ships (2003) Analysis of the statistics concerning with port calling and influence on Rock-Out of port in the western part of North America. Technical Note of National Institute for Land and Infrastructure Management No.145 (in Japanese)

Batik Life Cycle Assessment Analysis (LCA) for Improving Batik Small and Medium Enterprises (SMEs) Sustainable Production in Surakarta, Indonesia

Ghita Yoshanti and Kiyoshi Dowaki

Abstract Batik is one of Indonesian traditional fabrics with substantial values, honored by UNICEF as intangible cultural heritage in 2009, and has been exist since long time ago. Previously, the local wisdom of using natural dyes preserved the environment during batik production. Recently, due to the high demand, economic motives, and the effort to get affordable raw materials for cost saving, many unsustainable common practices are conducted by the SMEs during their production as the majority producers of Indonesian batik. It occurs most especially due to several limitations and challenges faced by the SMEs. This research provides ideas to improve batik production by the SMEs with environmentally conscious designs and sustainability through innovation in batik product life cycle design. By using LCA cradle to gate methodology with CML-IA method, the environmental burden occurred during batik stamp production is estimated. The estimation results are transformed into several recommendations for more sustainable batik life cycle design by SMEs in Surakarta, Indonesia.

Keywords Batik stamp production • Small and medium enterprises (SMEs) • LCA • CML-IA

1 Introduction

Textile industry is one of the industries with unpleasant environmental records from various points of views; the energy, raw material, and the waste treatment resulted after the production. Batik as a part of the textile industry also experiences the same condition. The traditional process of production requires less electricity and water if compared to the bigger textile industry. However, since the production is normally

G. Yoshanti (✉)
School of Business Management, Institut Teknologi Bandung, West Java, Indonesia
e-mail: ghita.yoshanti@sbm-itb.ac.id

K. Dowaki
Department of Industrial Administration, Tokyo University of Science, Chiba, Japan

M. Matsumoto et al. (eds.), *Sustainability Through Innovation in Product Life Cycle Design*, EcoProduction, DOI 10.1007/978-981-10-0471-1_68

conducted by small and medium enterprises, the concern toward the sustainable life cycle design during production is often ignored. SMEs also face several limitations, which causes environmental aspect not to be in the top tier of their priority during the production process.

To get a closer look at batik, the word derives from the Javanese "*amba*" (to write) and "*tik*" (to dot or point). Batik was previously used in traditional ceremonies and rituals, but currently, batik is largely used in daily life through various kinds of product from shirts, dresses, bed sheets, trousers, tablecloths, and other products. According to Indonesia's Ministry of Trade, batik is classified as a part of the creative industry focuses in creativity, ideas, and innovation as the core of its process. Batik, based on Indonesian National Standard Planning, is defined as a craft resulting from dyeing by using hot wax as the cover before the dyeing process, with *canting* and stamps as the main tools to create certain meaningful patterns. This definition also confirms that any textile products with batik motifs, which are not made using hot wax mechanism with canting or stamp, such as printed batik, actually could not be acknowledged as a batik product but only as a common textile product with printed batik motifs. Unfortunately, the consumers are not fully aware of this definition and thus prefer the cheaper product of batik textile rather than the original traditional batik. This condition hit the performance of traditional batik production, which is mostly SMEs. Many of batik SMEs cannot survive the competition with the printed batik textile producers. If they could, they are generally using cheaper raw materials and conduct several common practices, which are actually harmful to the environment. The motive is to gain a competitive price to their product and to maintain the production. The local wisdom of using environmental-friendly materials is then replaced by some chemical substances, which are more affordable and easier to use for cost saving and competing with the cheaper batik products in the marketplace.

After gaining international recognition, Indonesian government tried to re-brand and promote batik as one of Indonesian prominent commodities. Batik is a potential export commodity for Indonesia with high demand from the customers in the USA, Belgium, Germany, England, Japan, and South Korea [1]. The export destination countries are mostly the countries which have strict environmental regulations. Therefore, if the Indonesian government is still interested to make batik to be not only popular in domestic market but also in the international market, the environmental and sustainability aspect should be their priority.

2 Literature Review

2.1 Batik SMEs in Surakarta

In Indonesia, SMEs represent the backbone of industrial infrastructure by constituting more than 80 % of the companies (over 30 million enterprises) [2]. Most of batik enterprises are SMEs; the number reaches 55.573 industries or about 99.39 % from the total of batik enterprises based on the data of Indonesian Department of Industry 2010 [3]. Surakarta is one of the prominent cities in Java where batik history rooted since 1500s. Surakarta is located between the slopes of Mount *Merapi* and Mount *Lawu*, approximately 92 m above the sea level and covers an area of 44 km^2.

Batik from Surakarta is rich of the designs from the old kingdom's cultural heritage also other cultural and art combinations. There are 3400 batik designs in Surakarta; however there are only 900 designs that have patented by the regional government until 2008 [4]. Surakarta is an important center of batik production as much as Yogyakarta and Pekalongan. The production of batik in Surakarta has been a part of the traditional economy of the population in this area. There are several reliable manufacturers and commercial batik clusters in Surakarta completed by the long history since fourth and fifth centuries ago. This rich history and values motivated Surakarta government to brand its city to be the capital city of batik. The reputation and overwhelming history of batik in Surakarta should be balanced with the capacity of this city to preserve its environment during the production of batik. Unfortunately, the fact is that the environmental condition in Surakarta is getting worse, proved by the pollution in the river, odor around the production area, and the poor health condition of the people around the batik villages. Since most of the batik producers are SMEs, hence the production process is conducted in small to medium scale around the residential areas in Surakarta.

2.2 SMEs Limitation in Batik Production

The degradation of environmental condition around batik villages is caused by several limitation owned by batik producers. As for the scale of its business, batik SMEs are generally more reluctant to expand their business to a larger scale than their current condition, because the larger the enterprise, the more they need to conduct the environmental assessment due to what is stated in the regulations. If the batik producers maintain their business to remain in small and medium scales, it is not necessary for them to conduct the environmental assessment. SMEs are easier to start up their business and it causes a tighter competition, especially with batik from other areas in Indonesia and printed batik textile products from China. This pressure forces these SMEs to ignore the needs to consider environmental aspects during their production process.

Deteriorated environmental condition around batik villages is motivated by several unsustainable common practices during production. They are high quantity of excess fabrics which damage the environment, the use of chemical substances (by not properly storing and labeling the container), inadequate facility of waste and water treatment, improper safety equipment used by the workers, large amount of waxing liquid waste, and the use of unsustainable energy by burning woods and kerosene.

The SMEs possess several limitations behind the common unsustainable practices. Environmental consciousness is not within their top priority because it is still a challenge for them to survive among the similar businesses. Another limitation is that these SMEs also face the fluctuation of raw materials price. Mori fabric is the main raw materials for batik and it is made of cotton. Indonesia is not the producer of cotton; instead, Indonesia produces *kapuk*. Hence, to produce mori fabric, these SMEs need to import the cotton from India, China, the USA, or Australia. The fluctuation of the raw material price is really depending on the cotton market overseas. Batik SMEs with limited budget would not be able to afford to buy huge amount of mori. They buy mori in small amount to avoid the shrinkage and to limit their warehouse cost. Thus, the fluctuation of mori price is in the top priority for the batik SMEs. Outdated and low efficiency of technology and the lack of preventive maintenance also become the limitation for Batik SMEs. The batik SMEs practice the traditional mechanisms as easy as possible and tend to neglect the further long-term impact to the environment. Batik SMEs also suffer from improper production management, inefficient use of raw materials, and lack of profit due to the tight competition with the cheaper and massive product of textile with batik motifs from local and foreign printed textile industries.

2.3 Environmental Analysis in Textile and Batik Sector

Some previous researches have already addressed the impact of textile production to the environment like Defra, 2010 [5] and McGill, 2009 [6] with both of studies claimed that the production and also the use of textile possess a high impact toward environment. The researchers also compared several kinds of textile materials from natural to man made. The natural fabrics have an environmental burden in the growing phase for planting and irrigating since they require huge amount of water and fertilizers, while man-made fabrics require a lot of energy due to the longer production process than the natural fibers. Chapman, 2010 [7] compared several LCA studies from some kinds of clothing from T-shirt, tracksuits, blouse, flax shirt, cotton shirt, and many other kinds of clothing. The result of this study claimed that the two stages with the largest contribution to the environments are the production stage and usage stage, as the storage, transport, and retail factors are not really contributing to the environment.

In batik sector, the researches are limited to some Indonesian researchers. They mostly analyze the environmental burden by conducting qualitative studies like Arianti Gunawan (2013) [8] and Nawangpalupi (2012) [9]. Arianti Gunawan (2013) claimed that from ten of the batik entrepreneurs' location as the object of her study, not including Surakarta, none of the batik SMEs conduct green initiatives in their production. Nawangpalupi (2012) concluded that the environmental awareness in SMEs is influenced by the education of the SMEs owners and the access of support, which the SMEs get from the government. The external factors are the role of government and the suppliers, which do not support the environmental awareness positively. The perception of environmental awareness in SMEs is still low.

Batik LCA was also conducted by Puspita Sari et al. (2012) [10] using batik in Pekalongan as the object of the study. Puspita Sari calculated the eco-efficiency and eco-cost of chemical dyes in batik. The result showed that the chemical dyes batik is not sustainable with high global warming impact. It is, however, affordable for the producers. Rinawati and Dyah Ika et al. (2013) [11] also conducted LCA for batik in Pekalongan for batik Puspa Kencana. The study used a value stream mapping from batik process and estimated the eco-efficiency by using LCA. The result showed that in Puspa Kencana batik, there are four factors of inefficiency: defect, inappropriate processing, overproduction, and waiting. There were Rp 98.734.748,41 eco-cost and eco-efficiency index (EEI) 0.756 as well as eco-efficiency rate (EER), 88.1 %. Supartini et al. (2013) [12] also conducted the LCA analysis but for the natural dyes, and it is known that chemical dyes, however affordable, is not sustainable. The use of natural dyes requires longer processing time and is a complex process as well as costly. However, based on her analysis, batik with natural dyes is actually sustainable and affordable with EEI 1.14. In order to increase the sustainability, the batik producers should choose the raw material with cheaper price and low environmental impact.

The scope of this study is different from previous studies which generally only utilizing qualitative analysis; this research will also use LCA with CML-IA method. If the previous studies put individual batik producer for the object of the analysis using LCA, this study will highlight the SMEs in Surakarta as a whole. The data was gathered from direct interviews with batik producers, as the representatives of some prominent batik clusters in Surakarta. Therefore, the life cycle design as the recommendation from this research will hopefully fit the context of batik SMEs in Surakarta.

3 Life Cycle Assessment (LCA) in Batik Stamp Production

Life cycle assessment (LCA) is a methodology conducted to achieve the more sustainable production and consumption patterns, examining the implications of a product or service, in its whole supply chain, usage, waste management, or entire life cycle [13]. LCA is based on ISO 14040 and 14044: 2006. The steps are goal and scope definition, life cycle inventory, life cycle impact assessment, and interpretation.

3.1 Goal and Scope

The goal of the LCA analysis is helping to uncover the environmental profile of batik production especially batik stamp under several assumptions generated during the interview to strictly gain the same standard among several batik producers in Surakarta. CML-IA is the method used from the Leiden University in the Netherlands (2001). The method is commonly used in several LCA papers in textile industries. CML-IA presented a set of impact categories in problem-oriented approach (midpoint). Based on the environmental profile result from CML-IA, the hot spot analysis is counted and several recommendations for sustainable design of life cycle batik product are summarized. The LCA analysis in this paper covers cradle to gate analysis due to limited sources of data. The use phase and waste assessment after the use phase will be elaborated in further research.

Due to the wide variety of batik, the scope of analysis in this study is based on the functional unit of an enterprise to produce 100 pieces of short-sleeved batik shirt with standard L size. The material uses cotton from Surakarta supplier (for counting the transportation environmental burden) in *Pasar Klewer* and *Solo Raya*. Cotton is processed in PT Sri Tex and the raw cotton comes from India and is shipped to PT Sri Tex. The batik method is stamp batik and two colors are used with the same chemical dyes but having different colors from *Napthol*. The calculation involves the production process and excluded retailer transportation, use phase from consumers' point of view, and waste treatment after the consumer use phase.

3.2 Life Cycle Inventory (LCI)

The inventories are collected through interviews with three representatives of batik producer. They are interviewed by using the same assumption to produce 100 pieces of short-sleeved batik shirt in standard size with stamp method and using two colors. The table of materials needed is based on Fig. 1.

3.3 Life Cycle Impact Assessment (LCIA)

The environmental profile is calculated by using SimaPro P.hD 8.0.4.26 series with CML-IA method baseline V3.02/World 2000. The result is presented in Table 1.

The five higher environmental profiles in batik production are marine aquatic ecotoxicity, abiotic depletion (fossil fuel), human toxicity, global warming potential, and freshwater aquatic ecotoxicity.

Marine aquatic ecotoxicity is the impact of toxic substance on marine ecosystem, which contaminates the air, water, and soil. Abiotic depletion (fossil fuel) is related to extraction of minerals and fossil fuels due to input in the system. Human toxicity is concerning the effect of toxic substances on the fate, exposure, and effects of the toxic for an infinite time horizon; for CML-IA, the human toxicity does not include health risk exposure. Global warming potential (GWP) is related to the emission of green house gas (GHG) to the air with the time horizon in a hundred

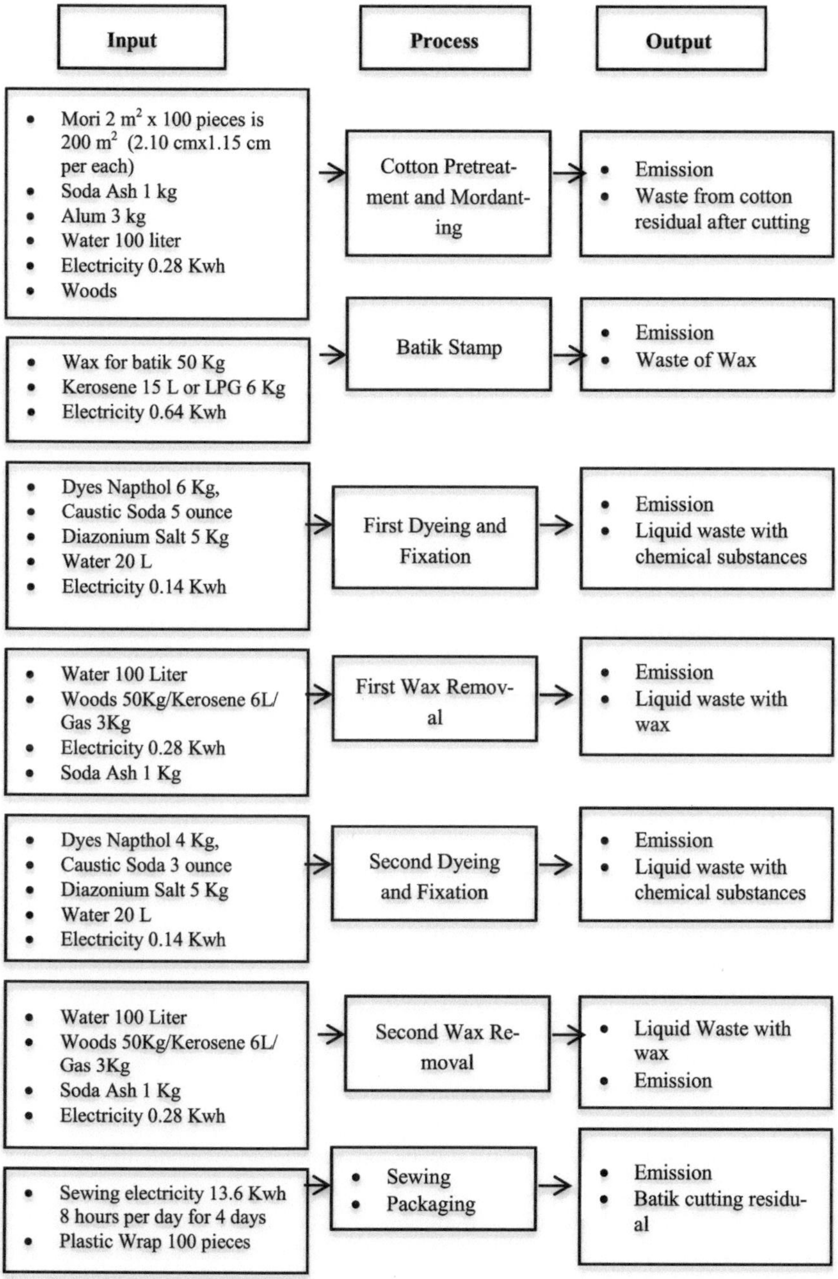

Fig. 1 Input, process, and output inventory in Batik stamp

Table 1 CML-IA environmental profile result

Impact category	Unit	Total
Marine aquatic ecotoxicity	kg 1,4-DB eq	525,966.4612
Abiotic depletion (fossil fuels)	MJ	8656.777622
Human toxicity	kg 1,4-DB eq	884.9247101
Global warming (GWP100a)	kg CO2 eq	405.2046583
Freshwater aquatic ecotoxicity	kg 1,4-DB eq	247.2605995
Terrestrial ecotoxicity	kg 1,4-DB eq	5.50603961
Acidification	kg SO2 eq	2.540122219
Eutrophication	kg PO4 – eq	1.181362281
Photochemical oxidation	kg C2H4 eq	0.126347316
Abiotic depletion	kg Sb eq	0.000478466
Ozone layer depletion (ODP)	kg CFC-11 eq	0.000285856

years. Freshwater aquatic ecotoxicity refers to the impact on freshwater ecosystem as the result of using chemical compound; it pollutes air, water, and also soil.

4 Analysis and Interpretation

4.1 Environmental Profile Analysis

Based on the calculation from CML-IA method, most of the high impacts are dealing with chemical substances and the emission from the heating process of the energy. Several environmental profiles related to the use of chemical substances are marine aquatic ecotoxicity, human toxicity, and freshwater ecotoxicity. The environmental profiles related to energy are abiotic depletion (fossil fuel) and global warming potential. Reducing the use of chemical compound during the production can solve impact of toxicity. Currently, some of leading batik producers are reconsidering the local wisdom by using natural dyes. Although natural dyes consumes a lot of water for getting brighter color and require larger area of land to plant the natural plantation for making dyes paste, the wastes from natural dyes are safer than the chemical dyes. The treatment for natural dyes also does not require complex proper placement and labeling just like the chemical substances. For the condition in Surakarta, several natural dyes like *mahogany*, *jambal*, and *secang* are widely available. The types of natural dyes used for batik should also be adjusted according to the availability of the material. In Surakarta, mahogany kind of plant for natural batik making can easily be found, but in other areas, the existing natural substance of dyes can be suited accordingly.

The use of kerosene as the source of heating energy during production is causing abiotic depletion (fossil fuel); thus, a more sustainable energy can be used, like LPG. LPG is able to lessen abiotic depletion better than kerosene. For GWP, the use of kerosene and woods for burning is also unsustainable. LPG is a better source of energy as it is largely available in Surakarta and the price is not really a burden for the SMEs scale.

4.2 Hot-Spot Analysis

Environmental profile in hot-spot analysis are estimated on behalf of impact portion for each process of production. The result is based on Table 2, where sewing and packaging are the processes which oftentimes the highest in several environmental impacts. The highest point for sewing and packaging is for marine and aquatic ecotoxicity impact, about 67 %; this is caused by the use of a form of plastic for wrapping that is not sustainable. Cotton after mordanting also has a high impact in the ozone layer depletion 82 % and terrestrial ecotoxicity since the cotton production, during the process of cotton input materials, requires huge amount of energy. Ozone layer depletion is the reduction of ozone in stratosphere due to human activities of emitting gas (CFC), which is harmful to the ozone. Terrestrial ecotoxicity refers to the impact of toxic substances emitted to terrestrial ecosystem like tundra, taiga, tropical rain forest, and deserts. Batik stamp, impacts around 41 % to abiotic depletion due to the use of kerosene. Dyeing and fixation process is about 40 % of overall abiotic depletion due to the chemical dyes and additives used during the process of dyeing and fixation.

4.3 Comparing GWP Energy and Transportation

GWP for energy used during the process of batik stamp is compared to find out which energy possesses lesser impact of global warming between kerosene-woods and LPG. As for the transportation, it is also compared to find out whether

Table 2 Hot-spot analysis' highest impact percentage for each process

Impact	%	Batik process
Abiotic depletion	40 %	Dyeing and fixation
Abiotic depletion (fossil fuel)	41 %	Batik stamp
GWP	49 %	Sewing and packaging
Ozone layer depletion	82 %	Cotton after mordanting
Human toxicity	43 %	Dyeing and fixation
Freshwater aquatic ecotoxicity	53 %	Sewing and packaging
Marine aquatic ecotoxicity	67 %	Sewing and packaging
Terrestrial ecotoxicity	92 %	Cotton after mordanting
Photochemical oxidation	35 %	Sewing and packaging
Acidification	47 %	Sewing and packaging
Eutrophication	65 %	Sewing and packaging

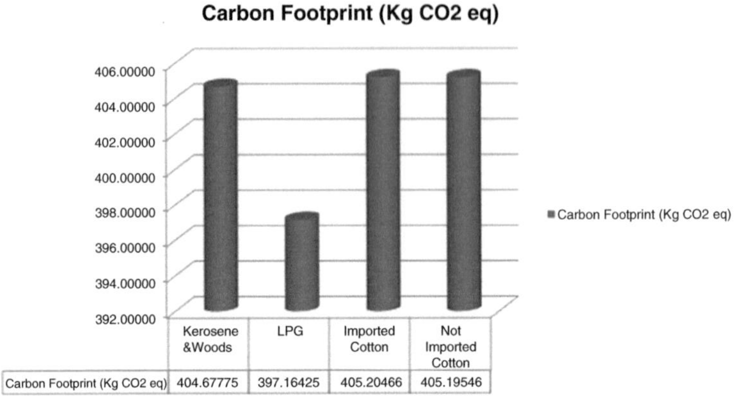

Fig. 2 Comparison GWP energy and transportation

importing the cotton will lead to higher global warming than not importing. The result is based on Fig. 2. LPG possesses fewer GWP if compared to kerosene and woods, the difference is about 7.5 Kg CO_2 eq. While for transportation between importing and not importing the cotton, the difference of the GWP is only 0.0092 Kg CO_2 eq.

5 Summary

The life cycle sustainable design of batik production for SMEs in Surakarta will be more sustainable if the energy used by the SMEs is substituted from kerosene and woods into LPG. If the batik producers are afraid of the increasing cost by using LPG, the regulator can be plugged into the LPG tube to control the amount of gas emitted from the tube. Creative energy community (CEC) in Surakarta makes this innovation and this is largely applicable by the SMEs. For imported cotton, it can be substituted to other fabrics available in Indonesia, but since the environmental burden is not really high, to use cotton would not be a poor option as long as the production management can be precisely designed to reduce defect and waste of cotton residuals. SMEs can also reduce the use of chemical substances by using natural dyes and natural additives like tapioca instead of chemical additives. The process will be more complex than using chemical dyes and additives; it is, however, more sustainable.

For the wax waste, the liquid of wax waste can be reused after usage, so it is better to filter the liquid waste and process the wax waste. Sewing process should be maintained so that it would not take a long time, and it is encouraged for the cable of the sewing machines to be unplugged after use. In packaging, since plastic wraps are not sustainable, they can be substituted with package from used papers; it will also give additional values to batik products for their consistency in preserving the environment up to the smallest detail like their packaging. The cotton or batik residual can be transformed into other products, which have added values, or can be used as the material for the label of the shirts.

The following are highlighted important:

To use the sustainable raw materials as much as possible
To reduce the use of chemical substances
To conduct water treatment before throwing the waste to the environment or to have a proper clean production mechanism
To use the existing raw materials near the production location
To equip the worker with several safety kits or protective gears when using chemical compounds
To possibly reuse the materials
To choose using the energy which has lesser emission to the environment

These life cycle designs will produce batik with environmental values and will possibly be accepted by the exporter countries that have strict environmental regulations.

References

1. Koran Sindo (2013) Dunia Makin Mencintai Batik. Retrieved from http://www.kemenperin.go.id/artikel/6827/Dunia-Makin-Mencintai-Batik, 8 Apr 2015
2. Switch Asia Program (2010) Switch Asia Background. Retrieved from http://www.cleanbatik.com/index.php?id=79, 7 Aug 2014
3. Indonesian Department of Industry (2010) Batik Sebagai motor penggerak ekonomi. Retrieved from http://ditjenpdn.kemendag.go.id/id/news/2014/08/09/batik-sebagai-motor-penggerak-ekonomi, 8 Apr 2015
4. Ministry of Trade Republic of Indonesia (2008) Handbook of Commodity Profile "Indonesian Batik: a Cultural Beauty" Retrieved from http://www.kemendag.go.id/files/pdf/2012/12/08/batik-id0-1354950532.pdf, 8 Feb 2015
5. Defra (2010) The role and business case for existing and emerging fibers in sustainable clothing. Retrieved from http://www2.wrap.org.uk/downloads/Emerging_fibres_- _summary.1cfb6183.10840.pdf, 17 Apr 2015
6. McGill M (2009) Carbon footprint analysis of textile reuse and recycling. Executive summary University Imperial College London, Center for Environmental Policy. Retrieved from https://workspace.imperial.ac.uk/environmentalpolicy/public/McGill09_textiles_Exec%20Summary.pdf, 17 Apr 2015
7. Chapman, Adrian (2010) LCA report of Clothing. Retrieved from http://www.oakdenehollins.co.uk/media/232/2010_mistra_review_of_life_cycle_assessments_of_clothing.pdf, 11 Feb 2015
8. Arianti Gunawan, Arien (2013) Ecopreneurship practice in Batik Industry: case study of batik small and medium enterprises in Ten Cities/Districts in Indonesia. In thesis of Master Science in Management SBM ITB, Bandung, Indonesia
9. Nawang Palupi (2012) Identifikasi dan Segmentasi Kesadaran Lingkungan Industri Kecil di Bidang Garmen dan Alas Kaki. In: Research Report, Universitas Katolik Parahyangan, Bandung, Indonesia

10. Puspita Sari D et al (2012) Pengukuran Tingkat Eko-Efisiensi Menggunakan Life Cycle Assessment untuk Menciptakan Sustainable Production di Industri Kecil Menengah Batik. J Teknik Ind 14(2):137–144
11. Rinawati DI (2013) Pengelolaan Produksi Menggunakan Pendekatan Lean and Green untuk Menuju Industri batik yang berkelanjutan Studi kasus UKM batik Puspa Kencana. J Teknik Ind VIII 1:43–50
12. Supartini, Sri Hartini, Dyah Ika Rinawati (2013) Pengukuran Tingkat Sustainabilitas Batik Tulis Pewarna Alam dengan Metode LCA Studi Kasus di Industri Batik mahkota Laweyan, Solo. Ind Eng Online J 2(4)
13. European Commission, Joint Research Centre, Institute for Environment and Sustainability (2010) International Reference Life Cycle Data System (ILCD) Handbook – General guide for Life Cycle Assessment – Detailed guidance. First edition March 2010. EUR 24708 EN. Luxembourg. Publications Office of the European Union

Eco-design and Life Cycle Assessment of Japanese Tableware from Palm-Melamine Bio-composites

Singh Intrachooto, Rattanawan Mungkung, and Kittiwan Kitpakornsanti

Abstract Palm waste upcycling has become a national policy due to the increase of palm oil plantations for alternative energy production. Because of the large waste volumes of 5.4 million tons per year, this study explores the potential use of palm wastes for innovative materials and eco-products as well as their commercial applications. The concept of eco-design was applied at the very beginning of designing phase by considering the use of waste as raw materials. Currently, there is no alternative application of palm wastes beyond being incinerated for heat or being turned into particleboards. The potential impacts from transport were minimized in the production of tableware by sourcing palm wastes from local palm mills at the nearest location. Processing procedure of fiber preparation included sundrying and grinding the palm fiber, followed by steam explosion (pressure at 18 bars, temperature at 200 °C, for 5 min). After that, the treated fibers were ground to the smallest size (lesser than 0.1 mm) to be mixed homogeneously with melamine compound. The ratio of palm fiber compound varied from 10, 20, and 30 % by weight. The goal of the new compound aims at maximum fiber content that complies with eco- or bio-based environmental product declaration. There are two sets of design: origami (inspired by a Japanese paper folding) and organic (inspired by mushroom forms). The new palm-melamine bio-composites were tested for their qualities in accordance with the required industrial standards: TISI 1245. Using palm fiber of 10 % of the total weight gave the best quality results, while the higher fiber content failed for water absorption as well as acid resistance

S. Intrachooto (✉)
Creative Center for Eco-Design, Department of Building Innovation, Faculty of Architecture, Kasetsart University, Bangkok, Thailand
e-mail: singhman@ku.th

R. Mungkung
Department of Environmental Technology and Management, Faculty of Environment, Kasetsart University, Bangkok, Thailand

Centre of Excellence on enVironmental strategy for GREEN business (VGREEN), Faculty of Environment, Kasetsart University, Bangkok, Thailand

K. Kitpakornsanti
Centre of Excellence on enVironmental strategy for GREEN business (VGREEN), Faculty of Environment, Kasetsart University, Bangkok, Thailand

M. Matsumoto et al. (eds.), *Sustainability Through Innovation in Product Life Cycle Design*, EcoProduction, DOI 10.1007/978-981-10-0471-1_69

tests. To facilitate the commercial application, the optimal production conditions at the lab scale were adjusted for actual industrial processes of a local melamine manufacture. Life cycle assessment (LCA) was performed to ensure that the use of palm-melamine bio-composite had the potential reduction on all impact categories. Prototypes were made and carbon footprinting was performed to support the environmental product declaration in terms of $kgCO_2e$ per piece. Through the integration of eco-design and LCA, the tableware from palm-melamine bio-composites has been developed and supported with environmental footprint information, so that customers can better engage in environmental practices.

Keywords Eco-design • Empty fruit bunch • Life cycle assessment • Palm-melamine bio-composites • Tableware • Upcycle

1 Introduction

Thailand has a national policy to increase palm oil plantation areas for alternative energy production. As a result, a huge amount of palm wastes are being generated from both the plantation as well as from processing of palm oil industry (Fig. 1). Of particular interest is the empty fruit bunch (EFB) because of their large waste volume approximately 5.4 million tons per year [1]. Previous studies have reported the use of palm wastes for fertilizer as it contains potassium and can be used for adjusting pH in soil. Moreover, palm wastes are used as biomass energy for steam or electricity production.

Having considered the chemical properties of EFB, it could be useful for natural fiber-reinforced polymer composite materials. The composite produced from these types of materials is of low density and low cost, has comparable specific properties, and most importantly is environmental friendly. This has led to the need to

Fig. 1 Empty fruit bunch (EFB)

explore the potential utilization of palm wastes for eco-product development in order to add more values. Market studies have shown that Japanese food and restaurants have enjoyed popularity in Thailand with its growth of 15–20 % since 2011, ranking fifth behind the USA, China, South Korea, and Taiwan, in terms of number of restaurants [2]. Its growth has toppled USD 600 millions in 2014 [3]. As a result, the study on exploring the potential use of EFB for natural fiber-reinforced polymer composite material to make an eco-tableware product development was developed and aimed to capture the growing trend of Japanese restaurants.

To anticipate the market trend of eco-products, life cycle thinking was applied, and life cycle assessment (LCA) was performed to ensure that the use of palm-melamine bio-composite would not cause higher impact than 100 % melamine. Carbon footprinting based on LCA was also conducted to prepare the industry to support the environmental product declaration. It was expected that the palm-fiber bio-composite and the resultant tableware will be commercially viable, which will eventually enhance sustainable consumption and production policy.

2 Methodology

2.1 *Palm Fiber Preparation and Palm-Melamine Composite*

EFB wastes were collected from the nearest local palm mill, 200 km from the studied site, and transported by a pickup car. They were sundried to reduce moisture and chopped into small pieces so that they can further be separated to bundle fiber. Then, they were treated by using steam explosion at temperature of 200 °C and 18 bar pressure for 5 min and then put into a blender with high-speed crushing to produce EFB powder. The EFB powders were then mixed with melamine compound. The palm-melamine composites were formed into flat circular disks: The ratio of palm fiber compound varied from 10, 15, 20, 25, and 30 % by weight and tested qualities according to Thai Industrial Standards Institute for melamine-ware (TISI 1245).

2.2 *Life Cycle Assessment (LCA)*

The LCA method according to the ISO14040 [4] was used to evaluate and compare the environmental performances of palm-melamine composite tableware with 100 % melamine compound tableware. The scope of the study was to assess the environmental performances of palm-melamine composite tableware from raw material extraction to final disposal or "cradle to grave" approach (Fig. 2). The evaluation assumed that the tableware is landfilled after the end of life. Primary data were primarily gathered from the industry such as input and output of palm fiber

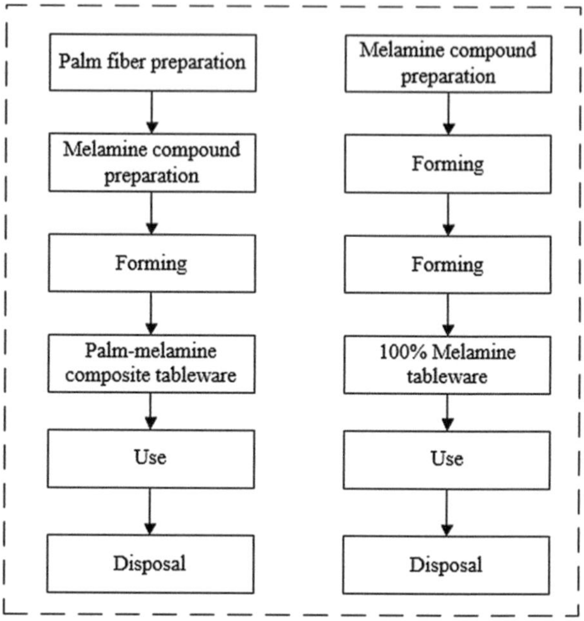

Fig. 2 System boundary of the study

preparation and tableware production processes. All relevant background data such as energy, transport, and waste disposal were taken from international databases when necessary. In this study, ReCiPe 2008 method was used as the impact assessment method. Five relevant environmental impacts for this study were climate change, human toxicity, terrestrial eco-toxicity, freshwater eco-toxicity, and marine eco-toxicity.

Carbon footprinting, based on the national guideline for carbon footprint of product [5], was also conducted. The emission factor (GHG emission) of palm waste was treated as zero in terms of material acquisition, but still took into account of the transport of EFB from palm mill to the studied site. Inputs and outputs associated with the palm fiber preparation and production processes of palm-melamine composite were collected, based on the primary data. The required background data (such as electricity, chemical, water production, etc.) were sourced from the national databases and supplemented by international databases when necessary.

2.3 *Eco-design*

The concept of eco-design was integrated with LCA to ensure sensible design practices. LCA is a tool to assess the potential environmental impacts of product along its life cycle. Along with applying eco-design concept, LCA results are used as the inputs for design from the very early stage (i.e., conceptual design). By doing so, the environmental aspects are captured and quantified for designers, so that the tableware will be environmental friendly. LCA of early prototypes also provides supporting data for the refinement of composite formula.

3 Results and Discussion

3.1 *Palm-Melamine Composite and Quality Test Results*

The new palm-melamine bio-composites of 10, 20, and 30 % were tested for their qualities in accordance with the required industrial standards for melamine-ware: TISI 1245. Using palm fiber of 10 % of the total weight gave the best quality results; the higher fiber content fails water absorption tests as well as acid resistance tests. The results of the palm-melamine composite of 10 % are shown in Table 1.

3.2 *LCA of Palm-Melamine Composite*

LCA was used to evaluate the potential impact reduction of using palm fiber at different ratios when mixed with melamine compound varying from 10, 20, and 30 % by weight. Using palm fiber of 30 % could reduce the climate change impact by about 21 % but merely 1.4 % if using palm fiber of 10 %. Similarly, the impact on

Table 1 Palm-melamine composite quality test results

Criteria	Assessment result	(TISI) for melamine-ware 2553-1245	Result
Molding ability		Easily removed from molds	PASS
Acid resistance		No visible crack or discoloration	PASS
Heat resistance		No visible crack or discoloration	PASS
Specific gravity	1.48	1.45–1.555	PASS
Boiling water absorption	0.55	<1.0 % wt	PASS
Water absorption	0.47	<0.8 % wt	PASS
Impact strength	4.04	>2 kJ/m^2	PASS
Evaporation residue (water)	0.0035	<30 mg/dm^3	PASS
Evaporation residue (4 % acetic acid)	0,0405	<30 mg/dm^3	PASS
Evaporation residue (n-heptane)	0.0047	<30 mg/dm^3	PASS
KMnO4	4.3	<10 mg/dm^3	PASS
Formaldehyde	Not found	Not found	PASS
Heavy metal	0.03	<1 mg/dm^3	PASS
Phenol	Not found	Not found	PASS

Table 2 LCA results of palm-melamine composite

	% palm fiber in palm-melamine composite			
Impact category	0 %	10 %	20 %	30 %
Climate change	3.5868	3.5362	3.1527	2.8314
Human toxicity	0.2194	0.2149	0.1934	0.1687
Terrestrial eco-toxicity	0.0015	0.0015	0.0014	0.0012
Freshwater eco-toxicity	0.0006	0.0006	0.0006	0.0005
Marine eco-toxicity	0.0034	0.0033	0.0030	0.0026

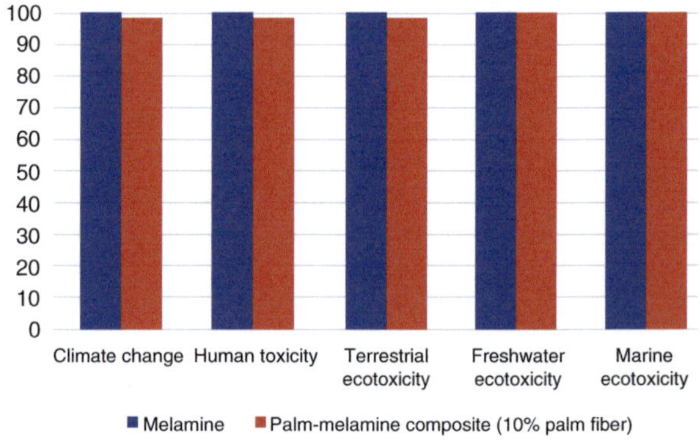

Fig. 3 Comparative LCA results of 100 % and palm-melamine composite (10 % palm fiber)

human toxicity could be reduced by about 23 % if using 30 % of palm fiber but only 2 % if using 10 % of palm fiber. The impacts on terrestrial eco-toxicity, freshwater eco-toxicity, and marine eco-toxicity were not significantly different (Table 2).

To optimize the environmental and quality performance, it was confirmed from the LCA results that the impacts of palm-melamine composite were not higher than that of 100 % melamine in all impact categories (Fig. 3). This information was used to support the decision on using palm fiber of 10 % to replace melamine since palm fiber helps reduce the impacts while still meeting all standard qualities. The use of more than 10 % palm fiber resulted in higher water absorption and failed to meet the standard quality requirements.

In addition, carbon footprinting was also performed to anticipate the market requirement on environmental product declaration (Table 3). This aimed at preparing the industry to get ready for the certification and verification for carbon footprint or even carbon footprint reduction label. For the carbon footprint label, the value of life cycle GHG emissions will be displayed. For the carbon footprint reduction label, the industry must implement the carbon footprint reduction measures to reduce the carbon footprint value by at least 2 % or to achieve the median carbon footprint values previously done and announced by the Thailand Greenhouse Gas Management (Public Organization).

Table 3 Carbon footprint values of palm-melamine tableware prototypes

Item	Carbon footprint values (gCO_2e per piece)
	105
	211
	335
	558
	493
	47

3.3 Eco-design of Tableware from Palm-Melamine Composite

The design of tableware from palm-melamine composite is aimed at the Japanese restaurant industry. There are two sets of tableware design: origami (inspired by Japanese paper folding) and organic (inspired by mushroom forms). Each set comprises of seven plates, bowls, and cups (Figs. 4 and 5).

The two sets of final designs were the result of design team's brainstorming with the material development team as well as the LCA team to keep the balance of design in terms of quality, environmental performances, and at the same time the design aesthetics. Although the team targeted at adding palm fiber as much as possible, the quality was an important issue, especially the water absorption aspect. Although we could solve this problem through the chemical treatment of palm fiber

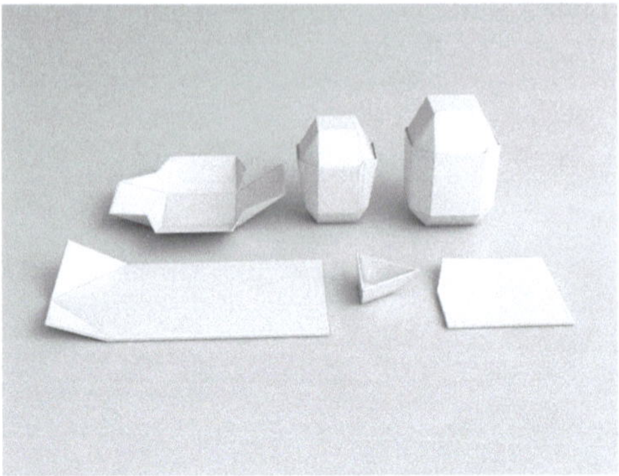

Fig. 4 Tableware made of palm-melamine composite (origami set)

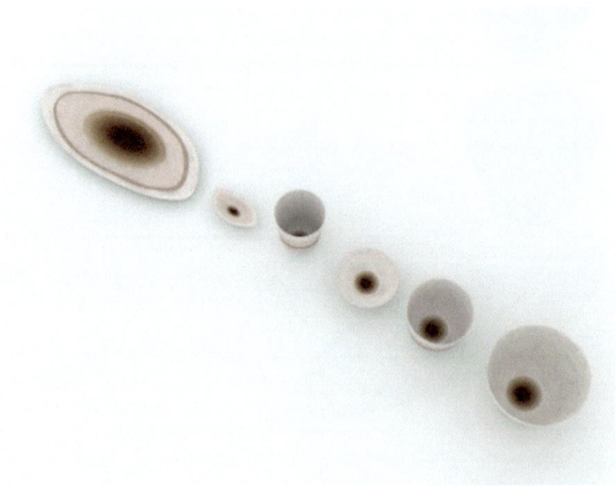

Fig. 5 Tableware made of palm-melamine composite (organic set)

to make it less hydrophilic, the potential impacts of chemical treatment had to be evaluated. As a consequence, the design and material development were adjusted to optimize quality and environmental performances by not applying further chemical treatment to palm fiber and keep the used pigments minimal.

3.4 Prototype Production

Molds of the tableware, origami and organic sets, were produced with technical difficulties. The molds of origami set required skillful mold makers, due to the complicated design with slopes and angles. The mold of organic set was slightly easier due to their shapes that can take out of the molds more conveniently. They were made in different shades by adding pigments (food grade) for the aesthetic purpose (Figs. 6 and 7).

Fig. 6 Prototype of the tableware made of palm-melamine composite (origami set) in different colors: (**a**) *white*, (**b**) *light brown*, and (**c**) *black*

Fig. 7 Prototype of the tableware made of palm-melamine composite (organic set)

In terms of production processes, the same process for manufacturing tableware from 100 % melamine was still applied for tableware made from palm-melamine bio-composite. Since this palm-melamine tableware passed the TISI standard, it can be safely inferred that its life span is equivalent to the widespread tableware made from 100 % melamine.

4 Summary

Palm-melamine tableware is an attempt to go beyond the development of a composite material from agro-wastes, i.e., palm fiber, but to also add creative design and assess environmental impact throughout its life cycle during product design process. The design and material were adjusted based on quality tests and LCA results. The lab-scale material development was also trialed at a manufacturing facility to ensure that the design with the new palm-melamine composite can readily be manufactured, hence creating commercial value from agro-wastes. Since the process of designing these tableware involved the recognition of market potentials, customer's preferences, as well as environmental evaluation and declaration, the resulted palm-melamine tableware will not only be an innovative composite material and design that is sensible to the environment but also a viable prototype for commercial production.

Further research could be conducted to increase the palm fiber content without shifting environmental impact to toxicity. With higher palm fiber content, designers could also explore other applications that are less restrictive than food-contact products. Plausible applications could be automotive interiors or wall covering for building interiors where water contact is minimal and being lightweight is advantageous.

Acknowledgments This research is financially supported by Thailand's Agricultural Research Development Agency (ARDA), and Design Catalyst and the Thai MFC Company are acknowledged for their effort in this R&D collaboration.

References

1. Office of Agriculture Economics (OAE) 2013 http://www.oae.go.th/download/prcai/farmcrop/palm.pdf. Accessed on 24 Mar 2014
2. Brandage (2008) Sushi-ramen: opportunity only, nocrisis. http://www.brandage.com/Modules/DesktopModules/Article/ArticleDetail.aspx?tabID=7&ArticleID=1597&ModuleID=701&GroupID=469. Accessed 17 Mar 2014
3. Kasikornthai (2010) Kasikornthai cooperated with organization to promote Japanese Restaurants Abroad (JOR) to support Japanese restaurant business in Thailand. http://www.kasikornbank.com/TH/WhatHot/Pages/Kbank_With_JRO.aspx. Accessed 28 Jun 2015
4. International Organization for Standardization (2006) ISO 14040: environmental management-life cycle assessment-principle and framework
5. Thailand Greenhouse Gas Management Organisation (TGO) (2015) National guideline of product carbon footprint, 35 pp

Consumer's Lifestyle and Its Impact on Eco-product Aesthetics

Chen-Fu Chen

Abstract This study is to explore the relationship between different consumer's lifestyle and its aesthetic preferences to eco-products. Thus, this study conducts the literature review, an expert interview, and an online pretest for constructing consumer lifestyle attributes as well as aesthetic attributes for the questionnaire survey with Likert scale. The lamp as one of three product samples used in the survey is chosen for analysis in this paper. The result of factor analysis indicates six eco-lifestyle factors: luxury life, up-to-date fashion, pursuing aesthetics for the living, protection of environment, ego-expression, and taste of craft. The result of cluster analysis shows four groups of consumer lifestyle: "pursuing product aesthetics for life but ignoring the environment," "pursuing luxury and fashion with less awareness of environmental protection," "simple and plain," and "green Lifestyles of Health and Sustainability (LOHAS)." Furthermore, the group of "pursuing product aesthetics for life but ignoring the environmental protection" emphasizes personal achievement and product design with aesthetics.

Keywords Consumer lifestyle • Eco-product • Aesthetic preference

1 Introduction

With the rapid development of industrialization, modernization, and living standard, individual consumption would further exacerbate the problems of resource depletion, ecosystem deterioration, and environmental pollution. And the current consumption pattern is still one that sacrifices natural resources and ecological environment and produces a series of environmental problems. So it is imperative to conduct sustainable consumption. Changing individual lifestyle into sustainable consumption may be a strategic initiative for sustainable development. However, different people may have different motivations in conducting sustainable consumption. This study tries to classify consumers into different groups with different

C.-F. Chen (✉)
Department of Product Design, Ming Chuan University, Taipei, Taiwan
e-mail: chenfu@mail.mcu.edu.tw

M. Matsumoto et al. (eds.), *Sustainability Through Innovation in Product Life Cycle Design*, EcoProduction, DOI 10.1007/978-981-10-0471-1_70

lifestyles, and then lifestyles of each group can be identified respectively so that product design can be focused based on the consumer lifestyle.

Moreover, the continuation of an eco-product's aesthetic appeal during its useful life needs to be considered by eco-product designers in the eco-product design process. However, the characteristics for environmental protection including consumer's perception and preference can affect the aesthetics of an eco-product.

The purpose of this paper is to classify consumer lifestyles and discuss their relationship according to perspectives of the price and product design in terms of product aesthetics. Questionnaire survey with the sample of lamps is conducted for item analysis, factor analysis, cluster analysis, and cross analysis.

2 Literature Review

2.1 *Sustainability and Eco-products*

Manufacturer responsibility or product stewardship is taking the manufacturing industry into a new era where manufacturers are considering the environmental impacts of the product at every stage of its life cycle from the extraction of raw materials to the end-of-life disposal. For example, office furniture manufacturers are making products from renewable and recycled materials and reducing toxins in chair frames, upholstery, carpet, flooring, and adhesives through the manufacturing process.

Some claimed eco-product examples produced by office chair manufacturers, such as recycled resin table, office chair, desk, partition, and lighting. Using recycled material has been emphasized commonly through the product development process, while CO_2 emission is more focused in recent years. For example, consumers may find the quantitative information of CO_2 emission or carbon footprint in the recent eco-product market. Typical environmental attributes are listed on various consumer guides, including recyclability, recycled content, fuel efficiency, toxic content reduction, and others such as carbon emission and environment-friendly packaging [10]. Many eco-products come with environmental labels that state product features to inform and appeal to consumers.

2.2 *Product Aesthetics and the Environment*

In the practice of eco-design, life cycle assessment (LCA) methodology provides the basic modeling framework and environmental criteria for evaluating the environmental load and impact throughout the entire product life cycle from material acquisition to disposal [8, 12]. It supports decision makers with quantified product or system-related information of energy and material streams and reveals initiation-related impacts to the ecosystem. However, the aesthetic effectiveness of an eco-product is particularly difficult to judge numerically, while the evaluation and final selection of a product's exterior (form) is frequently determined by a designer [4].

Charter and Tischner [2] propose that the lower aesthetic appeal of eco-products is one of the existing obstacles in buying eco-efficient products and aesthetics should be considered in eco-design. It is important to think about an aesthetic feature that supports long-term use. The continuation of an eco-product's aesthetic appeal during its useful life needs to be considered by eco-product designers in the eco-product design process. Moreover, Zafarmand et al. [14] propose a framework with seven aesthetic attributes and 12 aspects of product sustainability are established, which can be applied as guidelines for developing the aesthetics of sustainable products. For example, rasing product aesthetics would show its high correlation with modality, recovery, and recyclability, which is applied as part of the assessment attributes. For example, "aesthetic durability" involves designing a new eco-product that has the ability to change over time, which has strongly related to "user-product relation," "product life extension," "reduced material/energy use," and "reparability." Three aspects of product sustainability including "harmony with environment," "serviceability," and "renewability" are weakly related to "aesthetic durability." In other words, the product has to be made adaptable to changing technological possibilities and changing user preferences. "Aesthetic upgradability and modularity" may enable product lifetime optimization by keeping a strong relationship with these seven aspects of product sustainability.

2.3 Sustainable Marketing and Consumer Behavior

Imkamp [7] states that the interest of consumers in information about ecologically relevant product attributes is growing through several years of discussion in the mass media about ecological problems associated with consumer products and consumer behavior. However, eco-product marketing barriers may be laid by such reasons as higher price, lower quality, product/brand image, and lower reliability, due to unstable physical/chemical characteristics and lack of product information, communication, aesthetics, functionality, service, and corporate sustainable social responsibility.

The issue of aesthetic consumption in the modern society has been raised since some studies claim that effective manufacturing of products no longer holds a sufficient competitive advantage. Instead, the route to commercial success is supposed to be found in the artful creation of aesthetic offerings, of images, and of brands [3, 4, 6, 13, 14]. An aesthetic image serves as a stimulus, a sign, or a representation that drives cognition, interpretation, and preference. Aesthetics, taste, and design thus cannot be treated superficially, but they are drivers of consumption that are fundamental issues for understanding the path toward a sustainable future.

2.4 Sustainable Consumption and Consumer Lifestyle

As for the consumer's consumption behavior, Bret Leary et al. [1] found that environmental concern impacts sustainable consumption behavior. Tukker et al. [11] also pointed out that the studies on sustainable consumption from the perspective of internal factors like consumer's perception and attitude are very limited, which can be a key point for conducting studies on consumer lifestyles. The lifestyle of consumers is the reaction and recognition to everyday things for the individual or the group. Generally, there are two types of the lifestyle assessment such as activity, interest, opinion (AIO) and values, attitudes, and lifestyles (VALS) commonly applied for academia as well as practitioners to study and understand consumer lifestyle.

Hawkins et al. [5] propose that AIO should include personal attitude, viewpoint of values, activities and interests, demographics, the media used, and degree of usage of products, which has enlarged the range of lifestyle definition. VALS is applied with more focus on values which have formed an important part of lifestyle assessment. For exploring the consumer behavior and consumption to eco-products, it is imperative to analyze consumer's purchasing intention for eco-products through the consumer lifestyle.

3 Research Methodology

This study tries to explore the correlation between consumer lifestyle and eco-product aesthetics. A lifestyle assessment is constructed with attributes especially for both product aesthetics and sustainability as well as functionality. Then the questionnaire is conducted for factor analysis and cluster analysis, which defined the lifestyle and its characteristics. Cross analysis is applied to explore the preference of eco-product aesthetics correlated to different lifestyles of consumer groups.

3.1 Empirical Design

Eight lamps, eight hand bags, and eight chairs emphasizing sustainability are chosen from eco-product samples from the book *Green Design* [9]. This paper just adopts analysis for the lamp sample as an example. Eight lamp samples are illustrated in Table 1.

Questionnaire survey is conducted at IKEA shops around Taipei City area and 118 valid questionnaires are collected. Subjects who went to IKEA shops would be more interested to product design. Then, item analysis is applied for constructing

Table 1 Part of product survey samples

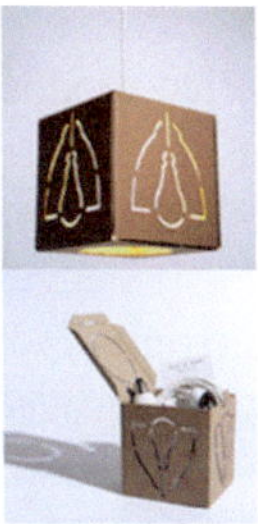			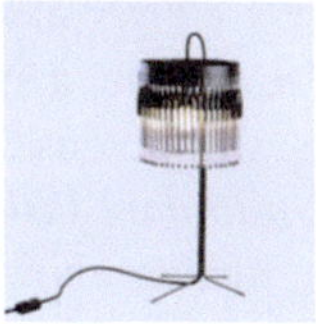
Name: Nat a Box	Name: oregami Lampshade	Name: Kina	Name: Volivik Lamp
Designer: David Graas	Designer: Matali Crasset	Designer: David Trubridge	Designer: en Pieza
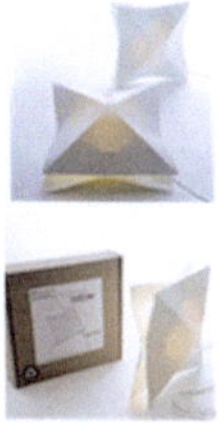	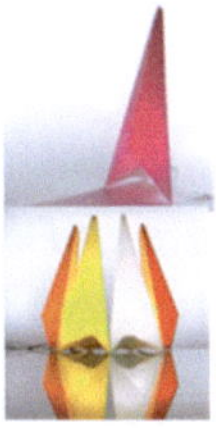		
Name: Hollow	Name: Stealth	Name: Firewinder -The Original Windlight	Name: Jacky Light
Designer: Blue Marmalade Studio	Designer: Blue Marmalade Studio	Designer: Tom Lawton	Designer: Tom Dixon

appropriate and reliable questionnaire survey questions for surveying eco-product aesthetics and consumer lifestyles.

Factor analysis is applied for naming common group characteristics and analysis. For classifying consumer groups with different opinions and life experiences on aesthetics and lifestyle, cluster analysis is used. Finally, cross analysis with t-test is applied for exploring the preference correlation between consumer groups and eco-product aesthetics.

3.2 *Factor Analysis for Eco-product Aesthetics and Consumer Lifestyles*

To facilitate the interpretation of a latent factor, the factor analysis with varimax rotation was carried out. The factor that has an eigenvalue of greater than one was extracted, and the six factors were obtained, including luxury life, up-to-date fashion, pursuing aesthetics for the living, protection of environment, ego-expression, and taste of craft.

Luxury life: a symbolic behavior from which consumers prefer high-price product with a sense of luxury for living more comfortable and meaningful.

Up-to-date fashion: fashion is an intermedia and changes fast from which consumers show their viewpoints of the value.

Pursuing aesthetics for the living: life is more sensitive to aesthetics from which consumers like the product design with aesthetics.

Protection of environment: more strong awareness of environmental protection from which consumers prefer simple living style.

Ego-expression: personal viewpoint of product aesthetics is unique with personal subjective declaration.

Taste of craft: good design craft can be durable and functional through generations from which consumers feel the sense of high living taste.

3.3 Cluster Analysis for Grouping Consumer Lifestyle Characteristics

This study applies two-stage clustering approach and starts with Ward's method and then K-means method. The statistical results for K-means method is shown in Table 2. The higher value of coefficient shows apparent lifestyle characteristics for each consumer group. The negative value of coefficient shows the opposite of the lifestyle characteristics for the consumer group. Characteristics for each cluster are explained as follows:

Cluster 1: preferring aesthetics for life and strong ego-expression but ignoring protection of environment. Eighteen subjects are in this group (15.3 %). This group of consumers ignores the protection of environment and shows very much interest to design. This group of consumers is named "pursuing product aesthetics for life but ignoring the environmental protection."

Cluster 2: pursuing much luxury life and up-to-date fashion with moderate awareness to the protection of the environment. Forty-three subjects (36.4 %) are in this group. This group of consumers likes sharing information about fashion while caring about the protection of environment and is named "pursuing luxury and fashion with less awareness of environmental protection."

Cluster 3: pursuing none of up-to-date fashion, ego-expression, and taste of craft. Nineteen subjects (16.1 %) are in this group. This group of consumers prefers simple life with few interest on design and is named "simple and plain."

Cluster 4: pursuing none of luxury life, moderate ego-expression, and high awareness of environmental protection. Thirty-eight subjects (32.2 %) are in this group. This group of consumers cares very much about environmental protection

Table 2 Factors for cluster analysis and their values of coefficient

Factors	Cluster 1	Cluster 2	Cluster 3	Cluster 4
Luxury life	0.02691	0.93191	−0.52327	−0.80563
Up-to-date fashion	−0.05872	0.26830	−0.78378	0.11610
Aesthetics for life	0.22814	−0.1118	0.14825	−0.05568
Protection of environment	−1.59576	0.33010	−0.31014	0.53742
Ego-expression	0.95509	−0.12277	−1.34266	0.35785
Taste of craft	−0.11446	0.12085	−0.52932	0.18212
No. (subject)	18	43	19	38
Percentage	15.3 %	36.4 %	16.1 %	32.2 %

as well as ego-expression. Thus, "green Lifestyles of Health and Sustainability (LOHAS)" is named.

Based on cluster analysis, four groups of consumers have illustrated different degree of awareness to environmental protection. Group of "green LOHAS" shows most of passion to the protection of environment while three groups as "pursuing luxury and fashion with less awareness of environmental protection," "simple and plain," and "pursuing product aesthetics for life but ignoring the environmental protection" are decreasing the awareness of environmental protection, respectively. Moreover, most of survey subjects prefer the lamp "Kina" for its aesthetics, durability, simple, and apparent form.

3.4 *Cross Analysis for Consumer Lifestyle and Eco-products*

Table 3 shows the significance for different perception to product sample "Nat a Box" by different clusters of the consumer lifestyle.

Table 4 shows sample of "Nat a Box" lamp that is highly chosen by group of "pursuing product aesthetics for life but ignoring the environmental protection" with 94.4 %, while group of "simple and plain" chose less with 47.54 %. Group of "simple and plain" may choose "Nat a Box" lamp with sustainable attributes of modularity, easy disassembly, and simple form.

Table 5 shows consumers' holistic preference to lamps and the top three lamps are Jacky Light (50 %), Kina (50.8 %), and Hollow (47.5 %), respectively. The sustainable characteristics including aesthetics of simple form, energy-saving lightbulb, recycled material, and green package are highly preferred by groups of "pursuing luxury and fashion with less awareness of environmental protection" and "green LOHAS." Groups of "pursuing product aesthetics for life but ignoring the environmental protection" and "simple and plain" prefer "o-Re-gami Lampshade" with 55.6 % and 52.6 % respectively. "o-Re-gami Lampshade" is designed by folding a single recycled leather into a lamp which has illustrated the unique form and aesthetics, which may attract these two groups of consumers.

Table 3 Chi-square test of lamps for each cluster

Cross content	Pearson Chi-square	df	p
Cluster *Nat a Box	7.979	3	0.046*
Cluster *JackyLight	7.457	3	0.059
Cluster *o-Re-gami Lampshade	3.007	3	0.390
Cluster *Volivik Lamp	2.978	3	0.395
Cluster *Kina	1.458	3	0.692
Cluster *Stealth	0.882	3	0.830
Cluster *Firewinder-The Original Windlight	2.437	3	0.487
Cluster *Hollow	0.298	3	0.960

$P < 0.05$

Table 4 Cross analysis for sample of "Nat a Box" and clusters of the consumer lifestyle

Consumer lifestyle	Item	Nat a Box		Total
		0	1	
"Pursuing product aesthetics for life but ignoring the environmental protection"	No.	17	1	18
	Observed value %	94.4 %	5.6 %	100.0 %
	"Nat a Box" %	20.2 %	2.9 %	15.3 %
	Whole %	14.4 %	0.8 %	15.3 %
"Pursuing luxury and fashion with less awareness of environmental protection"	No.	30	13	43
	Observed value %	69.8 %	30.2 %	100.0 %
	"Nat a Box" %	35.7 %	38.2 %	36.4 %
	Whole %	25.4 %	11.0 %	36.4 %
"Simple and plain"	No.	10	9	19
	Observed value %	52.6 %	47.4 %	100.0 %
	"Nat a Box" %	11.9 %	26.5 %	16.1 %
	Whole %	8.5 %	7.6 %	16.1 %
"Green LOHAS"	No.	27	11	38
	Observed value %	71.1 %	28.9 %	100.0 %
	"Nat a Box" %	32.1 %	32.4 %	32.2 %
	Whole %	22.9 %	9.3 %	32.2 %
Total	No.	84	34	118
	Observed value %	71.2 %	28.8 %	100.0 %
	"Nat a Box" %	100.0 %	100.0 %	100.0 %
	Whole %	71.2 %	28.8 %	100.0 %

Table 5 Cross analysis for consumer lifestyles and preferred lamps

Preferred lamp	Item	Consumer lifestyle (QCL) "Pursuing product aesthetics for life but ignoring the environmental protection"	"Pursuing luxury and fashion with less awareness of environmental protection"	"Simple and plain"	"Green LOHAS"	Total
Nat a Box	No	1	13	9	11	34
	Lamps %	2.9 %	38.2 %	26.5 %	32.4 %	
	QCL_1 %	5.6 %	30.2 %	47.4 %	28.9 %	
	Whole %	0.8 %	11.0 %	7.6 %	9.3 %	28.8 %
Jacky Light	No.	11	26	5	17	59
	Lamps %	18.6 %	44.1 %	8.5 %	28.8 %	
	QCL_1 %	61.1 %	60.5 %	26.3 %	44.7 %	
	Whole %	9.3 %	22.0 %	4.2 %	14.4 %	50.0 %
o-Re-gami Lampshade	No.	10	15	10	17	52
	Lamps %	19.2 %	28.8 %	19.2 %	32.7 %	
	QCL_1 %	55.6 %	34.9 %	52.6 %	44.7 %	
	Whole %	8.5 %	12.7 %	8.5 %	14.4 %	44.1 %
Volivik	No	3	3	4	4	14
	Lamps %	21.4 %	21.4 %	28.6 %	28.6 %	
	QCL_1 %	16.7 %	7.0 %	21.1 %	10.5 %	
	Whole %	2.5 %	2.5 %	3.4 %	3.4 %	11.9 %
Kina	No.	11	21	8	20	60
	Lamps %	18.3 %	35.0 %	13.3 %	33.3 %	
	QCL_1 %	61.1 %	48.8 %	42.1 %	52.6 %	
	Whole %	9.3 %	17.8 %	6.8 %	16.9 %	50.8 %
Stealth	No.	6	17	9	14	46
	Lamps %	13.0 %	37.0 %	19.6 %	30.4 %	
	QCL_1 %	33.3 %	39.5 %	47.4 %	36.8 %	
	Whole %	5.1 %	14.4 %	7.6 %	11.9 %	39.0 %
Firewinder	No.	3	13	4	13	33
	Lamps %	9.1 %	39.4 %	12.1 %	39.4 %	
	QCL_1 %	16.7 %	30.2 %	21.1 %	34.2 %	
	Whole %	2.5 %	11.0 %	3.4 %	11.0 %	28.0 %
Hollow	No.	9	21	8	18	56
	Lamps %	16.1 %	37.5 %	14.3 %	32.1 %	
	QCL_1 %	50.0 %	48.8 %	42.1 %	47.4 %	
	Whole %	7.6 %	17.8 %	6.8 %	15.3 %	47.5 %
Total	No.	18	43	19	38	118
	Whole %	15.3 %	36.4 %	16.1 %	32.2 %	100.0 %

Eventually, the most preferred lamp by all groups of consumers is “Kina” (50.8 %) which is designed with aesthetics of natural material and rhythm weaving form.

4 Conclusion

The technology for eco-product design has been well developing through recent years by both the industry and research institutions. It is time to disseminate sustainable consumption to consumers for reducing load to the environment. Eco-products should be designed to meet consumer needs from different perspectives. Being aware of the eco-product characteristics based on consumer lifestyle may help the product designer find the aesthetic characteristics for a certain group of consumers. For further study, the attributes for eco-product aesthetics and function can be classified according to product categories.

Acknowledgments The author would like to express appreciation to Mr. Yu-Wei Chuang as my graduate student for his help in this study. The author would also like to thank Ministry of Science and Technology, Taiwan for providing financial support (grant# MOST 103-2420-H-130-004-MY2) to present this research.

References

1. Bret L, Vann RJ, Mittelstaedt JD (2014) Changing the marketplace one behavior at a time: perceived marketplace influence and sustainable consumption. J Bus Res 67(9):1953–1958
2. Charter M, Tischner U (2001) Sustainable solutions: developing products and services for the future. Greenleaf Publishing, Sheffield
3. Davies A, Elliott R (2006) The evolution of empowered consumer. Eur J Mark 40:1106–1121
4. Dobers P, Strannegard L (2005) Design, lifestyle and sustainability, aesthetics consumption in a world of abundance. Bus Strateg Environ 14:324–336
5. Hawkins D, Roger I, Kenneth JB, Coney A (1995) Consumer behavior-implications for marketing strategy, 6th edn. IRWIN, New York, pp 331–341
6. Horie N, Iwamoto T, Okamoto S (2005) A study on consumer's conceptual image for an environmentally conscious product. In: Proceedings of the 4th international symposium on environmentally conscious design and inverse manufacturing, Tokyo, Japan, 12–14 December 2005, P-4
7. Imkamp H (2000) The interest of consumers in ecological product information is growing-evidence from two German surveys. J Consum Policy 23:193–202
8. Kobayashi H (2006) A systematic approach to eco-innovative product design based on life cycle planning. Adv Eng Inform 20:113–125
9. Lucas D (2011) Green design. Braun Publish, Arenenber
10. Otto H (2009) Visual analysis of information hierarchies in life cycle inventories. In: Proceedings of the 6th international symposium on environmentally conscious design and inverse manufacturing, Sapporo, Japan, 7–9 December 2009, pp 917–922
11. Tukker A, Emmert S, Charter M (2008) Fostering change to sustainable consumption and production: an evidence based view. J Clean Prod 16:1218–1225

12. Vezzoli C, Sciama D (2006) Life cycle design: from general methods to product type specific guidelines and checklists: a method adopted to develop a set of guidelines/checklist handbook for the eco-efficient design of NECTA vending machines. J Clean Prod 14:1319–1325
13. Walker S (1995) The environment, product aesthetics and surface. Des Issues Mass Inst Technol 11(3):15–27
14. Zafarmand SJ, Sugiyama K, Watanabe M (2003) Aesthetic and sustainability: the aesthetic attributes promoting product sustainability. J Sustain Prod Des 3:173–186

MIX
Papier aus verantwortungsvollen Quellen
Paper from responsible sources
FSC® C105338

If you have any concerns about our products,
you can contact us on
ProductSafety@springernature.com

In case Publisher is established outside the EU,
the EU authorized representative is:
Springer Nature Customer Service Center GmbH
Europaplatz 3, 69115 Heidelberg, Germany

Printed by Libri Plureos GmbH
in Hamburg, Germany